Toxicology of the Lung

Fourth Edition

TARGET ORGAN TOXICOLOGY SERIES

Series Editors

A. Wallace Hayes, John A. Thomas, and Donald E. Gardner

TOXICOLOGY OF THE LUNG, FOURTH EDITION
Donald E. Gardner, editor, 696 pp., 2006

TOXICOLOGY OF THE PANCREAS
Parviz M. Pour, editor, 720 pp., 2005

TOXICOLOGY OF THE KIDNEY, THIRD EDITION
Joan B. Tarloff and Lawrence H. Lash, editors, 1200 pp., 2004

OVARIAN TOXICOLOGY
Patricia B. Hoyer, editor, 248 pp., 2004

CARDIOVASCULAR TOXICOLOGY, THIRD EDITION
Daniel Acosta, Jr., editor, 616 pp., 2001

NUTRITIONAL TOXICOLOGY, SECOND EDITION
Frank N. Kotsonis and Maureen A. Mackey, editors, 480 pp., 2001

TOXICOLOGY OF SKIN
Howard I. Maibach, editor, 558 pp., 2000

TOXICOLOGY OF THE LUNG, THIRD EDITION
Donald E. Gardner, James D. Crapo, and Roger O. McClellan, editors, 668 pp., 1999

NEUROTOXICOLOGY, SECOND EDITION
Hugh A. Tilson and G. Jean Harry, editors, 386 pp., 1999

TOXICANT–RECEPTOR INTERACTIONS: MODULATION OF SIGNAL TRANSDUCTIONS AND GENE EXPRESSION
Michael S. Denison and William G. Helferich, editors, 256 pp., 1998

TOXICOLOGY OF THE LIVER, SECOND EDITION
Gabriel L. Plaa and William R. Hewitt, editors, 444 pp., 1997

FREE RADICAL TOXICOLOGY
Kendall B. Wallace, editor, 454 pp., 1997

(Continued)

ENDOCRINE TOXICOLOGY, SECOND EDITION
Raphael J. Witorsch, editor, 336 pp., 1995

CARCINOGENESIS
Michael P. Waalkes and Jerrold M. Ward, editors, 496 pp., 1994

DEVELOPMENTAL TOXICOLOGY, SECOND EDITION
Carole A. Kimmel and Judy Buelke-Sam, editors, 496 pp., 1994

IMMUNOTOXICOLOGY AND IMMUNOPHARMACOLOGY, SECOND EDITION
Jack H. Dean, Michael I. Luster, Albert E. Munson, and Ian Kimber, editors, 784 pp., 1994

NUTRITIONAL TOXICOLOGY
Frank N. Kotsonis, Maureen A. Mackey, and Jerry J. Hjelle, editors, 336 pp., 1994

OPHTHALMIC TOXICOLOGY
George C. Y. Chiou, editor, 352 pp., 1992

TOXICOLOGY OF THE BLOOD AND BONE MARROW
Richard D. Irons, editor, 192 pp., 1985

TOXICOLOGY OF THE EYE, EAR, AND OTHER SPECIAL SENSES
A. Wallace Hayes, editor, 264 pp., 1985

CUTANEOUS TOXICITY
Victor A. Drill and Paul Lazar, editors, 288 pp., 1984

Target Organ Toxicology Series

Toxicology of the Lung

Fourth Edition

Edited by

Donald E. Gardner

A CRC title, part of the Taylor & Francis imprint, a member of the Taylor & Francis Group, the academic division of T&F Informa plc.

Published in 2006 by
CRC Press
Taylor & Francis Group
6000 Broken Sound Parkway NW, Suite 300
Boca Raton, FL 33487-2742

Printed in the United States of America on acid-free paper
10 9 8 7 6 5 4 3 2 1

International Standard Book Number-10: 0-8493-2835-7 (Hardcover)
International Standard Book Number-13: 978-0-8493-2835-0 (Hardcover)
Library of Congress Card Number 2005047214

Cover illustration provided by Prof. Stephen Gardner from the Savannah College of Art and Design, Savannah, Georgia.

Library of Congress Cataloging-in-Publication Data

Toxicology of the lung / [edited by] Donald E. Gardner.-- 4th ed.
p. cm.
Third ed. published: Philadelphia, Pa. : Taylor and Francis, 1999.
Includes bibliographical references and index.
ISBN 0-8493-2835-7 (alk. paper)
I. Gardner, Donald E., 1931-

RC720.T695 2005
616.2'407--dc22 2005047214

Taylor & Francis Group
is the Academic Division of Informa plc.

Visit the Taylor & Francis Web site at
http://www.taylorandfrancis.com

and the CRC Press Web site at
http://www.crcpress.com

PREFACE

A strong response to the third edition of *Toxicology of the Lung* inspired this substantially updated and revised fourth edition. This fourth edition provides an excellent forum for addressing the recent growth in information in both basic and applied research in pulmonary toxicology since the third edition. It presents a comprehensive and critical synthesis of the latest advances, leading to a better understanding of how the body responds to airborne contaminants.

A number of new topic areas have been added to this edition, providing an insight into the most important current thinking regarding the (1) critical need for understanding the kinetics and dynamic interactions associated with toxic effects, (2) appropriateness of recent advances being made across disciplines to address human clinical testing and new emerging technology for using animal and *in vitro* models to detect adverse effects, (3) identification of how airborne substances can alter physiological, biochemical and morphological functioning of biological systems, and (4) the latest modeling approaches for predicting deposition and fate of inhaled particles. These 16 chapters address the important areas in toxicology as well as illustrate ongoing research efforts for better understanding the association between airborne substances and systemic effects.

All of the chapters in this new edition are written by a multidisciplinary team of authors, who are internationally recognized experts in their selected area and all except one are new contributors for this edition. Special attention is devoted to providing the reader with an up-to-date reference section for each topic area, a useful repository of comprehensive information that will direct the reader to additional resources.

This text is unique in that it presents the information from a target organ perspective, addresses critical issues and recent advances in toxicology research, and focuses on assessment of human effects through clinical human, animal, cellular toxicology, and modeling studies.

Donald E. Gardner

CONTRIBUTORS

Corrie B. Allen
Lone Wolf Bioscience Consulting
La Conner, Washington

Bahman Asgharian
CIIT
Centers for Health Research
Research Triangle Park, North Carolina

Paul H. Ayres
R.J. Reynolds Tobacco Company
Science & Regulatory Affairs Division
Bowman Gray Technical Center
Winston-Salem, North Carolina

David M. Bernstein
Consultant in Toxicology
Chemin de la Petite-Boissière
Geneva, Switzerland

Mitchell D. Cohen
New York University
School of Medicine
Tuxedo, New York

Geoffrey M. Curtin
R.J. Reynolds Tobacco Company
Science and Regulatory Affairs Division
Bowman Gray Technical Center
Winston-Salem, North Carolina

John Doull
University of Kansas Medical Center
Department of Pharmacology Toxicology and Therapeutics
Kansas City, Kansas

David J. Doolittle
R.J. Reynolds Tobacco Company
Science & Regulatory Affairs Division
Bowman Gray Technical Center
Winston-Salem, North Carolina

Mark Frampton
University of Rochester
School of Medicine
Rochester, New York

Shayne C. Gad
Gad Consulting Services
Cary, North Carolina

Werner Hofmann
Institute of Physics and Biophysics
University of Salzburg
Salzburg, Austria

Ilona Jaspers
University of North Carolina at Chapel Hill
CEMALB
Chapel Hill, North Carolina

Frederick J. Miller
Fred J. Miller and Associates, LLC
Cary, North Carolina

Paul Morrow
University of Rochester
School of Medicine
Rochester, New York

Jürgen Pauluhn
Bayer AG
Institut für Toxikologie
Wuppertal, Germany

Anthony P. Pietropaoli
University of Rochester
School of Medicine
Rochester, New York

Ryan J. Potts
R.J. Reynolds Tobacco Company
Science & Regulatory Affairs Division
Bowman Gray Technical Center
Winston-Salem, North Carolina

Christopher A. Reilly
University of Utah
Department of Pharmacology and Toxicology
Salt Lake City, Utah

Stephen I. Rennard
University of Nebraska Medical Center
Department of Internal Medicine
Nebraska Medical Center
Omaha, Nebraska

Keith Robinson
CTR
BioResearch Inc.
Senneville, Quebec, Canada

William L. Roth
Office of Food Additive Safety
College Park, Maryland

Karl K. Rozman
University of Kansas Medical Center
Department of Pharmacology Toxicology and Therapeutics
Kansas City, Kansas

James E. Swauger
R.J. Reynolds Tobacco Company
Science & Regulatory Affairs Division
Bowman Gray Technical Center
Winston-Salem, North Carolina

Mark Utell
University of Rochester
School of Medicine
Rochester, New York

Bellina Veronesi
U.S. Environmental Protection Agency
NHEERL
Research Triangle Park, North Carolina

André Viau
CTR
BioResearch Inc.
Senneville, Quebec, Canada

Jim C. Walker
Florida State University
Sensory Research Institute
Tallahassee, Florida

David B. Warheit
E.I. duPont de Nemours
Haskell Laboratory
Newark, Delaware

Hanspeter Witchii
University of California at Davis
ITEH
Davis, California

Gerold S. Yost
University of Utah
Department of Pharmacology and Toxicology
Salt Lake City, Utah

CONTENTS

EDITOR

Donald E. Gardner, Ph.D., Fellow, ATS, has over 40 years of experience in the field of toxicology. He earned B.S. and M.S. degrees from Creighton University, Omaha, Nebraska and a Ph.D. in environmental health from the University of Cincinnati.

Past employment has included the U.S. Environmental Protection Agency/ U.S. Public Health Service, where he was director of the Toxicology Division that was responsible for both the animal and human toxicology program that addressed the potential health risks associated with exposure to environmental chemicals. Following retirement from the EPA, he joined the staff of Northrop/ManTech Corp. as vice president and chief scientist. He has had adjunct professor appointments at Duke University, North Carolina State University and University of Massachusetts, Amherst.

Since 1995 he has been president of Inhalation Toxicology Associates, a consulting company that provides consulting services to governmental agencies including U.S. EPA, NIEHS, NASA, DOD and NIH and to private industry and law firms. He is active on numerous international advisory panels in the area of environmental health and toxicology.

He is a board certified Fellow in Toxicology and has chaired numerous expert panels for the National Academy of Science, National Research Council, Committee on Toxicology and has been vice chairman of the Committee on Toxicology. He is currently an associate editor of the *Journal of Immunotoxicology* and on the Editorial Board of *Toxicology Mechanisms and Methods.* He is coeditor of *Target Organ Toxicology Series,* editor of

Toxicology of the Lung and coeditor of *New Perspectives: Toxicology and the Environment.*

Throughout his career he has published over 250 research manuscripts and book chapters. He is the editor of the *Journal of Inhalation Toxicology* (1988-present).

1

METHODS FOR EVALUATING THE LUNG IN HUMAN SUBJECTS

Stephen I. Rennard and John R. Spurzen

CONTENTS

1.1 INTRODUCTION

The lung is frequently the target of environmental and systemic toxic exposures. A normal adult at rest takes in more than 7000 l of air daily (1). Gas exchange requires exposure of a surface area of approximately 80 m^2 of the lung to this inhaled air (2). This makes the lung uniquely exposed to airborne toxins. In addition, the lung receives the entire cardiac output from the right side of the heart, which places it at risk for systemic toxins. Finally, specific metabolic reactions in the lung can lead to the uptake and concentration of a variety of toxins in the lung. As a result, toxic exposures are a frequent cause of lung disease. Evaluation of the lung, therefore, is highly relevant in toxicological studies.

Of the parenchymal organs, the lung is relatively easy to evaluate, making the assessment of toxic exposures feasible in human subjects. The current chapter will provide an overview of the methods by which the lung can be evaluated, including physiological, imaging, and sampling techniques. The methods available include well-established ones that are widely available as well as developing technologies.

1.2 PHYSIOLOGY

1.2.1 Airflows

The most widely used assessment of lung functions are measures of airflow, particularly expiratory airflow (3). The most widely used method, spirometry, requires that a subject inhale to total lung capacity (TLC) and then exhale as rapidly and completely as possible (Figure 1.1). Expired-air volume is then measured as a function of time, which provides airflow as a function of time. The most important parameters obtained by this maneuver are the forced expiratory volume in 1 sec (FEV1) and the forced vital capacity (FVC). Consensus standards for the performance and interpretation of spirometry are available from both the American Thoracic Society (4,5) and the European Respiratory Society (6).

Vital capacity (VC), which is the volume of air that can be voluntarily moved in and out of the lungs, would, in theory, require an infinitely prolonged exhalation. In most individuals, the lungs are almost completely emptied after 6 sec, and FEV6 (forced expiratory volume after 6 sec) is often accepted as a surrogate for FVC. The maximal flow rate obtained during a forced maneuver occurs shortly after beginning the maneuver, and flow rates then decrease as lung volumes decrease. The peak expiratory flow (PEF) is fairly easily measured and is often used in monitoring subjects with asthma and in epidemiological studies.

It has been proved useful to plot the data generated by a spirometer as flow vs. volume rather than as volume vs. time. The resulting flow-volume

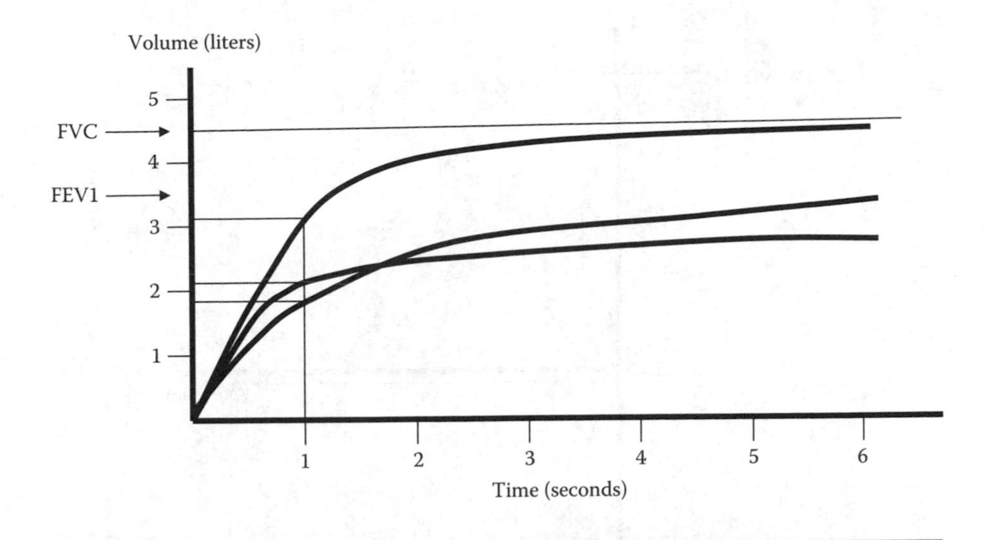

Figure 1.1 Spirogram. After inhaling as deeply as possible, a subject exhales as rapidly as possible. The volume of air (vertical axis) is plotted against time (horizontal axis). Airflow, as a function of time, is given by the slope at any point. Forced expiratory volume in 1 sec (FEV1) is the volume of air exhaled in 1 sec. It would be reduced in case of airway obstruction and airway restriction. Forced vital capacity (FVC) is the maximal volume of air exhaled. In obstruction, the FVC may be normal, but the subject may not be able to exhale long enough to make an accurate measure. In restriction, the FVC will be reduced. Peak expiratory flow rate (PEFR) is the steepest slope on the curve, and this occurs very near the beginning of the maneuver. FVC and FEV1 are indicated for a normal subject. (See the color version of this figure after page 18.)

curve has a characteristic shape and is often used in the diagnosis of pulmonary disease. Plotting data from inspiration, together with expiration data, results in a flow-volume loop (Figure 1.2). Data contained in the flow-volume loop is the same as that in the spirogram.

The increase in intrathoracic pressure generated in a forced maneuver can cause airways to collapse in some individuals, particularly those with lung disease. This can reduce the rate and volume of the exhaled air. As a result, FVC may differ from slow vital capacity (SVC), which is measured in a similar way except that after a subject inhales fully, he or she is allowed to exhale as completely as possible in a manner that is most comfortable.

1.2.2 Lung Volumes

In addition to airflow, measurement of lung volumes (3) provides information that can help distinguish the presence of air trapping or restrictive ventilatory defects (Figure 1.3). The first is common in obstructive disease,

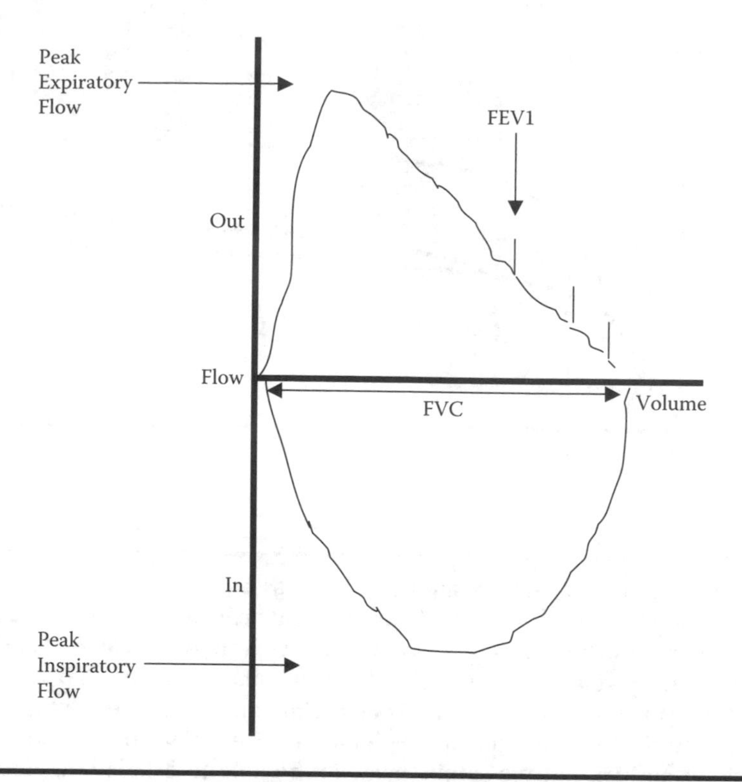

Figure 1.2 Flow-volume loop. Data from a spirogram is often plotted as flow vs. volume. This can be plotted for both expiration and inspiration, resulting in a loop. Peak inspiratory and expiratory flows, FEV1, and FVC can be read directly from the plot. The characteristically shaped triangular expiratory and semicircular inspiratory limbs of the curve are altered in disease.

and the latter is common with interstitial fibrosis. Although VC can be measured easily by spirometry, measurement of TLC is complicated by the fact that some air remains in the lung following maximal forced expiratory efforts. This volume, the residual volume (RV), can be estimated using Boyle's law. This requires the subject to sit inside a body plethysmograph. Although straightforward, this is a relatively expensive test that is not easily standardized across many centers.

RV can also be measured using a dilution tracer. Helium is most commonly used for this purpose. This method, unfortunately, is relatively inaccurate and becomes progressively more so in the presence of irregularities in ventilation, which are common in individuals with lung disease.

1.2.3 Challenge Tests

Smooth muscles surrounding the airways will constrict in response to a variety of stimuli, and this can be used in challenge tests. As the airways

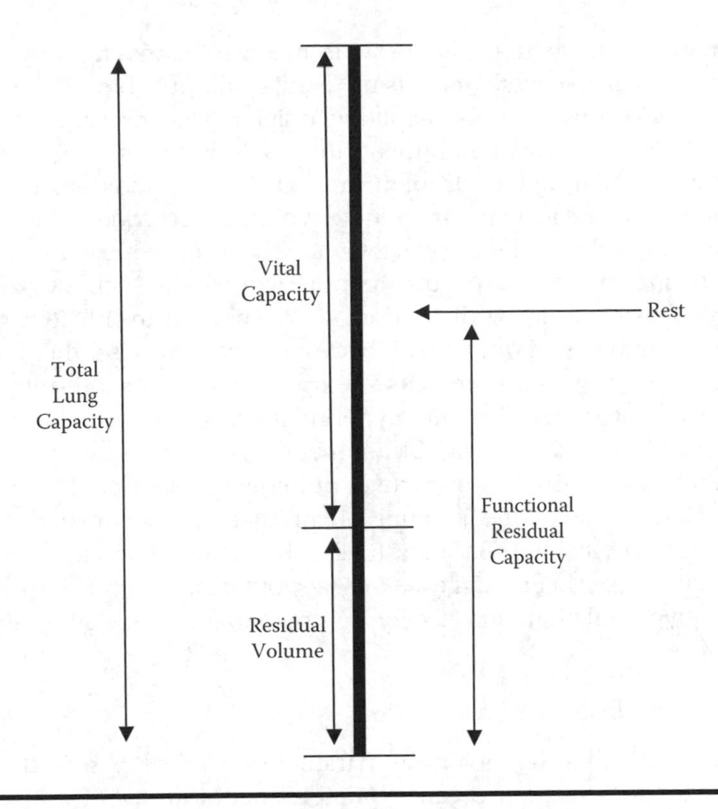

Figure 1.3 Lung volumes. Total lung capacity (TLC) is the total volume of air in the lungs at maximal inhalation. Vital capacity (VC) is the volume of air that can be exhaled from TLC. Residual volume (RV) is the volume of air that remains in the lung after a maximal exhalation. In the absence of any muscular activity, the lungs are partly expanded. This is the point at which normal inhalation begins and is termed *Functional Residual Capacity* (FRC). VC can be determined by spirometry. Measurements of TLC, RV, and FRC require other methodologies (see subsection titled "Lung Volumes").

in asthma are characteristically hyperresponsive, challenge tests have been used to diagnose and gauge the severity of asthma. Challenge tests using methacholine, histamine, cold dry air, nebulized distilled water, and specific allergens have been widely used (3,7). Other challenge tests have also been described. Although individuals with hyperreactive airways will generally respond to most stimuli, differences in response have been seen, and these may characterize different subsets of individuals with hyperreactive airways. Airway reactivity, although a feature of asthma, is not uniformly present in asthmatics (8). Moreover, airway reactivity may be present in individuals who appear to have no other features of asthma. Measurement of airway reactivity has been used in challenge studies of normal and asthmatic subjects exposed to a number of toxins.

Exercise challenge tests also provide means of assessing lung function in further detail (3). With increasing cardiac output, blood flow in the lung uses additional alveolar capillary units. In the presence of poorly ventilated areas of the lung, this can result in worsening ventilation perfusion matching and a drop in arterial PO_2. In addition, exercise is associated with an increase in minute volume occurring as a result of increased tidal volume and respiratory rate. This has been used in challenge studies to increase the exposure to potential toxins such as ozone (9). In the presence of airflow limitation, there may be insufficient time for complete emptying of the lungs because with exercise the increased respiratory rate results in a decreased expiratory time. As a result, exercise may be associated with dynamic hyperinflation (10). Recent physiological studies suggest that this is a significant mechanism for dyspnea, particularly in patients with chronic obstructive pulmonary disease (11). Dynamic hyperinflation reduces the amount of air that can be inhaled, i.e., the inspiratory capacity (IC). Because IC can be estimated quite easily using spirometry, it has been suggested as a potentially useful surrogate for dynamic hyperinflation, and it may be amenable to large-scale studies.

1.2.4 Other Tests

A number of other physiological parameters can be assessed. Airway conductance and resistance can be determined with a body plethysmograph. Measurement of lung compliance requires placement of an esophageal balloon. Measurement of blood gases (PO_2, PCO_2, and pH) provides an estimate of gas exchange and can also be regarded as a lung function test. Pulmonary hypertension often results from acute and chronic lung disease, and measurement of pressures in the pulmonary circulation is often performed. The most precise measures require right heart and pulmonary artery catheterization. Echocardiographic techniques, which estimate pulmonary pressures from the Doppler effect generated by the acceleration of blood in the pulmonary artery, have been developed, and they provide noninvasive estimates of peak pulmonary pressure (12). This method has also been applied to animal studies (13).

1.3 IMAGING

A variety of methods are available for imaging the lungs, the most widely used one being routine roentgenography. Posterior, anterior, and lateral views of the chest have been the mainstays of clinical diagnosis for a century. Compared to computed tomography, routine x-rays are relatively less sensitive to subtle changes. Such studies, however, have been widely

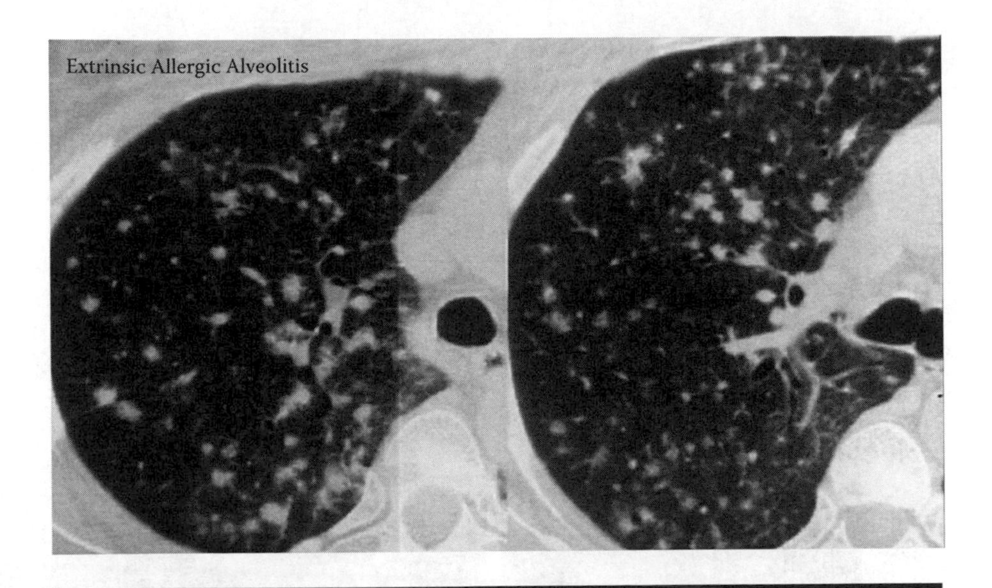

Figure 1.4 Computed tomography of the chest. Two axial sections of the right lung at the level of the carina (right) and more proximally in a subject with extrinsic allergic alveolitis are shown. The peribronchial location of the infiltrative lesions is readily recognizable. This could not have been ascertained by conventional radiographs. (Images courtesy of Prof. J. Gurney, University of Nebraska Medical Center, Omaha.)

used in the diagnosis of toxic exposures in both clinical and epidemiological studies.

Computed tomography of the chest provides much greater anatomic resolution than does routine roentgenography (Figure 1.4). In addition, algorithms for the quantitative assessment of the lungs using computed tomography have been developed. These can provide a measure of the severity and extent of emphysema (14–16) and airway disease (17). The development of high-speed scanners that use relatively low radiation exposures has made this test increasingly appropriate for research applications.

Magnetic resonance imaging has not been widely applied to the assessment of the lungs, primarily because conventional scanners image the protons present in water, and the large air content of the lungs makes them invisible by this method. It is possible, however, to utilize tracer gases that can be hyperpolarized to provide a magnetic signal (18,19). Both ^{3}He and ^{129}Xe have been used for this purpose. These methods can image air spaces of the lung at a level that permits investigation of abnormalities that cannot be detected by other means (Figure 1.5). Because attenuation of the magnetic signal depends upon its interactions with the

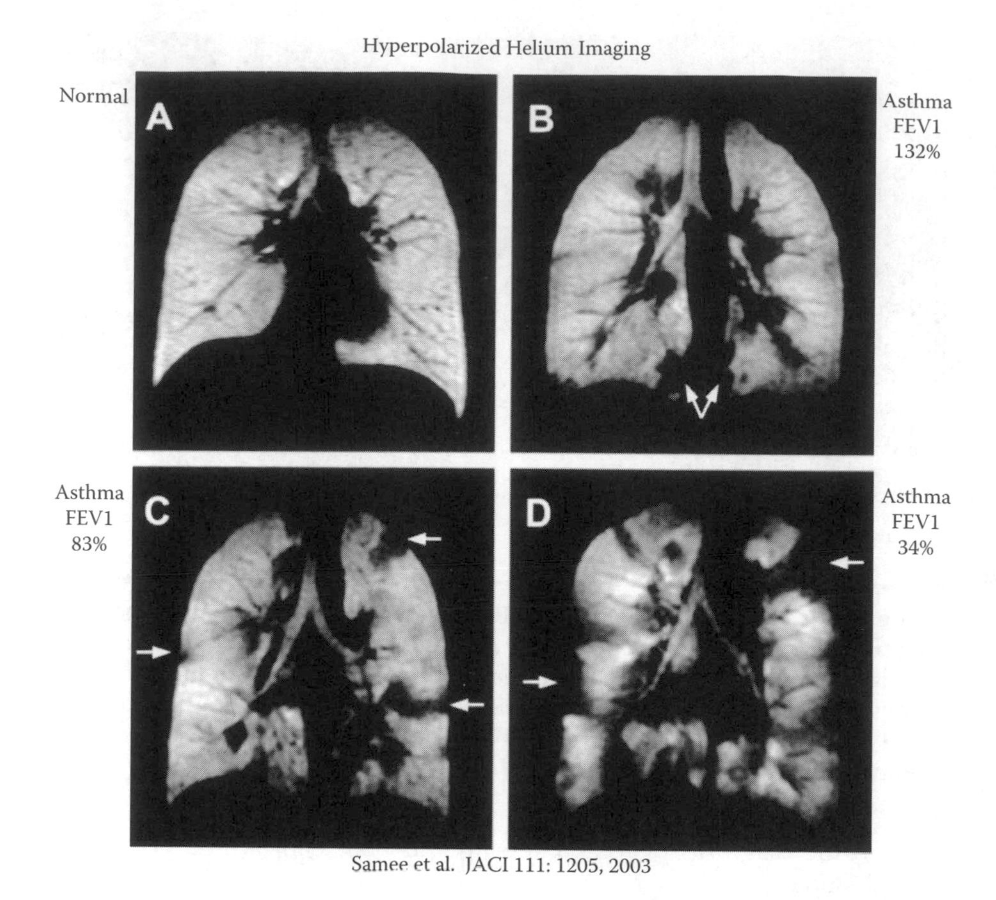

Figure 1.5 Hyperpolarized gas imaging of the lungs. Images were obtained following inhalation of ^{3}He in asthmatic subjects with varying lung function. The ability to image ventilation with a high degree of resolution makes possible the detection of abnormal areas even when lung function, as a whole, is normal. See Reference 99 for details. (From Samee, S., T. Altes, P. Powers, E.E. de Lange, J. Knight-Scott, G. Rakes, J.P. Mugler, 3rd, J.M. Ciambotti, B.A. Alford, J.R. Brookeman, and T.A. Platts-Mills: Imaging the Lungs in Asthmatic Patients by Using Hyperpolarized Helium-3 Magnetic Resonance: Assessment of Response to Methacholine and Exercise Challenge. *J Allergy Clin Immunol* 2003, 111: 1205–1211. With permission.)

surrounding space, it is possible to utilize magnetic methods to estimate the apparent diffusion coefficient within the lung, which provides an indirect measure of alveolar size (20) and can also be used in the assessment of small airway diameter (21). In addition to being a research tool, such studies may provide a noninvasive tool to assess the progression of emphysema.

1.4 SAMPLING THE LUNGS

Of the parenchymal organs, the lungs are among the most easily sampled. This is true for several reasons: First, the large volume of air that is inhaled is also exhaled. Contents in the exhaled gas that derive from the lungs can be collected and analyzed. Second, material present in the lungs may be expectorated in sputum, either spontaneously or in response to specific induction. Finally, it is fairly straightforward to introduce a flexible fiberoptic bronchoscope into the lower respiratory tract in order to collect samples by using a variety of techniques. Because samples of material from the lungs permit a variety of biochemical and cellular analyses, these methods have great potential in the evaluation of toxic exposures.

1.4.1 Exhaled-Breath Analysis

Exhaled breath contains a number of volatile components that derive from the lower respiratory tract, and their assessment has provided a means for evaluating the lung (22,23). The most widely studied component is nitric oxide (NO)(24). NO is produced endogenously by several enzyme systems, and it serves as an important mediator in most tissues. Inflammatory reactions are often associated with the induction of iNOS, an enzyme that generates large amounts of NO. Inflammatory lung diseases are associated with an induction of iNOS and a measurable increase in exhaled NO. Elevated NO production has been assessed in a large number of lung diseases. In asthma, measurement of NO has been suggested as a gauge of clinical severity and therapeutic effect (25). NO is produced throughout the respiratory tract. By analyzing NO concentrations at different flow rates, it is potentially possible to separate upper airway production from the distal lung production of nitric oxide (26).

Other gases present in exhaled air have also been measured. Carbon monoxide can be readily measured, and increased levels are exhaled by smokers and by individuals exposed to environmental CO (27–29). CO is also produced endogenously by two heme oxygenases, one of which, HOX-1, is associated with inflammation (30). Increased levels of CO in the exhaled air have been reported in some inflammatory lung diseases (22,23). Pentane is generated as a result of oxidation of cellular lipids. It has been quantified in exhaled air and has been suggested to be a marker of oxidative stress (31,32). Ethane and other alkanes may have similar significance (33,34).

1.4.2 Exhaled-Breath Condensate

By chilling exhaled breath, water vapor, together with other components present in the exhaled air, will condense and can be collected. Collection of exhaled breath condensate samples is fairly simple. A subject at rest breathes into a mouthpiece while tight seal is maintained with the mouth and a nose clip is in place. In a 15-min time frame, approximately 2 ml of material can be collected. Most studies have been performed with normal tidal breathing at rest. Although relatively few studies have addressed methods of sample collection, alterations in ventilation may affect the results (35). A major problem with this technique has been contamination of the exhaled breath with saliva and other oral contents. Several devices, however, have been developed that appear to address this problem (36,37).

A larger number of moieties have been found in exhaled breath condensate. Low-molecular-weight substances such as hydrogen peroxide are likely to be present as gases in the exhaled breath and may condense in chilled air. In addition, ions and macromolecules, including proteins, have been found, although the results are not uniformly reproducible and are controversial. Data using this method are, as yet, limited. Cigarette smokers have been described as having alterations of a number of parameters (Table 1.1).

Some controversy also exists regarding the mechanism by which nonvolatile substances reach the exhaled breath condensate. Aerosolization of microdroplets as a consequence of turbulent flow within the lungs has been suggested (38,39). Increases in various markers in the exhaled breath condensate, therefore, could represent alterations in lung anatomy or in the composition of the fluid or gaseous compartments of the lower respiratory tract. Although the significance of most of the markers present in thc lower respiratory tract is still incompletely defined, several have been evaluated in considerable detail. Hydrogen peroxide, for example, has been reported to be increased in various lung conditions (22,23). Alterations in the pH of exhaled breath condensate have also been reported (40,41). Because of the ease with which samples are obtained and the noninvasive nature of the test, analysis of exhaled breath condensates holds great promise for the assessment of biomarkers in the lower respiratory tract.

1.4.3 Sputum Analysis

Both spontaneously expectorated (42) and induced sputum (43) have been used to analyze components present in the lower respiratory tract. Sputum contains mucus produced by goblet cells that line the airways and by mucous glands. Both are much more common in the proximal airways (2), although goblet cell metaplasia in the distal airways is a common feature of many lung diseases. In addition to mucus, sputum also contains other components present within the airway lumen and within the lumen of the distal airways that frequently move proximally as a result of the

Table 1.1 Measures Reported to be Abnormal in the Exhaled Breath of Smokers as Compared to that of Nonsmokers

Analyte	*Reference*
4-HHE	100
4-HNE	100
Acrolein	100
H_2O_2 (μM)	101
Heptanal (nM)	100,102
Hexanal (nM)	102
IL-1 (pg/ml)	103
IL-6 (pg/ml)	104, 105
Isoprostane (pg/ml)	104
LTB4	105
Malondialdehyde (nM)	100, 102
NCA	103
Nitrite (μM)	103, 106
Nitrotyrosine	106
$NO_2 + NO_3$	106
NO_3 (μM)	107
Nonanal (nM)	100, 102
S-Nitrosothiols	106
TBARs (μM)	101
TNFα	103
Total protein (μg/ml)	103

mucociliary escalator. Sputum samples, therefore, contain both cellular and molecular components present within the lower respiratory tract.

Analysis of sputum generally requires processing, and a consensus panel has provided recommendations (44). A variety of techniques have been utilized to separate "true" sputum from saliva. Rinsing the specimens is less widely used than removing specimens of sputum manually with a forceps. The gel-like nature of sputum complicates its analysis. Ultracentrifugation can separate the "gel" phase from the "sol" phase. Alternatively, the structure of the gel is dependent on disulfide bonds within mucin molecules, and the samples can be solubilized by reduction. Dithiothriatol is most commonly used, but this, as with any form of processing, can potentially alter analytes of interest (45). Spontaneously produced sputum samples have the advantage of representing what is spontaneously present in the lower respiratory tract. Sputum induction, which likely serves as an irritant and, therefore, has the possibility of altering the composition of the contents of the lower respiratory tract, is more widely used for several reasons. First, 60 to 70% of normal individuals who do not spontaneously produce sputum will produce adequate samples following induction with hypertonic saline. Second, sputum induction permits collection of samples in a controlled laboratory setting where salivary contamination, recent food ingestion or smoking, and subject clinical state can be controlled. Finally, sputum induction permits evaluation of samples during collection, ensuring that adequate material is obtained.

Cells present in sputum can be readily evaluated. Alteration in cellular composition has been described in a number of clinical settings (46,47) as well as following challenge studies (48,49). Although the majority of these studies have evaluated asthma, exposure to toxins such as ozone has been performed, and the response assesed by induced sputum (50,51). Moreover, cells recovered in sputum may be viable, and they have been assessed functionally (52). Finally, recovery of cells in sputum permits their analysis using molecular methods (53).

1.4.4 Bronchoscopy

This is a method by which the lungs can be directly inspected and sampled. There are two basic techniques: rigid (54) and flexible bronchoscopy (55,56).

1.4.4.1 Rigid Bronchoscopy

This method was developed more than 100 yr ago. It utilizes a straight, hollow tube, which is inserted through the mouth, between the vocal cords, and into the trachea. Insertion of the instrument requires hyperextension of the neck. Rigid bronchoscopy is most commonly performed under general anesthesia. By means of a light and optical system, the carina, the proximal portions of the left mainstem bronchus, and the right middle and lower lobes may be inspected. Because the instrument is rigid, it is not easily manipulated around bends in the airways, and the ability to visualize the more-distal airways is limited. Rigid bronchoscopy has largely been supplanted by flexible bronchoscopy, which is more easily performed and which permits inspection of the more-distal airways. However, because the intraluminal diameter of the rigid bronchoscope is much larger than that of the flexible bronchoscope, the rigid instrument remains the procedure of choice in selected settings (57).

1.4.4.2 Flexible Bronchoscopy (55)

Fiber-optic elements are frequently included in the optical systems of the rigid bronchoscope and are sometimes termed flexible fiberoptic bronchoscopy.

The application of fiber optics has made the flexible bronchoscope possible. The most commonly used instrument for adult bronchoscopy is approximately 0.5 cm in diameter and is flexible. By means of a wire, the bronchoscopist can bend the bronchoscope tip up to 180°. By bending the tip and rotating the instrument, it is possible to insert the bronchoscope into all the major lobar and segmental airways in a normal adult lung. The limit of airways that may be inspected with the flexible bronchoscope is determined not by the flexibility of the instrument, but by its size: the instrument cannot be advanced beyond airways that have the same caliber as the external diameter of the bronchoscope.

The original flexible fiber optic bronchoscopes contained three fiber optic bundles. Two provided illumination without shadows. The third was used for inspection of the airways. Resolution of the image in the original bronchoscopes was determined by the number of fibers in the bundle. Current bronchoscopes have generally replaced the fiber optic inspection bundle with a wire connected to a high-resolution television camera small enough to be incorporated into the bronchoscope tip. The camera provides far superior resolution. In addition, the camera provides the possibility of digitally processing the recovered information. In fact, color images are generated from a black-and-white camera by using a series of images and filters. The application of similar methods to explore the biology and biochemistry of the airways is under investigation.

A number of modifications of the bronchoscope allow various types of inspection. A bronchoscope with a side-viewing camera has been developed (58). This provides relatively clear imaging of airway mucosa and a remarkable view of the airway vasculature, in contrast to the foreshortened images obtained with an end-viewing bronchoscope, in which the regions close to the bronchoscope tip are foreshortened. Bronchoscopes have been modified for florescence imaging and have been used to identify neoplastic cells with either florescent dyes (59) or endogenous florescence (60,61). Finally, bronchoscopes have been equipped with ultrasound probes (62,63). These permit imaging of bronchial wall thickness and of structures such as lymph nodes that are adjacent to airways.

Bronchoscopes are available in a variety of sizes (55). The most commonly used instruments for bronchoscopy in adults are approximately 5 mm in external diameter. The biopsy channel of such instruments is generally about 2 to 2.2 mm in diameter. Such an instrument can be routinely inserted into all the lobar segmental and subsegmental airways in normal adults. A pediatric instrument with an external diameter of about 3.5 mm and a sample channel diameter of 1.2 mm is routinely used. Obviously, the distance to which it can be inserted depends on the size of the child being bronchoscoped. An ultrathin bronchoscope 1.8 mm in diameter has been developed. This can be inserted through the lumen of a conventional bronchoscope and can be advanced into the peripheral airways (64). Although it does not have a sample channel, the ultrathin bronchoscope can be used to guide the collection of samples with a brush under direct visualization (65).

1.4.4.3 Performance of Bronchoscopy (55,56,66)

Prebronchoscopy evaluation should include electrocardiography, assessment of blood electrolytes, and coagulation studies, especially if biopsy procedures are planned. Subjects should refrain from using aspirin for one week prior to the procedure in order not to compromise platelet function. To minimize the risk of gastric aspiration, the stomach should be empty. This is routinely accomplished by keeping subjects NPO overnight prior to the procedure. The procedure may be performed either transnasally or transorally. Transoral procedures permit the use of an endotracheal tube. Once in place, this minimizes subject gagging, makes it possible to remove and reinsert the bronchoscope readily, and offers an additional measure of safety should bleeding occur. The transoral route also avoids contamination of airway samples with ciliated cells from the nasopharynx. The transnasal route, however, may be technically easier. Anesthesia of the nose (if necessary) is generally achieved with topical viscous lidocaine. Anesthesia of the posterior pharynx, hypopharynx, and larynx is also achieved with lidocaine administered via nebulization under direct vision. It is also common practice to administer lidocaine together with a beta agonist bronchodilator via nebulization prior to performing topical anesthesia as described in the preceding text (67). This may contribute to topical anesthesia in the upper airways, may reduce cough later in the procedure, and is believed to prevent bronchospasm. Subjects are routinely monitored with continuous oximetry. It is common practice to administer supplemental oxygen via a nasal cannula during and following the procedure. An intravenous line should be inserted in the event of emergency medications being required. An intravenous line can also facilitate administration of narcotics or sedatives as needed.

After insertion through the mouth or nose, the bronchoscope is inserted under direct vision between the vocal cords. The bronchoscope can then be advanced into the trachea. If desired, an endotracheal tube can be placed over the bronchoscope and the remainder of the procedure be performed through the in-place endotracheal tube. After inserting the bronchoscope, most bronchoscopists perform a full inspection of all airways prior to any other procedure. In addition to describing any lesions present, it is possible to evaluate airway inflammation. The Thompson Bronchitis Index is a scoring system that assesses erythema, edema, secretions, and friability systemically in the lobes of the lung (68). This scoring system has been utilized in research studies as it quantifies visible inflammation. Visible inflammation is sensitive to change, and reductions in this parameter have been reported in a variety of interventions (69–71). Interobserver agreement among trained bronchoscopists is very high (69). In addition, the bronchitis index can be scored on videotaped images, permitting independent and blinded assessment of visible airways inflammation, although color rendering of taped images and, in particular, of erythema may be somewhat problematic.

1.4.4.4 Sampling the Lungs Using Bronchoscopes

There are several methods for sampling the lungs using a bronchoscope. These include aspiration procedures in which intraluminal contents are aspirated through the sample channel and biopsy methods that use a cutting forceps, a brush, or a needle.

1.4.4.5 Bronchoalveolar Lavage

The simplest technique is simple aspiration, in which intraluminal contents are recovered by vacuum aspiration and collected in an in-line suction trap. This method is commonly used to remove secretions and has many clinical applications. A modification of this procedure involves the installation of 5 to 10 ml of sterile isotonic saline followed by the aspiration of the saline together with intraluminal airway contents. This is sometimes termed a *bronchial wash*, but it should not be confused with bronchoalveolar lavage (66,72), which represents a further modification of the aspiration procedure and allows routine sampling of both airway and alveolar spaces under controlled conditions. Bronchoalveolar lavage is most commonly performed by first "wedging" the bronchoscope. This is accomplished by advancing the bronchoscope gently into an airway that is approximately the same size as the outer diameter of the bronchoscope (Figure 1.6). Once in position, application of suction to the bronchoscope channel causes the airway to collapse. Instillation of saline through the bronchoscope can then be performed in a controlled manner such that only the portion of the lung in the airway distal to the bronchoscope is sampled. Because the volume of lung accessed by the bronchoscope is generally about 1% of the total lung volume, lavage can be performed with little physiological compromise. A decline in FEV1 of about 5 to 10% has been reported, and a similar decline occurs in mild asthmatics (73). Thus, bronchoalveolar lavage is generally well tolerated in individuals with relatively normal lung function. Patients with compromised lung function will be at increased risk for complications.

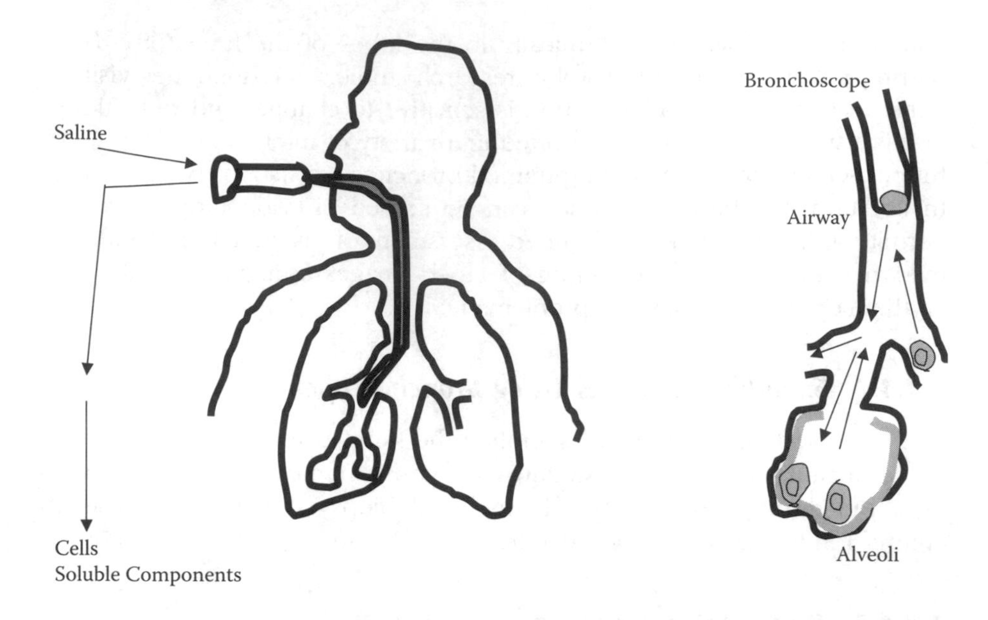

Figure 1.6 Schematic diagram of bronchoalveolar lavage. The bronchoscope is advanced to the "wedged" position, where the caliber of the airway is the same as that of the bronchoscope. An instilled fluid samples the intraluminal space in the airways and the alveoli distal to the bronchoscope. Samples of the cells and the fluid lining the alveoli and the airways are recovered by aspiration.

There are a number of "standardized" procedures for performing bronchoalveolar lavage (66,74). These vary, most prominently, in the volume of fluid infused per aliquot, the number of aliquots infused, the length of time the fluid is allowed to remain in the lung before aspiration (the "dwell" time), and the total volume of fluid infused. It is generally recommended that lavages of at least 100 ml be used in order to effectively assess the alveolar space. The most commonly used procedures are to infuse and sequentially aspirate four 60-ml aliquots at a single site or repeat five 20-ml aliquots at two or three different sites within the lung. Most commonly, a "zero dwell time" procedure is used in which the fluid is aspirated immediately following instillation.

The first fluid instilled reaches the proximal airways, but samples the alveolar spaces less efficiently (75). When sequential 20-ml aliquots are infused, the first aliquot is relatively enriched in "bronchial" contents as evidenced by a relatively high percentage of ciliated epithelial cells (76). Subsequent lavages are progressively enriched in alveolar contents as evidenced by an increase in alveolar macrophages as compared to columnar epithelial cells. Return from an initial 20-ml aliquot, which generally approaches 5 ml, may be processed separately in order to provide a somewhat selective evaluation of airways. Return from subsequent lavages progressively increases in volume, perhaps because of a "pump priming" effect. As a result, the four subsequent 20-ml lavages will yield 50 to 60 ml of return of the 80 ml infused. Some investigators pool all lavages, whereas others process the initial aliquot separately from the subsequent aliquots. Differences in methodologies may account for some differences in reported results. Use of larger-volume lavages does not permit separate analysis of the "bronchial" component (77).

Bronchoalveolar lavage fluid contains the molecular and cellular components present within the intraluminal space of the lung (66,72). A number of biochemical measures have been made and demonstrated to be abnormal in disease or following exposures (72,74,78). Several methods have been used to express the results of measures made on bronchoalveolar lavage fluid. The simplest is to express results as concentrations in the recovered lavage. However, because the fluid used to perform the lavage dilutes the intraluminal contents, there has been much interest in utilizing other denominators. Albumin is present in the epithelial-lining fluid of the lower respiratory tract at a concentration estimated to be about 10% that of plasma. BAL measures for a number of proteins, therefore, have been normalized to albumin concentrations (74,79). The concentration of albumin in the lower respiratory tract, however, can vary with disease, particularly inflammatory diseases that alter lung permeability. An alternative approach involves the measurement of urea. Because it is freely diffusible, urea concentrations in the epithelial-lining fluid of the lung are equal to those in plasma, which can be measured. Measurement of urea in recovered bronchoalveolar lavage fluid, therefore, permits estimation of the dilution engendered by the lavage procedure (80). Urea, however, diffuses quite rapidly. Therefore, the longer it takes to perform the lavage, the higher will be the amount of urea recovered and the less accurate the estimate of dilution (81,82). Nevertheless, this method has proved useful to estimate concentration of lower respiratory tract parameters, for example, to demonstrate that prostaglandin E and prostaglandin D are present in the lower respiratory tract at concentrations that are physiologically relevant (83,84).

Cellular components are also obtained by bronchoalveolar lavage (Figure 1.7) (66,74). These can be evaluated by a variety of histological and immunohistological methods. In addition, viable cells are routinely recovered, which permits their functional evaluation *ex vivo* (85). Finally, it is possible to use molecular methods, for example, to assess gene expression of recovered cells using microarray or other analytical methods.

Alterations in cellular populations of the lower respiratory tract are characteristic of a variety of disease states (66,74). In addition, exposures are often associated with acute recruitment into the lung of specific populations of cells. The ability to apply a variety of biochemical, histological, and molecular methods to samples recovered by bronchoalveolar lavage

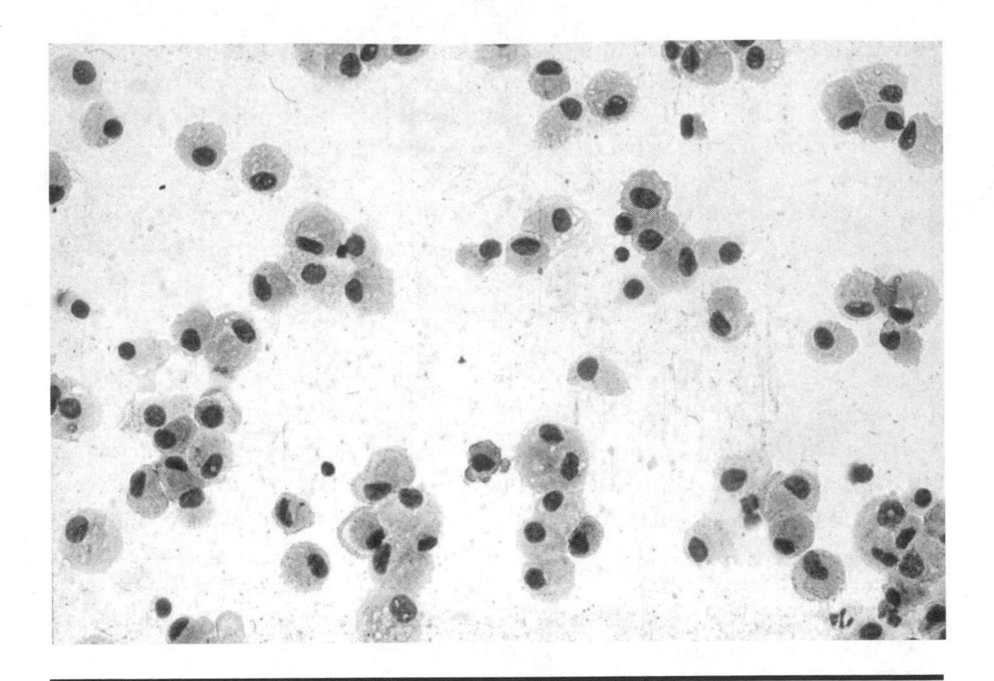

Figure 1.7 Bronchoalveolar lavage cells. A bronchoalveolar lavage sample from a normal subject was used to prepare a cytocentrifuge specimen that was stained by a modified Wright–Giemsa stain. The majority of the cells are alveolar macrophages. A ciliated epithelial cell, a neutrophil, and several lymphocytes are also present.

has permitted a great variety of studies, both in disease and following defined exposures (86), including environmental pollutants such as diesel exhaust (87) and ozone (88–90). It is also possible to administer exposure through the bronchoscope, and endobronchial challenge studies have been widely used in asthma (91). As noted earlier, bronchoalveolar lavage is easily performed, and there are a variety of acceptable variations on the procedure. These variations as well as variations in sample processing can contribute to differences in measurements. Although not generally a problem for the evaluation of clinical samples, such variations may be problematic in research studies and may account for variations in published results. Standardization of lavage procedures at a single site has proved easier than standardization of the procedure across several sites in multicenter studies.

1.4.4.6 Biopsy Procedures

The lower respiratory tract can also be sampled by several biopsy procedures. These include endobronchial biopsy, transbronchial biopsy, endobronchial brush biopsy, and transbronchial needle aspiration biopsy.

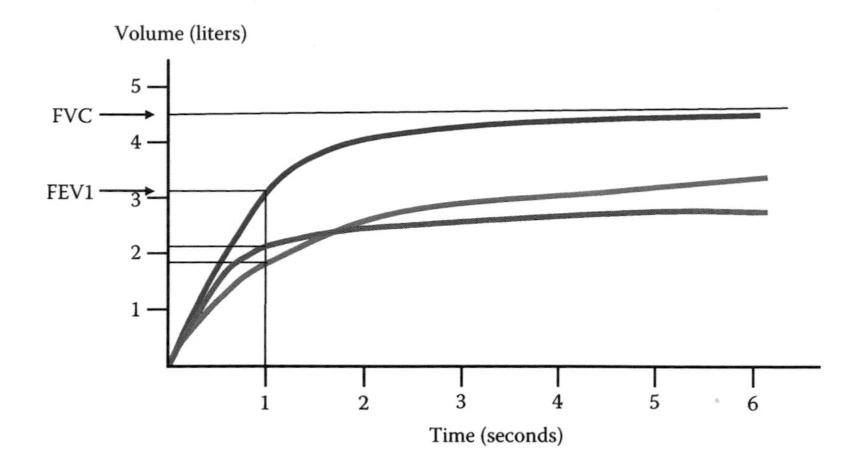

Color Figure 1.1 Spirogram. After inhaling as deeply as possible, a subject exhales as rapidly as possible. The volume of air (vertical axis) is plotted against time (horizontal axis). Airflow, as a function of time, is given by the slope at any point. Forced expiratory volume in 1 sec (FEV1) is the volume of air exhaled in 1 sec. It would be reduced in case of airway obstruction and airway restriction. Forced vital capacity (FVC) is the maximal volume of air exhaled. In obstruction, the FVC may be normal, but the subject may not be able to exhale long enough to make an accurate measure. In restriction, the FVC will be reduced. Peak expiratory flow rate (PEFR) is the steepest slope on the curve, and this occurs very near the beginning of the maneuver. FVC and FEV1 are indicated for a normal subject.

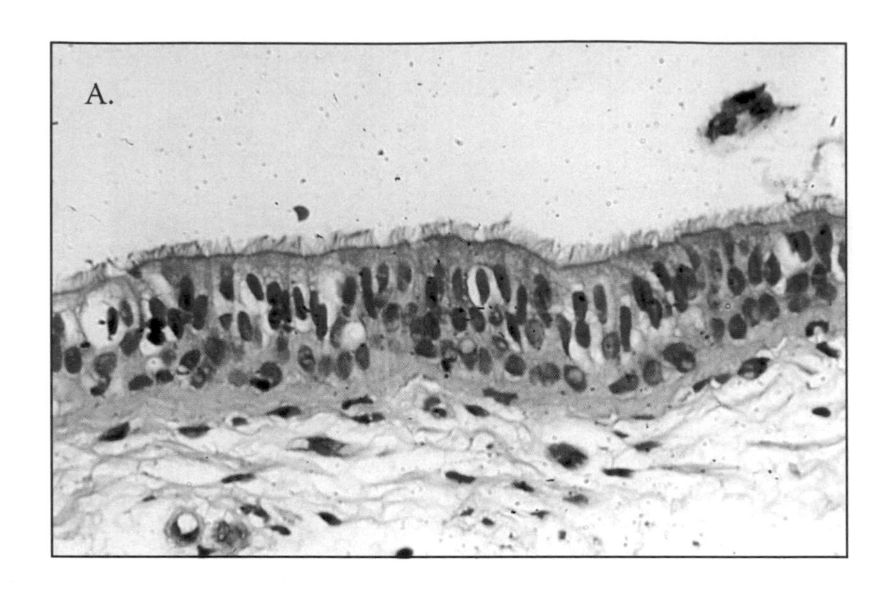

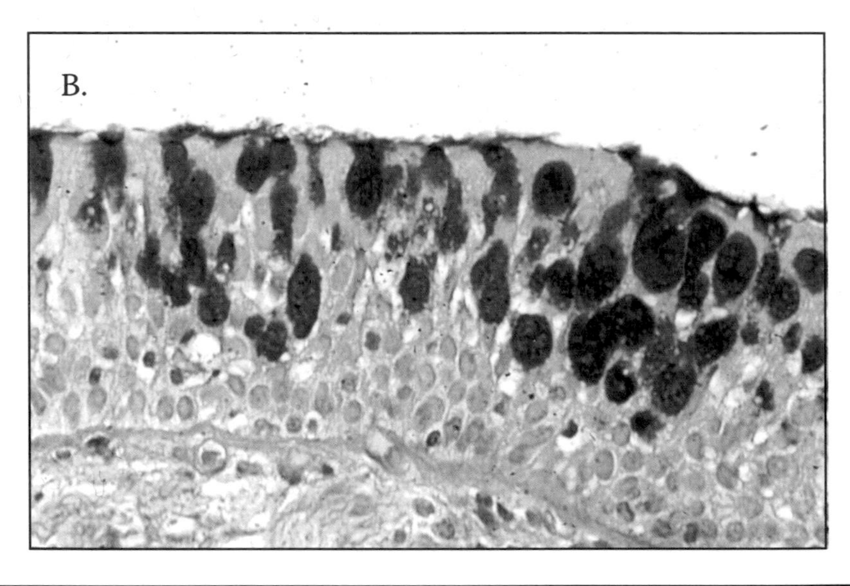

Color Figure 1.8 Endobronchial biopsy. Panel A: An endobronchial biopsy from a normal volunteer stained with hematoxylin-eosin. Note the pseudostratified columnar cells and the ciliated border. Panel B: an endobronchial biopsy from a "normal" smoker Stained with PAS. Note the hypercellularity, the partial loss of cilia, and the marked increase in mucin containing goblet cells.

1.4.4.6.1 Endobronchial Biopsy

In this procedure, a biopsy forceps is advanced through the sampling channel of the bronchoscope and, under direct vision, a biopsy is obtained of the bronchial wall, usually a branch point that fits nicely between the biopsy cups of the forceps. Cancers, which protrude into the airway lumen, can often be readily biopsied with this method. Biopsies of branch points of 1 to 2 mm in diameter can be routinely obtained from normal subjects and subjects with disease (Figure 1.8). Biopsy of the airway wall away from branch points is difficult as it is not easy to grasp the airway with forceps.

The biopsies are small, and the forceps can compress the tissue, possibly damaging the airway epithelium. Nevertheless, specimens obtained using this method by a skilled bronchoscopist usually yield readily available airway mucosa as well as submucosal tissue. Because the depth of the biopsy is much less than the thickness of the normal airway, there is little risk of pneumothorax. Bleeding may be a significant complication, although it is usually minimal in the absence of coagulation defects, such as renal failure or in the presence of abnormal blood vessels as may occur with cancer. Endobronchial biopsy, because of the relative ease of the procedure, has been easier to standardize in multicenter studies.

1.4.4.6.2 Transbronchial Biopsy

In this procedure, the biopsy forceps is advanced into distal regions of the lung. The biopsy forceps is used to cut through the small airway wall, and a biopsy of alveolar structures is obtained. The Zavala method has been suggested to minimize the risk of pneumothorax (92). With this technique, the subject is instructed to take a deep breath, and the biopsy forceps, in the closed position, is advanced under fluoroscopic control. The forceps is then retracted 2 to 3 cm and opened. The subject is then instructed to exhale, and the forceps is advanced approximately 1 cm. With the subject in forced exhalation, the forceps is closed and then retracted. The concept behind this procedure is that lung tissue is pressed into the biopsy forceps in a manner that prevents the forceps from reaching the pleural space, thus minimizing the risk of pneumothorax. The incidence of pneumothorax and significant bleeding following transbronchial biopsy is conventionally stated to be 3 to 5%, but better results have been reported (93,94). With transbronchial biopsy, both small airway and alveolar structure biopsies can be obtained. The technique is routinely used in selected clinical settings and has been applied to research studies in patients with asthma (95). The risk of pneumothorax

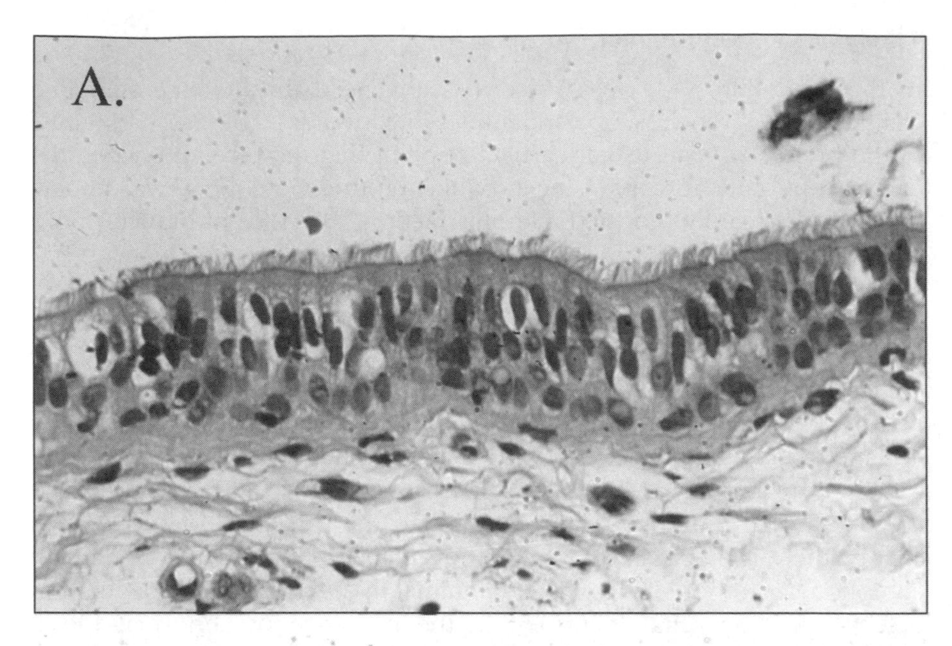

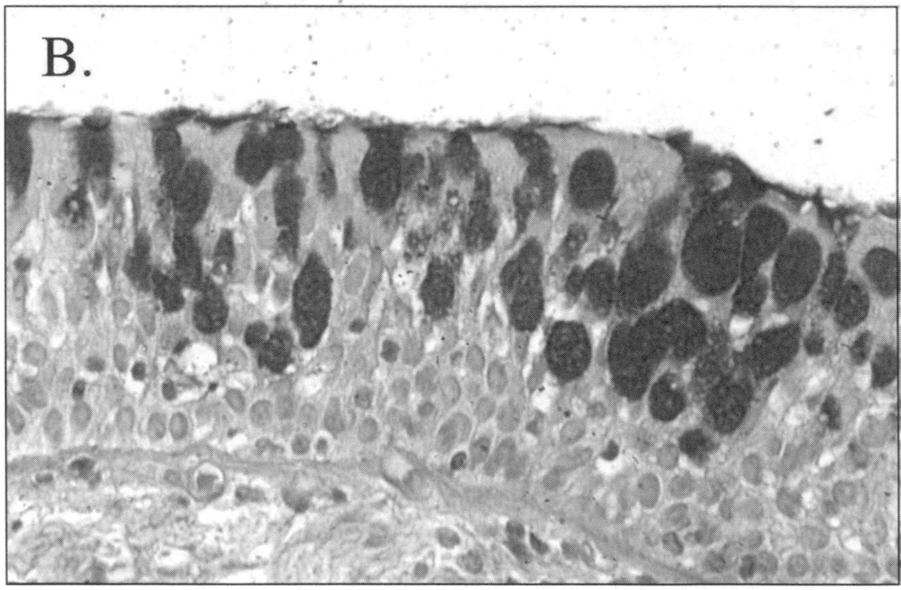

Figure 1.8 Endobronchial biopsy. Panel A: An endobronchial biopsy from a normal volunteer stained with hematoxylin-eosin. Note the pseudostratified columnar cells and the ciliated border. Panel B: an endobronchial biopsy from a "normal" smoker Stained with PAS. Note the hypercellularity, the partial loss of cilia, and the marked increase in mucin containing goblet cells. (See the color version of this figure after page 18.)

is believed to be greater in the presence of chronic obstructive pulmonary disease.

1.4.4.6.3 Endobronchial Brush Biopsy

With this method, an endobronchial brush is inserted through the bronchoscope. The brush, which has a distal tip approximately 1-cm long with bristles approximately 2 to 3 ml in length, is contained within a sheath. The sheath containing the brush is then advanced through the sampling channel of the bronchoscope, after which the brush is advanced out of the sheath. Under direct vision, the proximal airways can then be "brushed," i.e., superficial cells are scarped from the airway epithelium and are retained in the brush. The brush is then retracted into the sheath, and cells within the brush are removed either in a saline solution or by smearing on a slide. A variety of histochemical and molecular methods have been applied to the analysis of samples obtained by endobronchial brushings (96,97). The brush may also be advanced to more-distal regions in the lung in order to sample small airways. Peripheral airways are thus sampled, and the ultrathin bronchoscope permits sampling under direct vision (65). Fluoroscopic guidance may help minimize the risk for pneumothorax.

1.4.4.6.4 Transbronchial Needle Aspiration Biopsy

This method, developed by K.P. Wang, permits sampling of tissues outside the airways (98). A needle equipped with a stylette is inserted through the sampling channel of the bronchoscope. The needle is then used to penetrate the airway wall in the region of interest. The stylette is then withdrawn, and an aspiration sample can be obtained through the needle. This method has been utilized successfully to diagnose both neoplastic and infectious conditions in peribronchial tissues.

1.5 SUMMARY

The lung is frequently the target of toxic exposures. The lung, however, is also readily amenable to investigation. A number of well-established methods permit physiological assessment, imaging, and sampling of the lung. New methods are currently under investigation. Because of the ability to assess the lung, toxicologists have many options for the study of lung disease in humans. A variety of studies ranging from epidemiology to natural history and mechanisms of disease are possible with these methods. The application of these methods is likely to play an increasingly important role in toxicological studies.

REFERENCES

1. West, J.B. and P.D. Wragner: Ventilation, blood flow, and gas exchange, in *Textbook of Respiratory Medicine,* J.F. Murray et al., Ed. 2000, Saunders: Philadelphia. pp. 55–89.
2. Albertine, K.H., M.C. Williams, and D. Hyde: Anatomy of the lungs, in *Textbook of Respiratory Medicine,* J.F. Murray et al., Ed. 2000, Saunders: Philadelphia. pp. 3–33.
3. Gold, W.M.: Pulmonary function testing, in *Textbook of Respiratory Medicine,* J.F. Murray et al., Ed. 2000, Saunders: Philadelphia, 781–882.
4. American Thoracic Society: Lung function testing: selection of reference values and interpretative strategies. *Am Rev Respir Dis* 1991, 144: 1202–1218.
5. American Thoracic Society: Standardization of spirometry, 1994 update. *Am J Respir Crit Care Med* 1995, 152: 1107–1136.
6. European Respiratory Society, official statement: Standardized lung function testing. *Eur Respir J Suppl* 1993, 16: 1–100.
7. Cain, H.: Bronchoprovocation testing. *Clin Chest Med* 2001, 22: 651–659.
8. Pride, N.B.: Pulmonary physiology, in *Asthma and COPD,* P.J. Barnes et al., Ed. 2002, Academic Press: Amsterdam. pp. 43–56.
9. Gong, H., Jr., M.S. McManus, and W.S. Linn: Attenuated response to repeated daily ozone exposures in asthmatic subjects. *Arch Environ Health* 1997, 52: 34–41.
10. O'Donnell, D.E., S.M. Revill, and K.A. Webb: Dynamic hyperinflation and exercise intolerance in chronic obstructive pulmonary disease. *Am J Respir Crit Care Med* 2001, 164: 770–777.
11. O'Donnell, D.E.: Dyspnea in advanced chronic obstructive pulmonary disease. *J Heart Lung Transplant* 1998, 17: 544–554.
12. Chemla, D., V. Castelain, P. Herve, Y. Lecarpentier, and S. Brimioulle: Haemodynamic evaluation of pulmonary hypertension. *Eur Respir J* 2002, 20: 1314–1331.
13. Johnson, L.: Diagnosis of pulmonary hypertension. *Clin Tech Small Anim Pract* 1999, 14: 231–236.
14. Muller, N.L., C.A. Staples, R.R. Miller, and R.T. Abboud: "Density mask": an objective method to quantitate emphysema using computed tomography. *Chest* 1988, 94: 782–787.
15. Coxson, H.O., R.M. Rogers, K.P. Whittall, Y. D'Yachkova, P.D. Pare, F.C. Sciurba, and J.C. Hogg: A quantification of the lung surface area in emphysema using computed tomography. *Am J Respir Crit Care Med* 1999, 159: 851–856.
16. Archer, D.C., C.L. Coblentz, R.A. deKemp, C. Nahmias, and G. Norman: Automated in vivo quantification of emphysema. *Radiology* 1993, 188: 835–858.
17. Nakano, Y., N.L. Muller, G.G. King, A. Niimi, S.E. Kalloger, M. Mishima, and P.D. Pare: Quantitative assessment of airway remodeling using high-resolution CT. *Chest* 2002, 122: 271S–275S.
18. Oros, A.M. and N.J. Shah: Hyperpolarized xenon in NMR and MRI. *Phys Med Biol* 2004, 49: R105–53.
19. Altes, T.A. and M. Salerno: Hyperpolarized gas MR imaging of the lung. *J Thorac Imaging* 2004, 19: 250–258.
20. Mills, G.H., J.M. Wild, B. Eberle, and E.J. Van Beek: Functional magnetic resonance imaging of the lung. *Br J Anaesth* 2003, 91: 16–30.

21. Yablonskiy, D.A., A.L. Sukstanskii, J.C. Leawoods, D.S. Gierada, G.L. Bretthorst, S.S. Lefrak, J.D. Cooper, and M.S. Conradi: Quantitative in vivo assessment of lung microstructure at the alveolar level with hyperpolarized 3He diffusion MRI. *Proc Natl Acad Sci USA* 2002, 99: 3111–3116.
22. Kharitonov, S.A. and P.J. Barnes: Biomarkers of some pulmonary diseases in exhaled breath. *Biomarkers* 2002, 7: 1–32.
23. Kharitonov, S.A. and P.J. Barnes: Exhaled markers of inflammation. *Curr Opin Allergy Clin Immunol* 2001, 1: 217–224.
24. Deykin, A. and S.A. Kharitonov: Nitric oxide, in *Asthma and COPD,* P.B. Barnes et al., Ed. 2002, Academic Press: Amsterdam. pp. 307–314.
25. Dinakar, C.: Exhaled nitric oxide in the clinical management of asthma. *Curr Allergy Asthma Rep* 2004, 4: 454–459.
26. Hogman, M., N. Drca, C. Ehrstedt, and P. Merilainen: Exhaled nitric oxide partitioned into alveolar, lower airways and nasal contributions. *Respir Med* 2000, 94: 985–991.
27. Prignot, J.: Pharmacological approach to smoking cessation. *Eur Respir J* 1989, 2: 550–560.
28. Verhoeff, A.P., H.C. van der Velde, J.S. Boleij, E. Lebret, and B. Brunekreef: Detecting indoor CO exposure by measuring CO in exhaled breath. *Int Arch Occup Environ Health* 1983, 53: 167–173.
29. Hewat, V.N., E.V. Foster, G.D. O'Brien, and G.I. Town: Ambient and exhaled carbon monoxide levels in a high traffic density area in Christchurch. *N Z Med J* 1998, 111: 343–344.
30. Choi, A.M. and J. Alam: Heme oxygenase-1: function, regulation, and implication of a novel stress-inducible protein in oxidant-induced lung injury. *Am J Respir Cell Mol Biol* 1996, 15: 9–19.
31. Olopade, C.O., M. Zakkar, W.I. Swedler, and I. Rubinstein: Exhaled pentane levels in acute asthma. *Chest* 1997, 111: 862–865.
32. Olopade, C.O., J.A. Christan, M. Zakkar, C. Hua, W.I. Swedler, P.A. Scheff, and I. Rubinstein: Exhaled pentane and nitric oxide levels in patients with obstructive sleep apnea. *Chest* 1997, 111: 1500–1504.
33. Paredi, P., S.A. Kharitonov, and P.J. Barnes: Analysis of expired air for oxidation products. *Am J Respir Crit Care Med* 2002, 166: S31–7.
34. Paredi, P., S.A. Kharitonov, D. Leak, S. Ward, D. Cramer, and P.J. Barnes: Exhaled ethane, a marker of lipid peroxidation, is elevated in chronic obstructive pulmonary disease. *Am J Respir Crit Care Med* 2000, 162: 369–73.
35. McCafferty, J.B., T.A. Bradshaw, S. Tate, A.P. Greening, and J.A. Innes: Effects of breathing pattern and inspired air conditions on breath condensate volume, pH, nitrite, and protein concentrations. *Thorax* 2004, 59: 694–698.
36. Hunt, J.: Exhaled breath condensate: an evolving tool for noninvasive evaluation of lung disease. *J Allergy Clin Immunol* 2002, 110: 28–34.
37. Doniec, Z., D. Nowak, W. Tomalak, K. Pisiewicz, and R. Kurzawa: Passive smoking does not increase hydrogen peroxide (H2O2) levels in exhaled breath condensate in 9-year-old healthy children. *Pediatr Pulmonol* 2005, 39: 41–45.
38. Effros, R.M., K.W. Hoagland, M. Bosbous, D. Castillo, B. Foss, M. Dunning, M. Gare, W. Lin, and F. Sun: Dilution of respiratory solutes in exhaled condensates. *Am J Respir Crit Care Med* 2002, 165: 663–669.

39. Effros, R.M., M.B. Dunning, 3rd, J. Biller, and R. Shaker: The promise and perils of exhaled breath condensates. *Am J Physiol Lung Cell Mol Physiol* 2004, 287: L1073–80.
40. Hunt, J.F., K. Fang, R. Malik, A. Snyder, N. Malhotra, T.A. Platts-Mills, and B. Gaston: Endogenous airway acidification. Implications for asthma pathophysiology. *Am J Respir Crit Care Med* 2000, 161: 694–699.
41. Gessner, C., S. Hammerschmidt, H. Kuhn, H.J. Seyfarth, U. Sack, L. Engelmann, J. Schauer, and H. Wirtz: Exhaled breath condensate acidification in acute lung injury. *Respir Med* 2003, 97: 1188–1194.
42. Stockley, R.A. and D. Lomas: Assessment of soluble parameters in sputum. *Agents Actions Suppl* 1990, 30: 145–160.
43. Djukanovic, R., P.J. Sterk, J.V. Fahy, and F.E. Hargreave: Standardised methodology of sputum induction and processing. *Eur Respir J Suppl* 2002, 37: 1s–2s.
44. Efthimiadis, A., A. Spanevello, Q. Hamid, M.M. Kelly, M. Linden, R. Louis, M.M. Pizzichini, E. Pizzichini, C. Ronchi, F. Van Overvel, and R. Djukanovic: Methods of sputum processing for cell counts, immunocytochemistry and in situ hybridisation. *Eur Respir J Suppl* 2002, 37: 19s–23s.
45. Woolhouse, I.S., D.L. Bayley, and R.A. Stockley: Effect of sputum processing with dithiothreitol on the detection of inflammatory mediators in chronic bronchitis and bronchiectasis. *Thorax* 2002, 57: 667–671.
46. Pavord, I.D., P.J. Sterk, F.E. Hargreave, J.C. Kips, M.D. Inman, R. Louis *et al.*: Clinical applications of assessment of airway inflammation using induced sputum. *Eur Respir J Suppl* 2002, 37: 40s–43s.
47. Kips, J.C., M.D. Inman, L. Jayaram, E.H. Bel, K. Parameswaran, M.M. Pizzichini, I.D. Pavord, R. Djukanovic, F.E. Hargreave, and P.J. Sterk: The use of induced sputum in clinical trials. *Eur Respir J Suppl* 2002, 37: 47s–50s.
48. Campo, P., Z.L. Lummus, and D.I. Bernstein: Advances in methods used in evaluation of occupational asthma. *Curr Opin Pulm Med* 2004, 10: 142–146.
49. Beier, J., K.M. Beeh, D. Semmler, and R. Buhl: Sputum levels of reduced glutathione increase 24 hours after allergen challenge in isolated early, but not dual asthmatic responders. *Int Arch Allergy Immunol* 2004, 135: 30–35.
50. Vagaggini, B., M. Taccola, S. Cianchetti, S. Carnevali, M.L. Bartoli, E. Bacci, F.L. Dente, A. Di Franco, D. Giannini, and P.L. Paggiaro: Ozone exposure increases eosinophilic airway response induced by previous allergen challenge. *Am J Respir Crit Care Med* 2002, 166: 1073–1077.
51. Hiltermann, T.J., J. Stolk, P.S. Hiemstra, P.H. Fokkens, P.J. Rombout, J.K. Sont, P.J. Sterk, and J.H. Dijkman: Effect of ozone exposure on maximal airway narrowing in non-asthmatic and asthmatic subjects. *Clin Sci (Lond)* 1995, 89: 619–624.
52. Profita, M., G. Chiappara, F. Mirabella, R. Di Giorgi, L. Chimenti, G. Costanzo, L. Riccobono, V. Bellia, J. Bousquet, and A.M. Vignola: Effect of cilomilast (Ariflo) on TNF-alpha, IL-8, and GM-CSF release by airway cells of patients with COPD. *Thorax* 2003, 58: 573–579.
53. Germonpre, P.R., G.R. Bullock, B.N. Lambrecht, V. Van De Velde, W.H. Luyten, G.F. Joos, and R.A. Pauwels: Presence of substance P and neurokinin 1 receptors in human sputum macrophages and U-937 cells. *Eur Respir J* 1999, 14: 776–782.
54. Miller, J.I., Jr.: Rigid bronchoscopy. *Chest Surg Clin N Am* 1996, 6: 161–167.
55. Wang, K.P. and A.C. Mehta: *Flexible Bronchoscopy*. 1995, Cambridge, U.K.: Blackwell, p. 386.

56. Ono, R.: *Bronchoscopy*. 1994, Tokyo: Nakayama-Shoten. p. 182.
57. Helmers, R.A. and D.R. Sanderson: Rigid bronchoscopy. The forgotten art. *Clin Chest Med* 1995, 16: 393–399.
58. Tanaka, H., G. Yamada, T. Saikai, M. Hashimoto, S. Tanaka, K. Suzuki, M. Fujii, H. Takahashi, and S. Abe: Increased airway vascularity in newly diagnosed asthma using a high-magnification bronchovideoscope. *Am J Respir Crit Care Med* 2003, 168: 1495–1499.
59. Edell, E.S. and D.A. Cortese: Bronchoscopic localization and treatment of occult lung cancer. *Chest* 1989, 96: 919–921.
60. Lam, S., C. MacAulay, J.C. leRiche, and B. Palcic: Detection and localization of early lung cancer by fluorescence bronchoscopy. *Cancer* 2000, 89: 2468–2473.
61. McWilliams, A., J. Mayo, S. MacDonald, J.C. leRiche, B. Palcic, E. Szabo, and S. Lam: Lung cancer screening: a different paradigm. *Am J Respir Crit Care Med* 2003, 168: 1167–1173.
62. Ono, R.: *Ultrasonic Bronchoscopy*. 1997, Tokyo: Nakayama-Shoten. 181.
63. Falcone, F., F. Fois, and D. Grosso: Endobronchial ultrasound. *Respiration* 2003, 70: 179–194.
64. Tanaka, M., O. Kawanami, M. Satoh, K. Yamaguchi, Y. Okada, and F. Yamasawa: Endoscopic observation of peripheral airway lesions. *Chest* 1988, 93: 228–233.
65. Tanaka, M., H. Takizawa, M. Satoh, Y. Okada, F. Yamasawa, and A. Umeda: Assessment of an ultrathin bronchoscope that allows cytodiagnosis of small airways. *Chest* 1994, 106: 1443–1447.
66. Klech, H. and W. Pohl: Technical recommendations and guidelines for bronchoalveolar lavage (BAL). Report of the S.E.P. Task Group on BAL. *Eur Respir J* 1989, 2: 561–585.
67. Gove, R.I., J. Wiggins, and D.E. Stableforth: A study of the use of ultrasonically nebulized lignocaine for local anaesthesia during fiberoptic bronchoscopy. *Br J Dis Chest* 1985, 79: 49–59.
68. Thompson, A.B., G. Huerta, R.A. Robbins, J.H. Sisson, J.R. Spurzem, S. Von Essen et al.: The bronchitis index. A semiquantitative visual scale for the assessment of airways inflammation. *Chest* 1993, 103: 1482–1488.
69. Thompson, A.B., M. Mueller, and A.J. Heires et al.: Aerosolized beclomethasone in chronic bronchitis, improved pulmonary function and diminished airway inflammation. *Am Rev Respir Dis* 1992, 146: 389–395.
70. Rennard, S.I., D. Daughton, J. Fujita, M.B. Oehlerking, J.R. Dobson, M.G. Stahl, R.A. Robbins, and A.B. Thompson: Short-term smoking reduction is associated with reduction in measures of lower respiratory tract inflammation in heavy smokers. *Eur Respir J* 1990, 3: 752–759.
71. John, M., I. Fietze, A.C. Borges, U. Oltmanns, B. Schmidt, and C. Witt: Quantification of the effect of inhaled budesonide on airway inflammation in intermittent asthma by bronchitis index. *J Asthma* 2001, 38: 593–599.
72. Linder, J. and S.I. Rennard, *Atlas of Bronchoalveolar Lavage*. 1988, Chicago: American Society of Clinical Pathology Press. p. 196.
73. Rankin, J.A., P.E. Snyder, E.N. Schachter, and R.A. Matthay: Bronchoalveolar lavage. Its safety in subjects with mild asthma. *Chest* 1984, 85: 723–728.
74. Reynolds, H.Y.: State of the art: bronchoalveolar lavage. *Am Rev Respir Dis* 1987, 135: 250–263.
75. Lam, S., J.C. Leriche, K. Kijek, and K. Phillips: Effect of bronchial lavage volume on cellular and protein recovery. *Chest* 1985, 88: 856–859.

76. Rennard, S.I., M. Ghafouri, A.B. Thompson, J. Linder, W. Vaughan, K. Jones, R.F. Ertl, K. Christensen, A. Prince, M.G. Stahl, and R.A. Robbins: Fractional processing of sequential bronchoalveolar lavage to separate bronchial and alveolar samples. *Am Rev Respir Dis* 1990, 141: 208–217.
77. Davis, G.S., M.S. Giancola, M.C. Costanza, and R.B. Low: Analysis of sequential bronchoalveolar lavage samples from healthy human volunteers. *Am Rev Respir Dis* 1982, 126: 611–616.
78. Klech, H. and C. Hutter: Clinical guidelines and indications for bronchoalveolar lavage (BAL): Report of the European Society of Pneumology Task Group on BAL. *Eur Respir J* 1990, 3: 937–974.
79. Reynolds, H. and Newball, M.M., Analysis of proteins and respiratory cells obtained from human lungs by bronchial lavage. *J Lab Clin Med* 1974, 84: 559–573.
80. Rennard, S., G. Basset, D. Lecossier, K. O'Donnell, P. Martin, and R.G. Crystal: Estimation of volume of epithelial lining fluid recovered by lavage using urea as marker of dilution. *J Appl Physiol* 1986, 60: 532–538.
81. Marcy, T.W., W.W. Merrill, G.P. Naegel, J.A. Rankin, and H.Y. Reynolds: Limitations of the urea method for quantifying epithelial lining fluid in bronchoalveolar lavage. *Am Rev Respir Dis* 1987, 135: 1276–1280.
82. Kelly, C.A., J.D. Fenwick, P.A. Corris, A. Fleetwood, D.J. Hendrick, and E.H. Walters: Fluid dynamics during bronchoalveolar lavage. *Am Rev Respir Dis* 1988, 138: 81–84.
83. Borok, Z., A. Gillissen, R. Buhl, R.F. Hoyt, R.C. Hubbard, T. Ozaki, S.I. Rennard, and R.G. Crystal: Augmentation of functional prostaglandin E levels on the respiratory epithelial surface by aerosol administration of prostaglandin E. *Am Rev Respir Dis* 1991, 144: 1080–1084.
84. Liu, M.C., E.R. Bleecker, L.W. Lichtenstein, A. Kagey-Sobotka, Y. Niv, T.L. McLemore, S. Permutt, D. Proud, and W.C. Hubbard: Evidence for elevated levels of histamine, prostaglandin D2, and other bronchoconstricting prostaglandins in the airways of subjects with mild asthma. *Am Rev Respir Dis* 1990, 142: 126–132.
85. Rennard, S.I., G.W. Hunninghake, P. Bitterman, and R.G. Crystal: Production of fibronectin by the human alveolar macrophage: Mechanism for the recruitment of fibroblasts to sites of tissue injury in interstitial lung diseases. *Proc Natl Acad Sci USA* 1981, 78: 7147–7151.
86. Kavuru, M.S., R.A. Dweik, and M.J. Thomassen: Role of bronchoscopy in asthma research. *Clin Chest Med* 1999, 20: 153–189.
87. Salvi, S.S., C. Nordenhall, A. Blomberg, B. Rudell, J. Pourazar, F.J. Kelly, S. Wilson, T. Sandstrom, S.T. Holgate, and A.J. Frew: Acute exposure to diesel exhaust increases IL-8 and GRO-alpha production in healthy human airways. *Am J Respir Crit Care Med* 2000, 161: 550–557.
88. Balmes, J.R., R.M. Aris, L.L. Chen, C. Scannell, I.B. Tager, W. Finkbeiner, D. Christian, T. Kelly, P.Q. Hearne, R. Ferrando, and B. Welch: Effects of ozone on normal and potentially sensitive human subjects. Part I: Airway inflammation and responsiveness to ozone in normal and asthmatic subjects. *Res Rep Health Eff Inst* 1997, 1–37, discussion 81–99.
89. Frampton, M.W., P.E. Morrow, A. Torres, K.Z. Voter, J.C. Whitin, C. Cox, D.M. Speers, Y. Tsai, and M.J. Utell: Effects of ozone on normal and potentially sensitive human subjects. Part II: Airway inflammation and responsiveness to ozone in nonsmokers and smokers. *Res Rep Health Eff Inst* 1997, 39–72, discussion 81–99.

90. Frampton, M.W., J.R. Balmes, C. Cox, P.M. Krein, D.M. Speers, Y. Tsai, and M.J. Utell: Effects of ozone on normal and potentially sensitive human subjects. Part III: Mediators of inflammation in bronchoalveolar lavage fluid from non-smokers, smokers, and asthmatic subjects exposed to ozone: a collaborative study. *Res Rep Health Eff Inst* 1997, 73–9, discussion 81–99.
91. Frew, A.J., M.P. Carroll, C. Gratziou, and N. Krug: Endobronchial allergen challenge. *Eur Respir J Suppl* 1998, 26: 33S–35S.
92. Zavala, D.C.: Bronchoscopy, lung biopsy, and other procedures. In: Textbook of Respiratory Medicines 1988. Murray, J.F. and Nadel, S.A., Eds., WB Saunders, Philadelphia, pp. 562–595.
93. Hernandez Blasco, L., I.M. Sanchez Hernandez, V. Villena Garrido, E. de Miguel Poch, M. Nunez Delgado, and J. Alfaro Abreu: Safety of the transbronchial biopsy in outpatients. *Chest* 1991, 99: 562–565.
94. Clark, R.A., P.B. Gray, R.H. Townshend, and P. Howard: Transbronchial lung biopsy: a review of 85 cases. *Thorax* 1977, 32: 546–549.
95. Kraft, M., R. Djukanovic, J. Torvik, L. Cunningham, J. Henson, S. Wilson, S.T. Holgate, D. Hyde, and R. Martin: Evaluation of airway inflammation by endobronchial and transbronchial biopsy in nocturnal and nonnocturnal asthma. *Chest* 1995, 107: 162S.
96. Trapnell, B.C., C.S. Chu, P.K. Paakko, T.C. Banks, K. Yoshimura, V.J. Ferrans, M.S. Chernick, and R.G. Crystal: Expression of the cystic fibrosis transmembrane conductance regulator gene in the respiratory tract of normal individuals and individuals with cystic fibrosis. *Proc Natl Acad Sci USA* 1991, 88: 6565–6569.
97. Hackett, N.R., A. Heguy, B.G. Harvey, T.P. O'Connor, K. Luettich, D.B. Flieder, R. Kaplan, and R.G. Crystal: Variability of antioxidant-related gene expression in the airway epithelium of cigarette smokers. *Am J Respir Cell Mol Biol* 2003, 29: 331–343.
98. Wang, K.P., B.R. Marsh, W.R. Summer, P.B. Terry, Y.S. Erozn, and R.R. Baker: Transbronchial aspiration for diagnosis of lung cancer. *Chest* 1981, 80: 45.
99. Samee, S., T. Altes, P. Powers, E.E. de Lange, J. Knight-Scott, G. Rakes, J.P. Mugler, 3rd, J.M. Ciambotti, B.A. Alford, J.R. Brookeman, and T.A. Platts-Mills: Imaging the lungs in asthmatic patients by using hyperpolarized helium-3 magnetic resonance: assessment of response to methacholine and exercise challenge. *J Allergy Clin Immunol* 2003, 111: 1205–1211.
100. Andreoli, R., P. Manini, M. Corradi, A. Mutti, and W.M. Niessen: Determination of patterns of biologically relevant aldehydes in exhaled breath condensate of healthy subjects by liquid chromatography/atmospheric chemical ionization tandem mass spectrometry. *Rapid Commn Mass Spectrom* 2003, 17: 637–645.
101. Nowak, D., S. Kalucka, P. Bialasiewicz, and M. Krol: Exhalation of H2O2 and thiobarbituric acid reactive substances (TBARs) by healthy subjects. *Free Radic Biol Med* 2001, 30: 178–186.
102. Corradi, M., I. Rubinstein, R. Andreoli, P. Manini, A. Caglieri, D. Poli, R. Alinovi, and A. Mutti: Aldehydes in exhaled breath condensate of patients with chronic obstructive pulmonary disease. *Am J Respir Crit Care Med* 2003, 167: 1380–1386.
103. Garey, K.W., M.M. Neuhauser, R.A. Robbins, L.H. Danziger, and I. Rubinstein: Markers of inflammation in exhaled breath condensate of young healthy smokers. *Chest* 2004, 125: 22–26.

104. Montuschi, P., J.V. Collins, G. Ciabattoni, N. Lazzeri, M. Corradi, S.A. Kharitonov, and P.J. Barnes: Exhaled 8-isoprostane as an in vivo biomarker of lung oxidative stress in patients with COPD and healthy smokers. *Am J Respir Crit Care Med* 2000, 162: 1175–1177.
105. Carpagnano, G.E., S.A. Kharitonov, M.P. Foschino-Barbaro, O. Resta, E. Gramiccioni, and P.J. Barnes: Increased inflammatory markers in the exhaled breath condensate of cigarette smokers. *Eur Respir J* 2003, 21: 589–593.
106. Balint, B., L.E. Donnelly, T. Hanazawa, S.A. Kharitonov, and P.J. Barnes: Increased nitric oxide metabolites in exhaled breath condensate after exposure to tobacco smoke. *Thorax* 2001, 56: 456–461.
107. Corradi, M., A. Pesci, R. Casana, R. Alinovi, M. Goldoni, M.V. Vettori, and A. Cuomo: Nitrate in exhaled breath condensate of patients with different airway diseases. *Nitric Oxide* 2003, 8: 26–30.

2

HUMAN CLINICAL STUDIES OF AIRBORNE POLLUTANTS

Mark W. Frampton, Anthony P. Pietropaoli, Paul E. Morrow, and Mark J. Utell

CONTENTS

2.1 THE NEED FOR HUMAN STUDIES

The assessment of risk of acute or chronic inhalation of low-level environmental air pollutants is complex. Typically, the database for risk assessment arises from three separate investigative approaches: epidemiology, animal toxicology, and human inhalation studies. Each possesses advantages but also carries significant limitations. For example, the epidemiological approach examines exposures in the "real world" but struggles with the realities of conducting research in the community, where cigarette smoking, socioeconomic status, occupational factors, and inadequate exposure characterization are important confounders. Outcomes are most frequently evaluated by questionnaire, and sophisticated measures of physiological responses are often not practical. In contrast, inhalation studies in animals allow remarkable precision in quantifying exposure duration and concentration, measurement of a wide variety of physiologic, biochemical, and histological endpoints, and examination of extremes of the exposure–response relationship. Often, however, interpretation of these studies is constrained by the difficulty in extrapolating findings from animals to humans, especially at unrealistic exposure levels.

Carefully controlled, quantitative studies of exposed humans offer a third and complementary approach. Human clinical studies utilize laboratory atmospheric conditions, which can be designed to simulate specific ambient polluted atmospheres. The investigators then document the symptoms and health-related physiological effects that result from breathing these atmospheres. Advantage is taken of the highly controlled environment to identify responses to individual pollutants and to characterize exposure–response relationships. In addition, the controlled environment provides the opportunity to examine interactions among pollutants, or between pollutants and other environmental variables such as humidity, temperature, or exercise. Insofar as individuals with respiratory or cardiovascular diseases can participate in these exposure protocols, potentially susceptible populations can be studied. However, this approach has limitations as well. For practical and ethical reasons, studies must be limited to small groups that are presumably representative of larger populations, to short durations of exposure, and to pollutant concentrations that are expected to produce only mild and transient responses. End point assessment is largely restricted to minimally invasive measures, such as assessment of symptoms, pulmonary function, and behavioral and cardiovascular responses; more recently, a growing number of markers of pulmonary effects collected from exhaled breath or from lavage fluid from the nose or lung, and systemic markers measured in blood have been incorporated into such clinical studies. Finally, the usefulness of the acute, transient responses seen in clinical studies in predicting the health effects of chronic or repeated exposure is uncertain.

Although recent epidemiological studies have demonstrated adverse effects of exposure to ambient particulate matter (PM) at extremely low concentrations, the specific size or chemical composition that can cause these effects remains controversial. The U.S. Congress and the Environmental Protection Agency (EPA) have charged the National Research Council (NRC) with developing a framework to guide research in the health effects of PM. The first report of the NRC Committee on Research Priorities for Airborne Particulate Matter was published in 1998[1] and established 10 research priorities. Human clinical studies were identified as a key approach in addressing many of these priorities, including the effort to determine the mechanisms involved. Particles of interest for clinical studies include concentrated ambient particles, ultrafine particles (UFP; diameter $< 0.1\ \mu m$), acid aerosols, diesel exhaust, and metals.[2,3]

In the previous edition of this book,[4] we reviewed protocol design, methods for measuring health effects, and susceptible populations in human inhalation studies. Understanding the factors deciding the retention and the fate of inhaled pollutants is essential for understanding the health effects; this chapter reviews selected areas of dosimetry in human clinical studies. Ethical and safety issues are also reviewed. The significant advances in determining airway effects in clinical studies, including the use of exhaled-breath condensate analyses, as well as refinements in the more invasive technique of fiberoptic bronchoscopy, are updated. Recent epidemiological studies have established that exposure to low ambient concentrations of PM can have adverse effects on the cardiovascular system.[5] This reminds us that the lung is a portal to the rest of the body, and that inhalation effects can extend well beyond the lung. Clinical studies are becoming increasingly important in assessing these extrapulmonary effects, and their role will also be examined in this chapter. Interestingly, the definition of susceptible populations has broadened from those with respiratory disease to include people with cardiovascular disease or with diseases which predispose to cardiovascular disease such as diabetes. We will introduce the issue of clinical exposure studies in assessing responses in these susceptible populations.

2.2 STUDY DESIGN OPTIONS AND ISSUES

2.2.1 Recruitment and Characterization of Subjects

An important feature of the controlled clinical study is the opportunity to examine responses in both healthy volunteers and individuals with underlying pulmonary or cardiovascular diseases. Subjects are typically classified by age, gender, race, lung function, and severity of the underlying condition, if any. Healthy volunteers are often characterized by the absence of pulmonary or cardiovascular disease as judged by history, normal spirometry, and often in older subjects, a normal electrocardiogram (ECG). Within the rubric of healthy volunteers, adolescent, elderly normal, and smoker subgroups have been studied in various protocols. Healthy volunteers in many human exposure studies are able to perform vigorous exercise for extended periods and can usually tolerate the more invasive investigative techniques such as fiberoptic bronchoscopy. For these reasons, caution must be exercised in extrapolating findings from these young, fit subjects to the general population, especially those with underlying cardiac or pulmonary diseases.

2.2.2 Role of Exercise

Exercise performed either on a treadmill or bicycle ergometer is often a component of human exposure studies. Exercise enhances the pollutant dose both by increasing ventilation and by causing a switch from nasal to oral breathing. It is clear that minute ventilation during exposure is an important factor in determining the magnitude of change in selected physiological measures. The best studied example of this is with ozone. Exposures to ozone at rest have little effect on lung function at concentrations as high as 0.40 ppm. However, incorporation of heavy, prolonged exercise has elicited significant lung function decrements[6] and airway injury[7] at concentrations as low as 0.08 ppm. These findings contributed to the U.S. EPA's promulgation, in 1997, of a more restrictive air quality standard for ozone: 0.08 ppm for an 8-h average. Minute ventilation (V_E), concentration (C), and duration of exposure (T) are the key determinants of pulmonary function responses to short-term exposures to ozone, as illustrated by the following equation:

$$\text{Response} \sim V_E \cdot C \cdot T$$

Pulmonary function responses are known to decrease with age. McDonnell et al.[8] used data from 485 healthy subjects undergoing ozone exposure to derive a predictive model for decrements in the forced experatory volume in 1 sec (FEV_1). However, these authors found that much of the between-subject variability in responsiveness could not be explained by exposure parameters or subject age, suggesting the involvement of other, possibly genetic, determinants.

For some pollutants, oral breathing increases the concentration reaching the lower respiratory tract due to bypassing of the filtering effect of the nasal passages. Furthermore, exercise may alter the deposition efficiency of a pollutant. For example, in studies of exposure to ultrafine carbon particles (count median diameter 0.026 μm) in our laboratory, exercise increased the already high deposition efficiency of these particles.[9] In healthy subjects, the number of particles retained in the lung increased more than fourfold during exercise, as a consequence of both increased minute ventilation and deposition efficiency (see section titled "Dosimetry in Clinical Studies").

Exercise elicits other physiological changes that may be important in altering the response to pollutants. For example, vigorous exercise is accompanied by increases in circulating cytokines and enzymes associated with muscle injury and systemic inflammation.[10] There are also well-recognized changes in the hypothalamic–pituitary–adrenal axis which may influence the airway inflammatory effects of pollutant exposure. Plasma concentrations of interleukin-6 increase after vigorous exercise, initiating a systemic acute phase response with increases in plasma C-reactive protein.[10] The systemic acute phase response has been implicated as one of the mechanisms by which ambient PM exposure induces cardiovascular effects[11]; thus, exercise may enhance this effect of pollutants. The total white blood cell count and the number of blood neutrophils increase after exercise, largely as a consequence of reduced transit time of leukocytes through the pulmonary capillary bed. The surface expression of the CD11b/CD18 integrin complex on venous blood granulocytes increases after vigorous exercise, indicating that the cells that are marginated in the pulmonary capillary blood express higher levels of these adhesion molecules than circulating granulocytes.[12] Our own studies[13] suggest that inhalation of ultrafine carbon particles in humans may blunt or reverse exercise-induced increases in blood leukocyte expression of adhesion molecules, suggesting that these exposures modify leukocyte demargination during exercise.

2.2.3 The Double-Blind Crossover Design

For controlled clinical studies, the double-blind protocol that uses a crossover design is optimal. The control exposure generally consists of clean or purified air, although other control inhalations have been used, depending on the purpose of the study. For example, a sodium chloride aerosol of similar size distribution has been used as a control for exposures to acid aerosols.[14] The investigator, the subject, and the technicians performing outcome measures should be unaware of whether the pollutant or the control exposure is occurring. Such a strategy minimizes potential investigative bias and avoids relaying of cues to volunteers regarding anticipated responses. This is especially important when studying airway responses in subjects with asthma. A significant number of asthmatics respond with bronchoconstriction to psychological stimuli.[15] In the crossover design, where each subject receives both a pollutant and a control exposure, avoiding carryover effects from one exposure to the next is critical. Thus, a sufficient interval is needed between exposures for any physiological changes to return to baseline, whether they be the effects of the pollutant exposure, exercise, or the measurement techniques themselves. The presence of carryover effects may go unnoticed and could serve to obscure pollutant exposure effects. Statistical methods are available to test for carryover effects in clinical studies, but they require a balanced randomization scheme, with equal and adequate numbers of subjects in each exposure sequence. These analytical and statistical issues in study design have been reviewed.[16]

2.3 ETHICAL AND SAFETY ISSUES

Guidelines and principles have been developed for experimentation involving human subjects, and these have been put forward in the Belmont Report.[17] The basic ethical principles involved, and their applications, are summarized in Table 2.1. All studies involving human subjects should be reviewed by an independent body that adheres to the principles espoused

Table 2.1 The Belmont Report

Principles	*Applications*
Respect for persons	**Informed consent**
Treat people as autonomous agents	Information
Protect people with diminished autonomy	Comprehension
	Voluntariness
Beneficence	**Assessment of risks and benefits**
Maximize benefits	Justification of study
Minimize harm	Minimization risk
Justice	**Selection of subjects**
Equitable distribution of burdens and benefits	Individual justice
	Social justice
	Vulnerable subjects

in the Belmont Report. This review board has the important role of assuring the integrity of the consent process and ensuring that the potential benefits of the study outweigh the risks to the subjects.

Human clinical studies of air pollutants share design considerations with clinical trials of drugs or treatments but differ in one important aspect: with clinical studies of air pollution or toxicants, there is presumably no likelihood of personal benefit, whereas drug or treatment studies often carry the possibility of improving the treatment of the subjects' illness or condition. The hope with clinical exposure studies is that the findings will prove beneficial to others by improving risk assessment and future harm reduction from pollutant exposure. In conducting such studies, there must be no significant risk of irreversible organ dysfunction, and there must be an absence of severe pain or discomfort, due either to the exposure atmosphere or the procedures used to measure effects. The challenge in human clinical studies is to develop study designs and identify endpoints of exposure that provide insights into the potential adverse effects of exposure, even while protecting subject safety and health.

Adhering to these principles places obvious limitations on the questions that can be answered with human studies. Both safety and practicability often preclude studies of those most vulnerable to air pollution effects, such as infants, young children, pregnant women, and people with severe asthma, chronic obstructive pulmonary disease (COPD), or coronary artery disease. For example, people with severe COPD may be functioning close to the limits of gas ventilation or oxygenation imposed by their disease; even small reductions in lung function or oxygenation could initiate an exacerbation of their airway compromise, resulting in respiratory distress or even respiratory failure. Indeed, this is the hypothesis that has been put forward to explain the increases in mortality and morbidity seen with exposure to ambient particulate air pollution.

2.4 DOSIMETRY IN CLINICAL STUDIES

Clinical investigations into pollutant- or xenobiotic-induced biological responses following inhalation exposures are greatly enhanced by a dosimetric approach. By definition, this provides a quantitative basis for expressing dose–effect and dose–response data. These fundamental relationships are invaluable for examining potency, making species comparisons and intraspecies extrapolations, comparing normal and susceptible subjects, elucidating underlying mechanisms, and developing exposure guidelines.

Unfortunately, the use of the term "dose" has led to confusion, as it is often applied to a variety of parameters, ranging from a simple description of an administered concentration to an estimate of the time–concentration integral of a specific reactant at a target or receptor site. The meaning of the term dose, therefore, depends, *inter alia*, upon the biological level examined and the ability to quantify the material at that particular level.

In most clinical studies, it is important to establish, at a minimum, the administered dose. In inhalation studies, this can merely be a description of the inspired concentration or, more often, the concentration–time product. This simple dose description can be achieved by applying a quantitative aerosol sampling device (e.g., absolute filter) or an appropriate gas or vapor detector (e.g., infrared or oxidant gas analyzer) that quantitatively samples near, or reflects the airborne concentration accurately at, the breathing zone of the exposed subject.

Also, with the use of a particle-sizing device, such as a cascade impactor, an aerosol exposure can be better described than by mass concentration alone. Obtaining particle size data not only provides an additional quality control but permits the aerosol exposure to be interpreted and compared using any of several particle size–deposition models that have been published based on both theoretical[18,19] and experimental data.[20–22]

The foregoing methods for administered-dose estimations can be further enhanced by measuring the actual amount of material breathed during an exposure; in other words, the first type of dose estimation described is augmented by including a measurement of ventilation. Such a respiratory measurement is often achieved during an aerosol or gas exposure in a chamber by the use of an inductometric device worn on the thorax, such as the Respitrace (Noninvasive Systems, Miami Beach, FL), which provides unencumbered measurements of tidal volume and respiratory frequency and, hence, the minute ventilation.

In many controlled human investigations of inhaled materials, the subjects are connected to a face mask or mouthpiece unit and exposed individually. Under these conditions, it is possible, using respiratory valves, to distinguish between the inspired air concentration and the expired air concentration in a continuous manner. When coupled with measurements of respired volumes or airflows, this can yield a quantitative estimate of the amount of inhaled material that is deposited within the respiratory system over the exposure duration. Such an exposure ideally utilizes real-time analyses of the airborne concentrations of the gas, vapor, or particles in the inspired or expired air. A pneumotachograph is often incorporated in these systems for measuring the expired airflow after the gaseous component has been analyzed. In aerosol exposures, pneumotachography has been used especially when the aerosol component has first been removed (e.g., by electrostatic precipitation or filtration). In any case, electronic integration of pneumotachographic signals, which are based on airflow, is a well-established technology for obtaining volumetric ventilation data, e.g., Utell et al., 1998.[23]

All of these experimental methods have limitations, and their advantages and disadvantages vary. There are circumstances where the type of material to be studied can dictate the approach used; for example, when studying exposures to a radioactive material or a potent pharmacon it is desirable that a highly economic, well-contained system be used, i.e., with either a mouthpiece or face mask. There will be the concomitant problem of determining the incidental gas or aerosol losses in tubing and valves. Also, there are important considerations that must be given to pressure and resistance changes within the exposure system that are produced due to respiratory excursions, especially if exercise is included in the protocol. Finally, the type and duration of the inhalation exposure is often affected by concerns for the comfort and safety of the subject. In general, a large chamber, with the subject's being free to pursue normal activities, provides the most comfortable exposure conditions, even to the extent of permitting control of ambient temperature and relative humidity.

Dosimetry in its classical sense depends upon determining the amount of the test material that arrives at the target organ or receptor and, additionally, its retention time at the site(s) of interest. The experimental approaches that have just been described all fall short of accomplishing this final step, which depends on the determination of the clearance (removal) of the material from the target site by translocation, excretion, and metabolic processes. Unlike animal studies that usually depend upon invasive procedures or even animal sacrifice to obtain retention data, human studies can rarely quantify retention. The closest, noninvasive experimental approach applicable to humans is with the use of radioactive materials, allowing external counting of the subject during the postexposure period to provide a good estimate of regional or whole body retention times. In most clinical studies, the use of radioisotopes is not deemed ideal and often is not even possible; consequently, the most commonly used approach is to measure, by the least invasive method(s) possible, some index of retention. For example, with a gas or vapor exposure, postexposure measurements of expired air can give some insight into the material's clearance rate, especially from the respiratory system. Urine chemistry often can provide a measurement renal excretion of the test material or its metabolites and thereby give another index of clearance and metabolic rates. Blood levels usually provide important kinetic data on the body burden of the material and its metabolites and their rates of elimination. Pharmacokineticists regularly associate the total absorbed dose of a pharmacon by an area-under-the-curve (AUC) method, whereby the blood level over time is integrated usually by a trapezoidal approximation.[24]

These indices of retention are also very useful for developing pharmacokinetic models, which can be a very powerful tool in the comparisons of exposures to different materials in the same subject or of the same material among different subjects. Of course, the measurement and application of these kinds of indices will vary according to the material under investigation and existing knowledge about its pharmacokinetic behavior from animal experimentation.

There are additional complexities to inhalation dosimetry, such as variations in lung size; predisposing factors (e.g., age and disease) and consideration of regional doses, such as nasopharyngeal or tracheobronchial. Dosimetry is influenced by a host of physicochemical factors related to gaseous or particulate material, e.g., specific surface area, electrical charge, hygroscopicity, surface coatings, water and lipid solubility, and many kinds of transformations and interactions occurring *in vivo*, e.g., hydrolytic reactions, protein binding, and colloid formation. Consequently, there is the especially important matter of deciding what is the appropriate dosimetric. For example, with aerosols, there are compelling reasons why the dose expressed in terms of mass is not as relevant as that by particle number or surface area.[25] These diverse factors must always be considered, to a greater or lesser degree, in any inhalation exposure study, but they are beyond the scope of the generalized consideration of the subject in this chapter.

It is useful to review some actual experimental studies in which many of the aspects of inhalation dosimetry are illustrated. Two recent examples of human exposure studies undertaken in our laboratories will serve this purpose: a study of ultrafine carbon aerosol, a surrogate for environmental carbonaceous particulate pollution, utilizing a mouthpiece exposure system and normal subjects,[9] and a silicone vapor exposure study using both oral and nasal inhalation exposures of normal subjects.[23] Both studies used a double-blind crossover design and quantitative measurements of intake and deposition, both during rest and intermittent exercise.

Daigle et al.[9] exposed 12 subjects to both 10 and 25 μg/m^3 graphitic carbon (mean count median diameter 26 nm, with a geometric standard deviation of 1.6) and determined the deposition fraction (DF) for both particle number and mass by analyzing the change in the particulate number concentrations and particle sizes in the inspired and expired air. The DF measurement required condensation particle counters and particle size mobility analyzers. The inspired aerosol mass concentration was measured independently by a microbalance. Because the particle sizes of the inspired and expired aerosols were measured in 8 particle size ranges, viz., from 8.7 to 64.9 nm, it included more than 98% of the particles, a particle size vs. DF curve was constructed using the DF values, which ranged from 0.80 for 8.7 nm particles to 0.55 for 64.9 nm particles. The DFs were then compared to theoretical deposition curves for the same size range. The experimental data generally agreed with the predicted data under conditions of rest, but the DF was found to be significantly greater than predicted under conditions of exercise. No significant DF differences were seen between the sexes or between exposures of 10 and 25 μg/m^3.

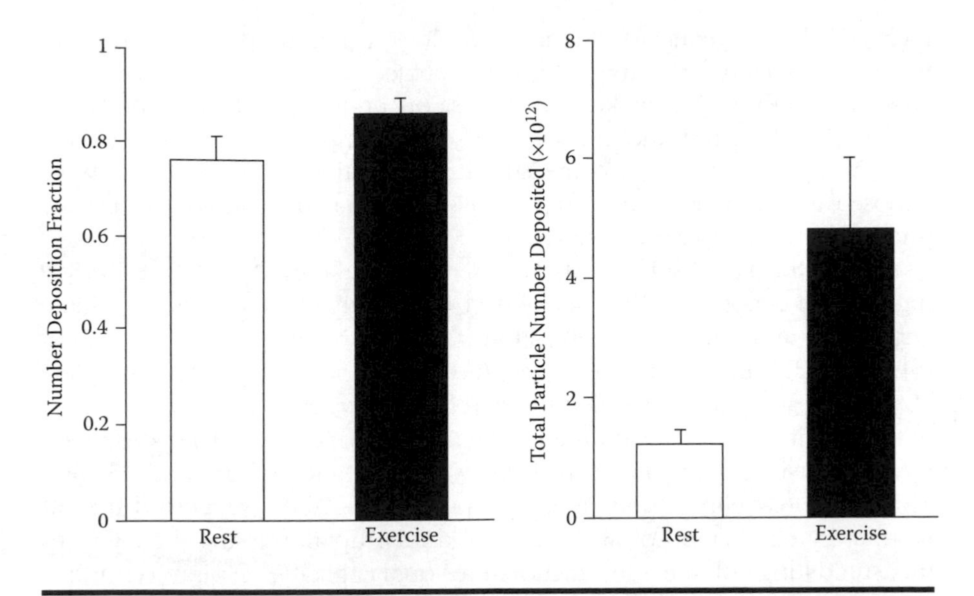

Figure 2.1 Particle number deposition fraction (DF; left panel) and total particle deposition over 1 h of exposure (right panel) at rest and exercise in 16 subjects with asthma.

A similar study was conducted in subjects with asthma,[26] allowing a comparison between healthy and asthmatic subjects. The total DF for healthy subjects was 0.66 ± 0.11 at rest and 0.83 ± 0.04 during exercise. Asthmatic subjects showed higher total DFs, viz., 0.76 ± 0.05 and 0.86 ± 0.04, respectively. Moreover, as shown in Figure 2.1, the combination of increased deposition and increased particle intake resulted in a marked increase in total particle deposition during exercise, compared with resting exposure, in subjects with asthma. Postexposure experimental measurements included symptoms, pulmonary function, blood markers of inflammation and coagulation, sputum inflammatory cells, and airway NO production. Inasmuch as the study was also designed to examine several possible cardiovascular effects, continuous 12-lead electrocardiography was undertaken to assess heart rate variability, repolarization, and arrhythmias in each subject. Thus, this approach allowed individual inhaled doses of the fully characterized UFP to be computed and correlated to health effects in healthy and susceptible groups.

A nasal and oral (mouthpiece) exposure study using a siloxane vapor (octamethylcyclotetrasiloxane, a.k.a D_4), a relatively inert, highly lipophilic silicone liquid, was reported by Utell et al.[23] in healthy subjects with a double-blind crossover design using an air control. At an average airborne D_4 concentration of 122 µg/l (10 ppm), inspiratory and expiratory concentrations were measured in real time. Expired air and plasma levels of D_4 were measured before (control), during, and after 1-h mouthpiece exposures, as were the expired air volumes. Twelve subjects were used in this D_4:air crossover study of D_4 intake and deposition under steady-state conditions during periods of rest and exercise; subsequently, postexposure D_4 elimination in expired air was also measured. An additional eight subjects were exposed to D_4 in order to compare nasal vs. oral deposition and uptake under the same exposure conditions.

Results for the 12 subjects showed a mean intake of D_4 of 137 ± 25 (SD) mg; a mean deposition efficiency of $0.74/(1 + 1.45V_E)$ where V_E is the minute ventilation; and a mean D_4 deposition fraction (inhaled-exhaled/inhaled) of 0.12. This DF indicated a 16.4 mg mean D_4 uptake assuming complete absorption of the deposited D_4 (adjusted for the postexposure D_4 elimination in the expired air). The expired D_4 elimination rate could be described by a biphasic exponential consistent with half-times of 0.6 and 13 min. This nonlinear elimination rate, extrapolated to 24 h, suggested a total respiratory elimination of about 24% of the D_4 uptake. Using the subject's measured lung volume and functional residual capacity, respiratory surface area was estimated and a mean D_4 mass transfer coefficient was calculated, viz., 5.7×10^{-5} cm/sec for transfer from respired air to blood.

In the nasal vs. oral exposure comparison, the mean D_4 deposition was also 0.12 for both exposures, but the estimated D_4 uptake was about 60% less for the nasal exposure, presumably reflecting a greater upper airway deposition which was not as completely absorbed.

An extension of the foregoing vapor exposure study in humans was conducted in an adjunctive study of six volunteers using a ^{14}C radiolabeled D_4.[27] This study, greatly enhancing the dosimetric approach, provided quantitative transport, retention, and metabolite data, and made it feasible for human and rat studies[28] to be compared and integrated into a unique physiologically based pharmacokinetic (PBPK) model for both species by Reddy et al.[29] The human inhalation PBPK model that was developed for D4 incorporated all available pharmacokinetic data for D4 deposition in humans following inhalation exposures.

2.5 OUTCOME MEASURES—THE LUNG AND BEYOND

Clinical studies of air pollution effects have traditionally focused on the lung. It is logical that inhalation effects should be largely limited to the respiratory tract; symptoms and changes in lung function can be measured reliably and safely using well-established methodologies. The pulmonary function effects of criteria air pollutants such as ozone, sulfur dioxide, and nitrogen dioxide (NO_2) have been studied extensively. Pulmonary function responses have been important determinants of the ambient-air-quality standards established by the U.S. EPA. However, studies using techniques for sampling the cells of the lower airways have shown that pollutant inhalation can cause airway inflammation. Furthermore, epidemiological studies over the past 10 yr have taught us that exposure to ambient PM, and perhaps other pollutants, increases cardiovascular mortality and morbidity. This section will review the traditional outcome measures of symptoms and pulmonary function and then review methods for measuring other pulmonary and cardiovascular effects.

2.5.1 Symptom Questionnaires

Questionnaires have been used to determine the presence of chronic respiratory conditions and assess short-term symptomatic responses in both clinical and epidemiological studies. Standardized respiratory questionnaires directed at chronic respiratory symptoms and diseases were first developed in the 1950s by British investigators for the purpose of investigating chronic bronchitis and chronic airflow obstruction.[30] The widely used questionnaire developed by the American Thoracic Society evolved from the earlier versions prepared by the British Medical Research Council[31] and is currently being revised and updated. Most investigators implementing either an epidemiological study or a clinical study have used the American Thoracic Society questionnaire or another standardized instrument for the purpose of characterizing the symptoms and illness status of participants. Although standardized instruments have long been available for chronic respiratory symptoms, similar standardized and validated questionnaires directed at acute symptoms have not been published, and investigators have typically developed their own instruments for individual studies.

Questionnaires utilized in clinical studies survey symptoms during exposure, immediately after exposure, and occasionally, 24 h or longer after exposure. Figure 2.2 provides an example of a questionnaire used to detect respiratory symptoms in response to irritant gases and particles. Typically, symptoms are identified as lower respiratory (e.g., cough, sputum production, wheeze, dyspnea, and chest pain), upper respiratory (e.g., throat irritation and nasal congestion) and nonrespiratory (e.g., headache, fatigue, and eye irritation). The symptoms are then scored by severity with higher scores representing greater severity (e.g., 0 = not present, 5 = incapacitating). Data can be expressed and scored on a continuous scale by having subjects indicate the symptom level on a continuous line.

In expressing symptoms as scores, it should be emphasized that the questionnaire responses need not necessarily correlate with the quantitative measurements obtained by physiological assessment. The order of scores

Symptom Questionnaire rev 10/04

Code:

Date: Time:

Place a mark on the line showing how you feel right now.

1 2 3 4 5 6 7 8 9 10

(a) Cough

(b) Sputum

(c) Shortness of Breath

(d) Chest Pains on Deep Inspiration

(e) Throat Irritation

(f) Nasal Congestion or Discharge

(g) Headache

(h) Fatigue (beyond that attributable to exercise)

(i) Nausea

(j) Wheeze

(k) Chest Tightness

(l) Eye Irritation

(m) Anxiety

(n) Fast Heart Beat or Pounding Heart

(o) Irregular Heart Beat, Skipped Beats

Could you smell or taste anything unusual about the air you were breathing ? Yes No

Do you think you were exposed to the pollutant today ? Yes No

General Comments:

Figure 2.2 Symptom questionnaire for human clinical studies.

is meaningful for any one symptom, with a higher score representing greater severity than a lower score; however, a score of 4 does not necessarily imply twice the severity of a score of 2. Summation of the individual symptom scores provides an overall numerical index of symptom severity, although it must be recognized that ambiguity may be introduced by adding scores from different symptoms.

Symptoms may or may not correlate with other outcome measures. Nonsmoking subjects experienced variable degrees of mild cough during and following a 4-h exposure to 0.22 ppm of ozone,[32] but the correlation with decrements in FEV_1 was weak ($r = .46$, $p < .001$). Other symptoms were even less predictive of lung function changes. In this study, smokers reported fewer respiratory symptoms than did nonsmokers following ozone exposure, but recovery of polymorphonuclear leukocytes (PMN) in bronchoalveolar lavage (BAL) at 18 h after exposure was similar in smokers and nonsmokers[33] and did not correlate with the increase in cough ($r = .14$, $p = .42$) or shortness of breath ($r = .01$, $p = .94$). Thus, induction of respiratory symptoms may involve pathways independent of those leading to airway inflammation.

2.5.2 Pulmonary Function Measures

2.5.2.1 Pulmonary Mechanics

Most studies of the effects of inhaled pollutants on the lung have measured FEV_1 or airway resistance (or its reciprocal, airway conductance). Although more sensitive methods may exist for detecting mild obstruction or peripheral airway changes, they often suffer from lack of reproducibility and uncertainty regarding their interpretation.

A variety of measurements of resistance to respiratory flow are now accepted as important clinical and investigative tools. Both airway resistance and total thoracic resistance measurements have been used in air pollution studies. Measurement of airway resistance is the most widely applied method to assess resistance to airflow in the lungs. The technique is rapid, noninvasive, and requires the subject to pant at 1 to 2 cycles per second with a nose clip in place. It also permits the recording of thoracic gas volume at the same time and thereby permits correction for the possible effects of changes in lung volume on resistance. The measurement of airway resistance requires a body plethysmograph. These measurements have proven to be reproducible, rapid, and sensitive indicators of mild changes such as those produced by bronchodilators and bronchoconstrictors. The strengths and limitations of these techniques have been discussed in a previous monograph.[4]

Certainly, the most commonly evaluated parameters in controlled exposure studies are measurements of forced vital capacity (FVC), FEV_1, and the maximal flow–volume curve. The reasons for selecting these tests are obvious. They are simple to perform, and small changes in FVC or FEV_1 can be detected in the same subject by repeating the test on the same day or consecutive days. One study on the daily intrasubject variability of FVC and FEV_1 in normal subjects and in patients with chronic bronchitis indicated that a change of more than 5% in the normal subjects and of

15% in patients with airflow limitation was required before it could be considered significant.[34] These measurement techniques, the equipment, and the interpretation have been reviewed extensively.[35]

One potentially confounding problem with spirometry is that the test itself may alter the parameters being tested. For instance, the deep inspiration, which precedes measurements of forced expiratory flow will cause transient bronchodilation and reduce induced bronchoconstriction in the normal individual but may cause bronchoconstriction in the asthmatic. The importance of these changes in airway tone in pulmonary function testing maneuvers remains to be determined.

Some general recommendations should improve the comparability of data between laboratories: The simplest tests using the simplest apparatus are often the most reproducible; calibration should be checked before each experiment; when studies are scheduled, the subject should refrain from smoking, inhaling other irritants, exercising, or using bronchodilator therapy 6 to 12 h prior to testing; tests that do not require deep inhalation should always be done first; and, finally, all studies should be repeated at that same time of the day and with the subject in the same position.

2.5.2.2 Airway Responsiveness

An important and sensitive technique involves the measurement, following pollutant exposure, of the responsiveness of the airways to increasing doses of known bronchoconstricting agents such as methacholine, carbachol, or histamine. Similarly, isocapnic cold-air hyperventilation and inhalation of increasing concentrations of SO_2 have been used to evaluate pollutant-induced airway hyperresponsiveness. Although these techniques may detect alterations in airway function in asymptomatic volunteers, their quantification also relies on measurements of pulmonary function. The American Thoracic Society has published guidelines for measuring airway responsiveness.[36]

The bronchoconstrictor response is generally measured using one of two methods: change in airway resistance and change in FEV_1. Both have shortcomings. The plethysmographic measurement of airway resistance includes the resistance of the larynx, so an increase in the measured resistance may reflect laryngeal narrowing or smooth muscle contraction. Unfortunately, the deep inhalation that precedes measurements of expiratory flow causes transient bronchodilatation and lessens the bronchoconstriction in a healthy subject, whereas it may increase bronchoconstriction in a subject with asthma. Nevertheless, the bronchoconstrictor response is usually measured as the change in FEV_1. Increasing doses of an aerosol are inhaled to construct a dose–response curve and the results are expressed as the provocation concentration (PC) necessary to produce a decrease in FEV_1 of 20% (PC_{20}). The PC_{20} is obtained from the log-dose-response curve by linear extrapolation of the last two points; the lower the PC_{20}, the greater the responsiveness. PC_{20} has been found to correlate closely with methacholine and histamine concentrations in the aerosol.[37]

The testing of nonspecific airway hyperresponsiveness has proved to be highly useful for assessing airway responses to low concentrations of environmental airway pollutants. Even after the return to baseline lung function following removal from acute NO_2[38] or sulfuric acid aerosol[39] exposure, asthmatics demonstrated increased airway responsiveness to cold air and hyperventilation or to carbachol aerosols, respectively. In addition, baseline nonspecific airway responsiveness has been examined as a predictor of responsiveness to pollutant exposure in normal and asthmatic volunteers. Utell et al.[40] found a relationship between baseline airways responsiveness assessed by carbachol in asthmatic subjects and responsiveness to an inhaled sulfuric acid aerosol. In contrast, airways responsiveness to methacholine was not predictive of the FEV_1 decrement in response to ozone exposure in either smokers or nonsmokers.[32]

2.5.3 Direct Sampling of Airways

Pulmonary function testing is generally available, safe, easy, and reproducible. However, these tests do not detect all respiratory effects of clinical importance. The development of fiber optic technology, digital imaging, and an increased understanding of cellular and molecular physiology has made feasible the assessment of biological changes in the airway following pollutant exposure. Some examples of the techniques and approaches developed to investigate cell and tissue responses in clinical studies are listed in Table 2.2. Table 2.3 lists the advantages and disadvantages of the techniques used to sample the respiratory tract of human volunteers.

2.5.3.1 Fiber Optic Bronchoscopy

Development of the technique of fiberoptic bronchoscopy in the 1970s revolutionized the diagnostic evaluation of patients with pulmonary disease, and the relative safety of the procedure is now well established.[41] This technique is frequently used to sample the respiratory tract for research purposes and has become an important tool in human clinical studies on the effects of air pollution. The fiber optic bronchoscope has also been used to instill particles into a single subsegmental airway in human subjects in order to examine localized epithelial responses without

Table 2.2 Measuring Biological Effects of Pollutant Exposure in Human Clinical Studies

Airway Effects	*Systemic Effects*
Mucociliary clearance	Nonspecific immune responsiveness
Epithelial permeability	Serum antibody responses to infectious challenge
Markers of inflammation	Neurobehavioral performance
Markers of tissue injury/repair	Cardiac responses (see Table 2.6)
Airway responses to challenge with infectious organisms	Vascular endothelial function
Cellular antimicrobial functions	Blood hemoglobin and leukocytes
Mucin composition	Blood markers of inflammation and the acute phase response
Surfactant composition and function	Blood coagulation
Diffusing capacity for CO and NO	

Note: IHD, ischemic heart disease; COPD, chronic obstructive pulmonary disease.

exposing the whole lung.[42] This approach is similar to segmental allergen challenge in studies of the pathophysiology of asthma.

BAL, the serial instillation and removal of fluid through a bronchoscope wedged in a distal airway, has provided an opportunity to sample the epithelial lining fluid and cells of the distal airways. Evidence suggests that cells recovered by BAL reflect those present in the pulmonary parenchyma in disease states such as sarcoidosis and idiopathic pulmonary fibrosis.[43] However, cells from the pulmonary interstitium, some of which may have a key role in the pulmonary immune response,[44] are not well sampled using this technique.

One problem in comparing the results of various clinical studies using BAL is the lack of a standardized protocol for the procedure, and variations in the technique itself or the manner in which the fluid is processed can significantly alter the findings. For example, the number and volume of the aliquots of fluid instilled varies between different centers. Increasing the number of aliquots instilled increases the total cell yield[45] but may result in the sampling of a different, more adherent, population of alveolar macrophages. The cells in the fluid obtained following instillation of the first aliquot may represent the more proximal conducting airways, whereas subsequent aliquots more closely reflect alveolar sampling.[46] Consequently, some investigators discard, or analyze separately, the first aliquot returned, whereas others pool and analyze the entire return. Finally, the differential cell count differs depending on the method used for preparing the samples. Lymphocytes are less adherent than alveolar macrophages and will tend

Table 2.3 Techniques for Sampling of the Respiratory Tract in Clinical Studies

Procedures	*Advantages*	*Disadvantages*
Nasal lavage and biopsy	Minimally invasive Easy to perform Allows serial measurements	High within- and between-subject variability Poor correlation with changes in lower airways
Proximal airway lavage	Samples large conducting airway separately from alveolar space	Relatively invasive Difficult to perform Uncomfortable for subject No serial measurements Limited experience
Bronchoalveolar lavage	Samples distal airway and alveolar space Large database Relatively reproducible findings	Relatively invasive Difficult to perform Uncomfortable for subject No serial measurements
Bronchial brush biopsy	Samples relatively pure population of airway epithelial cells	Low viability and limited number of recovered cells Relatively invasive Difficult to perform Uncomfortable for subject
Endobronchial biopsy	Samples epithelium *in situ* Provides tissue for histology, immunocytochemistry, or *in situ* hybridization	Limited to major conducting airways Small tissue samples Relatively invasive Difficult to perform Uncomfortable for subject
Induced sputum	Noninvasive Only minor discomfort Can be safely performed in presence of respiratory compromise Fluid and cells representative of lower airways	Relatively high between-subject variability Limited cell recovery Not all subjects produce sputum Questionable correlation with BAL
Markers in exhaled air and breath condensate	Noninvasive No discomfort	Technical challenges Limited experience Significance of changes often unclear

to be lost preferentially during processing of BAL samples.[47] Differential cell counts in smears prepared using BAL fluid that has not been processed or supplemented with serum protein appears to most closely reflect the actual percentage of lymphocytes, which is approximately 12% in normal volunteers.[48] The reproducibility of findings on BAL in normal volunteers is an important consideration in clinical studies, where there is a need to compare the findings following the exposure to both pollutant and control atmospheres. Ettensohn et al.[49,50] reported considerable variability in the total cell numbers when individual subjects were repeatedly lavaged on separate occasions. Differential cell count was the most consistent parameter observed. Nevertheless, isolated elevations of one or more cell type were occasionally observed in these healthy volunteers. It is possible that the variability in BAL findings in part reflects responses to various environmental exposures, such as sidestream cigarette smoke or allergens, and this variability is an important confounding factor in clinical studies using BAL. A number of different types of pollutants and aerosols have been studied using BAL, including ozone, NO_2, SO_2, acid aerosols, diesel exhaust, acid aerosols, and ambient PM.

Bronchoscopy and BAL have established that inhalation of environmental pollutants causes airway inflammation. However, the various pollutants differ in their potential for inducing airway inflammation. For example, exposure to 0.4 ppm ozone for 2 h, with exercise, resulted in an 8-fold increase in PMN recovered by BAL and was accompanied by increased recovery in lavage fluid of a variety of mediators of inflammation and injury, including fibronectin, elastase, plasminogen activator, and prostaglandin E_2.[51] The recovery of PMN was greater in the first lavage aliquot than in the remaining aliquots, suggesting that ozone exposure causes an acute bronchiolitis. Extended (7-h) exposures to levels of ozone as low as 0.1 ppm resulted in a 4.8-fold increase in PMN in BAL.[7] These and other findings caused the EPA to promulgate a more stringent National Ambient Air Quality Standard for ozone of 0.08 ppm. Exposure to concentrated ambient-air PM also caused mild airway inflammation in healthy subjects.[52]

In contrast, exposure to sulfuric acid aerosols (acid rain) fails to induce an airway inflammatory response, even at concentrations that are orders of magnitude higher than that found in ambient air. Healthy volunteers exposed for 2 h to a sulfuric acid aerosol at 1000 $\mu g/m^3$, with exercise, showed no increase in recovery of inflammatory cells by BAL 18 h after exposure.[53] In addition, there were no effects on the glycoprotein composition of mucins collected by bronchial wash at the time of bronchoscopy.[54]

An important question is whether noninvasive measurements such as pulmonary function or airway responsiveness testing reflect the inflammatory changes detected by BAL. With regard to ozone exposure, the answer appears to be in the negative. Two recent studies have shown that decrements in FEV_1 or FVC following exposure to ozone did not correlate with the airway inflammatory response found by BAL 18 h after exposure.[33,55] Nonspecific airways responsiveness was not predictive of the airways' functional or inflammatory response to ozone. Measures of function of the small airways may be correlated with markers of epithelial injury or repair. Weinmann et al.[56] exposed 8 healthy subjects to 0.35 ppm ozone for 130 min with intermittent exercise and found that reductions in isovolume flow at intermediate to low lung volumes (a presumptive measure of small airways function) correlated strongly with increases in fibrinogen in the first aliquot of BAL fluid 24 h after exposure ($r = .91$, $p = .001$). However, because there was no significant correlation with other markers of injury and inflammation, including inflammatory cells, albumin, prostaglandins, or kinins, interpretation of this very strong correlation with fibrinogen is difficult.

BAL has also been used to determine whether NO_2 induces inflammation of the lower airways. Exposure to 2 ppm NO_2 for 6 h, with intermittent exercise, resulted in a small but significant influx of PMNs into the airways, both at 1 and 18 h after exposure.[57] In a subsequent study, 3 h exposure to 1.5 ppm NO_2 induced mild airway inflammation and also caused small reductions in blood hemoglobin concentration.[58] These findings suggest that NO_2 exposure may have subtle airway and systemic effects; the clinical significance of these observations remains unclear at this time.

2.5.3.2 Proximal Airway Lavage

Bronchoscopic techniques have been developed to evaluate the more proximal conducting airway epithelium that is not sampled by BAL. One such technique involves the placement into the airway of a catheter equipped with two balloons which, on inflation, temporarily isolates a segment of a main-stem bronchus.[59] Small aliquots of fluid are then instilled and withdrawn through a port in the catheter. This technique can sample bronchial epithelial cells, inflammatory cells, and epithelial lining fluid following exposure for comparison with the findings on BAL. Care must be taken during the procedure to avoid prolonged balloon inflation, as some subjects may experience significantly decreased arterial oxygen tension during occlusion of the airway. A modification of this technique has been used to study the effects of ozone exposure in healthy and asthmatic subjects.[55]

2.5.3.3 Bronchoscopic Biopsy

Bronchial biopsy via the fiber optic bronchoscope provides a method for sampling the bronchial mucosa. Forceps biopsies provide specimens for histopathology and characterization of gene expression using immunofluorescence staining and *in situ* hybridization. This methodology has been safely performed in many studies on healthy and asthmatic subjects. Although bleeding can occur at the biopsy site, it is almost never clinically significant. Samples are small (only a few millimeters) and may show crush artifacts, but this method is important because findings on BAL or sputum induction (see next section) may not reflect changes in the bronchial mucosa. For example, Jörres et al.[60] studied the effects of 4 repeated daily exposures to 0.2 ppm ozone in healthy subjects, using lung function, BAL, and bronchial biopsy. Repeated exposures to ozone attenuated the decrease in FEV_1 and the increase in BAL inflammatory cells seen with a single exposure, as expected from previous studies. However, inflammatory cells in airway tissue increased with repeated exposure to ozone (Figure 2.3), indicating that inflammation and injury at the tissue level persist following repeated exposures, even as functional changes return to baseline. Other studies[61] have used bronchial biopsy to demonstrate that ozone exposure increases lung endothelial expression of the vascular adhesion molecules P-selectin and intercellular adhesion molecule-1 (ICAM-1). Similar findings were seen in healthy subjects exposed to diesel exhaust.[62]

Bronchial brush biopsies can be used to harvest bronchial epithelial cells for primary culture or *ex vivo* experiments. For example, our laboratory used bronchial brush biopsy to harvest airway epithelial cells from healthy subjects after exposure to 0.6 ppm NO_2, 1.5 ppm NO_2, or air for 3 h.[58] Bronchial epithelial cells in culture were then challenged with respiratory syncytial virus. Epithelial cells harvested following exposure to 1.5 ppm NO_2 showed greater injury (as determined by release of lactic dehydrogenase) than those with prior air exposure, suggesting that NO_2 exposure enhanced cell susceptibility to viral injury.

These and other studies show that the use of multiple bronchoscopic sampling methods, including proximal airway lavage, BAL, and bronchial biopsy, can enhance the understanding of the airway effects of single or repeated exposures to pollutants.

2.5.3.4 Nasal Lavage and Challenge

It has been postulated that changes in the nose may reflect inflammation in the lower airways. The nose is more accessible than the lower airways, and nasal inflammation can be assessed using nasal lavage (NL) with quantitation of inflammatory cells and mediators in the fluid returned, similar to BAL. Ozone has the potential to irritate airways throughout the respiratory tract because of its physical properties of high reactivity and relative insolubility and therefore can be expected to elicit changes in

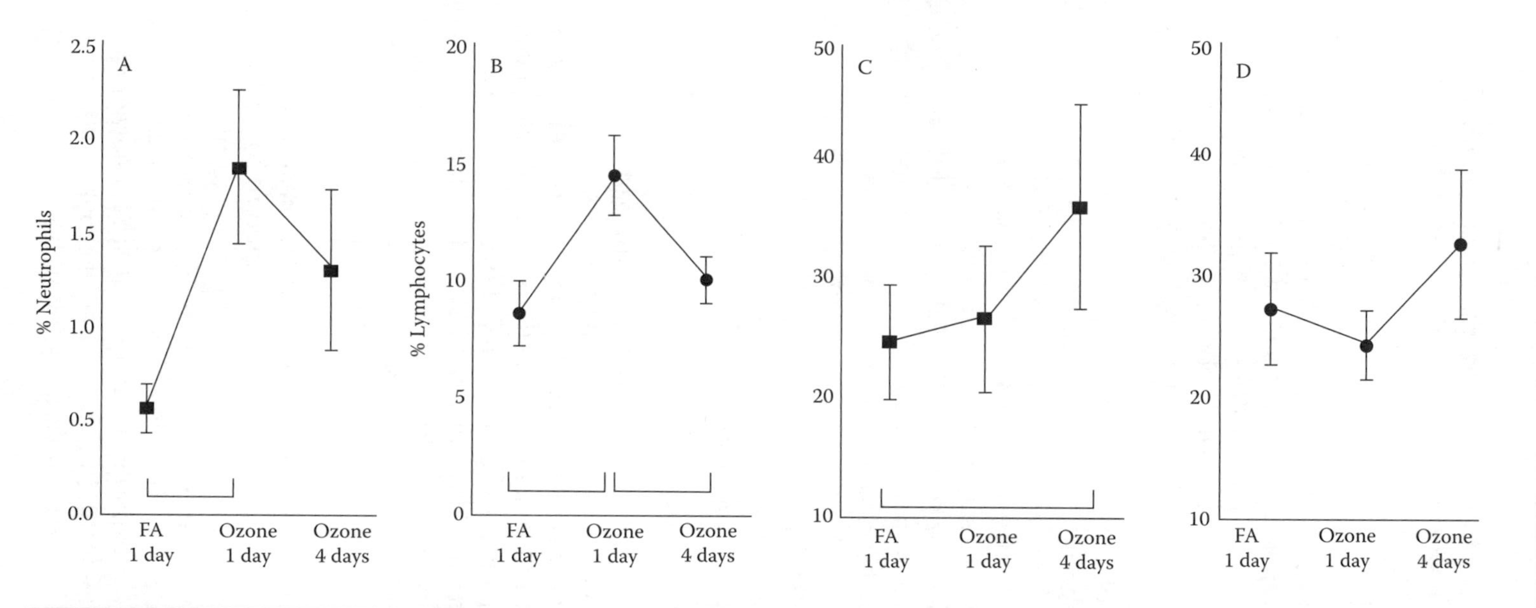

Figure 2.3 **Granulocytes and lymphocytes in bronchoalveolar lavage fluid (A and B) and in mucosal biopsies (C and D) 20 h after exposure. Exposures were to filtered Air (FA), 0.2 ppm ozone for 4 h on 1 d, and 4 h daily for 4 d. (Modified from Jörres, R.A., Holz, O., Zachgo, W., Timm, P., Koschyk, S., Müller, B., Grimminger, F., Seeger, W., Kelly, F.J., Dunster, C., Frischer, T., Lubec, G., Waschewski, M., Niendorf, A., and Magnussen, H., The effect of repeated ozone exposures on inflammatory markers in broncho-alveolar lavage fluid and mucosal biopsies. *Am J Respir Crit Care Med*, 161, 1855, 2000.)[60]**

nasal epithelium as well as the lower airways. Graham and Koren[63] found that exposure to 0.4 ppm ozone for 2 h resulted in a 6-fold increase in PMN recovered by NL 18 h after exposure. However, there was no significant correlation between the recovery of PMN in NL and BAL fluid after ozone exposure among the 10 subjects studied ($r = .4$, $p = .24$). Furthermore, subsequent studies have also failed to show a significant nasal inflammatory response to more environmentally relevant levels of ozone exposure.[33] Although investigation of changes in the nasal mucosa may provide insights into epithelial responses to pollutants, NL does not serve as a substitute for sampling of the lower airways following ozone exposure.

The accessibility of the nasal mucosa makes it useful in modeling allergic responses. Exposure to traffic-related PM, especially diesel exhaust particles (DEP), has been associated with asthma and allergic disease. Nasal challenge studies have provided evidence that DEP can enhance IgE production in atopic subjects[64] and can enhance the specific IgE response to ragweed in ragweed-sensitive subjects.[65] DEP also appear to induce nasal sensitization to a neoantigen.[66] However, the relatively high dose of DEP used in these nasal challenge studies may not be relevant to outdoor ambient exposures, and the responses also appear to be dependent on the source of DEP.

2.5.4 Other Measures of Respiratory Effects

2.5.4.1 Mucociliary Clearance

Mechanical clearance of inhaled microorganisms or particles that deposit on the respiratory mucous layer is an important element of respiratory host defense and is known to be altered in cigarette smokers.[67] Mucociliary clearance can be assessed in humans using timed clearance after inhalation of radiolabeled inert particles; careful control of particle size can determine the regional deposition and thus target the region of the lung to be studied. Schlesinger[68] reviewed the techniques used for measuring mucociliary clearance. Aerosol probes and radiolabeled aerosol techniques are also potential noninvasive tools for determining airway and airspace sizes in controlled clinical studies.

The most convincing evidence for the effects of inhaled pollutants on mucociliary clearance comes from studies of exposure to acid aerosols. Leikauf et al.[69] demonstrated an increase in the clearance rate of radiolabeled aerosols following 1-h exposures to H_2SO_4 aerosols at 100 $\mu g/m^3$ and a slowing of clearance following exposure to 1000 $\mu g/m^3$. Spektor et al.[70] found that a 2-h exposure to 100 $\mu g/m^3$ H_2SO_4 slowed clearance similar to 1-h exposures to 1000 $\mu g/m^3$. These data confirmed observations in animals[68] and revealed a possible mechanism linking exposure to acid aerosols to chronic bronchitis.[71] However, the concentrations of acid aerosols shown to have effects on mucociliary clearance far exceed outdoor ambient levels.

Exposure to ozone has been linked to an increase in mucociliary clearance when measured immediately after single exposures to a level of 0.4 ppm.[72] However, Gerrity and colleagues[73] found no effect of exposure to 0.4 ppm ozone for 1 h on mucociliary clearance when measured between 2 and 5 h after exposure. This suggests that the apparent increase in clearance in the prior study may have been caused by cough due to the irritant effects of the exposure.

2.5.4.2 *Epithelial Permeability*

Exposure to very high levels of oxidants such as NO_2 result in a marked increase in paracellular permeability of the alveolar epithelium, leading to pulmonary edema, as exemplified by silo filler's disease.[74] Measurement of the concentration of large molecular weight serum proteins in epithelial lining fluid sampled by BAL is used to assess change in epithelial permeability. Exposure to 0.4 ppm ozone for 2 h resulted in twofold increase in albumin and IgG in BAL fluid when assessed 18 h after exposure.[51] In contrast, exposures to NO_2 at concentrations as high as 2 ppm for 6 h, did not alter the levels of total protein or albumin in BAL fluid.[75,57]

Another technique used to assess changes in epithelial permeability following pollutant exposure is inhalation of technetium 99m (^{99m}Tc)-labeled diethylene triamine pentacetic acid (^{99m}Tc-DTPA). The rate of transfer of this molecule into the circulation is a sensitive measure of pulmonary epithelial permeability. Use of this technique has confirmed that exposure to ozone at levels sufficient to cause decrements in pulmonary function (0.4 ppm for 2 h) also results in increased epithelial permeability.[76] Interestingly, exposure of dogs to 5 ppm NO_2 for 20 min appeared to decrease the uptake of ^{99m}Tc-DTPA,[77] suggesting that other factors besides epithelial permeability may influence the results obtained using this technique. These findings also suggest that fundamental differences exist in the effects of the two oxidants, ozone and NO_2, on pulmonary alveolar epithelium.

2.5.4.3 *Sputum Induction*

Inhalation of nebulized hypertonic saline has been used to obtain expectorated sputum in the setting of respiratory infections. In recent years, sputum induction has emerged as a useful research tool for sampling the cells and epithelial lining fluid of the lower respiratory tract in humans. Hargreave et al.[78] in Canada and Fahy et al.[79] in the U.S., among other groups, have shown the reproducibility and utility of these techniques. Sputum can be reliably induced in asthmatic subjects as well as most healthy subjects using nebulized 3 to 5% saline. The recovered material derives from the proximal conducting airways, and not the peripheral airways,[80] and provides a measure of inflammation at that level. Asthmatic subjects had more eosinophils and PMNs in their sputum than healthy subjects. Sputum inflammatory cells increased with asthma exacerbations[81,82] and allergen challenge[83] and decreased with prednisone therapy.[84] The results appear to be qualitatively similar to the findings in the first (bronchial) aliquot of BAL fluid.[85] Sputum had higher numbers of nonsquamous cells and higher concentrations of eosinophil cationic protein and albumin in the soluble fraction compared with BAL fluid. However, eosinophil recovery in sputum correlated closely with the first aliquot of BAL fluid.[85]

The process of sputum induction itself appears to induce a mild airway inflammatory response. For example, when two inductions are performed 24 h apart, the second induction has a higher recovery of PMN.[86,87] Thus, caution must be used when designing studies with serial sputum inductions. Sputum induction is well tolerated even in severe asthma, but some asthmatic subjects may experience bronchoconstriction in response to the nebulized saline. When this happens it is transient; saline inhalation does not cause a late-phase response as can be seen with allergen challenge. Pretreatment with bronchodilators is effective in preventing the aerosol-induced bronchoconstriction. Because of this risk of bronchoconstriction in asthmatic subjects, monitoring of lung function is an important part of the procedure. In a study assessing safety in 64 patients with asthma of varying severity,[88] the procedure had to be stopped because of reductions in FEV_1 or other effects in 11.6% of patients with severe asthma, although none of the changes were considered severe.

Some subjects are unable to produce sputum following induction. Approximately two-thirds of healthy subjects and more than 90% of asthmatic subjects respond to the procedure. Individuals who are able to produce sputum do so on subsequent challenges also, and the findings on repeat sputum inductions are quite consistent (r = .80) among individuals.[89] Reference values have been published[90] for cell recovery in healthy subjects.

Fahy et al.[91] have demonstrated that sputum induction detects airway inflammation following exposure to 0.4 ppm ozone for 2 h in healthy subjects. In contrast, exposure to 0.3 ppm NO_2 for 1 h, with moderate intermittent exercise, did not alter cell recovery in induced sputum in subjects with mild asthma or COPD.[92] These findings are consistent with the findings in BAL fluid following exposures to ozone and to NO_2. Sputum induction therefore holds promise as a noninvasive method for assessing airway inflammation in clinical and field studies.

2.5.4.4 Markers in Exhaled Air

Some biochemical processes involving airway epithelial cells and the epithelial lining fluid release gaseous and chemical products that can be detected in trace amounts in the exhaled breath. These biochemical reactions may be influenced by various airway disease states characterized by airway inflammation (like asthma) and by inhalation of reactive pollutants in the ambient air. Investigators have sampled exhaled air for a variety of volatile and nonvolatile substances as markers of either airway inflammation or injury.

Nonvolatile substances are measured by analyzing the exhaled-breath condensate collected during rapid cooling of expired air. Most of the condensate consists of exhaled water vapor,[93] but a small fraction consists of aerosolized fluid droplets from lower respiratory surfaces, potentially containing chemicals and mediators of interest. The proportion of such aerosols in the exhalate increases with faster and more turbulent air flow and with lower surface tension of the respiratory tract lining fluid.[94] The basic technical procedure is simple. The study subject performs tidal breathing for 5 to 60 min through a one-way nonrebreathing valve attached to tubing directed through a cooling chamber and connected to a condensation chamber at its distal end. The most commonly measured mediators and their pathophysiologic significance are listed in Table 2.4.

Advantages of the technique are its noninvasiveness, simplicity, and repeatability. The most important disadvantage is the lack of standardization of collection techniques and measurement procedures. Recently, Effros and colleagues[96] have demonstrated the importance of controlling for the enormous and variable dilution of these mediators with exhaled water vapor. These investigators further showed that ammonia gas originating in the mouth forms high concentrations of ammonium cation (NH_4^+) in the exhaled condensate, a potential confounder when interpreting condensate pH measurements.[95] By lyophilizing the exhaled condensate, Effros et al. developed a method for removing the NH_4^+.[96] The sum of the remaining condensate cations is measured, and should equal the sum of cations in serum if there is no dilution. Dilution, if present, can be calculated by dividing the sum of cation concentrations in serum by the sum of cation concentrations in the lyophilized condensate.

Finally, Effros et al. demonstrated that measurement of condensate conductivity equals the sum of the cations. Because fluid conductivity is technically easier to measure than each of the individual cations, this final step adds simplicity to a novel technique, which should markedly improve the accuracy and reproducibility of these measurements. One problem is the uncertainly regarding the origin of collected substances: Do they arise from the alveoli or the conducting airways?[94] This problem is similar to that initially encountered with exhaled nitric oxide (NO) measurements (see the following text).

Table 2.4 Nonvolatile Mediators in Exhaled Air

Mediator	*Pathophysiologic Significance*	*Association with Disease State*	*Reference*
Hydrogen Peroxide	Reactive oxygen species	Increased in smokers	182
		Increased in asthmatics, decreases after inhaled steroid therapy	183
		Increased in bronchiectasis	184
		Increased in COPD	185
		Increased in ARDS	186
Thiobarbituric acid-reactive (TBARS)	lipid peroxidation metabolite	Increased in asthma	183
Nitrosothiol	Nitric oxide metabolite	Increased in asthma	187
8-isoprostane	Membrane phosholipid peroxidation metabolite	Increased in asthma	188
		Increased in patients at risk of ARDS	189
		Increased in COPD	190
Leukotrienes	Inflammatory mediator	Increased in asthma	191
Nitrite	Nitric oxide metabolite	Increased in cystic fibrosis	192
		Increased in asthma	193
		Increased in COPD	187
pH	Postulated to reflect inflammatory cell enzyme activity	Decreased in asthma	194
		Decreased in COPD	195
		Decreased in bronchiectasis	195

Studies evaluating the effects of airborne pollutants on the measurement of nonvolatile mediators in exhaled-breath condensates are limited. Two-hour exposure to 0.1 ppm ozone induced a significant rise in the markers of oxidative stress (8-isoprostane and thiobarbituric acid-reactive substances (TBAR) and inflammation (leukotriene-B4 or LTB-4) in a group of normal subjects possessing a genotype associated with increased susceptibility to ozone.[97] Montuschi et al.[98] demonstrated a rise in 8-isoprostane levels after 2 h of exposure to 0.4 ppm ozone with intermittent exercise in 9 healthy subjects. This rise was not affected by pretreatment with inhaled budesonide.

Volatile substances in exhaled air can also be measured, and it is now possible to measure over 100 different volatile chemicals in exhaled breath; this has potential applications in field studies of environmental exposures.[99] These volatile substances include NO,[100] carbon monoxide (CO),[101] isoprene,[102] ethane,[103] and pentane.[104]

Measurements of exhaled NO have attracted wide interest as a means of detecting lung inflammation.[100] NO is produced by the action of nitric oxide synthase (NOS) on L-arginine. A variety of cells, including airway epithelial and endothelial cells, express one or more of the three isoforms of NOS. Concentrations of NO are increased in the exhaled air of people with asthma compared with healthy subjects.[105] Mild asthmatics not requiring inhaled or oral corticosteroids have NO levels in exhaled breath that are sevenfold greater than normal subjects.[106] NO levels increase further with clinical exacerbations, correlate with the degree of airway hyperresponsiveness in steroid-naive asthmatics,[107] and decrease following therapy with corticosteroids.[108,109]

Over the past decade, exhaled NO measurement techniques have arguably attracted the greatest interest, and NO has been the best characterized of all exhaled volatile markers. With refinement of chemiluminescence technology, NO analyzers are now capable of measuring NO concentrations of 1 ppb.[110] The chemiluminescence method involves a reaction between the sample NO and ozone generated within the instrument. This reaction produces electronically excited NO_2, which emits light in proportion to the amount of NO present in the sample. The quantity of light, in turn, is detected by the instrument's photomultiplier tube.

Early on, published recommendations advocated NO measurement during exhalation at a single expiratory flow rate, without an initial breathhold maneuver.[111,110] However, in 1997 it was found that exhaled NO concentrations were directly related to expiratory flow rate.[112,113] This observation catalyzed the development of two-compartment models of pulmonary NO exchange by a number of research groups.[114–118] These models all describe exhaled NO as arising from two compartments — the alveoli and the conducting airways. By measuring the exhaled NO concentration at many different constant expiratory flow rates, several flow-independent NO exchange parameters can be determined. These parameters, in turn, uniquely characterize NO exchange in the alveoli vs. the conducting airways.

These models hold promise as a means of separately characterizing perturbations in NO exchange in these different lung regions in response to disease or environmental exposures. The details of these models and their parameters are beyond the scope of this chapter, but the reader is referred to a recent review article for more information on this topic.[119] When interpreting clinical studies that have not utilized these techniques, it is important to note that exhaled NO sampled at faster expiratory flow rates (e.g., > 200 ml/sec) is more reflective of the alveolar compartment, whereas NO sampled at slower expiratory flow rates (e.g., 50 ml/sec) is more reflective of the conducting airway compartment.

A number of studies have measured exhaled NO at single expiratory flow rates after exposure to airborne pollutants. No changes occurred in exhaled NO (measured at an expiratory flow rate of approximately 80 to 100 ml/sec) after 2 h of exposure to 0.4 ppm ozone, despite alterations in other measurements such as ozone-induced changes in spirometry, methacholine reactivity, cellularity of induced sputum, and myeloperoxidase in exhaled-breath condensate.[120] On the other hand, Nickmilder et al.[121] did observe higher exhaled NO levels in children exposed to ambient ozone concentrations greater than 135 $\mu g/m^3$ during summer camp. Expiratory flow rates during these NO measurements were not reported. In another study, exposure to PM with aerodynamic diameter < 2.5 μm ($PM_{2.5}$) was associated with elevation in exhaled NO levels measured off-line in 6- to 13-yr-old children.[122] In this study, subjects were coached to maintain constant expiratory flow, although the actual flow rates were not measured. Steerenberg et al. reported that exhaled NO levels measured at a slow expiratory flow rate (8.3 ml/sec) were independently and directly correlated with ambient levels of CO, NO_2, and $PM_{2.5}$ in children.[123] Finally, Sundblad et al.[124] reported that exhaled NO measured at a single expiratory flow rate of approximately 100 ml/sec increased significantly in swine-house workers after 3 h of exposure to the organic dusts in that environment.

The inconsistent results in these studies are likely related, in part, to lack of standardization in expiratory flow rates during NO measurement. For example, if airborne pollutants predominantly affect NO dynamics in the conducting airways and exhaled NO is measured at fast expiratory flow rates, these changes are unlikely to be detected. Similarly, if airborne pollutants predominantly affect NO dynamics in the alveolar compartment and exhaled NO is measured at slow expiratory flow rates, these changes are also unlikely to be detected. Clearly, standardization of expiratory flow rate will be an important first step in improving the utility of exhaled NO measurements in clinical studies of airborne pollutants. Moreover, determination of alveolar and conducting airway NO exchange parameters potentially allows more precise determination of the location of perturbations in NO metabolism. These measurements have been effectively used to determine the predominant site of derangement in pulmonary NO dynamics in a variety of disease states. These include scleroderma,[125] cystic fibrosis,[126] asthma,[117,127] COPD,[127] and allergic rhinitis.[127]

We have applied our two-compartment model of NO exchange to healthy subjects who inhaled elemental carbon UFP in concentrations up to 50 μg/m^3 vs. filtered air.[128] Alveolar NO exchange parameters differed slightly after inhalation of UFP (50 μg/m^3) but consistent patterns of change were not evident. Further studies are under way to explore these findings.

CO is another volatile substance that can be measured in exhaled breath.[129] Exhaled CO may be a reflection of pulmonary oxidative stress and inflammation.[130] Endogenous CO is formed when erythrocytic heme is oxidatively degraded by the enzyme heme oxygenase (see Figure 2.4). Induction of heme oxygenase-1 (HO-1) may serve an important protective function during periods of oxidant stress.[131] Elevations in exhaled CO have been reported in a variety of pulmonary disorders including asthma,[101,129,132] cystic fibrosis,[133,134] allergic rhinitis,[135] and upper respiratory infections.[136]

Studies measuring exhaled CO concentrations in response to airborne pollutants are limited. Exhaled CO did not change after 0.4 ppm ozone exposure for 2 h.[120] However, in healthy subjects exposed to diesel exhaust particles (PM_{10}, 200 μg/m^3) vs. air for 2 h,[137] exhaled CO levels increased 1 h after particle exposure (4.4 ± 0.3 ppm vs. 2.9 ± 0.2 ppm) in conjunction with slight elevations of neutrophils and myeloperoxidase in sputum induced 4 h after exposure.

Unfortunately, the origin of the exhaled CO are incompletely characterized to date,[138] and the measurement techniques are not well standardized.[139] Zetterquist and colleagues[140] carefully studied healthy controls and patients with allergic rhinitis, asthma, and cystic fibrosis using two independent CO measurement techniques. In contrast to previous observations (see earlier citations), there were no differences in exhaled CO between normal subjects and patients with any of these respiratory disorders. These investigators further demonstrated that, unlike NO, exhaled CO levels do not depend on expiratory flow rate, and that breath-holding prior to exhalation increases the exhaled CO concentration. The authors state that these findings suggest an alveolar, rather than conducting airway, origin of exhaled CO and conclude that exhaled CO measurements will be of limited value in assessing disease or inflammation of the conducting airways. Not surprisingly, these findings have generated considerable controversy.[141,142] These measurements are further complicated by the fact that high ambient CO levels increase exhaled CO levels.[143] Indeed, exhaled CO was initially used to determine whether research subjects were compliant with antismoking programs, as the CO in cigarette smoke raises exhaled CO levels.[144,140] Finally, the subjects' degree of airflow obstruction also affects the concentration of exhaled CO.[145] Clearly, additional studies

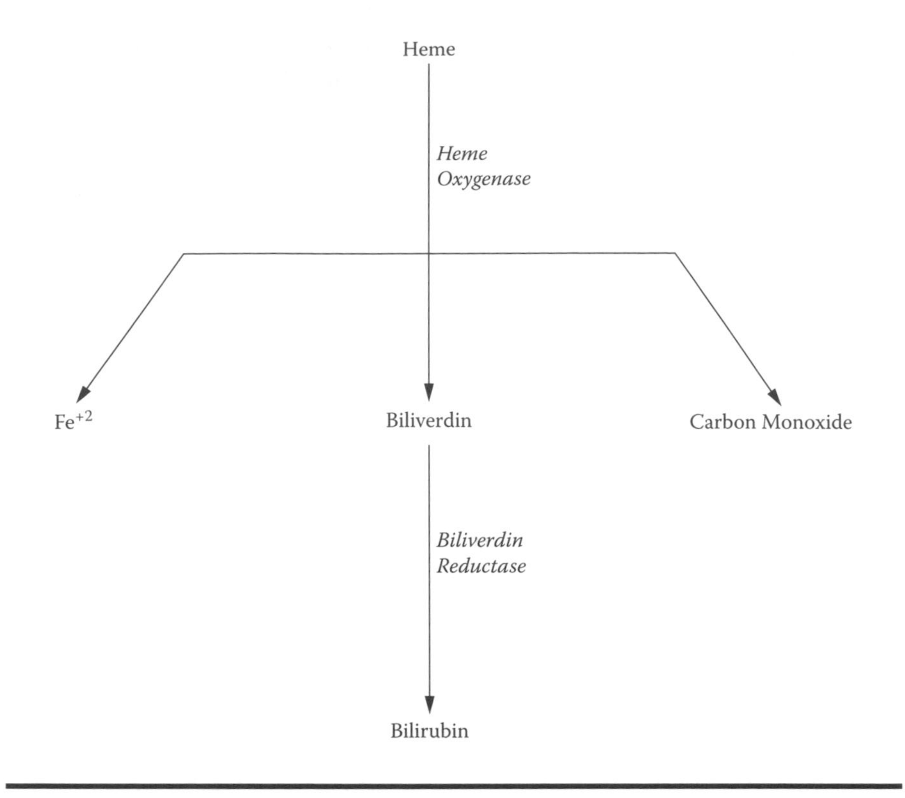

Figure 2.4 The heme oxygenase pathway. Erythrocyte heme is oxidatively metabolized to biliverdin by heme oxygenase. The inducible form of this enzyme is heme oxygenase-1 (HO-1). Byproducts of this reaction are ferrous iron (Fe^{2+}) and carbon monoxide; the latter is measured in exhaled air. Biliverdin is then reduced to bilirubin by biliverdin reductase. Bilirubin has antioxidant effects, and this is one of the mechanisms whereby HO-1 is considered to play a cytoprotective role.

of exhaled CO are needed to clarify its origins in the respiratory tract, identify and correct for confounding variables that affect exhaled concentration, and standardize measurement techniques.

Measurement of volatile hydrocarbons in exhaled air has also attracted interest.[130] Ethane and pentane are produced during oxidant-induced lipid peroxidation, and can be measured in the exhaled air with gas chromatography.[146] Elevated levels of these hydrocarbons have been detected in a variety of pulmonary disorders.[147,133,148] Foster et al.[102] found small but significant increases in exhaled isoprene levels in healthy subjects following exposure to ozone at 0.15 to 0.35 ppm for 130 min. However, whether exhaled hydrocarbons can be considered as accurately reflecting oxidative stress and lipid peroxidation must await further study.

In conclusion, measurement of volatile and nonvolatile organic molecules in exhaled air show promise as noninvasive markers of pollutant effects, but considerable work remains to be done to clarify the significance of the observed changes.

2.5.5 Cardiovascular Effects

Epidemiological studies provide convincing evidence that exposures to increasing concentrations of ambient PM triggers adverse cardiac events. Clinical studies on the cardiovascular effects of air pollution were the subject of a recent workshop.[149] The American Heart Association has acknowledged the role of air pollution in cardiovascular disease in a recently published statement[150] reviewing the proposed mechanistic pathways and providing advice for clinicians. Human clinical studies have begun to test specific hypotheses regarding the cardiovascular effects of air pollution exposure.[151] For obvious safety reasons, human clinical studies are not designed to include outcomes such as myocardial infarction, serious arrhythmias, or congestive heart failure. Furthermore, clinical studies cannot answer questions about long-term effects such as worsening of atherosclerosis; however, they can utilize our growing understanding of atherosclerotic cardiovascular disease as an inflammatory process, manifested by dysfunction of vascular endothelium, to identify markers of cardiovascular effects. Table 2.5 lists some of the methods available for measuring the cardiovascular effects of air pollution in clinical studies. Here we will review three methods: electrophysiologic monitoring, measurement of vascular and endothelial function, and monitoring of cardiovascular hemodynamics.

2.5.5.1 Electrophysiologic Monitoring

Continuous recordings of the ECG for periods of 24 h or more have been used extensively in clinical cardiology and are now being used in both clinical and panel studies to investigate the effects of air pollution. Newer recording systems allow monitoring of all 12 leads of a standard ECG rather than just the three leads traditionally used in clinical monitoring. This allows measurement of focal changes in the ST segment (ischemia) and of heterogeneity of ventricular repolarization.[152] Measurement of both time- and frequency-domain variation in heart rate (HRV) has prognostic value in patients with heart disease. Generally, reductions in HRV are seen with impaired cardiac function and indicate an increased likelihood of adverse events.[153,154]

Clinical studies have begun to utilize continuous ECG monitoring to study the cardiac effects of PM. Devlin et al.[155] exposed healthy elderly

Table 2.5 Procedures Available for Assessing Cardiac Functional Responses

Procedure	*Measurement*	*Significance*
Noninvasive		
Continuous ECG	Heart rate variability Heart rate turbulence ST segment elevation Myocardial repolarization Arrhythmias	Provides data on parasympathetic influence on cardiac function, baroreflex sensitivity, ischemia, and rhythm disturbance; significance of transient changes in HRV unclear
Echocardiography	Ejection fraction Chamber size Ventricular wall motion Pulmonary artery pressure (estimation)	Quality of results dependent on body habitus; well-established method for assessing ventricular function
Scintigraphic scanning	Myocardial perfusion Ventricular wall motion Ejection fraction	Requires intravenous injection of radioisotope; time consuming
Impedance cardiography	Cardiac output Systemic vascular resistance Systolic time intervals	Results heavily dependent on operator experience and placement of electrodes
Invasive		
Right heart catheterization	Pulmonary artery pressure Left ventricular filling pressure Cardiac output Pulmonary vascular resistance Systemic vascular resistance	"Gold standard" for measurement of pulmonary artery pressure and cardiac output
Left heart catheterization	Left ventricular ejection fraction and wall motion Coronary artery anatomy and flow	"Gold standard" for assessment of left ventricular function and coronary arteries; invasive and costly.

subjects between the ages of 60 and 80 yr to concentrated outdoor PM and to clean air for 2 h. Changes in HRV were measured immediately before, immediately following, and 24 h after exposure. Elderly subjects experienced significant decreases in HRV in both time and frequency domains immediately following exposure. Some of these changes persisted for at least 24 h. These data were compared with HRV data collected from young healthy volunteers exposed to concentrated ambient particles (CAPS) in a previous study in which no CAPS-induced changes in HRV were found. These concentrated ambient-air pollution particle-induced changes in heart rate variability in a controlled human exposure study extend similar findings reported in recent panel studies and indicate the potential mechanisms by which PM may induce adverse cardiovascular events.

2.5.5.2 Vascular Function

Abnormal endothelial function, defined as an impaired vasodilator response to appropriate stimuli, is clearly and closely linked with vascular disease and cardiac events. Endothelial function is most often measured in the systemic arterial circulation using flow-mediated dilatation of the forearm blood vessels. The response of the forearm circulation to flow-mediated dilatation is via endothelial release of NO. Depressed flow-mediated dilatation is considered to be secondary to failure of shear-stress-induced NO release because of endothelial cell injury and/or NO inactivation by reactive oxygen species. In patients with risk factors for atherosclerosis, there is evidence for reduced vascular NO bioavailability and for increased production of superoxide anion. Measures which increase NO bioavailability, such as treatment with L-arginine, improve endothelial function.[156] Patients with reduced flow-mediated dilatation are at increased risk of adverse cardiovascular events, and measures that improve endothelial function may also reduce cardiovascular risk. There is evidence that changes in flow-mediated dilatation of the forearm vessels correlate with changes in endothelial function in the coronary circulation.[157] A finding of transient reduction in flow-mediated dilatation following air pollution would suggest an increased cardiovascular risk especially for people with coronary artery disease, and this would be compatible with the adverse cardiovascular effects of repeated or long-term exposures.

The test is fairly simple:[158,159] a blood pressure cuff is inflated around the forearm to at least 50 mmHg above the systolic blood pressure, making the forearm ischemic for 5 min. The cuff is then suddenly released. Ultrasound imaging is used to precisely measure brachial artery diameter and flow before and after inflation of the cuff. Figure 2.5 shows an ultrasound image of the brachial artery before and after the ischemia-induced increase in blood flow. The increase in diameter of the artery is

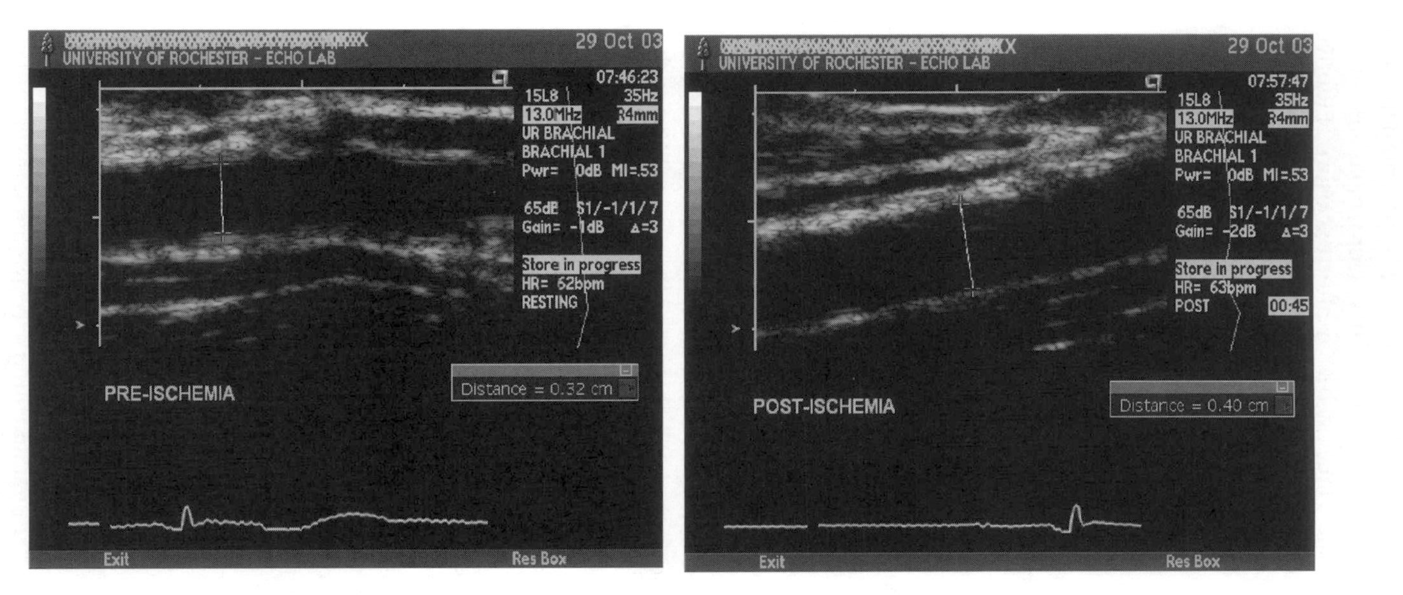

Figure 2.5 Flow-mediated dilatation: ultrasound images of the brachial artery before (left panel) and after (right panel) forearm ischemia. The brachial artery diameter increased by 8 mm after ischemia. Courtesy of Karl Schwarz MD, University of Rochester Medical Center.

"flow-mediated," and primarily due to NO release by endothelial cells. Endothelium-independent dilatation can then be measured by administering nitroglycerine.

Endothelial function is a dynamic process and is sensitive to a variety of influences, including diet, temperature, exercise, emotional stress, the menstrual cycle in females, body position, and exposure to cigarette smoke. These variables need to be controlled to the extent possible. Reproducibility is highly dependent on the ability of the technician to image the brachial artery at the same site and at the same angle, with careful standardization of other aspects of testing.

Brook et al.[160] demonstrated that a 2-h exposure to both ozone 0.12 ppm and concentrated ambient PM decreased the diameter of the brachial artery. Preliminary work in our laboratory[128] suggests that inhalation of elemental carbon UFP blunts the improvement in vascular responsiveness seen following exercise.

NO plays a critical role in vascular function and dysfunction.[161–163] Measurement of NO metabolites in blood has potential as a means of assessing vascular function.[164] Arbak and colleagues[165] reported elevated nitrite and nitrate levels in highway toll workers exposed to diesel exhaust compared to an office-based control group. We observed changes in venous nitrate levels after carbon UFP exposure that appeared to parallel changes in forearm blood flow,[128] but these findings are preliminary and require confirmation. Currently, data regarding endothelial dysfunction after airborne pollutant exposure remain sparse, and this is a fertile area for future investigation.

2.5.5.3 Hemodynamic Monitoring

Measurement of cardiovascular hemodynamics provides information on intrinsic cardiac function but generally involves the invasive technique of cardiac catheterization, with its attendant risk, complexity, and cost. Key indicators include measurements of cardiac output, stroke volume, ventricular ejection fraction, left ventricular end-diastolic pressure, pulmonary artery pressure, pulmonary and systemic vascular resistance, and central venous pressure. There have been few studies of pollutant effects on these endpoints. However, the increasing evidence for cardiovascular effects of exposure to PM has increased interest in such studies.

Gong et al.[166] performed catheterization of the right heart and radial artery in 10 hypertensive and 6 healthy men undergoing exposures to 0.3 ppm ozone or air for 3 h, with intermittent exercise. Ozone exposure caused increases in the blood pressure–heart rate product and in the alveolar–arterial oxygen gradient, indicating increased myocardial work and impaired gas exchange in response to ozone exposure. However, there were no effects on intrinsic cardiac function.

Drechsler-Parks[167] used a different technique to assess changes in cardiac output: noninvasive impedance cardiography. Eight older adults (56 to 85 yr of age) were exposed to 0.60 ppm NO_2, 0.45 ppm ozone, and a combination of 0.60 ppm NO_2 and 0.45 ppm ozone for 2 h, with intermittent exercise. The exercise-induced increase in cardiac output was smaller with the NO_2 + ozone exposures than with the filtered air or O_3 exposures alone. There were no significant differences in minute ventilation, heart rate, or cardiac stroke volume, although the mean stroke volume was lower for NO_2 + ozone than for air. The author speculated that chemical interactions between ozone and NO_2 at the level of the epithelial lining fluid led to the production of nitrite, leading to vasodilatation, with reduced cardiac preload and cardiac output. This finding has not been confirmed.

One previous study[168] reported small but statistically significant reductions in blood pressure after exposure to 4 ppm NO_2 for 75 min, a finding consistent with systemic vasodilatation in response to the exposure. However, many subsequent studies at concentrations generally less than 4 ppm have not reported changes in blood pressure in response to NO_2 exposure. It therefore remains uncertain whether NO_2 exposure has an effect on cardiovascular hemodynamics in humans.

2.6 SUSCEPTIBLE POPULATIONS IN CLINICAL STUDIES

The U.S. Clean Air Act mandated that the National Ambient Air Quality Standards for the criteria pollutants were to be set low enough to protect the health of all susceptible groups within the population. The term "susceptible" has been most often applied to these with one or more diseases or characteristics that place them at increased risk compared to people without these characteristics. Examples of groups considered to have increased susceptibility to air pollution, along with the potential mechanisms and health effects, are shown in Table 2.6. Individuals at the extremes of age are considered to have increased risk. The increased mortality associated with exposure to particulate air pollution occurs mostly in the elderly,[169] possibly related to lung or heart disease or the other frailties that often accompany aging. However, few clinical studies have examined age-based susceptibility. There are few studies on children due to ethical and technical reasons. Lung function decrements in response to ozone exposure are known to decrease with increasing age; however, we know little about the airway inflammatory response to pollutants in elderly subjects.

The diseases and conditions associated with increased health risks from air pollution have been reviewed.[151] That the presence of lung disease

Table 2.6 Groups Considered Susceptible to Inhaled Air Pollutants

Population	*Potential Mechanisms*	*Health Effects*
Infants	Immature defense mechanisms of the lung; impaired lung growth	Increased risk of respiratory infection; asthma symptoms; reduced lung function for age
Elderly	Impaired respiratory defenses; reduced functional reserve; underlying cardiovascular disease	Increased risk of mortality; cardiac arrhythmias, infection, myocardial infarction, congestive heart failure
Asthmatics	Increased airways responsiveness; reduced lung function; unequal ventilation	Increased risk of exacerbation of respiratory symptoms; reduced lung function
Cigarette smokers	Impaired defense and clearance; lung injury	Increased damage through synergism
People with COPD	Reduced lung function; impaired clearance; unequal ventilation; airway inflammation	Increased risk of clinically significant effects on lung function
People with cardiovascular disease	Vascular inflammation; hypoxia; vasoconstriction; endothelial dysfunction; enhanced coagulation	Increased risk of myocardial infarction, congestive heart failure, stroke, cardiac arrhythmias
People with diabetes	Vascular disease increases risk	Increased risk of pulmonary and cardiac effects of exposure to particulate matter

confers increased risk seems intuitively obvious. To the degree that air pollution exposure decreases lung function, people who already have impaired lung function and reduced reserve will suffer greater consequences from a given reduction in function, compared with healthy people. A transient 25% decline in FEV_1 after exercising outdoors on a high-ozone day may have little consequence for a healthy person, but for the severe asthmatic who is already dyspneic on minimal exertion, a 25% decline in FEV_1 could prove disastrous or precipitate a hospital visit, even though the actual volume of lost function is relatively less.

In clinical studies, asthmatics exhibit exaggerated responses to SO_2, acidic aerosols, and perhaps NO_2, whereas individuals with COPD may have increased responsiveness to NO_2 but have generally not shown similar responsiveness to ozone. The most striking effect of acute exposure to SO_2 at concentrations of 1 ppm or below is the induction of bronchoconstriction in asthmatics after exposures lasting only 5 min.[170] In contrast, inhalation of concentrations in excess of 5 ppm causes only small decrements in airway function in normal subjects. Similarly, clinical studies have identified exercising adolescent asthmatics[171] and adult asthmatics[172] as being susceptible to sulfuric acid aerosols at high ambient concentrations — levels that do not affect healthy volunteers. Although several controlled human studies have found asthmatics responsive to low levels of NO_2, the findings have not been consistent.[173] The conflicting results among these studies are probably related to differences in subject selection and exposure protocols.

Factors other than impaired reserve likely contribute to the increased susceptibility of people with respiratory disease. Clinical studies have demonstrated that people with asthma and COPD may have enhanced fractional deposition of inhaled particles compared with healthy subjects.[26,174,175] Moreover, ventilation in diseased lungs is uneven, resulting in airway "hot spots" in which particle deposition may be several-fold higher than in other lung regions. Thus, the dose of inhaled particles reaching some parts of the airway epithelium may be markedly increased in people with obstructive disease. Asthma and COPD are characterized by underlying airway inflammation. Epithelial denudation and loss of cilia reduce clearance of inhaled particles, and may increase the likelihood of particle translocation to the interstitium.

Two studies have suggested that mild asthmatics may experience a neutrophilic airway inflammatory response to ozone exposure that is more intense than in healthy subjects.[176,177] Recent studies have also shown that, in asthmatics, ozone exposures at concentrations sufficient to induce an airway inflammatory response increase responsiveness to a subsequent allergen challenge.[178] Clinical investigations therefore suggest that the mechanisms for exacerbations of asthma following ozone exposure include bronchoconstriction, worsening of airway inflammation, and increased responsiveness to allergen challenge.

Asthmatics may also be more sensitive to combinations of pollutants. Frampton et al.[14] examined the effects of prior exposure to low-level sulfuric acid aerosol on the airway response to ozone in healthy and asthmatic subjects. Exposure–response relationships were examined using three levels of ozone, 0.08, 0.12, and 0.18 ppm. Ozone exposures were preceded 24 h earlier by exposure to a 100 μg/m^3 H_2SO_4 or NaCl aerosol. The acidic aerosol and oxidant exposures were 3 h in duration. Thirty healthy and an equal number of allergic asthmatics were studied. The findings revealed an interactive relationship between the ozone exposure concentration and sulfuric acid or sodium chloride aerosol preexposure in asthmatics, but this was not seen in healthy subjects. For the asthmatic subjects, ozone concentration-related differences in lung function were observed with H_2SO_4 preexposure but not for NaCl preexposure. These effects were observed for both FVC and FEV_1 immediately after and 4 h after ozone exposure. These data suggest that preexposure to sulfuric acid aerosols may alter responses to ozone in exercising asthmatics.

A number of epidemiological studies have suggested that emergency room and hospital visits for asthma are increased on high-ozone days. It is therefore surprising that controlled clinical studies have generally not found striking differences in lung function responses to ozone in the asthmatic as compared with healthy subjects. Several possible explanations exist. In contrast to studies with healthy volunteers, studies of asthmatic subjects have not been performed using prolonged exposures or repeated daily exposures. Furthermore, few studies with asthmatic subjects have incorporated multiple periods of moderate to intense exercise, a factor that contributes to changes in airway function with low-level ozone exposure in healthy volunteers.

One study has addressed some of these issues. Horstman and colleagues (Horstman et al., 1995) exposed 17 subjects with clinically active asthma and 13 healthy subjects to 0.16 ppm ozone for 7.6 h, with multiple prolonged periods of mild exercise. As shown in Table 2.6, asthmatic subjects had significantly larger decrements in FEV_1 and in FEV_1/FVC, despite occasional use of bronchodilators before and during the exposure. This study suggests that people with clinically active asthma may be at increased risk of bronchoconstriction following prolonged exposures to environmentally relevant concentrations of ozone.

Effects of inhaled pollutants in COPD have not been extensively examined. COPD patients exposed to ozone levels up to 0.30 ppm[179] showed similar or less response than healthy nonsmokers. To determine if low-level NO_2 induces changes in pulmonary function, Morrow et al.[180] investigated responses to inhalation of 0.3 ppm NO_2 for 4 h in 20 COPD subjects (mean age of 60 yr) with a history of cigarette smoking. These subjects were compared with 20 elderly healthy subjects of comparable age. Criteria for inclusion included dyspnea on exertion, airways obstruction ($FEV_1/FVC = 0.58 \pm 0.09$ [SD]), and a lack of response to inhaled bronchodilators. During intermittent light exercise, COPD subjects demonstrated progressive decrements in FVC and FEV_1 compared to baseline on exposure to NO_2 but not with air. Subgroup analyses suggested that responsiveness to NO_2 decreased with the severity of the COPD. In the cohort of elderly healthy subjects, the NO_2-induced reduction in FEV_1 was greater among smokers than nonsmokers. A comparison of the COPD and elderly healthy subjects also revealed distinctions in NO_2-induced responsiveness: no changes in lung function were observed in the elderly healthy group.

Studies of low-level exposure to CO have focused on subpopulations with ischemic heart disease and peripheral vascular disease. In patients with exertional angina, early onset of angina pectoris and ST segment depression have been consistently observed at carboxyhemoglobin (COHb) levels of 2 to 4% by several investigative teams. In the largest of these studies, the Health Effects Institute Multicenter CO Study,[181] 5 and 12% decreases in the time to onset of ST segment depression were observed at COHb levels of 2 and 4%, respectively. Significant decreases in time to onset of angina were also demonstrated at these COHb levels. These endpoints are remarkably consistent, and are compatible with the hypothesis that an elevated COHb level impairs the response of the myocardium to increased metabolic demands.

Other subpopulations (susceptible populations) that have been recruited to participate in clinical studies include subjects with asthma, individuals with allergies or acute upper respiratory infections, and subjects with COPD or coronary artery disease. Asthmatics are often characterized by their responsiveness to methacholine or carbachol, presence or absence of allergy (skin tests or IgE levels), use of medications, severity of symptoms, and degree of airway obstruction assessed by pulmonary function tests. Subjects with severe asthma have rarely participated in controlled clinical studies.

2.7 CONCLUSIONS

Controlled clinical studies provide a means for examining responses to air pollutants, especially those identified from epidemiological studies. Well-characterized exposures have been performed either in environmental chambers or by mouthpiece. Responses have traditionally been assessed by changes in respiratory mechanics, but more recent studies have involved direct sampling of the airways using fiberoptic bronchoscopy. The search continues for noninvasive markers of lower airway effects, and promising techniques include sputum induction and measurements of gases in exhaled air. Clinical studies with air pollutants have identified susceptible populations, characterized exposure–response relationships, and examined "lowest effect" levels. In brief, clinical studies have become increasingly sophisticated and should permit better definition of dosimetry as well as cellular and molecular mechanisms of pollutant effects. With the interest of government and industry in understanding ambient-air particle effects on human health, clinical studies will continue to provide data important in understanding air pollution health risks and their mechanisms.

REFERENCES

1. National Research Council. Research Priorities for Airborne Particulate Matter: I. Immediate Priorities and a Long-Range Research Portfolio, Commission on Life Sciences, Commission on Geosciences, Environment, and Resources, National Research Council, 1998.
2. Utell, M.J., and Drew, R. Summary of the workshop on clinical studies and particulate matter. *Inhal Toxicol*, 10, 625, 1998.
3. Lippmann, M., Frampton, M., Schwartz, J., Dockery, D., Schlesinger, R., Koutrakis, P., and Froines, J., Special report: the EPA's particulate matter (PM) health effects research centers program: a mid-course (2 year) report of status, progress, and plans, *Environ Health Perspect*, 111, 1074, 2003.
4. Frampton, M.W., and Utell, M.J., Clinical studies of airborne pollutants, in *Toxicology of the Lung*, Gardner, E.E., Crapo, J.D., and McClellan, R.O., Eds., 3rd ed., Taylor and Francis, Philadelphia, PA, 1999, p. 455.
5. Bonner, F.T., and Hughes, M.N., No lack of NO activity, *Science,* 260, 145, 1993.
6. McDonnell, W.F., Kehrl, H.R., Abdul-Salaam, S., Ives, P.J., Folinsbee, L.J., Devlin, R.B., O'Neil, J.J., and Horstman, D.H., Respiratory response of humans exposed to low levels of ozone for 6.6 hours. *Arch Environ Health*, 46, 145, 1991.
7. Devlin, R.B., McDonnell, W.F., Mann, R., Becker, S., House, D.E., Schreinemachers, D., and Koren, H.S., Exposure of humans to ambient levels of ozone for 6.6 hours causes cellular and biochemical changes in the lung. *Am J Respir Cell Mol Biol*, 4, 72, 1991.
8. McDonnell, W.F., Stewart, P.W., Andreoni, S., Seal, E., Jr., Kehrl, H.R., Horstman, D.H., Folinsbee, L.J., and Smith, M.V., Prediction of ozone-induced FEV_1 changes. *Am J Respir Crit Care Med*, 156, 715, 1997.
9. Daigle, C.C., Chalupa, D.C., Gibb, F.R., Morrow, P.E., Oberdörster, G., Utell, M.J., and Frampton, M.W., Ultrafine particle deposition in humans during rest and exercise. *Inhal Toxicol*, 15, 539, 2003.
10. Cannon, J.G., Meydani, S.N., Fielding, R.A., Fiatorone, M.A., Meydani, M., Farhangmehr, M., Orencole, S.F., Blumberg, J.B., and Evans, W.J., Acute phase response in exercise. II. Associations between vitamin E, cytokines, and muscle proteolysis. *Am J Physiol*, 260, R1235–R1240, 1991.
11. Seaton, A., MacNee, W., Donaldson, K., and Godden, D., Particulate air pollution and acute health effects. *Lancet*, 345, 176, 1995.
12. van Eeden, S.F., Granton, J., Hards, J.M., Moore, B., and Hogg, J.C., Expression of the cell adhesion molecules on leukocytes that demarginate during acute maximal exercise. *J Appl Physiol*, 86, 970, 1999.
13. Frampton, M.W., Azadniv, M., Chalupa, D., Morrow, P.E., Gibb, F.R., Oberdörster, G., Boscia, J., and Speers, D.M., Blood leukocyte expression of LFA-1 and ICAM-1 after inhalation of ultrafine carbon particles. *Am J Respir Crit Care Med*, 163, A264, 2001.
14. Frampton, M.W., Morrow, P.E., Cox, C., Levy, P.C., Condemi, J.J., Speers, D., Gibb, F.R., and Utell, M.J., Sulfuric acid aerosol followed by ozone exposure in healthy and asthmatic subjects. *Environ Res*, 69, 1, 1995.
15. Spektor, S., Luparello, T.J., Kopetzky, M.T., Souhrada, J., and Kinsman, R.A., Response of asthmatics to methacholine and suggestion. *Am Rev Respir Dis,* 113, 43, 1976.

16. Jones, B., and Kenward, M.G., *Design and Analysis of Cross-Over Trials,* Chapman and Hall, New York, 1989.
17. National Institutes of Health, The Belmont Report: Ethical Principles and Guidelines for the Protection of Human Subjects of Research. Bethesda, MD, National Commission for the Protection of Human Subjects of Biomedical and Behavioral Research, 1979.
18. Findeisen, W., Uber das absetzen kleiner in der luft suspendierte teilchen in der menschlichen lunge bei der atmung. *Pflügers Arch Ges Physiol,* 236, 367, 1935.
19. Landahl, H.D., On the removal of airborne droplets by the human respiratory tract—I. The lung. *Bull Math Biophys,* 12, 43, 1950.
20. Albert, R.E., and Arnett, L.C., Clearance of radioactive dust from the human lung. *Arch Ind Health,* 12, 99, 1955.
21. Lippmann, M., and Albert, R.E., Clearance of radioactive dust from the human lung. *Am Ind Hyg Assoc J,* 30, 257, 1969.
22. Heyder, J., Gebhart, J., Heigwer, G., Roth, C., and Stahlhofen, W., Experimental studies of total deposition of aerosol particles in the human respiratory tract. *J Aerosol Sci,* 4, 191, 1973.
23. Utell, M.J., Gelein, R., Yu, C.P., Kenaga, C., Geigel, E., Torres, A., Chalupa, D., Gibb, F.R., Speers, D., Mast, R.W., and Morrow, P.E., Quantitative exposure of humans to an octametylcyclotetrasiloxane (D4) vapor. *Toxicol Sci,* 44, 206, 1998.
24. Bolton, S., Ed., Data graphics, in *Pharmaceutical Statistics: Practical and Clinical Applications,* Marcel Dekker, New York, 1998, p. 32.
25. Oberdörster, G., Gelein, R.M., Ferin, J., and Weiss, B., Association of particulate air pollution and acute mortality: involvement of ultrafine particles? *Inhal Toxicol,* 7, 111, 1995.
26. Chalupa, D.C., Morrow, P.E., Oberdörster, G., Utell, M.J., and Frampton, M.W., Ultrafine particle deposition in subjects with asthma. *Environ Health Perspect,* 112, 879, 2004.
27. Utell, M.J., Absorption, Kinetics, and Elimination of ^{14}C-octamethylcyclotetrasiloxane (C-14 D_4) in Humans After a One Hour Respiratory Exposure. Report to the Dow Corning Corporation, 2000.
28. Plotzke, K.P., Crofoot, S.D., Ferdinandi, E.S., Beattie, J.G., Reitz, R.H., McNett, D.A., and Meeks, R.G., Disposition of radioactivity in fischer 344 rats after single and multiple inhalation exposure to [(14)C]Octamethylcyclotetrasiloxane ([(14)C]D(4)). *Drug Metab Dispos,* 28, 192, 2000.
29. Reddy, M.B., Andersen, M.E., Morrow, P.E., Dobrev, I.D., Varaprath, S., Plotzke, K.P., and Utell, M.J., Physiological modeling of inhalation kinetics of octamethylcyclotetrasiloxane in humans during rest and exercise. *Toxicol Sci,* 72, 3, 2003.
30. Samet, J.M., A historical and epidemiologic perspective on respiratory symptoms questionnaires. *Am J Epidemiol* 108, 435, 1978.
31. Ferris, B.G., Epidemiology standardization project. *Am Rev Respir Dis* 118 (6, part 2), 1, 1978.
32. Frampton, M.W., Morrow, P.E., Torres, A., Cox, C., Voter, K.Z., and Utell, M.J., Ozone responsiveness in smokers and nonsmokers. *Am J Respir Crit Care Med,* 155, 116, 1997.
33. Torres, A., Utell, M.J., Morrow, P.E., Voter, K.Z., Whitin, J.C., Cox, C., Looney, R.J., Speers, D.M., Tsai, Y., and Frampton, M.W., Airway inflammation in smokers and nonsmokers with varying responsiveness to ozone. *Am J Respir Crit Care Med,* 156, 728, 1997.

34. Hruby, J., and Butler, J., Variability of routine pulmonary function tests. *Thorax* 31, 548, 1975.
35. Boushey, H.A. Jr., and Dawson, A., Spirometry and flow-volume curves, in *Pulmonary Function Testing-Guidelines and Controversies: Equipment, Methods, and Normal Values*, Clausen, J.L., Ed., Academic Press, New York, 1982, p. 61.
36. American Thoracic Society, Guidelines for methacholine and exercise challenge testing–1999. *Am J Respir Crit Care Med*, 161, 309, 2000.
37. Hargreave, F.E., Dolovich, J., and Boulet, L.P., Inhalation provocation tests. *Respir Med* 4: 224–236, 1983.
38. Bauer, M.A., Utell, M.J., Morrow, P.E., Speers, D.M., and Gibb, F.R., Inhalation of 0.30 ppm nitrogen dioxide potentiates exercise-induced bronchospasm in asthmatics. *Am Rev Respir Dis*, 134, 1203, 1986.
39. Utell, M.J., Morrow, P.E., and Hyde, R.W., Airway reactivity to sulfate and sulfuric acid aerosols in normal and asthmatic subjects. *J Air Pollut Control Assoc*, 34, 931, 1984.
40. Utell, M.J., Morrow, P.E., Speers, D.M., Darling, J., and Hyde, R.W., Airway responses to sulfate and sulfuric acid aerosols in asthmatics: an exposure-response relationship. *Am Rev Respir Dis*, 128, 1983.
41. Fulkerson, W.J., Fiberoptic bronchoscopy. *N Engl J Med*, 311, 511, 1984.
42. Ghio, A.J., and Devlin, R.B., Inflammatory lung injury after bronchial instillation of air pollution particles. *Am J Respir Crit Care Med*, 164, 704, 2001.
43. Hunninghake, G.W., Kawanami, O., Ferrans, V.J., Young Jr., R.C., Roberts, W.C., and Crystal, R.G., Characterization of the inflammatory and immune effector cells in the lung parenchyma of patients with interstitial lung disease. *Am Rev Respir Dis*, 123, 407, 1981.
44. Nicod, L.P., and El Habre, F., Adhesion molecules on human lung dendritic cells and their role for T-cell activation. *Am J Respir Cell Mol Biol*, 7, 207–213, 1992.
45. Hunninghake, G.W., Gadek, J.E., Kawanami, O., Ferrans, V.J., and Crystal, R.G., Inflammatory and immune processes in the human lung in health and disease: evaluation by bronchoalveolar lavage. *Am J Pathol*, 97, 149–206, 1979.
46. Rennard, S.I., Ghafouri, M.O., Thompson, A.B., Linder, J., Vaughan, W., Jones, K., Ertl, R.F., Christensen, K., Prince, A., Stahl, M.G., and Robbins, R.A., Fractional processing of sequential bronchoalveolar lavage to separate bronchial and alveolar samples. *Am Rev Respir Dis*, 141, 208, 1990.
47. Saltini, C., Hance, A.J., Ferrans, V.J., Basset, F., Bitterman, P.B., and Crystal, R.G., Accurate quantification of cells recovered by bronchoalveolar lavage. *Am Rev Respir Dis*, 130, 650–658, 1984.
48. Willcox, M., Kervitsky, A., Watters, L.C., and King, T.E. Jr., Quantification of cells recovered by bronchoalveolar lavage. Comparison of cytocentrifuge preparations with the filter method. *Am Rev Respir Dis*, 138, 74, 1988.
49. Ettensohn, D.B., Jankowski, M.J., Duncan, P.G., and Lalor, P.A., Bronchoalveolar lavage in the normal volunteer subject. I. Technical aspects and intersubject variability. *Chest*, 94, 275–280, 1988.
50. Ettensohn, D.B., Jankowski, M.J., Redondo, A.A., and Duncan, P. G., Bronchoalveolar lavage in the normal volunteer subject. 2. Safety and results of repeated BAL, and use in the assessment of intrasubject variability. *Chest*, 94, 281–285, 1988.
51. Koren, H.S., Devlin, R.B., Graham, D.E., Mann, R., McGee, M.P.D., Horstmann, H., Kozumbo, W.J., Becker, S., House, D.E., McDonnell, W.F., and Bromberg, P.A., Ozone-induced inflammation in the lower airways of human subjects. *Am Rev Respir Dis* 139, 407–415, 1989.

52. Ghio, A.J., Kim, C., and Devlin, R.B., Concentrated ambient air particles induce mild pulmonary inflammation in healthy human volunteers. *Am J Respir Crit Care Med,* 162, 981, 2000.
53. Frampton, M.W., Voter, K.Z., Morrow, P.E., Roberts, N.J., Jr., Culp, D.J., Cox, C., and Utell, M.J., Sulfuric acid aerosol exposure in humans assessed by bronchoalveolar lavage. *Am Rev Respir Dis,* 146, 626, 1992.
54. Culp, D.J., Chen, Y., Frampton, M.W., Gibb, F.R., Speers, D.M., and Utell, M.J., Sulfuric acid inhalation and biochemical characteristics of human airway mucous glycoproteins, *Am Rev Respir Dis,* 141, A77, 90.
55. Balmes, J.R., Chen, L.L., Scannell, C., Tager, I., Christian, D., Hearne, P.Q., Kelly, T., and Aris, R.M., Ozone-induced decrements in FEV_1 and FVC do not correlate with measures of inflammation. *Am J Respir Crit Care Med,* 153, 904, 1996.
56. Weinmann, G.G., Liu, M.C., Proud, D., Weidenbach-Gerbase, M., Hubbard, W., and Frank, R., Ozone exposure in humans: inflammatory, small and peripheral airway responses. *Am J Respir Crit Care Med,* 152, 1175, 1995.
57. Azadniv, M., Utell, M.J., Morrow, P.E., Gibb, F.R., Nichols, J., Roberts, N.J., Jr., Speers, D.M., Torres, A., Tsai, Y., Abraham, M.K., Voter, K.Z., and Frampton, M.W., Effects of nitrogen dioxide exposure on human host defense. *Inhal Toxicol,* 10, 585, 1998.
58. Frampton, M.W., Boscia, J., Roberts, N.J.J., Azadniv, M., Torres, A., Cox, C., Morrow, P.E., Nichols, J., Chalupa, D., Frasier, L.M., Gibb, F.R., Speers, D.M., Tsai, Y., and Utell, M.J., Nitrogen dioxide exposure: effects on airway and blood cells. *Am J Physiol,* 282, L155–L165, 2002.
59. Eschenbacher, W.L., and Gravelyn, T.R., A technique for isolated airway segment lavage. *Chest,* 92, 105, 1987.
60. Jörres, R.A., Holz, O., Zachgo, W., Timm, P., Koschyk, S., Müller, B., Grimminger, F., Seeger, W., Kelly, F.J., Dunster, C., Frischer, T., Lubec, G., Waschewski, M., Niendorf, A., and Magnussen, H., The effect of repeated ozone exposures on inflammatory markers in bronchoalveolar lavage fluid and mucosal biopsies. *Am J Respir Crit Care Med,* 161, 1855, 2000.
61. Stenfors, N., Pourazar, J., Blomberg, A., Krishna, M.T., Mudway, I., Helleday, R., Kelly, F.J., Frew, A.J., and Sandstrom, T., Effect of ozone on bronchial mucosal inflammation in asthmatic and healthy subjects. *Respir Med,* 96, 352, 2002.
62. Salvi, S., Blomberg, A., Rudell, B., Kelly, F., Sandstrom, T., Holgate, S.T., and Frew, A., Acute inflammatory responses in the airways and peripheral blood after short-term exposure to diesel exhaust in healthy human volunteers. *Am J Respir Crit Care Med,* 159, 702, 1999.
63. Graham, D.E., and Koren, H.S., Biomarkers of inflammation in ozone-exposed humans. Comparison of the nasal and bronchoalveolar lavage. *Am Rev Respir Dis,* 142, 152, 1990.
64. Diaz-Sanchez, D., Dotson, A.R., Takenaka, H., and Saxon, A., Diesel exhaust particles induce local IgE production in vivo and alter the pattern of IgE messenger RNA isoforms. *J Clin Invest,* 94, 1417, 1994.
65. Diaz-Sanchez, D., Tsien, A., Fleming, J., and Saxon, A., Combined diesel exhaust particulate and ragweed allergen challenge markedly enhances human in vivo nasal ragweed-specific IgE and skews cytokine production to a T helper cell 2-type pattern. *J Immunol,* 158, 2406, 1997.

66. Diaz-Sanchez, D., Garcia, M.P., Wang, M., Jyrala, M., and Saxon, A., Nasal challenge with diesel exhaust particles can induce sensitization to a neoallergen in the human mucosa. *J Allergy Clin Immunol,* 104, 1183, 1999.
67. Vastag, E., Matthys, H., Zsamboki, G., Kohler, D., and Daikeler, G., Mucociliary clearance in smokers. *Eur J Respir Dis,* 68, 107–113, 1986.
68. Schlesinger, R.B., The interaction of inhaled toxicants with respiratory tract clearance mechanisms. *Crit Rev Toxicol,* 20, 257–286, 1990.
69. Leikauf, G., Yeates, D.B., Wales, K.A., Spektor, D., Albert, R.E., and Lippmann, M., Effects of sulfuric acid aerosol on respiratory mechanics and mucociliary particle clearance in healthy nonsmoking adults. *Am Ind Hyg Assoc J,* 42, 273–282, 1981.
70. Spektor, D.M., Yen, B.M., and Lippmann, M., Effect of concentration and cumulative exposure of inhaled sulfuric acid on tracheobronchial particle clearance in healthy humans. *Environ Health Perspect,* 79, 167–172, 1989.
71. Lippmann, M., Progress, prospects, and research needs on the health effects of acid aerosols. *Environ Health Perspect,* 79, 203–205, 1989.
72. Foster, W.M., Costa, D.L., and Langenback, E.G., Ozone exposure alters tracheobronchial mucociliary function in humans. *J Appl Physiol,* 63, 996–1002, 1987.
73. Gerrity, T.R., Bennett, W.D., Kehrl, H., and Dewitt, P.J., Mucociliary clearance of inhaled particles measured at 2 h after ozone exposure in humans. *J Appl Physiol,* 74, 2984–2989, 1993.
74. Douglas, W.W., Hepper, N.G.G., and Colby, T.V., Silo-filler's disease. *Mayo Clin Proc,* 64, 291, 1989.
75. Frampton, M.W., Finkelstein, J.N., Roberts, N.J.Jr., Smeglin, A.M., Morrow, P.E., and Utell, M.J., Effects of nitrogen dioxide exposure on bronchoalveolar lavage proteins in humans. *Am J Respir Cell Mol Biol,* 1, 499, 1989.
76. Kehrl, H.R., Vincent, L.M., Kowalsky, R.J., Horstman, D.H., O'Neill, J.J., McCartney, W.H., and Bromberg, P.A., Ozone exposure increases respiratory epithelial permeability in humans. *Am Rev Respir Dis,* 135, 1124, 1987.
77. Oberdörster, G., Utell, M.J., Morrow, P.E., Hyde, R.W., Weber, D.A., and Drago, S.R., Decreased lung clearance of inhaled ^{99m}Tc-DTPA aerosols after NO_2 exposure: indication of lung epithelial permeability change? *J Aerosol Sci,* 17, 320, 1986.
78. Hargreave, F.E., Popov, T., Kidney, J., and Dolovich, J., Sputum measurements to assess airway inflammation in asthma. *Allergy,* 48, 81, 1993.
79. Fahy, J.V., Liu, J., Wong, H., and Boushey, H.A., Cellular and biochemical analysis of induced sputum from asthmatic and from healthy subjects. *Am Rev Respir Dis,* 147, 1126, 1993.
80. Alexis, N.E., Hu, S.-C., Zeman, K., Alter, T., and Bennett, W.D., Induced sputum derives from the central airways. Confirmation using a radiolabeled aerosol bolus delivery technique. *Am J Respir Crit Care Med,* 164, 1964, 2001.
81. Pin, I., Gibson, P.G., Kolendowicz, R., Girgis-Gabardo, A., Denburg, J.A., and Dolovich, J., Use of induced sputum cell counts to investigate airway inflammation in asthma. *Thorax,* 47, 25, 1992.
82. Fahy, J.V., Kim, K.W., Liu, J., and Boushey, H.A., Prominent neutrophilic inflammation in sputum from subjects with asthma exacerbation. *J Allergy Clin Immunol,* 95, 843, 1995.

83. Pin, I., Freitag, A.P., O'Byrne, P.M., Girgis-Gabardo, A., Watson, R.M., Dolovich, J., Denburg, J.A., and Hargreave, F.E., Changes in the cellular profile of induced sputum after allergen-induced asthmatic responses. *Am Rev Respir Dis*, 145, 1265, 1992.
84. Claman, D.M., Boushey, H.A., Liu, J., Wong, H., and Fahy, J.V., Analysis of induced sputum to examine the effects of prednisone on airway inflammation in asthmatic subjects. *J Allergy Clin Immunol*, 94, 861, 1994.
85. Fahy, J.V., Wong, H., Liu, J., and Boushey, H.A., Comparison of samples collected by sputum induction and bronchoscopy from asthmatic and healthy subjects. *Am J Respir Crit Care Med*, 152, 53, 1995.
86. Holz, O., Richter, K., Jorres, R.A., Speckin, P., Mucke, M., and Magnussen, H., Changes in sputum composition between two inductions performed on consecutive days. *Thorax*, 53, 83, 1997.
87. Nightingale, J.A., Rogers, D.F., and Barnes, P.J., Effect of repeated sputum induction on cell counts in normal volunteers. *Thorax*, 53, 87, 1998.
88. de la Fuente, P.T., Romagnoli, M., Godard, P., Bousquet, J., and Chanez, P., Safety of inducing sputum in patients with asthma of varying severity. *Am J Respir Crit Care Med*, 157, 1127, 1998.
89. Pizzichini, E., Pizzichini, M.M.M., Efthimiadis, A., Evans, S., Morris, M.M., Squillace, D., Gleich, G.J., Dolovich, J., and Hargreave, F.E., Indices of airway inflammation in induced sputum: reproducibility and validity of cell and fluid-phase measurements. *Am J Respir Crit Care Med*, 154, 308, 1996.
90. Spanevello, A., Confalonieri, M., Sulotto, F., Romano, F., Balzano, G., Migliori, G.B., Bianchi, A., and Michetti, G., Induced sputum cellularity: reference values and distribution in normal volunteers. *Am J Respir Crit Care Med*, 162, 1172, 2000.
91. Fahy, J.V., Wong, H., Liu, J., and Boushey, H.A., Analysis of induced sputum after air and ozone exposures in healthy subjects. *Environ Res*, 70, 77, 1995.
92. Vagaggini, B., Paggiaro, P.L., Giannini, D., Franco, A.D., Cianchetti, S., Carnevali, S., Taccola, M., Bacci, E., Bancalari, L., Dente, F.L., and Giuntini, C., Effect of short-term NO_2 exposure on induced sputum in normal, asthmatic and COPD subjects. *Eur Respir J*, 9, 1852, 1996.
93. Effros, R.M., Wahlen, K., Bosbous, M., Castillo, D., Foss, B., Dunning, M., Gare, M., Lin, W., and Sun, F., Dilution of respiratory solutes in exhaled condensates. *Am J Respir Crit Care Med*, 165, 663, 2002.
94. Mutlu, G.M., Garey, K.W., Robbins, R.A., Danziger, L.H., and Rubinstein, I., Collection and analysis of exhaled breath condensate in humans. *Am J Respir Crit Care Med*, 164, 731, 2001.
95. Hyde, R.W., I don't know what you guys are measuring but you sure are measuring it! A fair criticism of measurements of exhaled condensates? *Am J Respir Crit Care Med*, 165, 301, 2002.
96. Effros, R.M., Biller, J., Foss, B., Hoagland, K., Dunning, M.B., Castillo, D., Bosbous, M., Sun, F., and Shaker, R., A simple method for estimating respiratory solute dilution in exhaled breath condensates. *Am J Respir Crit Care Med*, 168, 1500, 2003.
97. Corradi, M., Alinovi, R., Goldoni, M., Vettori, M.V., Folesani, G.M.P., Cavazzini, S., Bergamaschi, E., Rossi, L., and Mutti, A., Biomarkers of oxidative stress after controlled human exposure to ozone. *Toxicol Lett*, 134, 219, 2002.
98. Montuschi, P., Nightingale, J.A., Kharitonov, S.A., and Barnes, P.J., Ozone-induced increase in exhaled 8-isoprostane in healthy subjects is resistant to inhaled budesonide. *Free Radic Biol Med*, 33, 1403, 2002.

99. Blaser, L., Measured breath. *Environ Health Perspect*, 104, 1292, 1996.
100. Barnes, P.J., and Kharitonov, S.A., Exhaled nitric oxide: a new lung function test. *Thorax*, 51, 233, 1996.
101. Zayasu, K., Sekizawa, K., Okinaga, S., Yamaya, M., Ohrui, T., and Sasaki, H., Increased carbon monoxide in exhaled air of asthmatic patients. *Am J Respir Crit Care Med*, 156, 1140, 1997.
102. Foster, W.M., Jiang, L., Stetkiewicz, P.T., and Risby, T.H., Breath isoprene: temporal changes in respiratory output after exposure to ozone. *J Appl Physiol*, 80, 706, 1996.
103. Arterbery, V.E., Pryor, W.A., Jiang, L., Sehnert, S.S., Foster, W.M., and Risby, T., Breath ethane generation during clinical total body irradiation as a marker of oxygen-free-radical mediated lipid peroxidation: a case study. *Free Radic Biol Med*, 17, 569, 1994.
104. Euler, D.E., Dave, S.J., and Guo, H., Effect of cigarette smoking on pentane excretion in alveolar breath. *Clin Chem*, 42, 303, 1996.
105. Alving, K., Weitzberg, E., and Lundberg, J.M., Increased amount of nitric oxide in exhaled air of asthmatics. *Eur Respir J*, 6, 1368, 1993.
106. Kharitonov, S.A., Chung, K.F., Evans, D., O'Connor, B.J., and Barnes, P.J., Increased exhaled nitric oxide in asthma is mainly derived from the lower respiratory tract. *Am J Respir Crit Care Med*, 153, 1773, 1996.
107. Dupont, L.J., Rochette, F., Demedts, M.G., and Verleden, G.M., Exhaled nitric oxide correlates with airway hyperresponsiveness in steroid-naive patients with mild asthma. *Am J Respir Crit Care Med*, 157, 894, 1998.
108. Djukanovic, R., Homeyard, S., Gratziou, C., Madden, J., Walls, A., Montefort, S., Peroni, D., Polosa, R., Holgate, S., and Howarth, P., The effect of treatment with oral corticosteroids on asthma symptoms and airway inflammation. *Am J Respir Crit Care Med*, 155, 826, 1997.
109. Kharitonov, S.A., Yates, D.H., and Barnes, P.J., Inhaled glucocorticoids decrease nitric oxide in exhaled air of asthmatic patients. *Am J Respir Crit Care Med*, 153, 454, 1996.
110. American Thoracic Society, Recommendations for standardized procedures for the online and offline measurement of exhaled lower respiratory nitric oxide and nasal nitric oxide in adults and children. *Am J Respir Crit Care Med*, 160, 2104, 1999.
111. Kharitonov, S., Alving, K., and Barnes, P.J., Exhaled and nasal nitric oxide measurements: recommendations. *Eur Respir J*, 10, 1683, 1997.
112. Hogman, M., Stromberg, S., Schedin, U., Frostess, C., Hedenstierna, G., and Gustaffson, L.E., Nitric oxide from the human respiratory tract efficiently quantified by standardised single breath measurements. *Acta Physiol Scand*, 159, 345, 1997.
113. Silkoff, P.E., McClean, P.A., Slutsky, A.S., Furlott, H.G., Hoffstein, E., Wakita, S., Chapman, R., Szalai, J.P., and Zamel, N., Marked flow-dependence of exhaled nitric oxide using a new technique to exclude nasal nitric oxide. *Am J Respir Crit Care Med*, 155, 260, 1997.
114. Tsoukias, N.M., and George, S.C., A two-compartment model of pulmonary nitric oxide exchange dynamics. *J Appl Physiol*, 85, 653, 1998.
115. Pietropaoli, A.P., Perillo, I.B., Torres, A., Perkins, P.T., Frasier, L.M., Utell, M.J., Frampton, M.W., and Hyde, R.W., Simultaneous measurement of nitric oxide production by conducting and alveolar airways of humans. *J Appl Physiol*, 87, 1532, 1999.

116. Hogman, M., Drca, N., Ehrstedt, C., and Merilainen, P., Exhaled nitric oxide partitioned into alveolar, lower airways and nasal contributions. *Respir Med*, 94, 985, 2000.
117. Silkoff, P.E., Sylvester, J.T., Zamel, N., and Permutt, S., Airway nitric oxide diffusion in asthma: role in pulmonary function and bronchial responsiveness. *Am J Respir Crit Care Med*, 161, 1218, 2000.
118. Jörres, R.A., Modelling the production of nitric oxide within the human airways. *Eur Respir J*, 16, 555, 2000.
119. George, S.C., Hogman, M., Permutt, S., and Silkoff, P.E., Modeling pulmonary nitric oxide exchange. *J Appl Physiol*, 96, 831, 2004.
120. Nightingale, J.A., Rogers, D.F., Chung, K.F., and Barnes, P.J., No effect of inhaled budesonide on the response to inhaled ozone in normal subjects. *Am J Respir Crit Care Med*, 161, 479, 2000.
121. Nickmilder, M., Carbonnelle, S., deBurbure, C., and Bernard, A., Relationship between ambient ozone and exhaled nitric oxide in children. *JAMA*, 290, 2546, 2003.
122. Koenig, J.Q., Jansen, K., Mar, T.F., Lumley, T., Kaufman, J., Trenga, C.A., Sullivan, J., Liu, L.J., Shapiro, G.G., and Larson, T.V., Measurement of offline exhaled nitric oxide in a study of community exposure to air pollution. *Environ Health Perspect*, 111, 1625, 2003.
123. Steerenberg, P.A., Bischoff, E.W., de Klerk, A., Verlaan, A.P., Jongbloets, L.M., van Loveren, H., Opperhuizen, A., Zomer, G., Heisterkamp, S.H., Hady, M., Spieksma, F.T., Fischer, P.H., Dormans, J.A., and van Amsterdam, J.G., Acute effect of air pollution on respiratory complaints, exhaled NO and biomarkers in nasal lavages of allergic children during the pollen season. *Int Arch Allergy Immunol*, 131, 127, 2003.
124. Sundblad, B.M., Larsson, B.M., Palmberg, L., and Larsson, K., Exhaled nitric oxide and bronchial responsiveness in healthy subjects exposed to organic dust. *Eur Respir J*, 20, 426, 2002.
125. Girgis, R.E., Gugnani, M.K., Abrams, J., and Mayes, M.D., Partitioning of alveolar and conducting airway nitric oxide in scleroderma lung disease. *Am J Respir Crit Care Med*, 165, 1587, 2002.
126. Shin, H.W., Rose-Gottron, C.M., Sufi, R.S., Perez, F., Cooper, D.M., Wilson, A.F., and George, S.C., Flow-independent nitric oxide exchange parameters in cystic fibrosis. *Am J Respir Crit Care Med*, 165, 349, 2002.
127. Hogman, M., Homkvist, T., Wegener, T., Emtner, M., Andersson, M., Hedenstrom, H., and Merilainen, P., Extended NO analysis applied to patients with COPD, allergic asthma, and allergic rhinitis. *Respir Med*, 96, 24, 2002.
128. Pietropaoli, A.P., Delehanty, J.M., Perkins, P.T., Utell, M.J., Oberdörster, G., Hyde, R.W., Frasier, L.M., Speers, D.M., Chalupa, D.C., and Frampton, M.W., Venous nitrate, nitrite, and forearm blood flow after carbon ultrafine particle exposure in healthy human subjects, *Am J Respir Crit Care Med*, 169, A883, 2004.
129. Horvath, I., Loukides, S., Wodehouse, T., Kharitonov, S.A., Cole, P.J., and Barnes, P.J., Increased levels of exhaled carbon monoxide in bronchiectasis: a new marker of oxidative stress. *Thorax*, 53, 867, 1998.
130. Paredi, P., Kharitonov, S.A., and Barnes, P.J., Analysis of expired air for oxidation products. *Am J Respir Crit Care Med*, 166, S31–S37, 2002.
131. Choi, A.M., and Alam, J., Heme oxygenase-1: function, regulation, and implication of a novel stress-inducible protein in oxidant-induced lung injury. *Am J Respir Cell Mol Biol*, 15, 9, 1996.

132. Yamaya, M., Sekizawa, K., Ishizuka, S., Monma, M., and Sasaki, H., Exhaled carbon monoxide levels during treatment of acute asthma. *Eur Respir J*, 13, 757, 1999.
133. Paredi, P., Shah, P.L., Montuschi, P., Sullivan, P., Hodson, M.E., Kharitonov, S.A., and Barnes, P.J., Increased carbon monoxide in exhaled air of patients with cystic fibrosis. *Thorax*, 54, 917, 1999.
134. Antuni, J.D., Kharitonov, S.A., Hughes, D., Hodson, M.E., and Barnes, P.J., Increase in exhaled carbon monoxide during exacerbations of cystic fibrosis. *Thorax*, 55, 138, 2000.
135. Monma, M., Yamaya, M., Sekizawa, K., Ideda, K., Suzuki, N., Kikuchi, T., Takasaka, T., and Sasaki, H., Increased carbon monoxide in exhaled air of patients with seasonal allergic rhinitis. *Clin Exp Allergy*, 29, 1537, 1999.
136. Yamaya, M., Sekizawa, K., Ishizuka, S., Monma, M., Mizuta, K., and Sasaki, H., Increased carbon monoxide in exhaled air of subjects with upper respiratory tract infections. *Am J Respir Crit Care Med*, 158, 311, 1998.
137. Nightingale, J.A., Maggs, R., Cullinan, P., Donnelly, L.E., Rogers, D.F., Kinnersley, R., Chung, K.F., Barnes, P.J., Ashmore, M., and Newman-Taylor, A., Airway inflammation after controlled exposure to diesel exhaust particulates. *Am J Respir Crit Care Med*, 162, 161, 2000.
138. Horvath, I., MacNee, W., Kelly, F.J., Dekhuijzen, P.N.R., Phillips, M., Doring, G., Choi, A.M.K., Yamaya, M., Bach, R.H., Wilis, D., Donnelly, L.E., Chung, K.F., and Barnes, P.J., Haemoxygenase-1 induction and exhaled markers of oxidative stress in lung diseases, summary of the ERS research seminar in Budapest, Hungary, September, 1999. *Eur Respir J*, 18, 420, 2001.
139. Chapman, J.T., and Choi, A.M.K., Exhaled monoxides as a pulmonary function test. *Clin Chest Med*, 22, 817, 2001.
140. Zetterquist, W., Marteus, H., Johannesson, M., Nordvall, S.L., Ihre, E., Lundberg, J.O.N., and Alving, K., Exhaled carbon monoxide is not elevated in patients with asthma or cystic fibrosis. *Eur Respir J*, 20, 92, 2002.
141. Paredi, P., Kharitonov, S.A., and Barnes, P.J., Exhaled carbon monoxide is not elevated in patients with asthma or cystic fibrosis. *Eur Respir J*, 21, 197, 2003.
142. Zetterquist, W., Marteus, H., Johannesson, M., Nordval, S.L., Ihre, E., Lundberg, J.O., and Alving, K., Exhaled carbon monoxide is not elevated in patients with asthma or cystic fibrosis. *Eur Respir J*, 21, 197, 2003.
143. Deveci, S.E., Deveci, F., Acik, Y., and Ozan, A.T., The measurement of exhaled carbon monoxide in healthy smokers and non-smokers. *Respir Med*, 98, 551, 2003.
144. Jarvis, M.J., Russell, M.A.H., and Saloojee, Y., Expired air carbon monoxide: a simple breath test of tobacco smoke intake. *Br Med J*, 2, 484, 1980.
145. Togores, B., Bosch, M., and Agusti, A.G.N., The measurement of exhaled carbon monoxide is influenced by airflow obstruction. *Eur Respir J*, 15, 177, 2000.
146. Zarling, E.J., and Clapper, M., Technique for gas-chromatographic measurement of volatile alkanes from single-breath samples. *Clin Chem*, 33, 140, 1987.
147. Paredi, P., Kharitonov, S.A., Leak, D., Ward, S., Cramer, D., and Barnes, P.J., Exhaled ethane, a marker of lipid peroxidation, is elevated in chronic obstructive pulmonary disease. *Am Ann Deaf Am J Respir Crit Care Med*, 162, 369, 2000.
148. Olapade, C.O., Zakkar, M., Swedler, W.I., and Rubinstein, I., Exhaled pentane levels in acute asthma. *Chest*, 111, 862, 1997.
149. Utell, M.J., Frampton, M.W., Zareba, W., Devlin, R.B., and Cascio, W.E., Cardiovascular effects associated with air pollution: potential mechanisms and methods of testing. *Inhal Toxicol*, 14, 1231, 2002.

150. Brook, R.D., Franklin, B., Cascio, W., Hong, Y., Howard, G., Lipsett, M., Luepker, R., Mittleman, M., Samet, J., Smith, S.C.J., and Tager, I., Air pollution and cardiovascular disease: a statement for healthcare professionals from the expert panel on population and prevention science of the American Heart Association. *Circulation*, 109, 2655, 2004.
151. Frampton, M.W., Utell, M.J., and Samet, J.M., Cardiopulmonary consequences of particle inhalation, in *Particle-Lung Interactions*, Gehr, P. and Heyder, J., Eds., Marcel Dekker, New York, 2000, p. 653.
152. Zareba, W., Moss, A.J., and Badilini, F., Dispersion of repolarization: a noninvasive marker of nonuniform recovery of ventricular excitability, in *Noninvasive Electrocardiology: Clinical Aspects of Holter Monitoring*, Moss, A.J. and Stern, S., Eds., WB Saunders, London, 1995, p. 405.
153. Stys, A., and Stys, T., Current clinical applications of heart rate variability. *Clin Cardiol*, 21, 719, 1998.
154. Lombardi, F., Frequency-domain heart rate variability, in *Noninvasive Electrocardiology in Clinical Practice*, Zareba, W., Maison-Blanche, and Locati, E.H., Eds., Futura Publishing Company, Armonk, NY, 2001, p. 163.
155. Devlin, R.B., Ghio, A.J., Kehrl, H., Sanders, G., and Cascio, W., Elderly humans exposed to concentrated air pollution particles have decreased heart rate variability. *European Respir J-Supplement*, 40, 76s, 2003.
156. Quyyumi, A.A., Does acute improvement of endothelial dysfunction in coronary artery disease improve myocardial ischemia? A double-blind comparison of parenteral D- and L-arginine. *J Am Coll Cardiol*, 32, 904, 1998.
157. Anderson, T.J., Uehata, A., Gerhard, M.D., Meredith, I.T., Knab, S., Delagrange, D., Lieberman, E.H., Ganz, P., Creager, M.A., Yeung, A., and Selwyn, A.P., Close relation of endothelial function in the human coronary and peripheral circulations. *J Am Coll Cardiol*, 26, 1235, 1995.
158. Corretti, M.C., Anderson, T.J., Benjamin, E.J., Celermajer, D., Charbonneau, F., Creager, M.A., Deanfield, J., Drrexler, H., Gerhard-Herman, M., Herrington, D., Vallance, P., Vita, J., and Vogel, R., Guidelines for the ultrasound assessment of endothelial-dependent flow-mediated vasodilation of the brachial artery. *J Am Coll Cardiol*, 39, 257, 2002.
159. Aeschlimann, S.E., Mitchell, C.K.C., and Korcarz, C.E., Ultrasound brachial artery reactivity testing: technical considerations. *J Am Soc Echocardiogr*, 17, 697, 2004.
160. Brook, R.D., Brook, J.R., Urch, B., Vincent, R., Rajagopalan, S., and Silverman, F., Inhalation of fine particulate air pollution and ozone causes acute arterial vasoconstriction in healthy adults. *Circulation*, 105, 1534, 2002.
161. Ignarro, L.J., Buga, G.M., Wood, K.S., Byrns, R.E., and Chaudhuri, G., Endothelium derived relaxing factor produced and released from artery and vein is nitric oxide. *Proc Natl Acad Sci U S A*, 84, 9265, 1987.
162. Palmer, R.M.J., Ferrige, A.G., and Moncada, S., Nitric oxide release accounts for the biological activity of endothelium-derived relaxing factor. *Nature*, 327, 524, 1987.
163. Schechter, A.N., and Gladwin, M.T., Hemoglobin and the paracrine and endocrine functions of nitric oxide. *N Engl J Med*, 348, 1483, 2003.
164. McMahon, T.J., Moon, R.E., Luschinger, B.P., Darraway, M.S., Stone, A.E., Stolp, B.W., Gow, A.J., Pawloski, J.R., Watke, P., Singel, D.J., Piantadosi, C.A., and Stamler, J.S., Nitric oxide in the human respiratory cycle. *Nat Med*, 8, 711, 2002.

165. Arbak, P., Yavuz, O., Bukan, N., Balbay, O., Ulger, F., and Annakkaya, A.N., Serum oxidant and antioxidant levels in diesel exposed toll collectors. *J Occup Health*, 46, 281, 2004.
166. Gong, H.J., Wong, R., Sarma, R.J., Linn, W.S., Sullivan, E.D., Shamoo, D.A., Anderson, K.R., and Prasad, S.B., Cardiovascular effects of ozone exposure in human volunteers. *Am J Respir Crit Care Med*, 158, 538, 1998.
167. Drechsler-Parks, D.M., Cardiac output effects of O_3 and NO_2 exposure in healthy older adults. *Toxicol Ind Health*, 11, 99, 1995.
168. Linn, W.S., Solomon, J.C., Trim, S.C., Spier, C.E., Shamoo, D.A., Venet, T.G., Avol, E.L., and Hackney, J.D., Effects of exposure to 4 ppm nitrogen dioxide in healthy and asthmatic volunteers. *Arch Environ Health*, 40, 234, 1985.
169. Health Effects Institute, Special Report: Revised Analyses of Time-Series Studies of Air Pollution and Health. 2003.
170. Sheppard, D., Saisho, A., Nadel, J.A., and Boushey, H.A., Exercise increases sulfur dioxide-induced bronchoconstriction in asthmatic subjects. *Am Rev Respir Dis*, 123, 486, 1981.
171. Koenig, J.Q., Covert, D.S., and Pierson, W.E., Effects of inhalation of acidic compounds on pulmonary function in allergic adolescent subjects. *Environ Health Perspect*, 79, 173, 1989.
172. Morrow, P.E., Utell, M.J., Bauer, M.A., Speers, D.M., and Gibb, F.R., Effects of near ambient levels of sulfuric acid aerosol on lung function in exercising subjects with asthma and chronic obstructive pulmonary disease. *Ann Occup Hyg*, 38(Suppl. 1), 933, 1994.
173. Utell, M.J., and Frampton, M.W., Oxides of nitrogen, in *Toxicology of the Respiratory System*, Sipes, I.G., McQueen, C.A., and Gandolfi, A.J., Eds., Elsevier Science, Oxford, England, 1997, p. 303.
174. Brown, J.S., Zeman, K.L., and Bennett, W.D., Ultrafine particle deposition and clearance in the healthy and obstructed lung. *Am J Respir Crit Care Med*, 166, 1240, 2002.
175. Kim, C.S., and Kang, T.C., Comparative measurement of lung deposition of inhaled fine particles in normal subjects and patients with obstructive airway disease. *Am J Respir Crit Care Med*, 155, 899, 1997.
176. Scannell, C., Chen, L., Aris, R.M., Tager, I., Christian, D., Ferrando, R., Welch, B., Kelly, T., and Balmes, J.R., Greater ozone-induced inflammatory responses in subjects with asthma. *Am J Respir Cell Mol Biol*, 154, 24, 1996.
177. Basha, M.A., Gross, K.B., Gwizdala, C.J., Haidar, A.H., and Popovich, J., Jr., Bronchoalveolar lavage neutrophilia in asthmatic and healthy volunteers after controlled exposure to ozone and filtered purified air. *Chest*, 106, 1757, 1994.
178. Jorres, R., Nowak, D., and Magnussen, H., The effect of ozone exposure on allergen responsiveness in subjects with asthma or rhinitis. *Am J Respir Crit Care Med*, 153, 56, 1996.
179. Kehrl, H.R., Hazucha, M.J., Solic, J.J., and Bromberg, P.A., Responses of subjects with chronic obstructive pulmonary disease after exposures to 0.3 ppm ozone. *Am Rev Respir Dis*, 131, 719, 1985.
180. Morrow, P.E., Utell, M.J., Bauer, M.A., Smeglin, A.M., Frampton, M.W., Cox, C., Speers, D.M., and Gibb, F.R., Pulmonary performance of elderly normal subjects and subjects with chronic obstructive pulmonary disease exposed to 0.3 ppm nitrogen dioxide. *Am Rev Respir Dis*, 145, 291, 1992.

181. Allred, E.N., Bleecker, E.R., Chaitman, B.R., Dahms, T.E., Gottlieb, S.O., Hackney, J.D., Pagano, M., Selvester, R.H., Walden, S.M., and Warren, J., Short-term effects of carbon monoxide exposure on the exercise performance of subjects with coronary artery disease. *N Engl J Med*, 321, 1426, 1989.
182. Nowak, D., Antczak, A., Krol, M., Pietras, T., Shariati, B., Bialasiewicz, P., Jeczkowski, K., and Kula, P., Increased content of hydrogen peroxide in the exhaled breath of cigarette smokers. *Eur Respir J*, 9, 652, 1996.
183. Antczak, A., Nowak, D., Shariati, B., Krol, M., Piasecka, G., and Kurmanowska, Z., Increased hydrogen peroxide and thiobarbituric acid-reactive products in expired breath condensate of asthmatic patients. *Eur Respir J*, 10, 1235, 1997.
184. Loukides, S., Horvath, I., Wodehouse, T., Cole, P.J., and Barnes, P.J., Elevated levels of expired breath hydrogen peroxide in bronchiectasis. *Am J Respir Crit Care Med*, 158, 991, 1998.
185. Dekhuijzen, P.N., Aben, K.K., Dekker, I., Aarts, L.P., Wielders, P.L., van Herwaarden, C.L., and Bast, A., Increased exhalation of hydrogen peroxide in patients with stable and unstable chronic obstructive pulmonary disease. *Am J Respir Crit Care Med*, 154, 813, 1996.
186. Wilson, W.C., Swetland, J.F., Benumof, J.L., Laborde, P., and Talor, R., General anesthesia and exhaled breath hydrogen peroxide. *Anesthesiology*, 76, 703, 1992.
187. Corradi, M., Montuschi, P., Donnelly, L.E., Pesci, A., Kharitonov, S.A., and Barnes, P.J., Increased nitrosothiols in exhaled breath condensate in inflammatory airway diseases. *Am J Respir Crit Care Med*, 163, 854, 2001.
188. Montuschi, P., Corradi, M., Ciabattoni, G., Nightingale, J., Kharitonov, S.A., and Barnes, P.J., Increased 8-isoprostane, a marker of oxidative stress, in exhaled condensate of asthma patients. *Am J Respir Crit Care Med*, 160, 216, 1999.
189. Carpenter, C.T., Price, P.V., and Christman, B.W., Exhaled breath condensate isoprostanes are elevated in patients with acute lung injury or ARDS. *Chest*, 114, 1653, 1998.
190. Montuschi, P., Collins, J.V., Ciabattoni, G., Lazzeri, N., Corradi, M., Kharitonov, S.A., and Barnes, P.J., Exhaled 8-isoprostane as an in vivo biomarker of lung oxidative stress in patients with COPD and healthy smokers. *Am J Respir Crit Care Med*, 162, 1175, 2000.
191. Hanazawa, T., Kharitonov, S.A., and Barnes, P.J., Increased nitrotyrosine in exhaled breath condensate of patients with asthma. *Am J Respir Crit Care Med*, 162, 1273, 2000.
192. Yang, R.-B., Mark, M.R., Gray, A., Huang, A., Xie, M.H., Zhang, M., Goddard, A., Wood, W.I., Gurney, A.L., and Godowski, P.J., Toll-like receptor-2 mediates lipopolysaccharide-induced cellular signalling. *Nature*, 395, 284, 1998.
193. Hunt, J., Byrns, R.E., Ignarro, L.J., and Gaston, B., Condensed expirate nitrite as a home marker for acute asthma. *Lancet*, 346, 1235, 1995.
194. Hunt, J.F., Fang, K., Malik, R., Snyder, A., Malhotra, N., Platts-Mills, T.A.E., and Gaston, B., Endogenous airway acidification. *Am J Respir Crit Care Med*, 161, 694, 2000.
195. Kostikas, K., Papatheodorou, G., Ganas, K., Psathakis, K., Panagou, P., and Loukides, S., pH in expired breath condensate of patients with inflammatory airway diseases. *Am J Respir Crit Care Med*, 165, 1364, 2002.

3

NASAL IRRITATION—CURRENT UNDERSTANDING AND FUTURE RESEARCH NEEDS

James C. Walker

CONTENTS

3.1 BACKGROUND, PERSPECTIVES, AND SCOPE

For the majority of individuals, most of the air taken in with each inhalation passes through the nasal cavity. Under normal environmental conditions, the air is warmed and humidified as it passes through the upper airway—the nasopharynx and the trachea. A small fraction of the chemical stimuli in the incoming airstream is deposited onto patches of mucus overlying dendrites of the olfactory receptor neurons and endings of the trigeminal nerve. A fraction of these deposited molecules act as stimuli that trigger action potentials in the olfactory nerve (cranial nerve I), branches of the trigeminal nerve (cranial nerve V), or both. An excellent explanation of the anatomy and innervation of the nasal cavity, from the standpoint of chemosensitivity, is provided by Silver and Finger (1991). If both nerves are intact, as is the case in the vast majority of individuals, it is unlikely that nasal irritation (NI) would be experienced in the absence of odor sensation. Thus, a given inhalation and the consequent chemosensory stimulation are likely to produce one of three outcomes: no sensation, odor sensation but no NI, or both odor sensation and NI. The third case is the subject of this chapter.

With both odor and NI, there are at least four aspects or dimensions that should be noted. First, ideally, each should be operationally defined to minimize or eliminate ambiguity, and straightforward procedures should be employed to generate quantitative stimulus–response functions. Once it is demonstrated that a given type of stimulus is perceived, the simplest way to express its potency is in terms of the minimum concentration needed to elicit the sensation of interest. This assumes, of course, precise control of the stimulus parameters, collection of a sufficient number of samples for each stimulus condition, adequate accounting for the different sources of variation in the responses taken as measures of perception, and clear operational definitions of threshold concentration. Some recent progress in these areas may be noted (Kendal-Reed et al., 1998; Walker et al., 1999; Kendal-Reed et al., 2001; Walker et al., 2003). Second, one should be able to measure the increase in the magnitude of the sensation with increases in stimulus concentration. In addition to these experimental factors of importance in detection, reliable tools must be developed and validated for use by subjects in communicating clearly the magnitude of the sensation that they experience. Two somewhat recent innovations in this area are the labeled magnitude scale (LMS) approach (Green et al., 1996) and the isoresponse technique (Kendal-Reed et al., 1998). Third, and still more experimentally challenging, is the assessment of what may be termed the character or quality of the sensation. In the case of NI, only a small number of kinds of sensation have been noted: stinging, scratching, itching, burning, and the somewhat vague descriptor "painful." Despite great difficulties, ingenious approaches have yielded valuable new information in this area (e.g., Laska et al., 1997). With odor, this assessment is even more daunting, but, again, some impressive efforts have been made in quantifying odor discrimination capability (Olsson and Cain, 2000; Laska and Grimm, 2003), and these will likely form the foundation for future breakthroughs. Finally, an important, though little-studied, element of NI is the hedonic or affective impact of these sensations. Considerable individual differences in both the use of rating scales and the experiences that help determine hedonic impact (Distel et al., 1999) must be grappled with in the effort to derive reliable stimulus–impact relationships. In addition, understandable difficulties are encountered when experimental subjects are asked to judge the hedonic impact of, for example, an odor sensation when NI is also present.

NI is defined here as any combination of stinging, scratching, itching, or burning sensations that are localized to the nasal cavity and are caused by the introduction of airborne chemical stimuli into this cavity. Researchers, it is argued, should generate an understanding of the stimulus, organismic and response variables that give rise to NI in actual environments. In terms of environmental impact, NI has a small effect on a given individual but has a large aggregate impact because a very large number of individuals experience this (generally aversive) sensation each day in "real-world" environments. As discussed by Walker et al. (2001b), responses to short-term airborne chemical exposures have not been studied in depth even though such effects are extremely costly in terms of total morbidity caused (Fisk and Rosenfeld, 1997).

NI is of far greater concern in the developed countries, and one may (using the United States as an example) divide the environments that can cause NI into two categories: one is the industrial or manufacturing setting and the other is the office environment. Up to the early 1970s, NI as a problem was confined largely to exposures in the industrial or manufacturing sector. Exposures usually occurred in the course of healthy workers performing their duties, the range of chemicals to which workers were exposed was relatively narrow, and the concentrations were typically high enough to be sensed by real-time instrumentation. The individual most often given the responsibility of protecting workers and monitoring exposures was one with industrial hygiene or toxicology training. In some cases, especially when exposures were high or the consequences were severe, an occupational medicine specialist would be involved and the adverse effects were treated using conventional medical procedures. The office environment is quite different. Roughly 30 yr ago, an effort began in the U.S. to essentially seal buildings with the aim of conserving energy. This has combined with the long-lasting trend in the economy toward an increasingly large information-processing workforce. The net result is that a large fraction of the U.S. workforce spends the majority of its waking hours working in an indoor environment where there is little influx of fresh air (see Fisk and Rosenfeld, 1997).

In relation to the first category, the following factors may be said to typify the office environment:

1. In the office setting, variation among individuals is much greater, and workers are not selected with any regard to how well they will deal with environmental chemical exposures.
2. The duration of each exposure may well be an entire workday. This raises the question of the role of cumulative time of exposure as a factor in determining effects. That is, one logically needs to take into account that some adverse effects, such as NI, may develop over the course of a workday. Clearly, this has a bearing on the work of those charged with investigating NI. Because, for example, it is extremely difficult to partition a normal office workday into exposed and nonexposed (to chemical stimuli) periods, and because office workers tend to be much more stationary over the course of the workday, developing reliable stimulus–response relationships in the office setting becomes extremely challenging, yet critically important.
3. The author is unaware of any systematic comparison of the air chemistry of office and manufacturing environments but offers the plausible suggestion that these two types of exposure settings are distinctly different. Although the numbers of different chemicals present in the two categories may be roughly similar, a reasonable assumption at present is that factory, but not office, environments are at least occasionally characterized by moderate-to-high concentrations of a few chemicals. These few chemicals may predominate in the factory setting to such a degree that most of the effects observed, some of which may be of significant medical concern, may be attributable to a small number of compounds.
4. Given the preceding points, the effects of airborne chemical exposures tend to be much more subtle in the office than in the factory or plant environment. The generally less-severe effects seen in the office environment are more susceptible to being altered by the greater social interactions present in the office environment and/or by cognitive biases and expectations (Distel and Hudson, 2001; Dalton et al., 1997; Lees-Haxley and Brown, 1992; Knasko et al., 1990).
5. In view of the lesser severity of effects in the office setting, it seems reasonable to suggest that the relative importance of NI is likely to be somewhat less. For example, if NI is only slight, then odor is likely to be given more weight as a determinant of the overall air quality than if the irritation were severe.

To address these complexities, the ideal scientific approach will have to be one that is more multidisciplinary than was necessary when the issue was primarily one of industrial hygiene and exposures in the factory or chemical plant setting. Research to understand short-term environmental effects must also deal with low-level complex mixtures, take into account psychological factors to a greater extent, place greater relative emphasis on organismic variables and interindividual variation than in the past, and clarify the role of the duration of exposure.

In view of the need to bring rigorous, multidisciplinary science to bear on the question of NI as it presents in actual environments, experiments were evaluated for this chapter based, in part, on the degree to which the following ideal features were present:

1. Stimuli were generated in a manner that was methodologically sound and repeatable in other laboratories, and real-time measurements to verify concentration and other parameters such as relative humidity and temperature of the stimulus airstream were feasible. (Note: It is understood that the degree to which stimulus parameters may be specified precisely will vary with the identity of the chemical(s) and the range of concentrations used.)
2. Stimuli were presented in a natural way. That is, the individuals received the stimuli by inhaling. Further, the interface between the apparatus and the participants was such that the experimenter could be certain that individuals were exposed to the stimulus concentrations that were generated.
3. The chemicals and the concentrations at which they were presented had some reasonable relation to the kind and intensity of stimuli encountered in actual environments.
4. Similarly, the duration of exposure was varied (ideally, from a few seconds up to an hour or so) so that the experimenter could evaluate possible effects of cumulative exposure on various measures of NI.
5. Among the response measures used was self-report of the presence or absence of NI and, ideally, sensation magnitude so that with each individual tested, the limits of sensitivity to NI had to be determined. (Note: It is recognized that this may be difficult or impossible to do with longer durations of exposure, but it can certainly be done in a straightforward way with short-term exposures of only the nose.) Information on sensitivity was collected in a manner that furthers our understanding of variation, both within and between individuals.

These features may seem somewhat obvious to the reader, but the rather nascent and disparate nature of research in the NI area warrants a clear statement regarding the need for improved experimental rigor and more comprehensive conceptualization of interrelated questions.

The purposes of this chapter are to: (1) highlight some of the key findings (on which there is general agreement) regarding the environmental stimuli giving rise to NI, (2) explain what is known, and what is being debated, concerning the neural underpinnings of NI, (3) discuss additional research issues requiring resolution, and (4) propose some potentially useful research for the future.

3.2 CATEGORIZING RESEARCH ON NI

As noted earlier, two basic experimental approaches have been used to study NI. After some mention of the very limited work on NI using environmental chambers for prolonged exposures, a discussion is provided of two key lines of research on NI in response to very brief exposures.

3.2.1 Chamber Studies

The practice of placing individuals in an environmental chamber for the purpose of measuring NI (and perhaps other effects) in response to whole-body exposures of extended duration is clearly of great value in the effort to understand the effects of everyday environmental contaminants. This approach allows one to conduct exposures in ways that mimic most of the features that occur in actual environments and it can provide data that serve as a link between the findings based on very brief "nasal-only" exposures in the laboratory and those in actual environments (Walker et al., 2001b). Although second-hand smoke (e.g., Weber et al., 1979; Cain et al., 1987; Walker et al., 1997) and some other pollutants have been studied in this way, relatively few chamber studies of NI in response to well-characterized single chemicals or mixtures have been published. These are summarized briefly in an excellent, and very recent, critical review (Doty et al., 2004, pp. 105–106).

In one sense, the paucity of work in this area may be appropriate and fortunate because, as noted in the text that follows, current ambiguity as to the neural basis of NI hinders sound research. As a practical example, normosmic individuals (with intact olfactory and trigeminal systems) would be of dubious value in such studies if future work were to reveal that valid measures of NI are possible only if they exclude the involvement of the olfactory system. Based on these considerations, a discussion of chamber studies of NI may well be premature.

3.2.2 NI in Response to Very Brief Exposures

Research in this area can be grouped into four categories. Two of these are not covered in any detail in this chapter. The first is an impressive set of studies, primarily by Hummel and colleagues (e.g. Hummel, 2000; Hummel and Livermore, 2002) that involves electrophysiological measurements, from the nose and/or scalp, of candidate correlates of NI. Stimulation involves a modern variant of the blast injection technique (see Prah et al., 1995). The participant is taught to perform a velopharyngeal closure maneuver in which the nasal cavity is sealed off from the oral cavity by an upward movement of the soft palate, and the participant breathes orally. Through a sophisticated annular vacuum system, a very brief (~0.2 sec) pulse of chemical vapor is discharged from the end of a thin tube inserted into the nose. Electrical changes from the nasal mucosa and/or scalp are related to this stimulation and to self-reports of NI. This line of work exhibits outstanding technical rigor, and there is little doubt that many of the results will ultimately be of great value in understanding the neurobiology of NI. However, maximal use of these findings will require a number of bridge or linking experiments to generate the transfer functions needed to apply the results to other studies and everyday life, in which stimuli are delivered to individuals in the course of normal breathing. In addition, maximally informed interpretation of the findings from these electrophysiological studies must await resolution of the question of possible olfactory contribution to NI. In some ways, this resolution may also be aided by electrophysiological studies. For these reasons, a discussion of this impressive series of studies was not included in this chapter.

A second category of studies has also been largely omitted. These are aimed at understanding how NI sensitivity is modulated by conditions in the nasal cavity or by the subjects' beliefs or expectations (whether preexisting or "planted" by the experimenter). Shusterman (2002) has recently provided an excellent review of the research on the former case in the context of understanding variation (primarily interindividual) in NI sensitivity. With regard to the important and oft-neglected role of "top-down" influences in chemosensory psychophysical data, Dalton (e.g., Dalton et al., 1997) is due much credit for enlivening this area with a great deal of new data and clever experimental procedures. As with the electrophysiological studies of Hummel and others, the topic of cognitive influences is not dealt with in any detail in this chapter. Maximal value for this creative line of work must await clarification of the operational definition of NI, and this will require new research on the neural basis of NI as experienced in everyday life.

Following are brief summaries of two lines of NI research. These two types of studies focus on a set of very important experimental issues that are considered more fundamental than the evaluation of either the electrophysiological correlates of NI perception or the variables that can be shown to modify (at least initially) the reporting of NI. Implicit or explicit in the research programs discussed in the following section are several key decisions or assumptions that the experimenters have made as to the neural basis of NI, the need for careful generation and presentation of stimuli so that concentrations in the nasal cavity are known, the possible relationship between NI and various breathing parameters, the quantity of data needed for a given stimulus–individual combination, the kind of response data collected, and the strategies and tools for data analysis and interpretation. These issues are discussed as they appear.

3.3 MODELING OF NASAL TRIGEMINAL POTENCY OF CHEMICALS

Over the last decade or so, a considerable number of studies have been published (e.g., Cometto-Muñiz and Cain, 1990; Cometto-Muñiz et al., 1998; Abraham et al., 2001; Cometto-Muñiz et al., 2001) on the topic of the chemical determinants of NI. This QSAR (quantitative structure–activity relationship) research program is included in a recent review of sensory irritation (Doty et al., 2004, pp. 103–104). Also included in this summary are some details concerning the regression modeling used to predict NI (often termed *nasal pungency* by this group) using five chemical parameters: excess molar refraction, dipolarity/polarizability, overall or effective hydrogen bond acidity, overall or effective hydrogen bond basicity, and the solute gas-liquid partition coefficient on hexadecane at 298°K. Organismic, stimulus, and response variables (see Walker et al., 2001b) for this line of research are considered in the following subsections.

3.4 WORK BY COMETTO-MUÑIZ, ABRAHAM, AND COLLEAGUES

3.4.1 Organismic Variables

In much of these researchers' work, experimental participants are restricted to those considered to have anosmia, a condition conceptually defined as an absence of connection between the olfactory receptors in the nose and the olfactory bulb. In practice, individuals with anosmia are often identified based on performance on odor tests; of course, this largely unavoidable approach reflects a degree of circularity in that somewhat arbitrary assumptions are made as to the degree of loss of function that can be taken as indicating complete loss of structure. The emphasis on anosmics is based on the assumptions that NI is solely a matter of trigeminal activation and that olfactory input leads to some combination of confusion and bias with a tendency to overreport NI. Often cited in support of this view is the observation (e.g., Cometto-Muñiz and Cain,

1994, 1996; Cometto-Muñiz, 2001; Walker et al., 1990c) that the NI thresholds of normosmics are lower than those of anosmics, although this difference declines as NI intensity grows with increases in concentration (see Kendal-Reed et al., 1998). Because the vast majority of individuals exposed to environmental chemicals have both intact olfactory and trigeminal nerves, reliance on only anosmics carries the implication that much of the NI that people believe they experience in everyday life is not "real" NI. At least two facts should be considered when evaluating the assumption that NI is solely a matter of nasal trigeminal activation. First, the author is aware of no experimental evidence supportive of the notion that, in everyday life, the sensation of NI derives entirely from trigeminal activation. Second, the vast majority of individual–exposure combinations in everyday life will involve odor sensation of weak to moderately strong intensity, with little or no NI. Thus, although odor may often occur in the absence of NI, NI will seldom (perhaps never) occur in the absence of odor. Research approaches resting on the assumption that NI experienced at the same time as odor either cannot be measured or can only be measured using methods (e.g., nasal localization) that exclude the possibility of an olfactory contribution are therefore of questionable predictive value. Additional discussion of the issue of neural mediation is included in the section titled "Suggestions for future research."

Anosmia was defined as a score of 0 to 1.75 on a combined threshold and 7-item odor-identification test where the maximum possible score was 7 (Cain, 1989). This is a reasonable operational definition, but it is not clear if this can be confidently taken as evidence of the complete lack of olfactory input to the brain. This reflects the somewhat primitive state of science in olfactory neurology; that is, there is currently no procedure that can provide independent verification of the completeness of olfactory loss. Quite understandably, given the rarity of individuals with no olfactory function, these researchers have allowed both age and interval between anosmia-producing event and testing to vary widely. Stevens et al. (1982) reported a decline in trigeminal sensitivity with age. Although the results of a small electrophysiological study by Hummel et al. (1996) are open to various interpretations, one interpretation is that trigeminal sensitivity *per se* is lowered when olfactory input to brain is removed. In sum, it is possible that these time-related variables may have added statistical noise to the NI sensory measurements of Cometto-Muñiz and colleagues.

A second approach to the question of how to measure trigeminal stimulation in ways that are not influenced by the presence of the olfactory system, and one taken by Cometto-Muñiz and colleagues in some of their work (e.g., Cain and Cometto-Muñiz, 1996), is to use the nasal localization technique established by prior investigators (von Skramlik, 1925; Kobal et al., 1989). In brief, the person being tested (whether normosmic or anosmic) receives chemical stimulation into only one nostril, with the other receiving only clean air. The lowest concentration at which the individual can reliably report which nostril receives stimulation is taken as a measure of nasal trigeminal sensitivity. Preliminary indications are that this measure yields data from normosmics that are in generally good agreement with those derived from testing anosmics (Cometto-Muñiz et al., 1998; Cometto-Muñiz and Cain, 1996). If future work supports this view, the nasal localization approach could perhaps be useful in understanding olfactory–trigeminal interactions. At present, it cannot be stated that nasal localization provides an objective measure of NI; it simply appears to be a useful gauge of nasal trigeminal sensitivity.

3.4.2 Stimulus Selection, Generation, and Presentation

The work by Cometto-Muñiz, Abraham, and colleagues is to be applauded for the large variety of stimuli that have been used. This is necessary in order to collect a sufficient amount of data on a number of different compounds to develop predictive models of trigeminal sensitivity. This variety has provided a sufficient range in the values of each of the five chemical parameters so that the resultant models are statistically robust. There is, of course, a trade-off in that the focus on large numbers of stimuli, in practice, imposes limits on the time and effort that can be expended in refining the precision with which stimuli are presented or increasing the amount of sampling done for a given individual–stimulus combination. As a result, there may be considerable uncertainty regarding the concentration of irritant taken into the nasal cavity during a given trial. In some of these studies (e.g., Cometto-Muñiz et al., 2002), stimuli were presented using a puff-bottle technique in which the participant has to coordinate his or her inhalation with the expulsion of air from a small vapor bottle. This is a simple approach but one must take into account that there will be variation in the proportion of air entering the nose that comes from the stimulus bottle (with the remainder of the inhaled volume being ambient air). In other studies (e.g., Cometto-Muñiz et al., 2001), an effort was made to use dilution by liquid solvents and a static delivery system for stimulus presentation. With this approach, the force for pulling air over the liquid odorant or irritant is supplied by the inhalation of the person being tested. This change represents an improvement over the puff-bottle approach in that most or all of the airflow entering the nose comes from the odorant/irritant delivery device. Two problems remain that prevent one from assuming that the device meets the requirement of an ideal delivery system, i.e., one that would deliver known concentrations on each trial regardless of the inhalation behavior of the person being tested. First, as explained by Prah et al. (1995), liquid dilution is not satisfactory as a means of confidently generating known vapor-phase concentrations. Second, with this approach there is the likelihood that, at higher inhalation flow rates, there will be a drop in the expected concentration because of a progressive (and unknown) depletion of the headspace over the liquid during the course of the inhalation.

Some attempt has been made to address the issue of using data from two different stimulation approaches (Cometto-Muñiz et al., 2000). However, although one might be able to estimate some average offset, it is not possible to predict how data collected with the puff-bottle technique would have been different had the data been collected using the glass vessel approach. Similarly, it is not possible to mathematically remove uncertainty concerning headspace depletion from glass vessel data and convert the data to what would have been obtained had the depletion problem been eliminated by the use of air dilution olfactometry. The net effect of uncertainty regarding stimulus concentration is that the threshold and other response data from this line of work cannot be integrated in a straightforward manner with information from other studies.

3.4.3 Response Measurement

Most of this work employed the 2-AFC (two-alternative forced choice) technique to estimate threshold. In a given trial, the participant received a dose from each of two puff-bottles (only one of which contained the irritant chemical) and then stated which one had the greater NI intensity. Typically, an ascending concentration series was employed, with each step representing a 3-fold increase. Correct responses led to the same concentration being presented again, and errors led to the concentration being raised by one step. This procedure was continued for a given combination of individual and stimulus until a concentration was encountered that elicited five consecutive correct responses. This concentration was defined as the threshold. The criterion used may be considered somewhat conservative because the likelihood that the level of performance used to define threshold would be expected to occur by chance alone is 0.03. As indicated earlier, this group has also used the nasal localization technique as a means of estimating trigeminal sensitivity in normosmic (having both olfactory and trigeminal inputs to the brain) individuals. The procedure was very similar to the 2-AFC detection technique used with normosmics; the only change was that the subject was asked to judge which nostril was stimulated as opposed to which puff-bottle or glass vessel elicited the greater intensity of NI.

Just as a strong focus on collecting data on large numbers of chemicals can significantly lessen stimulus control, such an emphasis leads to a small amount of data being collected for each individual–stimulus combination. Implicit in the idea of obtaining a threshold concentration estimate (whether based on simple detection by anosmics or nasal localization by normosmics) is the notion that sufficient sampling has occurred to minimize uncertainty in the concentration estimate. It seems unlikely that the very limited sampling in the Cometto-Muñiz work minimized uncertainty. In some cases, a single instance of five correct responses was considered sufficient to define the NI threshold for a given individual-chemical combination. Therefore, it is not possible to take data from this experimental approach to develop estimates of intraindividual variability in NI sensitivity. Of course, it is not possible to gain a sound estimate of interindividual variation until intraindividual variation has been accounted for (Walker et al., 1999).

This line of work is to be applauded for its survey of the approximate stimulatory effectiveness of a wide variety of systematically selected chemicals for the human nasal trigeminal system. It appears that data were collected on a sufficient number of chemicals to allow the development of a reasonably robust QSAR model (see Abraham et al., 1998) of NI thresholds in anosmics. This model accounts for ~95% of the variance in NI thresholds in anosmics, requires only five physicochemical parameters as inputs, and is sufficiently robust to predict results for several chemical classes. Ideally, future work in this area will include additional tests (see Abraham et al., 2001) of this model's ability to predict responses to previously untested stimuli.

It should perhaps be emphasized that reservations about the assumptions regarding the neural mediation of NI should not be construed as suggesting that trigeminal input is unimportant. As will be discussed in more detail in subsequent parts of the chapter, it is clear that trigeminal activation is critically important for moderate to severe intensities of NI. Given the value of information on the chemosensitivity of human nasal trigeminal afferents, the impressive QSAR models produced by Cometto-Muñiz and colleagues are potentially of much value. The precedent for this regression model can be found in the studies done to better understand the respiratory rate depression response of mice. This bioassay was established largely through the efforts of Alarie (e.g., see Alarie, 1973) who has also contributed to an understanding of the possible pharmacology underlying what is normally termed the RD50 test (Alarie et al., 1995; Alarie et al., 1998). Many regulatory groups have relied heavily on the RD50 (concentration causing 50% depression in respiratory rate) as a parameter that can predict the potency of a compound for causing short-term adverse effects in humans. For this and other reasons, it would seem desirable that there should be a direct within-compound comparison of RD50 and anosmic detection (or nasal localization) values. Similarly, it would be valuable to evaluate, as predictors of human nasal trigeminal chemosensitivity, more direct measures of trigeminal nerve activation. This could be done using behavioral (e.g., Walker et al., 1979; Walker et al., 1986) and/or electrophysiological (e.g., Tucker, 1971) methods. Although it has been recently stated (Doty et al., 2004) that, due to differences in nasal anatomy, animal models are not appropriate for predicting human nasal trigeminal sensitivity, Silver and Moulton (1982) have presented data suggesting the opposite. As in many other aspects of NI-related research, there is a paucity of solid information on this point.

3.5 COMBINING AIR DILUTION OLFACTOMETRY, A PSYCHOBIOLOGICAL PERSPECTIVE AND A FOCUS ON ACCOUNTING FOR VARIATION

The second category of studies in the NI area is that of Walker and colleagues. This has been a much more modest effort than that by Cometto-Muñiz, Abraham, and their colleagues. In comparison to that line of work, the studies by Walker and colleagues have: (1) not been driven to any significant degree by an interest in understanding the physicochemical determinants of trigeminal stimulatory effectiveness; (2) emphasized precise control over all stimulus airstream parameters; (3) emphasized quantitative comparisons between normosmic and anosmic individuals in terms of NI sensitivity; (4) focused on evaluating candidate nonverbal correlates of NI in normosmic and anosmic individuals; (5) sought to understand NI as reported by normosmic individuals in everyday life, with no presuppositions as to the neural mediation of NI; (6) focused at least as much on suprathreshold scaling of NI as on threshold determinations; and (7) included some effort to quantify intra- and interindividual variation in NI sensitivity in both normosmics and anosmics.

In terms of instrumentation, the work by Walker and colleagues is based on a computerized air dilution olfactometer (Walker et al., 1990b) that has been calibrated using several different chemicals (Walker et al., 1990a; Maiolo et al., 1996; Maiolo et al., 2000). This system provides for completely automated conduct of test sessions in which perceptual and breathing responses are measured during at least ten trials, presented at each of four or five concentration levels, as well as an equal number of clean air control trials. A short series of exploratory studies (Walker et al., 1990c; Warren et al., 1992; Warren et al., 1994) was conducted which employed stimuli that varied considerably in terms of known or presumed relative stimulatory effectiveness for the olfactory and nasal trigeminal nerves. This limited amount of work indicated the following:

1. In agreement with work from other laboratories, normosmics may report NI at concentrations not detected by anosmics.
2. To a first approximation, odor thresholds for commonly used odorants are about 1 ppm ± an order of magnitude. Trigeminal sensitivity appears to vary more for different compounds, and thresholds in human anosmics are reasonably well predicted by behavioral studies in anosmic pigeons (Walker et al., 1979, Walker et al., 1986) and electrophysiological studies on animals (Tucker, 1971; Silver and Moulton, 1982).
3. For a given compound, the degree to which the growth of NI intensity lags the growth of odor intensity is reasonably well predicted by the size of the gap between olfactory and trigeminal threshold concentrations as determined by human or animal studies.
4. In tests with normosmics, odorants that are poor or completely ineffective as trigeminal stimuli (e.g., *n*-amyl, phenethyl alcohol) caused only very modest levels of NI and did not reliably alter breathing or nasal cross-sectional area. In the same individuals, the potent trigeminal stimulus acetic acid caused comparable odor intensity, but the much higher NI ratings were associated with clear reductions in inhaled volume. With very high NI intensity, nasal cross-sectional area was also significantly reduced.

This pattern of observations, in combination with the need to understand nested sources of variation, led to a comprehensive comparison of normosmic and anosmic responses. Using the same air dilution olfactometric approach cited earlier, 31 normosmic and 4 anosmic individuals were each tested with the same range of four concentrations of propionic acid during four 75-trial sessions. As in prior work, the system employed precise breathing measurements and synchronized stimulus delivery with exhalation onset; this practice allowed stimulus concentration in the delivery mask to reach an asymptotic value before the onset of the next inhalation. In addition, quantification of breathing during stimulus presentation allowed various parameters to be evaluated as correlates of NI perception. Collection of psychophysical response data was straightforward. Following each trial, the individual entered the perceived intensities of odor and NI into a computer; sensation magnitude was reported relative to the strongest each could recall prior to participating in the study.

As described in Kendal-Reed et al. (1998), an isoresponse technique was developed and used so that estimates could be developed of the (individually determined) concentrations needed to equate four levels of perceived intensity among individuals. Expressing sensitivity to odor and NI in units of absolute chemical concentration allowed straightforward comparisons between the normosmic and anosmic groups. This method of expressing responses, when combined with the repeated measures taken in the study, also allowed a rigorous analysis of several sources of variation (Kendal-Reed et al., 2001). The following was learned in terms of perceptual responses:

1. Although only four anosmics were used, the data allow the tentative conclusion that intra- and interindividual variation in NI perception is far less in anosmics than in normosmics.
2. Consistent with prior work, NI sensitivity was greater in normosmics than in anosmics. There was essentially no overlap between the two groups at any of the four perceptual magnitudes.
3. The clearest evidence for normosmic–anosmic differences was seen with the lowest (likely the most representative of real-world intensities) of the four perceptual magnitudes, and this level exhibited the greatest dispersion among normosmics. With progressively higher NI intensities, differences between the two groups lessened.

Finally, the possible value of one or more aspects of breathing as a reliable correlate of NI was evaluated (Walker et al., 2001a). For both normosmics and anosmics, two approaches were used. With one, respiratory behavior was analyzed on a breath-by-breath basis, as the percentage change from the prestimulus period, before being combined within each group (normosmic or anosmic) to arrive at group averages. With the second approach, the cumulative inhaled volume (CIV) was calculated within each individual–concentration combination. These values were then normalized by being expressed, for each individual, as a proportion of the mean volume inhaled during the first 2 sec of the stimulus period (beginning with the onset of the first inhalation after stimulus arrival) for clean air trials. Relating both ways of expressing breathing behavior to NI responses showed that:

1. Within each group, the degree of decline in CIV tracked the rise in NI. There was greater inhibition in normosmics and the minimum concentration needed to inhibit breathing was lower with normosmics than with anosmics. Thus, the apparent olfactory nerve contribution to NI perception is also reflected in the greater responsiveness seen in breathing responses.
2. Figures 3.1 and 3.2 (from Walker et al., 2001a) provide some insights into the respiratory mechanics of the CIV changes and illustrate the

relationships between perception and breathing in both groups. In normosmics (Figure 3.1), the rise in NI over the range of concentrations too low to be detected by anosmics (see Figure 3.2) is accompanied by a graded decline in inhalation volume but no change in inhalation duration. Thus, the CIV decline in normosmics is initially attributable entirely to a decline in the volume of inhalations. Based on the pattern of results seen in prior work, and discussed earlier, the increasing odor magnitudes over this concentration range are interpreted as simply a measure of olfactory nerve activation that, only in the presence of trigeminal activation, contributes to both NI and an inhibition of breathing. With the highest concentration, changes in NI and breathing in normosmics and anosmics are qualitatively similar, but the olfactory nerve contribution to both endpoints remains evident.

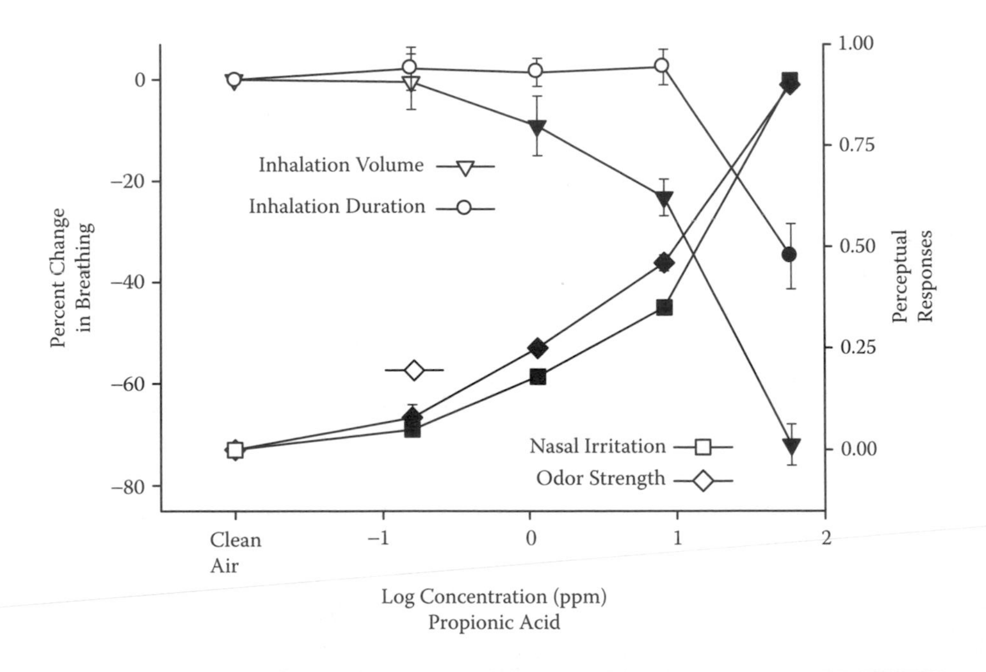

Figure 3.1 Summary of odor and NI magnitude ratings, and two respiratory measures, in 20 normosmic individuals. Filled symbols denote statistically significant changes ($p < .05$) relative to clean air trials. (From Walker et al. (2001a). With the permission of Oxford University Press.)

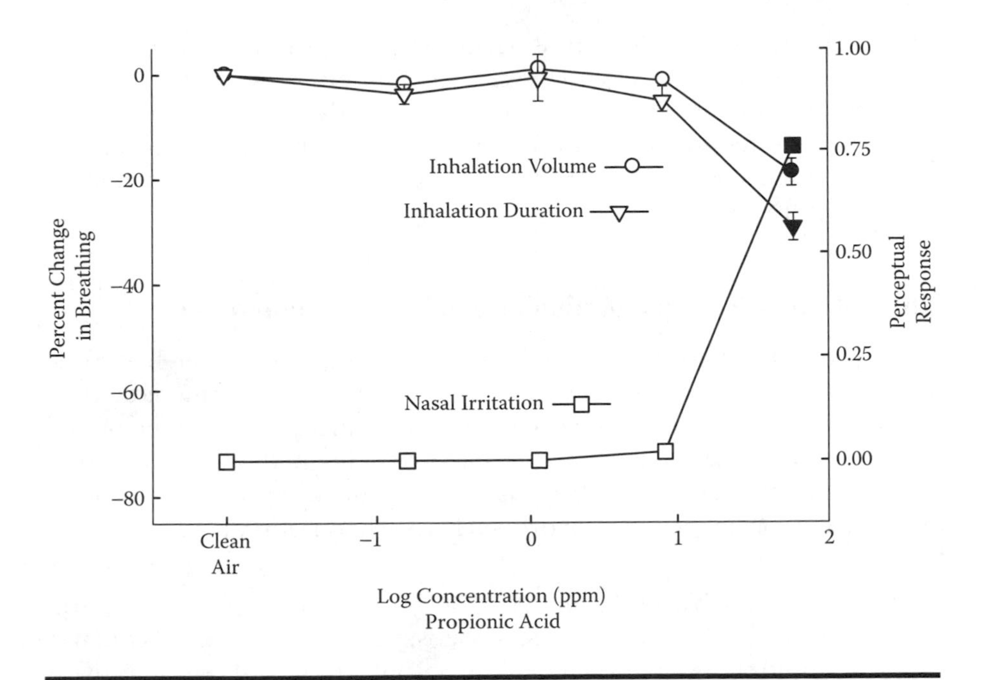

Figure 3.2 Summary of NI magnitude ratings, and two respiratory measures, in four anosmic individuals. Filled symbols denote statistically significant changes ($p < .05$) relative to clean air trials. (From Walker et al. (2001a). With the Permission of Oxford University Press.)

3.6 SUGGESTIONS FOR FUTURE RESEARCH

The discussion in this chapter is not intended to provide a survey of all research done thus far on NI. Given the major issues that have been highlighted, it was felt that such a cataloging of findings would not be useful. The aim of this brief discussion has been to illustrate the state of the science. Resolution of the issue of the neurobiology of NI would seem to be a prerequisite for any major progress on basic and applied issues regarding NI. There are currently two views. One, which could be termed the "trigeminal-only" approach, postulates that the only legitimate NI is that which is mediated solely by the trigeminal nerve. The alternate view of the neurobiology of NI, and the current working hypothesis of the author, has been stated in a series of reports (Kendal-Reed et al., 1998; Kendal-Reed et al., 2001; Walker et al., 2001a). This hypothesis holds that, with levels of stimulation of the trigeminal system that are too low to support conscious perception by anosmics, NI is a joint outcome of this stimulation and greater levels of olfactory activation. By this view, the relative importance of olfactory input in NI progressively declines as the trigeminal detection threshold is exceeded, the evidence for which is NI intensity increases in anosmics. Throughout the dynamic range of NI for normosmics, differences in both perception and breathing (between these individuals and anosmics) are attributed to the presence of a functioning olfactory system.

3.6.1 Addressing the Question of Neural Mediation of NI

One's presuppositions and assumptions influence how research is conducted, how data are interpreted and how findings are communicated. It is suggested that the persistence of ambiguity on the question of the neural mediation of NI will continue to be a problem in designing research studies and interpreting data until it is satisfactorily resolved through sound experimentation. A plan to achieve such resolution is proposed.

1. Recruit a substantial number of subjects who are presumed to be anosmics. For each, measure sensitivity to propionic acid and categorize as completely anosmic only those whose sensitivity closely matches those of the anosmics tested by Kendal-Reed et al. (1998).
2. Once 10 such individuals have been identified, recruit 20 or so age- and gender-matched normosmics.
3. With each individual, quantify perceptual (NI) and breathing responses to a range of precisely controlled concentrations of six or more chemicals (e.g., hydrogen sulfide, ethyl vanillin, acetic acid, propionic acid, amyl acetate, and phenethyl alcohol) that are known to differ widely in terms of odor pleasantness, odor potency, and in the difference between their olfactory and trigeminal potencies. Ideally, the same range of concentrations should be tested in all individuals and should extend from sub- or perithreshold for odor in normosmics to that eliciting moderate to severe NI in all anosmics. For compounds that are detected weakly or not at all by anosmics, the concentration range tested might extend as much as ~10,000-fold above the odor threshold for normosmics.
4. Examine data with the following key questions in mind:
   a. For each chemical that is detectable by anosmics, is there a reasonably constant gap or offset between the concentration that is just detectable and the point at which NI intensity reaches some criterion value for normosmics?
   b. For each compound, is there a range of concentrations, that lie clearly below the anosmic detection threshold, in which

the growth in odor intensity is accompanied by a rise in NI magnitude?

c. Across compounds, can the lowest concentration that causes a reliable drop in inhalation duration in normosmics be used as a reliable indicator of anosmic detection threshold?
d. Across compounds, can the lowest concentration that causes a reliable drop in inhalation volume be used as a physiological correlate of NI in both normosmics and anosmics?

To the degree that the answers to these questions are in the affirmative, it would seem reasonable to conclude that there is an olfactory contribution to NI. Were this the case, it would be appropriate for future research efforts to factor this important detail into experimental designs and data interpretation. Conversely, if the opposite pattern of answers is obtained to the research questions outlined above, it would seem wise for future work to focus entirely on the trigeminal nerve as the neural basis of NI and its physiological correlates.

3.6.2 Suggested Follow-Up Work

Once the fundamental question of neural mediation is put to rest, subsequent follow-up studies would do well to focus on such questions as:

1. Does the apparent utility of breathing parameters as predictors of NI hold up even for chemicals that do not (as is expected to be the case for propionic acid) stimulate only polymodal nociceptors (see Silver and Finger, 1991). Similarly, does the apparent utility of the breathing response as a correlate of NI hold up when mixtures of chemicals are employed?
2. Based on the information yet to be learned concerning the neural basis of NI, the use of the QSAR model put forward by Cometto-Muñiz and colleagues and, possibly, information on animal or human olfactory and trigeminal responsiveness, can models be developed to predict the NI responses of normosmic humans to previously untested single chemicals or mixtures of chemicals? Based on recent work on odor detectability (Walker et al., 2003), it may be best to develop such models first on an individual basis and then, having understood the nature of individual responses, extend the modeling effort to the group level.
3. What principles are needed for taking the responses of normosmics, based on very brief (a few seconds) stimulations, and making predictions as to how NI perception and its physiological correlates change over time with exposures of up to an hour or more?

4. Finally, what principles or rules are needed to understand the way that nasal inputs (whether trigeminal only or olfactory plus trigeminal) are combined with ocular chemosensory (and possibly other) inputs to determine the response of the human organism to airborne chemical exposures that cover a wide range of durations?

ACKNOWLEDGMENTS

The author was supported in part by Philip Morris USA Inc. and *Philip Morris International* during the preparation of this chapter.

REFERENCES

Abraham, M.H., Kumarsingh, R., Cometto-Muñiz, J.E., Cain, W.S. 1998. An algorithm for nasal pungency thresholds in man. *Archives of Toxicology* 72: 227–232.

Abraham, M.H., Gola, J.M.R., Cometto-Muñiz, J.E., Cain, W.S. 2001. The correlation and prediction of VOC thresholds for nasal pungency, eye irritation and odour in humans. *Indoor Built Environment* 10: 252–257.

Alarie, Y. 1973. Sensory irritation by airborne chemicals. *CRC Critical Reviews in Toxicology* 2: 299–363.

Alarie, Y., Nielsen, G.D., Andonian-Haftvan, J., Abraham, M.H. 1995. Physicochemical properties of nonreactive volatile organic chemicals to estimate RD_{50}: alternatives to animal studies. *Toxicology and Applied Pharmacology* 134: 92–99.

Alarie, Y., Schaper, M., Nielsen, G., Abraham, M.H. 1998. Structure-activity relationships of volatile organic chemicals as sensory irritants. *Archives of Toxicology* 72: 125–140.

Cain, W.S., Tosun, T., See, L.C., Leaderer, B. 1987. Environmental tobacco smoke: sensory reactions of occupants. *Atmospheric Environment,* 21(2): 347–353.

Cain, W.S. 1989. Testing olfaction in a clinical setting. *Ear, Nose, and Throat Journal* 68: 316–328.

Cain, W.S., Cometto-Muñiz, J.E. 1996. Sensory irritation potency of VOCs measured through nasal localization thresholds. In Yoshizawa, S., Kimura, K., Ikeda, K., Tanabe, S., and Iwata, T. (Eds.), *Indoor Air '96 (Proceedings of the 7th International conference on Indoor Air Quality and Climate),* Nagoya, Japan, 1: 167–172.

Cometto-Muñiz, J.E., Cain, W.S. 1990. Thresholds for odor and nasal pungency. *Physiological Behavior* 48: 719–725.

Cometto-Muñiz, J.E., Cain, W.S. 1994. Sensory reactions of nasal pungency and odor to volatile organic compounds: the alkylbenzenes. *American Industrial Hygiene Association Journal* 55: 811–187.

Cometto-Muñiz, J.E., Cain, W.S. 1996. Nasal localization thresholds in normosmics mirror nasal pungency thresholds in anosmics, for homologous N-alcohols. Paper presented at the *Annual Meeting of the Association for Chemoreception Sciences,* Sarasota, FL, April 17–21.

Cometto-Muñiz, J.E., Cain, W.S., Abraham, M.H., Kumarsingh, R. 1998. Trigeminal and olfactory chemosensory impact of selected terpenes. *Pharmacology Biochemistry and Behavior* 60: 765–770.

Cometto-Muñiz, J.E., Cain, W.S., Hiraishi, T., Abraham, M.H., Gola, J.M.R. 2000. Comparison of two stimulus-delivery systems for measurement of nasal pungency thresholds. *Chemical Senses* 25: 285–291.

Cometto-Muñiz, J.E., Cain, W.S., Abraham, M.H., Gola, J.M.R. 2001. Ocular and nasal trigeminal detection of butyl acetate and toluene presented singly and in mixtures. *Toxicological Sciences* 63: 233–244.

Cometto-Muñiz, J.E. 2001. Physicochemical basis for odor and irritation potency of VOCs. In Spengler, J.D., Samet, J., and McCarthy, J.F. (Eds.), *Indoor Air Quality Handbook.* McGraw-Hill, New York, pp. 20.1–20.21.

Cometto-Muñiz, J.E., Cain, W.S., Abraham, M.H., Gola, J.M.R. 2002. Psychometric functions for the olfactory and trigeminal detectability of butyl acetate and toluene. *Journal of Applied Toxicology* 22: 25–30.

Dalton, P., Wysocki, C.J., Brody, M.J., Lawley, H.J. 1997. The influence of cognitive bias on the perceived odor, irritation and health symptoms from chemical exposure. *International Archives of Occupational Environmental Health* 69: 407–417.

Distel, H., Ayabe-Kanamura, S., Martínez-Gómez, M., Schicker, I., Kobayakawa, T., Saito, S., Hudson, R. 1999. Perception of everyday odors—correlation between intensity, familiarity and strength of hedonic judgment. *Chemical Senses* 24: 191–199.

Distel, H., Hudson, R. 2001. Judgement of odor intensity is influenced by subjects' knowledge of the odor source. *Chemical Senses* 26: 247–251.

Doty, R.L., Cometto-Muñiz, J.E., Jalowayski, A.A., Dalton, P., Kendal-Reed, M., Hodgson, M. 2004. Assessment of upper respiratory tract and ocular irritative effects of volatile chemicals in humans. *Critical Reviews in Toxicology* 34(2): 85–142.

Fisk, W.J., Rosenfeld, A.H. 1997. Estimates of improved productivity and health from better indoor environments. *Indoor Air* 7: 158–172.

Green, B.G., Dalton, P., Cowart, B., Shaffer, G., Rankin, K., Higgins, J. 1996. Evaluating the "labeled magnitude scale" for measuring sensations of taste and smell. *Chemical Senses* 21: 323–334.

Hummel, T., Barz, S., Lotsch, J., Roscher, S. 1996. Loss of olfactory function leads to a decrease of trigeminal sensitivity. *Chemical Senses* 21: 75–79.

Hummel, T. 2000. Assessment of intranasal trigeminal function. *International Journal of Physiology* 36: 147–155.

Hummel, T., Livermore, A. 2002. Intranasal chemosensory function of the trigeminal nerve and aspects of its relation to olfaction. *International Archives of Occupational and Environmental Health* 75: 305–313.

Kendal-Reed, M., Walker, J.C., Morgan, W.T., LaMachio, M., Lutz, R.W. 1998. Human responses to propionic acid. I. Quantification of within- and between-participant variation in perception by normal and anosmic subjects. *Chemical Senses* 23: 71–82.

Kendal-Reed, M., Walker, J.C., Morgan, W.T. 2001. Investigating sources of response variability and neural mediation in human nasal irritation. *Indoor Air* 11(2): 185–191.

Knasko, S.C., Gilbert, A.N., Sabini, J. 1990. Emotional state, physical well-being and performance in the presence of feigned ambient odor. *Journal of Applied Social Psychology* 20: 1345–1357.

Kobal, G., Van Toller, S., Hummel, T. 1989. Is there directional smelling? *Experientia* 45: 130.

Laska, M., Distel, H., Hudson, R. 1997. Trigeminal perception of odorant quality in congenitally anosmic subjects. *Chemical Senses* 22: 447–456.

Laska, M., Grimm, N. 2003. The substitution-reciprocity method for measurement of odor quality discrimination thresholds: replication and extension to nonhuman primates. *Chemical Senses* 28(2): 105–111.

Lees-Hayley, P.R., Brown, R.S. 1992. Biases in perception and reporting following a perceived toxic exposure. *Perceptual and Motor Skills* 75: 531–544.

Maiolo, K.C., Walker, J.C., Ogden, M.W. 1996. Method for calibrating olfactometer output. Part 1. Acetic and propionic acids. *Analytical Communications* 33: 199–202.

Maiolo, K.C., Walker, J.C., Ogden, M.W. 2000. Method for calibrating olfactometer output. Part 2. Amyl acetate. *Analyst* 125: 1295–1298.

Olsson, M.J., Cain, W.S. 2000. Psychometrics of odor quality discrimination: method for threshold determination. *Chemical Senses* 25: 493–499.

Prah, J.D., Sears, S.B., Walker, J.C. 1995. Modern approaches to air-dilution olfactometry. In R.L. Doty (Ed.), *Handbook of Olfaction and Gustation*, Marcel Dekker, New York, pp. 227–255.

Shusterman, D.J. 2002. Individual factors in nasal chemesthesis. *Chemical Senses* 27: 551–564.

Silver, W.L., Moulton, D.G. 1982. Chemosensitivity of rat nasal trigeminal receptors. *Physiology and Behavior* 28: 927–931.

Silver, W.L., Finger, T.E. 1991. The trigeminal system. In Getchell, T.V., Doty, R.L., Bartoshuk, L.M., Snow, J.B., Jr. (Eds.), *Smell and Taste in Health and Disease.* New York: Raven, pp. 97–108.

Stevens, J.C., Plantinga, A., Cain, W.S. 1982. Reduction of odor and nasal pungency associated with aging. *Neurobiology of Aging* 3: 125–132.

Tucker, D. 1971. Nonolfactory responses from the nasal cavity: Jacobson's organ and the trigeminal system. In L.M. Beidler (Ed.), *Handbook of Sensory Physiology,* Vol. IV, *Chemical Senses, Part 1, Olfaction.* Springer-Verlag, Berlin, New York, pp. 151–181.

Von Skramlik, E. 1925. Uber die Lokalisation der Empfindungen bei den niederen Sinnen. *Z. Sinnesphysiol.* 56: 69.

Walker, J.C., Tucker, D., Smith, J.C. 1979. Odor sensitivity mediated by the trigeminal nerve in the pigeon. *Chemical Senses and Flavor* 4: 107–116.

Walker, J.C., Walker, D.B., Tambiah, C.R., Gilmore, K.S. 1986. Olfactory and nonolfactory odor detection in pigeons: elucidation by cardiac acceleration paradigm. *Physiology and Behavior* 38: 575–580.

Walker, J.C., Kurtz, D.B., Shore, F.M., Ogden, M.W., Reynolds, J.H., IV. 1990a. Apparatus for the automated measurement of the responses of humans to odorants. *Chemical Senses* 15(2): 165–177.

Walker, J.C., Kurtz, D.B., Shore, F.M. 1990b. Apparatus for Assessing Responses of Humans to Stimulants. U.S. Patent 4,934,386. June 19.

Walker, J.C., Reynolds, J.H., IV, Warren, D.W., Sidman, J.D. 1990c. Responses of normal and anosmic subjects to odorants. In Green, B.G., Mason, J.R., Kare, M.R. (Eds.), *Chemical Senses, Vol.* 2: *Irritation.* New York, Marcel Dekker, 1990, 95–117.

Walker, J.C., Nelson, P.R., Cain, W.S., Utell, M.J., Joyce, M.B., Morgan, W.T., Steichen, T.J., Prichard, W.S., Stancill, M.W. 1997. Perceptual and psychophysiological responses of non-smokers to a range of environmental tobacco smoke concentrations. *Indoor Air* 7: 173–188.

Walker, J.C., Kendal-Reed, M., and Morgan, W.T. 1999. Accounting for several related sources of variation in chemosensory psychophysics. In G.A. Bell and A. Watson (Eds.), *Tastes and Aromas: The Chemical Senses in Science and Industry*. University of New South Wales Press, Sydney, pp. 105–113.

Walker, J.C., Kendal-Reed, M., Morgan, W.T., Polyakov, V.V., Lutz, R.W. 2001a. Human responses to propionic acid. II. Quantification of breathing responses and their relationship to perception. *Chemical Senses* 26: 351–358.

Walker, J.C., Kendal-Reed, M., Utell, M.J., Cain, W.S. 2001b. Human breathing and eye blink rate responses to airborne chemicals. *Environmental Health Perspectives* 109: 507–512.

Walker, J.C., Hall, S.B., Walker, D.B., Kendal-Reed, M.S., Hood, A.F., Niu, X.-F. 2003. Human odor detectability: new methodology used to determine threshold and variation. *Chemical Senses* 28: 817–826.

Warren, D., Walker, J.C., Drake, A.F., Lutz, R. 1992. Assessing the effects of odorants on nasal airway size and breathing. *Physiology and Behavior* 51: 425–430.

Warren, D., Walker, J.C., Drake, A.F., Lutz, R. 1994. Effects of odorants and irritants on respiratory behavior. *Laryngoscope* 104: 623–626.

Weber, A., Fischer, T., Grandjean, E. 1979. Passive smoking in experimental and field conditions. *Environmental Research* 20: 205–216.

4

IN VITRO MODELS FOR LUNG TOXICOLOGY

Corrie B. Allen

CONTENTS

4.1 INTRODUCTION

In vitro is literally translated to mean "in glass," undoubtedly referring to experiments performed in the ubiquitous glass test tube. *In vitro* experiments are more broadly defined as being "in an artificial environment outside the living organism" (1). *In vitro* models are widely used in multiple fields of biological inquiry. A comprehensive description of all the variations of *in vitro* models described in the literature would consume multiple volumes. The goal of this discussion is to present descriptions of several popular models used in lung toxicology along with pertinent references that can be consulted for additional information. *In vitro* models used in this field of study range from experiments in molecular biology to the examination of responses of multiple organ systems:

Molecular biology and biochemistry experiments are used to examine the direct reaction of a toxin on chemical reactions at the molecular level.

Cultured primary and transformed cells can be used to explore the intracellular processes occurring in response to a chemical challenge.

Airway explants have been widely used for assessing responses to inhaled chemicals. The intact mosaic of epithelial, interstitial, and muscular cells comprising the explanted tissue allows for a wide variety of experimental approaches.

Explants of lung parenchymal tissue have been used to assess the release of chemical messengers for the variety of cell types present.

Isolated perfused lungs (IPL) have been used to measure a wide variety of lung functions in the absence of confounding systemic effects. These preparations can be maintained for several hours and have been used in a wide variety of biochemical and pharmacological studies.

4.2 WHY STUDY *IN VITRO* MODELS?

Two advantages of performing experiments using *in vitro* models come immediately to mind. First, *in vitro* experiments allow the investigator to study the individual components of the network of pulmonary reactions to inhaled toxins. Exposure of animals to airborne toxins may initiate a cascade of reactions across several cell types in a number of organs. Some of these reactions to chemical agents are short lived, whereas other responses persist for some time or perhaps may even be permanent. Simpler experimental models are required to understand the individual reactions that combine in producing a more complex overall response.

A second reason for performing *in vitro* experiments is to address questions of lung toxicology in human lung cells. Epidemiological studies certainly provide valuable information concerning possible associated effects of environmental toxins on human lungs. However, there are limitations to the scope of questions that may be addressed using epidemiological techniques. Some of the questions that are inappropriate for epidemiological examination may be addressed using *in vitro* experiments with human cells.

In vitro models have limitations. At every level of biological organization, whether biochemical, cellular, tissue, organ, or organismic, isolated study of a single element neglects the rich interplay between elements at the same and at different levels. In the living animal, biochemical reactions do not occur in isolation from one another nor do cellular functions of one cell type occur without affecting the behavior of other cell types. Organ reactions to an environmental insult affect the function of other organs and, indeed, responses of individual animals may be affected by the presence of other animals in the exposure environment.

Airway and alveolar cells are rarely exposed directly to environmental toxins. These agents must first pass through the nose or mouth and then through any airway structures proximal to the cells under study. Such prior contact will affect the *in vivo* response of the cell in a fashion that may not be seen in the *in vitro* cell culture preparation. Likewise, such upstream effects will not contribute to the responses of an isolated perfused lung (IPL) preparation.

Several spatially diffuse systems participate in an organismic response to an environmental toxin. Although the immune system plays a particularly significant role in integrating bodywide responses to such challenges, the nervous and hematopoietic systems also make important functional contributions. In addition, behavioral responses of the exposed animal can modify the cellular exposure levels along with the biological consequences of a chemical exposure. Changes in breathing patterns can alter the distribution of a toxic agent within the lung (2–4). Furthermore, grooming behavior (5), along with behaviors such as food and water consumption (6,7), may be altered by exposure to some toxins. Such behavioral changes may contribute to the final effect of those toxins on the respiratory system.

Many of the responses of the intact animal to a toxic challenge change over the span of an exposure period or during the time between intermittent exposures. Adaptation to the toxin may occur, or sensitization may result in a more robust response to a subsequent exposure. The limited lifespan of an *in vitro* preparation prevents the study of such responses exclusively in the *in vitro* setting.

Animals comprise a massive network of interacting systems that integrate physiological function. Conclusions derived from the data collected from *in vitro* preparations in isolation from these integrating systems must be cautiously extrapolated when considering responses of the intact animal.

4.3 COMMON CONSIDERATIONS FOR ALL MODELS

4.3.1 Short-Term vs. Long-Term Experiments

Studies of the effect of chemical exposure on specific cells, tissues, or organs are generally of two types. In the first, immediate or short-lived responses to chemical exposures are studied. In these experiments, the apparatus used for the *in vitro* preparation must maintain the cells of interest in a physiologically relevant state while at the same time allowing application of the agent of interest and the study of consequent biological or chemical effects. The second type of experiment is designed either to study the biological effects of a prolonged exposure to the chemicals or to examine the effects of the chemical exposure at significantly delayed time points. Such effects may persist or be short term. Often, an *in vitro* preparation cannot be maintained for the duration required for these types of experiments. Cultured cell lines can be maintained indefinitely. However, lung parenchymal and airway explants can be studied for a few days, whereas IPLs can only be maintained for a few hours. When the duration of a proposed exposure exceeds the lifespan of the *in vitro* preparation to be studied, procedures and apparatus must be designed to allow exposure of the intact animal to the agent of interest. A separate apparatus is then needed for the study of that chemical's effects on the cells, tissues, or organs of interest. The instrumentation needed to assess cellular function is still required, but there is less of a need for the apparatus to incorporate the chemical exposure capacity.

Taken together, five general patterns of experiments can be envisioned:

1. Short-term experiments with concurrent chemical exposure and sampling
2. Short-term experiments with sampling or tests at the end of chemical exposure
3. Brief chemical exposure followed by a delay after which testing is performed
4. Prolonged experiments with concurrent chemical exposure and intermittent testing
5. Prolonged chemical exposure followed by testing at the end of the experiment

4.3.2 Delivery Path of Inhaled Agents

Airborne chemicals travel to the target cells in the body along a variety of paths. The agents may contact the target cells directly, pass through a single liquid layer (as with the epithelial cells of the lung), or arrive at the target cells after passing across a complex sequence of liquid and membranous barriers (e.g., endothelial cells of the pulmonary vasculature). Depending on the chemical nature of the agent of interest, passage through the various liquid and cellular barriers en route to the target tissue of interest may produce changes in either the chemical itself or the barrier materials. Chemical changes in the toxic agent may subsequently alter the downstream response of the target cell. For example, ascorbate levels in bronchoalveolar lavage (BAL) fluid are reduced in the lungs of rats exposed to ozone (8) and may represent scavenging of some of the reactive oxygen species involved in cell signaling. The decline in ascorbate levels may also affect the antioxidant capacity of the lung. Chemical alterations of the barrier materials caused by exposure to the airborne agent may produce secondary reaction products that alter cell function. This is exemplified by the production of toxic lipid products from lung surfactants exposed to ozone (9).

4.3.3 Humidity

Most cell types exposed to airborne chemical compounds *in vivo* are kept moist both by an overlying layer of fluid and by moisture delivered to the basolateral surface from the interstitial spaces. Explant cultures and isolated perfused organ preparations also require aggressive humidity control. If this care is not taken, cell desiccation and increases in concentration of media components likely will confound the results of any chemical exposures. High gas flow velocities above the media can accelerate the loss of water from both the cells and media.

4.3.4 pH

A number of buffer systems act *in vivo* to maintain appropriate pH levels around cells in the body. Changes in these pH levels can profoundly affect cell function. Once cells or tissues are removed from the physiological buffering systems, care must be taken to ensure that appropriate pH levels are maintained for the cell types under study. Special attention should be paid to the selection of experimental buffering systems. The use of the more robust bicarbonate buffering system may make some experiments more difficult (e.g., measurement of respiratory cell metabolism using ^{14}C-labeled substrates such as glucose and glutamine). Furthermore, peroxynitrite formed from the reaction of nitric oxide with superoxide ion can react with aqueous CO_2 to form nitrating, nitrosating, and oxidizing intermediate products (10). At the elevated levels of CO_2 used in bicarbonate buffer systems (5 to 10%), the reactions of biotargets with these products will predominate strongly over those with peroxynitrite. Other buffering systems are available but cannot be used without caution. Some buffers such as HEPES have been shown to increase endothelial permeability to albumin in isolated rabbit lungs (11) and contribute to the production of toxic oxygen radicals in cultured endothelial cells (12). Weak buffers may be unable to maintain pH in the face of rapid metabolic production of CO_2.

4.3.5 Temperature

Temperatures of *in vitro* preparations are usually maintained at the normal body temperature of the source organism. The main consequence of changes in temperatures results from changes in cell metabolism. In cell culture systems, evaporative cooling of cells near the interface between the culture media and a gas phase that is not water saturated may lead to a reduction in the temperature to which the cells are exposed. In addition, gas passing over the cells from an external source should be prewarmed before entering the exposure chamber to prevent cooling. Some experimental protocols include changes in the experimental temperature. This is usually done so that differences in temperature-dependent chemical reactions can be observed.

4.3.6 Appropriate Agent Concentrations

The toxicity of airborne compounds depends both on the nature of the chemical compound being used, along with the biological target and end point being studied. In order for a study to provide meaningful results, the concentration of the experimental agent should be related in some fashion to those concentrations seen by the cell, tissue, or organ under *in vivo* conditions. Excessive doses might be used to elicit a positive effect but cannot be used to prove that the compound has such an effect *in vivo* or under naturally attainable environmental conditions. When calculating the concentration of an experimental compound, those calculations should account for reactions with biological structures "upstream" from the cellular target being tested. In most cases, such contact will diminish the amount of the agent reaching the target cells. In some situations, however, reaction of the test compound with other biological compounds may produce secondary toxic products.

4.3.7 Coreactants (CO_2, Proteins, Serum, Antioxidants, Nutrients, and Plastics)

In vitro preparations consist of more than just the cells or tissues being directly examined. They also include the liquid and gas phases surrounding the biological material, along with the structural materials used to maintain the *in vitro* environment (chambers, tubing, etc.) Many components of *in vitro* media and apparatus will likely react with the experimental agents. This can result in a reduction of the amount of the agent available for reactions with the cells, a decline in the concentration of some critical media component, or perhaps even the production of novel reaction products. Investigators need to consider any peripheral reactions that may confound the findings of the planned study.

4.4 END POINTS

Many end points have been studied in numerous *in vitro* models. The following discussion addresses only a few of the many combinations of toxic compounds, target cells, and measured end points reported in the literature.

Exposure of cells to toxic agents often affects the release or processing of chemical *messenger compounds* such as cytokines, chemokines, prostaglandins, and leukotrienes. For example, cultured rat pleural mesothelial cells exposed to amosite asbestos particles were found to release increased amounts of intercellular adhesion molecule-1 and monocyte chemoattractant protein-1 into the culture media (13), whereas RAW 264.7 cells (a mouse macrophage cell line) exposed to ambient air particles were shown to release increased amounts of tumor necrosis factor-α (TNF-α) (14). The content or release of catecholamines or serotonin from some cells may also be affected by the toxin exposure. Rats exposed to 0.5 ppm ozone for 5 d exhibited a reduction in norepinephrine turnover in the heart, but not in the lungs (15). A cultured mast cell line (RBL-2H3) was shown to release serotonin when injured by high levels of ozone (16), whereas IPLs from rats exposed to 0.7-ppm ozone 20 h/d for 20 d exhibited a 32%

decrease in serotonin clearance (17). A further example would be the decline in serotonin uptake by calf aortic endothelial cells observed when the cultured cells were exposed to 95% oxygen for 20 h (18).

Metabolic effects of a toxic agent on an *in vitro* preparation can be identified by monitoring such indices as the cellular ATP content, oxygen consumption, glucose consumption, and glutamine consumption, along with lactate production. Measurement of oxygen consumption requires a closed system to prevent replacement of the consumed oxygen by atmospheric oxygen. Although oxygen consumption is often measured using polarographic electrodes or fiber-optic fluorescence oxygen probes, glucose, glutamine, and lactate measurements are usually made using simple spectrophotometric methods. Enzymes involved in energy metabolism are often measured to determine the site of biochemical disruption caused by a toxic agent.

The bioenergetic consequences of toxin exposure have been studied for several agents in a number of cell types. Mustafa and coworkers reported that both lung mitochondria from rats exposed to ozone *in vivo* and lung mitochondria isolated from rats and then exposed to ozone *in vitro* displayed a significant decline in the oxygen consumption, ADP:O ratios, and respiratory control ratios (19,20) Similar findings of ozone-induced respiratory impairment of lung mitochondria were reported by Zychlinski (21). Interestingly, Mustafa (22) found that mitochondrial oxygen consumption in the lungs of rats exposed to ozone rebounded above that of the controls during a postozone recovery period. Bassett described an increase in glucose consumption in IPLs of rats exposed to 0.6-ppm ozone for 3 d (23). Beach and Harmon found that glucose consumption by isolated lung mitochondria was inhibited after *in vitro* exposure to either benzene, naphthalene, acenaphthene, or 1-chloronaphthalene (24). Morin and coworkers found that exposure of lung slices to diesel exhaust markedly decreased ATP levels (25).

Other workers have studied the effects of toxin exposures on single cell types from the lung. For example, Castronova reported that exposure of alveolar macrophages (AM) to either carbon tetrachloride, toluene, or xylene resulted in a decline in the cellular oxygen consumption (26). Ritter and coworkers exposed cells from a human lung fibroblast line (Lk004) and from a human bronchial epithelial line (HFBE-21) to either ozone or NO_2 and found reduced ATP/ADT ratios. Gabridge and Gladd demonstrated a significant decline in ATP content in cells of a human lung fibroblast line exposed to sulfur dioxide (25). They also found that nitrogen dioxide exposure resulted in a smaller decline in ATP content in these cells.

Cellular damage can result from exposure to a toxic agent. Examples of this are mitochondrial DNA damage, nuclear DNA damage, protein nitrosylation, and lipid peroxidation. Some agents may also affect repair mechanisms normally acting in cells. Without causing direct damage, these agents may cause cellular injuries to accumulate through effects on these repair processes.

Gene expression may be altered by exposure to toxic agents. A typical example would be the upregulation of CYP1A1 gene expression caused by binding of dioxin to the aryl hydrocarbon receptor (27,28). Changes in gene expression can be monitored using such techniques as northern blotting and real-time PCR, whereas related changes in relevant protein production can be studied using western blots. The popularization of chip technology has enabled the study of changes in a far greater spectrum of genes and their proteins than was previously possible.

AM and polymorphonuclear leukocytes (PMN) are among the first cells responding to many foreign agents or organisms entering the lung. Furthermore, these cells both release and respond to chemical messengers (cytokines, NO, etc.) that serve to coordinate different aspects of the *immune cell function*. The ability of these cells to appropriately respond to inhaled hazards can be affected by toxic compounds (29,30). Several aspects of AM and PMN function are amenable to testing in *in vitro* models, including adhesion, chemotaxis and phagocytosis, and bactericidal activity. In addition, production of the chemical messengers and cellular response to chemical messengers can be measured as indices of immune cell function.

One of the first steps in the response of the mobile cells of the immune system is *adhesion* to the blood vessel wall prior to extravasation. This adhesion may be affected by toxin exposure. For example, Shigehara et al. reported that sulfite activates PMN to adhere to fibronectin by way of the β2-integrin Mac-1 (31). In another study, inflammatory lung cells (mostly macrophages) isolated from rats exposed to ozone showed greater adherence to cultured lung epithelial cells (ARL-14) than did the harvested cells from air-exposed rats (32).

Both macrophages and PMN can be induced into *chemotaxis*, i.e., to migrate toward the source of a chemotactic chemical. Some toxins may promote the release of such a signal. For example, rabbit AM exposed to 0.3- or 1.2-ppm ozone *in vitro* release factors into their media that stimulate chemotaxis in PMN (33).

Phagocytosis of particulate and cellular debris along with living infectious organisms is an important activity of both AM and PMN. This activity may also be affected by chemical exposure. AM harvested from ad-lib-fed rats after exposure to ozone exhibited a decline in phagocytosis of latex beads while AM from calorie-restricted rats actually increased phagocytosis. This was correlated with a prolonged bacterial infection when inoculated with *Streptococcus zooepidemicus*. The bacteria were rapidly cleared in the calorie-restricted animals (34). Sherwood and coworkers exposed rats to hexachlorophene aerosols for various time periods and measured phagocytosis along with other indices of host defense (35). They found that alveolar macrophage phagocytosis of labeled red blood cells was increased after 4 d, but not after 1 or 16 d of exposure to the hexachlorophene aerosol.

Reactive oxygen and nitrogen species (e.g., hydrogen peroxide, superoxide, nitric oxide, and peroxynitrite) are released by activated macrophages and neutrophils for cell killing and as intracellular messenger compounds. Inappropriate release of these compounds may damage the host, also. Some environmental toxins affect oxidant release. Becker and coworkers found that exposure of human macrophages to residual oil fly ash (ROFA) caused increased oxidant production (36). In contrast, Nguyen and coworkers showed that the production of nitrating and chlorinating oxidants by stimulated human PMNs was inhibited after treatment of the PMNs with cigarette smoke (37). Furthermore, *in vitro* exposure of human neutrophils to the insecticide dieldrin at nonnecrotic concentrations resulted in an increased production of superoxide (38).

Some investigators have studied the effect of inhaled chemicals on the ability of the lung's immune system to kill bacterial cells. The majority of studies of this issue have addressed the processing of bacterial infections under *in vivo* conditions in chemical-exposed animals. In one study, mice were challenged with *S. aureus* or *P. mirabilis*, and then exposed to NO_2 at a number of concentrations for 4 h. Pulmonary antibacterial defenses against *S. aureus* were suppressed at NO_2 levels of 4.0 ppm and greater, whereas 10 ppm NO_2 was required to suppress antibacterial activity against the *P. mirabilis* (39). In another *in vivo* study, significantly decreased pulmonary bactericidal activity was observed in mice after single exposures to 500, 250, 100, and 2.5 ppm toluene, and after 5 daily 3-h exposures to 1.0 ppm of toluene (40). In one of the relatively rare *in vitro* studies of bactericidal activity, rabbits were first exposed to *Listeria monocytogenes* aerosols. Macrophages were harvested, 10 to 48 d after infection, from the lower respiratory tract of the rabbits and found to express an increased ability to ingest and kill both the original infecting organism and unrelated organisms, when compared to normal AM (41).

Environmental chemicals may affect the release of one or more of the multitude of intracellular messenger compounds. Human AM exposed to ozone *in vitro* showed no effect on spontaneous release of TNF, IL-1, and IL-6, whereas LPS-stimulated release of these compounds was reported to be reduced compared to that of the ozone-naive cells (42). However, human alveolar macrophage IL-6, TNF, and MCP-1 production were induced by exposure to ambient air particles while phagocytosis and oxidant production were inhibited. Particle size and endotoxin contamination modulated some of these effects (43). In another report, human

AM incubated with air particulate matter (PM) greater than 10-μm diameter (PM10) produced TNF-α in a dose-dependent fashion (44). Huang et al. reported that TNF-α production in RAW 264.7 cells was stimulated to a greater degree by larger (2.5 to 10 μm particles) than by fine (<2.5 μm) particles (14). Furthermore, van Eeden et al. found that macrophages incubated with a collected ambient air particulate mixture (EHC-93) produced a broad spectrum of proinflammatory cytokines, including IL-6, IL-1-β, macrophage inflammatory protein-1 (MIP-1), and granulocyte-macrophage colony-stimulating factor (GM-CSF) (45). Pelletier has shown that sodium sulfite increases the release of IL-8 but not TNF-α in human neutrophils (46).

It is interesting to note that Thomassen et al. found that IL-10 administration can inhibit the LPS-stimulated release of TNF, IL-1, IL-6, and IL-8 from human AM *in vitro* (47). This suggests that IL-10 may then act to moderate some of the aforementioned cytokine production. It should also be noted that some PMN functions are altered as time in culture increases (48). PMN integrity declines after 18 h under *in vitro* conditions. Furthermore, neutrophils cultured for 24 h show reduced phagocytosis capacity, while exhibiting unchanged ability to produce hydrogen peroxide and superoxide. However, the superoxide production in response to PMA was reduced at 24 h in culture. These findings illustrate the need to design experiments with such time-dependent effects in mind.

A number of inhaled toxins can initiate the process of programmed cell death known as *apoptosis*. Several assays have been developed for the identification and study of the apoptotic process in cells and tissues.

TUNEL (terminal deoxyribonucleotide-transferase-mediated dUTP nick end-labeling) identifies DNA strand nicks and strand breaks that occur during apoptosis. TUNEL staining has been used to study asbestos-induced apoptosis in alveolar epithelial cell lines (49) and to study apoptosis in lung slices exposed to diesel exhaust (50).

DNA cleavage by actuated endonucleases is one of the hallmarks of the apoptotic process. Endonucleases break the DNA strand at the regularly spaced internucleosomal junctions producing a ladder-like appearance of the DNA when separated electrophoretically on an agarose gel. This assay has been used in studies of urban particle-induced apoptosis in human lung macrophages (51) and in the MRC-5 human lung fibroblast cell line (52).

Caspase activation is studied as an indicator of activation of the apoptotic process. Several inhaled compounds have been found to participate in the activation of the caspase cascade, including ozone, silica, and asbestos (9,53,54).

Chromatin cleavage indicated by an increase in the presence of cytoplasmic nucleosomes can precede plasma membrane breakdown by several hours (55). These nucleosomes can be detected in cytoplasmic fractions using ELISA techniques. Such assays have been used in a number of studies to address the role of particle exposure on macrophage apoptosis induced by silica, volcanic ash, ROFA, and urban air PM samples (SRMs 1648 and 1649) (51,56).

Cell proliferation may also be affected by inhaled toxins. Cell proliferation can be assessed using a number of techniques, including the incorporation of either ^{3}H-thymidine or bromodeoxyuridine into cellular DNA along with flow cytometric analysis of cell cycle position.

4.5 GENERATING SYSTEMS FOR AIRBORNE AGENTS

A number of systems can be used to generate airborne agents to *in vitro* preparations. A summary of some of the more commonly used systems is presented in Table 4.1.

In many cases, the fashion in which the generating system is interfaced with the *in vitro* apparatus is more art than science (e.g., connecting a running diesel engine to a plate of cultured cells) (57).

4.5.1 Gas Mixing (e.g., Ozone, NO_2, and CO)

Some gases used in exposure studies are quite stable and can be purchased compressed in cylinders (e.g., O_2 and NO_2). The compressed gas is then mixed with air (or air + 5 to 10% CO_2) to achieve the experimental

Table 4.1 Generating Systems

Class	*Example Agent*	*Generating System*
Gases	Ozone	Gas generator
	NO_2, NO, SO_2, vinyl chloride	Compressed gas
Liquid	Toluene, methylethylketone, xylene	Nozzle atomizer Ultrasonic atomizer
Particles	Asbestos, silica, carbon black, TiO2	Particle disperser
	Ambient air particles	Concentrated ambient air particle system
Complex mixtures	Cigarette smoke	Cigarette smoking machines
	Combustion engine exhaust	Combustion engine

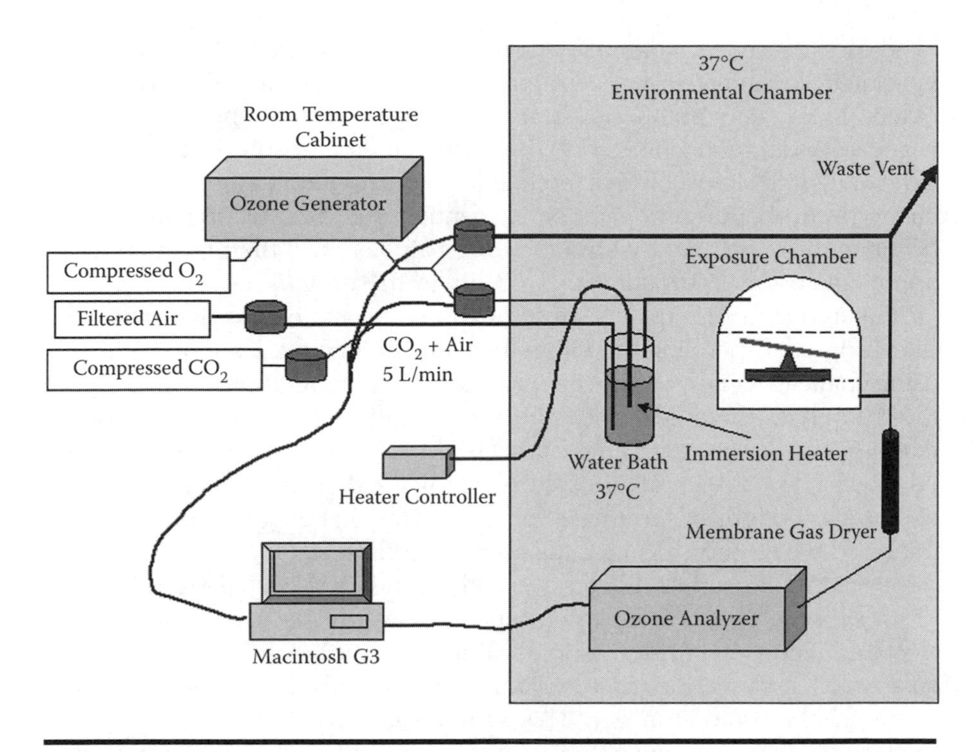

Figure 4.1 ***In vitro*** **ozone exposure system Schematic. This schematic presents the components of a system containing only a single ozone exposure chamber. In the system presented in the text, there are two additional exposure chambers and a control chamber, each with a dedicated set of mass flow controllers and water bath humidifiers.**

concentrations desired. However, some gases are unstable and must be generated continuously during an exposure (e.g., ozone). The mixing can be performed using systems as simple as two needle-valve flowmeters or as complex as computer-controlled electronic mass flow controllers. In some systems, such as some involving ozone exposure, feedback regulation of the gas mixing is used to closely regulate the concentration of gas to which the cells are exposed (Figure 4.1) (58).

4.5.2 Aerosols

Many chemicals that are normally liquids are transported in air either as vapor in the gas phase or in small suspended droplets as aerosols. In some cell culture systems that include a small volume of apical fluid, the agent can simply be dispersed or dissolved in that apical fluid, analogous to the *in vivo* deposition of the chemical in airway lining fluids. Complex exposure systems for cultured cells have been used in which continuously generated aerosols are passed across the surface of the cell monolayer. Often, these use authentic sources such as diesel engines (57) or cigarette-smoking machines (59) to generate their aerosols. In experiments using IPLs, it is almost always more appropriate to deliver the experimental chemicals to the lung preparation suspended in the ventilating gas. Spatial distribution of the agent within the lung's airways will then most correctly approximate the distribution that occurs in the lung *in vivo*.

Cultured cells or explants can be exposed to a variety of PM by simply dissolving or suspending the PM in the tissue culture media. In experiments using epithelial cells on tissue culture inserts, it is more appropriate to apply the PM only to the apical side of the cells and in a solution comparable to that existing *in vivo* (e.g., lung surfactant). In addition, investigators should consider the possibility that insoluble PM will either float to the surface or sink to the bottom of the tissue culture media, changing the local concentration to which the cells or explants are exposed.

Isolated perfused lungs can be used to study not only the direct effect of an airborne toxin on lung cell function, but can be used to investigate the distribution of PM deposition within the ventilating lung. PM can be suspended in an appropriate liquid and an aerosol solution generated using an ultrasonic atomizer. This solution is passed by the inlet of the trachea so that the generated aerosol is drawn into the lung during inspiration. Airborne dust has been generated in inhalation studies with mice using a powder disperser (60). Such devices are available from multiple sources and can certainly be used to provide airborne PM to IPLs.

Several types of materials are used for *in vitro* PM exposures. These may be acquired from commercial suppliers, private sources, or government agencies. Examples of such materials include crocidolite, crysotile, and amosite asbestos; silica; ROFA; latex particles; titanium dioxide; and carbon black. Complex mixtures of airborne PM have been collected by baghouse filtration of ambient air or by collection of filtered PM from the exhaust of diesel engines. The National Institute of Standards and Testing (NIST; www.nist.gov) has a number of environmental standard reference materials (SRM) available. SRM-1648 is an urban PM standard collected from urban PM collected in a baghouse located in St. Louis, MO. SRM-1649a is called "urban dust." This material was collected in a baghouse facility in Washington, DC. Also available from NIST are two diesel exhaust PM standards, SRM-1650a and SRM-2975. The former was collected from the exhaust of several four-cycle diesel engines, whereas the latter was collected from the exhaust of diesel-powered forklifts. An organic solution extracted from SRM-2975 is available as SRM-1975. Finally, SRM-1597 is a complex mixture of polycyclic aromatic hydrocarbons from coal tar delivered as a solution in toluene.

Concentrated ambient particle (CAP) systems represent a relatively new technology that has been continually refined over the past decade (61,62). The ambient particle concentrators are comprised of a series of virtual impactors that sequentially increase the concentration of the collected particles within the device's airstream. Up to 150-fold increases in the airstream particle concentration have been reported (63). The particle-containing airstreams from these devices have been used to study *in vivo* responses to the concentrated particles in rodent (64,65) and canine (66) as well as human (67–72) subjects. Furthermore, *in vitro* exposure to concentrated particles has been achieved in human AM (73,74) and in two human epithelial-like cell lines [Calu-1 (75), and A549 (76)] using these devices.

4.6 INDIVIDUAL MODELS

4.6.1 Cultured Cells

Primary cell cultures are derived either from cells that migrate out from tissue samples or from tissue samples that have been disaggregated mechanically, chemically, or enzymatically. Once separated and purified, these cells can be grown for a number of passages in either suspension or adherent culture conditions. Cells can be directly examined and the surrounding media tested for released autocrine or paracrine mediators along with cell metabolites. Care must be taken to ensure that the cells are exposed to the experimental compounds in a fashion similar to that which occurs *in vivo*.

Primary cells harvested directly from animals for *in vitro* study respond more like the normal cells *in vivo* than do cell lines immortalized by various means (77). In a specific example, McDonald and Usachencko reported that primary human bronchial epithelial cells were resistant to injury when exposed to neutrophils for 2 h after 60 min of ozone exposure. In contrast, the virus-transformed human bronchial epithelial cell line BEAS-2B exhibited significant cytolysis when subjected to the same protocol (78). The use of primary epithelial cells harvested from human airways allows for the study of chemical effects that would be unethically studied in lungs *in situ*. Unfortunately, the number of primary cells that can be harvested from a single animal is limited, and these cells differentiate (or dedifferentiate) after a few passages. Primary human airway epithelial cells also gradually lose their ability to divide after several passages and become senescent. These growth-arrested cells are no longer appropriate models for the normal cells under *in vivo* conditions.

AM play a significant role in moderating the inflammatory response of the lung to foreign compounds (79–81). AM can be harvested from both animal and human subjects by BAL can be used to study the effect of airborne chemicals on primary cultures of these cells. A number of aspects of AM biology have been used to determine chemical effects, including phagocytic capacity (82,83), adhesion (84,85), DNA damage (86,87), and release of chemical messenger compounds (43,47,80,88–92). Many millions of AM can be harvested from the lungs of larger animals (93). However, less than 2 million AM can be collected from smaller rodents (guinea pigs, rats, hamsters, and mice) (93,94). The reduced yield from these species must be considered when designing experiments with BAL AM.

The number of cells needed to answer specific scientific questions varies widely. For example, microscopic analysis can be used to examine the morphology of single AM. For flow cytometric analysis, 10^4 cells are usually recommended to assure statistically relevant data (95). Measurement of stimulated nitric oxide production from macrophage-like RAW 264.7 cells requires nearly a million cells, a small volume of media (~1 ml) and several hours of incubation when using the Griess reagent for measurement (96). Many AM products that are released into the surrounding cell culture media are measured as indices of cellular function. The rate of production and the sensitivity of laboratory detection methods used to measure these macrophage products determine cellular requirements. Furthermore, the number of AM that can be exposed to environmental agents under *in vitro* conditions may be limited by the growth surface area available when studying adherent cells, especially when using tissue culture inserts.

This is a significant problem with transgenic mouse models when considering the small macrophage yield available from mice in general. In the majority of studies using AM, the animals are exposed to the experimental compound, and the AMs are subsequently harvested by BAL for study. In a remarkably small number of studies, BAL-harvested AMs have been subsequently exposed to experimental compounds under *in vitro* conditions (97–101).

4.6.2 Continuous Cell Lines

Continuous cell lines are useful tools for toxicological studies because they can be maintained indefinitely under tissue culture conditions without displaying changes in phenotype or senescence. Unfortunately, this immortality comes with a price to the investigator. Processes similar to those involved in the transformation process may be those affected by the toxic agents being studied. When using these continuous cell lines, the investigator must keep in mind the nature of the transformation and the impact that this difference in cell biology might have on the experimental outcome. Some transformed cells used in *in vitro* experiments result from naturally occurring mutations in cells from the region of interest. For example, the THP-1 cell line is a human monocytic leukemia cell line

(102) while the NR8383.1 alveolar macrophage cell line was derived from normal rat alveolar macrophage cells cultured for several months in the presence of gerbil lung cell conditioned medium. Both of these cell lines have been used as models for AM. Cells may also be transformed by viral infection. The AMJ2-C11 line resulted from the *in vitro* infection of mouse AM with the J2 retrovirus (103), and the RAW 264.7 macrophage cell line came from tumors produced in a mouse infected *in vivo* with the Abelson leukemia virus (104). Transformed airway epithelial cells, smooth muscle cells, and fibroblasts have also been used as experimental models.

4.7 COCULTURE EXPERIMENTS

The biological effects of airborne chemicals in the airways and lung result from the complex reactions both of the agent with individual cell types and from changes the chemical produces in the manner of interaction between the numerous cell types the agent contacts. In some cases, investigators may be specifically interested in the change in those interactions. Coculture experiments in which two or more cell types are maintained together can be designed to identify changes in the cell interactions in the absence of confounding input from other cell types and from other systemic processes. For example, Ning and coworkers have examined the role of neutrophils in moderating the release of IL-8 from Calu-1 cells exposed to concentrated ambient air particles (75). Cheek et al. have described the effect of neutrophils cocultured with rat alveolar type II cells in studies of ozone injury and repair (105). To identify possible cellular interactions between human bronchial epithelial cells and human umbilical vein endothelial cells in ozone exposure experiments, Mogel et al. incubated the two cell types on opposite sides of filter supports and then measured IL-6 and IL-8 release, along with endothelial ICAM-1 expression, in response to ozone exposure (106). Jujii and coworkers placed freshly prepared human AM directly on confluent human bronchial epithelial cell monolayers and studied cell viability and bone marrow growth factor release after 24 h of exposure to EHC-93 particles (107). It is important that the media selected for coculture experiments provides the appropriate nutritional environment for the combination of cells being studied. In some cases, optimal media for one cell type will not support normal function in a second cell type in the system. For example, one of the media preferred for growth of human microvascular endothelial cells (MCDB-131) usually contains 5% fetal bovine serum. An attempt to coculture these endothelial cells with primary human airway epithelial cells in this media may fail because proliferation of human airway epithelial cells has been found to decline in the presence of small amounts of serum (108). Interestingly, bovine bronchial epithelial cell growth is stimulated by the presence of serum (109). This point illustrates the need for careful review of the literature with special attention paid to cell type and species differences.

4.7.1 State of Differentiation of Primary Cell Culture

Primary mammalian cells slowly differentiate after harvest and eventually reach a state of senescence accompanied by growth arrest and death. Cellular responses to toxic agents may be affected by the level of differentiation. It is important when conducting experiments to compare treated cells at similar states of differentiation.

4.7.2 Level of Confluence

Spatial density of cultured cells affects multiple aspects of cell function. Cell density influences measured indices of bioenergetics including oxygen consumption (110–112), glucose consumption (113,114), lactate production (115,116), mitochondrial membrane potential (117), and cellular ATP content (115,116). These observations have been made in multiple cell types. Further, pyridine nucleotide redox balance in CRL1606 hybridoma cells (118), insulin-like growth factor binding protein secretion by diploid fibroblasts (119), and low-density lipoprotein secretion by bovine endothelial cells (120) have all been reported to be affected by cell density. Clearly, these and many other effects of cell density on cell function are strongly dependent on the cell type studied.

Toxic compounds may directly affect several aspects of cell function while at the same time producing additional effects indirectly through changes in cell density. Careful experiments performed using a range of cell seeding densities will help the investigator discriminate between the two mechanisms.

4.7.3 Choice of Tissue Culture Media and Conditions

A wide variety of tissue culture media formulations has been described. Many of these media formulations have been used in lung toxicology studies (Table 4.2). Some of these media have been designed for specific cell lines. Three particular media come to mind when considering the culture of human lung cells. MCDB-131 was originally optimized for growth of human microvascular endothelial cells (121). It is one of the preferred media for support of endothelial cells from both large and small blood vessels (77) and is the basis for a number of the proprietary formulations commercially available. The alpha modification of Eagle's Minimum Essential Medium (α-MEM) was originally formulated for optimum support of fibroblast growth (122) and has been found to be an

Table 4.2 Media Used for *In Vitro* Cell Culture in Lung Toxicology Studies

Compound	*Cell Type*	*Base Medium*	*% Serum*	*Reference*
95% oxygen	A-549 human adenocarcinoma cells	F12-K	10	251
Asbestos	Human alveolar macrophage	Medium 1990	10	252
CAPS	Calu-1 human lung epidermoid carcinoma cells	McCoy's 5a medium	10	75
Carbon monoxide	RAW 264.7 mouse macrophage cell line	DMEM	10	253
Cigarette smoke	HFBE-21 human bronchial epithelial cell line	RPMI-1640	0	142
Cigarette smoke extract	NCI-H292 human pulmonary carcinoma cells	RPMI-1640	10	254
Diesel exhaust particles	Human bronchial epithelial cells	Medium 199	2.5[a]	255
Diesel motor exhaust	HFBE-21 human bronchial epithelial cell line	RPMI-1640	0	57
EHC-93	Human alveolar macrophages	RPMI-1640	10	44
Ozone	A-549 human adenocarcinoma cells	F12-K	10	256
Ozone	BEAS-2B human bronchial epithelial line	KGM[a]	0	90
Ozone	Human bronchial epithelial cells	1:1 F12/ DMEM	1[c]	257
Ozone	Human nasal epithelial cells	1:1 F12/ DMEM	10[a]	258
Ozone	Human nasal epithelial cells	1:1 F12/ DMEM	10[a]	132

(*continued*)

Table 4.2 Media Used for *In Vitro* Cell Culture in Lung Toxicology Studies (Continued)

Compound	*Cell Type*	*Base Medium*	*% Serum*	*Reference*
Ozone	Primate bronchial epithelial cells	1:1 F12/ DMEM	10[a]	132
Silica	Human alveolar macrophage	RPMI-1640	10	81
Silica	Rat alveolar macrophages	DMEM	0	259
Sulfur dioxide	Human nasal epithelial cells	1:1 F12/ DMEM	10[a]	133
Nitrogen dioxide	RLE - rat lung type II epithelial cell line	1:1 F12/ DMEM	0.5	260
Residual oil fly ash	BEAS-2B human bronchial epithelial line	KGM[b]	0	261

Note: CAPS = Concentrated ambient particles

[a] Used Nu-Serum (Becton-Dickenson, San Jose, CA).

[b] Keratinocyte Growth Medium—essentially MCDB 153 medium plus several supplements; product of Clonetics (San Diego, CA).

[c] Used 1% low protein serum replacement (Sigma, St Louis, MO).

excellent medium for the growth and maintenance of both alveolar and peritoneal macrophages (123–128). LHC-9 medium was developed by Lechner and LaVeck specifically for the culture of human bronchial epithelial cells (129). Other media can be used to support a broad spectrum of cell types. Several resources are available to guide the selection of media for a particular cell type. Freshney's *Culture of Animal Cells: a Manual of Basic Technique* (77) and the American Type Culture Collection Website (www.ATCC.org) are particularly comprehensive. Many of the commercial suppliers of primary human cells also market specific media for the individual cell types. It must be noted, however, that the formulations of these media are often proprietary so investigators wishing to examine the effects of specific media components might well consider other media whose formulae are publicly available.

Media volumes should be chosen so that the cultured cells will not deplete the media of an essential nutrient during the course of the study. This can be troublesome in studies involving the exposure of cultured cells to gases that are soluble in the culture media. The rate of equilibration of the media with the experimental gas is dependent on both the solubility of the gas with the media and on the thickness of the fluid layer (130). Investigators electing to use small media volumes to accelerate gas equilibration may find that the cultured cells deplete the media of glucose, glutamine, or some other critical nutrient before the end of the experimental period. In contrast, luxuriant media volumes prevent nutrient deficiencies, while delaying the equilibration of poorly soluble gases for several hours (131).

Growth of isolated mammalian cells has become a valuable tool in many aspects of biological inquiry. This widespread popularity belies the difficulties overcome by early investigators. "Recipes" are now available for the culture of many mammalian cell types. However, the popular use of antibiotics, fungicides, and mycoplasma detection kits in cell culture laboratories illustrates the continuing difficulty in maintaining clean cultures of mammalian cells. Investigators should keep in mind that the antibiotic and antifungal agents used in their cell culture experiments may alter the effects of the chemicals to which the cells are exposed.

4.7.4 Apical Fluid Issues

Many investigators have recognized the need to expose cultured cells directly to airborne test compounds when conducting *in vitro* experiments. However, the somewhat conflicting need to maintain cellular hydration and nutrition was also recognized. A number of techniques have been used to simultaneously address both of these needs.

In some studies, tissue culture plates were slowly rotated at an inclined angle (~15°) so that the cells on the plates would be alternately exposed directly to the gas phase overlying the cells and then to the tissue culture media that only partially covered the growth surface when tilted. Rotation of the plates was achieved in some experiments by mounting an inclined platform inside a stationary exposure chamber and slowly rotating that platform with an external drive motor (132–134). In other systems, the tissue culture plates were placed in a tilted exposure chamber, and the entire exposure chamber rotated within a warm incubator.

As an alternative to the rotating apparatus described earlier, Tarkington and coworkers have described a rocking platform to move liquid across airway explant cultures (135,136). In this system, the exposure chambers are mounted on a motorized rocker that cycled at a rate of 22/min. When using systems similar to this, care must be taken to ensure that the gas flow lines used to deliver the experimental agent to the exposure chambers do not become disconnected or crimped. Furthermore, cyclical strain on most gas fittings will inevitably result in mechanical failure unless monitored closely.

Roller bottles have been used to alternately expose cells to experimental airborne agents and to the tissue culture media (137,138). This technique allows a large number of cells to be exposed directly to test chemicals with a reduced risk of desiccation. However, morphological examination of the cells *in situ* and studies of live-cell function are difficult.

Cell culture inserts (e.g., Transwell or Millicell units) can be used to maintain cultured cells at the gas–liquid interface. Adequate hydration and nutrition can be maintained if the gas phase above the cells is water saturated. Cultured cells studied in these systems most appropriately model cells exposed to chemicals *in vivo*. However, *in vivo*, there is usually a thin layer of liquid between the apical surface of the cells and the ventilating air. Conducting airway cells are covered with a layer of mucus-containing lining liquid while alveolar epithelial cells whereas reside under a layer of surfactant-rich fluid. Depending on the goals of the experiment and the cell types used, this overlying fluid may be included by the investigator as part of the experiment or produced by the cultured cells as a normal cell product. In these situations, maintaining a thin layer of the apical liquids may be difficult due to the effect of fluid surface tension. Often times, the fluid develops a meniscus, with the thinnest layer of apical fluid at the center of the growth matrix and the thickest layer present at the periphery of the membrane against the plastic insert housing. Rocking of these inserts has little effect on this meniscus formation.

Nutritional support of cells maintained on the tissue culture inserts is an important consideration for investigators. The amount of media that can be used with cells exposed at the air–media interface is quite limited. For example, the 30 mm tissue culture inserts can only contain about 1 ml of basolateral media without flooding the apical surface of the cells. Prolonged experiments with these small media volumes may result in depletion of media components such as glucose and glutamine. Potentially toxic waste products such as lactic acid may also accumulate in the small media volumes during extended experiments. Furthermore, the experimental environment that the cells are exposed to may cause chemical changes to the media that will subsequently affect cell function. The effect of the nutrient depletion and waste accumulation can partially be mitigated by gently mixing the media throughout an exposure experiment to prevent the formation of chemical gradients within the media. This has been achieved at one facility by slowly rocking the 6-well tissue culture plates holding the tissue culture inserts (58). The rocking platform is placed within the exposure chamber to simplify gas connections, and motion is imparted to the device by alternating the force delivered by electromagnets mounted on the outside of the exposure chamber. Other investigators have achieved similar goals by placing the entire exposure chamber on a rocking platform within the tissue culture incubator.

In a perfect *in vitro* system, alveolar cells would be exposed to environmentally relevant levels of a test compound delivered in a warm, water-saturated gas mixture containing low levels of CO_2, as occurs in the lung under *in vivo* conditions. It is difficult to achieve this goal in the real world. Investigators need to critically consider specific aspects of various existing exposure systems when deciding on a system to use in their experiments. A number of exposure systems have been described that incorporate the use of tissue culture inserts and polystyrene tissue culture plates in environmental exposure systems (58,139–141). Some test compounds may react with the polystyrene material to produce unexpected and undesired artifactual responses in the cells under study. Furthermore, the media volume that can be maintained under the tissue culture membrane of these inserts is quite small when used in conjunction with their respective multiwell plate. This limits the duration of possible experiments. Finally, direct access of the experimental compound to the media itself may produce unintended reactions.

An alternative to the tissue culture insert–multiwell plate combination has been developed, and its use in studying a number of environmental toxins has been described in a number of papers (57,59,140,142–145). In this system, the tissue culture inserts are held within glass chambers that are constructed such that there is no contact of the apical gas phase with the basolateral media. The basolateral chambers are plumbed to allow exchange of the media without removing the apical gas. In this fashion, media nutrient depletion and reactions between the environmental agents and the tissue culture media can be avoided. However, the issue previously addressed concerning the reaction of polystyrene with the test compounds in the apical gas also exists here. In addition, the media chambers are connected to the media reservoir and drains by silicone tubing. In this design, bicarbonate-buffering systems cannot be used. Any dissolved carbon dioxide arriving in the media from the apical gas phase will rapidly diffuse out of the basolateral media through the silicone tubing. Furthermore, the exposure gas is delivered without humidification. These investigators state that the apical surface of the culture cells in the Cultex system is kept moist by a "microclimate" of humid air formed by evaporation of liquid moving from the basolateral to the apical surface of the cell layer. Such evaporation would seem to result in an accumulation of media salts on the apical surface of the cells that would likely affect cell function and cell responses to the environmental challenge. As described in the literature, the Cultex system overcomes some of the limitations characteristic of other systems but introduces a new set of concerns. Investigators need to consider all of the aspects of the exposure system needed for their particular experiment and then choose, modify, or invent a system that achieves their experimental goals while minimizing the potential for artifacts.

4.7.5 Airway Explant Culture

Experiments involving the use of explants can generally be divided into two designs. In the first design, the intact animal is exposed to the experimental compound and the explant then harvested at a later time point for study. In the second design, the explant is harvested from the unexposed animal. Then, under *in vitro* conditions, the explant is exposed to the experimental compound and subsequently studied.

A broad spectrum of methods has been used to study the effects of a number of agents on airway cell function in airway explants. Some of these methods involve simple examination of the exposed tissue either under an optical or electron microscope. Fisher and Placke used morphological methods to study tracheal explants from hamsters in studies of benzo(a)pyrene and nickel exposure (146). These authors also describe the morphological effects of silica exposure on peripheral lung explants.

Other more technically challenging preparations using airway explants have been used to address specific experimental questions. Rings and strips of airway have been used to examine smooth muscle contractility. Microscopic examination of ciliary motion has demonstrated the role of chemical exposure on mucociliary clearance processes.

4.7.6 Airway Smooth Muscle Responses

Assessment of airway smooth muscle responses to airborne toxins can be performed with rings or strips of trachea or bronchi. Complete experimental systems for these experiments can be purchased commercially or built from individual components by the investigator. An excellent description of a typical system is provided by Gelfand et al. (147). The ring or strip is suspended between a fixed lower hook and a second hook connected to an isometric strain gauge transducer. The explant is submerged in a physiological salt solution within a heated water bath (Figure 4.2). A mechanical preload is placed on the tissue by way of a micrometer-positioning device holding the strain gauge transducer. Electrical field stimulation can be applied through electrodes placed in the bath solution while direct electrical stimulations can be applied directly to the explant through electrodes placed in the tissue or through the hooks suspending the explant. In most cases, the cross-sectional area of the ring or strip is determined so that the contractile forces measured can be normalized to the amount of smooth muscle present. These forces are often presented as g/cm^2.

Smooth muscle responses have been evaluated in airways from a number of species, including humans (148,149), dogs (150,151), cats (152), rabbits (147), guinea pigs (153–157), rats (158), and mice (159–163). Contraction or relaxations have been induced using EFS (150–154,156) or

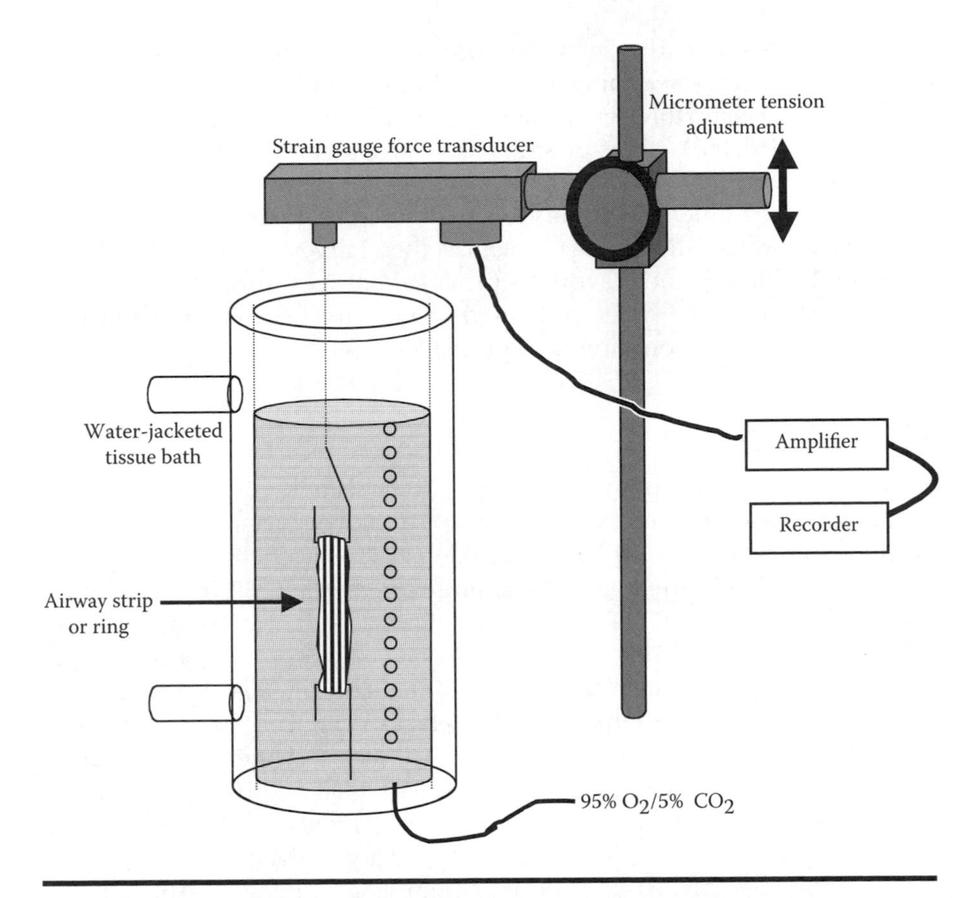

Figure 4.2 Airway ring/strip apparatus. Strips or rings or airway tissue are suspended between a lower fixed hook and an second upper hook attached to a strain gauge force transducer. A preload tension can be placed on the specimen by micrometer adjustment of the force transducer height. The water-jacketed tissue bath is maintained at 37°C for mammalian tissue. The tissue bath is filled with a physiological salt solution bubbled with a gas mixture usually containing 95% O_2 and 5% CO_2.

chemical stimulation with acetylcholine (150,151,153–155,158), substance P (149,155,164), neurokinin A (164), histamine (149), methacholine (157), or serotonin (148,152,155,165). A variety of receptor antagonists has been used to isolate specific mechanisms involved in the contractile responses observed.

Studies addressing the effects of inhaled gases on airway contractile function have been conducted using two distinct exposure schemes. In the first, the live animal inhales the experimental gas for a fixed time period, and the airway then is harvested for subsequent study

(150–153,155,156). In the second experimental design, airways are harvested from unexposed animals and flushed with air containing the experimental gas mixture while submerged in a physiological salt solution (149,154,157,164,166). Rings or strips are then prepared from the exposed airway segment.

In other experiments, animals are exposed to a toxin added to either their feed or water. The airway explant is then harvested and studied. For example, Michielsen and coworkers provided rats feed supplemented with hexachlorobenzene for 6, 14, or 21 d, then used tracheal explants to study that chemical's effect on airway hyperreactivity (165).

4.7.7 Ciliary Motion

Mucociliary clearance is an important mechanism by which biological debris and environmental particles can be removed from the airways of the lung. Ciliary motion has been studied in airway epithelial cells exposed to a number of airborne agents including nitric oxide (167,168), acetaldehyde (169), ozone (170,171), sulfuric acid (170,171), sulfur dioxide (172,173), cadmium (174), and nickel (175). In one series of studies, a stroboscopic light source was used to illuminate tracheal ring explants during microscopic examination (170,171,174,175). The strobe rate was adjusted until ciliary motion appeared to cease and the corresponding frequency read from the stroboscopic device. Other investigators have used a photoelectric technique to assess ciliary motion (168,173,176,177). In this technique, light reflected from the ciliary surface is measured by a photomultiplier tube and attendant recording apparatus. The rate of variation of the light intensity detected by the photomultiplier tube is used to reflect ciliary beat frequency. Analysis of videotape recording of the microscopic images of airway epithelial cells has also been used to quantify ciliary activity (169,178).

4.7.8 Microscopy Systems for Cultured Cells and Explants

Microscopic examination of culture cells and tissue explants can be used to examine the effect of airborne chemicals on several aspects of cell physiology. For the study of more prolonged effects of chemical exposure, the animals are exposed to the airborne agent and then the cells harvested at a subsequent time point for study. In studies addressing more immediate or transient effects of chemical exposure, cells or tissues are harvested from the animals and then exposed to the chemicals in systems that allow real-time measurements of cell function and morphology while under microscopic examination. Antibodies targeting specific cell molecules are conjugated to a variety of fluorescent compounds to visually determine the location and motion of those target molecules in the cell. Other compounds, such as phalloidin, naturally bind to interesting cellular targets (i.e., actin) and can be conjugated to fluorochromes to localize those targets. A cornucopia of fluorescent probes has become available to study nearly every biological target conceivable. Many of these either fluoresce more brightly or exhibit a quenching of fluorescent intensity upon binding to their respective target. The spatial distribution of other compounds is determined by the electrical potential existing across various cell membranes. Others react with reactive oxygen and nitrogen compounds, resulting in a change in their fluorescent character.

Although this wondrous assortment of probes is widely used to expand the frontiers of cell biology, caution must be exercised when examining the effects of toxic chemicals on cell systems. Some of these compounds may react directly with the toxin under study, giving the false impression that the amount of the assumed target of a probe has changed. Furthermore, some of these compounds (e.g., rhodamine 123) can produce singlet oxygen when illuminated (179,180). Although this type of reaction is a boon for those working with photodynamic therapy in the treatment of cancer, such a reaction will confound the interpretation of the observations using such compounds as probes of cellular function. A further concern when using fluorescent probes is the effect of the multidrug resistance (MDR) process on the cellular levels of the fluorescent compound. This process involves a number of distinct ATP-dependent drug efflux "pumps" (181,182). Many compounds are extruded from cells by the MDR process, including some of the fluorescent probes used to assess mitochondrial membrane potential (183–186) or nucleic acid content (186–189). Furthermore, the acetoxymethyl ester (AM) derivatives of a number of fluorescent probes are extruded by the MDR pumps. These include AM derivatives of the cell viability probe calcein (190) along with the AM derivatives of the calcium probes fura-2 (191), indo-1 (191,192), and fluo-3 (192). Any changes in the MDR process caused by the toxic exposure will affect fluorescent intensities of these probes.

Live-cell microscopy systems require a heating device to maintain the cells at their normal body temperature. These systems must also allow for delivery of fresh media to prevent nutrient depletion during extended studies. When using volatile chemical agents in these systems, the cell chamber should be sealed both to prevent a decline in the concentration of the experimental agent in the media and to prevent exposure of the investigator to potentially toxic chemicals. Water-immersion objective lenses can be used if the experimental agent remains in the media with little evaporation. This allows ready access to the media and cells for additions and manipulations.

Reaction of the experimental compound directly with the live-cell apparatus may produce artifacts. Materials selected for construction of these systems must be compatible with the biological system being used as well as unreactive to the experimental agents.

4.7.9 Lung Parenchymal Explants

The uninflated lung collapses with distortion of the normal cellular architecture. The collapse of the lung will cause changes in cell shape along with alterations in cell–cell and cell–matrix relationships. Such changes have been shown to affect a variety of cellular functions (193–202). A number of investigators have attempted to minimize these kinds of effects by inflating the lungs with agarose and then sectioning 0.5- to 2-mm slices for subsequent experimentation (203–221). Such preparations can be studied for several weeks (206,212,213). In these preparations, the type of agarose used is chosen to take advantage of the hysteresis in the liquid-to-gel transition. The type of agarose is chosen so that it melts at temperatures above those usually used in cell culture but forms a gel upon cooling to temperatures below cell culture temperatures. Once the agarose solidifies, the lung can be rewarmed to normal cell culture temperatures without remelting of the agarose.

Several end points have been studied using agarose-filled lung explants. These include ATP release (203), airway smooth muscle reactivity (204,210,214–219,221), mucociliary function (208), mitogen release (207), and protein synthesis (205).

4.7.10 Isolated Perfused Lung

A number of parameters of function can be measured in IPL. Vasomotor tone can be measured by estimating vascular resistance to movement of the perfusion fluid through the lung. Changes in elastic properties of the lung have been assessed by monitoring changes in airway pressure during the ventilatory cycle. Furthermore, IPLs have been used to study the production, destruction, and modification of a variety of biological compounds by the lung [e.g., angiotensin (222), serotonin (223), interleukin-6 (224), and interferon-gamma (225)]. The production of free radicals within IPLs has been studied using chemiluminescent probes and specialized monitoring equipment (226,227). Strain gauge transducers can be used to measure the weight of the heart–lung ensemble in order to monitor the development of pulmonary edema.

IPLs have been extensively used to study the effects of ozone (17,228–237), cigarette smoke (224,225,238), and airborne PM (239–244). In many of these experiments, the animal was first exposed to the experimental toxin, and the lung was then isolated and perfused for subsequent functional studies (224,225,228,238,242–244). In others, the IPL itself was exposed to the toxic substance. IPLs have also been used to assess hypoxic pulmonary vasoconstriction in a number of species (245–250). For logistical reasons, IPLs are usually prepared from larger rodents (i.e., rats, ferrets, rabbits). A wide variety of IPL preparations have been described that differ in method of ventilation, perfusion pressure control, and perfusate composition. The method chosen depends on the specific goals of the experiment. Both positive and negative pressure systems have been used to ventilate the IPL. (Figure 4.3 and Figure 4.4, respectively)

Positive pressure ventilation systems are easier to set up and allow more aggressive control of airway pressures. A peristaltic or roller pump is used to move the perfusate solution through the lung, usually by way of a cannula placed in an incision in the right ventricle of the heart. A small animal ventilator maintains cyclic gas flow into the airways by way of a cannula placed in the trachea. Pressure transducers are used to monitor fluid perfusion pressure while another may be used to measure airway pressure in the trachea. A warm-water-jacketed bath is usually used to surround the heart–lung ensemble to maintain physiological temperatures.

In comparison, negative pressure systems require a closed compartment surrounding the IPL to contain the oscillating negative transpleural pressure. These systems are technically more complicated than the positive pressure versions. The heart–lung ensemble is placed in the artificial thorax chamber and the trachea connected to a line passing to the outside of the chamber. The chamber is then sealed and ventilated at negative pressures relative to that in the trachea. As the movement of gases is driven by changes in transpleural pressure, the distribution of airflow within the lung is more similar to that occurring in the lung *in vivo*. Furthermore, experimental agents can be introduced into the airways at ambient atmospheric pressures. However, adjustments of cannula position or other physical manipulations are more difficult with the lungs located inside the artificial thorax.

Intravascular pressures in the lung are normally quite low *in vivo*. Likewise, IPLs are normally perfused at low pressures (only a few mmHg pressure). However, changes in pulmonary vascular resistance occur in a variety of experimental settings and may result in changes in the intravascular pressure in the IPL. IPL systems can be constructed that regulate either intravascular pressure to maintain a steady flow rate or that regulate the perfusate flow rate to maintain a stable intravascular pressure. Because the lung *in vivo* is physically obligated to receive the entire cardiac output, many investigators feel that the most appropriate perfusate control system is one that maintains constant flow while allowing vascular-remarks resistance-dependent changes in vascular pressure.

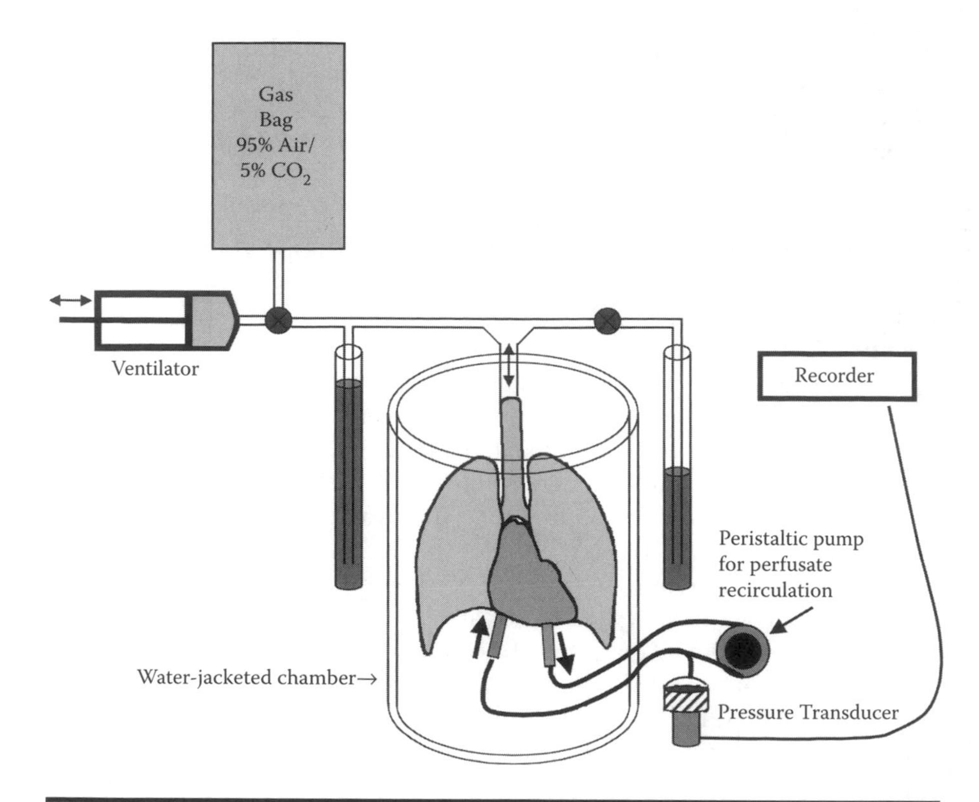

Figure 4.3 Positive pressure isolated perfused lung system. The heart-lung ensemble is suspended within a water-Jacketed organ chamber maintained by water circulation at 37°C. The lungs are ventilated through a cannula placed in the trachea while the pulmonary vasculature is perfused by way of a cannula placed in the right ventricle of the heart. Venous effluent is collected from a second cannula placed in the left ventricle of the heart. The perfusate can be used for only one pass through the lung or it may be recirculated. In this particular design, positive end-inspiratory pressure and positive end-expiratory pressure are maintained by appropriately plumbed water columns. pulmonary resistance can be assessed by monitoring changes in pressure within the pulmonary arterial cannula. Pulmonary compliance changes can be assessed by monitoring pressure changes within the tracheal cannula.

IPLs have been perfused with a variety of solutions, ranging from simple salt solutions to autologous whole blood. In experiments in which bicarbonate-buffered salt solutions are used, the lungs are usually ventilated with 95% oxygen and 5% carbon dioxide to provide both pH control and for adequate oxygenation of the tissue. In addition, the salt solutions used in IPL systems often contain an oncotic support compound such as albumin or Ficoll to prevent the development of edema. The span of time

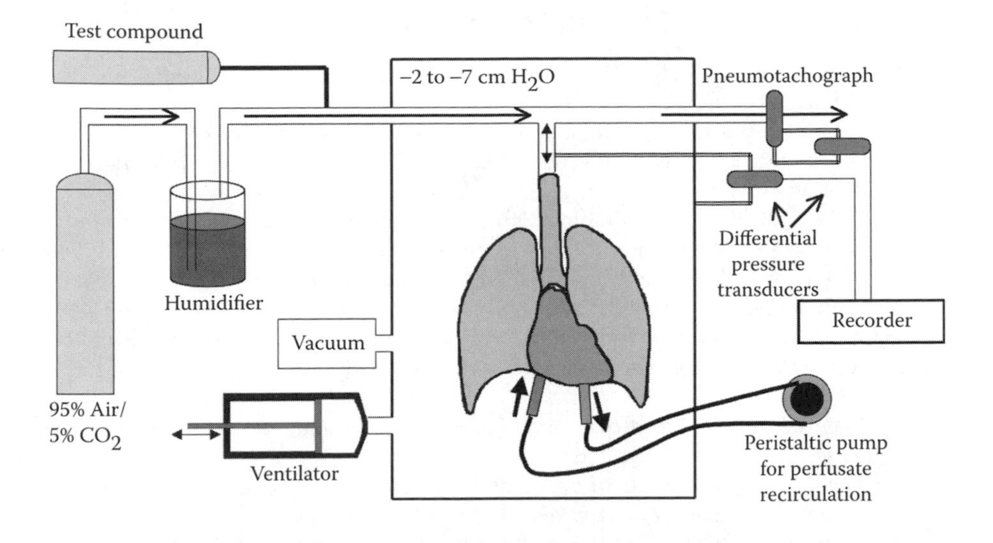

Figure 4.4 Negative pressure isolated perfused lung system. the heart–lung ensemble is suspended within a closed artificial thorax chamber. Slight negative pressure is maintained within the artificial thorax by application of an external vacuum. Ventilation is achieved using a small animal ventilator connected to the artificial thorax. Humidified gas for lung ventilation is passed through a cannula placed in the trachea. A pneumotachograph is used to measure ventilatory volumes while a differential pressure transducer is used to measure transpleural pressure. Perfusion of the pulmonary vasculature is achieved using a roller or peristaltic Pump connected to cannulae placed in the left and right ventricles of the heart. Experimental compounds are introduced to the airway ventilation line downstream from humidification bath to prevent their loss in water.

an IPL can be used for experiments is normally limited to 2 or 3 h. IPLs used for extended time often develop edema and become unresponsive to experimental stimuli.

4.8 CLOSING REMARKS

In vitro models are valuable tools in lung toxicology. Several of these models have been described here. Variations of these models have been implemented in different laboratories in different fashions in order to best address the specific questions being addressed. Other organs within the body are affected by inhaled toxins. Many of the same *in vitro* techniques that are used to study lung insults can also be used to explore the effects of toxins on these distal organs. In addition, other *in vitro* models have been developed that are specific for study of these organs.

REFERENCES

1. Taber, C.W., *Taber's Cyclopedic Medical Dictionary*. 15th ed., Thomas, C., Ed., F.A. Davis Company, Philadelphia 1985.
2. Selgrade, M.K. et al., Evaluation of effects of ozone exposure on influenza infection in mice using several indicators of susceptibility. *Fundam Appl Toxicol*, 11, 169, 1988.
3. Kabel, J.R. et al., Longitudinal distribution of ozone absorption in the lung: comparison of nasal and oral quiet breathing. *J Appl Physiol*, 77, 2584, 1994.
4. Mautz, W.J., Exercising animal models in inhalation toxicology: interactions with ozone and formaldehyde. *Environ Res*, 92, 14, 2003.
5. Musi, B. et al., Effects of acute and continuous ozone (O3) exposure on activity/exploration and social behavior of CD-1 mice. *Neurotoxicology*, 15, 827, 1994.
6. Umezu, T. et al., Effect of ozone toxicity in the drinking behavior of rats. *Arch Environ Health*, 42, 58, 1987.
7. Umezu, T. et al., Effects of ozone and nitrogen dioxide on drinking and eating behaviors in mice. *Environ Res*, 61, 51, 1993.
8. Vincent, R. et al., Sensitivity of lungs of aging Fischer 344 rats to ozone: assessment by bronchoalveolar lavage. *Am J Physiol*, 271, L555, 1996.
9. Uhlson, C. et al., Oxidized phospholipids derived from ozone-treated lung surfactant extract reduce macrophage and epithelial cell viability. *Chem Res Toxicol*, 15, 896, 2002.
10. Squadrito, G.L. and Pryor, W.A., Oxidative chemistry of nitric oxide: the roles of superoxide, peroxynitrite, and carbon dioxide. *Free Radic Biol Med*, 25, 392, 1998.
11. Douglas, G.C. et al., HEPES buffer perfusate alters rabbit lung endothelial permeability. *J Appl Physiol*, 75, 1423, 1993.
12. Bowman, C.M. et al., HEPES may stimulate cultured endothelial cells to make growth-retarding oxygen metabolites. *In Vitro Cell Dev Biol*, 21, 140, 1985.
13. Hill, G.D. et al., Soluble ICAM-1, MCP-1, and MIP-2 protein secretion by rat pleural mesothelial cells following exposure to amosite asbestos. *Exp Lung Res*, 29, 277, 2003.
14. Huang, S.L. et al., Contribution of endotoxin in macrophage cytokine response to ambient particles *in vitro*. *J Toxicol Environ Health A*, 65, 1261, 2002.
15. Cottet-Emard, J.M. et al., Long-term exposure to ozone alters peripheral and central catecholamine activity in rats. *Pflugers Arch*, 433, 744, 1997.
16. Peden, D.B. and Dailey, L., Modulation of mast cell functions by in vitro ozone exposure. *Am J Physiol*, 268, L902, 1995.
17. Gross, K.B. et al., The effect of ozone inhalation on metabolic functioning of vascular endothelium and on ventilatory function. *Toxicol Appl Pharmacol*, 109, 336, 1991.
18. Block, E.R. and Stalcup, S.A., Depression of serotonin uptake by cultured endothelial cells exposed to high O_2 tension. *J Appl Physiol*, 50, 1212, 1981.
19. Mustafa, M.G. and Cross, C.E., Effects of short-term ozone exposure on lung mitochondrial oxidative and energy metabolism. *Arch Biochem Biophys*, 162, 585, 1974.
20. Mustafa, M.G. et al., Ozone interaction with rodent lung. II. Effects on oxygen consumption of mitochondria. *J Lab Clin Med*, 82, 357, 1973.

21. Zychlinski, L. et al., Age-related difference in bioenergetics of lung and heart mitochondrial from rats exposed to ozone. *J Biochem Toxicol*, 4, 251, 1989.
22. Mustafa, M.G., Augmentation of mitochondrial oxidative metabolism in lung tissue during recovery of animals from acute ozone exposure. *Arch Biochem Biophys*, 165, 531, 1974.
23. Bassett, D.J. and Bowen-Kelly, E., Rat lung metabolism after 3 days of continuous exposure to 0.6 ppm ozone. *Am J Physiol*, 250, E131, 1986.
24. Beach, A.C. and Harmon, H.J., Additive effects and potential inhibitory mechanism of some common aromatic pollutants on in vitro mitochondrial respiration. *J Biochem Toxicol*, 7, 155, 1992.
25. Morin, J.P. et al., Development of a new in vitro system for continuous in vitro exposure of lung tissue to complex atmospheres: application to diesel exhaust toxicology. *Cell Biol Toxicol*, 15, 143, 1999.
26. Castranova, V. et al., Toxicity of metal ions to alveolar macrophages. *Am J Ind Med*, 1, 349, 1980.
27. Yoon, B.I. et al., Aryl hydrocarbon receptor mediates benzene-induced hematotoxicity. *Toxicol Sci*, 70, 150, 2002.
28. Santini, R.P. et al., Regulation of Cyp1a1 induction by dioxin as a function of cell cycle phase. *J Pharmacol Exp Ther*, 299, 718, 2001.
29. Gardner, D.E., Alterations in macrophage functions by environmental chemicals. *Environ Health Perspect*, 55, 343, 1984.
30. Girard, D., Activation of human polymorphonuclear neutrophils by environmental contaminants. *Rev Environ Health*, 18, 75, 2003.
31. Shigehara, T. et al., Sulfite induces adherence of polymorphonuclear neutrophils to immobilized fibrinogen through activation of Mac-1 beta2-integrin (CD11b/CD18). *Life Sci*, 70, 2225, 2002.
32. Pearson, A.C. and Bhalla, D.K., Effects of ozone on macrophage adhesion in vitro and epithelial and inflammatory responses in vivo: the role of cytokines. *J Toxicol Environ Health*, 50, 143, 1997.
33. Driscoll, K.E. and Schlesinger, R.B., Alveolar macrophage-stimulated neutrophil and monocyte migration: effects of in vitro ozone exposure. *Toxicol Appl Pharmacol*, 93, 312, 1988.
34. Dong, W. et al., Altered alveolar macrophage function in calorie-restricted rats. *Am J Respir Cell Mol Biol*, 19, 462, 1998.
35. Sherwood, R.L. et al., Effects of inhaled hexachlorobenzene aerosols on rat pulmonary host defenses. *Toxicol Ind Health*, 5, 451, 1989.
36. Becker, S. et al., Differential particulate air pollution induced oxidant stress in human granulocytes, monocytes and alveolar macrophages. *Toxicol In Vitro*, 16, 209, 2002.
37. Nguyen, H. et al., Cigarette smoke impairs neutrophil respiratory burst activation by aldehyde-induced thiol modifications. *Toxicology*, 160, 207, 2001.
38. Pelletier, M. et al., Activation of human neutrophils *in vitro* and dieldrin-induced neutrophilic inflammation in vivo. *J Leukoc Biol*, 70, 367, 2001.
39. Jakab, G.J., Modulation of pulmonary defense mechanisms by acute exposures to nitrogen dioxide. *Environ Res*, 42, 215, 1987.
40. Aranyi, C. et al., Effects of toluene inhalation on pulmonary host defenses of mice. *Toxicol Lett*, 25, 103, 1985.
41. Johnson, J.D. et al., Activation of alveolar macrophages after lower respiratory tract infection. *J Immunol*, 115, 80, 1975.

42. Becker, S. et al., Modulation of human alveolar macrophage properties by ozone exposure *in vitro. Toxicol Appl Pharmacol*, 110, 403, 1991.
43. Soukup, J.M. and Becker, S., Human alveolar macrophage responses to air pollution particulates are associated with insoluble components of coarse material, including particulate endotoxin. *Toxicol Appl Pharmacol*, 171, 20, 2001.
44. Mukae, H. et al., Phagocytosis of particulate air pollutants by human alveolar macrophages stimulates the bone marrow. *Am J Physiol Lung Cell Mol Physiol*, 279, L924, 2000.
45. van Eeden, S.F. et al., Cytokines involved in the systemic inflammatory response induced by exposure to particulate matter air pollutants (PM10). *Am J Respir Crit Care Med*, 164, 826, 2001.
46. Pelletier, M. et al., Activation of human neutrophils by the air pollutant sodium sulfite (Na(2)SO(3)): comparison with immature promyelocytic HL-60 and DMSO-differentiated HL-60 cells reveals that Na(2)SO(3) is a neutrophil but not a HL-60 cell agonist. *Clin Immunol*, 96, 131, 2000.
47. Thomassen, M.J. et al., Regulation of human alveolar macrophage inflammatory cytokine production by interleukin-10. *Clin Immunol Immunopathol*, 80, 321, 1996.
48. Curi, T.C. et al., Percentage of phagocytosis, production of O_2.-, H_2O_2 and NO, and antioxidant enzyme activities of rat neutrophils in culture. *Cell Biochem Funct*, 16, 43, 1998.
49. Aljandali, A. et al., Asbestos causes apoptosis in alveolar epithelial cells: role of iron-induced free radicals. *J Lab Clin Med*, 137, 330, 2001.
50. Le Prieur, E. et al., Toxicity of diesel engine exhausts in an *in vitro* model of lung slices in biphasic organotypic culture: induction of a proinflammatory and apoptotic response. *Arch Toxicol*, 74, 460, 2000.
51. Holian, A. et al., Urban particle-induced apoptosis and phenotype shifts in human alveolar macrophages. *Environ Health Perspect*, 106, 127, 1998.
52. Zhao, X.H. et al., Automobile exhaust particle-induced apoptosis and necrosis in MRC-5 cells. *Toxicol Lett*, 122, 103, 2001.
53. Panduri, V. et al., Mitochondrial-derived free radicals mediate asbestos-induced alveolar epithelial cell apoptosis. *Am J Physiol Lung Cell Mol Physiol*, 286, L1220, 2004.
54. Chao, S.K. et al., Cell surface regulation of silica-induced apoptosis by the SR-A scavenger receptor in a murine lung macrophage cell line (MH-S). *Toxicol Appl Pharmacol*, 174, 10, 2001.
55. Duke, R.C. and Cohen, J.J., IL-2 addiction: withdrawal of growth factor activates a suicide program in dependent T cells. *Lymphokine Res*, 5, 289, 1986.
56. Iyer, R. et al., Silica-induced apoptosis mediated via scavenger receptor in human alveolar macrophages. *Toxicol Appl Pharmacol*, 141, 84, 1996.
57. Knebel, J.W. et al., Exposure of human lung cells to native diesel motor exhaust—development of an optimized in vitro test strategy. *Toxicol In Vitro*, 16, 185, 2002.
58. Allen, C.B., An automated system for exposure of cultured cells and other materials to ozone. *Inhal Toxicol*, 15, 1039, 2003.
59. Aufderheide, M. et al., An improved in vitro model for testing the pulmonary toxicity of complex mixtures such as cigarette smoke. *Exp Toxicol Pathol*, 55, 51, 2003.

60. Bouthillier, L. et al., Acute effects of inhaled urban particles and ozone: lung morphology, macrophage activity, and plasma endothelin-1. *Am J Pathol*, 153, 1873, 1998.
61. Sioutas, C. et al., A technique to expose animals to concentrated fine ambient aerosols. *Environ Health Perspect*, 103, 172, 1995.
62. Ghio, A.J. and Huang, Y.C., Exposure to concentrated ambient particles (CAPs): a review. *Inhal Toxicol*, 16, 53, 2004.
63. Demokritou, P. et al., Development of a high-volume concentrated ambient particles system (CAPS) for human and animal inhalation toxicological studies. *Inhal Toxicol*, 15, 111, 2003.
64. Kodavanti, U.P. et al., Variable pulmonary responses from exposure to concentrated ambient air particles in a rat model of bronchitis. *Toxicol Sci*, 54, 441, 2000.
65. Clarke, R.W. et al., Urban air particulate inhalation alters pulmonary function and induces pulmonary inflammation in a rodent model of chronic bronchitis. *Inhal Toxicol*, 11, 637, 1999.
66. Wellenius, G.A. et al., Inhalation of concentrated ambient air particles exacerbates myocardial ischemia in conscious dogs. *Environ Health Perspect*, 111, 402, 2003.
67. Ghio, A.J. et al., Concentrated ambient air particles induce mild pulmonary inflammation in healthy human volunteers. *Am J Respir Crit Care Med*, 162, 981, 2000.
68. Devlin, R.B. et al., Elderly humans exposed to concentrated air pollution particles have decreased heart rate variability. *Eur Respir J Suppl*, 40, 76s, 2003.
69. Harder, S.D. et al., Inhalation of PM2.5 does not modulate host defense or immune parameters in blood or lung of normal human subjects. *Environ Health Perspect*, 109(Suppl. 4), 599, 2001.
70. Gong, H., Jr. et al., Controlled exposures of healthy and asthmatic volunteers to concentrated ambient particles in metropolitan Los Angeles. *Res Rep Health Eff Inst*, 1, 2003.
71. Holgate, S.T. et al., Health effects of acute exposure to air pollution. Part II: Healthy subjects exposed to concentrated ambient particles. *Res Rep Health Eff Inst*, 31, 2003.
72. Ghio, A.J. et al., Exposure to concentrated ambient air particles alters hematologic indices in humans. *Inhal Toxicol*, 15, 1465, 2003.
73. Imrich, A. et al., Insoluble components of concentrated air particles mediate alveolar macrophage responses in vitro. *Toxicol Appl Pharmacol*, 167, 140, 2000.
74. Imrich, A. et al., Lipopolysaccharide priming amplifies lung macrophage tumor necrosis factor production in response to air particles. *Toxicol Appl Pharmacol*, 159, 117, 1999.
75. Ning, Y. et al., Particle-epithelial interaction: effect of priming and bystander neutrophils on interleukin-8 release. *Am J Respir Cell Mol Biol*, 30, 744, 2004.
76. Stringer, B. et al., Lung epithelial cell (A549) interaction with unopsonized environmental particulates: quantitation of particle-specific binding and IL-8 production. *Exp Lung Res*, 22, 495, 1996.
77. Freshney, R.I., *Culture of Animal Cells: a Manual of Basic Technique*. 4th ed., John Wiley & Sons, New York, 2000.
78. McDonald, R.J. and Usachencko, J., Neutrophils injure bronchial epithelium after ozone exposure. *Inflammation*, 23, 63, 1999.

79. Koike, E. et al., Effect of ozone exposure on alveolar macrophage-mediated immunosuppressive activity in rats. *Toxicol Sci*, 41, 217, 1998.
80. Koike, E. et al., Mechanisms of ozone-induced inhibitory effect of bronchoalveolar lavage fluid on alveolar macrophage-mediated immunosuppressive activity in rats. *J Leukoc Biol*, 66, 75, 1999.
81. Hamilton, R.F., Jr. et al., Silica and PM1648 modify human alveolar macrophage antigen-presenting cell activity in vitro. *J Environ Pathol Toxicol Oncol*, 20(Suppl. 1), 75, 2001.
82. Becker, S. and Soukup, J.M., Decreased CD11b expression, phagocytosis, and oxidative burst in urban particulate pollution-exposed human monocytes and alveolar macrophages. *J Toxicol Environ Health A*, 55, 455, 1998.
83. Hadnagy, W. and Seemayer, N.H., Inhibition of phagocytosis of human macrophages induced by airborne particulates. *Toxicol Lett*, 72, 23, 1994.
84. Pendino, K.J. et al., Stimulation of nitric oxide production in rat lung lavage cells by anti-Mac-1beta antibody: effects of ozone inhalation. *Am J Respir Cell Mol Biol*, 14, 327, 1996.
85. Bhalla, D.K., Alteration of alveolar macrophage chemotaxis, cell adhesion, and cell adhesion molecules following ozone exposure of rats. *J Cell Physiol*, 169, 429, 1996.
86. Bermudez, E. et al., DNA strand breaks caused by exposure to ozone and nitrogen dioxide. *Environ Res*, 81, 72, 1999.
87. Bermudez, E., Detection of poly(ADP-ribose) synthetase activity in alveolar macrophages of rats exposed to nitrogen dioxide and ozone. *Inhal Toxicol*, 13, 69, 2001.
88. Becker, S. et al., Stimulation of human and rat alveolar macrophages by urban air particulates: effects on oxidant radical generation and cytokine production. *Toxicol Appl Pharmacol*, 141, 637, 1996.
89. Driscoll, K.E. et al., Alveolar macrophage cytokine and growth factor production in a rat model of crocidolite-induced pulmonary inflammation and fibrosis. *J Toxicol Environ Health*, 46, 155, 1995.
90. Devlin, R.B. et al., Ozone-induced release of cytokines and fibronectin by alveolar macrophages and airway epithelial cells. *Am J Physiol*, 266, L612, 1994.
91. Bhalla, D.K. et al., Modification of macrophage adhesion by ozone: role of cytokines and cell adhesion molecules. *Ann N Y Acad Sci*, 796, 38, 1996.
92. Ishii, Y. et al., Rat alveolar macrophage cytokine production and regulation of neutrophil recruitment following acute ozone exposure. *Toxicol Appl Pharmacol*, 147, 214, 1997.
93. Henderson, R.F., in *Toxicology of the Lung* (Gardner, D.E., Crapo, J.D., and Massaro, E.J., Eds.), 1st ed., Raven Press, New York, 1988, p. 239.
94. Henderson, R.F. et al., Comparative study of bronchoalveolar lavage fluid: effect of species, age, and method of lavage. *Exp Lung Res*, 13, 329, 1987.
95. Shapiro, H.M., *Practical Flow Cytometry*. 4th ed., Wiley-Liss, Hoboken, NJ, 2003.
96. Lorsbach, R.B. et al., Expression of the nitric oxide synthase gene in mouse macrophages activated for tumor cell killing. Molecular basis for the synergy between interferon-gamma and lipopolysaccharide. *J Biol Chem*, 268, 1908, 1993.
97. Zeidler, P.C. et al., Response of alveolar macrophages from inducible nitric oxide synthase knockout or wild-type mice to an in vitro lipopolysaccharide or silica exposure. *J Toxicol Environ Health A*, 66, 995, 2003.

98. Lundborg, M. et al., Human alveolar macrophage phagocytic function is impaired by aggregates of ultrafine carbon particles. *Environ Res*, 86, 244, 2001.
99. Dorger, M. et al., Comparison of the phagocytic response of rat and hamster alveolar macrophages to man-made vitreous fibers in vitro. *Hum Exp Toxicol*, 19, 635, 2000.
100. Quinlan, T.R. et al., Mechanisms of asbestos-induced nitric oxide production by rat alveolar macrophages in inhalation and in vitro models. *Free Radic Biol Med*, 24, 778, 1998.
101. Dandrea, T. et al., Differential inhibition of inflammatory cytokine release from cultured alveolar macrophages from smokers and non-smokers by NO_2. *Hum Exp Toxicol*, 16, 577, 1997.
102. Helmke, R.J. et al., From growth factor dependence to growth factor responsiveness: the genesis of an alveolar macrophage cell line. *In Vitro Cell Dev Biol*, 23, 567, 1987.
103. Palleroni, A.V. et al., Tumoricidal alveolar macrophage and tumor infiltrating macrophage cell lines. *Int J Cancer*, 49, 296, 1991.
104. Raschke, W.C. et al., Functional macrophage cell lines transformed by Abelson leukemia virus. *Cell*, 15, 261, 1978.
105. Cheek, J.M. et al., Neutrophils enhance removal of ozone-injured alveolar epithelial cells in vitro. *Am J Physiol*, 269, L527, 1995.
106. Mogel, M. et al., A new coculture-system of bronchial epithelial and endothelial cells as a model for studying ozone effects on airway tissue. *Toxicol Lett*, 96–97, 25, 1998.
107. Fujii, T. et al., Particulate matter induces cytokine expression in human bronchial epithelial cells. *Am J Respir Cell Mol Biol*, 25, 265, 2001.
108. Lechner, J.F. et al., Induction of squamous differentiation of normal human bronchial epithelial cells by small amounts of serum. *Differentiation*, 25, 229, 1984.
109. Beckmann, J.D. et al., Serum-free culture of fractionated bovine bronchial epithelial cells. *In Vitro Cell Dev Biol*, 28A, 39, 1992.
110. Smith, M.D. et al., Techniques for measurement of oxygen consumption rates of hepatocytes during attachment and post-attachment. *Int J Artif Organs*, 19, 36, 1996.
111. Bruttig, S.P. and Joyner, W.L., Metabolic characteristics of cells cultured from human umbilical blood vessels: comparison with 3T3 fibroblasts. *J Cell Physiol*, 116, 173, 1983.
112. Jorjani, P. and Ozturk, S.S., Effects of cell density and temperature on oxygen consumption rate for different mammalian cell lines. *Biotechnol Bioeng*, 64, 349, 1999.
113. Martin, M. et al., Energetic and morphological plasticity of C6 glioma cells grown on 3-D support; effect of transient glutamine deprivation. *J Bioenerg Biomembr*, 30, 565, 1998.
114. Kittlick, P.D. and Neupert, G., Hypoxia in fibroblast cultures. I. Changes in glucose consumption by 5 vol.-% oxygen concentration and reduced pH. *Exp Pathol*, 10, 109, 1975.
115. Fiorani, M. et al., Cell density-dependent regulation of ATP levels during the growth cycle of cultured Chinese hamster ovary cells. *Biochem Mol Biol Int*, 32, 251, 1994.
116. Bereiter-Hahn, J. et al., Dependence of energy metabolism on the density of cells in culture. *Cell Struct Funct*, 23, 85, 1998.

117. Sung, S.S. et al., Extracellular ATP perturbs transmembrane ion fluxes, elevates cytosolic [Ca^{2+}], and inhibits phagocytosis in mouse macrophages. *J Biol Chem*, 260, 13442, 1985.
118. Zupke, C. et al., Intracellular flux analysis applied to the effect of dissolved oxygen on hybridomas. *Appl Microbiol Biotechnol*, 44, 27, 1995.
119. Grigoriev, V.G. et al., Senescence and cell density of human diploid fibroblasts influence metabolism of insulin-like growth factor binding proteins. *J Cell Physiol*, 160, 203, 1994.
120. Kenagy, R. et al., Metabolism of low density lipoprotein by bovine endothelial cells as a function of cell density. *Arteriosclerosis*, 4, 365, 1984.
121. Knedler, A. and Ham, R.G., Optimized medium for clonal growth of human microvascular endothelial cells with minimal serum. *In Vitro Cell Dev Biol*, 23, 481, 1987.
122. Stanners, C.P. et al., Two types of ribosome in mouse-hamster hybrid cells. *Nat New Biol*, 230, 52, 1971.
123. Stewart, C.C., Nutrient utilization by peritoneal exudate cells. *J Reticuloendothel Soc*, 14, 332, 1973.
124. van der Zeijst, B.A. et al., Proliferative capacity of mouse peritoneal macrophages in vitro. *J Exp Med*, 147, 1253, 1978.
125. Reppun, T.S. et al., Isokinetic separation and characterization of mouse pulmonary alveolar colony-forming cells. *J Reticuloendothel Soc*, 25, 379, 1979.
126. Lin, H.S. et al., Effects of hydrocortisone acetate on pulmonary alveolar macrophage colony-forming cells. *Am Rev Respir Dis*, 125, 712, 1982.
127. Lin, H.S. and Hsu, S., Effects of dose rate and dose fractionation of irradiation on pulmonary alveolar macrophage colony-forming cells. *Radiat Res*, 103, 260, 1985.
128. Lin, H.S. et al., Clonal growth of hamster free alveolar cells in soft agar. *J Exp Med*, 142, 877, 1975.
129. Lechner, J.F. and LaVeck, M.A., A serum-free method for culturing normal human bronchial epithelial cells at clonal density. *J Tissue Cult Methods*, 9, 43, 1985.
130. Metzen, E. et al., Pericellular PO_2 and O_2 consumption in monolayer cell cultures. *Respir Physiol*, 100, 101, 1995.
131. Allen, C.B. et al., Limitations to oxygen diffusion and equilibration in in vitro cell exposure systems in hyperoxia and hypoxia. *Am J Physiol Lung Cell Mol Physiol*, 281, L1021, 2001.
132. Dumler, K. et al., The effects of ozone exposure on lactate dehydrogenase release from human and primate respiratory epithelial cells. *Toxicol Lett*, 70, 203, 1994.
133. McManus, M.S. et al., Human nasal epithelium: characterization and effects of in vitro exposure to sulfur dioxide. *Exp Lung Res*, 15, 849, 1989.
134. Valentine, R., An in vitro system for exposure of lung cells to gases: effects of ozone on rat macrophages. *J Toxicol Environ Health*, 16, 115, 1985.
135. Tarkington, B.K. et al., Ozone exposure of cultured cells and tissues. *Methods Enzymol*, 234, 257, 1994.
136. Tarkington, B.K. et al., In vitro exposure of tracheobronchial epithelial cells and of tracheal explants to ozone. *Toxicology*, 88, 51, 1994.
137. Baker, F.D. and Tumasonis, C.F., Modified roller drum apparatus for analyzing effects of pollutant gases on tissue culture systems. *Atmos Environ*, 5, 891, 1971.

138. Bolton, D.C. et al., An in vitro system for studying the effects of ozone on mammalian cell cultures and viruses. *Environ Res*, 27, 466, 1982.
139. Lee, J.G. et al., The use of the single cell gel electrophoresis assay in detecting DNA single strand breaks in lung cells in vitro. *Toxicol Appl Pharmacol*, 141, 195, 1996.
140. Aufderheide, M. et al., Novel approaches for studying pulmonary toxicity in vitro. *Toxicol Lett*, 140–141, 205, 2003.
141. Umstead, T.M. et al., In vitro exposure of proteins to ozone. *Toxicol Mech Meth*, 12, 1, 2002.
142. Ritter, D. et al., Exposure of human lung cells to inhalable substances: a novel test strategy involving clean air exposure periods using whole diluted cigarette mainstream smoke. *Inhal Toxicol*, 15, 67, 2003.
143. Aufderheide, M. et al., A method for the in vitro exposure of human cells to environmental and complex gaseous mixtures: application to various types of atmosphere. *Altern Lab Anim*, 30, 433, 2002.
144. Aufderheide, M. et al., A method for in vitro analysis of the biological activity of complex mixtures such as sidestream cigarette smoke. *Exp Toxicol Pathol*, 53, 141, 2001.
145. Ritter, D. et al., In vitro exposure of isolated cells to native gaseous compounds—development and validation of an optimized system for human lung cells. *Exp Toxicol Pathol*, 53, 373, 2001.
146. Fisher, G.L. and Placke, M.E., in *Toxicology of the Lung*, Gardner, D.E., Crapo, J.D., and Massaro, E.J., Eds., 1st ed., Raven Press, New York, 1988, p. 285.
147. Gelfand, A.S. et al., Effect of aspiration of milk on mechanisms of neural control in the airways of developing rabbits. *Pediatr Pulmonol*, 23, 198, 1997.
148. Roux, E. et al., Human isolated airway contraction: interaction between air pollutants and passive sensitization. *Am J Respir Crit Care Med*, 160, 439, 1999.
149. Ben-Jebria, A. et al., Effect of in vitro nitrogen dioxide exposure on human bronchial smooth muscle response. *Am Rev Respir Dis*, 146, 378, 1992.
150. Walters, E.H. et al., The responsiveness of airway smooth muscle in vitro from dogs with airway hyper-responsiveness in vivo. *Clin Sci (Lond)*, 71, 605, 1986.
151. Montano, L.M. et al., Effect of ozone exposure in vivo on response of bronchial rings in vitro: role of intracellular Ca^{2+}. *J Appl Physiol*, 75, 1315, 1993.
152. Takata, S. et al., Ozone exposure suppresses epithelium-dependent relaxation in feline airway. *Lung*, 173, 47, 1995.
153. Yoshida, M. et al., Ozone exposure may enhance airway smooth muscle contraction by increasing $Ca^{(2+)}$ refilling of sarcoplasmic reticulum in guinea pig. *Pulm Pharmacol Ther*, 15, 111, 2002.
154. Chitano, P. et al., In vitro exposure to nitrogen dioxide (NO_2) does not alter bronchial smooth muscle responsiveness in ovalbumin-sensitized guinea-pigs. *Pulm Pharmacol*, 7, 251, 1994.
155. Murlas, C.G. et al., O3-induced mucosa-linked airway muscle hyperresponsiveness in the guinea pig. *J Appl Physiol*, 69, 7, 1990.
156. Sommer, B. et al., Effect of different ozone concentrations on the neurogenic contraction and relaxation of guinea pig airways. *Fundam Clin Pharmacol*, 11, 501, 1997.
157. van Hoof, I.H. et al., Changes in receptor function by oxidative stress in guinea pig tracheal smooth muscle. *Cent Eur J Public Health*, 4(Suppl. 3), 1996.

158. Roux, E. et al., Human and rat airway smooth muscle responsiveness after ozone exposure in vitro. *Am J Physiol*, 271, L631, 1996.
159. Christ, M.J. et al., Amiloride-induced contraction of isolated guinea pig, mouse, and human fetal airways. *Am J Physiol*, 274, R209, 1998.
160. Garssen, J. et al., An isometric method to study respiratory smooth muscle responses in mice. *J Pharmacol Methods*, 24, 209, 1990.
161. Iwamoto, L.M. et al., Loop diuretics and in vitro relaxation of human fetal and newborn mouse airways. *Pediatr Res*, 50, 273, 2001.
162. Justice, J.P. et al., IL-10 gene knockout attenuates allergen-induced airway hyperresponsiveness in C57BL/6 mice. *Am J Physiol Lung Cell Mol Physiol*, 280, L363, 2001.
163. Sipahi, E.Y. et al., Nitric oxide and prostanoid-dependent relaxation induced by angiotensin II in the isolated precontracted mouse tracheal muscle and the role of potassium channels. *Pharmacol Res*, 42, 69, 2000.
164. Chitano, P. et al., Isotonic smooth muscle response in human bronchi exposed in vitro to nitrogen dioxide. *Eur Respir J*, 9, 2294, 1996.
165. Michielsen, C. et al., The environmental pollutant hexachlorobenzene causes eosinophilic and granulomatous inflammation and in vitro airways hyperreactivity in the Brown Norway rat. *Arch Toxicol*, 76, 236, 2002.
166. Chitano, P. et al., In-vitro exposure of guinea pig main bronchi to 2.5 ppm of nitrogen dioxide does not alter airway smooth muscle response. *Respir Med*, 89, 323, 1995.
167. Jain, B. et al., Modulation of airway epithelial cell ciliary beat frequency by nitric oxide. *Biochem Biophys Res Commn*, 191, 83, 1993.
168. Uzlaner, N. and Priel, Z., Interplay between the NO pathway and elevated $[Ca^{2+}]i$ enhances ciliary activity in rabbit trachea. *J Physiol*, 516 (Pt. 1), 179, 1999.
169. Sisson, J.H. et al., Acetaldehyde-mediated cilia dysfunction in bovine bronchial epithelial cells. *Am J Physiol*, 260, L29, 1991.
170. Grose, E.C. et al., Pulmonary host defense responses to inhalation of sulfuric acid and ozone. *J Toxicol Environ Health*, 10, 351, 1982.
171. Grose, E.C. et al., Response of ciliated epithelium to ozone and sulfuric acid. *Environ Res*, 22, 377, 1980.
172. Sakai, N. et al., Inhibitory effect on sulfur dioxide on ciliary motility in rabbit tracheal epithelium and its prevention by intracellular cyclic AMP. *Nihon Kyobu Shikkan Gakkai Zasshi*, 31, 733, 1993.
173. Tamaoki, J. et al., Effect of azelastine on sulphur dioxide induced impairment of ciliary motility in airway epithelium. *Thorax*, 48, 542, 1993.
174. Adalis, D. et al., Toxic effects of cadmium on ciliary activity using a tracheal ring model system. *Environ Res*, 13, 111, 1977.
175. Adalis, D. et al., Cytotoxic effects of nickel on ciliated epithelium. *Am Rev Respir Dis*, 118, 347, 1978.
176. Korngreen, A. and Priel, Z., Purinergic stimulation of rabbit ciliated airway epithelia: control by multiple calcium sources. *J Physiol*, 497(Pt. 1), 53, 1996.
177. Sakai, A. et al., Nitric oxide modulation of Ca^{2+} responses in cow tracheal epithelium. *Eur J Pharmacol*, 291, 375, 1995.
178. Jain, B. et al., TNF-alpha and IL-1 beta upregulate nitric oxide-dependent ciliary motility in bovine airway epithelium. *Am J Physiol*, 268, L911, 1995.
179. Saxton, R.E. et al., Dose response of human tumor cells to rhodamine 123 and laser phototherapy. *Laryngoscope*, 104, 1013, 1994.

180. Petrat, F. et al., NAD(P)H, a primary target of $1O_2$ in mitochondria of intact cells. *J Biol Chem*, 278, 3298, 2003.
181. Krishan, A. et al., Drug retention, efflux, and resistance in tumor cells. *Cytometry*, 29, 279, 1997.
182. Tang-Wai, D.F. et al., Human (MDR1) and mouse (mdr1, mdr3) P-glycoproteins can be distinguished by their respective drug resistance profiles and sensitivity to modulators. *Biochemistry (Mosc)*, 34, 32, 1995.
183. Altenberg, G.A. et al., Unidirectional fluxes of rhodamine 123 in multidrug-resistant cells: evidence against direct drug extrusion from the plasma membrane. *Proc Natl Acad Sci USA*, 91, 4654, 1994.
184. Nare, B. et al., Characterization of rhodamine 123 binding to P-glycoprotein in human multidrug-resistant cells. *Mol Pharmacol*, 45, 1145, 1994.
185. Ludescher, C. et al., Rapid functional assay for the detection of multidrug-resistant cells using the fluorescent dye rhodamine 123. *Blood*, 78, 1385, 1991.
186. Kessel, D. et al., Characterization of multidrug resistance by fluorescent dyes. *Cancer Res*, 51, 4665, 1991.
187. Lautier, D. et al., Detection of human leukemia cells with multidrug-resistance phenotype using multilabeling with fluorescent dyes. *Anticancer Res*, 13, 1557, 1993.
188. Lahmy, S. et al., Identification of multi-drug resistant cells in sensitive Friend leukemia cells by quantitative videomicrofluorimetry. *Cell Biochem Funct*, 10, 9, 1992.
189. Frey, T. et al., Dyes providing increased sensitivity in flow-cytometric dye-efflux assays for multidrug resistance. *Cytometry*, 20, 218, 1995.
190. Olson, D.P. et al., Detection of MRP functional activity: calcein AM but not BCECF AM as a multidrug resistance-related Protein (MRP1) substrate. *Cytometry*, 46, 105, 2001.
191. Nelson, E.J. et al., Fluorescence methods to assess multidrug resistance in individual cells. *Cancer Chemother Pharmacol*, 42, 292, 1998.
192. Brezden, C.B. et al., Constitutive expression of P-glycoprotein as a determinant of loading with fluorescent calcium probes. *Cytometry*, 17, 343, 1994.
193. Blum, J.L. and Wicha, M.S., Role of the cytoskeleton in laminin induced mammary gene expression. *J Cell Physiol*, 135, 13, 1988.
194. Li, M.L. et al., Influence of a reconstituted basement membrane and its components on casein gene expression and secretion in mouse mammary epithelial cells. *Proc Natl Acad Sci USA*, 84, 136, 1987.
195. Folkman, J. and Moscona, A., Role of cell shape in growth control. *Nature*, 273, 345, 1978.
196. Streuli, C.H. et al., Control of mammary epithelial differentiation: basement membrane induces tissue-specific gene expression in the absence of cell-cell interaction and morphological polarity. *J Cell Biol*, 115, 1383, 1991.
197. Lin, C.Q. and Bissell, M.J., Multi-faceted regulation of cell differentiation by extracellular matrix [see comments]. *FASEB J*, 7, 737, 1993.
198. Ingber, D.E. et al., Cellular tensegrity: exploring how mechanical changes in the cytoskeleton regulate cell growth, migration, and tissue pattern during morphogenesis. *Int Rev Cytol*, 150, 173, 1994.
199. Wang, N. and Ingber, D.E., Control of cytoskeletal mechanics by extracellular matrix, cell shape, and mechanical tension. *Biophys J*, 66, 2181, 1994.

200. Ingber, D.E. et al., Cell shape, cytoskeletal mechanics, and cell cycle control in angiogenesis. *J Biomech*, 28, 1471, 1995.
201. Chen, C.S. et al., Geometric control of cell life and death. *Science*, 276, 1425, 1997.
202. Ingber, D.E., The architecture of life. *Sci Am*, 48, 1998.
203. Hays, A.M. et al., Correlation between in vivo and in vitro pulmonary responses to jet propulsion fuel-8 using precision-cut lung slices and a dynamic organ culture system. *Toxicol Pathol*, 31, 200, 2003.
204. Wang, C.G. et al., In vitro bronchial responsiveness in two highly inbred rat strains. *J Appl Physiol*, 82, 1445, 1997.
205. Price, R.J. et al., Toxicity of 3-methylindole, 1-nitronaphthalene and paraquat in precision-cut rat lung slices. *Arch Toxicol*, 69, 405, 1995.
206. Placke, M.E. and Fisher, G.L., Adult peripheral lung organ culture—a model for respiratory tract toxicology. *Toxicol Appl Pharmacol*, 90, 284, 1987.
207. Vallan, C. et al., Release of a mitogenic factor by adult rat lung slices in culture. *Exp Lung Res*, 21, 469, 1995.
208. Kurosawa, H. et al., Mucociliary function in the mouse measured in explanted lung tissue. *J Appl Physiol*, 79, 41, 1995.
209. Fisher, R.L. et al., The use of human lung slices in toxicology. *Hum Exp Toxicol*, 13, 466, 1994.
210. Dandurand, R.J. et al., Responsiveness of individual airways to methacholine in adult rat lung explants. *J Appl Physiol*, 75, 364, 1993.
211. Warren, J.S. and Barton, P.A., In vitro analysis of pulmonary inflammation using rat lung organ cultures. *Exp Lung Res*, 18, 55, 1992.
212. Siminski, J.T. et al., Long-term maintenance of mature pulmonary parenchyma cultured in serum-free conditions. *J Appl Physiol*, 262, L105, 1992.
213. Kinnard, W.V. et al., Regulation of alveolar type II cell differentiation and proliferation in adult rat lung explants. *Am J Respir Cell Mol Biol*, 11, 416, 1994.
214. Martin, C. et al., Videomicroscopy of methacholine-induced contraction of individual airways in precision-cut lung slices. *Eur Respir J*, 9, 2479, 1996.
215. Martin, C. et al., Differential effects of the mixed ET(A)/ET(B)-receptor antagonist bosentan on endothelin-induced bronchoconstriction, vasoconstriction and prostacyclin release. *Naunyn Schmiedebergs Arch Pharmacol*, 362, 128, 2000.
216. Martin, C. et al., Cytokine-induced bronchoconstriction in precision-cut lung slices is dependent upon cyclooxygenase-2 and thromboxane receptor activation. *Am J Respir Cell Mol Biol*, 24, 139, 2001.
217. Martin, C. et al., Effects of the thromboxane receptor agonist U46619 and endothelin-1 on large and small airways. *Eur Respir J*, 16, 316, 2000.
218. Wohlsen, A. et al., The early allergic response in small airways of human precision-cut lung slices. *Eur Respir J*, 21, 1024, 2003.
219. Wohlsen, A. et al., Immediate allergic response in small airways. *Am J Respir Crit Care Med*, 163, 1462, 2001.
220. Williams, P.P. and Gallagher, J.E., Preparation and long-term cultivation of porcine tracheal and lung organ cultures by alternate exposure to gaseous and liquid medium phases. *In Vitro Cell Dev Biol*, 14, 686, 1978.
221. Held, H.D. et al., Characterization of airway and vascular responses in murine lungs. *Br J Pharmacol*, 126, 1191, 1999.

222. Jackson, R.M. et al., Lung MK 351A uptake after hypoxia adaptation and subsequent hyperoxia exposure. *Lung*, 166, 209, 1988.
223. Block, E.R. and Fisher, A.B., Depression of serotonin clearance by rat lungs during oxygen exposure. *J Appl Physiol*, 42, 33, 1977.
224. Pessina, G.P. et al., Pulmonary catabolism of interleukin-6 evaluated by lung perfusion of normal and smoker rats. *J Pharm Pharmacol*, 48, 1063, 1996.
225. Pessina, G.P. et al., Pulmonary catabolism of interferon-gamma evaluated by lung perfusion of both normal and smoke-exposed rats. *J Interferon Cytokine Res*, 15, 225, 1995.
226. Archer, S.L. et al., Detection of activated O_2 species in vitro and in rat lungs by chemiluminescence. *J Appl Physiol*, 67, 1912, 1989.
227. Archer, S.L. et al., Simultaneous measurement of O_2 radicals and pulmonary vascular reactivity in rat lung. *J Appl Physiol*, 67, 1903, 1989.
228. Joad, J.P. et al., Effects of ozone and neutrophils on function and morphology of the isolated rat lung. *Am Rev Respir Dis*, 147, 1578, 1993.
229. Joad, J.P. et al., Effect of respiratory pattern on ozone injury to the airways of isolated rat lungs. *Toxicol Appl Pharmacol*, 169, 26, 2000.
230. Joad, J.P. et al., The local C-fiber contribution to ozone-induced effects on the isolated guinea pig lung. *Toxicol Appl Pharmacol*, 141, 561, 1996.
231. Joad, J.P. et al., Ozone effects on mechanics and arachidonic acid metabolite concentrations in isolated rat lungs. *Environ Res*, 66, 186, 1994.
232. Postlethwait, E.M. et al., Three-dimensional mapping of ozone-induced acute cytotoxicity in tracheobronchial airways of isolated perfused rat lung. *Am J Respir Cell Mol Biol*, 22, 191, 2000.
233. Postlethwait, E.M. et al., Determinants of inhaled ozone absorption in isolated rat lungs. *Toxicol Appl Pharmacol*, 125, 77, 1994.
234. Delaunois, A. et al., Interactions between cytochrome P-450 activities and ozone-induced modulatory effects on endothelial permeability in rabbit lungs: influence of gender. *Inhal Toxicol*, 11, 999, 1999.
235. Delaunois, A. et al., Comparison of ozone-induced effects on lung mechanics and hemodynamics in the rabbit. *Toxicol Appl Pharmacol*, 150, 58, 1998.
236. Pino, M.V. et al., Functional and morphologic changes caused by acute ozone exposure in the isolated and perfused rat lung. *Am Rev Respir Dis*, 145, 882, 1992.
237. Bassett, D.J. et al., Rat lung recovery from 3 days of continuous exposure to 0.75 ppm ozone. *J Toxicol Environ Health*, 25, 329, 1988.
238. Joad, J.P. et al., Perinatal exposure to aged and diluted sidestream cigarette smoke produces airway hyperresponsiveness in older rats. *Toxicol Appl Pharmacol*, 155, 253, 1999.
239. Moller, L. et al., Metabolism of the carcinogenic air pollutant 2-nitrofluorene in the isolated perfused rat lung and liver. *Carcinogenesis*, 8, 1847, 1987.
240. Tornquist, S. et al., Bioavailability of benzo(a)pyrene deposited in the lung. Correlation with dissolution from urban air particulates and covalently bound DNA adducts. *Drug Metab Dispos*, 16, 842, 1988.
241. Gotze, J.P. et al., Effects of induction and age-dependent enzyme expression on lung bioavailability, metabolism, and DNA binding of urban air particulate-absorbed benzo[a]pyrene, 2-nitrofluorene, and 3-amino-1,4-dimethyl-5H-pyridol-(4,3)-indole. *Environ Health Perspect*, 102(Suppl. 4), 147, 1994.

242. Morgan, D.D. et al., The pharmacokinetics of benzo[alpha]pyrene in the isolated perfused rabbit lung: the influence of benzo[alpha]pyrene, n-dodecane, particulate, or sulfur dioxide. *Toxicology*, 33, 275, 1984.
243. Warshawsky, D. et al., Influence of airborne particulate on the metabolism of benzo[a]pyrene in the isolated perfused lung. *J Toxicol Environ Health*, 11, 503, 1983.
244. Schoeny, R. and Warshawsky, D., Mutagenicity of benzo(a)pyrene metabolites generated on the isolated perfused lung following particulate exposure. *Teratog Carcinog Mutagen*, 3, 151, 1983.
245. Becker, P.M. et al., Effects of oxygen tension and glucose concentration on ischemic injury in ventilated ferret lungs. *J Appl Physiol*, 75, 1233, 1993.
246. Brower, R.G. et al., Locus of hypoxic vasoconstriction in isolated ferret lungs. *J Appl Physiol*, 63, 58, 1987.
247. Tucker, A. et al., Pulmonary vascular responsiveness in lungs isolated from altitude-exposed and altitude recovery sympathectomized rats. *Fed Proc*, 45, 1031, 1986.
248. Tucker, A. et al., Altered vascular responsiveness in isolated perfused lungs from aging rats. *Exp Lung Res*, 3, 29, 1982.
249. McMurtry, I.F. et al., Inhibition of hypoxic pulmonary vasoconstriction by calcium antagonists in isolated rat lungs. *Circ Res*, 38, 99, 1976.
250. McMurtry, I.F. et al., Blunted hypoxic vasoconstriction in lungs from short-term high-altitude rats. *Am J Physiol*, 238, H849, 1980.
251. Allen, C.B. and White, C.W., Glucose modulates cell death due to normobaric hyperoxia by maintaining cellular ATP. *Am J Physiol*, 274, L159, 1998.
252. Holian, A. et al., Asbestos- and silica-induced changes in human alveolar macrophage phenotype. *Environ Health Perspect*, 105(Suppl. 5), 1139, 1997.
253. Sarady, J.K. et al., Carbon monoxide modulates endotoxin-induced production of granulocyte macrophage colony-stimulating factor in macrophages. *Am J Respir Cell Mol Biol*, 27, 739, 2002.
254. Richter, A. et al., Autocrine ligands for the epidermal growth factor receptor mediate interleukin-8 release from bronchial epithelial cells in response to cigarette smoke. *Am J Respir Cell Mol Biol*, 27, 85, 2002.
255. Bayram, H. et al., The effect of diesel exhaust particles on cell function and release of inflammatory mediators from human bronchial epithelial cells in vitro. *Am J Respir Cell Mol Biol*, 18, 441, 1998.
256. Jaspers, I. et al., Ozone-induced IL-8 expression and transcription factor binding in respiratory epithelial cells. *Am J Physiol*, 272, L504, 1997.
257. Cromwell, O. et al., Expression and generation of interleukin-8, IL-6 and granulocyte-macrophage colony-stimulating factor by bronchial epithelial cells and enhancement by IL-1 beta and tumour necrosis factor-alpha. *Immunology*, 77, 330, 1992.
258. Beck, N.B. et al., Ozone can increase the expression of intercellular adhesion molecule-1 by human nasal epithelial cells. *Inhal Toxicol*, 6, 345, 1994.
259. Spech, R.W. et al., Surfactant protein A prevents silica-mediated toxicity to rat alveolar macrophages. *Am J Physiol Lung Cell Mol Physiol*, 278, L713, 2000.
260. Persinger, R.L. et al., Nitrogen dioxide induces death in lung epithelial cells in a density-dependent manner. *Am J Respir Cell Mol Biol*, 24, 583, 2001.
261. Oortgiesen, M. et al., Residual oil fly ash and charged polymers activate epithelial cells and nociceptive sensory neurons. *Am J Physiol Lung Cell Mol Physiol*, 278, L683, 2000.

5

DOSIMETRY OF PARTICLES IN HUMANS: FROM CHILDREN TO ADULTS

Bahman Asgharian, Werner Hofmann, and Frederick J. Miller

CONTENTS

5.1 INTRODUCTION

Suspended particles in the air may enter the human body by inhalation. Once inhaled, airborne particles travel through various parts of the respiratory tract and may deposit preferentially in specific regions as a result of the external forces exerted on them (Hatch and Gross, 1964; Lippmann and Altshuler, 1976; Brain and Valberg, 1979; Medinsky et al., 1997). Particles that originate in the environment and land on lung surfaces may have deleterious effects (Pope and Dockery, 1992), whereas therapeutic effects can be seen for particles delivered to the lung as aerosolized medication (e.g., insulin, interferon, and albuterol) to treat illnesses such as diabetes, cancer, cystic fibrosis, and asthma (Niven, 1995; Smith and Bernstein, 1996).

The deposition and fate of particles in the lungs have long been the subjects of extensive research on human health and associated health risk assessments (ICRP, 1994; NCRP, 1997). Children have been identified as a sensitive subpopulation for the determination of regulatory standards for air pollution (Browner, 1996). Epidemiology studies have found a link between current levels of airborne particulate matter pollutants and a growing list of adverse health effects in children (Pope and Dockery, 1992; Beyer et al., 1998; Conceição et al., 2001; Gauderman et al., 2002). Recently, Somers et al. (2004) showed that exposure to airborne pollutants leads to heritable DNA mutations in male mice. A direct link between mutations and health effects was not established in the study, and whether the findings could be extrapolated to humans was not clear. Samet et al. (2004) provided a possible explanation for the cell mutations resulting from exposure to airborne particles but noted that the findings of Somers et al. (2004) should be interpreted with caution.

Administration of therapeutic drugs to people by inhalation has received increased attention in recent years owing to the feasibility of direct and fast targeting of specific sites within the lung and delivery to the systemic circulation while bypassing the stomach, intestine, and other organs in which undesired reactions can occur with subsequent adverse side effects. This interest in drug delivery via inhalation has created an area of research that raises challenges in formulation and production of therapeutic aerosols with desired physicochemical properties for optimal deposition and prolonged clearance in a cost-effective and efficient manner (Edwards et al., 1997, 1998; Le Souef, 2002).

Both hazardous and beneficial effects of aerosol inhalation require knowledge of particle transport, deposition, and clearance in the lungs. Most efforts in identifying the fate of inhaled particles in the lung have focused on adults. Few deposition or clearance studies in children are available even though some biological outcomes may be the result of accumulating dose originating in childhood. As a sensitive subpopulation, children present a special challenge because the deposited dose may be drastically higher than that of adults due to smaller lung geometry and larger minute ventilation compared with lung size as shown by Overton and Graham (1989) for ozone. In addition, children may be more sensitive than adults to a given dose of particles due to incomplete development of body defense or repair mechanisms.

Knowledge of particle deposition in the lung is a first step in understanding the biological mechanisms that lead to adverse health effects. As recognized by the National Research Council (NRC, 1998), information on the deposition of inhaled particles in the lungs is plentiful for human adults but is severely lacking in children. The same shortcoming prevails in the pharmaceutical industry, in which administration of proper doses of aerosolized drugs to children is desired (Thomas et al., 1993). As a result, most assessments have been performed with reference to adult lungs. Despite the need for additional research, little national attention has been devoted to particle deposition in children.

In this chapter, the current literature on the fate of particles in the lungs of children (and adults, as a means of comparison) is examined to stimulate interest in additional research and to identify research needs and the path forward. First, the information on morphometric measurements of various airway parameters of growing lungs is described along with available lung geometry models. Next, mechanisms of particle deposition in the lung are described, followed by a section on deposition measurements in various sites of the respiratory tract. In addition, a brief overview of deposition models and predicted results in children is presented. Finally, data gaps and future research directions are identified by means of examining existing information.

5.2 STRUCTURE AND FUNCTION OF THE RESPIRATORY TRACT

The respiratory tract begins at the nasal and oral airways, which together form the upper respiratory tract (URT) and merge to connect to the trachea, where the lower respiratory tract (LRT) begins. The LRT may be defined as a network of branching airways beginning with the trachea and proceeding to the main bronchi, lobar bronchi, and segmental bronchi to which three right lobes and two left lobes are attached. The branching structure continues in each lobe, extending beyond the terminal bronchioles, in which an increasing number of alveoli appear on airway surfaces. The definition of airway generations in the LRT is based on the bifurcating pattern of the airways in which a parent airway branches into two (or three, on some occasions) daughter airways, which, in turn, are parents to their daughter branches. The trachea is commonly termed as *generation 0*, the main bronchi as *generation 1*, their four daughters as *generation 2*, and so forth.

The respiratory tract can also be subdivided into three main regions according to function: the extrathoracic (ET) region (from the nose and mouth down to and including the larynx); the tracheobronchial (TB) tree (trachea through terminal bronchioles); and the pulmonary region (respiratory bronchioles to terminal alveolar sacs). Among disciplines, different nomenclature is often used for these regions. For example, URT is the same as the ET region; pulmonary (P), alveolar (A), and acinar are used interchangeably. Also, the TB region is often referred to as the conducting airways. The LRT is comprised of the TB and A regions and is also referred to as the thoracic region. A synopsis of key aspects of the morphology, cytology, histology, function, and structure of the respiratory tract of humans is presented in Table 5.1.

The anatomical or structural features of the lung are complete at birth for the ET and TB regions (Reid, 1984; Burri, 1985) and develop to completion by about the age of 8 for the alveolar region (Dunnill, 1962; Reid, 1984). The airways, however, continue to grow in size until chest wall growth ceases during young adulthood (Reid, 1977). In the subsections that follow, the structure and function of each region are briefly discussed for adults and children.

5.2.1 ET Structure and Function

In the ET region, inhaled air is conditioned in respect to temperature and humidity (Proctor and Anderson, 1982), and there is an active mucociliary system for the removal of foreign solid material that lands on nasal airway surfaces (Phipps, 1981). Four types of epithelium cover the surface of the ET region: squamous, transitional, respiratory, and olfactory (Harkema, 1992). The structure of the nasal passages is shown in Figure 5.1, in which the geometry is constructed by the assembly of successive, equally distanced airway coronal cross sections (in transparent gray) extending from the nares to nasal pharynx. Selected cross sections (in black) are also shown in Figure 5.1 to illustrate the complexity of the geometry and changes in

Table 5.1 Key Aspects of the Structure and Function of the Respiratory Tract of Humans

Functions	Cytology (Epithelium)	Histology (Walls)	Anatomy	Zones (air)	Location	
Air Conditioning: Temperature and Humidity, and Cleaning; Fast Particle Clearance; Air Conduction	Epithelium Types: Squamous Transitional Respiratory Olfactory Cell Types: Ciliated Cells Nonciliated Cells: Goblet Cells Mucous (secretory) Cells Serous Cells Brush Cells Endocrine Cells Basal Cells Intermediate Cells	Mucous Membrane, Respiratory Epithelium (Pseudostratified, Ciliated, Mucous), Glands Mucous Membrane, Respiratory or Stratified Epithelium, Glands	Anterior Nasal Passages Nose Mouth Pharynx Posterior Larynx Esophagus 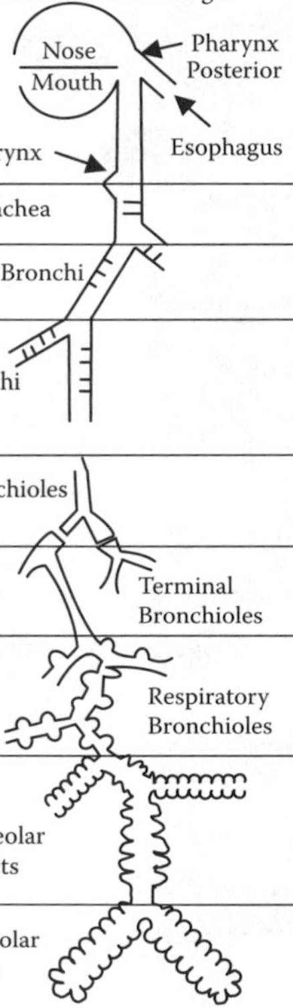	Conduction	Extrathoracic	
		Mucous Membrane, Respiratory Epithelium Cartilage Rings, Glands	Trachea Main Bronchi		Thoracic	Tracheobronchial
		Mucous Membrane, Respiratory Epithelium Cartilage Plates, Smooth Muscle Layer, Glands	Bronchi			
	Respiratory Epithelium with Clara Cells (no goblet cells) Cell Types Ciliated Cells Nonciliated Cells: Clara (secretory) Cells	Mucous Membrane, Respiratory Epithelium, No Cartilage, No Glands, Smooth Muscle Layer	Bronchioles			
		Mucous Membrane, Single-Layer Respiratory Epithelium, Less Ciliated, Smooth Muscle Layer	Terminal Bronchioles			
Air Conduction; Gas Exchange; Slow Particle Clearance	Respiratory Epithelium Consisting Mainly of Clara Cells (Secretory) and Few Ciliated Cells	Mucous Membrane, Single-Layer Respiratory Epithelium of Cuboidal Cells, Smooth Muscle Layer	Respiratory Bronchioles	Gas Exchange Transitory		Alveolar
Gas Exchange; Very Slow Particle Clearance	Squamous Alveolar Epithelial Cells (Type I), Covering 94% of Alveolar Surface Areas Cuboidal Alveolar Epithelial Cells (Type II, Surfactant-Producing), Covering 6% of Alveolar Surface Areas Alveolar Macrophages	Wall Consists of Alveolar Entrance Rings, Squamous Epithelial Layer, Surfactant	Alveolar Ducts			
		Interalveolar Septa Covered by Squamous Epithelium, Containing Capillaries, Surfactant	Alveolar Sacs			
			Lymphatics			

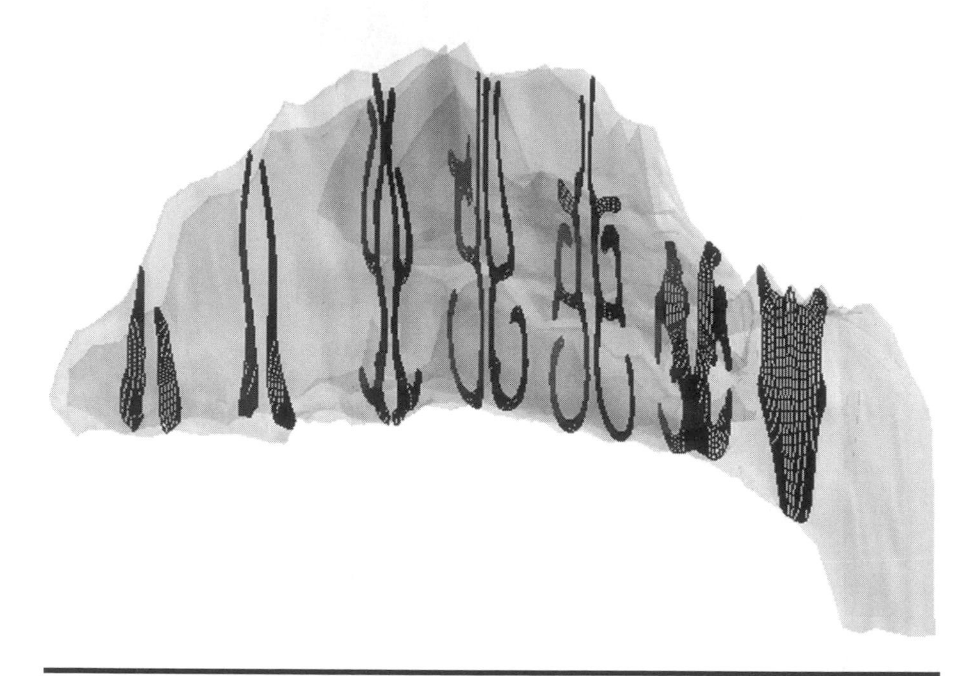

Figure 5.1 Geometry of the human nasal passages. The geometry is constructed by the assembly of successive, equally distanced airway coronal cross sections extending from the nares to nasal pharynx.

the cross section along the nasal pathway. Bulk airflow patterns are influenced by these complex structures such that major medial and lateral airflow streams are formed in the human nose (Hahn et al., 1993; Subramaniam et al., 1998).

The structure of the ET region is complete at birth and grows in dimension only with age. At any age, the size of ET airways is proportional to the tracheal diameter (Prahl-Anderson et al., 1979). Morphometric studies of the ET airways from cadavers have been conducted in adults (Montgomery et al., 1979) and children (Bosma, 1986). Other studies have used MRI or CT scan data to examine the ET airway anatomy of children for use in the production of hollow-cast models (Swift, 1991; Swift et al., 1994; Janssens et al., 2001). In these studies, a 4-week-old infant (Swift et al., 1994), a 6-week-old infant (Swift, 1991), and a 9-month-old child (Janssens et al., 2001) were investigated. Only the model of Janssens and colleagues (2001) included a larynx.

5.2.2 TB Structure and Function

The main function of the TB region is to deliver oxygen efficiently to the alveolar region, where gas exchange (oxygen for carbon dioxide) occurs.

The types of cells comprising the TB epithelium are similar to those contained in the transitional and respiratory epithelium of the ET region.

The structure of the TB region at the gross level can be thought of as a complex branching system of tubes or pipes. The anatomical depiction of the TB airways of humans (Table 5.1) corresponds to a symmetric branching system denoted as regular and dichotomous because each branching parent tube gives rise to two daughter tubes of the same diameter. This corresponds to the lung model developed by Weibel (1963) and represents the lung structure that has been most often used in human dosimetry models for both gases and particles. In actuality, the conducting airways of humans exhibit an irregular bipodial and even tripodial branching pattern (Crapo et al., 1990). Despite this, when Martonen (1983) compared results obtained using symmetric TB geometries (Weibel, 1963; Soong et al., 1979) or the asymmetric model of Horsfield et al. (1971), he found that the symmetric models gave better agreement with the available human experimental deposition data than did the asymmetric model.

The airways in each lobe of the human lung require anywhere from 9 to 22 branchings to reach a terminal bronchiole (Yeh and Harkema, 1993). Reports on the average number of conducting airway generations are contradictory. Weibel (1963) proposed 17 generations, which was later revised by Haefeli-Bleuer and Weibel (1988) to 15 or 16. Yeh and Schum (1980) and Horsfield and Cumming (1968) also reported a value of 16.

Several studies have reported TB airway measurements in adults (Weibel, 1963; Horsfield and Cumming, 1968; Horsfield et al., 1971; Raabe et al., 1976; Yeh and Schum, 1980; Phalen et al., 1985) and children (Mortensen et al., 1983, 1988; Phalen 1985) from replica casts prepared from excised lungs. A few measurements are also available in living subjects but have been confined to the trachea due to limitations in the resolution of the procedures used (Morgan and Steward, 1982; Kawakami et al., 1986; Griscom and Wohl, 1985). Although the number of conducting airways is of the order of 60,000, no more than about 8,000 have been measured (Raabe et al., 1976). Measurements typically include all airways in the first few airway generations (Mortensen et al., 1983) followed by selected airways along several paths down the bronchial tree (Weibel, 1963; Phalen et al., 1985).

There is general agreement that the full number of bronchial airways is present at birth and that airways only grow in size until full maturity. Beech et al. (2000) estimated the number of terminal bronchioles in 14 subjects younger than 1 yr old to be between 15,000 and 61,000, which corresponds to 15 to 17 generations and is consistent with that of adults. Several measurements of bronchial airway dimensions exist in the literature (Engel, 1913; Wood et al., 1971; Hieronymi, 1961; Johnson et al., 1978; Griscom and Wohl, 1985; Hislop et al., 1972; Phalen et al., 1985). Airway measurements are often taken along a single path or within the first several

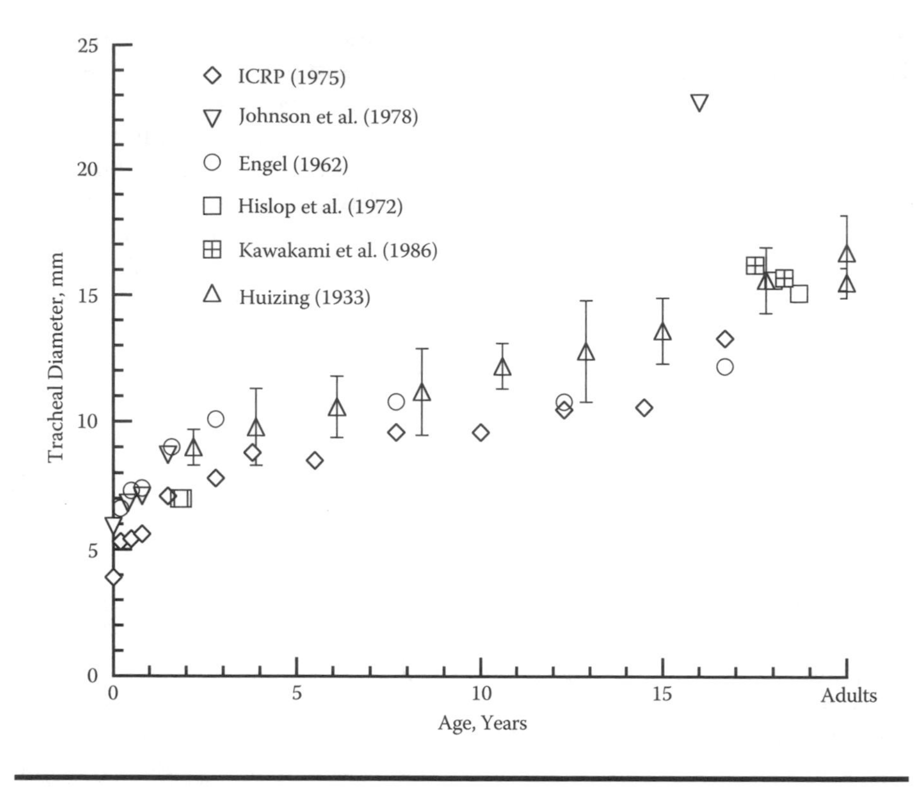

Figure 5.2 Reported measurements of the tracheal diameter as a function of age.

airway generations. Disagreements among various measurements could be due to intersubject variability or errors associated with the selected technique. Data collected by several investigators on tracheal diameter growth are shown in Figure 5.2. There is a slight variability in the reported measurements. However, the agreement is remarkable given vast differences in measurement techniques and time of measurements in the breathing cycle. This agreement has been attributed to the stiffness of the trachea (Ménache and Graham, 1992).

There are contradictory findings regarding the growth pattern of the airways, particularly central versus peripheral airways (or large airways vs. small airways). Hislop et al. (1972) claimed that all airways grew in proportion to the change in lung volume, whereas the measurements of Cudmore et al. (1962) indicated that proximal airways grew faster than the distal airways. Horsfield et al. (1987) also arrived at this conclusion. However, Hogg et al. (1970) reported an opposite trend in the growth of distal airway diameters compared with central airway diameters up to the age of 5 yr. Yet Hieronymi (1961) showed that the proximal and distal airways grew equally during infancy but that distal airways grew faster than proximal airways after the age of 1 yr. The disagreements among various studies have been attributed to the use of different measurement techniques, intersubject variability, and sampling problems associated with the large numbers and structural variations of lung airways (Phalen et al., 1985).

There is the possibility that children have more conducting airway generations at birth than adults and that conducting airways are converted to respiratory bronchioles through retrograde alveolarization (Boyden, 1965, 1967; Burri, 1985). Horsfield et al. (1987) also claimed evidence of retrograde alveolarization in their study. Hislop and Reid (1974) presented a similar statement suggesting several possibilities, including terminal bronchioles transforming into respiratory bronchioles and respiratory bronchioles transforming into alveolar ducts.

5.2.3 P Structure and Function

The alveolar region is where gas exchange occurs. The alveolar region includes respiratory bronchioles, alveolar ducts, and alveolar sacs (Table 5.1). The respiratory bronchioles are different from terminal bronchioles in that they have pockets of alveoli attached to their walls. The average number of generations of respiratory bronchioles in adults is identified as three (Pinkerton et al., 2000; Weibel, 1963; Yeh and Schum, 1980; Horsfield and Cumming, 1968). Haefeli-Bleuer and Weibel (1988) also reported an average number of three respiratory bronchiole generations but found that number to be quite variable. Respiratory bronchioles are followed by approximately five generations of alveolar ducts in which the entire surface of each duct is composed of alveoli (Hafeli-Bleuer and Weibel, 1988). The last generation of alveolar or acinar region comprises alveolar sacs that cover the peripheral end of the terminal alveolar duct. The reported number of alveoli in adult humans varies considerably. Dunnill (1982) reported between 200 and 600 million. In a study of six adult lungs, Ochs et al. (2004) reported a 37% coefficient of variation associated with a range of 274 to 790 million alveoli and a mean alveolar number of 480 million. Their mean is close to the value of 486 million reported by Mercer et al. (1994) for four adults, together with a standard error of 37 million.

The liquid layer lining the alveolar epithelium is comprised of surfactant, which contains a number of surface-active materials, primarily phospholipids. Surfactant lowers the work of breathing by lowering surface tensions, thereby stabilizing alveoli and preventing them from collapsing. The surfactant layer is nonuniform in that there is a thin film (<0.01 μm thick) on a hypophase approximately 10 times thicker, but there is significant pooling of surfactant in corners (pockets) of alveoli during expiration.

Unlike the ET and TB region, the morphometric structure of the alveolar region is incomplete at birth. Although major airway growth in the alveolar region takes place within the first 2 yr (Gehr, 1987; Zeltner et al., 1987; Zeltner and Burri, 1987), the growth is completed only by the age of about 8 (Dunnill, 1962; Reid, 1984). The number of respiratory airway generations is not complete at birth. As in adults, there are three generations of respiratory bronchioles (Hislop and Reid, 1974; Reid, 1984). Hislop and Reid (1974) reported four generations of alveolar ducts by the age of two months and six generations at the age of 7 yr. Data on the diameter of respiratory bronchioles in children that indicated a diameter increase with age are given by Hislop and Haworth (1989) and Weibel (1963). There is no information on the length of respiratory airways in children. Limited information is available regarding alveolar duct dimensions (Hislop et al., 1986; Weibel, 1963).

A number of studies have attempted to estimate the number of alveoli as a function of age based on limited measurements (Dunnill, 1962, 1982; Weibel, 1963; Thurlbeck, 1975; Polgar and Weng, 1979; Langston et al., 1984; Thurlbeck, 1988). The common consensus is that most alveoli are added after birth (Thurlbeck, 1988). Estimates of the number at birth vary between 17 and 71 million (Dunnill, 1962), which is attributed to biological variability and difficulties in identifying the alveoli.

5.3 LUNG GEOMETRIC MODELS: ADULTS AND CHILDREN

A complete representation of the lung structure and dimensions (geometry) does not exist even for a typical lung. The complexity of the lung structure has made complete airway measurements cumbersome and formidable. Lung geometry models with various degrees of complexity have been constructed from the limited measurements. Early efforts in lung geometry modeling focused on human adults. Three general models of lung geometry have been proposed: typical-path symmetric, lobar-symmetric, and asymmetric structures (Figure 5.3). Each geometric model is selected according to the desired level of deposition predictions. Although symmetric geometry is sufficient for regional predictions of particle deposition, detailed local predictions can be obtained only from an asymmetric lung geometry. Different geometry models are briefly described in the following text.

The early lung geometry model of Findeisen (1935) consisted of only nine generations that included both the conducting and pulmonary airways with the last section comprising 5.2×10^7 alveolar sacs. A description of the airway structure was developed by Weibel (1963) based on measurements

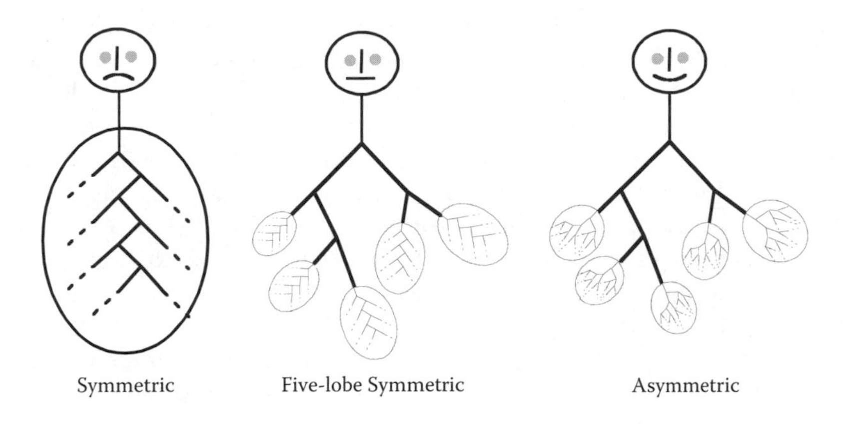

Figure 5.3 Schematic diagram of three different geometric models of the lung.

from a lung cast prepared from an excised lung of a young male. The airways in this model were assumed to be a dichotomous, branching system comprised of 23 generations. Beginning with the 17th generation, increasing numbers of alveoli were present on distal airway walls, and the last three generations were completely alveolated. The Weibel model has been widely used to predict gas and particle deposition in the lung because of its simplicity. However, as the lung is really asymmetric, the Weibel model overestimates the number of airways. Still, it produces reasonable average particle deposition predictions for the TB and alveolar regions of the lung.

Efforts to develop an asymmetric lung model include the work of Horsfield and Cumming (1968), but only measurements of the conducting airways of their cast were reported. A complete asymmetric lung model was later proposed by Olson et al. (1970) that closely resembled the Weibel model. However, there were considerably fewer airways in the deep lung. A different representation of the lung was reported by Yeh and Schum (1980) that included both symmetric, typical-path, and five-lobe symmetric but structurally different lung models. The five-lobe model of Yeh and Schum (1980) included branching and inclination angles in addition to airway length and diameter, which made it attractive for particle deposition calculations.

The preceding lung models were either too simplistic (symmetric) or lacked enough information to provide accurate predictions of particle deposition in the lung. There is also considerable intersubject variability of airway dimensions among the population. Efforts in the development of a realistic description of lung geometry have moved toward the development of stochastic lung geometries that can be reproduced to create lungs of different size. Koblinger and Hofmann (1985) developed a stochastic morphometric model of the human TB tree. The model was based on data describing the distribution of airway morphometric parameters such as length, diameter, branching angle, cross-sectional area of the daughter tubes, gravity angle, and correlations between these parameters as a function of airway generation. Koblinger and Hofmann (1990) combined the bronchial tree of Koblinger and Hofmann (1985) with the asymmetric dichotomous alveolar branching structure of Haefeli-Bleuer and Weibel (1988) to develop a complete stochastic model of the adult human lung. The stochastic model was also used to create a complete lung geometry containing all pathways of the adult TB airways (Asgharian et al., 2001). The advantage of the model is that more than one lung geometry can be reproduced stochastically to reflect, to some extent, intersubject variability (Hofmann et al., 2002).

Because early studies used an adult male lung to represent a typical lung, little attention was given to gender, race, age, and other variables. Consequently, limited efforts were spent on the growing lung. Early models of the lung geometry of children assumed that the lung of a child is a miniature version of an adult lung because some studies (e.g., Hislop et al., 1972) showed that airway dimensions in children were proportional to the change in lung volume. Hofmann et al. (1979) formulated an age-dependent lung model based on the simple anatomical model of Landahl (1950) and Jacobi (1965). During postnatal growth, airways increased in proportion to the size of the adult lung. The number of airways remained the same in the TB region during growth but changed with age in the alveolar region. Crawford (1980, 1982) also assumed that lungs of children were simply scaled-down versions of the adult lung model of Weibel (1963), with only the pulmonary airways changing in number until the age of eight. For ages less than 8 yr, the dimensions of individual airways in the pulmonary region remained the same, with the pulmonary volume changing only in response to the increase in the number of airways.

Children's lungs are not miniature versions of adult lungs as previously thought. Hofmann (1982a) used the anatomical structure of the dichotomous Model A of Weibel (1963) as the basis for structural development with age. The number of airways was assumed constant for generations 0 to 16, and the change in airway parameters was represented by the same analytical function that was based on experimental data for the main bronchi. Assuming that the number of airways in the alveolar region followed that of the respiratory airways of Dunnill (1962), growth of airway length and diameter in the alveolar region was assumed proportional to the cube root of the pulmonary airway volume divided by the number of airways in a given generation. In all, a combination of curve fits to experimental data and theoretically derived assumptions were used to obtain airway parameters as a function of age.

In a manner similar to Hofmann (1982a), Xu and Yu (1986) based their lung geometry on the Weibel (1963) model and derived slightly different expressions for airway length and diameter in the TB and alveolar regions. Airway growth curves in the TB region were found by fitting to the measurements of the main bronchi and terminal bronchioles. Alveolar region volumes were also slightly different from those of Hofmann (1982a). Hofmann et al. (1989) and Martonen et al. (1989) modified the earlier model of Hofmann (1982a) by allowing the airway parameters in an airway generation belonging to the TB region to grow at different rates. The airway growth rates of Phalen et al. (1985) were used in constructing the TB region. The main improvement in the model was that the airway growth rate was a function of height and was different in each airway generation. Other existing lung models (e.g., Martonen and Zhang, 1993; Musante and Martonen, 2000) were a variation of previous ones.

To date, most of the lung geometry models of children are symmetric with typical-path representation similar to that of Weibel for adults. Similar to the adult model, such a geometry is sufficient for predicting average deposition per lung region. Detailed, site-specific deposition information either cannot be obtained or is not reliable. Airway parameters are identical for all airways of the same generation in a symmetric lung structure and grow with age. Figure 5.4 provides airway diameter at different generations of the lungs of children ages 3 months, 23 months, 8 yr, and 14 yr, and adults (Ménache et al., 2005). Airway diameter decreases with increasing generation number and increases with age.

The first attempt to develop a lobar-specific geometric representation of children's lungs was made by Ménache et al. (2005). In this model, lobar-specific symmetric models of airway geometry for the TB and alveolar airways of children were developed based on the Utah Biomedical Test Laboratory (UBTL) measurements of airway dimensions in, for example, generations 0 through 9 (Mortensen et al., 1983) and limited information on dimensions of respiratory airways published in the peer-reviewed literature (Weibel, 1963). The number of airway generations was first estimated from published information (Beech et al., 2000; Haefeli-Bleuer and Weibel, 1988; Horsfield and Cumming, 1968). At each age group, nonlinear functions were then obtained using airway parameters for generations 0 through 9 and published data on terminal bronchioles, respiratory bronchioles, alveolar ducts, and alveolar sacs to estimate airway dimensions for the remaining airway generations. Eleven dichotomous, symmetric branching lungs for ages from 3 months to 21 yr were developed. Each lung geometry consisted of five symmetric lobes that were different structurally.

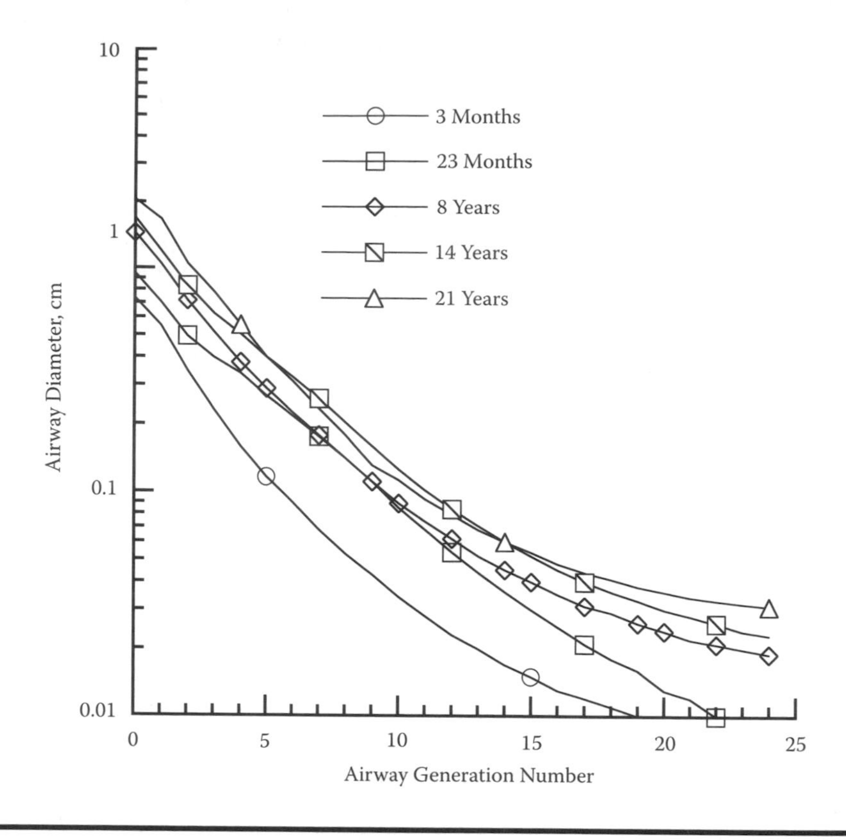

Figure 5.4 Airway diameter in different generations of a symmetric lung geometry for children. Reported in Ménache, M.G., Hofmann, B., Asgharian, B., Miller, F.J. Airway Geometry Models of Children's Lungs for Use in Dosimetry Modeling, 2005.

5.4 MECHANISMS OF DEPOSITION

Inhaled tidal air carries suspended particles to various sites within the respiratory tract, in which particles may deposit by different deposition mechanisms. Because airflow velocity reduces with each bifurcation of the airways, airflow rate drops very quickly in the TB region and approaches fully developed laminar flow after a few generations. The flow resembles creeping flow throughout the alveolar region. Even at flow rates 10 to 15 times greater than needed for normal respiration, convection is only 12% as important as diffusion for gas transport (Davidson and Fitz-Gerald, 1974).

Laws of physics govern the transport of particles entrained in the air. Particle transport and physical properties of the particles combine to yield mechanisms by which particles are removed from the airstream. The nature of the major mechanisms of deposition and significance in various sites within the lung will be described prior to discussing lung deposition measurements and predictions.

The major mechanisms by which noncharged particles deposit are inertial impaction, sedimentation, and Brownian diffusion. These mechanisms are illustrated schematically in Figure 5.5. Electrostatic attraction may be an important mechanism for the deposition of particles in some workplace exposure settings if the processes being used to generate charged particles and workers are in close proximity to the source of these particles. However,

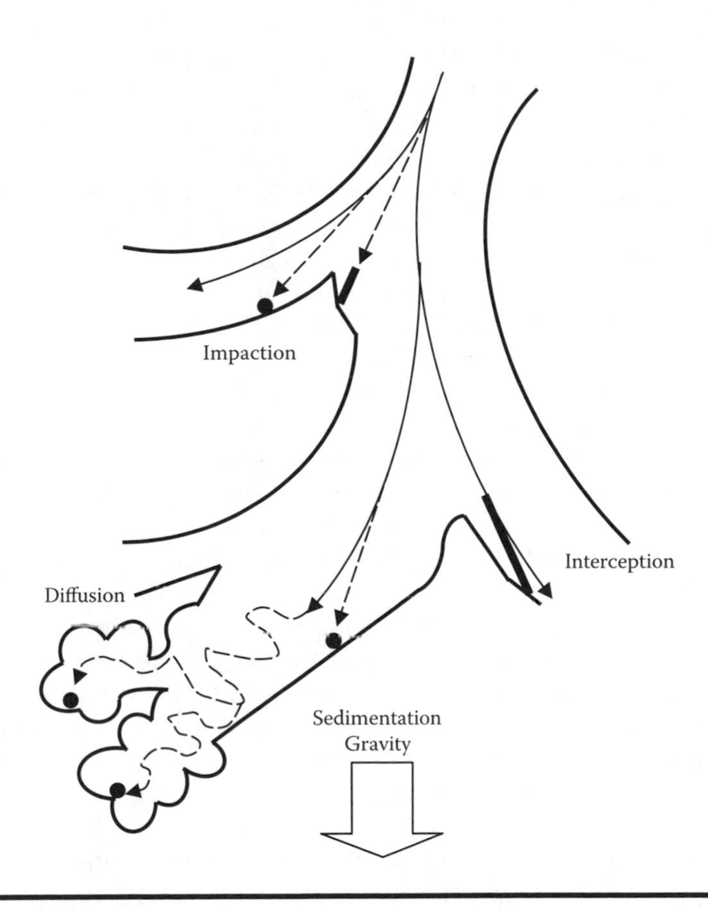

Figure 5.5 Primary mechanisms of particle deposition of inhaled noncharged particles in the respiratory tract. Solid lines, airflow streamline; dashed lines, particle trajectory. (Modified from McClellan, R.O., Miller, F.J. An overview of EPA's proposed revision of the particulate matter standard. *CIIT Activities* 1997, 17: 1–23.)

electrostatic attraction is not an important deposition mechanism for ambient exposures to particles because there is ample time for such aerosols to come to Boltzman equilibrium prior to inhalation. Inertial impaction is the process by which the inertia of the particle makes it unable to follow changes in airstream direction or air velocity streamlines. Sedimentation refers to the settling out of particles from the airstream due to gravitational forces. The random displacement motion of particles resulting from constant bombardment by air molecules (Brownian diffusion) can result in particles coming into close proximity with airway surfaces. Depending on particle size and airflow rate, diffusion, sedimentation, and impaction do not necessarily act as independent processes. Once particles are brought into close proximity with the airway walls, they are removed from the airstream by interception with the walls. Because objects with a relatively long length-to-diameter ratio have an increased probability of the ends intercepting the wall of an airway, interception is often considered by some as a separate deposition mechanism for fibers (Figure 5.5).

The geometric (physical or diffusion equivalent) and aerodynamic equivalent diameters of a particle are important determinants of the relative importance of the preceding mechanisms in the deposition of particles. Aerodynamic diameter (d_{ae}) takes into account the size, shape, and density of a particle and is defined as the diameter of a unit density (1 g/cm^3) sphere having the same terminal settling velocity as the particle. Because particles of different sizes and density can have the same d_{ae}, they can be deposited in the same locations within the respiratory tract.

Various deposition mechanisms are active or dominant depending on the location within the respiratory tract, the size of the particle, and the depth and route of breathing. In the ET region of humans, the torturous nasal passages result in inertial impaction being the predominant mechanism by which inhaled particles larger than about 5 μm are deposited. In addition, ultrafine particles (<0.1 μm in diameter) are effectively removed in the nose via diffusion as the surfaces of the nasal turbinates are large compared to the cross-sectional area and are in close proximity to the bulk airflow streams.

The site of particle deposition in the lung is a complex interaction of the respiratory tract anatomy with particle aerodynamic properties and is influenced by lung physiological parameters. Figure 5.6 illustrates regions of the lung where particle deposition occurs by various loss mechanisms. Impaction is also the predominant mechanism for particle deposition in the TB airways for particles greater than 2.5 μm d_{ae}. Because mean flow rate and residence time influence impaction, TB deposition is even greater for activities that increase minute ventilation. Enhanced deposition of particles larger than 2.5 μm during inspiration occurs at airway bifurcations (Martonen and Hofmann, 1986; Schlesinger, 1988). With expiration, there is a tendency for increased deposition on the walls of the parent tube within a distance of about one diameter of the

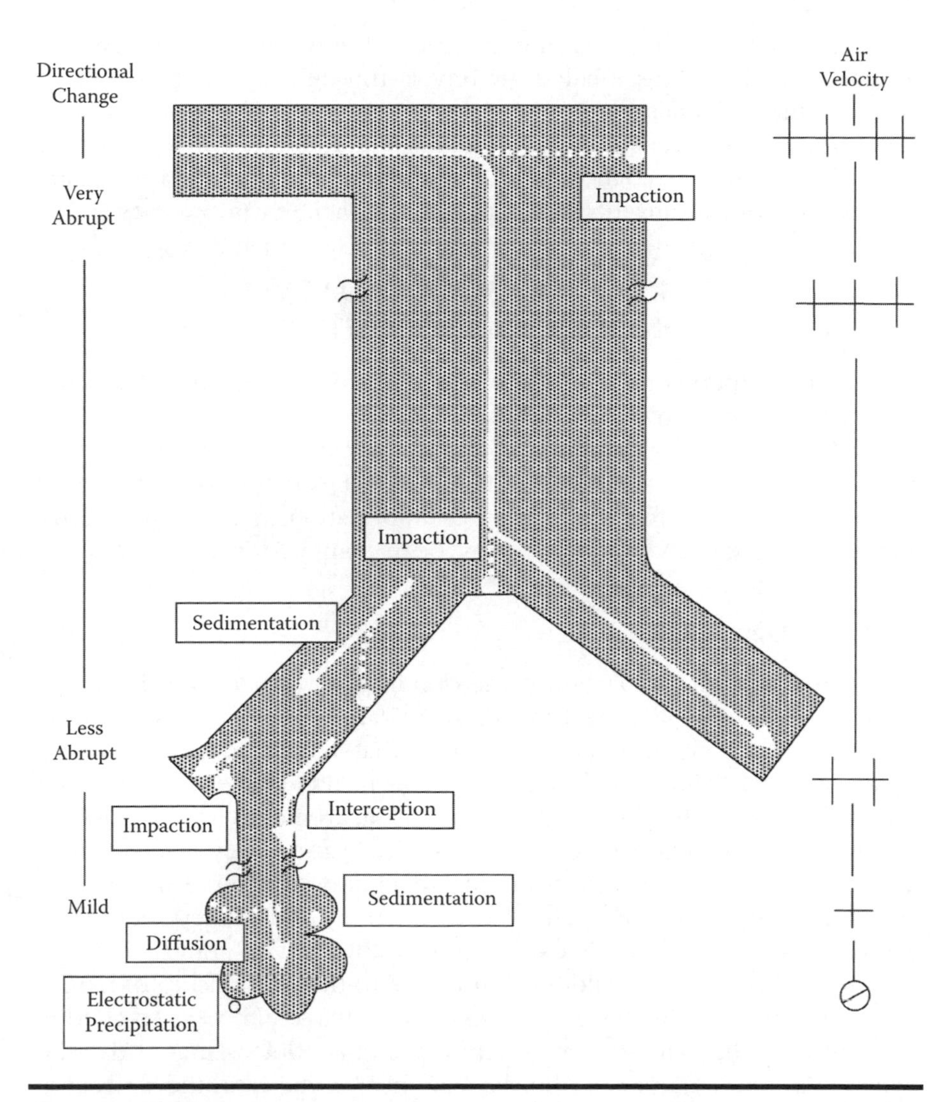

Figure 5.6 Schematic representation of major mechanisms of deposition of particles in the respiratory tract. Airflow is signified by the arrows and particle trajectories by the dashed lines. (Modified from U.S. Environmental Protection Agency. Dosimetry of Inhaled Particles in the Respiratory Tract. In *Air Quality for Particulate Matter*, Vol. II, pp. 10–18. Research Triangle Park, NC: National Center for Environmental Assessment, 1996.)

airway size. Also, the turbulence created by the laryngeal jet tends to result in enhanced deposition of ultrafine particles (<0.1 μm in geometric diameter) in the trachea and larger TB airways due to diffusion (Cohen, 1987; Cohen et al., 1990).

As a result of an increasing number of airways, the airflow rate continues to decline as inhaled air travels through the lung. Slow flow increases the residence time of particles in the small airways, resulting in deposition by sedimentation for particles > 1 μm d_{ae} and diffusion for particles < 0.5 μm. Impaction is no longer present as particles with slow velocity do not acquire the inertia necessary for impaction losses.

5.5 EXPERIMENTAL DATA ON REGIONAL DEPOSITION

A number of experimental studies on the regional and total lung deposition of particles have been conducted using healthy adult subjects. The reader may refer to several articles for a detailed review of deposition data in human adults (ICRP, 1994; Miller, 1999, 2000*). We do not repeat adult data here but rather focus on available information in children, because these data are far fewer and have not been compiled in one report.

5.5.1 ET Deposition

As discussed before, impaction is the dominant deposition mechanism in the ET region of particles 1 μm in size or larger. Particle loss efficiency is a function of an impaction parameter that is related to the particle Stokes number (Hinds, 1999). Despite the health concern for children exposed to particles, limited deposition measurements are available in children. Experimental studies in children have used only a few particle sizes, and, as a result, one cannot establish a deposition curve with particle inertia as in the case for adults. However, the trend of deposition with particle inertia in children is expected to follow that of adults.

A greater concern in children is in regard to the relationship of particle deposition with age as this is relevant to potential health risks to children from inhaling the same exposure atmosphere as adults. Limited data are available in the literature on the deposition of particles in the ET region. A study by Becquemin, Swift et al. (1991) and one by Bennett et al. (1997) are the only published reports on particle deposition in the nasal and oral passages of children, respectively. Nasal deposition losses were plotted in Figure 5.7a against the impaction parameter $\rho d^2 Q$, where ρ is particle mass density, Q is mean inspiratory flow rate, and d is particle diameter. Becquemin, Swift et al. (1991) found that ET deposition in the nasal passages increased with particle inertia and age for coarse particles (Figure 5.7a). Figure 5.7b shows oral deposition fraction data of

* Also see Errata (*Inhal. Toxicol.* 12:1257–1259, 2000), particularly for corrections to interspecies dosimetric comparisons contained in these publications.

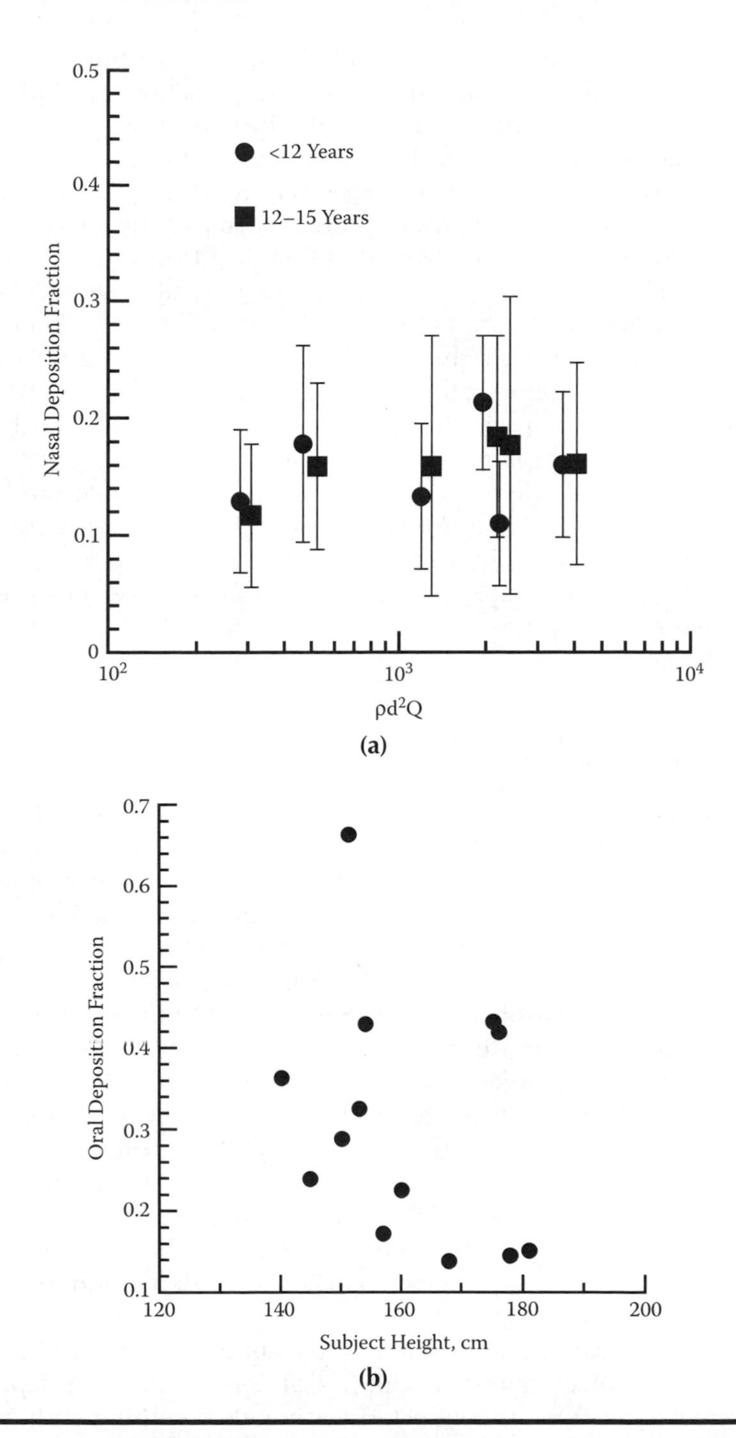

Figure 5.7 Deposition fraction of inhaled particles in the ET region of children. (a) nasal airways; (b) oral airways.

Bennett et al. (1997) plotted against subject height. Bennett et al. (1997) suggested enhanced oral deposition of coarse particles in children compared with adults. They also found that the increase was correlated with decreasing height of the subject. The data sets in these studies are very limited because only limited particle sizes were considered. Although some fine and coarse particle sizes were studied, no measurements were conducted for ultrafine particle losses by diffusion. In addition, study protocols were not designed to separate losses during inhalation and exhalation due to difficulties in making the necessary measurements. Ultrafine and coarse particles may have significant deposition in the head region. The study of particle deposition in the lung requires a knowledge of ET filtering capability during inhalation; ample data exist for adults in regard to this capability but there are limited data on children.

In the absence of adequate data, it is recommended that predictions of particle deposition in the ET region of children be calculated from expressions for adults following the scaling procedure suggested by Swift (1989). The procedure accounts for the effect of the subject's body size on nasal deposition of particles in the aerodynamic size range (ICRP, 1994).

5.5.2 Lung Deposition

Deposition measurements in the lung are typically reported for the TB and P regions. However, these values are obtained through indirect measurements. Experimental protocols for adults employ radiolabeled particles, and the amount of activity retained in the lung is measured as a function of time. TB deposition is derived from the fast-decay component of the time-activity curve with alveolar deposition being estimated from the slow-decay portion of the curve. Such a separation is based on the assumption of fast clearance by the mucociliary escalator and slow clearance by macrophage phagocytosis and translocation. In children, however, only nonradiolabeled particles have been used due to expected higher sensitivity and increased possible health risk from exposure to radiation. Particle concentration before entering and after exiting the respiratory tract were measured and used to calculate the amount deposited in the lung. Consequently, time history measurement of deposited dose in the lung was not possible, and separation of TB and P deposition cannot be obtained by this technique.

The data on particle deposition in the lungs of children are quite limited. Due to uncertainty regarding nasal deposition, lung deposition measurements have been conducted via mouth breathing with particle sizes between 1 and 3 μm, for which oral deposition is negligible.

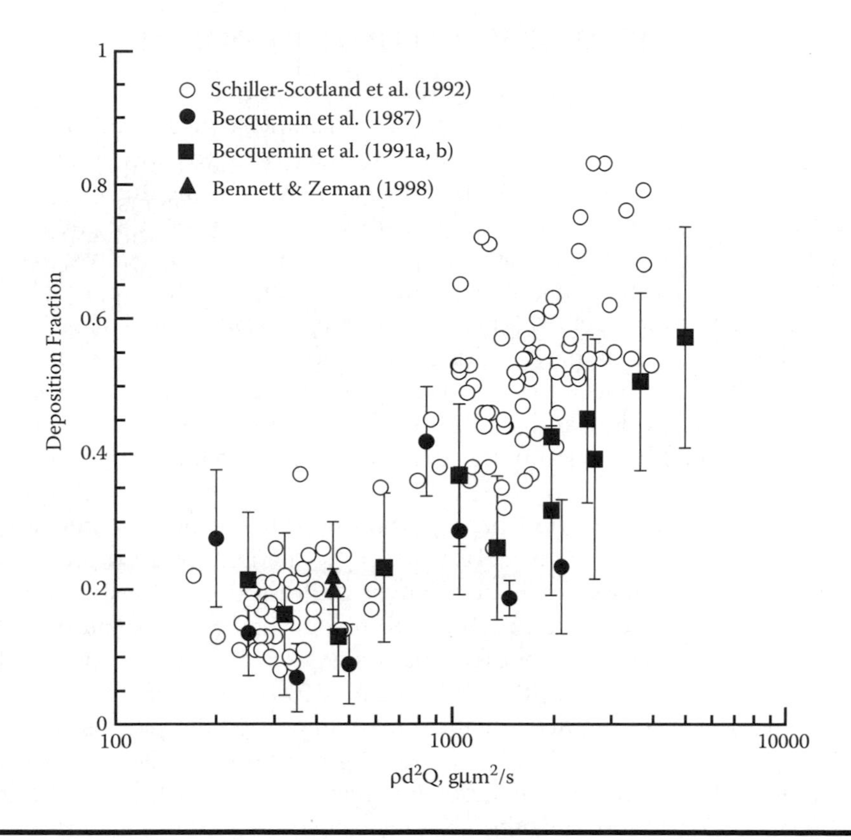

Figure 5.8 Measured deposition fraction of inhaled particles in the lungs of children.

Particle losses in this size range are primarily by impaction in the upper airways and sedimentation in the lower airways of the lung. As these loss mechanisms depend on the impaction parameter, $\rho d^2 Q$, deposition measurements are plotted against this parameter. Figure 5.8 shows measurements from studies by Schiller-Scotland et al. (1992), Becquemin et al. (1987), Becquemin, Swift et al. (1991), Becquemin, Yu et al. (1991), and Bennett and Zeman (1998). Only the data set of Schiller-Scotland contained one-to-one information on the deposition value and corresponding particle size and breathing parameters. Other data sets reported the mean plus standard deviation as reflected in the figure. Overall, lung deposition increased with increasing values of the impaction parameter. However, some scattering of the reported data, aside from uncertainties regarding the measurements, arose from intersubject variability.

5.6 PREDICTIVE DEPOSITION MODELS

Mathematical models to predict particle deposition in the lung should contain three major components: first, a lung geometry prescribed at a level of detail commensurate with the desired accuracy of deposition predictions; second, a model of airflow transport to carry particles in the lung; and, finally, a mathematical treatment of particles traveling in the lung that computes the deposition fraction of particles based on the conservation of particle mass by marching through individual lung airways in sequential and parallel fashion. Deposition fractions are typically calculated during a breathing cycle that may include inhalation, pause, and exhalation. Although the mechanisms of particle transport and deposition remain similar between children and adults, lung growth and changes in breathing physiology combine to modify the deposition and distribution of particles in the lung as a function of age.

As in the previous section on reporting deposition measurements, we will focus our attention on available models in children because predictive models for adults have been adequately addressed previously. Several articles exist in the literature to which the reader may refer for information on predictive models of particle deposition in the lungs of adults (e.g., Stöber et al., 1993; ICRP, 1994; Miller, 1999). Furthermore, we will not discuss details of the mathematical modeling of particle deposition but will only provide reference to and a brief description of existing studies in children. The mathematical modeling may be different in each study, and the reader is referred to the original source for further information.

There is a fair degree of confidence in recorded breathing parameters of children at different ages due to the availability of noninvasive techniques for measuring these parameters (Hofmann, 1982b). Reported values of the breathing parameters at selected ages during different physical activities are given in Table 5.2. Although breathing frequency decreases with age, inhaled tidal volume increases, resulting in an increase in minute ventilation with age.

There is uncertainty related to growth of both the ET and thoracic airways. Realistic predictive models of particle deposition in the lung need to address this variability. Limited lung dosimetry models for predicting airway-specific deposition in children currently exist, primarily because of the lack of detailed geometry models. Lung deposition models were originally designed for a standard man, defined as a young adult male weighing nearly 70 kg with a height of about 170 cm. The age-dependent lung geometry model of Hofmann et al. (1979) assumed that only the bronchial airway tree of a child was a miniature version of an adult lung during postnatal growth (Hislop et al., 1972). The deposition and retention model of ICRP (1975) was then used to calculate regional

Table 5.2 Ventilation Parameters of Children and Adults by Age

	Sedentary Activity[a]		*Maximal Activity*[b]	
Age (y)	*Frequency (breaths/min)*	*Tidal Volume (ml)*	*Frequency (breaths/min)*	*Tidal Volume (ml)*
0.58	35	42	70	165
1.83	28	84	70	328
4	22	152	68	570
8.17	18	266	60	1190
30	14	500	40	3050

[a] Based on equations proposed in Hofmann, W. Mathematical model for the postnatal growth of the human lung. *Respir Physiol* 1982b, 49: 115–129.

[b] Values reported in International Commission on Radiological Protection (ICRP). *Task Group on Reference Man.* ICRP Publication 23. Pergamon Press: Oxford, 1975.

and total lung deposition. The dosimetric model consisted of a nasopharyngeal region that was accounted for by a deposition probability expression of adults. Doses to the TB and P regions were found to depend strongly on age.

In a follow-up study, Hofmann (1982a) used the more detailed Weibel geometry, which permits the calculation of anatomical parameters such as length and diameter with age. The new model was also used by other investigators (Martonen and Zhang, 1993; Musante and Martonen, 2000). Using the deposition model of ICRP (1975), Hofmann (1982a) found that the dose to the lung decreased with increasing age. The maximum dose was calculated around the age of six. This finding was shared by Crawford (1982), who normalized the adult model of particle nasal deposition of Pattle (1961) for children using the relationship given by Johnson et al. (1975). Regional deposition was determined following the methodology of Landahl (1950), which incorporated the effects of pulmonary mixing. Crawford further found that age effects were more evident for ultrafine particles.

Hofmann et al. (1989) employed the deposition model of Hofmann (1982a) with an improved age-dependent lung geometry in which the growth of bronchial airway dimensions was based on the measurements of Phalen et al. (1985). Excluding nasopharyngeal deposition, they found that deposition predictions were highly sensitive to the assumed lung geometry, stressing the need for a morphometrically accurate model of the entire respiratory tract.

The deposition models used by Hofmann et al. (1979), Crawford (1982), Hofmann (1982a), and Hofmann et al. (1989) used the Weibel (1963) lung geometry that was based on the compartmental models of Findeisen (1935) and Landahl (1950). Xu and Yu (1986) argued that the predictions from the compartmental model deviate from the experimental data of Lippmann and Altshuler (1976). These authors proposed a continuous model (called the *trumpet* model) originally developed for adults (Yu and Diu, 1983) to be used for predictions of particle deposition in the lungs of children (Xu and Yu, 1986; Yu and Xu, 1987). The geometry input to the deposition model included a naso- and oropharyngeal region scaled from adults by the tracheal diameter at a given age and a growing lung geometry based on available morphometric measurements in the literature. The results suggested larger ET and total deposition in children than in adults. The regional deposition in children was found to be higher or lower depending on particle size.

As is the case for adults (Rudolf et al., 1986, 1990), empirical expressions have been derived based on available measurements to predict regional particle deposition in the lungs of children. Yu et al. (1992) assumed the lung to be comprised of a series of two filters representing the TB and P regions and obtained relationships for the deposition fraction in each region as a function of particle geometric diameter, breathing period, tidal volume, and the subject's age. ICRP (1994) also proposed empirical expressions for the TB and P regions by introducing age-dependent scaling factors to correct for the measured deposition fractions in adults. There were no assumptions regarding lung geometry, airflow pattern and distribution, and particle deposition mechanisms. The results were applicable for the particle size range in the measurements, as well as the range of the other parameters in the equations.

The deposition models described in the preceding text used simplified typical-path lung structures that were obtained from limited airway measurements in children's lungs (Hofmann, 1982b; Phalen et al., 1985). These models have been validated in adults for regional deposition fractions but have proved inadequate for site-specific predictions (Asgharian et al., 2001). The human lung has an asymmetric branching structure that affects both airflow and particle deposition. Asgharian et al. (2004) developed a lobar-specific symmetric model capable of predicting regional deposition within each lobe of the lung. The model used the lung geometry proposed by Menache et al. (2005). Figure 5.9 gives the predicted deposition in the lungs of an 8-yr-old child. The predicted lobar deposition was highest in the lower lobes, followed by that in the upper lobes. Deposition fraction of particles was smallest in the right middle lobe. Lobar deposition fractions were found to depend on the volume of the lobe, in which the lower

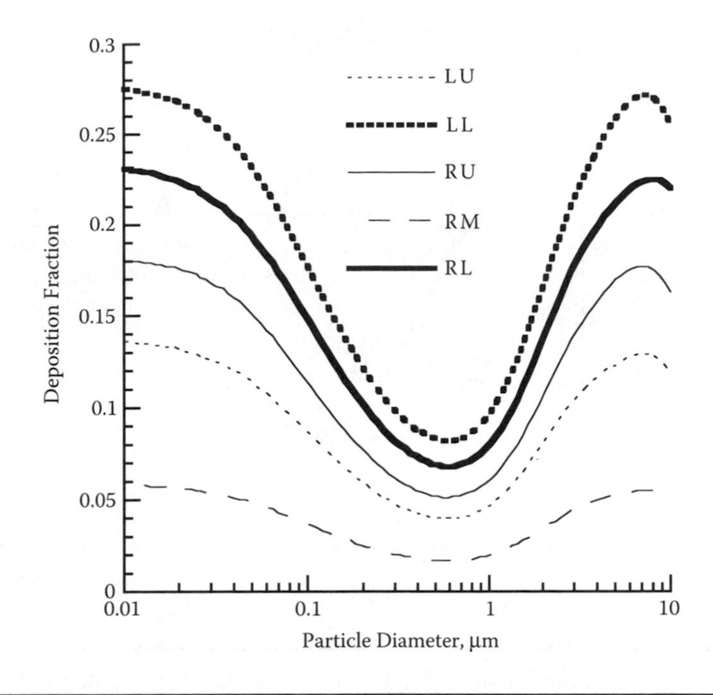

Figure 5.9 Predicted lobar deposition fraction of particles in an 8-yr-old child.

lobes with the highest volume gave the largest deposition fractions. Furthermore, deposition fractions for any region of the lung (e.g., TB, P, total) were related to the volume of that region of the lung. In fact, when deposition fractions were divided by the respective volumes, practically the same deposition curves were obtained.

Figure 5.10 gives the total lobar deposition fractions in children ages 3 months, 23 months, 8 yr, and 14 yr, and in a 21-yr-old adult. The 3-month-old child had the highest deposition fraction, which is similar to those of other ages, except for the 23-month-old, which, surprisingly, gives the lowest predictions. The figure shows that there is no clear relationship between particle deposition and age, which may be due in part to intersubject variability among similar ages.

As particle deposition fraction can be normalized with respect to lung volume, the ratio can be used to compare lung deposition among different ages. The adjusted deposition shown in Figure 5.11 was greatest for infants and decreased with age, which was in agreement with the findings of Hofmann (1982a) and Xu and Yu (1986). Model predictions were also shown to be consistent with empirically measured deposition in children.

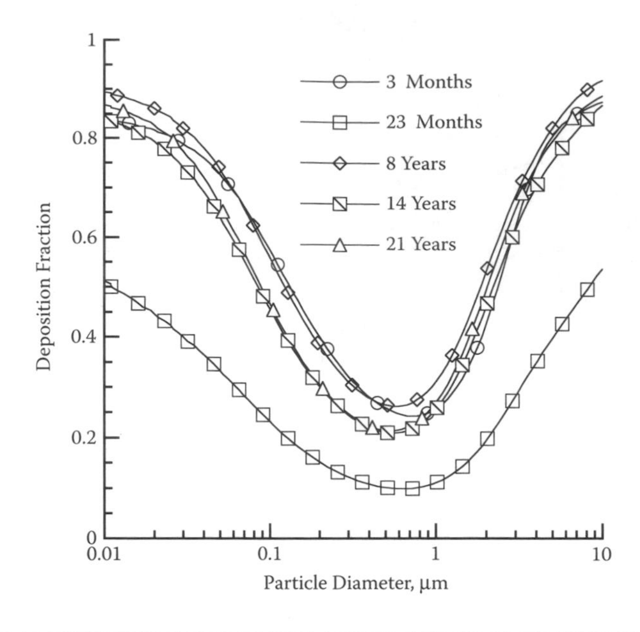

Figure 5.10 Deposition fraction of different size particles in the lungs of children.

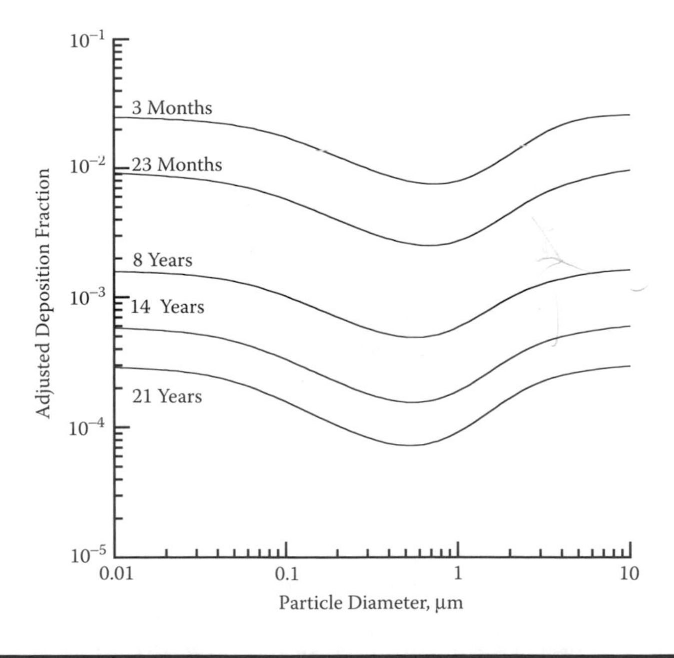

Figure 5.11 Predicted adjusted deposition fraction of inhaled particles in the lung as a function of age. Adjusted deposition fraction is deposition fraction divided by lung volume.

5.7 PARTICLE CLEARANCE FROM THE LUNG

Particles depositing on lung airway surfaces initiate a response by lung defense mechanisms that proceed to remove the particles from the lung. Particle clearance takes place gradually over time and depends on the initial site of deposition. Prolonged residence of the particles in the lung may lead to injury and biological responses. Long-term consequences of deposited particles in the lung are mainly a concern in adults. However, the onset of exposure quite often occurs in childhood, and thus the study of particle clearance in children is also needed. Many childhood lung disorders such as asthma occur in the TB region of the respiratory tract. Furthermore, very little information exists on alveolar clearance during the early stages of life when lung structure and defense mechanisms are developing.

Two major and distinct clearance mechanisms are observed throughout the respiratory tracts of adults and children. Particles in the upper airways of the LRT of healthy individuals are cleared mainly by the mucociliary escalator in a time period of approximately 24 h (Albert and Arnett, 1955; Albert et al., 1969, 1973). Particles removed from lung airways can be transported by constant flow of the mucous layer covering airway surfaces to the epiglottis, where they are swallowed and enter the gastrointestinal tract. The mucous layer protects the tissue from direct exposure to inhaled pollutants and is composed of an epiphase (gel phase) and an underlying hypophase (sol phase) in which cilia beat in a manner that propels mucus to the glottis where it is swallowed. For information on the comparative biochemistry of this layer, the reader is referred to Hatch (1992).

Detailed airway structure is also required to properly capture TB clearance processes. Mucous velocities in all fast-clearing airways of the lung (TB airways) are required to accurately assess particle removal. Mucous velocity measurements for all airways are a challenging, if not impossible, endeavor. However, velocities in all airways are connected; the mucous blanket can be thought of as a continuous flow of mucus in succeeding airways similar to the way air travels through the lumen. Therefore, measurement of mucous velocity in the trachea plus the knowledge of mucous transport in the succeeding airways should be sufficient to estimate mucous velocities in all subsequent airways.

The thickness of the mucociliary layer varies as a function of location within the TB region. Table 5.3 gives available data on the thickness of the liquid lining layer in the TB and ET regions and on mucous velocities in adult humans. The mean mucous velocities given in Table 5.3 have relatively large standard deviations associated with them. As noted by Wolff (1992), mucous clearance is not necessarily as relentless and uniform as some might presume, particularly in view of the fact that there are preferential routes of clearance. Thus, some deposited particles may be retained in areas of

Table 5.3 Extrathoracic and Tracheobronchial Liquid Lining Layer Thickness and Mucus Velocities in Humans

Location	*Thickness*[a] *(μm)*	*Mucus Velocity (mm/min)*	*Reference*
Nose		5.2 ± 2.3 to 8.4 ± 4.8	van Ree and van Dishoeck (1962)
			Andersen et al. (1971)
Trachea		3.6 ± 1.5 to	Yeates et al. (1975)
		21.5 ± 5.5	Santa Cruz et al. (1974)
Bronchi (main stem)	8.3[b]		Mercer et al. (1992)
Bronchi (segmental)	6.9[b]		
Bronchioles	1.8[b]		

[a] Unless specified, values apply only to the epiphase above cilia tips.

[b] Thickness of epiphase plus hypophase.

slow clearance for extended periods of time (Stalhofen, 1989; Stalhofen et al., 1994). Observations (Goodman et al., 1978) that mean tracheal mucus velocities decline with increasing age have important implications for assessing the risks of relatively insoluble particles, particularly if comparable reductions in mucociliary clearance also occur in smaller TB airways.

The second major clearance mechanism occurs in the alveolar region, where the mucous blanket gives way to a thin surfactant layer containing macrophages. Macrophage activity accounts for most insoluble particle removal in the alveolar region. Macrophages engulf free particles that land on the airway surface, move upward in the direction of the trachea to reach the TB airways, and are then cleared by the mucociliary escalator. Clearance of insoluble particles in the alveolar region is typically on the order of months to years.

5.8 FUTURE DIRECTIONS AND DATA GAPS

To assess the health risk from exposure to pollutants, the dose to the lungs of children must be determined to establish a linkage between exposure and potential biological responses. Despite children having possibly enhanced health risks from exposure when compared with adults, the information on the fate of particles in the lungs of children is quite limited. Major data gaps exist in the current database and include several dosimetry aspects that are briefly discussed in the following subsections.

5.8.1 Particle Inhalability

Not all particles in the environment are inhalable. When inhaling a volume of air, particles approaching the ET region may escape deposition due to their inertia. Particle inhalability is defined as that fraction of particles in an aerosol that can enter the ET region upon inhalation (Ménache et al., 1995). Inhalability affects the number of particles that enter the lung and are available for deposition. Thus to study the toxic effects of particles, inhalability must be accounted for. Particle inhalability in children may be different from that of adults due to differences in ET geometry and physical activity. Several studies have been done on the inhalability of particles in adult humans (Aitken et al., 1999; Breyesse and Swift, 1990; Erdal and Esmen, 1995; Hsu and Swift, 1999; Kennedy and Hinds, 2002; Ogden and Birkett, 1977; Vincent et al., 1990; Vincent and Armbruster, 1981) but virtually none in children. Inhalability measurements in adults are often performed in manikins that face incoming airflow at various wind speeds and directions. Ménache et al. (1995) used available information on particle inhalability in adult humans to predict the inhalability fraction as a function of particle diameter. However, inhalability is more appropriately a function of particle inertia as it also depends on breathing rates. Ideally, inhalability should be related to the impaction parameter, $\rho d^2 Q$, as suggested by Asgharian et al. (2003). In addition to having different physical activities and breathing rates from adults, children have a smaller size ET geometry (i.e., smaller size ET opening) that is inversely proportional to age. The combined effects of ET size and breathing rates may lead to different inhalability values compared with adults. Thus, evaluation of particle dose in the lungs of children requires information on particle inhalability as a function of age. Using the information on breathing rates and particle characteristics, measured inhalable fractions of particles in children must be fit to a parametric expression of inhalability as a function of particle inertia (Stokes number).

5.8.2 ET Deposition

Information on the number of particles deposited in the ET region is necessary to assess the amount that enters the lung. This data set is incomplete in children, mainly because there are limitations regarding conducting deposition studies in children. To obtain the much-needed data while keeping experiments in children to a minimum, computational studies along with *in vitro* measurements can be performed in nasal and oral casts of children. Scanned images of the ET region of children at various ages admitted to hospitals throughout the U.S. for reasons not related to their respiratory system could be used. Modern technology allows the ET scans to be reconstructed on the computer and also to be used to create nasal molds through stereolithography and other techniques (Deitz, 1990; Kimbell and Greenwood, 1995; Guilmette and Gagliano, 1994). One can then computationally study particle deposition in the ET region and determine parameters that influence deposition to obtain a relationship between deposition and variables controlling it (Kimbell et al., 2003). In addition, nasal molds can be constructed and used in *in vitro* studies designed to measure the deposition of particles in the ET region of children at various ages (Cheng et al., 1995; Janssens et al., 2001) and adults (Cheng et al., 1993, 1988; Guilmette et al., 1994; Itoh et al., 1985; Kelly et al., 2004a, 2004b; Montassier et al., 1992; Swift, 1991) and to establish a functional relationship between ET deposition and age.

Computational results, along with *in vitro* laboratory validations, will improve our understanding of particle deposition in the ET region of children by identifying the mechanisms controlling deposition and determining the influence of age, activity level, and other variables on particle deposition in this region. These studies will also determine the amount that penetrates the ET passages to reach the lungs of children. However, computational and *in vitro* studies cannot replace *in vivo* studies entirely because of the high sensitivity of the nasal mucosa to ambient air features and frequent nasal cycling. *In vivo* data are required for the final confirmation of model predictions. As discussed earlier, the data set by Becquemin, Swift et al. (1991) and Bennett et al. (1997) are the only *in vivo* data available in the literature on deposition in nasal and oral airways. These data sets are incomplete as measurements were performed for only a few particle sizes in the fine and coarse size range. There is no information on ultrafine particles in the ET region in spite of recent studies suggesting that ultrafine particles have a greater health impact on people (Utell and Frampton, 2000; Oberdörster, et al., 1995; Ibald-Mulli et al., 2002). In addition, the measurements of Becquemin, Swift et al. (1991) were taken for resting and moderately exercising breathing conditions. No measurements of ET deposition were available at other activity levels. As a result, a functional relationship of deposition fraction with particle size and activity level could not be established, even in the impaction-dominated region. Moreover, the study design did not allow separation of deposition during inhalation and exhalation. To find the number of particles that enter the lung, determining fractional deposition during inhalation is an *a priori* requisite. The limited information in existing databases is not conclusive, and the trend of deposition with the child's age, particle size, and other parameters is difficult to determine.

Some modeling efforts exist on ET deposition in children. Predictive particle loss models available only for impaction losses argue that the airway structure of the head is complete at birth and increases only in size with age (Xu and Yu, 1986). Assuming the size of airways in the head to be proportional to the diameter of the trachea (Prahl-Anderson et al., 1979), loss formulas in adults have been extended to children by applying a correction factor based on tracheal diameter (Xu and Yu, 1986; Yu and Xu, 1987). The predicted head deposition calculations showed enhanced deposition in children compared with adults, particularly for ages below 10 yr. This finding is in contradiction with the measurements of Becquemin, Swift et al. (1991), who showed greater nasal deposition in adults compared with children. The data on oral deposition, on the other hand, suggest that children have enhanced oral deposition when compared with adults. These contradictions hint toward inadequacies of existing predictive models and the inconclusiveness of various data sets. Additional measurements, whether *in vivo* or *in vitro*, are necessary to demonstrate the trend of deposition with age and to determine the parameters controlling ET deposition. Once collected, the data set should be used to obtain a predictive model for the filtering efficiency of the ET region during inhalation and exhalation through the oral and nasal routes as a function of parameters affecting deposition.

5.8.3 Oronasal Distribution of Ventilation

Breathing route is an important determinant of deposited dose of inhaled particles to the lungs and changes with increased physical activity. People breathing through their noses will augment their nasal ventilation by additional breathing through their mouths (oronasal breathing). Thus, determination of the oronasal switching point is critical in assessing the deposition and fate of inhaled particles in the lungs of children and adults.

The relationship between physical activity and the oronasal switching point has been investigated in a number of studies for adults (Becquemin et al., 1999; Bennett et al., 2003; Chadha et al., 1987; James et al., 1997; Leiter and Baker, 1989; Niinimaa et al., 1980, 1981; Patrick and Sharp, 1969; Saibene et al., 1978; Schultz and Horvath, 1989; Wheatley et al., 1991) The underlying mechanisms for switching from nasal to oronasal breathing are not clear; nasal airway pressure drop and airflow resistance probably play a role as switching allows the delivery of the same airflow to the lung without an increase in work of the lung. Because ET geometry and breathing rates are different between children and adults, the switching point changes with age.

A few studies have attempted to measure oronasal breathing in children in recent years (Becquemin et al., 1999; James et al., 1997; Leiberman et al., 1990). Due to ethical considerations regarding the use of children in *in vivo* studies, the number of children in the studies was very limited. James et al. (1997) found that the switching point from nasal to oronasal breathing was more variable in children during exercise compared with adults. The study was unable to find a relationship between nasal and oral partitioning of airflow with age. Becquemin et al. (1999) found that children breathe more often orally both at rest and during exercise than do adults. The highest oral fractions were also found in the youngest children. Because of the limited number of children studied and also insufficient selection of the number of ventilation rates critical in identifying the precise switching point, these studies did not provide conclusive results. The extent of nasal to oronasal switching remains largely unknown in children, and the underlying mechanisms causing the switching point are not understood. Additional studies are required to shed light on nasal–oronasal airflow partitioning to determine accurately the amount of particles reaching the lung, particularly at higher activity levels.

5.8.4 Lung Geometry of Children

To assess the quantity and distribution of dose to the lungs of children, a complete airway morphometry structure including the conducting airways, transitional airways (respiratory bronchioles), and alveolar airways is required. Asgharian et al. (2004) showed that lung geometry and volume were major determinants in predicting particle deposition in the lungs of children. Information on the lung geometry of children is limited. Inadequate measurement techniques, the tedious nature of morphometry measurements, and the large number of airways in the human lung have contributed to the lack of complete measurements of the lengths, diameters, and branch and gravity angles in a single lung, let alone in several.

Additional morphometric airway measurements are needed to describe the entire LRT of children at different stages of growth. Before setting out to collect additional morphometric measurements, the existing database should be fully utilized. Mortensen and colleagues (1983, 1988) at the UBTL made complete measurements of TB airway lengths, diameters, and branch angles for the first ten generations of the lungs of 11 children aged between 3 months and 21 yr. This database is fairly extensive and can be used to create the TB region of the lungs of children at ages specified in the database. To create individual airway geometries for children of the ages of the subjects in the UBTL database, Ménache et al. (2005) used this data set combined with limited published information on terminal bronchioles, respiratory airways, and alveoli in children as well as the method used by Weibel (1963) to develop his single-path model for the adult lung. Although the development of this lung geometry is an improvement over previous geometry models that used other morphometric data to create typical-path models of the lung geometry, more realistic data are needed if the interest is in local or site-specific deposition predictions. Morphometric measurements indicate that human lung geometry is asymmetric. The asymmetric feature of the lung influences the way airflow and particles are transported to various locations, resulting in a heterogeneous distribution in the lung. As a result, particle deposition at different locations in the lung but at the same distance from the trachea may vary substantially. The abundance of airway measurements in the Utah data set should allow for statistical analyses of the available data to obtain distribution functions describing various airway parameters similar to what Koblinger and Hofmann (1985) did for adults. Stochastic lungs can be generated as a function of age in the same way as for adults (Asgharian et al., 2001). A number of other issues such as retrograde alveolarization are not currently well understood and need to be further investigated. There is also a need for further investigation into airway growth in the alveolar region as little information is available on alveolar airways during the first few years of life when airways increase in size and number.

5.8.5 Lung Deposition Measurements

As mentioned before, few studies focused on particle deposition in the lungs of children. Clearly, additional studies are necessary. Ideally, studies that measure deposition in the ET and lung region should be combined because regional deposition fractions of particles are related, and also experiments in children should, from an ethical perspective, be kept to a minimum. The database for lung deposition suffers from the same shortcomings as that for the ET region. To date, the number of particle sizes studied in the fine and coarse ranges are very limited. The number of data points is not adequate to establish a functional relationship between deposition and deposition parameters (sedimentation, impaction, and diffusion). Also, data must be collected on the deposition of ultrafine particles in the lungs of children at different ages.

Lung deposition data suffer an additional shortcoming compared with data for the ET region. Because of indirect measurements of deposition as particles enter and exit the lung, measuring regional and lobar deposition of particles is not possible. Thus, only total lung deposition measurements can be obtained. Information on lung deposition is useful but hardly sufficient to study disease outcome as lung diseases are probably initiated locally at sites with the largest accumulated doses. At a minimum, studies should be designed to allow regional deposition measurements. This means that measurements of particle deposition in the lung must be made. New techniques should be explored for *in vivo* measurements. These techniques should be employed to measure particle deposition in finer detail such as in a region, lobe, or, ideally, a specific airway of the lung.

5.8.6 Lung Deposition Modeling

Currently, limited lung dosimetry models exist to predict airway-specific deposition in children, primarily because of the lack of detailed geometry models. The available models are either compartmental or typical-path and are based on idealized geometry of the lung obtained from limited airway measurements in children's lungs (Hofmann, 1982a; Phalen et al., 1985). Such models are useful for regional prediction of deposition. These models have recently been improved. Asgharian et al. (2004) developed a lobar-specific symmetric model capable of predicting regional deposition within each lung lobe. The model uses the lung geometry of Ménache et al. (2005) described previously.

Because biological responses are initiated at a specific site as opposed to an entire region of the lung, the precise dose, not the average regional dose (as predicted by typical-path models), at the sites of action is required. The development of asymmetric lung geometry models proposed as a future research goal calls for an improved deposition model that is capable of taking advantage of the geometry to calculate detailed transport and deposition of particles in the lung. Such models would be extremely powerful for assessing potential risks to children from exposure to urban aerosols.

5.8.7 Particle Clearance from the Lungs of Children

A number of studies have measured mucous velocity in the trachea of adults using different techniques. The reader can refer to the ICRP (1994) for a compilation of these studies. Such information is missing in children. Knowledge of mucous velocities in children compared with adults would be of great value. Some information is available in animals that can be used to calculate tracheal mucous velocities. Wolff (1992) reported the tracheal mucous velocities in beagle dogs increased to a maximum value in young adults and then declined with age. He argued that age-related changes in canine lung function are most similar to those in humans. In addition, the influence of physical activity, if any, on mucous transport would be useful because children have an activity pattern that is considerably different from that of adults.

Radiolabeled particle exposures in adults have been conducted to study the time course of particle transport and removal in the lung following deposition on airway surfaces (ICRP, 1994). Children are a sensitive subpopulation and thus have not been exposed to radiolabeled particles. Studies in children have included exposure to inert particles followed by measurement of particle concentration at the nasal or oral port of entry on inhalation and exhalation. These concentrations can be used to compute deposition in the lungs of children, but no information regarding clearance can be obtained from these studies. Due to a lack of information on clearance in children, adult clearance models have been extended to children (Hofmann, et al., 1979, 1982a; Crawford, 1982). Currently, there is no way to assess whether the adult model is a good indicator of clearance in children or whether the risk from exposure is underpredicted or overpredicted. Virtually no information exists on particle clearance in the P region of the lungs of children. Thus, there is a need to devise methods and techniques for the clearance measurements of particles in the lungs of children without jeopardizing their safety. A comparison of health risk from particle exposure between children and adults must include the long-term fate of particles in the lung.

ACKNOWLEDGMENTS

This research was supported in part by the Long Range Research Initiative of the American Chemistry Council and the Commission of the European Communities, Contract No. FIGH-CT-2000-00053. The authors would also like to thank Dr. Barbara Kuyper for her editorial assistance in manuscript preparation.

REFERENCES

1. Aitken, R.J., Baldwin, P.E.J., Beaumont, C.G., Kenny, L.C., Maynard, A.D. Aerosol inhalability in low air movement environments. *J Aerosol Sci* 1999, 30(5): 613–626.
2. Albert, R.E., Arnett, L.C. Clearance of radioactive dust from the human lung. *AMA Arch Ind Health* 1955, 12: 99–106.
3. Albert, R.E., Lippmann, M., Briscoe, W. The characteristics of bronchial clearance in humans and the effects of cigarette smoking. *Arch Environ Health* 1969, 18: 738–755.
4. Albert, R.E., Lippmann, M., Peterson, H.T., Jr., Berger, J., Sanborn, K., Bohning, D. Bronchial deposition and clearance of aerosols. *Arch Intern Med* 1973, 131: 115–127.
5. Andersen, I., Lundqvist, G.R., Proctor, D.F. Human nasal mucosal function in a controlled climate. *Arch Environ Health* 1971, 23: 408–420.
6. Asgharian, B., Hofmann, W., Bergmann, R. Particle deposition in a multiple-path model of the human lung. *Aerosol Sci Technol* 2001, 34: 332–339.
7. Asgharian, B., Kelly, J.T., Tewksbury, E.W. Respiratory deposition and inhalability of monodisperse aerosol in Long-Evans rats. *Toxicol Sci* 2003, 71: 104–111.
8. Asgharian, B., Ménache, M.G., Miller, F.J. Modeling age-related particle deposition in humans. *J Aerosol Med* 2004, 17(3): 213–224.
9. Becquemin, M.M., Bertholon, J.F., Bouchiki, A., Malarbet, J.L., Roy, M. Oronasal ventilation partitioning in adults and children: effect on aerosol deposition in airways. *Radiat Prot Dosimetry* 1999, 81(3): 221–228.

10. Becquemin, M.H., Roy, M., Bouchikhi, A., Teillac, A. Deposition of inhaled particles in healthy children. In Hoffman, W., Ed. *Deposition and Clearance of Aerosols in the Human Respiratory Tract.* Vienna, Austria: Facultas, 1987: 22–27.
11. Becquemin, M.H., Swift, D.L., Bouchikhi, A., Roy, M., Teillac, A. Particle deposition and resistance in the noses of adults and children. *Eur Respir J* 1991, 4(6): 694–702.
12. Becquemin, M.H., Yu, C.P., Roy, M., Bouchikhi, A. Total deposition of inhaled particles related to age: comparison with age-dependent model calculations. *Radiat Prot Dosimetry* 1991, 38(1/3): 23–28.
13. Beech, D.J., Sibbons, P.D., Howard, C.V., van Velzen, D. Terminal bronchiolar duct ending number does not increase post-natally in normal infants. *Early Hum Dev* 2000, 59(3): 193–200.
14. Bennett, W.D., Zeman, K.L. Deposition of fine particles in children spontaneously breathing at rest. *Inhal Toxicol* 1998, 10(9): 831–842.
15. Bennett, W.D., Zeman, K.L., Jarabek, A.M. Nasal contribution to breathing with exercise: effect of race and gender. *J Appl Physiol* 2003, 95: 497–503.
16. Bennett, W.D., Zeman, K.L., Kang, C.W., Schechter, M.S. Extrathoracic deposition of inhaled, coarse particles (4.5 μm) in children versus adults. *Ann Occup Hyg* 1997, 41(Suppl. 1): 497–502.
17. Beyer, U., Franke, K., Cyrys, J., Peters, A., Heinrich, J., Wichmann, H.E., Brunekreef, B. Air pollution and respiratory health of children: the PEACE panel study in Hettstedt and Zerbst, Eastern Germany. *Eur Respir Rev* 1998, 8: 61–69.
18. Bosma, J.F. *Anatomy of the Infant Head.* Baltimore:The John Hopkins University Press, 1986.
19. Boyden, E.A. The terminal air sacs and their blood supply in a 37-day infant lung. *Am J Anat* 1965, 116: 413–428.
20. Boyden, E.A. Notes on the development of the lung in infancy and early childhood. *Am J Anat* 1967, 121: 749–761.
21. Brain, J.D., Valberg, P.A. Deposition of aerosol in the respiratory tract. *Am Rev Respir Dis* 1979, 120: 1325–1373.
22. Breyesse, P.N., Swift, D.L. Inhalability of large particles into the human nasal passages: in vivo studies in still air. *Aerosol Sci Technol* 1990, 13: 459–464.
23. Browner, C.M. *Environmental Health Threats to Children.* Washington, D.C.: U.S. Environmental Protection Agency, 1996.
24. Burri, P.H. Development and growth of the human lung. In Fishman, A.P., Sec. Ed., Macklem, P.T., Mead, J., Vol. Eds. *Handbook of Physiology. Section 3: The Respiratory System Vol. 1.* Bethesda, MD: American Physiology Society, 1985: 1–46.
25. Chadha, T.S., Birch, S., Sackner, M.A. Oronasal distribution of ventilation during exercise in normal subjects and patients with asthma and rhinitis. *Chest* 1987, 92: 1037–1041.
26. Cheng, Y.S., Smith, S.M., Heh, H.C., Cheng, K.H., Swift, D.L. Deposition of ultrafine aerosols and thoron progeny in replicas of nasal airways of young children. *Aerosol Sci Technol* 1995, 23(4): 541–552.
27. Cheng, Y.S., Su, Y.F., Yeh, H.C., Swift, D.L. Deposition of thoron progeny in human head airways. *Aerosol Sci Technol* 1993, 18: 359–375.
28. Cheng, Y.S., Yamada, Y., Yeh, H.C., Swift, D.L. Diffusional deposition of ultrafine aerosols in a human nasal cast. *J Aerosol Sci* 1988, 19(6): 741–751.

29. Cohen, B.S. Deposition of ultrafine particles in the human tracheobronchial tree: a determinant of the dose from radon daughters. In Hopke, P.H., Ed. *Radon and Its Decay Products.* Washington, D.C.: American Chemical Society, 1987: 475–486.
30. Cohen, B.S., Sussman, R.G., Lippmann, M. Ultrafine deposition in a human tracheobronchial cast. *Aerosol Sci Technol* 1990, 12: 1082–1091.
31. Conceição, G.M.S., Miraglia, S.G.E., Kishi, H.S., Saldiva, P.H.N., Singer, J.M. Air pollution and child mortality: a time-series study in São Paulo, Brazil. *Environ Health Perspect* 2001, 109(Suppl. 3): 347–350.
32. Crapo, J.D., Chang, Y.L., Miller, F.J., Mercer, R.R. Aspects of respiratory tract structure and function important for dosimetry modeling: Interspecies comparisons. In Gerrity, J.R., Henry, J.C., Eds. *Principle of Route-to-Route Extrapolation for Risk Assessment.* New York: Elsevier, 1990: 15–32.
33. Crawford, D.J. A generalized age-dependent lung model with applications to radiation standards. 1980, ORNL/TM-7454.
34. Crawford, D.J. Identifying critical human sub-populations by age groups: radioactivity and the lung. *Phys Med Biol* 1982, 27: 539–552.
35. Cudmore, R.F., Emery, J.L., Mithal, A. Postnatal growth of bronchi and bronchioles. *Arch Dis Child* 1962, 37: 481–484.
36. Davidson, M.R., Fitz-Gerald, J.M. Transport of O2 along a model pathway through the respiratory region of the lung. *Bull Math Biol* 1974, 36: 275–303.
37. Deitz, D. Stereolithography automates prototyping. *Mech Eng* 1990, 112(2): 34–39.
38. Dunnill, M.S. Postnatal growth of the lung. *Thorax* 1962, 17: 329–333.
39. Dunnill, M.S. The problem of lung growth (editorial). *Thorax* 1982, 37: 561–563.
40. Edwards, D.A., Ben-Jebria, A., Langer, R. Recent advances in pulmonary drug delivery using large porous inhaled particles. (Invited Review) *J Appl Physiol* 1998, 84: 379–385.
41. Edwards, D.A., Hanes, J., Caponetti, G., Hrkach, J.S., Ben-Jebria, A., Eskew, M.L., Mintzes, J., Deaver, D., Lotan, N., Langer, R. Large porous aerosols for pulmonary drug delivery. *Science* 1997, 276: 1868–1871.
42. Engel, S. Form, lage und Lageveranderungen des bronchialbaumes im kindersalter. *Arch Kinderheilkd* 1913, 60: 267–288.
43. Engel, S. *Lung Structure.* Springfield, IL: Thomas, 1962.
44. Erdal, S., Esmen, N.A. Human head model as an aerosol sampler: calculation of aspiration efficiencies for coarse particles using an idealized human head model facing the wind. *J Aerosol Sci* 1995, 26(2): 253–272.
45. Findeisen, W. Über das absetzen kleiner, in der luft suspendierter. Teilchen in der menschlichen lunge bei der atmug. *Pflügers Arch Ges Physiol* 1935, 236: 367–379.
46. Gauderman, W.J., Gilliland, G.F., Vora, H., Avol, E., Stram, D., McConnell, R., Thomas, D., Lurmann, F., Margolis, H.G., Rappaport, E.B., Behane, K., Peters, J.M. Association between air pollution and lung function growth in southern California children: results from a second cohort. *Am J Respir Crit Care Med* 2002, 166: 76–84.
47. Gehr, P. Inhalation pathway in infants and children. In Gerber, G.B., Métivier, H., Smith, H., Eds. *Age-Related Factors in Radionuclide Metabolism and Dosimetry.* (Proceedings of a workshop held in Angers, France, November 26–28, 1986). Dordrecht, The Netherlands: Martinus Nijhoff, 1987: 67–78.

48. Goodman, R.M., Yergin, B.M., Landa, J.F., Golivanaux, M.H., Sackner, M.A. Relationship of smoking history and pulmonary function tests to tracheal mucous velocity in nonsmokers, young smokers, ex-smokers, and patients with chronic bronchitis. *Am Rev Respir Dis* 1978, 117: 205–214.
49. Griscom, N.T., Wohl, M.E. Dimensions of the growing trachea related to body height. *Am Rev Respir Dis* 1985, 131: 840–844.
50. Guilmette, R.A., Cheng, Y.S., Yeh, H.C., Swift, D.L. Deposition of 0.005-12 μm monodisperse particles in a computer-milled, MRI-based nasal airway replica. *Inhal Toxicol* 1994, 6(Suppl. 1): 395–399.
51. Guilmette, R.A., Gagliano, T.J. Construction of a model of human nasal airways using in vivo morphometric data. *Ann Occup Hyg* 1994, 38(Suppl. 1): 69–75.
52. Haefeli-Bleuer, B., Weibel, E.R. Morphometry of the human pulmonary acinus. *Anat Rec* 1988, 220: 401–414.
53. Hahn, I., Scherer, P.W., Mozell, M.M. Velocity profiles measured for airflow through a large-scale model of the human nasal cavity. *J Appl Physiol* 1993, 75: 2273–2287.
54. Harkema, J.R. Epithelial cells of the nasal passages. In Parent, R.A., Ed. *Comparative Biology of the Normal Lung.* Boca Raton, FL: CRC Press, 1992: 27–36.
55. Hatch, G.E. Comparative biochemistry of airway lining fluid. In Parent, R.A., Ed. *Comparative Biology of the Normal Lung.* Boca Raton, FL: CRC Press, 1992: 617–632.
56. Hatch, T.F., Gross, P. *Pulmonary Deposition and Retention of Inhaled Aerosols.* New York: Academic Press, 1964.
57. Hieronymi, G. Uber den Durch das alter bedingten formwandel menschlicher lungen. *Ergeb Allg Pathol Pathol Anat* 1961, 41: 1–62.
58. Hinds, W.C. *Aerosol Technology.* New York: John Wiley & Sons, 1999: 121.
59. Hislop, A.A., Haworth, S.G. Airway size and structure in the normal fetal and infant lung and the effect of premature delivery and artificial ventilation. *Am Rev Respir Dis* 1989, 140: 1717–1726.
60. Hislop, A.A., Muir, D.C.F., Jacobsen, M., Simon, G., Reid, L. Postnatal growth and function of the pre-acinar airways. *Thorax* 1972, 27: 265–274.
61. Hislop, A.A., Reid, L. Development of the acinus in the human lung. *Thorax* 1974, 29: 90–94.
62. Hislop, A.A., Wigglesworth, J.S., Desai, R. Alveolar development in the human fetus and infant. *Early Hum Dev* 1986, 13: 1–11.
63. Hofmann, W. Dose calculations for the respiratory tract from inhaled natural radioactive nuclides as a function of age—II. Basal cell dose distribution and associated lung cancer risk. *Health Phys* 1982a, 43(1): 31–44.
64. Hofmann, W. Mathematical model for the postnatal growth of the human lung. *Respir Physiol* 1982b, 49: 115–129.
65. Hofmann, W., Asgharian, B., Winkler-Heil, R. Modeling intersubject variability of particle deposition in human lungs. *J Aerosol Sci* 2002, 33(2): 219–235.
66. Hofmann, W., Martonen, T.B., Grahm, R.C. Predicted deposition of nonhygroscopic aerosols in the human lung as a function of subject age. *J Aerosol Med* 1989, 2(1): 49–68.
67. Hofmann, W., Steinhäusler, F., Pohl, E. Dose calculations for the respiratory tract from inhaled natural radioactive nuclides as a function of age — I. Compartmental deposition, retention and resulting dose. *Health Phys* 1979, 37: 517–532.

68. Hogg, J.C., Williams, J., Richardson, J.B., Macklem, P.T., Thurlbeck, W.M. Age as a factor in the distribution of lower airway conductance and in the pathologic anatomy of obstructive lung disease. *N Engl J Med* 1970, 282: 1283–1287.
69. Horsfield, K., Cumming, G. Morphology of the bronchial tree in man. *J Appl Physiol* 1986, 24: 373–383.
70. Horsfield, K., Dart, G., Olson, D.E., Filley, G.F., Cumming, G. Models of the human bronchial tree. *J Appl Physiol* 1971, 31: 207–217.
71. Horsfield, K., Gordon, W.I., Kemp, W., Phillips, S. Growth of the bronchial tree in man. *Thorax* 1987, 42: 383–388.
72. Hsu, D.J., Swift, D. The measurements of human inhalability of ultralarge aerosols in calm air using mannikins. *J Aerosol Sci* 1999, 30(10): 1331–1343.
73. Huizing, E. Ueber die weite und das wachstum des brochialbaumes. *Z Hals Nasen Ohren Heilkd* 1933, 33: 546–558.
74. Ibald-Mulli, A., Wichmann, H.E., Kreyling, W., Peters, A. Epidemiological evidence on health effects of ultrafine particles. *J Aerosol Med* 2002, 15(2): 189–201.
75. ICRP (International Commission on Radiological Protection). *Task Group on Reference Man*. ICRP Publication 23. Pergamon Press: Oxford, 1975.
76. ICRP (International Commission on Radiological Protection). *Human Respiratory Tract Model for Radiological Protection*. ICRP Publication 66, Pergamon Press: Oxford. *Ann ICRP* 1994, 24: 272.
77. Itoh, H., Smaldone, G.C., Swift, D.L., Wagner, H.N. Mechanisms of aerosol deposition in a nasal model. *J Aerosol Sci* 1985, 16(6): 529–534.
78. Jacobi, W. Die natürliche Strahlenwirkung auf den Atemtrakt. *Biophysics* 1965, 2: 282–300.
79. James, D.S., Lambert, W.E., Mermier, C.M., Stidley, C.A., Chick, T.W., Samet, J.M. Oronasal distribution of breathing at different ages. *Arch Environ Health* 1997, 52: 118–123.
80. Janssens, H.M., de Jongste, J.C., Fokkens, W.J., Robben, S.G., Wouters, K., Tiddens, H.A. The Sophia anatomical infant nose-throat (SAINT) model: a valuable tool to study aerosol deposition in infants. *J Aerosol Med* 2001, 14(4): 433–441.
81. Johnson, J.R., Isles, K.D., Muir, D.C.F. Inertial deposition of particles in human branching airways. *Inhaled Part* 1975, 1: 61–73.
82. Johnson, T.R., Moor, W.M., Jeffries, J.E., Eds. *Children are Different: Developmental Physiology*. Columbus, OH: Ross Laboratories, 1978.
83. Kawakami, Y., Kusaka, H., Nishimura, M., Abe, S. Trachea and lung dimensions in nonsmoking twins: morphological and functional studies. *J Appl Physiol* 1986, 61: 495–499.
84. Kelly, J.T., Asgharian, B., Kimbell, J.S., Wong, B.A. Particle deposition in human nasal airway replicas manufactured by different methods. Part I: Inertial regime aerosols. *Aerosol Sci Technol* 2004a (accepted for publication).
85. Kelly, J.T., Asgharian, B., Kimbell, J.S., Wong, B.A. Particle deposition in nasal airway replicas manufactured by different methods. Part II: Diffusion regime aerosols. *Aerosol Sci Technol* 2004b (accepted for publication).
86. Kennedy, N.J., Hinds, W.C. Inhalability of large solid particles. *J Aerosol Sci* 2002, 33(2): 237–255.

87. Kimbell, J.S., Asgharian, B., Wong, B.A., Segal, R.A., Schroeter, J.D., Southall, J.P., Brace, G., Dickens, C.J., Miller, F.J. Characterization of particle deposition and penetration from current nasal spray devices. Part II—using a computational fluid dynamics (CFD) model of the human nasal passages. *J Aerosol Med* 2003, 6(2): 05.
88. Kimbell, J.S., Greenwood, D.D. Use of stereolithography to construct a three-dimensional physical model of the anterior F344 rat nasal passages. *Toxicologist* 1995, 15: 4–5.
89. Koblinger, L., Hofmann, W. Analysis of human lung morphometric data for stochastic aerosol deposition calculations. *Phys Med Biol* 1985, 30: 541–556.
90. Koblinger, L., Hofmann, W. Monte Carlo modeling of aerosol deposition in human lungs. Part I: simulation of particle transport in a stochastic lung structure. *J Aerosol Sci* 1990, 21(5): 661–674.
91. Landahl, H.D. On the removal of air-borne droplets by the human respiratory tract—I. The lung. *Bull Math Biophys* 1950, 12: 43–56.
92. Langston, C., Kida, K., Reed, M., Thurlbeck, W.M. Human lung growth in late gestation and in the neonate. *Am Rev Respir Dis* 1984, 129: 607–613.
93. Le Souef, P.N. Drug delivery. *Med J Aust* 2002, 177(Suppl. 6): S69–S71.
94. Leiberman, A., Ohki, M., Forte, V., Fraschetti, J., Cole, P. Nose/mouth distribution of respiratory airflow in "mouth-breathing" children. *Acta Otolaryngol* 1990, 109: 454–460.
95. Leiter, J.C., Baker, G.L. Partitioning of ventilation between nose and mouth: the role of nasal resistance. *Am J Orthod Dentofacial Orthop* 1989, 95: 432–438.
96. Lippmann, M., Altshuler, B., Aharaonson, A.E.F., Ben-David, A., Klinberg, M.A., Eds. Regional deposition of aerosols. In *Air Pollution and the Lung*. New York: Halstead Press-Wiley, 1976. pp. 25–48.
97. Martonen, T. On the fate of inhaled particles in the human: a comparison of experimental data with theoretical computations based on a symmetric and asymmetric lung. *Bull Math Biol* 1983, 45: 409–424.
98. Martonen, T.B., Graham, R.C., Hofmann, W. Human subject age activity level: factors addressed in a biomathematical deposition program for extrapolation modeling. *Health Phys* 1989, 57(Suppl. 1): 49–59.
99. Martonen, T., Hofmann, W. Factors to be considered in a dosimetry model for risk assessment of inhaled particles. *Radiat Prot Dosimetry* 1986, 15(4): 225–231.
100. Martonen, T., Zhang, Z. Deposition of sulfade acid aerosols in the developing human lung. *Inhal Toxicol* 1993, 5: 165–187.
101. McClellan, R.O., Miller, F.J. An overview of EPA's proposed revision of the particulate matter standard. *CIIT Activities* 1997, 17: 1–23.
102. Medinsky, M.A., Asgharian, B., Schlosser, P.M. Toxicokinetics: inhalation exposure and absorption of toxicants. In Sipes, G., McQueen, C.A., Gandolfi, A.J., Eds. *Comprehensive Toxicology,* Vol. 1. Oxford: Pergamon Press, 1997.
103. Ménache, M.G., Graham, R.C. Issues that must be addressed when constructing anatomical models of the developing lung. In Crapo, J.D., Smolko, E.D., Miller, F.J., Graham, J.A., Hayes, A.W., Eds. *Extrapolation of Dosimetric Relationships for Inhaled Particles and Gases*. New York: Academic Press, 1992, pp. 257–272.
104. Ménache, M.G., Hofmann, B., Asgharian, B., Miller, F.J. Airway geometry models of children's lungs for use in dosimetry modeling. 2005, submitted to Clinical Anatomy.
105. Ménache, M.G., Miller, F.J., Raabe, O.G. Particle inhalability curves for humans and small laboratory animals. *Ann Occup Hyg* 1995, 39(3): 317–328.

106. Mercer, R.R., Russell, M.L., Crapo, J.D. Mucous lining layers in human and rat airways. *Am Rev Respir Dis* 1992, 145: A335.
107. Mercer, R.R., Russell, M.L., Crapo, J.D. Alveolar septal structure in different species. *Am Physiol Soc* 1994, 477(3): 1060–1066.
108. Miller, F.J. Dosimetry of particles in laboratory animals and humans. In Gardner, D.E., Ed. *Toxicology of the Lung*. London: Taylor and Francis, 1999, pp. 513–555.
109. Miller, F. Dosimetry of particles: critical factors having risk assessment implications. *Inhal Toxicol* 2000, 12(Suppl. 3): 389–395.
110. Montassier, N.N., Karpen-Hayes, K., Hopke, P.K., Swift, D.L. The penetration of ultrafine particles of 218Po through human nasal and oral cast models. *Radiat Prot Dosimetry* 1992, 45(1-4): 665–667.
111. Montgomery, W.M., Vig, P.S., Staab, E.V., Matteson, S.R. Computed tomography: a three-dimensional study of the nasal airway. *Am J Orthod* 1979, 76: 363–375.
112. Morgan, G., Steward, D. Linear airway dimensions in children including those with cleft palate. *Can Anaesth Soc J* 1982, 29: 1–8.
113. Mortensen, J.D., Schaap, R.N., Bagley, B., Stout, L., Young, J.D., Stout, A., Burkart, J.A., Baker, C.D. Final Report: A Study of Age Specific Human Respiratory Morphometry, Technical Report TR 01525-010, University of Utah Research Institute, UBTL Division, 1983.
114. Mortensen, J.D., Stout, L., Bagley, B., Burkart, J.A., Schaap, R.N. Age-related morphometric analysis of human lung casts. In Crapo, J., Miller, F.J., Graham, J., Hayes, A., Eds. *Extrapolation of Dosimetric Relationships for Inhaled Particles and Gases*. New York: Academic Press, 1988, pp. 59–68.
115. Musante, C.J., Martonen, T.B. Computer simulations of particle deposition in the developing lung. *J Air Waste Manag Assoc* 2000, 50: 1426–1432.
116. NCRP (National Council on Radiological Protection Measurements). Deposition, Retention, and Dosimetry of Inhaled Radioactive Substances. NCRP Report 125, Bethesda, MD, 1997.
117. Niinimaa, V., Cole, P., Mintz, S., Shephard, R.J. The switching point from nasal to oronasal breathing. *Respir Physiol* 1980, 42: 61–71.
118. Niinimaa, V., Cole, P., Mintz, S., Shephard, R.J. Oronasal distribution of respiratory flow. *Respir Physiol* 1981, 43: 69–75.
119. Niven, R.W. Delivery of biotherapeutics by inhalation aerosol. *Crit Rev Ther Drug Carrier Syst* 1995; 12(2–3): 151–231.
120. NRC (National Research Council). *Research Priorities for Airborne Particulate Matter. I: Immediate Priorities and a Long-Range Research Portfolio*. Washington, D.C.: National Academy Press, 1998.
121. Oberdörster, G., Gelein, R.M., Ferin, J., Weiss, B. Association of particulate air pollution and acute mortality: involvement of ultrafine particles? *Inhal Toxicol* 1995, 7: 111–124.
122. Ochs, M., Nyengaard, J.R., Jung, A., Knudsen, L., Voigt, M., Wahlers, T., Richter, J., Gundersen, H.J.G. The number of alveoli in the human lung. *Am J Respir Crit Care Med* 2004, 169: 120–124.
123. Ogden, T.L., Birkett, J.L. The human head as a dust sampler. In Walton, W.H., Ed. *Inhaled Particles IV*. Oxford: Pergamon Press. 1977, pp. 93–105.
124. Olson, D.E., Dart, G.A., Filley, G.F. Pressure drop and fluid flow regime of air inspired into the human lung. *J Appl Physiol* 1970, 28(4): 482–494.
125. Overton, J.H., Graham, R.C. Predictions of ozone absorption in human lungs from newborn to adult. *Health Phys* 1989, 57(Suppl. 1): 29–36.

126. Pattle, R.R. The retention of gases and particles in the human nose. In Davis, C.N., Ed. *Inhaled Particles and Vapours.* Oxford: Pergamon Press, 1961, pp. 302–311.
127. Patrick, G.A., Sharp, G.R. Oronasal distribution of inspiratory flow during various activities. *J Physiol* (London) 1969, 71: 546–551.
128. Phalen, R.F., Oldham, M.J., Beaucage, C.B., Crocker, T.T., Mortensen, J.D. Postnatal enlargement of human tracheobronchial airways and implications for particle deposition. *Anat Rec* 1985, 212: 368–380.
129. Phipps, R.J. The airway mucociliary system. In Widdocombe, J.G., Ed. *Respiratory Physiology III.* Baltimore, MD: University Park Press, 1981, pp. 213–260.
130. Pinkerton, K.E., Green, F.H., Saike, C., Vallyathan, V., Plopper, C.G., Gopal, V., Hung, D., Bahne, E.B., Lin, S.S., Ménache, M.G., Schenker, M.B. Distribution of particulate matter and tissue remodeling in the human lung. *Environ Health Perspect* 2000, 108: 1063–1069.
131. Polgar, G., Weng, T.R. The functional development of the respiratory system from the period of gestation to adulthood. *Am Rev Respir Dis* 1979: 120: 625–695.
132. Pope, C.A., Dockery, D.W. Acute health effects of PM_{10} pollution on symptomatic and asymptomatic children. *Am Rev Respir Dis* 1992, 145: 1123–1128.
133. Prahl-Anderson, B., Kowalski, C.J., Heydendael, P.H.J.M. *A Mixed-Longitudinal Interdisciplinary Study of Growth and Development.* New York: Academic Press, 1979.
134. Proctor, D.F., Anderson, I., Eds. *The Nose, Upper Airway Physiology and the Atmospheric Environment.* New York: Elsevier, 1982.
135. Raabe, O.G., Yeh, H.C., Schum, G.M., Phalen, R.F. Tracheobronchial Geometry: Human, Dog, Rat, Hamster. Technical Report LF-53, Lovelace Foundation for Medical Education and Research, 1976.
136. Reid, L. The lung: its growth and remodeling in health and disease. *Am J Roentgenol* 1977, 129: 777–788.
137. Reid, L. Lung growth in health and disease. *Br J Dis Chest* 1984, 78: 113–134.
138. Rudolf, G., Gebhart, J., Heyder, J., Schiller, Ch.F., Stahlhofen, W. An empirical formula describing aerosol deposition in man for any particle size. *J Aerosol Sci* 1986, 17(3): 350–355.
139. Rudolf, G., Köbrich, R., Stahlhofen, W. Modelling and algebraic formulation of regional aerosol deposition in man. *J Aerosol Sci* 1990, 21(Suppl. 1): S403–S406.
140. Saibene, F., Mognoni, P., Lafortuna, C.L., Mostardi, R. Oronasal breathing during exercise. *Pflügers Arch* 1978, 378: 65–69.
141. Samet, J.M., DeMarini, D.M., Malling, H.V. Biomedicine. Do airborne particles induce inheritable mutations? *Science* 2004, 304: 971–972.
142. Santa Cruz, R., Landa, J., Hirsch, J., Sackner, M.A. Tracheal mucous velocity in normal man and patients with obstructive lung disease: effects of terbutaline. *Am Rev Respir Dis* 1974, 109: 458–463.
143. Schiller-Scotland, Ch.F., Hlawa, R., Gebhart, J., Wönne, R., Heyder, J. Total deposition of aerosol particles in the respiratory tract of children during spontaneous and controlled mouth breathing. *J Aerosol Sci* 1992, 23(Suppl. 1): S457–S460.
144. Schlesinger, R.B. Biological disposition of airborne particles: basic principles and application to vehicular emissions. In Watson, A.Y., Bates, R.R., Kennedy, D., Eds. *Air Pollution, the Automobile, and Public Health.* Washington, D.C.: National Academy Press, 1988, pp. 239–298.

145. Schultz, E.L., Horvath, S.M. Control of extrathoracic airway dynamics. *J Appl Physiol* 1989, 66: 2839–2843.
146. Smith, S.J., Bernstein, J.A. Therapeutic uses of lung aerosols. *Lung Biol Health Dis* 1996, 94: 233–269.
147. Somers, C.M., McCarry, B.E., Malek, F., Quinn, J.S. Reduction of particulate air pollution lowers the risk of heritable mutations in mice. *Science* 2004, 304: 1008–1010.
148. Soong, T.T., Nicolaides, P., Yu, C.P., Soong, S.C. A statistical description of the human tracheobronchial tree geometry. *Respir Physiol* 1979, 37161–172.
149. Stahlhofen, W. Human lung clearance following bolus inhalation of radioaerosols. In *Extrapolation of Dosimetric Relationships for Inhaled Particles and Gases.* Washington, D.C.: Academic Press, 1989, pp. 153–166.
150. Stahlhofen, W., Scheuch, G., Baily, M.R. Measurement of the tracheobronchial clearance of particles after aerosol bolus inhalation. In Dodgson, J., McCallum, R.I., Eds. *Inhaled Particles VII (Proceedings, International Symposium on Inhaled Particles Organized by the British Occupational Hygiene Society,* September 16–22, 1991), *Ann. Occup. Hyg.* 1994: 189.
151. Stöber, W., McClellan, R.O., Morrow, P.E. Approaches to modeling disposition of inhaled particles and fibers in the lung, In Gardner, D.E., Ed. *Toxicology of the Lung.* New York: Raven Press, 1993, pp. 527–601.
152. Subramaniam, R.P., Richardson, R.B., Morgan, K.T., Kimbell, J.S., Guilmette, R.A. Computational fluid dynamics simulations of inspiratory airflow in the human nose and nasopharynx. *Inhal Toxicol* 1998, 10: 91–120.
153. Swift, D.L. Age-related scaling for aerosol and vapor deposition in the upper airways of humans. *Health Phys* 1989, 57(Suppl. 1): 293–297.
154. Swift, D.L. Inspiratory inertial deposition of aerosols in human nasal airway replicate casts: implications for the proposed NCRP lung model. *Radiat Prot Dosimetry* 1991, 38(1–3): 29–44.
155. Swift, D.L., Cheng, Y.S., Su, Y.F., Yeh, H.C. Ultrafine aerosol deposition in the human nasal and oral passages. *Ann Occup Hyg* 1994, 38(Suppl. 1): 77–81.
156. Thomas, S.H., Batchelor, S., O'Doherty, M.J. Therapeutic aerosols in children. *Br Med J* 1993, 307: 245–247.
157. Thurlbeck, W.M. Postnatal growth and development of the lung. *Am Rev Respir Dis* 1975, 111, 803–844.
158. Thurlbeck, W.M. Quantitative anatomy of the lung. In Thurlbeck, W.M., Ed. *Pathology of the Lung.* Stuttgart: Thieme, 1988. pp. 51–55.
159. U.S. Environmental Protection Agency. Dosimetry of inhaled particles in the respiratory tract. In *Air Quality for Particulate Matter,* Vol. II, pp. 10–18. Research Triangle Park, NC: National Center for Environmental Assessment, 1996.
160. Utell, M.J., Frampton, M.W. Acute health effects of ambient air pollution: the ultrafine particle hypothesis. *J Aerosol Med* 2000: 13(4), 355–359.
161. van Ree, J.H.L., van Dishoeck, H. Some investigations on nasal ciliary activity. *Pract Otorhinolaryngol (Basel)* 1962, 24: 383–390.
162. Vincent, J.H., Mark, D., Miller, B.G., Armbruster, L., Ogden, T.L. Aerosol inhalability at higher windspeeds. *J Aerosol Sci* 1990, 21(4): 577–586.
163. Vincent, J.H., Armbruster, L. On the quantitative definition of the inhalability of airborne dust. *Ann Occup Hyg* 1981, 24(2): 245–248.
164. Wheatley, J.R., Amis, T.C., Engel, L.A. Oronasal partitioning of ventilation during exercise in humans. *J Appl Physiol* 1991, 71: 546–551.

165. Weibel, E.R. *Morphometry of the Human Lung*. Berlin: Springer-Verlag, 1963.
166. Wolff, R.K. Mucociliary function. In Parent, R.A., Ed. *Comparative Biology of the Normal Lung*. Lewis, Boca Raton, 1992, pp. 659–680.
167. Wood, L.D.H., Prichard, S., Weng, T.R., Kruger, K., Bryan, A.C., Levison, H. Relationship between anatomic dead space and body size in health, asthma, and cystic fibrosis. *Am Rev Respir Dis* 1971, 104: 215–222.
168. Xu, G.B., Yu, C.P. Effects of age on deposition of inhaled aerosols in the human lung. *Aerosol Sci Technol* 1986, 5: 349–357.
169. Yeates, D.B., Aspin, N., Levison, H., Jones, M.T., Bryan, A.C. Mucociliary tracheal transport rates in man. *J Appl Physiol* 1975, 39(3): 487–495.
170. Yeh, H.-C., Harkema, J.R. Gross morphometry of airways. In Gardner, D.L., Crapo, J.D., McClellan, O.R., Eds. *Toxicology of the Lung,* 2nd ed. New York: Raven Press, 1993, pp. 55–79.
171. Yeh, H.C., Schum, G.M. Models of human lung airways and their application to inhaled particle deposition. *Bull Math Biol* 1980, 42: 461–480.
172. Yu, C.P., Diu, C.K. Total and regional deposition of inhaled aerosols in humans. *J Aerosol Sci* 1983, 14(5): 599–609.
173. Yu, C.P., Xu, G.B. Predicted deposition of diesel particles in young humans. *J Aerosol Sci* 1987, 18: 419–429.
174. Yu, C.P., Zhang, L., Becquemin, M.H., Roy, M., Bouchikhi, A. Algebraic modeling of total and regional deposition of inhaled particles in the human lung of various ages. *J Aerosol Sci* 1992, 23(1): 73–79.
175. Zeltner, T.B., Burri, P.H. The postnatal development and growth of the human lung. II. Morphometry. *Respir Physiol* 1987, 67: 269–282.
176. Zeltner, T.B., Caduff, J.H., Gehr, P., Pfenninger, J., Burri, P.H. The postnatal development and growth of the human lung. I. Morphometry. *Respir Physiol* 1987, 67: 247–267.

6

INFLUENCES OF DYNAMICS, KINETICS, AND EXPOSURE ON TOXICITY IN THE LUNG

Karl K. Rozman, William L. Roth, and John Doull

CONTENTS

6.1 INTRODUCTION TO A THEORY OF TOXICOLOGY

Whelan J. (Jack) Hayes, one of our most illustrious colleagues, lamented in his book the lack of accepted theories of toxicology.[1] He compared and contrasted toxicology with physics, which has many theories amenable to exceptionally accurate testing. On the other hand, toxicology has ancient (50-yr-old-or-more) protocols, which were designed at a time when the volume of knowledge on toxicology was but a small fraction of the current level. The novel informational content of standard protocols, be it acute, subchronic, chronic, developmental, reproductive, or, for that matter, inhalation studies, has long ago been exhausted for any type of theoretical consideration. Kinetics is routinely ignored in study design, and even though the asserted goal of toxicologic studies is to characterize the dynamics of toxic effects, it is done in a haphazard, cookbook fashion that is completely unsuitable for epistemology. Not surprisingly, statistics plays a very important role in analyzing experimental data that are largely uninterpretable. The fundamental characteristic of any scientific discipline is to conduct hypothesis testing and thereby accumulate a critical volume of knowledge that an emerging theory then simplifies and makes available to others in a greatly abbreviated manner. Toxicology is still in the developmental stage of hypothesis testing, with the focus of research moving from biochemistry to cell biology to molecular biology to genetics and back in the hope of finding the "magic bullet"—the solutions to contradictory and, to a large extent, uninterpretable data provided by standard protocols. This may not happen. If not for a theory, or at least partial theories, forcing a radical rethinking of the standard protocols, there will be no tangible progress in this fascinating discipline, which, in our view, perhaps holds the key to understanding the molecular basis of life.[2]

6.2 DOSE AND TIME AS FUNDAMENTAL VARIABLES OF TOXICITY

Toxicity (T) is a function of exposure (E), and E is a function of dose (c) and time (t), i.e., $T = f[E(c,t)]$. *Toxicity* is the manifestation of an interaction between molecules constituting some form of life and molecules of exogenous chemicals or physical insults. Consequences of such molecular interactions or physical insults may propagate through causality chains all the way to the manifestation of toxicity at the organismic level. There are two fundamental ways to view this interaction: (1) what an organism does to a chemical and (2) what a chemical does to an organism. Dealing with the first question led to the development of the discipline of pharmacokinetics, which was later incorporated into some toxicity studies, and therefore in that context it is often called *toxicokinetics* (K). The second question was addressed by the discipline of pharmacology through pharmacodynamic experiments, which again in the context of toxicity, would be more properly termed as *toxicodynamics* (*D*). The use of the prefixes *pharmaco* and *toxico* in the context kinetics and dynamics is problematic because both involve value judgements not compatible with the unbiased interpretation of the laws of nature. Therefore, in this text, we will be referring to kinetics and dynamics without any prefixes.

Toxicity may also be described as a function of dynamics, kinetics, and exposure.

$$T = f(D,K,E) \tag{6.1}$$

This functional relationship can be defined mathematically by a simple differential equation using the chain rule as

$$\frac{\mathrm{d}T}{\mathrm{d}E} = \frac{\mathrm{d}T}{\mathrm{d}D}\frac{\mathrm{d}D}{\mathrm{d}K}\frac{\mathrm{d}K}{\mathrm{d}E} \tag{6.2}$$

for stable and metabolized gases. When rate-determining steps or rate-limiting steps do not originate in kinetics but in dynamics, this relationship simplifies to

$$\frac{\mathrm{d}T}{\mathrm{d}E} = \frac{\mathrm{d}T}{\mathrm{d}D}\frac{\mathrm{d}D}{\mathrm{d}E} \tag{6.3}$$

for highly reactive gases. In this case, exposure drives toxicity directly via a dynamic process. Further discussion of these and other limiting cases is given on page 210–212.

A definition of toxicity, according to Rozman and Doull,[3] is as follows: "[toxicity] is the accumulation of injury over short or long periods of time, which renders an organism incapable of functioning within the limits of adaptation or other forms of recovery." This definition implies that toxicity is a function of time and dose. This concept was already recognized by Paracelsus 500 yr ago, although time was not explicitly depicted as a variable of toxicity. A closer scrutiny of the definition of toxicity indicates that the relationship between toxicity, dose, and time is a complex one because kinetic processes are themselves dose and time dependent, i.e., $K = f(c,t)$, as are dynamic processes, i.e., $D = f(c,t)$. It should be noted that the various time dependencies seldom have the same timescale.

Conceptually, kinetics may also be viewed as a function of the shifting of the momentary equilibrium between absorption (*Abs*) and elimination (*El*)

$$K = f(Abs,\ El) \tag{6.4}$$

because it is the ratio between the entry rate (absorption) and exit rate (elimination) that determines the time course of a compound in an organism. A true equilibrium exists only under conditions of kinetic steady state when the rate of absorption equals the rate of elimination, which is the case after the elapse of 6.64 kinetic half-lives for intravenous infusion and continuous inhalation. In the simplest case of an intravenous bolus injection (instantaneous absorption), the time course is determined by the rate of elimination alone for a compound obeying a one-compartment model. Usually, absorption is faster than elimination, making, in most instances processes related to elimination (distribution, biotransformation, and excretion) rate determining or limiting.

By analogy, dynamics may be viewed as a function of a shifting of the momentary equilibrium between injury (*I*) and recovery (*R*)

$$D = f(I,\ R) \tag{6.5}$$

because it is the ratio of injury to recovery that determines the time course of an adverse effect in an organism. A true equilibrium exists only under conditions of dynamic steady state when the rate of recovery is the same as the rate of injury, which is the case after the elapse of 6.64 dynamic half-lives. The simplest case for such an injury would be when an organism recovers from injury in accordance with a one-compartment model. Again, processes related to recovery are usually slower than the rate of injury. Therefore, more often, recovery (adaptation, repair, and reversibility) will be rate determining or limiting.

Most often, compounds in an organism do not behave according to a one-compartment model. The reason for this is that elimination from the systemic circulation can itself be a function of excretion (*Ex*), distribution (*Dist*), and biotransformation (*Bio*).

$$El = f(Ex,\ Dist,\ Bio) \tag{6.6}$$

When any or all of these processes become rate limiting, two- or multi-compartment models are needed.

Again, in analogy to kinetics, recovery (*R*) in a dynamic model may not be a simple function of, for example, reversibility (*Rv*) but could also require repair (*Rp*). In addition, adaptation (*Adp*) may also occur.

$$R = f(Rv,\ Rp,\ Adp) \tag{6.7}$$

In such instances, two- or multicompartment dynamic analyses would be needed to describe the toxicity of a compound that affects any or all of these processes. Absorption and injury can be thought of as being analogous manifestations of kinetics and dynamics, respectively. Absorption is a function of site (S) and mechanism (M), as is injury.

$$Abs = f(S, M) \tag{6.8}$$

$$I = f(S, M) \tag{6.9}$$

This analysis can be continued all the way to the molecular level. It is clear that any rate-determining or rate-limiting steps, originating at the level of molecular interactions, will then propagate through causality chains to the levels depicted in Figure 6.1, which is a schematic illustration of this concept.

Each of the processes depicted in Figure 6.1 may be dose and time dependent, although past experiments often failed to demonstrate this because they were conducted with preponderant emphasis on one factor or the other, e.g., dynamics was mainly studied as a function of dose, and kinetics mainly as a function of time.

Time has always been an important factor in designing toxicological experiments; yet, time has been afforded very little attention as an explicit variable of toxicity. It is interesting that after Warren[4] was severely criticized by Ostwald and Dernoscheck[5] for his analogy of $c \times t = k$ to $P \times V = k$ of ideal gases, the entire issue was forgotten. Even though $c \times t = k$ kept surfacing repeatedly in many studies, an analogy to thermodynamics was not contemplated again (at least not to our knowledge).[6–9] When Rozman[10] "rediscovered" the $c \times t = k$ concept in another context (delayed acute oral toxicity), reevaluation of the role of time in toxicology in both a historical context and as an independent variable was required.

Ostwald and Dernoscheck's[5] analogy of toxicity to an adsorption isotherm is problematic because adsorption entails processes that are far from ideal conditions. Much more reasonable is Warren's[4] analogy to $P \times V = k$ for ideal gases as a comparison for ideal conditions in toxicology. Reducing the volume of a chamber containing a given number of molecules or atoms of an ideal gas will decrease the time for any given molecule or atom to collide with the walls of the chamber, leading to increased pressure, which is simply an attribute of the increased number of molecules per unit volume, which is concentration. Thus, $c \times t = k$ and $P \times V = k$ are compatible with each other mechanistically. Ostwald and Dernoscheck's[5] comparison of toxicity to an adsorption isotherm is much closer to the real-life situation of toxicology in which the most frequent finding is that $c \times t^x = k$ or $c^y \times t = k$.

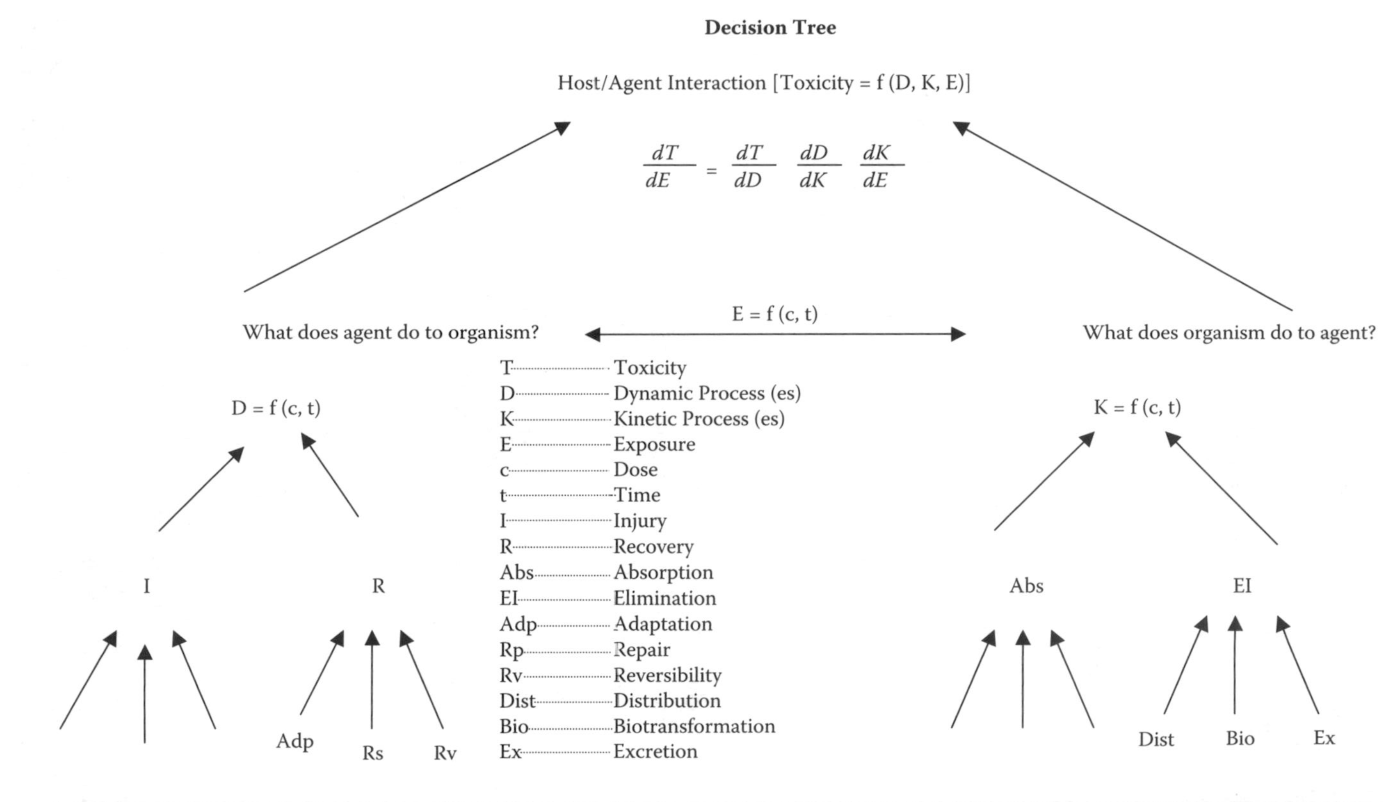

Figure 6.1 Conceptual outline of the decision tree approach.

A controversy still exists among toxicologists and statisticians as to where the exponent belongs, even though the equivalency of the two statements was already established in 1940 by Bliss.[11]

These considerations and some discussions led to the recognition that toxicologists and thermodynamicists worked in opposite ways. Instead of starting out with the simplest model (an ideal gas in thermodynamics corresponds to ideal conditions in toxicology experiments) and building into it step by step the increasing complexity of the real world, toxicologists tried to extrapolate conclusions from one complex situation to another. In addition, time was largely ignored, although it is one of the two fundamental variables of toxicity.[12] It is unlikely that a better understanding of biological processes at the molecular level alone will lead to improved risk predictions in toxicology as long as the experimental designs of toxicological studies provide the wrong reference points for departure from ideal to real conditions. For example, the standard inhalation toxicity protocols (6 h/5 d/week) cannot yield $c \times t = k$ because after 6 h of intoxication, there are up to 18 h of recovery, and on weekends there are up to 66 h of recovery. This would require at least two additional functions to correct for the departure from steady state. A real-life situation is even more complex because departures from the ideal condition (steady state) are highly irregular. Nevertheless, it is reasonable to expect that risk predictions will be possible for even the most irregular exposure scenarios once the reference points are established as dose and time responses under ideal conditions (dynamic or kinetic steady state).

In 25 years of studying the toxicity of TCDD and related compounds, the concept of $c \times t = k$ did not emerge in any other experimental context except in several recent subchronic or chronic studies, which were conducted under conditions of kinetic steady state[13–15] or under conditions of geometrically well-defined (monotonic) departure from it as shown by Table 6.6.[16] Nevertheless, a general interest in the role of time in toxicology pervaded our thinking for many years.[3,12,13,17] Most toxicologists are familiar with Haber's rule of inhalation toxicology and its applicability to war gases and some solvents.[6,13] Much less attention has been given to Druckrey's work, which extended the $c \times t$ concept to lifetime cancer studies using oral rather than inhalation exposure. Also, there is very little cross-referencing of the $c \times t = k$ data generated by entomologists[11,19,20] and those established by toxicologists. History demonstrates that a fundamental relationship in science keeps reappearing in different contexts, as is the case with $c \times t = k$. During this 25-yr period, many apparent exceptions with no satisfactory explanation were found. Attempts at generalization usually fail until a commonality is detected among all experiments, as is this case among those that yielded $c \times t = k$. This commonality is a kinetic steady state or some degree of irreversibility of an effect, which leads to a dynamic steady state (see Equation 6.3), which of course can be driven by kinetics as shown by Equation 6.2. Anesthesia, similar to intravenous infusion, leads to rapid and sustained steady state for compounds of short kinetic half-life. Most anesthetics and solvents do have short kinetic half-lives and many obey Haber's rule, except when measurements are taken while an adaptive process such as induction of a protein is underway. Druckrey[38] and the ED_{01}-Study[8] used feeding as a route for exposure, which yields a better steady state for compounds of intermediate half-life than does gavage, for example. However, the exponent x in Druckrey's general formula increases above 1 rapidly as the half-life of compounds becomes shorter because there is intermittent recovery between bouts of feeding. Most of the entomology studies were related to fumigation, which often, but not always, resulted in fairly rapid steady state in the fumigation chamber. Also, HpCDD, which has a half-life of 314 d[14] in female rats, yields virtual steady state for a 70-d observation period after any route of administration, but TCDD, with a half-life of 20 d, does not. However, when TCDD's toxicity was studied under steady-state conditions, its subchronic or chronic toxicity also occurred according to $c \times t = k$.[17] An example of a long dynamic half-life of an effect resulting in a dynamic steady state of the DNA adducts that drives the $c \times t$ product, rather than its kinetics, is represented by 2-acetaminofluocrene-induced bladder or liver cancer.[21]

6.3 DEFINITION OF DOSE AND TIME

Before analyzing dose–time relationships further, it is useful to establish clear definitions of these fundamental variables of toxicity. Historically, neither dose nor time has been defined with clarity as variables of kinetics and dynamics. It is customary to use the terms *acute dose* and *acute effect* as if the two were interchangeable. In fact, an acute dose can lead to chronic effects,[22] and multiple doses can trigger a fulminant episode of toxicity.[23] In risk (safety) assessment, it is always the total dose delivered that is of concern, although in therapeutics the daily dose rate is often referred to simply as the *dose*. Therefore, a useful definition of dose in toxicology would be

$$Dose = \sum_{n=1}^{n} Dose\ Rates$$

According to this definition, a single acute dose would represent the limiting case of the dose rate equaling the dose. This definition would be valid for any kind of irregularity in the dosing regimens and is analogous to the definition of dose in radiation biology. Also, this definition would not limit dose rate to a constant daily dose rate; it could be any dose rate per unit of time, such as a drug taken three times a day, inhalation of a volatile agent for 6 or 8 h per day, or for any other fractional exposure times.

Dose and dose rate were traditionally expressed in units of mg/kg or mg/kg/day, although often mole/kg or mole/kg/day were also used, which is certainly more appropriate from a thermodynamic point of view. It was suggested by Rozman et al.[13] that plotting dose responses as log of the number of molecules vs. effect is useful to establish a global perspective for all dose reponses because the intersection of the abscissa and the ordinate is the same for all dose responses on this scale, *viz.*, 1 molecule, the logarithm of which is zero. Waddell[24] demonstrated the usefulness of this method for a large number of carcinogenic dose responses.

Humans have always struggled with the notion of time. It is not possible to predict what influence the concept of toxicological time will have on our perception of time. Suffice it to say at this juncture that it is not possible to think of toxicity without the implicit presence of time as a variable, although time has received only semiquantitative designations (acute, subacute, subchronic, and chronic) in toxicity studies. In fact, one could view organisms as instruments exquisitely sensitive to time. Important to toxicology is the concept that the time course of a toxicant in an organism (kinetics) is very often different from the time course of toxicity (dynamics). The underlying biological processes (absorption, distribution, elimination, injury, adaptation, and recovery) have their own timescales depending on the molecular events behind each process (e.g., enzyme induction or receptor regulation either directly or via gene expression). Thus, in toxicology, the dose is a pure variable, but there are many different processes that occur on different timescales, yielding different $\int cdt$ integrals and leading to complex interactions, which can be described as $c \times t^x$. In spite of this complexity, science can deal with toxicology in a traditional, analytical fashion. Because only the knowledge of rate-limiting steps is required to accurately describe toxicity, this will most often reduce the complexity to manageable proportions.

6.4 DOSE AND TIME RELATIONSHIPS

Toxicity is a function of exposure, and exposure is a function of dose and time, i.e., $T = f[E(c, t)]$. Consequences of interactions between a toxic agent and an organism at the molecular level propagate through dynamic or kinetic–dynamic causality chains all the way to the manifestation of toxicity at the organismic level (Figure 6.1). If the recovery (consisting of adaptation, repair, and reversibility) half-life of an organism is longer than the kinetic half-life of the causative agent in the organism, then dynamics become rate determining (one-compartment model) or rate limiting (multi-compartment model). If the kinetic half-life of the compound is longer than the recovery half-life, then kinetics will be rate determining or limiting, in which case the kinetic AUC (area under the curve) will be identical to the dynamic AUC. There are three limiting conditions for $c \times t = k$ to emerge when the causality chain propagates through either dynamic or kinetic–dynamic processes:

Dynamics

1. In case of no recovery (no reversibility, repair, or adaptation), linear accumulation of injury will occur according to a triangular geometry ($c \times t/2 = k$) after repeated dose rates, or according to a rectangular geometry after a single dose ($c \times t = k$), provided the $c \times t$ lifetime threshold has been exceeded, which occurs when $c_{threshold} \times t_{lifespan} = k$ (lifetime threshold).
2. After dynamic (reversibility, repair, and adaptation) steady state has been reached, injury will occur according to a rectangular geometry ($c \times t = k$) after exceeding the $c \times t$ lifetime threshold.

Kinetics

1. In case of no elimination, linear accumulation of a compound (and as a result injury) will occur according to a triangular geometry ($c \times t/2 = k$) after repeated doses, or according to rectangular geometry after a single dose ($c \times t = k$) above the $c \times t$ lifetime threshold.
2. After kinetic (and, as a consequence, dynamic) steady state has been reached, injury will occur above the $c \times t$ lifetime threshold according to a rectangular geometry ($c \times t = k$).

Exposure Frequency or Duration:

As the kinetic and dynamic half-lives become shorter, the distinction between elimination and recovery half-lives becomes less important, because another time dependence, that of the frequency of exposure, starts dominating.

1. Compounds having very short kinetic or dynamic half-lives will reach steady state rapidly and yield $c \times t = k$ upon continuous exposure, according to a rectangular geometry above the $c \times t$ lifetime threshold, provided adaptation and repair are also at steady state.

2. Other types of geometries can certainly be created by elaborate but regular dosing regimens. These scenarios are less likely to play a practical role in toxicology, although they may be of theoretical interest in the establishment of model parameters for predicting toxicity after irregular dosing regimens.

It should be kept in mind that the mathematics of first-order processes, when appropriate, are valid for bimolecular reactions (e.g., receptor binding), which result in the propagation of the causality chain to the level of observation (Figure 6.1). Therefore, 90% of a dynamic steady state will not be reached after 3.32 recovery half-lives have elapsed. Thus, Haber's rule will be obeyed only if the observation period is greater than about four recovery half-lives or if recovery is a zero-order process.

Thus, the various $c \times t = k$ scenarios represent limiting conditions. The magnitude of the $c \times t$ product is a function of the potency of the compound, the susceptibility of the organism, and the deviation from ideal conditions, and $c \times t^x = k$ is obtained for nonlimiting conditions. It should be recognized that the dose (c) does not have inherent exponential properties, but time (t) might have such properties, because under non-ideal conditions toxicity is a function of at least two independent timescales, one being the half-life of the rate-determining step (dynamic or dynamic–kinetic) of the intoxication (an intrinsic property of a compound or organism), the other being the frequency (which includes duration) of exposure, which is independent of both the compound and the organism.

In conclusion, these data and considerations, based on a significant body of evidence accumulated over the last 100 yr, suggests that $c \times t = k$ is part of a fundamental law of toxicology, and possibly of biology in general, which can be seen only under ideal conditions. If confirmed in other classes of compounds and the herein-described ideal conditions, then Paracelsus' famous statement might have to be supplemented to read "Dosis et tempus fiunt (faciunt) venenum" (Dose and time together make the poison). Implications for risk assessment are that the margin of exposure (MOE) must be defined in terms of both dose and time. This can be done by relating the real-life (discontinuous) exposure scenario to that of ideal (continuous) exposure condition:

$$MOE = \frac{c \times t^x}{c \times t}$$

The margin of safety and its reciprocal, the margin of risk, can be determined when MOE exceeds the $c \times t$ lifetime threshold.

Figure 6.1 could also be used as a decision tree to identify critical steps needed for modeling to predict toxicity. It is important to note that dynamic steady state is not a frequently seen phenomenon in toxicology, although it can be seen any time when the observation period is much shorter than the recovery half-life. In real-life situations, there are usually at least two or three rate-limiting steps in kinetics and possibly as many in dynamics. It must be emphasized, though, that multiple kinetic compartmental models do not necessarily require multiple dynamic models, and *vice versa*. Therefore, a practical approach would be to conduct experiments at dynamic steady state (which would require a preexisting kinetic steady state in many instances) as a point of reference clearly defined by $c \times t = k$. Then, experiments would need to be carried out for different compounds with different half-lives to establish model parameters, which describe departures from a kinetic or dynamic steady state of increasing frequency and irregularity.

Current regulatory practices of automatically including uncertainty factors for reasons of both kinetic and dynamic inter- and intraspecies differences are not consistent with these conceptual considerations because there can be no uncertainty associated with non-rate-determining (limiting) steps. In very rare instances, when there are several different rate-limiting processes occurring on different timescales in kinetics and several different rate-limiting processes taking place on several different overlapping timescales in dynamics, the scenario would represent a formidable computational task for a theoretical treatise on risk assessment, including the application of uncertainty factors to account for both kinetic and dynamic differences.

In summary, $c \times t = k$ represents the most efficient (a kind of worst-case) exposure scenario for producing an effect, namely, continuous exposure until manifestation of an effect. Experimentally, this condition is often met by continuous inhalation exposure of compounds with short half-lives or daily oral administration of chemicals that have dynamic or kinetic half-lives of a few days or longer. It must be emphasized that any departure from the worst-case scenario will result in the change of $c \times t = k$ to $c \times t^x = k (x > 1)$. Departures are represented by regular or irregular interruptions of exposure or intermittent recovery from injury. The larger the departure, the larger will x, be indicating that increasing x is equivalent to decreasing toxicity. This is entirely logical; increasing interruptions of exposure or injury will result in longer periods of time needed to cause equivalent toxicity compared with that of continuous exposure, because of increasing intermittent recovery. To express this more clearly, we can write

$$c \times t^x = k \quad \text{or} \quad c \times t \times t^{x-1} = k$$

Thus, t^{x-1} is a simple transforming factor that changes the slope of the log c vs. log t plot back to unity. It may be viewed as the toxicological timescale of recovery that runs counter to the timescale of toxicity, thereby reducing it.

A limiting condition for first-order processes will be reached when exposure occurs outside of 6.64 kinetic or dynamic half-lives, because 99% of the elimination or recovery will have occurred at that time. Under such conditions (which are closest to real-life situations for most compounds), toxicity will be less dependent on dose and kinetic or dynamic time and mainly the frequency of exposure will determine x. If x is then determined experimentally, for, say, 1, 2, 4, 8, 16 and 32 d for a compound with a kinetic or dynamic half-life less than 3.6 h after continuous or intermittent exposure under isoeffective conditions, then plotting of this data will allow extrapolation to any exposure scenario outside of 6.64 half-lives (which, in this case, corresponds to 1 d). Most dietary constituents fall in this category. For zero-order processes, two half-lives are needed for elimination or recovery. It should be kept in mind that the half-life of zero-order processes (unlike that of first-order processes) is concentration dependent.

A recent series of articles explored how other disciplines deal with complex systems.[25–28] Goldenfeld and Kandanoff[25] made some important observations that are relevant for toxicology. Simple laws of physics give rise to enormous complexity when the number of actors is very large. We have the same paradox in toxicology in that the $c \times t$ concept is very simple, but the real-world manifestation of toxicity is very complicated because of the trillions of cells in a complex organism. Another observation that is equally relevant is "Use the right level of description to catch the phenomena of interest. Don't model bulldozers with quarks."[25] The decision tree approach shown in Figure 6.1 was developed to aid toxicologists and modelers to identify the appropriate phenomena and the right level of modeling. Toxicologists can avoid much of the unnecessary experimentation by using this top-to-bottom approach rather than the bottom-to-top approach that is in use today.

6.5 ANALOGY TO THERMODYNAMICS

In physics, Boyle's law of ideal gases gave rise to thermodynamics, and molecular or mechanistic considerations led to a theory of gas reactions. The former is based on using the minimum number of fundamental variables that can describe the simplest possible dynamic systems ($P \times V = k$ for ideal gases). The latter requires a great deal of knowledge about the mechanism of chemical reactions (wall reaction, activation energy, etc.). Both these approaches have been attempted in toxicology with as-yet-limited success, as we shall see in subsequent discussions. The reason for the lack of advance in theoretical toxicology is probably the fact that unlike thermodynamicists, toxicologists did not start out by defining the simplest possible toxicological conditions with a minimum number of variables as a point of departure toward more complexity, although experiments were coincidentally conducted under such ideal conditions and, in every such instance, Haber's rule proved to be applicable (e.g., Gardner et al.[29]) even though experimenters may have failed to notice it.[30]

The lack of conceptualization of the three variables of toxicity resulted in arbitrary study designs, which further eroded the predictability from one experiment to another. A thinking analogous to that in thermodynamics might help to alleviate this problem by optimizing study design and eventually by building a theory of toxicology. Thermodynamics, as with toxicology, has three fundamental variables i.e., in thermodynamics *P*, *V*, and *T* and *c*, *t*, and *W* in toxicology).

W (*Wirkung* in German) will be used to represent effect because *E* is used to represent exposure, elimination, and excretion in English. Before the development of a comprehensive theory of thermodynamics, it was clear to scientists that to study the relationship between an independent and a dependent variable, a third variable, or other variables, had to be kept constant. This was not done in toxicology, although most dose–response studies were conducted at constant time (isotemporal). However, to study the relationship between time and effect, the dose needs to be kept constant (isodosic) and to examine the relationship between dose and time, the effect must be kept constant (isoeffective). The $c \times t$ product will not emerge from the equation of ergodynamics (Figure 6.2) until after elucidation of the relationship between specific effect at constant time and specific effect at constant dose. In other words, we must learn more about *k*

Ergodynamics
(Wirkungslehre)

$$dW = \left(\frac{\partial W}{\partial c}\right)_t dc + \left(\frac{\partial W}{\partial t}\right)_c dt$$

dW = 0 isoeffective

$$-\left(\frac{\partial W}{\partial c}\right)_t dc = \left(\frac{\partial W}{\partial t}\right)_c dt$$

dt = 0 isotemporal

$$dW = \left(\frac{\partial W}{\partial c}\right)_t dc$$

dc = 0 isodosic

$$dW = \left(\frac{\partial W}{\partial t}\right)_c dt$$

Figure 6.2 Conceptualization of ergodynamics (*Wirkungslehre*) at constant action (*Wirkung*), time, and dose.

before significant theoretical advance is possible. As mentioned before, most experiments (14-d, 90-d, and 104-week studies) were conducted in the past under isotemporal conditions, which is appropriate for dose–response studies. However, an arbitrary choice of these time points and the inexactitude of diagnosis led to a great deal of confusion, for example, in the 14-d studies because different dose responses (meaning different mechanisms) were often lumped together. Experiments in toxicology have frequently been conducted under isoeffective conditions, with the endpoint usually being 100% of an effect (mortality, or cancer). However, systematic investigation of $c \times t = k$ has not been carried out, for example, at 20 or 80% of an effect. Finally, few experiments were conducted under isodosic conditions, because this requires that the concentration be kept constant at the site of action. The only experimental condition other than aquatic and *in vitro* toxicity is continuous inhalation exposure that keeps the concentration at the site of action constant. For example, Gardner et al.[29] have reported such data after continuous inhalation exposure of experimental animals to benzene and SO_2 when the end point (chronaxy, leukopenia) in question was measured immediately after termination of exposure. However, when the end point of measurement was not immediately done (streptococcal-infection-related mortality) after cessation of NO_2 exposure, the time–response curve started flattening out.[31] A systematic investigation of these issues has been carried out for HpCDD after oral administration with delayed acute toxicity as the end point of measurement.[10] More recently, these experiments have been extended to chronic toxicities, including cancer.[16] These data provide support to the suggestion by Rozman et al.[13] that dose–time–response may be viewed as a three-dimensional surface area similar to, but conceptually distinctly different from, the traditional model proposed by Hartung.[32] Experiments conducted under isoeffective conditions (planes parallel to the dose–time plane) yield hyperbolas that represent Haber's rule $c \times t = k$. Studies carried out under isotemporal conditions (planes parallel to the time–effect plane) yield S-shaped dose–response curves along with $c \times t = k \times W$, whereas isodosic investigations (planes parallel to the dose–effect plane) produce S-shaped time–response curves along with $c \times t = k \times W$. Indeed, plotting of the $c \times t$ product vs. W (effect) for HpCDD doses causing about 10 to 90% wasting or hemorrhage yielded a straight line of high correlation ($r^2 = 0.96$).[10]

This may form the core of a theory of toxicology analogous to $P \times V = k$ corresponding to isotherms and $P \times V = k \times T$ corresponding to isobars or isochors in thermodynamics. In thermodynamics $k = n \times R$, but in toxicology such a relation has not been found yet. In analogy to thermodynamics, it is possible that the toxiological k also consists of a universal constant multiplied by the molar potency of each toxicant. Moreover, we do not have convenient and exact measurements of

Ergodynamics

c × t = k × Effect (wirkung)

W............Effect (action/ wirkung)
c............Dose
t............Time

Thermodynamics

P × V = n × R × T

P............Pressure
V............Volume
T............Temperature
n............Number of moles
R............Universal gas constant

Physics

Action (wirkung) = Energy × Time

Toxicology

Effect (wirkung) = Dose × Time

Figure 6.3 Analogy between ergodynamics and thermodynamics. (From *Handbook of Pesticide Toxicology,* Academic Press, 2001. With permission.)

effects in toxicology, as provided by temperature readings in physics. The dimension of $P \times V$ is energy, whereas the dimension of $c \times t$ is energy × time, i.e., action, which is called *effect* in toxicology (Figure 6.3).

One example in which a correlation between the thermodynamic properties of gases and toxicological properties has been worked out in some detail is the relationship of gas solubility and reactivity to the extent of respiratory tract absorption and consequential effects, which may be local or systemic.

Gases can be broadly classified into four categories on the basis of solubility and reactivity (Table 6.1 adapted from Dahl[33]). Reactivity classes are defined as follows:

1. *Stable molecules* ($\Delta F > 3$ kcal/mol): These are defined as chemical compounds or elements whose free energy [$\Delta F = -RT\ln(K)$] of a chemical reaction that occurs in a physiological system ($A + B \leftrightarrow C + D$; $K = [C][D]/[A][B]$) is greater than 3 kcal/mol. At thermodynamic equilibrium of the chemical compound with its reaction products, $K = 0.0077$. Hence, less than 1% of the total concentration of a stable gas will be in the form of products.
2. *Reactive gases* ($\Delta F < -3$ kcal/mol; $t_{1/2} < 10$ min): These are gases that can react with physiological components rapidly without catalysts. They will probably not attain substantial systemic concentrations

Table 6.1 Thermodynamic–Toxicologic Property Correlations for Gases

Class	*Solubility*	*Reactivity (F)*	*Expected Absorption or Effects*
I	High	Reactive	Nasal and tracheobronchial absorption or reaction ~100%.
II	Low	Metabolizable	Slow absorption, delayed systemic effects
III	High	Metabolizable	Rapid absorption, rapid systemic effects
IV	Low	Stable	Limited absorption, physical effects only (anesthesia)

Note: Reactive: $\Delta F < -3$ kcal/mol, $t_{1/2} < 10$ min, metabolizable: $\Delta F \geq -3$ kcal/mol, $t_{1/2} > 10$ min, stable: $\Delta F > 3$ kcal/mol.

relative to their concentration at the portal of entry (POE), i.e., the nasal epithelia, trachea, and lungs. At thermodynamic equilibrium, less than 1% ($K = 130$) of the reactive gas will be in the form of the parent compound.

3. *Metabolizable gases* ($\Delta F \leq -3$ kcal/mol; $t_{1/2} > 10$ min): These are gases whose free energy of reaction is favorable, but they are stable in the absence of physiological catalysts. Metabolism may lead to the formation of products that damage normal physiological cycles.

If a gas is stable, only physical or reversible effects such as asphyxiation or anesthesia are expected. Reactive gases (methylphosphonic difluoride, formaldehyde, etc.) that are quickly converted into nonvolatile products in the mucus have primarily local effects at the POE. Rapidly metabolizable gases (such as volatile epoxides) can become systemic toxicants. Slowly metabolizable gases may be eliminated (primarily exhaled) largely unchanged, or they may accumulate metabolites that are potential toxicants. In the latter case, it is the dose of the metabolite, and not the parent compound, that is relevant to any toxic effects.

Dose–response relationships are as follows:

Gas concentration: Effects such as anesthesia may be related directly to the concentration [A] of stable gases.

Amount inhaled: For reactive gases, effects at the POE may be related to the inhaled dose because equilibrium favors quantitative conversion to products at the POE (inhaled dose = $V_{minute} \int [A] dt$).

Amount retained: Systemic effects may be observed and correlated with the amount of gas retained in tissues (not rapidly exhaled) or converted to metabolites (a full set of differential equations must be solved for [A], [B], [C], and [D]).

The action (effect) of a chemical may also be defined as

$$A = A_{NS} + A_S \quad \text{where} \quad A_S = \int c \times \mathrm{d}t$$

The total action (A) consists of a nonspecific action (A_{NS}) and a specific action (A_S). The nonspecific action is comparable to heat in thermodynamics that dissipates without being converted to work by an expanding gas. A chemical may have several specific dynamic actions such as enzyme induction, porphyria, or liver cancer. At the same time, the organism may have specific or nonspecific actions on the chemical. The portion of a dose that is not converted to the effect of interest must be viewed as nonspecific action with regard to this particular effect. For example, a chemical that is rapidly converted to a metabolite that is much less toxic and eliminated will have very little specific action. In case of metabolic activation, only the portion of the dose that is converted to a more toxic metabolite will constitute the specific action, and the remaining portion must be viewed as nonspecific action. For example, an above-threshold dose of TCDD will be very efficiently converted to specific toxic action because very little biotransformation takes place and because kinetics drive the toxicity of TCDD. Dynamics (binding to DNA) is the driving force for the toxicity of nitrosamines, and they require metabolic activation, which is just one of several possible metabolic pathways for this class of compounds. Therefore, nitrosamines are less efficiently converted into toxic action than is TCDD.

Substituting k/t for c or k/c for t and integrating $c\mathrm{d}t$ between c_1 and c_2 or t_1 and t_2 yields another logarithmic form of Haber's rule for isoeffective conditions:

$$\ln \frac{c_2}{c_1} = \ln \frac{t_1}{t_2} \tag{6.11}$$

An analogy to this relation exists in thermodynamics, i.e., the work (A) performed by the isothermic and reversible compression of an ideal gas:

$$A = n\mathrm{R}T \int_{V_1}^{V_2} \frac{\mathrm{d}V}{V} = n\mathrm{R}T \ln \frac{V_2}{V_1} = n\mathrm{R}T \ln \frac{P_1}{P_2} \tag{6.12}$$

$$\ln \frac{V_2}{V_1} = \ln \frac{P_1}{P_2}$$

6.6 DISCRIMINATION AMONG AGENTS WHOSE EFFECTS ARE DRIVEN BY DYNAMICS, KINETICS, AND EXPOSURE

In a standard textbook of toxicology[34] authored by Casarett et al., Table 4.3, titled "EPA/FIFRA Requirements for Hazard Evaluation of Pesticides," lists 15 requirements, beginning with acute studies and ending with kinetics. This is wrong, and is the reason why standard protocols do not give answers that are conducive to further the development of a theory of toxicology. At a time when the standard protocols were developed, the first textbook on pharmacokinetics was just published.[35] Pharmacokinetics became fully conceptualized in the next 20 yr.[36] However, for reasons the analyses of which are beyond the scope of this chapter, toxicologists paid little attention to this important scientific development, paying lip service to its existence in standard documents, almost never using it to justify standard settings, and completely ignoring it in the by-then-outdated standard protocols.

A theory of toxicology as outlined in the first part of this chapter would require a dramatic change in the previously mentioned table. In fact, kinetics should be the first entry in the table, that is, the very first study for hazards evaluation should be an IV and oral kinetic experiment establishing bioavailability and the kinetic half-life of any as-yet-unstudied chemical entity. Classical compartmental analysis or PB-PK modeling should accompany this effort. Any further study design would depend on the outcome of this first, and for decision making, most important study. There are three limiting conditions to be considered for rate determination (limitation), as discussed in the following subsections.

6.6.1 Very Long Kinetic Half-Life

If the kinetic half-life of a compound is very long, then the rate-determining steps (one-compartment models) or rate-limiting steps (multicompartment models) occurring on overlapping timescales will arise as a result of kinetics, and the dynamics of the toxic effect will be driven entirely by kinetics. Determination of the half-life of a chemical will allow one to calculate loading dose rates and maintenance dose rates (both oral and IV) for acute experiments so that experiments can be conducted at kinetic steady state. In companion studies, the acute toxicity (IV, oral, dermal, or inhalation) must be determined under conditions of kinetic steady state as well as after a single-dose administration. This provides critical information about reversibility or irreversibility of the effects driven either by kinetic or dynamic recovery, because at kinetic steady state all variables except the toxic effect are kept constant and, therefore, this is the only true and unequivocal measure of toxicity. If steady state is not maintained as is usually the case after administration of a single dose, recovery will occur at the same time that toxicity develops and the combination of two variables will be measured. If the half-life of a chemical is not known, such a single-dose study will be uninterpretable. At the same time, this is the major cause of large interlaboratory variations in the determination of dose responses for acute toxicity.

For compounds of very long kinetic half-life, there will be very little, if any, difference in the dose responses between various dosing regimens if the observation period is short compared to the half-life of a compound because a single dose creates a pseudo–steady state for, say, 10 to 20% of one half-life, which can be quite prolonged (50 d and longer).

6.6.2 Very Long Dynamic Half-Life

If the dynamic half-life turns out to be very long, with a short kinetic half-life, another set of considerations takes precedence. After a single dose, elimination of the chemical will be rapid, and slow recovery will dominate the dose dependence. There will also be little, if any, difference in toxicity between single-dose administrations and multiple-dose-rate administrations (kinetic steady state). This will become apparent from a comparison of the acute studies under conditions of kinetics steady state and single-dose administration.

6.6.3 Short Kinetic–Dynamic Half-Life

With these experiments, the rate-determining (limiting) steps have been reduced to the third and final possibility—when both the kinetic and dynamic half-lives are short. In this case, kinetic steady state will be reached rapidly if exposure is uninterrupted, and rapid recovery will dominate the single-dose exposure.

A comparison of the slopes of the dose–response experiments under conditions of dynamic and kinetic–dynamic steady state provides a quantitative estimate for kinetic or dynamic recovery. All slopes under conditions of dynamic or kinetic–dynamic steady state are very steep (a factor of less than two, possibly less than 1.5), whereas slopes away from dynamic or kinetic–dynamic steady state flatten out in direct proportion to the dynamic or kinetic half-life of a compound. Thus, a slope of 10 after a single dose would indicate a highly reversible effect, in which the predominant variable measured would be recovery and not toxicity, although this fact has been ignored or misinterpreted since the beginnings of modern toxicology.

If both the kinetic and dynamic half-lives are short, attainment of either kinetic or kinetic–dynamic steady state is very difficult by any means other than intravenous infusion (which is impractical in toxicity studies of longer duration), dermal application for compounds that have moderate to good bioavailability by this exposure route, and by inhalation exposure for most volatile or volatilizable compounds. Continuous inhalation of such compounds represents the only experimental design that measures toxicity alone without at least some recovery and, in many instances, a great deal of recovery such as in the case of standard 6 h/5 d/week inhalation toxicity protocols. Again, a comparison of the slope of the dose–response curve between a continuous and discontinuous inhalation exposure paradigm would provide a quantitative measure of the degree of recovery.

6.7 TOXICITY DRIVEN BY THE DYNAMICS OF EFFECT

For compounds causing effects such that the recovery half-life is longer (slower recovery) than the kinetic half-life, the former will dominate the dynamics of the effect. These are the "hit-and-run" type of poisons, which can be eliminated very rapidly and yet produce an effect that accords with $c \times t = k$.

Example 1: Very Long Dynamic Half-Life

There are few examples of recovery taking place on a timescale of years or longer. Chemical neuropathies are the closest examples that come to mind. In the case of the nervous system, it is the prolonged recovery period of damaged nervous tissue, rather than the prolonged elimination period of the toxic chemical, that is rate limiting. Because of the enormous reserve capacity and plasticity of the nervous system, it is difficult to conduct conclusive studies in this area. If the damage is highly irreversible, as is the case with ginger jake paralysis, which is caused by tris(o-tolyl) phosphate[37] or methanol's damage to the retina or optic nerve, accumulation of injury will occur according to a triangular geometry after repeated above-threshold exposures in spite of the short kinetic half-lives of these compounds. If an essentially irreversible injury is very severe, it is very difficult to titrate the dose and injury to assure reproducible survival, which is necessary when studying time-to-effect phenomena.

Example 2: Intermediate Dynamic Half-Life

Diethylnitrosamine (DENA) has a short kinetic half-life of about 10 min in rats.[38] Feeding rats a diet with different daily dose rates resulted in a reasonably good $c \times t = k$ relationship, with cancer as the endpoint of toxicity (Table 6.2). The daily dose rate was a poor surrogate for dose (cumulative) because of the vastly different life spans of the animals receiving the various daily dose rates. It is worthwhile to note that $c \times t = k$ is less variable under isoeffective conditions than when departure from such conditions takes place. Again, with a kinetic half-life of 10 min for DENA and two bouts of feeding per day, no satisfactory $c \times t = k$ relationship should be expected if kinetics are rate determining, because very rapid kinetic recovery occurs after completion of absorption of each daily dose. However, in general, the half-life of DNA adducts of potent carcinogens is on the order of weeks to months, which provides the key to understanding the reasonably good $c \times t = k$ obtained. DENA itself is not at kinetic steady state, but the DNA adducts are at steady state, which is the reason for the good $c \times t = k$ relationship.

Another example is benzene. An enormous amount of effort has gone into understanding the mechanism of toxicity of benzene in terms of metabolic activation. However, the half-life

Table 6.2 Average (Total) Dose and Induction Time in Diethylnitrosamine Carcinogenesis (mg/kg Body Weight, BDII Rats)

Dose (mg/kg)	*Yield of Carcinomas (Survivors/Number)*	*Induction Time (d)*	*$c \times t$ (mg/kg/d)*
1000	5/5	68	68 000
963	25/25	101	97 263
660	25/25	137	96 420
460	34/34	192	88 320
285	36/36	238	67 830
213	49/49	355	75 615
137	67/67	458	62 609
91	27/30	609	55 419
64	5/7	840	53 760

Source: Modified from Druckrey, H. et al., Erzeugung von Krebs durch eine einmalige Dosis von Methylnitroso-Harnstoff und verschiedenen Dialkylnitrosaminen an Ratten, *Zeitsch. f. Krebsforschg.* 66, 1, 1964.

of benzene is about 8 h in humans and not much different in animals;[39,40] whereas the hematopoietic system replenishes erythrocytes on a timescale of about 120 d, i.e., four half-lives.[41] In equilibrium, this must be the rate of maturation of erythrocytes from stem cells. The propagation of any lesion at any stage of this process will be subject to this timescale. Therefore, the dynamics of benzene toxicity will be dominated by the dynamics of the lesions and not by kinetics. Consequently, risk assessment of benzene exposure should be driven by the frequency of exposure above the toxic effect threshold and not by low-level continuous environmental exposure. According to these considerations, once-a-month exposure to very high ambient concentrations of benzene would result in accumulation of residual damage (according to $c \times t$ after reaching steady state) until the individual's aplastic anemia (or leukemia on yet another timescale) threshold has been exceeded. Therefore, preventive measures regarding benzene toxicity should focus on reducing the peak concentrations, and frequency of exposure (ideally, peak-exposure frequency should be reduced to less than one in 120 d). If safety considerations were to be based erroneously on kinetics, the conclusion would be that continuous exposure to lower levels of benzene are more dangerous (larger kinetic AUC) than intermittent exposures to high-peak concentrations. It is indeed known that workers intermittently exposed to high concentrations of benzene (cleaners of reaction vessels) develop aplastic anemia and leukemia, but workers inhaling low levels of benzene for 8 h per day for decades did not.

Another example is arsine, which produces hemolysis. The recovery half-life from arsine-induced hemolysis is identical to the production half-life of erythrocytes. At very high doses, when the time to death is less than a fraction of the recovery half-life, injury will accumulate linearly (zero-order process). However, at lower doses, when animals do not die, hemolysis will reach a steady state between injury and recovery, and if the dosing interval is within 120 d, chronic consequences might ensue.

Example 3: Short Dynamic Half-Life

The kinetic half-lives of soman, sarin, and tabun are in the range of 10 min to 1 h in all species studied, but the recovery half-life from intoxication is about 12 h.[43,49] The ratio between the Lintern et al.[49] observation period (6 h) and the kinetic half-life of these organophosphates (10 to 60 min) suggests extremely unfavorable conditions for $c \times t = k$ to occur because of virtually complete kinetic recovery by the end of the observation period. In fact, there is an excellent $c \times t = k$ relationship for at least those organophosphates that show slow recovery, because it is the dynamic half-life of the effect and not the kinetic half-life of the compound that determines $c \times t = k$.

The kinetic half-life of isocyanates is short, but recovery half-life from depression of respiratory rate is much slower (a day or two). Isocyanates are highly reactive compounds that bond avidly to any nucleophile, including macromolecules, in an organism. Because of this reactivity, their kinetic half-life is expected to be very short (about 1 h), which was confirmed by Brorson et al.[44] Isocyanates exert several well-known effects on the respiratory tract, from functional impairment to death. Recovery from the nonlethal effects is slow (e.g., respiratory rate) to extremely slow (sensitization). 2,4-toluene diisocyanate (TDI) was selected as another example illustrating the application of the theory outlined in the first section of this chapter because inhalation of TDI is a major occupational health hazard and because its study by Sangha and Alarie[45] was well conducted and described in detail. The time course of change in respiratory rate was measured in mice exposed to various concentrations of TDI. Log of effect was plotted vs. time on an arithmetic scale. Log of time vs. effect is its inverse function and as such is entirely symmetrical, and would have yielded identical information. The latter would have been more suitable for analysis if the percentage of change instead of percentage of controls was plotted, because the curves would, in that case, have had an ascending shape (similar to kinetic steady state). Change in respiratory rate is a graded response and, therefore, will approach a different maximum value for each dose rate. The isoeffective dose rates of RD 50 (50% decrease of respiratory rate) were near this maximum rate after about 180 min (Figure 6.3, Sangha and Alarie[45]). Therefore, the longest daily exposure period for this effect was appropriately chosen at 180 min. The recovery half-life for the respiratory rate seems to be longer than the kinetic half-life, as judged from Figures 4 to Figure 6 of Sangha and Alarie.[45] Therefore, the rate-determining step in this toxicity is most likely due to dynamic processes driven by exposure. It is, therefore, in line with the analysis shown in

Figure 6.1 that the effect is somewhat cumulative upon repeated exposure, even though complete kinetic recovery occurs (about 21 half-lives) before each additional exposure. It takes about 4 d of 3-h exposures to reach steady state in terms of effect, indicating a recovery half-life of about 1 to 2 d. Figure 1 of Sangha and Alarie[45] also demonstrated the development of the effect (injury) during the 3-h exposure period. It appears that the effect develops as a result of more than one process because multiple compartments were needed for modeling. Curve-fitting was not performed to reveal the true number of compartments, and in the absence of the original data, curve-peeling cannot be conducted to establish the half-life of the processes underlying this dynamic behavior. The authors did perform linear regression arbitrarily on portions of the curve, which is a semiquantitative indicator of the time course of each dynamic process. However, the authors failed to evaluate the $c \times t$ product under isoeffective conditions. Again, in the absence of the original data, it is possible to do only a limited analysis of Figure 1 of Sangha and Alarie.[45] Using a ruler, the $c \times t$ product at 70% of controls for dose rates ranging from 0.078 to 0.82 ppm can be estimated at 5.7 ± 1.0 ppm · min (n = 7). Thus, it is likely that all isocyanates will display $c \times t$ products for most effects via the dynamics of the respective underlying processes. Whether or not deviation from the $c \times t$ product will occur under isoeffective conditions depends on whether adaptation, repair, and reversibility take place during the observation period. Regarding the respiratory rate, the ratio of recovery (days) to observation period (hours) is favorable for the $c \times t$ product to emerge. For chronic effects, this may not be the case if exposure is discontinuous, and instead, $c \times t^x = k$ might be needed to describe those effects for which significant recovery would occur between repeated exposure episodes.

6.8 TOXICITY DRIVEN BY KINETICS OF CHEMICALS

If the kinetic half-life of a compound is longer than the dynamic half-life of the effect, the kinetics of the compound will dominate the overall process of toxicity.

Example 1: Very Long Kinetic Half-Life

HpCDD yielded $c \times t = k$ in terms of delayed acute toxicity,[10] whereas TCDD did not.[46] The half-life of TCDD in female

Sprague-Dawley rats is about 20 d and that of HpCDD is about 300 d.[15] Rats die as a result of wasting for up to 70 d after treatment with a single dose (rate) of HpCDD, but no rat dies after 30 d when treated with a single dose (rate) of TCDD. The reason for this difference is that 70 d represents about 20% of the half-life of HpCDD, whereas 30 d amounts to 1.5 half-lives of TCDD. Consequently, there is minimal departure from steady state (about 11%) regarding the body burden of HpCDD, but a major departure (62%) in TCDD. As a result, single-dose experiments with HpCDD were being conducted under nearly ideal conditions (near kinetic steady state), allowing for little recovery to occur during 70 d after dosing, whereas in TCDD-treated animals, a sufficient amount of chemical was removed from the organism for significant recovery to occur during 30 d after dosing. When TCDD was administered under isoeffective conditions to female rats, with a loading dose rate followed by maintenance dose rates every 4 d, the $c \times t = k$ relationship emerged with clarity in the case of this dioxin congener[43] also. It should be noted that this experiment was conducted under isoeffective conditions, which represent the ideal conditions to study the relationship between dose and time. It is apparent that variability is larger in this experiment than in the one with HpCDD.[10] The reason for this is that the rats received a different number of maintenance dose rates, and different doses were administered to different groups because of the different times to death. The lesson from these and other experiments (with lower doses) is that the slightest departure from ideal conditions (from kinetic steady state, other than monotonic departure from it) has a major impact on $c \times t = k$, making it disappear in most toxicological experiments. It appears that there are no exceptions to the $c \times t = k$ relationship under ideal conditions and that most toxicological experiments do have uncontrolled (concealed) variables. For example, the kinetics of chemicals have virtually never been controlled in toxicological experiments unless they were naturally controlled by long kinetic or dynamic half-lives.

Some might argue that this finding was just coincidental for another short-term effect and merely a coincidental confirmation of Haber's rule. Table 6.3 shows the major chronic toxic effects of HpCDD, in continuation of the experiment that addressed its delayed acute toxicity in a previous publication.[10] Animals surviving the acutely toxic insult (including an acute NOAEL of 2.5 mg/kg) developed anemia, lung cancer, liver

Table 6.3 Causes of Death in Rats Chronically Intoxicated with HpCDD

Anemia
Squamous cell carcinoma of the lungs
Hepatocellular carcinoma and cholangiocarcinoma
Various kidney pathologies
Others

cancer, various kidney pathologies, and a variety of low-incidence pathologies toward the end of the animal's natural life span in the sequence as listed. Squamous cell carcinoma of the lungs in Table 6.3 had the highest incidence among these effects. Table 6.4 demonstrates the chronic time response to various single doses of HpCDD in terms of all causes of death. Clearly, this chronic time response, which predicts with great accuracy the shortening of the animals' life span due to HpCDD doses, is truncated at the 3.8-mg/kg dose by the delayed acute time response having killed most or all animals in higher-dose groups. At least two of the effects (anemia and lung cancer) are present

Table 6.4 Time to Death in HpCDD-Treated Rats and Controls

Single Dose Rate (mg/kg)	*Time (d)*	*n*
0.0	540 ± 28.2[a]	36
2.5	438 ± 32.4	30
2.8	421.2 ± 18.9	55
3.1	410 ± 37.3	20
3.4	382 ± 49.1	10
3.8	286 ± 52.0	5

[a]Mean ± SE.

Source: Modified from Rozman, K.K., Approaches for using toxicokinetic information in assessing risk to deployed U.S. forces, in *Workshop Proceedings*: *Strategies to Protect the Health of Deployed U.S. Forces*, National Academy Press, Washington, D.C., 2000, 113.

Table 6.5 The $c \times t$ Product for all Causes of Chronic Toxicity of HpCDD

Single Dose Rate (mg/kg)	*$c \times t$ (mg/kg/d)*	*n*
0.0	N/A	36
2.5	1094 ± 81[a]	30
2.8	1184 ± 53.0	55
3.1	1277 ± 117	20
3.4	1313 ± 169	10
3.8	1074 ± 195	5
Average	1183 ± 48	

[a] Mean ± SE.

Source: Modified from Rozman, K.K., Approaches for using toxicokinetic information in assessing risk to deployed U.S. forces, in *Workshop Proceedings: Strategies to Protect the Health of Deployed U.S. Forces*, National Academy Press, Washington, D.C., 2000, 113.

concurrently in many animals, which makes separation of these two effects difficult other than by the time to occurrence of the effects in the first subject to be affected. However, if each effect occurs according to $c \times t = k$ then the sum of all effects must also be $c \times t = k$ for all subjects. This was indeed confirmed by Rozman et al.[16] in a study that depicted the $c \times t$ product for all effects combined (Table 6.5). The remarkably low variability is comparable to data obtained for delayed acute toxicity.[10] It should be noted that these chronic effects developed while the animals were eliminating a significant portion of their body burden, allowing for increasing recovery to occur. With a half-life of about 300 d, 75% was eliminated by day 600, a period during which most chronic deaths occurred. To elucidate the role of recovery in the toxicity of HpCDD, the compound was also administered as a loading dose rate (2.8 mg/kg) followed by biweekly maintenance dose rates(0.085 mg/kg) to maintain kinetic steady state. Table 6.6 shows the difference between the single-dose (rate) and the multiple-dose-rate experiments. This table shows that the dosing regimen had no impact, either acute or chronic, on the $c \times t$ product. Even though the multiple-dosing regimen led to a considerably larger dose of

Table 6.6 Comparison of the Average Acute and Chronic $c \times t$ Products after Single or Multiple Doses of HpCDD to Female Sprague-Dawley Rats

	Single Dose	*Multiple Dose*
Acute	106 ± 6.1[a]	120
Chronic	1212 ± 43	1230 ± 129

[a] Mean ± SE of the average $c \times t$ product; From Rozman, K.K., Delayed acute toxicity of 1,2,3,4,6,7,8-heptachloro-dibenzo-p-dioxin (HpCDD) after oral administration obeys Haber's Rule of inhalation toxicology, *Toxicol. Sci.* 49, 102, 1999.

HpCDD, the $c \times t$ product remained constant because of an entirely predictable shortening of the life span of the animals. Another important message is depicted in Table 6.7. Regardless of dosing regimen, the lung cancer dose response to HpCDD developed according to the dose–time paradigm. Lung cancer was the major chronic toxic effect of HpCDD and the major toxic effect after TCDD.[47] Yet, regulatory action on dioxins was and is based on liver toxicity or cancer, which is a low-incidence, late-life effect.

Example 2: Intermediate Kinetic Half-Life

Monochloroacetic acid (MCA) has a half-life of about 2 h in male rats and a time to effect (coma or death) of the same order of magnitude. Therefore, in the course of development of toxicity, a significant portion of a dose will be eliminated, allowing for some recovery to occur given while toxicity is developing. As with TCDD, MCA does not obey Haber's rule when administered as a single dose, although it comes close to it after subcutaneous injection[48] because of the slow release of MCA in this procedure. To create ideal conditions, MCA had to be infused by osmotic minipumps because repeated administrations of maintenance dose rates would have severely disturbed the animals, triggering convulsions (and death) and thereby offsetting their normal time schedule of going into coma according to $c \times t$. Rozman[42] demonstrated that this experimental modification resulted in a reasonably good $c \times t = k$ relationship

Table 6.7 Prevalence of Squamous Cell Carcinoma of the Lungs and Liver Cancer in Female Sprague-Dawley Rats Dosed with Various Dosing Regimens of HpCDD

Dose (mg/kg)	*Lung Cancer (%)*
0.0[a]	2.8
0.0[b]	3.3[c]
1.0[d]	0.0
2.1[e]	16.6
2.8[d]	30.9
3.1[d]	73.3
3.1[f]	66.6
3.4[d]	60.0
3.8[d]	66.0
4.6[g]	62.5

[a] Single dose of vehicle (4 ml/kg).

[b] Biweekly dose rate of vehicle (4 ml/kg).

[c] Adenocarcinoma.

[d] Single doses.

[e] 1.0 mg/kg + 0.03 mg/kg biweekly.

[f] 2.5 mg/kg + 0.63 mg/kg as a single additional dose rate.

[g] 2.8 mg/kg + 0.086 mg/kg biweekly.

Source: Modified from Rozman, K.K., et al.[16]

for this compound also and for another endpoint of toxicity (coma). It must be recognized that in single-dose experiments in which steady state is not carefully controlled, it is not the absolute timescale, but the ratio of the half-life of a compound and the observation period, that determines whether or not $c \times t = k$ will manifest. If the observation period is much shorter than the half-life of a compound, then $c \times t = k$ will always be observable unless, for example, adaptation introduces a concealed variable. However, the more unfavorable this ratio becomes, the more will be of the recovery occurring concurrently with the development of toxicity. This will not only introduce an uncontrolled variable but will also flatten the dose and time responses.

Example 3: Very Short Kinetic Half-Life

Methylene chloride (MCI) is an example of such compounds, having an estimated kinetic half-life of 5 to 40 min. The following considerations, however, also apply to compounds of even shorter half-lives, such as ozone; although for ozone, the dynamics of recovery may be rate determining. For compounds of very short kinetic or dynamic half-lives, the distinction between kinetics and dynamics becomes less important because rapid elimination or recovery reduces the time dependencies of toxicity. Such compounds will be more concentration dependent in any type of discontinuous exposure regimen, and only continuous exposure until the actual occurrence of the effect will yield $c \times t = k$. This is probably the origin of the notion that Haber's rule applies only to inhalation toxicology, whereas, in fact, inhalation happens to be the only practical way (intravenous infusion for days to months is clearly not practical) to provide continuous exposure for compounds having very short kinetic or dynamic half-lives. Here, the need for continuous exposure, rather than exposure by inhalation, is the key requirement. Intermittent inhalation exposure (i.e., 6 h/d) to kinetically acting compounds will yield $c \times t = k$ only if their kinetic half-lives are at least 1 d to allow for a reasonable steady state with little kinetic recovery (elimination) occurring between exposure episodes. If intermittent exposure is further fragmented by weekends, the half-life has to be correspondingly longer.

6.9 SPECIAL CASES

The outcome of an experiment conducted under conditions of kinetic steady state will be in accordance with Haber's rule unless some concealed variables impede the outcome. *In vitro* studies, experiments involving the use of aquatic species (fish, pond snails, etc.), or when continuous exposure is an implicit part of the experimental design will result in experiments showing simple or complex forms of the $c \times t = k$ relationship unless kinetic or dynamic adaptations occur during the measurement of toxicity.[49] In fact, Table 6.8 provides indications of the accuracy of the $c \times t = k$ relationship using sea urchin sperm motility as the endpoint of toxicity and divalent cations as toxicants.

Table 6.8 Effect of Hg Cl_2, $CdCl_2$ and Be Cl_2 on Sea Urchin Sperm Motility

Compound	*Dose (μM)*	*Time to EC_{50}*	*c × t (μM min)*
$HgCl_2$	1.0	13.3	13.3
	1.5	9.9	14.8
	2.0	8.4	16.8
			15.0 ± 1[a]
$CdCl_2$	2.5	15.0	37.6
	3.5	10.0	34.9
	4.5	6.6	29.7
			34.1 ± 2.3
$BeCl_2$	90	4.0	361.9
	120	4.6	550.9
	150	4.2	636.2
			516.3 ± 81.0

[a] Mean ± SE.

Source: Rozman, K. K.[50]

6.10 CONCLUSIONS

In this chapter, a theory of toxicology is outlined using a large existing toxicology database and interpreting it analogously to thermodynamics. Continuous inhalation exposure to toxicants represents the most efficient exposure scenario that comes closest to ideal conditions in toxicology experiments, which is the reason for finding confirmation of Haber's rule of inhalation toxicology most frequently in this particular field. However, any time a departure from Haber's rule was observed, it was considered an exception. A systematic analysis of the toxicology literature along the guidelines depicted in Figure 6.1 using theoretical considerations of toxicology revealed the conditions under which Haber's rule is observable. Therefore, the many presumed exceptions to Haber's rule are not really exceptions but are due to uncontrolled variables (usually involving time) of which the investigators are unaware.

REFERENCES

1. Hayes, W.J., *Handbook of Pesticide Toxicology*, Eds., W.J. Hayes, Jr., E.R. Laws, Jr., San Diego, Academic Press, 1991.
2. Rozman, K.K., Doull, J., and Hayes, W.J., Jr., Dose, time, and other factors influencing toxicity, in *Handbook of Pesticide Toxicology*, R. Krieger, Ed. in Chief, 2nd ed., Vol. 1, Academic Press, San Diego, 2001, p. 1.

3. Rozman, K.K. and Doull, J., General principles of toxicology, in *Environmental Toxicology*, J. Rose, Ed., Gordon and Breach Science, Amsterdam, 1998, 1.
4. Warren, E., On the reaction of Daphnia magna to certain changes in its environment, *Quart. J. Microsc. Sci.* 43, 199, 1900.
5. Ostwald, W. and Dernoscheck, A., Über die Beziehung zwischen Adsorption und Giftigkeit, *Kolloid-Zeitschr* 6, 297, 1910.
6. Flury, F. and Wirth, W., Zur Toxikologie der Lösungsmittel. *Archiv f. Gewerbepath. u. Gewerbehyg.* 5, 1, 1934.
7. Druckrey, H. and Küpfmüller, K., Quantitative Analyse der Krebsentstehung, *Zeitschr. f. Naturforschg.* 36, 254, 1948.
8. Littlefield, N.A., Farmer, J.H., and Gaylor, D.W., Effects of dose and time in a long-term, low-dose carcinogenic study, in *Innovation in Cancer Risk Assessment.* (ED_{01},-Study). *J. Environ. Pathol. Toxicol.* 3, 1980, 17.
9. Peto, R., Gray, R., Brantom, P., and Grasso, P., Effects on 4080 rats of chronic ingestion of N-nitrosodiethylamine or N-nitrosodimethylamine: a detailed dose-response study, *Cancer Res.* 51, 6415, 1991.
10. Rozman, K.K., Delayed acute toxicity of 1,2,3,4,6,7,8-heptachloro-dibenzo-p-dioxin (HpCDD) after oral administration obeys Haber's Rule of inhalation toxicology, *Toxicol. Sci.* 49, 102, 1999.
11. Bliss, C.I., The relation between exposure time, concentration and toxicity in experiments on insecticides, *Ann. Entomol. Soc. Am.* 33, 721, 1940.
12. Rozman, K.K., Quantitative definition of toxicity: a mathematical description of life and death with dose and time as variables, *Med. Hypotheses* 51, 175, 1998.
13. Rozman, K.K. et al., A toxicologist's view of cancer risk assessment, *Drug. Metab. Rev.* 28, 29, 1996.
14. Viluksela, M. et al., Subchronic/chronic toxicity of 1,2,3,4,6,7,8-heptachlorodibenzo-p-dioxin (HpCDD) in rats. Part 1. Design, general observations, hematology and liver concentrations, *Toxicol. Appl. Pharmacol.* 146, 207, 1997.
15. Viluksela, M. et al., Subchronic/chronic toxicity of a mixture of four chlorinated dibenzo-p-dioxins in rats, I. Design, general observations, hematology and liver concentrations, *Toxicol. Appl. Pharmacol.* 151, 57, 1998.
16. Rozman, K.K., Lebofsky, M., and Pinson, D.M., Chronic toxicity and carcinogenicity of 1,2,3,4,6,7,8-heptachlorodibenzo-p-dioxin displays a distinct dose/time toxicity threshold ($c \times t = k$) and a life-prolonging subthreshold effect. *Food Chem. Toxicol.* 43, 729–740, 2005.
17. Rozman, K.K. et al., Relative potency of chlorinated dibenzo-p-dioxins (CDDs) in acute, subchronic and chronic (carcinogenicity) toxicity studies: implications for risk assessment of chemical mixtures, *Toxicology* 77, 39, 1993.
18. Haber, F., On the history of gas warfare, in *Five Lectures from the Years* 1920–1923, Julius Springer, Berlin (in German), 1924, p. 76.
19. Peters, G. and Ganter, W., Zur Frage der Abtötung des Kornkäfers mit Blausäure, *Zeitschr. f. Angew. Entomol.* 21, 547, 1935.
20. Busvine, J.R., The toxicity of ethylene oxide to *Calandra oryzae, C. granaria, Tribolium castaneum,* and *Cimex lectularius. Ann. Appl. Biol.* 25, 605, 1938.
21. Rozman, K.K. and Doull, J., The role of time as a quantifiable variable of toxicity and the experimental conditions when Haber's $c \times t$ product can be observed: implications for therapeutics, *J. Pharmacol. Exp. Ther.* 296, 663, 2001.

22. Druckrey, H. et al., Erzeugung von Krebs durch eine einmalige Dosis von Methylnitroso-Harnstoff und verschiedenen Dialkylnitrosaminen an Ratten, *Zeitsch. f. Krebsforschg.* 66, 1, 1964.
23. Garrettson, L.K., Lead. In *Clinical Management of Poisoning and Drug Overdose*, Eds., L.M. Haddad and J.F. Winchester, 2nd ed., Saunders, Philadelphia, 1983, 1017.
24. Waddell, W.J., Thresholds in chemical carcinogenesis: what are animal experiments telling us? *Toxicol. Pathol.* 31, 260, 2003.
25. Goldenfeld, N., and Kandanoff, L.P., Simple lessons from complexity, *Science* 284, 87, 1999.
26. Whitesides, G.M. and Ismagilov, R.F., Complexity in chemistry, *Science* 284, 89, 1999.
27. Weng, G., Bhalla, U.S., and Iyengar, R., Complexity in biological signaling systems, *Science* 284, 92, 1999.
28. Koch, Ch. and Laurent, G., Complexity and the nervous system, *Science* 284, 96, 1999.
29. Gardner, D.E. et al., Role of time as a factor in the toxicity of chemical compounds in intermittent and continuous exposure, Part 1: Effects of continuous exposure, *J. Toxicol. Environ. Health* 3, 811, 1977.
30. Sivan, P.S., Hoskins, B., and Ho, I.K., An assessment of comparative acute toxicity of diisopropylfluorophosphate, taban, sarin and soman in relation to cholinergic and GABA ergic enzyme activities in rats, *Fundam. Appl. Toxicol.* 4, 531, 1984.
31. Gardner, D.E. et al., Influence of exposure mode on the toxicity of NO_2, *Environ. Health Perspect.* 30, 23, 1979.
32. Hartung, R., Dose response relationships, in *Toxic Substances and Human Risk,* Eds., R.G. Tardiff and J.V. Rodricks, Plenum, New York, 1987, p. 29.
33. Dahl, A., Dose concepts for inhaled gases and vapors, *Toxicol. Appl. Pharmacol.* 103, 185, 1990.
34. Casarett, L.J., Doull, J., and Klaassen, C.D., *Casarett and Doull's Toxicology: The Basic Science of Poisons,* Ed., Curtis D. Klaassen, New York, McGraw-Hill Medical Pub Division, 2001.
35. Dost, F.H., *Der Blutspiegel,* Thieme Verlag, Leipzig, 1953.
36. Gibaldi, M., and Perrier, D., Trapezoidal rule, in *Pharmacokinetics.* Marcel Dekker, New York, 1975, p. 293.
37. Morgan, J.P., Penovich, P., Jamaica Ginger paralysis, *Arch. Neurol.* 35, 530, 1978.
38. Druckrey, H.R. et al., Organotrope carcinogen Wirkungen bei 65 verschiedenen N-Nitroso-Verbindunger an BD-Ratten, *Zeitsch. Krebsforschg.* 69, 103, 1967.
39. Brugnone, F. et al., Reference values for blood benzene in the occupationally unexposed general population. *Int. Arch. Occup. Environ. Health* 64, 179, 1992.
40. ATSDR, *Toxicological Profile for Benzene,* U.S. Department of Health and Human Services, Public Health Service, 1997.
41. Wintrobe, M.M. and Lee, G.R., *Hematologic Alterations in Harrison's Principles of Internal Medicine,* Ed., Wintrobe, M.M. et al., McGraw-Hill, New York, 1974, p. 289.
42. Lintern, C.M., Wetherell, J.R., and Smith, M.E., Differential recovery of acetylchinesterare in guinea pig muscle and brain regions after soman treatment, *Hum. Exp. Toxicol.* 17(3), 157, 1998.
43. Rozman, K.K., Approaches for using toxicokinetic information in assessing risk to deployed U.S. forces, in *Workshop Proceedings: Strategies to Protect the Health of Deployed U.S. Forces,* National Academy Press, Washington, D.C., 2000, 113.

44. Brorson, T. et al., Test atmospheres of diisocyanates with special reference to controlled exposure of humans, *Int. Arch. Occup. Environ. Health* 61, 495, 1989.
45. Sangha, G.K. and Alarie, Y., Sensory irritation by toluene idiisocyanate in single and repeated exposures, *Toxicol. Appl. Pharmacol.* 50, 533, 1979.
46. Stahl, B.U., Kettrup, A., and Rozman, K.K., Comparative toxicity of form chlorinated dibenzo-p-dioxins and their mixture, Part I. Acute toxicity and toxic equivalency factors (TEFs), *Arch. Toxicol.* 66, 471, 1992.
47. Kociba, R.J. et al., Results of a two-year chronic toxicity and oncogenicity study of 2,3,7,8-tetrachlorodibenzo-p-dioxin in rats. *Toxicol. Appl. Pharmacol.* 46, 279, 1978.
48. Hayes, F.D., Short, T.D., and Gibson, J.E., Differential toxicity of monochloro-acetate, monofluoroacetate and monoiodoacetate in rats, *Toxicol. Appl. Pharmacol.* 26, 93, 1973.
49. Verhaar, H.J.M. et al., An LC_{50} vs. time model for the aquatic toxicity of reactive and receptor-mediated compounds, consequences for bioconcentration kinetics and risk assessment, *Environ. Sci. Technol.* 33, 758, 1999.
50. Rozman, K.K., The role of time in toxicology or Haber's $c \times t$ product. *Toxicology* 149, 35–42, 2000.

7

SAFETY ASSESSMENT OF THERAPEUTIC AGENTS ADMINISTERED BY THE RESPIRATORY ROUTE

Shayne C. Gad

CONTENTS

7.1 INTRODUCTION

Drugs and medicinal agents administered by the inhalation route include gaseous and vaporous anesthetics, coronary vasodilators, aerosols of bronchodilators, corticosteroids, mucolytics, expectorants, antibiotics, and an increasing number of peptides and proteins, in all of which there is significant nasal absorption (Cox et al. 1970; Williams 1974; Paterson et al. 1979; Hodson et al. 1981; Lourenco and Cotromanes 1982). Concerns with the environmental effects of chlorofluorocarbons has also led to renewed interest in dry powder inhalers (DPIs), which have additionally shown promise for better tolerance and absorption of some new drugs. Recent advances have also led to new nasal delivery systems, such as those in Table 7.1. Excessive inhalation of a drug into the pulmonary system during

Table 7.1 Nasal Delivery Systems

Liquid nasal formulations	
Instillation and rhinyle catheter	Drops
Unit-dose containers	Squeeze bottle
Metered-dose pump sprays	Airless and preservative-free sprays
Compressed-air nebulizers	
Powder dosage forms	
Insufflators	Monodose powder inhaler
Multidose dry powder system	
Pressurized MDIs (dose inhalers)	
Nasal gels	

therapy or manufacturing may result in adverse local and systemic effects, or both. Consequently, safety assessment of medicinal preparations delivered via respiratory routes with respect to local tissue toxicity, systemic toxicity, and the therapeutic-to-toxicity ratio is essential. The data generated is essential for charting the course of evaluation and development of a potential therapeutic agent, the general course of which is summarized in Figure 7.1.

Currently, respiratory-route drugs represent less than 1% of marketed therapeutics. However, advances in pharmaceutical technology and the introduction of new structural classes of therapeutic entities (particularly peptides and highly reactive small molecules such as the nitrous oxide analogs) have led to renewed interest in the use of respiratory routes of delivery to treat a range of diseases, including respiratory (asthma, SARS, and cystic fibrosis), CNS (pain), and systemic (such as diabetes) diseases.

7.2 PULMONARY DELIVERY OF THERAPEUTIC INHALED GASES AND VAPORS

Pulmonary dynamics, the dimension and geometry of the respiratory tract, and the structure of the lungs, together with the solubility and chemical reactivity of therapeutic inhalants, greatly influence the magnitude of penetration, retention, and absorption of inhaled gases, vapors (Dahl 1990), and aerosols (Phalen 1984), and, therefore, the therapeutic potential of the agents. The quantity of an inhalant effectively retained in the pulmonary system constitutes the inhaled dose that leads to both therapeutic and toxic (dose-limiting) responses.

Highly reactive and soluble gaseous or vaporous drugs react and dissolve readily in the mucosal membrane of the nasopharynx and the upper respiratory tract (URT), thereby exerting pharmacological effects

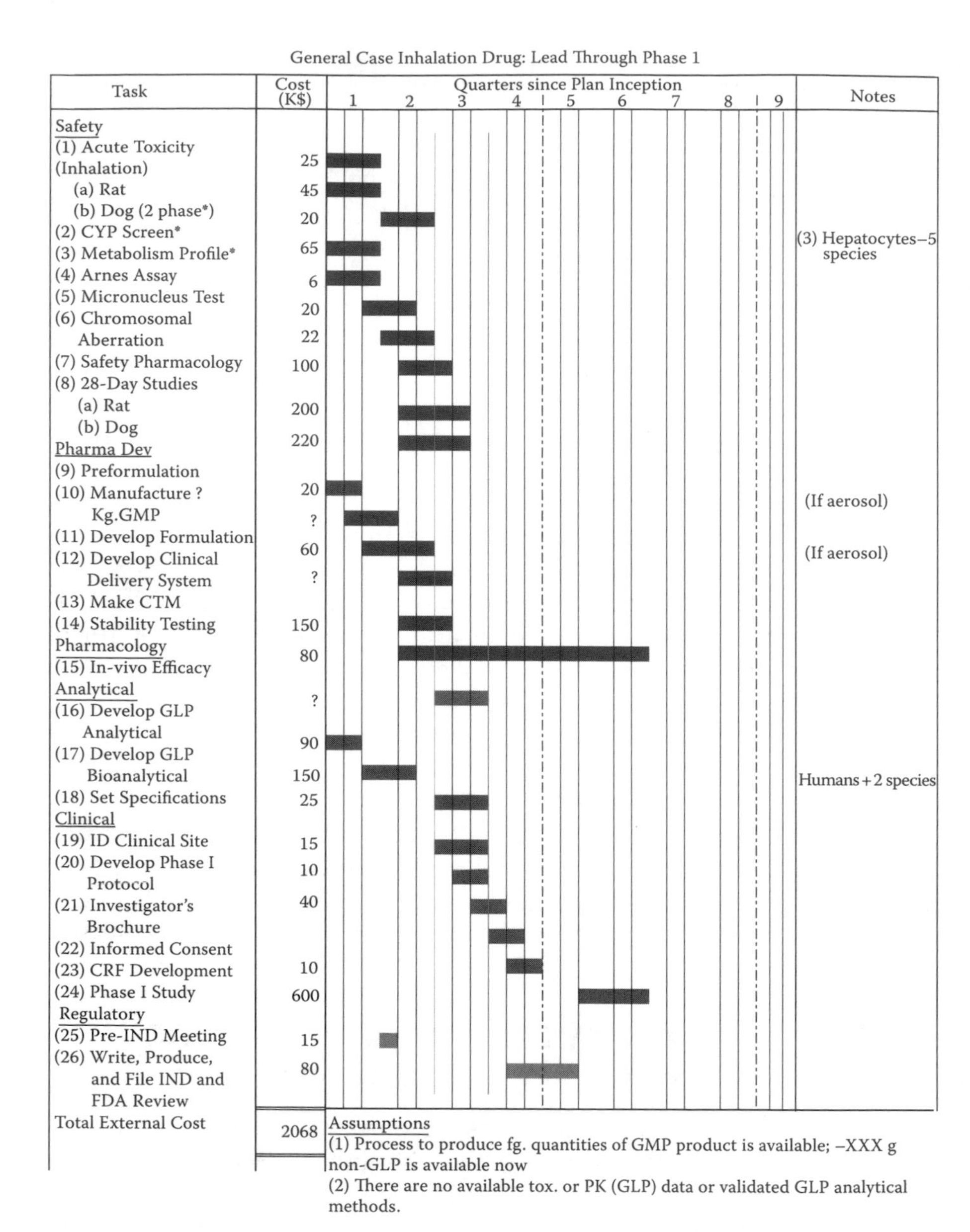

Figure 7.1 Typical pharmaceutical development activities to first-in-man (Phase I) studies for respiratory-route drugs. All the tasks involved in developing a potential drug from lead designation to the completion of an initial (first-in-man [FIM]) clinical study, usually performed on normal, healthy volunteers. (From Gad 2002.)

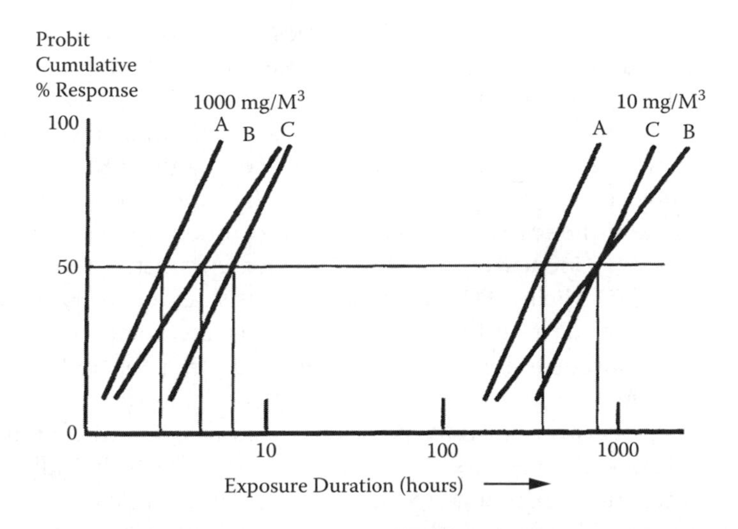

Figure 7.2 Dose–response is plotted in terms of the probit of cumulative percentage response to logarithm of the exposure concentrations, where A, B, and C are agents acting by different mechanisms or kinetics.

or causing local irritation or adverse effects, or both, on the ciliated, goblet, brush border columnar, and squamous cells of the epithelium. The dissolved drug is also absorbed into the bloodstream and transported to a target organ in which it can exert systemic effects. Less-reactive and less-soluble gaseous or vaporous drugs are likely to penetrate beyond the URT and reach the bronchial and alveolar regions, causing local and systemic effects. The unabsorbed gases or vapors are then exhaled. For example, ammonia gas generated from 10% ammonia water may be inhaled for reflex respiratory stimulation purposes (Budavari 1989). Ammonia is extremely soluble in water at a concentration of 715 ml of ammonia per ml of water (Phalen 1984) and is readily solubilized in the mucous lining, causing URT irritation. By contrast, oxygen is only sparingly soluble in water (0.031 cc of oxygen per ml of water) and capable of penetrating deeply into the alveoli, in which gas exchange takes place. Oxygen that binds reversibly with the hemoglobin of erythrocytes is unloaded at the target tissues, whereas the unbound oxygen is exhaled. Inhalation of properly humidified oxygen is life supporting, but inhalation of unhumidified oxygen may cause a reduction in the mucociliary clearance of secretions and other substances deposited in the trachea of animals (Pavia 1984) and humans (Lichtiger et al. 1975; Gamsu 1976). Gases or vapors of low lipid solubility are also poorly absorbed in the lungs, with much of the inhaled vapor exhaled. Other pharmacological gases and vapors, such as the anesthetics (nitrous oxide, halothane, enflurane, isoflurane, etc.) and coronary vasodilators (amyl nitrite), likewise affect the epithelium of the respiratory tract and lungs. The absorbed drugs exert local effects on various types of epithelial cells of the respiratory tract, and on type I and type II cells and the alveolar macrophages (AMs) in the alveoli. Repeated inhalation of some halogenated hydrocarbon anesthetics will result in the accumulation of vapors and systemic toxicity (Chenoweth et al. 1972). By contrast, vapors such as fluorocarbons (FC 11 and FC 12), which are used extensively as propellants for bronchodilator and corticosteroid aerosols, are absorbed rapidly but are not accumulated in the body even upon repeated inhalation (Aviado 1981; Aviado and Micozzi, M.S., 1981).

In general, dissolved gases or vapors at a nontoxic concentration are absorbed and metabolized locally by the lungs and systemically by the liver. The unchanged parent drug and its metabolites may be excreted to some extent via exhalation but mainly via the renal system. A dissolved gas or vapor at a toxic concentration, however, is likely to exert local effects such as altering the surface tension of the alveoli linings (surfactant) or disrupting the normal functions of the epithelial cells, the pneumocytes, and the AMs. The disrupted AMs in turn release their intracellular enzymes, potentially causing destruction of the alveolar septa and contributing to histopathologic changes of the respiratory tract and the lungs. Again, the magnitude of the adverse effects is dependent on pulmonary dynamics and the solubilities of the inhalants in the mucous membrane of the URT and in the plasma or lipids of the erythrocytes. The pharmacologic and toxicologic aspects of this are discussed later.

7.3 PULMONARY DELIVERY OF INHALED AEROSOLS

For inhaled aerosols, particle size is a major factor affecting the penetration, deposition, and hence the dose and site of pharmacological action (Dautrebande 1962a, 1962b; Agnew 1984). Particle size is expressed in terms of *aerodynamic diameter* (AD), defined as the diameter of a spherical particle of unit density (lg/cm^3) that has the same terminal settling velocity as the particle in question, regardless of its shape and density (Marple and Rubow 1980). The unit for AD is micrometers (μm). A sample of aerosol particles having ADs within a narrow size range is considered to be a monodisperse aerosol, whereas a sample of aerosols with a wide range of ADs is a heterodisperse, (or polydisperse) aerosol. The pattern of particle-size distribution is usually bell shaped, with smaller and larger particles on both sides of the mean AD. An aerosol sample with a high proportion of particles of similar size has a narrow particle-size distribution, or small geometric standard deviation (GSD). An aerosol sample with a

GSD of less than 2 is considered to be a monodisperse aerosol. Therefore, both the AD and GSD of 2 or less is considered to be optimal for pulmonary penetration and distribution in the respiratory tract and the lungs. For example:

In nose breathing, aerosol particles with ADs > 15 μm are likely to be trapped in the nasopharynx (extrathoracic, or head region) by filtration and impaction. Particles deposited in the nasopharynx are considered to be noninhalable (Lippmann 1970; Miller et al. 1979).

In mouth breathing, only 10 to 15% of 15-μm particles penetrate through the larynx to the intrathoracic tracheobronchial (TB) region. Particles reaching the TB region are considered to be inhalable (Lippmann 1970; Miller et al. 1979).

In natural nose and mouth breathing, only a negligible proportion of aerosol particles of AD > 10 μm reach the lungs (Swift and Proctor 1982). Aerosol particles of 3 to 4 μm in AD are considered to be of optimal size for TB deposition. The mechanisms of deposition are by impaction along the trachea, at bronchial branchings at which the direction of airflow changes, and by gravity settlement in the fine airways in amounts proportional to the particle-settling velocity and the time available for settlement (Hatch and Gross 1964; Heyder et al. 1980). Aerosol particles of 1 to 2 μm in AD, however, decrease in TB deposition because the particles are too small for effective impaction and sedimentation (Lippmann 1977; Chan and Lippmann 1980; Stahlhofen et al. 1980). Consequently, the majority of the very fine particles are exhaled. However, the deposition of the ultrafine particles of approximately 0.5 μm in AD on the walls of the finest bronchioles and the alveoli increases again due to molecular diffusion processes. Even so, about 90% of the inhaled 0.5-μm particles will still be exhaled during quiet tidal breathing and much more under forced exhalation (Davis et al. 1972; Taulbee et al. 1978). Those fine particles reaching the finest bronchioles and alveoli are considered to be respirable (Lippmann 1970).

In general, particles of AD > 10 μm deposit mainly in the URT, whereas particles of 1 to 5 μm AD, with a GSD of less than 2, are likely to reach the lower respiratory tract, which includes the TB region and the alveoli, with small oropharyngeal loss.

The proportion of an aerosol sample suitable for inhalation can also be determined on the basis of mass median aerodynamic diameter (MMAD), which is defined as the AD in which 50% (by weight) of an aerosol sample is less than or equal to the stated median AD. For example, a sample with an MMAD of 5 μm means that 50% by weight of that sample has ADs of 5 μm and smaller. The MMAD is, therefore, a good index for determining the proportion of an aerosol sample that is noninhalable, inhalable, or respirable. An aerosol sample with an MMAD of

5 μm and a GSD of less than 2 is considered to be optimal for pulmonary deposition and retention (Task Group on Lung Dynamics 1966).

In addition to AD and GSD, the pulmonary dynamics of a subject also greatly influence the distribution of aerosol particles in various regions of the respiratory tract (Agnew 1984). For example, the velocity of airflow in the respiratory tract significantly influences the pattern of TB deposition. An increase in airflow velocity in the airways (possibly due to exercise or work) increases the effectiveness of particle impaction at the bifurcations of the large airways (Dennis 1961; Hatch and Gross 1964; Parent 1991). As a result, spots impacted with a high concentration of particles (hot spots) are frequently present at the carina and the bifurcations of the airways (Lee and Wang 1977; Bell 1978; Stahlhofen et al. 1981). Furthermore, the depth of each breath (tidal volume) also influences the distribution of aerosols. A small tidal volume permits greater impaction in the proximal conducting airways and less sedimentation in the distal airways.

In general, slow, deep inhalation followed by a period of breath holding increases the deposition of aerosols in the peripheral parts of the lungs, whereas rapid inhalation increases the deposition in the oropharynx and in the large central airways. Thus, the frequency of respiration (the flow velocity) and the depth of breath (tidal volume) influence the pattern of pulmonary penetration and deposition of inhaled aerosols. Therefore, a therapeutic aerosol of ideal size will penetrate deeply into the respiratory tract and the lungs, producing an ideal effect only when the aerosols are inhaled in the correct manner.

7.4 ABSORPTION AND CLEARANCE OF INHALED AEROSOLS

Soluble therapeutic aerosols deposited on the epithelial linings of the respiratory tract are absorbed and metabolized in the same way as soluble gases and vapors. Insoluble medicinal aerosols are few in number. Sodium chromoglycate (SCG) is probably the only insoluble powder to be administered as a prophylactic antiasthmatic (Wanner 1979). Insoluble particles deposited on the ciliated linings of the URT are removed by a mucociliary clearance mechanism. Particles deposited on a terminal airway devoid of ciliated cells may be endocytosed into the epithelial cells. At a toxic concentration, the cells die and the debris is then phagocytized and transported into the interstitial space for removal via the lymph or vascular drainages, or reenters the ciliated zone of the airway. Particles deposited on the alveolar walls may be phagocytized by the AMs or neutrophils (PMNs) and transported from the low-surface-tension surfactant in the alveolar lining to the high-surface-tension bronchial fluid of the ciliate airways for elimination by the mucociliary clearance mechanism (Laurweryns et al. 1977). The particle sizes optimal for phagocytosis are 2 to 3 μm, whereas particles smaller than 0.26 μm are less effective in activating the macrophages (Holma 1967). In any case, AMs can phagocytose only a small fraction of a large number of deposited particles. The nonphagocytosed particles are translocated to the lymphatic system or vascular circulation (as discussed in the preceding text) for elimination (Ferin 1977).

Similar to the inhaled gases or vapors, soluble and insoluble aerosol particles can directly exert desirable and undesirable local effects at the site of deposition and systemic effects after solubilization, absorption, and metabolization.

7.5 PHARMACOTOXICITY OF INHALED AEROSOLS, GASES, AND VAPORS

The inhalation route for administering drugs into the pulmonary system for treatment of respiratory diseases eliminates many bioavailability problems such as plasma binding and "first-pass" metabolism, which are encountered in parenteral or oral administration. Consequently, a small inhalation dose is adequate for achieving the desirable therapeutic dose and response without inducing many undesirable side effects. Furthermore, the direct contact of the drug with the target site ensures rapid action. Nevertheless, the effects from inhaled drug aerosols, gases, and vapors also depend on the pharmacological properties of these substances and the location of their deposition in the respiratory system. For example, the classic experiments on bronchodilation drugs (Dautrebande 1962a, 1962b) showed that fine aerosol particles of isoproterenol penetrate deeply into the lower respiratory airways (LRA). In this way, an effective concentration of the drug aerosol can reach the beta-adrenergic receptors of the bronchial smooth muscles. Stimulation of the receptors causes relaxation of the smooth muscle fibers and results in bronchodilation (Weiner 1984; McFadden 1986). Such rapid bronchial responses can be produced in both healthy and asthmatic subjects without inducing any adverse cardiac effects. In contrast, the same delivered dose of isoproterenol of large particle sizes deposits mainly along the URT, with a minimal amount reaching the smooth muscles of the LRA. The drug is quickly absorbed into the tracheal and bronchial veins and delivered immediately to the left ventricle of the heart. A high plasma concentration of the drug in the heart causes prominent cardiovascular effects such as tachycardia and hypertension. Other aerosols of beta-adrenergic drugs, such as epinephrine, isoprenaline, terbutaline, and salbutamol, induce bronchodilation effects in animals and humans (Pavia 1984) via inhalation and stimulate ciliary beat frequency and mucus production at the site of deposition in the trachea (Wanner 1981). Such stimulation of mucus production can be detrimental to the breathing process, especially in asthmatics and cystic fibrosis patients. Thus, the TB mucociliary clearance mechanism is also stimulated. By contrast, anticholinergic bronchodilators, such as atropine and ipratropium bromide, cause mucus retention in the lungs (Pavia, Sutton, Agnew et al. 1983; Pavia, Sutton, Lopez-Vidriero et al. 1983). Therefore, in pharmacological or safety assessments of inhalant beta-adrenergic bronchial dilatation drugs, aerosols should be of small particle sizes suitable for deposition in the peripheral airways to minimize side effects. However, anticholinergic agents should be of larger particle sizes suitable for deposition in the large airways (Ingram et al. 1977; Hensley et al. 1978).

Other therapeutic aerosols, such as beclomethasone dipropionate, betamethasone valerate, and budesonide corticosteroid (Williams 1974); the carbenicillin and gentamicin antibiotics (Hodson et al. 1981); the 2-mercaptoethane-sulfonate (Pavia, Sutton, Lopez-Vidriero 1983) and *n*-acetylcysteine mucolytics; and even vaccines for the prevention of influenza and tuberculosis (Lourenco and Cotromanes 1982), are active by inhalation and oral administration. When these drugs are produced so as to be administered as aerosols, certain particle sizes may be targeted to a specific region or to multiple regions of the pulmonary system, depending on the therapeutic target sites. In any case, when aerosols are delivered as fine particles, the rate of absorption is increased because of an increase in the distribution area per unit mass of the drug. Thus, an effective aerosol dose of corticosteroid for treatment of asthma and bronchitis is merely a fraction of an oral dose (Williams 1974). An aerosol of disodium chromoglycate (DSG) dry powder, a prophylactic dose for preventing the onset of bronchoconstriction in asthmatic attacks (Cox et al.1970), is effective mainly by local inhibition of the release of chemical mediators from mast cells in bronchial smooth muscle. Therefore, DSG particle sizes should be approximately 2 μm in AD for the most effective penetration into the bronchial regions (Godfrey et al. 1974; Curry et al. 1975). Likewise, therapeutic aerosols of local anesthetics and surfactants may require appropriate particle sizes to be targeted to a specific region of the pulmonary system.

Other than undesirable pharmacological effects, toxic concentrations of soluble insoluble aerosol particles may lead to adverse physiological and histophathologic responses. For example, irritating aerosols cause dose-related reflex depression of the respiratory rate (Alarie 1966, 1981a), whereas phagocytosed particles cause chemotaxis of AMs and neutrophils to the site of deposition (Brain 1976). The maximum response usually occurs at 24-h postexposure and returns to normal in approximately 3-d postexposure (Kavet et al. 1978). Furthermore, a toxic quantity of phagocytosed particles may interact with the lysosomal membrane within a macrophage, releasing cytotoxic lysosomal enzymes, proteases, and free radicals that in turn damage the adjacent lung tissue (Hocking and Golde 1979).

In general, a specific category of drug delivered to a specific site of the pulmonary system will exert a specific pharmacological or toxicological action locally or systemically. Therefore, in safety assessments of inhalants, a drug should be delivered to the target sites of the pulmonary system according to the toxicological information required.

Finally, there are many drugs in the categories of amphetamines, anorectics, antihistamines, antipsychotics, tricyclic antidepressants, analgesics and narcotics, and beta-adrenergic blocking agents that are known to accumulate in the lung (Wilson 1982), even though these drugs are not administered via the inhalation route. Therefore, in safety assessments of these drugs, their pulmonary toxicity should also be evaluated. There are also *in vitro* techniques that are proposed for use in evaluating inhalation effects on respiratory tissues (Agu et al. 2002) that may serve as useful screening tools for sets of potential therapeutics.

7.6 NASAL DELIVERY OF THERAPEUTICS

The biotechnology revolution in therapeutics has led to renewed interest in the nasal route as a means of safely delivering an effective dose of many therapeutics. This, in turn, has led to investigations of new approaches to drug administration by this route that have proved very rewarding (Smaldone 1997; Sharma et al. 2001; Aldridge 2003; Lawrence 2002). Table 7.2 presents a comparison of this route vs. others that point out the advantages. These advantages have led to the development (and successful marketing) of an impressive number of new drugs with both local (Table 7.3) and systemic (Table 7.4) therapeutic uses. At the same time, this has also led to new guidelines on the subject being promulgated by the FDA (CDER 2002) and concerns about end-use safety (Kannisto et al. 2002). These guidelines and concerns will directly impact the development of new drugs, such as those in Table 7.5.

The target for nasal administration is the nasal cavity, with a volume (in adults) of only 20 ml but a total surface area (in humans) of ~180 cm^2. The cavity surface is covered with a 2 to 4 mm thick nasal mucosa, composed of both respiratory and olfactory components.

There are three separate mechanisms for transmucosal transport of potential drugs and other substances including toxicants: (1) simple diffusion—a nonsaturable mechanism with no carrier or energy involvement, (2) facilitated transport—saturable, with carrier involvement but no

Table 7.2 Factors Influencing Selection of the Nasal Route

Major Considerations	*Routes of Administration*					
	Oral	*IV*	*IM/SC*	*Transdermal*	*Nasal*	*Pulmonary*
Delivery interface to blood	Indirect; absorbed through GI system	Direct bolus administration into vein	Indirect; absorbed from muscular/ subcutaneous tissue	Indirect; absorbed through relatively impermeable skin	Indirect; absorbed through the highly permeable nasal mucosa	Indirect, but drug delivered to a large highly permeable epithelia
Delivery issues and concerns	Subject to digestive process, first-pass metabolism	Requires administration by healthcare professional	Painful injection, may require administration by healthcare professional	Highly variable, slow delivery; potential for skin reactions	Self administration. Requirement of high solubility	Requires deep, slow inhalation of small aerosol particles
Patient convenience	High	Low	Low	Moderate	Moderate to high	Moderate to high
Onset of action	Slow	Rapid	Moderate	Slow	Rapid	Moderate to rapid
Delivery of macromolec-ules	No	Yes	Yes	No	Yes	Yes
Bioavailability	Low to high	Reference standard	Moderate to high	Low	High	Moderate to high
Dose control	Moderate	Good	Moderate	Poor	Moderate	Moderate to good

Source: From Moren, F. (1993). Aerosol dosage forms and formulations. In *Aerosols in Medicine*. 2nd ed. (S. Moren, M.B. Dolovich, M.T. Newhouse, S.P. Newman, Eds.). Elsevier, Amsterdam, pp. 329–336; Durham, S.R. (2002). The ideal nasal corticosteroid: balancing efficacy, safety and patient preference. *Clin Exp Allergy Rev* 2: 32–37; Wills and Greenstone 2001.

Table 7.3 Marketed Nasal Products (for Topical Activity)

Product	*Drug*	*Indication*	*Manufacturer*
Astelin® Nasal Spray	Azelastine hydrochloride	Treatment of seasonal allergic rhinitis	Wallace Laboratories
Beconse® AQ Nasal Spray	Beclomethasone dipropionate monohydrate	Symptomatic treatment of seasonal and perennial allergic rhinitis	Allen and Hanbury's/Glaxo Wellcome, Inc.
Vancenase® AQ Nasal Spray	Beclomethasone dipropionate monohydrate	Symptomatic treatment of seasonal and perennial allergic rhinitis	Schering Plough Corp.
Rhinocort® Nasal Inhaler	Budesonide	Management of symptoms of seasonal and perennial allergic rhinitis and nonallergic perennial rhinitis	Astra USA, Inc.
Nasalcrom® Nasal Solution	Cromolyn sodium	Symptomatic prevention and treatment of seasonal and perennial allergic rhinitis	Sandoz Pharmaceutical Coop
Adrenalin® Chloride	Epinephrine hydrochloride	Nasal decongestant	Parke Davis
Nasalide® Nasal Solution	Flunisolide	Treatment of seasonal and perennial allergic rhinitis	Dura Trading Co., Ltd.
Flonase® Nasal Spray	Fluticasone propronate	Symptomatic treatment of seasonal and perennial allergic rhinitis	Glaxo SmithKline, Inc.
Atrovent® Nasal Spray	Ipratropium bromide	Symptomatic relief of rhinitis	Boehringer Ingelhem Pharmaceuticals, Inc.

(*continued*)

Table 7.3 Marketed Nasal Products (for Topical Activity) (Continued)

Product	*Drug*	*Indication*	*Manufacturer*
Livostin® Nasal Spray	Levocabastine	Treatment of allergic rhinitis	Janssen Research FDN Division of Johnson and Johnson
Privine® Nasal Spray, Nasal Solution and Nasal Drops	Naphazoline hydrochloride	Prompt and prolonged relief of nasal congestion due to common colds, sinusitis	Ciba Consumer Pharmaceuticals
Flunisolide Nasal Solution	Flunisolide	Nasal decongestant	Bausch & Lomb
Afrin® Nasal Spray	Oxymetazoline hydrochloride	Temporary relief of nasal congestion associated with colds, hay fever, and sinusitis	Schering Plough Healthcare Products
Vicks® Sinex ®Regular Decongestant Nasal Spray and Ultra Fine Mist	Phenylephrine hydrochloride	Temporary relief of nasal congestion due to colds, hay fever, upper respiratory allergies or sinusitis	Procter and Gamble
Vick® Vapor Inhaler *OTC	1-Des-oxyephedrine	Nasal decongestant	Procter and Gamble
Nasonex®	Mometasone	Treatment of seasonal and perennial nasal allergy symptoms	Schering Corp.
Nasacort®	Triamcinole acetonide	Treatment of seasonal and perennial allergic rhinitis	Aventis

Table 7.4 Marketed Nasal Products (for Systemic Activity)

Product	*Drug*	*Indication*	*Manufacturer*
Stadol NS® Nasal Spray	Butorphanol tartrate	Management of pain including migraine headache pain	Bristol-Myers Squibb
Miacalcin® Nasal Spray	Calcitonin-salmon	Treatment of hypercalcemia and osteoporosis	Novartis
DDAVP® Nasal Spray	Demopressin acetate	Diabetes insipidus	Aventis Pharmaceuticals
Migranal® Nasal Spray	Dihydroergota mine mesylate	Treatment of migraine	Novartis
Medihaler-ISO® Spray	Isoproterenol sulfate	Treatment of bronchospasm	3M Pharmaceuticals, Inc.
Nitrolingual® Spray	Nitroglycerin	Prevention of angina pectoris due to coronary artery disease	G Pohl Boskamp GMBH and Co.
Synarel® Nasal Solution	Nafarelin acetate	Central precocious puberty, endometriosis	Roche Laboratories
Nicotrol® Inhalation	Nicotine	Smoking cessation	Pharmacia
Syntocinon® Nasal Spray	Oxytocin	Promote milk ejection in breast feeding mothers	Novartis
Imitrex® Nasal Spray	Sumatriptan	Migraine	Glaxo SmithKline
Relenza® Powder for Inhalation	Zanamivir	Treatment of uncomplicated acute illness due to influenza A and B	Glaxo SmithKline

energy directly expended, and (3) active transport—a saturable mechanism that involves both a carrier and energy expenditure.

7.6.1 Nose-to-Brain Delivery

- Transport of substances from nasal cavity to CNS has been recorded since the mid-1900s when it was observed that viruses could move from nose to brain via olfactory pathways.

Table 7.5 Drugs in Development

Product	*Indication*	*Manufacturer*	*Development*
Nasal Nicotine Spray (NNS)	Smoking cessation	Pharmacia Corporation	Phase III trials
Formoterol, Oxis Turbuhaler®	Asthma	Astra Zenca PLC	Phase III trials
Zomig (zolmitriptan)®	Migraine therapy	Astra Zenca PLC	NDA filed
Ciches onide	Bronchial–respiratory	Aventis Pharmaceuticals	Phase III trials
FluMist	Vaccine	Aiviron	Approval recommended
Inhaled Insulin	Diabetes therapy	Eli Lilly & Co.	Phase II trials
GW	Bronchial–respiratory	Glaxo Smith Kline	Phase II trials
INS 37217 Intranasal	Bronchial–respiratory	Inspire Pharmaceuticals	Phase II trials
Beclomethasone	Bronchial–respiratory	Mediera Pharmaceuticals	Clinical
Salbutanol	Bronchial–respiratory	Mediera Pharmaceuticals	Clinical
VLA-4 antagonist	Bronchial–respiratory	Merck & Co., Inc.	Phase II trials
PT-141	Reproductive system therapy	Palatin Technologies, Inc.	Phase II trials
PA-1806	Bronchial–respiratory	PathoGenesis Corporation	Clinical
Exubera-inhaled insulin	Diabetes therapy	Pfizer	Phase III trials
NNS	Smoking cessation aid	Pharmacia Corporation	Phase III trials
Iloprost	Cardiovascular agent	Schering Plough Corporation	Phase III trials
Salloutanol	Bronchial–respiratory	Sheffield Pharmaceutical	Phase II trials
Pulmicort	Bronchial–respiratory	Astra Zenca PLC	New Labeling Approval
Serevent Dis Kus	Bronchial–respiratory	Glaxo Smith Kline	New Indication Approval

Table 7.5 Drugs in Development (Continued)

Product	*Indication*	*Manufacturer*	*Development*
Nicotrol	Smoking cessation aide	Pharmacia	New Formation Approval
Flunisolide	Severe asthma and lung disease	Bausch & Lomb Pharma	Approved
Cromolyn Sodium	Persistent asthma	Novex Pharma	Approved

- A number of studies have also reported the transport of heavy metals from nose to brain via olfactory pathways.
- Studies with tracer materials such as potassium ferricyanide, horseradish peroxidase, colloid gold, and albumin have shown transport of these substances from nose to brain.
- Various low-molecular-weight drugs such as estradiol, cephalexin, cocaine, and certain peptides have been shown to reach the cerebrospinal fluid (CSF), olfactory bulb, and some other parts of brain after nasal administration.

What has increased the utility of the nasal route is the development of strategies for enhancing nasal delivery of drugs. Two major categories or strategies that have increased such utilization are: (1) manipulation of formulation (either by coadministration with an enzyme inhibitor or an absorption enhancer or by the use of a bioadhesive system) and (2) structural modification of the drug molecule (that is, a prodrug approach) (Ugwoke et al. 2001). Each of these is discussed in the following text.

7.6.2 Advantage of Formulations

- Increase the permeability of the nasal mucosa by interaction of the formulation components with the nasal membrane in a safe, effective, and reversible manner
- Increase in drug solubility and protection against enzymatic degradation
- Increase in the residence time of the drug in the nasal cavity

Commonly used formulations approaches include:

Liquid formulations

- Aqueous solutions
  - Synthetic surfactants
  - Bile salts
  - Phospholipids
  - Cyclodextrins
- Micelles
- Liposomes
- Emulsions

Polymeric microspheres

7.7 EXCIPIENTS USED IN FORMULATION

A wide range of excipients and excipient technologies have become available for use with nasally administered products since the 1980s (Tamulinas, C.B. and Leach, C.L., 2000). A broad overview of these follows.

7.7.1 Surfactants

The effects of surfactants on drug absorption across nasal mucosa has been studied since the 1970s.

In 1981, Hirai et al. compared nonionic, anionic, amphoteric, synthetic surfactants and natural anionic surfactants for *in vivo* nasal absorption of insulin. These authors reported that:

- A majority of the surfactants enhanced insulin absorption relative to the extraction of membrane components.
- Enhancing effects correlate with the extraction efficiency of the membrane components by the surfactants.
- Alkyl glycosides constitute a novel class of sugar-derived surfactants used in cosmetics.
- Maltoside derivatives with an alkyl chair length between 12 and 14 enhance insulin absorption at low surfactant concentrations.
- Mechanism involved the loosening of tight junctions and increasing the paracellular transport.
- Enhancement effect produced by surfactant monomers is related to the ability of the surfactant molecules to penetrate and fluidize the lipid bilayers.
- About carboxymethyl cellulose (CMC), monomers aggregate into micelles that can solubilize components, particularly cholesterol and phospholipids.

7.7.2 Bile Salts

- Enhancement of insulin absorption but with milder effects on the biomembrane.
- Additionally, the absorption-promoting effects appear to arise from an inhibitory effect on the enzyme degradation of insulin.
- Bile salts appeared to be the most promising and effective absorption enhancers of peptides and proteins and still remain widely used as permeation enhancers.

7.7.3 Cyclodextrins

- Cyclodextrins are cyclic oligosaccharides containing a minimum of 6-D-glycopyranose units attached by an alpha 1, 4 linkage.
- Produced by enzymatic conversion of prehydrolized starch.
- Natural cyclodextrins are designated by the Greek letters α, β, and γ. The form is the most soluble of the three.
- The ring structure resembles a truncated cone with the characteristic cavity volume.
- The internal surface of cavity has slight hydrophobic properties whereas the outer surface is hydrophilic.
- Cyclodextrins form inclusion complexes with lipophilic molecules.
- Generally appear to be less irritating to the nasal mucosa than bile salts and surfactants.
- Large increases in solubility through complex formation and protection from enzymatic degradation improve nasal absorption of lipophilic drugs.
- Very potent absorption enhancers for hydrophilic peptides that are not complexed.
- Shao et al. (1992) evaluated several cyclodextrins for their absorption-promoting effect on insulin. The best results for absorption-promoting effect on insulin were obtained with dimethyl-β-cyclodextrin.
- Promoting order correlated well with the extent of nasal mucosal perturbation.

7.7.4 Mixed Micelles

- Nasal absorption of insulin in the presence of sodium glycocholate (NaGC) and linoleic acid increased relative to the increase by NaGC and linoleic acid alone.
- Mixed micelles of bile salts and fatty acids appear to have a synergistic effect on the absorption of peptides.

- Maximal nasal absorption enhancement of [D-Arg] kyotorphin has been observed with mixed micelles of sodium glycocholate and linoleic acid, with an effect greater than that with glycocholate alone.

7.7.5 Formulation and Potential Mucosal Damage

- Improved absorption involves interactions with the mucosal membrane.
- Proposed enhancement mechanisms are:
  - Extraction of membrane components
  - Penetration and fluidization of membrane
  - Loosening of tight junctions
  - Perturbation of nasal mucociliary clearance system
  - Simultaneous transport of environmental toxins

- Adverse effects have to be of short duration, mild, and rapidly reversible.
- Kinetics of lipid and protein extraction from the membrane are measures of the extent of damage evaluated by the measuring activity of membrane marker enzymes:
  - Lactate dehydrogenase: cytosolic enzyme, related to intracellular damage
  - 5'-nucleotidase: membrane-bound enzyme, indicator of membrane perturbations
  - Alkaline phosphatase: membrane-bound enzyme, related to membrane damage

- Ideal characteristics of absorption enhancers include:
  - Pharmacological inertness
  - Nonirritant, nontoxic, and nonallergenic
  - Effect on nasal mucosa should be transient and completely reversible
  - Potent in low concentrations
  - Compatible with other adjuvants
  - Has no offensive odor or taste
  - Inexpensive and readily available

- The factors influencing mucosal damage include:
  - Drug administration
  - Dose
  - Frequency
  - Interspecies difference
  - Sensitivity toward absorption enhancers

- Clinical signs of nasal irritation studies in rats include:
  - In studies of less than 90 d:
    Struggling, sneezing, salivation, head shaking, and nose rubbing
  - In studies of more than 90 d:
    Histological signs of nasal irritation including inflammation of septal and turbinate mucosal surfaces, epithelial and submucosal infiltration of inflammatory cells, purulent exudates, and mucosal hyperplasia

Zhang and Jiang (2001) have recently characterized specific approaches for reduction of the local tissue nasal toxicity of drugs and should be consulted for these.

7.7.6 Methods to Assess Irritancy and Damage

- Erythrocytes
  Used to study the membrane activity of absorption enhancers
- Histology
  Histological studies of nasal membranes
- Intracellular protein release
  Index of cellular damage due to exposure to absorption enhancers
- Tolerability
  These are subjective (double-masked) studies in which individuals report any effects due to the use of enhancers in the formulation.
- Cilia A function
  Cilia beat frequency is obtained from tissue samples at sacrifice using video capture systems.
  Tissues used for ciliary function studies include chicken embryo trachea, cryopreserved human mucosa taken from sphenoidal sinus, rat nasal mucosa, and, recently, human nasal epithelial cells.

7.7.7 Reported Nasal Irritancies Include

- Local irritation, burning, and stinging upon both acute and/or chronic administration were reported for Laureth-9, bile salts, and sodium taurodihydrofusidate.
- Slight nasal itch was reported when dimethyl-cyclodextrin was used for nasal insulin delivery
- Nasal burning and sinusitis were reported during studies involving nasal insulin delivery with glycocholate and methylcellulose

7.8 DELIVERY FORMS

There are a wide range of approaches to clinical drug products and delivery—that is, the actual final form of the drug as administered to patients (Mercer, 1981). The key points on the major classes of these are summarized in the following subsections. It should be noted that if a delivery device is used, it must also be reviewed and approved by the U.S. Food and Drug Administration.

7.8.1 Liquid Nasal Formulation

- These are the most widely used dosage forms in clinical practice.
- Mainly based on aqueous formulations.
- Humidifying effect is convenient because of the drying of mucous membranes owing to allergic and chronic diseases.
- Major drawback is the limitation on microbiological stability.
- Reduced chemical stability of the drug and short residence time in the nasal cavity are other disadvantages.
- Deposition site and deposition pattern are dependent on delivery device, mode of administration, and physicochemical properties of formulation.
- Preparations depend on whether administered for local or systemic applications.
- Patient compliance, cost effectiveness, and risk assessment.

7.8.2 Instillation and Rhinyle Catheter

- Catheters are used for delivery to a defined region.
- The combination of an instillation catheter to a Hamilton threaded plunger syringe has been used to compare the deposition of drops, nebulizers, and sprays in rhesus monkeys.
- These are used only for experimental studies and not for commercial clinical products.

7.8.3 Drops

- This is one of the oldest delivery systems.
- Low-cost devices utilized.
- Easy to manufacture.
- Disadvantages related to microbiological and chemical stability.
- Delivered volume cannot be controlled.
- Formulation can be easily contaminated by pipette (delivery device).

7.8.4 Powder Dosage Forms

- Dry powders are less frequent in nasal drug delivery.
- Major advantages include lack of need for preservatives and improved drug stability.
- Prolonged retention times in nasal region when compared to solutions.
- Addition of bioadhesive excipients results in further decreased clearance rates.
- Nasal powders may increase patient compliance especially for children if smell and taste of drug are otherwise unacceptable.

7.8.5 Insufflators and Monodose Powder Inhaler

- Many insufflators work with predosed powder doses in capsules.
- The use of gelatin capsules enables the filling and application of different amounts of powder.
- In a monodose powder inhaler, pushing a piston results in the precompression of air in a chamber.
- The piston pierces a membrane, and the expanding air expels air into the nostrils.

7.8.6 Pressurized MDIs

- They are manufactured by suspending the drug in liquid propellants with the aid of surfactants.
- Physicochemical compatibility between the drug and propellants must be evaluated.
- Phase separation, precipitation, crystal growth, polymorphism, dispersibility, and adsorption of the drug influence the drug particle size, dose distribution, and deposition pattern.
- Their advantages include portability, small size, availability over a wide dose range, dose consistency and accuracy, and protection of contents.
- Their disadvantages include nasal irritation by propellants and depletion of ozone layer by CFCs.

7.8.7 Nasal Gels

- Nasal administration of gels can be achieved by precompression pumps.
- The deposition of gel in the nasal cavity depends on the mode of administration because of its viscosity and poor spreading properties.
- Nasal gel containing vitamin B12 for systemic administration is available in the market.

7.8.8 Patented Nasal Formulations

- West Pharma developed its nasal technology (ChiSys) based on the use of chitosan as an absorption enhancer.
- Chitosan is a natural polysaccharide with bioadhesive properties.
- It prolongs the retention time of the formulation in the nasal cavity.
- It may facilitate absorption through promoting paracellular transport.

7.9 METHODS FOR SAFETY ASSESSMENT OF INHALED THERAPEUTICS

Methods for evaluation of inhalation toxicity should be selected according to the pharmacological and toxicological questions asked, and the design of experiments should specify the delivery route of a drug to the target sites in the pulmonary system. For example, if an immunologic response of the lungs to a drug is in question, then the lymphoid tissues of the lungs should be the major target of evaluation. The following are some of the physiological, biochemical, and pharmacological tests that are applicable for the safety assessment of inhaled medicinal gases, vapors, or aerosols.

URT irritation can occur from inhalation of a medicinal gas (nitrous oxide), vapor (salicylates), or aerosol (virtually any using a surfactant). For assessing the potential of an inhalant to cause URT irritation, the mouse body plethysmographic technique (Alarie 1966, 1981a, 1981b) has proven to be extremely useful. This technique operates on the principle that respiratory irritants stimulate the sensory nerve endings located at the surface of the respiratory tract from the nose to the alveolar region. The nerve endings, in turn, stimulate a variety of reflex responses (Alarie 1973; Widdicombe 1974) that result in characteristic changes in inspiratory and expiratory patterns and, most prominently, depression of the respiratory rate. Both the potency of irritation and the concentration of the irritant are positively related to the magnitude of respiratory rate depression. The concentration response can be quantitatively expressed in terms of RD_{50}, defined as the concentration (in logarithmic scale) of the drug in the air that causes a 50% decrease in respiratory rate. The criteria for positive URT irritation in intact mice exposed to the drug atmosphere are depression in breathing frequency and a qualitative alteration of the expiratory patterns. Numerous experimental results have shown that the responses of mice correlated almost perfectly with those of humans (Alarie et al. 1980; Alarie and Luo 1984). Thus, this technique is useful for predicting the irritancy of airborne medicinal compounds in humans. From the drug-formulating point of view, an inhalant drug with URT-irritating properties indicates the need for an alternate route of administration. From the industrial hygiene point of view, the recognition of the irritant properties is very important. If a chemical gas, vapor, or aerosol irritates, it has a *warning property*. With an adequate warning property, a worker will avoid inhaling damaging amounts of the airborne toxicant; without such a warning property, a worker may unknowingly inhale a harmful amount of the toxicant. However, warning properties are not very reliable for alerting individuals to potential health risks because a person can become tolerant or adapt to the smell or irritation properties of toxicants.

Respiratory tract irritation can alter absorption of a therapeutic agent in at least two ways. If irritation causes cell death, loss, or both, such damage will act to improve the extent of absorption. If, however, irritation leads to increased secretion of mucus, it can serve to act as an increased barrier to absorption and, therefore, decreased systemic drug availability.

Inhalation of a cardiovascular drug such as an aerosol of propranolol (a beta-adrenergic receptor agonist) may affect the respiratory cycle of a subject. For evaluating the cardiopulmonary effects of an inhalant, the plethysmograph technique using a mouse or a guinea pig model is useful. The criteria for a positive response in intact mice or guinea pigs are changes in the duration of inspiration and expiration, and the interval between breaths (Schaper 1989).

Pulmonary sensitization may occur from inhalation of drug vapors such as enflurane (Schwettmann and Casterline 1976) and antibiotics such as spiramycin (Davies and Pepys 1975) and tetracycline (Menon and Das 1977). To detect pulmonary sensitization from inhalation of drug and chemical aerosols, the body plethysmographic technique using a guinea pig model has been shown to be useful (Patterson and Kelly 1974; Karol 1988; Karol et al. 1989; Thornea and Karol 1989). The criteria for positive pulmonary sensitization in intact guinea pigs are changes in breathing frequency and their extent, and the time of onset of an airway-constrictive response after induction and after a challenge dose of the test drug (Karol et al. 1989).

The mucociliary transport system of the airways can be impaired by respiratory irritants, local analgesics, and anesthetics, and parasympathetic stimulants (Pavia 1984). Any one of these agents will retard the beating frequency of the cilia and the secretion of the serous fluid of the mucous membranes. As a result, the propulsion of the inhaled particles, bacteria, or endogenous debris toward the oral pharynx for expectoration or swallowing will be retarded. Conversely, inhalation of adrenergic agonists increases the activity of the mucociliary transport system and facilitates the elimination of noxious material from the pulmonary system. Laboratory evaluation of the adverse drug effects on mucociliary transport in animal models can be achieved by measuring the velocity of the linear flow of mucus in the trachea of surgically prepared animals (Rylander 1966). Clinically, the transportation of markers placed on the tracheal epithelium of normal human subjects can also be observed using a fiber-optic bronchoscopic technique (Pavia et al. 1980; Mussatto et al. 1988). The criteria of a positive response are changes in the transport time over a given distance of markers placed on the mucus, or changes in the rate of mucus secretion (Davis et al. 1976; Johnson et al. 1983, 1987; Webber and Widdicombe 1987). More comprehensive discussion on mucociliary clearance can be found in several reviews (Last 1982; Pavia 1984; Clarke and Pavia 1984).

Cytological studies on the bronchial alveolar lavage fluid (BALF) permit the evaluation of the effects of an inhaled drug on the epithelial lining of the respiratory tract. This fluid can be obtained from intact animals or from excised lungs (Henderson 1984, 1988, 1989). Quantitative analyses of fluid constituents such as neutrophils, antibody-forming lymphocytes, and antigen-specific IgG provide information on the cellular and biochemical responses of the lungs to the inhaled agent (Henderson 1984; Henderson et al. 1985, 1987). For example, BALF parameters were found to be unperturbed by the inhalation of halothane (Henderson and Loery 1983). The criteria of a positive response are increase in protein content, increase in the number of neutrophils and macrophages for inflammation; increase in the number of lymphocytes and alteration of lymphocyte profiles for immune response; increase in cytoplasmic enzymes (lactate dehydrogenase) for cell lysis (Henderson 1989); and the presence of antigen-specific antibodies for specific immune responses (Bice 1985).

Morphological examination of the cellular structure of the pulmonary system is the foundation of most inhalation toxicity studies. Inhalation of airborne drug vapors or aerosols at harmful concentrations results mainly in local histopathologic changes in the epithelial cells of the airways, of which there are two types: nonciliated and ciliated cells. The nonciliated cells are the Clara cells, which contain secretory granules with smooth endoplasmic reticulum (SER); the secretory granules that lack SER; and the brush cells, which have stubby microvilli and numerous cytoplasmic fibers on their free surfaces. If the concentration gradient of the drug in the lung is high enough to reach the alveoli, the type I alveoli cells will also be affected (Evans 1982). Drugs that affect the lungs via the bloodstream, such as bleomycin (Aso et al. 1976), cause changes to the endothelial cells of the vascular system that result in diffuse damage to the alveoli. The criteria of cellular damage are loss of cilia, swelling, and necrosis and sloughing of cell debris into the airway lumina. Tissues recovering from injuries are characterized by increases in the number of dividing progenitor cells, followed by increases in intermediate cells that eventually differentiate into normal surface epithelium.

Pulmonary drug disposition studies are essential in research and development of new inhalant drugs. Inhaled drugs are usually absorbed and metabolized to some extent in the lungs because the lungs, similar to the liver, contain active enzyme systems. A drug may be metabolized to an inactive compound for excretion or to highly reactive toxic metabolites that cause pulmonary damage. In most pulmonary disposition studies, a gas or vapor is delivered via whole-body exposure (Paustenbach et al. 1983) or head-only exposure (Hafner et al. 1975). For aerosols, over 90% of a dose administered by mouth breathing is deposited in the oropharynx and swallowed. Consequently, the disposition pattern reflects that of ingestion in combination with a small contribution from pulmonary metabolism. For determining the disposition of inhaled drugs by the pulmonary system alone, a dosimetric endotracheal nebulization technique (Leong et al. 1988) is useful. In this technique, microliter quantities of a radiolabeled drug solution can be nebulized within the trachea using a miniature air–liquid nebulizing nozzle. Alternatively, a small volume of liquid can be dispersed endotrachially using a microsyringe. In either technique, an accurate dose of a labeled drug solution is delivered entirely into the respiratory tract and lungs. Subsequent radioassay of the excreta thus reflects only the pulmonary disposition of the drug without complication from aerosols deposited in the oropharyngeal regions as would be the case if the drug had been delivered by mouth inhalation. For example, in a study of the antiasthmatic drug lodoximide tromethamine, the urinary metabolites produced by beagle dogs after receiving a dose of the radiolabeled drug via endotracheal nebulization showed a high percentage of the intact drug. However, metabolites produced after oral administration were mainly nonactive conjugates. The differences were due to the drug's escape from first-pass metabolism in the liver when it was administered through the pulmonary system. The results thus indicated that the drug had to be administered by inhalation to be effective. This crucial information was extremely important in the selection of the most effective route of administration and formulation of this antiasthmatic drug (Leong et al. 1988).

Cardiotoxicity of inhalant drugs should also be evaluated. For example, adverse cardiac effects may be induced by inhaling vapors of fluorocarbons, which are used extensively as propellants in drug aerosols. Inhalation of vapors of anesthetics has also been shown to cause depression of the heart rate and alteration of the rhythm and blood pressure (Merin 1981; Leong and Rop 1989). More importantly, inhalation of antiasthmatic aerosols of beta-receptor agonists delivered in a fluorocarbon propellant have been shown to cause marked tachycardia, electrocardiogram (ECG) changes, and sensitization of the heart to arrhythmia (Aviado 1981; Balazs 1981). Chronic inhalation of drug aerosols can also result in cardiomyopathy (Balazs 1981). For detection of cardiotoxicity, standard methods of monitoring arterial pressures, heart rate, and ECGs of animals during inhalation of a drug or at frequent intervals during a prolonged treatment period should be useful in safety assessments of inhalant drugs.

Because the inhalation route is just a method for administering drugs, other nonpulmonary effects, such as behavioral effects (Ts'o et al. 1975), and renal and liver toxicity, should also be evaluated. In addition, attention should be given to drugs such as bleomycin, which are not administered via the inhalation route but which accumulate in the lungs, where they cause pulmonary damage (Wilson 1982).

7.10 PARAMETERS OF TOXICITY EVALUATION

Paracelsus stated over 400 yr ago that "All substances are poison. The right dose differentiates a poison and a remedy." Thus, in safety assessments of inhaled drugs, the dose, or magnitude of inhalation exposure, in relation to the physiological, biochemical, cytological, or morphological responses must be determined. Toxicity information is essential to establishing guidelines to prevent the health hazards of acute or chronic overdosage during therapy or of unintentional exposure to bulk drugs and their formulated products during manufacturing and industrial handling.

7.10.1 The Inhaled Dose

Most drugs are designed for oral or parenteral administration in which the dose is calculated in terms of drug weight in milligrams (mg) divided by the body weight in kilograms (kg):

$$dose = \frac{drug\ weight\,(\mathrm{mg})}{body\ weight\,(\mathrm{kg})} = \mathrm{mg/kg}$$

For inhalant drugs, the inhaled dose has been expressed in many mathematical models (Dahl 1990). However, the practical approach is based on exposure concentration and duration rather than on theoretic concepts. Thus, an inhaled dose is expressed in terms of the exposure concentration (C) in milligrams per liter (mg/l) or milligrams per cubic meter (mg/m^3), or, less commonly, parts per million (ppm) parts of air, the duration of exposure (t) in minutes, the ventilatory parameters including the respiratory rate (R) in number of breaths per minute, and the tidal volume (Tv) in liters per breath, and a dimensionless retention factor α (alpha), which is related to the reactivity and the solubility of the drug. The product of these parameters divided by the body weight in kilograms gives the dose:

$$dose = \frac{C \cdot t \cdot R \cdot Tv \cdot \alpha}{body\,weight} = \text{mg/kg}$$

In critical evaluation of the effect of a gas, vapor, or aerosol inhaled in to the respiratory tract of an animal, the dosimetric method has been recommended (Oberst 1961). However, due to the complexity of measuring the various parameters simultaneously, only a few studies on gaseous drugs or chemicals have employed the dosimetric method (Weston and Karel 1946; Leong and MacFarland 1965; Landy et al. 1983; Stott and McKenna 1984; Dallas et al. 1986, 1989). For studies on liquid or powdery aerosols, modified techniques such as intratracheal instillation (Brain et al. 1976) or endotracheal nebulization (Leong et al. 1988) were used to deliver an exact dose of the test material into the lower respiratory tract (LRT) while bypassing the URT and ignoring the ventilatory parameters. These methods deliver a bolus to the lung which does not mimic the exposure and distribution pattern achieved by actual inhalation.

In routine inhalation studies, it is generally accepted that the respiratory parameters are relatively constant when the animals are similar in age, sex, and body weight. This leaves only C and t to be the major variables for dose consideration.

$$dose = C \cdot t = \text{mg} \cdot \text{min/l}$$

The product Ct is not a true dose because its unit is mg · min/l rather than mg/kg. Nevertheless, Ct can be manipulated as though it were a dose—an approximated dose (MacFarland 1976).

The respiratory parameters of an animal will dictate the volume of air inhaled and hence the quantity of test material entering the respiratory system. Commonly used parameters for a number of experimental species and man are given in Table 7.6 to illustrate this point and include the alveolar surface area because this represents the target tissue for most inhaled materials. It can be seen that by taking the ratios of these parameters and comparing the two extremes, i.e., mouse and man, that (McNeill 1964) a mouse inhales approximately 30 times its lung volume in one min, whereas a man at rest inhales approximately the same volume as that of his lung. This can increase with heavy work up to the same ratio as the mouse but is not sustained for long periods. This means that the dose per unit lung volume is up to 30 times higher in the mouse than man at the same inhaled atmospheric concentration

Table 7.6 Respiratory Parameters for Common Experimental Species and Humans

Species	*Body Weight (kg)*	*Lung Volume (ml)*	*Minute Volume (ml min–1)*	*Alveolar Surface Area (m2)*	*Lung Volume % Surface Area*	*Minute Volume % Lung Volume*	*Minute Volume % Surface Area*
Mouse	0.023	0.74	24	0.068	10.9	32.4	353
Rat	0.14	6.3	†84	0.39	16.2	13.3	215
Monkey	3.7	184	694	13	14.2	3.77	53
Dog	22.8	1501	2923	90	16.7	1.95	33
Human	75	7000	6000	82	85.4	0.86	73

Source: From Altman (1974).

(Touvay and Le Mosquet 2000). The minute volume of the mouse is in contact with five times less alveolar surface area than man; hence, the dose per unit area is up to five times greater in the mouse. (Akoun 1989) The lung volume in comparison with the alveolar surface area in experimental animals is less than in humans, meaning that the extent of contact of inhaled gases with the alveolar surface is greater in experimental animals.

Although it is possible, and common, to refer to standard respiratory parameters for different species to calculate inhaled dose and deposited dose with time, the inhaled materials usually influence the breathing patterns of test animals. The most common examples of this are irritant vapors, which can reduce the respiratory rate by up to 80%. This phenomenon results from a reflexive pause during the breathing cycle due to stimulation by the inhaled material of the trigeminal nerve endings situated in the nasal passages. The duration of the pause and hence the reduction in the respiratory rate are concentration related, permitting concentration–response relationships to be plotted. This has been investigated extensively by Alarie (1981a) and forms the basis of a test screen for comparing quantitatively the irritancy of different materials; it has found application in assessing appropriate exposure limits for human exposure when respiratory irritancy is the predominant cause for concern.

Although irritancy resulting from this reflex reaction is one cause of altered respiratory parameters during exposure, there are many others. These include other types of reflex responses, such as bronchoconstriction, the narcotic effects of many solvents, the development of toxic signs as exposure progresses, or simply a voluntary reduction in respiratory rate by the test animal due to the unpleasant nature of the inhaled atmosphere. The extent to which these affect breathing patterns, and hence inhaled dose, can only be assessed by actual measurement.

By simultaneous monitoring of tidal volume and respiratory rate or minute volume, and the concentration of an inhaled vapor in the bloodstream and the vapor in the exposure atmosphere, pharmacokinetic studies on the $C \cdot t$ relationship have shown that the effective dose was nearly proportional to the exposure concentration for vapors such as 1,1, 1-trichloroethane, which has a saturable metabolism, and found that the steady-state plasma concentrations were disproportionally greater at higher exposure concentrations.

Acknowledging the possible existence of deviations, this simplified approach of using C and t for dose determination provides that basis for dose–response assessments in practically all inhalation toxicological studies.

7.10.2 The Dose–Response Relationship

However, Figure 7.2 illustrates the classic viewpoint of dose response. The reader will recognize, as a starting point, that at least three dimensions are present, and all must be considered to understand a biologic system response. As dose increases:

- Incidence of responders in an exposed population increases (population incidence).
- Severity of response in affected individuals increases (severity).
- Time to occurrence of response or of progressive stage of response decreases (lag time).

The oldest principle of dose–response determination in inhalation toxicology is based on Haber's rule, which states that responses to an inhaled toxicant will be the same under conditions where C varies in a complementary manner to t (Haber 1924), for example, if $C \cdot t$ elicits a specific magnitude of the same response; that is, $Ct = K$, where K is a constant for the stated magnitude of response. This was first developed for use with war gases, and it holds up reasonably well for shorter exposure to such agents.

This rule also maintains fairly well when C or t varies within a narrow range for acute exposure to a gaseous compound (Rinehart and Hatch 1964) and for chronic exposure to an inert particle (Henderson et al. 1991). Excursion of C or t beyond these limits will cause the assumption $Ct = K$ to be incorrect (Adams et al. 1950, 1952; Sidorenko and Pinigin 1976; Andersen et al. 1979; Uemitsu et al. 1985). For example, an animal may be exposed to 1000 ppm of diethyl ether for 420 min or 1400 ppm for

300 min at a constant rate without incurring any anesthesia. However, exposure to 420,000 ppm for 1 min will surely cause anesthesia or even death of the animal. Furthermore, toxicokinetic study of liver enzymes affected by inhalation of carbon tetrachloride (Uemitsu 1985), which has a saturable metabolism in rats, showed that $Ct = K$ does not correctly reflect the toxicity value of this compound. Therefore, the limitations of Haber's rule must be recognized when it is used in interpolation or extrapolation of inhalation toxicity data, and is not recommended for most inhalation situations.

7.10.3 Exposure Concentration vs. Response

In certain medical situations (e.g., a patient's variable exposure duration to a surgical concentration of an inhalant anesthetic, or the repeated exposures of surgeons and nurses to subanesthetic concentrations of an anesthetic in the operating theater) it is necessary to know the duration of safe exposure to a drug. Duration safety can be assessed by determining a drug's median effective time (Et_{50}) or median lethal time (Lt_{50}). These statistically derived quantities represent the duration of exposure required to affect or kill 50% of a group of animals exposed to a specified concentration of an airborne drug or chemical in the atmosphere.

The graph in Figure 7.3 is the probit plot of cumulative percentage response to logarithm of exposure duration. It shows the 1000 mg/m^3 for 10 h to 10 mg/m^3 for 1000 h, each with a Ct (an approximated dose) of ~10,000 h mg/m^3. Similar to concentration–response graphs, the slopes indicate the differences in the mechanism of action and the margins of safe exposure of the three drugs. The ratio of the ET_{50} or LT_{50} of two drugs indicate their relative toxicity, and the ratio of ET_{50} over LT_{50} of the same drug is the therapeutic ratio.

7.10.4 Product of Concentration and Duration (*Ct*) vs. Responses

To evaluate inhalation toxicity in situations in which workers are exposed to various concentrations and durations of a drug vapor, aerosol, or powder during manufacturing or packaging in the work environment, a more comprehensive determination of $E(Ct)_{50}$ or $L(Ct)_{50}$ values is used. The $E(Ct)_{50}$ and $L(Ct)_{50}$ values are statistically derived values that represent the magnitude of exposure, expressed as a function of the product of C and t, that is expected to affect or kill approximately 50% of the animals exposed. The other curve represents exposures that kill 50% or >50% of each group of animals (Irish and Adams 1940).

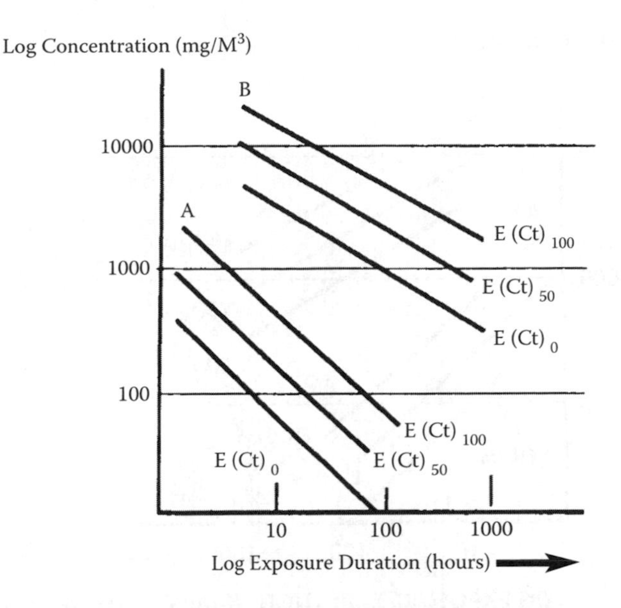

Figure 7.3 Dose–response plotted in terms of the probit plot of cumulative percentage response to logarithm of exposure duration.

The graph in Figure 7.4 illustrates inhalation exposures to a drug using various combinations of C and t that kill 50% of the animals. For example, 50% mortality occurs when a group of animals is exposed to drug A at a concentration of 1000 mg/m^3 for a duration of approximately 2 h, or at a concentration of 100 mg/m^3 for a duration of approximately 20 h. Furthermore, the graph also illustrates that the inhalation toxicity of drug A is more than one order of magnitude higher than that of drug B. For example, an exposure to drug A at the concentration of 100 mg/m^3 for 100 h kills 100% of the animals, whereas an exposure to drug B at the concentration of 1000 mg/m^3 for 100 h does not kill any.

7.10.5 Units for Exposure Concentration

For therapeutic gases and vapors, exposure concentrations are traditionally expressed in parts per million (ppm). The calculation for the ppm of a gas or vapor in an air sample is based on Avogadro's law, which states that equal volumes contain equal numbers of molecules under the same temperature and pressure. In other words, under standard temperature and pressure (STP), one gram molecular weight (mole) of any gas under a pressure of one atmosphere (equivalent to the height of 760 mm mercury) and a temperature of 273 K has the same number of molecules

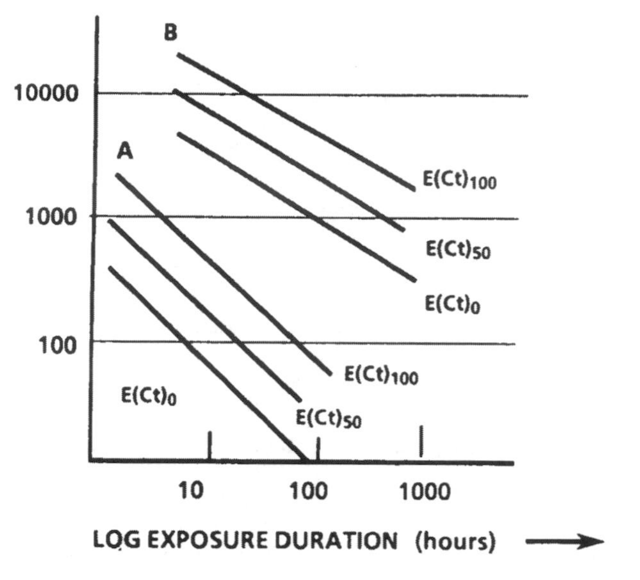

Figure 7.4 Dose–response plotted in terms of logarithms of drug exposure concentration and durations.

and occupies the same volume of 22.4 l. However, under ambient conditions, the volume of 22.4 l has to be corrected to a larger volume based on Charles' law, which states that the volume of a gas varies directly with the absolute temperature at constant pressure. Thus, at a room temperature of 25°C, one mole of a gas occupies a volume of 24.5 l.

$$22.4\,\text{l} \times \frac{298\ \text{K}}{273\ \text{K}} = 24.5\,\text{l}$$

Further correction of volume for an atmospheric pressure deviation from one atmosphere may be done by applying Boyle's law, which states that the volume of a gas without change of temperature varies inversely with the pressure applied to it.

$$24.5\,\text{l} \times \frac{758\ \text{mm Hg}}{760\ \text{mm Hg}} = 24.4\,\text{l}$$

In practice, atmospheric pressure in most animal experimental environments usually varies only a few mm of Hg, so little or no correction is required.

Using the aforementioned principles, the volume of a vapor generated from a given weight of a liquid can be calculated. For example, 1 mole of water weighs 18 g, whereas 1 mole of ethanol weighs 46 g. When 1 mole of each liquid is totally vaporized, each will occupy the same volume of 24.5 l at room temperature (25°C) and pressure (760 mmHg). In an inhalation experiment, if the volume of test liquid and the rate of airflow being mixed in the animal exposure chamber are known, the vapor concentration in the chamber atmosphere can be calculated in parts per million or milligrams per liter. A conversion table published by the U.S. Bureau of Mines enables quick conversion between parts per million and milligrams per liter for compounds with molecular weights up to 300 g (Fieldner et al. 1921; Patty 1958).

For aerosols of nonvolatile liquid and powder pharmaceutical compounds, the concentration of the mist or dust atmosphere must be expressed in terms of milligrams per liter or milligrams per cubic meter (mg/m^3) of air. With advances in biotechnology, many pharmacological testing techniques are based on specific receptor bindings, in which the ratio of the number of molecules to those of the receptors are considered, in which case the exposure concentration may be more appropriately expressed in micromoles per unit volume of air ($\mu mol/m^3$).

7.11 RESPIRATORY SAFETY PHARMACOLOGY

Pharmaceuticals differ from industrial and environmental chemicals in that the scope of concern for their adverse safety effects on the respiratory system extend both to reversible functional degradations and to effects on the respiratory system functionality due to systemically distributed agents administered by routes other than direct respiratory. This is the realm of the relatively new field of safety pharmacology.

As early as 1964, it became apparent that β-adrenergic blocking agents could lead to bronchoconstriction (and possible death) in asthmatics (McNeill 1964). Since then, many similar adverse effects have been identified. These known effects of drugs from a variety of pharmacologic/therapeutic classes on the respiratory system are summarized in Tables 7.7 to Table 7.9. Resulting worldwide regulatory requirements (Table 7.10 and Table 7.11) require the conduct of prescribed respiratory evaluations prior to drugs in humans. The objective of such studies is to evaluate the potential for drugs to cause unintended pharmacologic or toxicologic effects that influence respiratory function. Changes in respiratory function can result either from alterations in the pumping apparatus that controls the pattern of pulmonary ventilation or from changes in the mechanical

Table 7.7 Drugs Known to Cause Pulmonary Disease

Chemotherapeutic	*Analgesics*
	Heroin[a]
Cytotoxic	
Azathioprine	Methadone[a]
Bleomycin[a]	Noloxone[a]
Busulfan	Ethchlorvynol[a]
Chlorambucil	Propoxyphene[a]
Cyclophosphamide	Salicylates[a]
Etoposide	
Melphalan	
	Cardiovascular
Mitomycin[a]	Amiodarone[a]
Nitrosoureas	Angiotensin-converting enzyme inhibitors
Procarbazine	
Vinblastine	Anticoagulants
Ifosfamide	Beta-blockers[a]
	Dipyridamole
Noncytotoxic	
Methotrexate[a]	Fibrinolytic agents[a]
Cytosine arabinoside[a]	Protamine[a]
Bleomycin[a]	Tocainide
Procarbazine[a]	
	Inhalants
	Aspirated oil
Antibiotic	
Amphotericin B[a]	Oxygen[a]
Nitrofurantoin	
Acute[a]	
	Intravenous
Chronic	Blood[a]
Sulfasalazine	Ethanolamine oleate (sodium morrhuate)[a]
Sulfonamides	Ethiodized oil (lymphangiogram)
Pentamidine	Talc
	Fat emulsion
Antiinflammatory	

Table 7.7 Drugs Known to Cause Pulmonary Disease (Continued)

Chemotherapeutic	*Analgesics*
Acetylsalicyclic acid[a]	
Gold	
	Miscellaneous
Methtrexate	Bromocripitine
Nonsteroidal anti-inflammatory agents	Dantrolene
	Hydrochlorothiazide[a]
Penicillamine[a]	Methysergide
	Oral contraceptives
	Tocolytic agents[a]
Immunosuppressive	Tricyclics[a]
	L-Tryptophan
Cyclosporin	Radiation
Interleukin-2[a]	Systemic lupus erythematosus (drug-induced) [a]
	Complement-mediated leukostasis[a]

[a]Typically causes acute or subacute respiratory insufficiency.

Source: From Touvay, C. and LeMosquet, B. (2000); Akoun, G.M. (1989). Natural history of drug-induced pneumonitis. In *Drug Induced Disorders Volume 3*: *Treatment Induced Respiratory Disorders* (Akoun, G.M., White J.P., Eds.). Elsevier, New York, pp. 3–9; Dorato, M.A. (1994). Toxicological evaluation of intranasal peptide and protein drugs. *Drugs Pharm Sci* 62: 345–381; Lalej-Bennis, D. (2001). Six-month administration of gelified intranasal insulin in 16 type 1 diabetic patients under multiple injections: efficacy vs. subcutaneous injections and local tolerance. *Diabetes Metab* 27(3): 372–377; Mauderly, J.L. (1989). Effects of inhaled toxicants on pulmonary function. In *Concepts in Inhalation Toxicology* (McClellan, R.O., Henderson, R.F., Eds.). New York, Hemisphere Publishing, pp. 347–401; Rosnow, E.C. (1992). Drug-induced pulmonary disease: an update. *Chest* 102: 239–250.

properties of the lung that determine the transpulmonary pressures (work) required for lung inflation and deflation.

Under ICH (S7A) and FDA guidelines, all new drugs (with limited exceptions; see Gad 2003) must be evaluated for pharmacologic safety in three core organ systems (the central nervous system, cardiovascular

Table 7.8 Drugs That Adversely Affect Respiratory Function

Drugs Known to Cause or Aggravate Bronchospasm	*Agents Associated with Pleural Effusion*
Vinblastine	Chemotherapeutic agents
Nitrofurantoin (acute)	Nitrofurantoin (acute)
Acetylsalicyclic acid	Bromocriptine
Nonsteroidal anti-inflammatory agents	Dantrolene
Interleukin-2	Methysergide
Beta-blockers	L-Tryptophan
Dipyridamole	Drug-inducing systemic lupus erythematosus
Protamine	Tocolytics
Nebulized pentamidine, beclomethasone, and propellants	Amiodarone
Hydrocortisone	Esophageal variceal sclerotherapy agents
Cocaine	Interleukin-2
Propafenone	
Agents Associated with Acute-Onset Pulmonary Insufficiency[a]	**Agents That Cause Subacute Respiratory Failure**
Bleomycin plus O_2	Chemotherapeutic agents
Mitocycin	Nitrofurantoin (chronic)
Bleomycin[b]	Amiodarone
Procarbazine[b]	L-Tryptophan
Methotrexate[b]	Drug-inducing systemic lupus erythematosus
Amphotericin B	
Nitrofurantoin (acute)[c]	
Acetylsalicyclic acid [c]	
Interleukin-2[c]	
Heroin and other narcotics[c]	
Epinephrine[c]	
Ethchlorvynol[c]	
Fibrinolytic agents	
Protamine	
Blood products[c]	
Fat emulsion	
Hydrochlorothiazide	
Complement-mediated leukostasis	

Table 7.8 Drugs That Adversely Affect Respiratory Function (Continued)

Drugs Known to Cause or Aggravate Bronchospasm	*Agents Associated with Pleural Effusion*
Hyskon (dextran-70)[c]	
Tumor necrosis factor[c]	
Intrathecal methotrexate	
Tricyclic antidepressants[c]	
Amiodarone plus O_2	
Naloxone	

[a] Onset at less than 48 h

[b] Associated with hypersensitivity with eosinophilia

[c] Usually reversible within 48 to 72 h, implying noncardiac pulmonary edema rather than inflammatory interstitial pneumonitis.

Source: From McNeill, R.S. (1964). Effect of a β-adrenergic blocking agent, propranol, on asthmatics. *Lancet.* November 21: 1101–1102; Borison, H.L. (1977). Central nervous system depressants: control-systems approach to respiratory depression. *Pharmacol Ther B* 3: 211–226; Tattersfield, A.E. (1986). Beta adrenoreceptor antagonists and respiratory disease. *J Cardiovasc Pharmacol* 8 (Suppl. 4): 535–539; Illum, L. and Davis, S.S. (1992); Shao, Z. and Mitra, A. K. (1992); Shao, Z. and Mitra, A.K. (1992). Nasal membrane and intracellular protein and enzyme release by bile salts and bile salt fatty-acid mixed micelles-correlation with facilitated drug transport. *Pharm Res* 9: 1184–1189; Fariba et al. (2002).

system, and respiratory system). Table 7.10 presents the required determinations under these regulations for mandated respiratory system evaluations.

An example set of data for some standard positive controls is shown in Table 7.12. Note that each of the different agents has a separate pattern of effects.

The respiratory system is responsible for generating and regulating the transpulmonary pressures needed to inflate and deflate the lung. Normal gas exchange between the lung and blood requires breathing patterns that ensure appropriate alveolar ventilation. Ventilatory disorders that alter alveolar ventilation are defined as *hypoventilation* or *hyperventilation syndromes.* Hyperventilation results in an increase in the partial pressure of arterial CO_2 above normal limits and can lead to acidosis, pulmonary hypertension, congestive heart failure, headache, and disturbed sleep. Hypoventilation results in a decrease in the partial pressure of arterial CO_2 below normal limits and can lead to alkalosis, syncope, epileptic attacks, reduced cardiac output, and muscle weakness (Indans, I., 2002).

Table 7.9 Drugs Known to Influence Ventilatory Control

Depressants	*Stimulants*
Inhaled anesthetics	Alkaloids
Barbiturates	Nicotine
Benzodiazepines	Lobeline
Diazepam	Piperdine
Temazapan	Xanthine analogs
Chlordiazepoxide	Theophyline
Serotonin analogs	Caffeine
Methoxy-(dimethyl)-tryptamine	Theobromine
Dopamine analogs	Analeptics
Apomorphine	Doxapram
Adenosine analogs	Salicylates
2-Chloroadenosine	Progesterone analogs
R-Phenylisopropyl-adenosine	Almitrine
(R-PIA)	Glycine analogs
N-Ethylcarboxamide (NECA)	Strychnine
B-Adrenergic antagonists	GABA antagonists
Timolol maleate	Picrotoxin
GABA analogs	Bicuculline
Muscimol	Serotonin synthesis inhibitors
Baclofen	p-Chlorophenylalanine
Opiates	Reserpine
Morphine	
Codeine	
Methadone	
Meperidine	
Phenazocine	
Tranquilizers/Analgesics	
Chlorpromazine	
Hydroxyzine	
Rompum (xylazine)	
Nalorphine	

Table 7.10 Required Respiratory System Safety Pharmacology Evaluation

Respiratory Functions
Measurement of rate and relative tidal volume in conscious animals
Pulmonary function
Measurement of rate, tidal volume, and lung resistance and compliance in anaesthetized animals

Table 7.11 Regulatory Documents Recommending Respiratory Function Testing in Safety Pharmacology Studies

U.S.	FDA Guideline for the Format and Content of the Nonclinical Pharmacology/Toxicology Section of an Application (Section IID, p. 12, February 1987)
Japan	Ministry of Health and Welfare Guidelines for Safety Pharmacology Studies Required for the Application for Approval to Manufacture (Import) Drugs. Notification YAKUSHIN-YAKU No. 4, January 1991.
Australia	Guidelines for preparation and presentation of Applications for Investigational Drugs and Drug Products Under the Clinical Trials Exemption Scheme (STET 12, 15).
Canada	RA5 Exhibit 2, Guidelines for Preparing and Filing Drug Submissions (p. 21).
U.K.	Medicines Act 1968, Guidance Notes on Applications for Product Licenses (MAL 2, p. A3F-1).

Normal ventilation requires that the pumping apparatus provide both adequate total pulmonary ventilation (minute volume) and the appropriate depth (tidal volume) and frequency of breathing. The depth and frequency of breathing required for alveolar ventilation are determined primarily by the anatomic deadspace of the lung. In general, a rapid shallow breathing pattern (tachypnea) is less efficient than a slower, deeper breathing pattern that achieves the same minute volume. Thus, any change in minute

Table 7.12 Functional Respiratory Responses to Standard Pharmacologic Agents

Parameters	*Theophyline 10 mg/kg PO*	*Pentobarbital 35 mg/kg IP*	*Diazepam 35 mg/kg IP*	*Codeine 100 mg/kg IP*
F(breaths/min)	+ + +	– – –	– – –	No change
TV (ml)	No change	No change	No change	-
Ti (s)	– –	+ +	+ +	+
Te (s)	– –	+ + +	+ +	-
PIF (ml/s)	+ +	-	-	-
PEF (ml/s)	+ +	No change	+	-

Note: Where + is an increase and – a decrease, and s are seconds. F is respiratory rate, TV tidal volume, Ti inhalation time or duration, Tc exhalation time, PIF the pulmonary inhalation rate, and PEF the pulmonary exhalation rate.

Source: From Touvay, C. and Le Mosquet, B. (2000). Systeme respirataire et pharmacolgie de securite. *Therapie* 55: 71–83.

volume, tidal volume, or the rate of breathing can influence the efficiency of ventilation (Milic-Emili 1982). The inspiratory and expiratory phases of individual breath rates of airflow and durations that are distinct and independently controlled (Boggs 1992). Thus, by characterizing changes in the airflow rate and duration of each of these phases, mechanisms responsible for changes in tidal volume or respiratory rate can be identified (Milic-Emili 1982). For example, a decrease in airflow during inspiration (the active phase) is generally indicative of a decrease in the respiratory drive, whereas a decrease in airflow during expiration (the passive phase) is generally indicative of an obstructive disorder.

Mechanisms of ventilatory disorders can also be characterized as either central or peripheral. Central mechanisms involve the neurologic components of the pumping apparatus that are located in the central nervous system and include the medullary central pattern generator (CPG) as well as integration centers located in the medulla, pons, hypothalamus, and cortex of the brain that regulate the output of the CPG (Boggs 1992). The major neurologic inputs from the peripheral nervous system that influence the CPG are the arterial chemoreceptors (Boggs 1992). Many drugs stimulate or depress ventilation by selective interaction with the central nervous system (Eldridge and Millhorn 1981; Keats 1985; Mueller 1982) or arterial chemoreceptors (Heymans 1955; Heymans 1958).

Defects in the pumping apparatus are classified as hypo- or hyperventilation syndromes and are best evaluated by examining ventilatory parameters in a conscious animal model. The ventilatory parameters include respiratory rate, tidal volume, minute volume, peak (or mean) inspiratory flow, peak (or mean) expiratory flow, and fractional inspiratory time. Defects in mechanical properties of the lung are classified as obstructive or restrictive disorders and can be evaluated in animal models by performing flow-volume and pressure-volume maneuvers, respectively. The parameters used to detect airway obstruction include peak expiratory flow, forced expiratory flow at 25 and 75% of forced vital capacity, and a timed forced expiratory volume, whereas the parameters used to detect lung restriction include total lung capacity, inspiratory capacity, functional residual capacity, and compliance. Measurement of dynamic lung resistance and compliance, obtained continuously during tidal breathing, is an alternative method for evaluating obstructive and restrictive disorders, respectively, and is used when the response to drug treatment is expected to be immediate (within minutes, postdose). The species used in the safety pharmacology studies are the same as those generally used in toxicology studies (rats and dogs) because pharmacokinetic and toxicologic/pathologic data are available in these species. These data can be used to help select test measurement intervals and doses and to aid in the interpretation of functional change. The techniques and procedures for measuring respiratory function parameters are well established in guinea pigs, rats, and dogs (Murphy 1994; Amdur and Mead 1958; Diamond and O'Donnell 1977; King 1966; Mauderly 1974).

The key questions in safety pharmacology of the respiratory system are:

- Does the substance effect the mechanisms of respiratory control (central or peripheral) leading to hypoventilation (respiratory depression) or hyperventilation (respiratory stimulation)?
- Does the substance act on a component of the respiratory system to induce, for example, bronchospasm, obstruction, or fibrosis?
- Does the substance induce acute effects, or can we expect chronic effects?
- Are the observed effects dose dependent or independent?

7.11.1 Plethysomography

The classic approach to measuring respiratory function in laboratory animals is called *plethysomography*. It has two basic governing principles (Boggs 1992; O'Neil and Raub 1984; Palecek 1969; Brown and Miller 1987).

- The animal (mice, rat, or dog), anaesthetized or not, restrained or not, is placed in a chamber (single or double) with pneumotachographs.
- The variations of pressure in the chamber at the time of inspiration and expiration make it possible to obtain the respiratory flow of the animal.

There are three main types of body plethysmographs: constant volume, constant pressure, and pressure–volume. The constant volume body plethysmograph is a sealed box that detects volume change by measurement of pressure changes inside the box. While inside the plethysmograph, inhalation of room air (from outside the plethysmograph) by the test animal induces an increase in lung volume (chest expansion) and thus an increase in the plethysmograph pressure. On the other hand, exhalation to the atmosphere (outside the plethysmograph) induces a decrease in the plethysmograph pressure. The magnitude of lung volume change can be obtained via measurement of the change in plethysmograph pressure and the appropriate calibration factor. The plethysmograph is calibrated by injecting or withdrawing a predetermined change in box pressure. To avoid an adiabatic artifact, the rate of air injection or withdrawal is kept the same as that of chest expansion, indicated by the same dP/dt (change in pressure over time).

The constant pressure body plethysmograph is a box with a pneumotachograph port built into its wall. This plethysmograph detects volume change via integration of the flow rate, *flow*, which is monitored by the pneumotachograph port. There is an outward flow (air moving from the plethysmograph to the atmosphere) during inspiration and inward flow during expiration. Alternatively, in place of a pneumotachograph, a spirometer can be attached to the constant pressure plethysmograph to detect volume changes. For detection of plethysmograph pressure and flow rate, sensitive pressure transducers are usually employed. It is important that the transducer be capable of responding to volume changes in a linear fashion within the volume range studied. The plethysmograph should have only negligible leaks, and the temperature should not change during the respiratory maneuvers. The plethysmograph should also have linear characteristics with no hysteresis. Dynamic accuracy requires an adequate frequency response. A fast integrated-flow plethysmograph, with a flat amplitude response for sinusoidal inputs up to 240 Hz, has been developed for rats, mice, and guinea pigs (Sinnet 1981). Similar plethysmographs can also be provided for use with large mammals.

A third type of pressure–volume plethysmograph has the mixed characteristics of the two types of body box mentioned earlier. For a constant pressure plethysmograph, the change in volume at first is associated with gas compression or expansion. This fraction of the volume change can be corrected by electronically adding the plethysmograph pressure change to the volume signal. Therefore, the combined pressure–volume plethysmograph has excellent frequency-response characteristics and a wide range of sensitivities (Leigh and Mead 1974).

If volume, flow rate, and pressure changes are detected at the same time, several respiratory variables can be derived simultaneously from the raw signals. The whole-body plethysmograph method can then be used to measure most respiratory variables, such as tidal volume, breathing frequency, minute ventilation, compliance, pulmonary resistance, functional residual capacity, pressure–volume characteristics, and maximal expiratory flow-volume curves. Table 7.12 defines the parameters that are typically determined by these methods.

Selection of the proper reference values for interpretation of findings is essential (American Thoracic Society 1991; Drazen 1984).

7.11.2 Design of Respiratory Function Safety Studies

The objective of a safety pharmacology evaluation of the respiratory system is to determine whether a drug has the potential to produce a change in respiratory function. Because a complete evaluation of the respiratory function must include both the pumping apparatus and the lung, respiratory function safety studies are best designed to evaluate both these functional components. The total respiratory system is first evaluated by testing for drug-induced changes in ventilatory patterns of intact conscious animals. This is followed by an evaluation of drug-induced effects on the mechanical properties of the lung in anesthetized/paralyzed animals. Together, these evaluations are used to determine (McNeill 1964) if drug-induced changes in the total respiratory system have occurred and (Touvay and Le Mosquet 2000) whether these changes are related to pulmonary or extrapulmonary factors.

The time intervals selected for measuring ventilatory patterns following oral administration of a drug should be based on pharmacokinetic data. The times selected generally include the time to reach peak plasma concentration of the drug (T_{max}), at least one time before and one after T_{max}, and one time that is approximately 24 h after dosing to evaluate possible delayed effects. If the drug is given as a bolus IV injection, ventilatory parameters are monitored for approximately 5-min predose and continuously for 20 to 30 min postdose. Also, 1, 2, 4, and 24-h time intervals are monitored to evaluate possible delayed effects. If administered by inhalation or intravenous (IV) infusion, ventilatory parameter would generally be monitored continuously during the exposure period and at 1, 2, 3, and 24-h time intervals after dosing.

The time interval showing the greatest ventilatory change is selected for evaluating lung mechanics. However, if no ventilatory change occurred, the T_{max} would be used. If the mechanical properties of the lung need to be evaluated within 30 min after dosing, then dynamic measurements of compliance and resistance are performed. Measurements include a predose baseline and continuous measurements for up to approximately 1 h postdose. If the mechanical properties of the lung need to be measured at 30 min or longer after dosing, then a single time point is selected, and the pressure–volume and flow-volume maneuvers are performed.

Supplemental studies, including blood gas analysis, end-tidal CO_2 measurements, or responses to CO_2 gas and NaCN, can be conducted to supplement after the ventilatory and lung mechanical findings have been evaluated. In general, these would be conducted as separate studies.

7.11.3 Capnography

The measurements of rates, volumes, and capacities provided by plethysmograph measurements have a limited ability to detect and evaluate some ventilatory disorders (Murphy 1994) that markedly affect blood gases.

Detection of hypo- or hyperventilation syndromes requires measurement of the partial pressure of arterial CO_2 (P_{CO2}). In humans and large animal models, this can be accomplished by collecting arterial blood with a catheter or needle and analyzing for $PaCO_2$ using a blood gas analyzer. In conscious rodents, however, obtaining arterial blood samples by needle puncture or catheterization during ventilatory measurements is generally not practical. An alternative and noninvasive method for monitoring $PaCO_2$ is the measurement of peak-expired (end-tidal) CO_2 concentrations. This technique has been successfully used in humans (Nuzzo and Anton 1986) and has recently been adapted for use in conscious rats (Murphy 1994). Measuring end-tidal CO_2 in rats requires the use of a nasal mask and a microcapnometer (Columbus Instruments, Columbus, OH) for sampling air from the mask and calculating end-tidal CO_2 concentrations. End-tidal CO_2 values in rats are responsive to ventilatory changes and accurately reflect changes in $PaCO_2$ (Murphy 1994).

A noninvasive procedure in conscious rats has been developed for use in helping distinguish between the central and peripheral nervous system effects of drugs on ventilation. Exposure to CO_2 gas stimulates ventilation primarily through a central mechanism (Borison 1977). In contrast, a bolus injection of NaCN produces a transient stimulation of ventilation through a mechanism that involves selective stimulation of peripheral chemoreceptors (Heymans and Niel 1958). Thus, to distinguish central from peripheral nervous system effects, our procedure measures the change in ventilatory response (pretreatment vs. posttreatment) to both a 5-min exposure to 8% CO_2 gas and a bolus IV injection of 300 μg/kg of NaCN. In this paradigm, a central depressant (e.g., morphine sulfate) inhibits the CO_2 response and has little effect on the NaCN response.

The species selected for use in safety pharmacology studies should be the same as those used in toxicology studies. The advantages of using these species (rat, dog, or monkey) is that (McNeill 1964) the pharmacokinetic data generated in these species can be used to define the test measurement intervals and (Touvay and Le Mosquet 2000) acute toxicity data can be used to select the appropriate high dose. Further, the toxicologic/pathologic findings in these species can be used to help define the mechanism of functional change. Rats are the primary choice because they are readily available, and techniques for measuring pulmonary function are well established in this species.

7.12 INHALATION EXPOSURE TECHNIQUES FOR THERAPEUTIC AGENTS

Many inhalation exposure techniques, such as the whole-body, nose-only, mouth-only, or head-only technique (Drew and Laskin 1973; MacFarland 1976; Leong et al. 1981; Phalen 1984), the intranasal exposure technique

(Elliott and De Young 1970), the endotracheal nebulization technique (Leong et al. 1985, 1988; Schreck et al. 1986), and the body plethysmographic techniques (Alarie 1966; Thorne and Karol 1989), have been developed for inhalation toxicity studies. Table 7.13 provides a summary of the advantages and disadvantages of each of the major inhalation exposure methodologies.

The main criteria for the design and operation of any dynamic (as opposed to static) inhalation exposure system are:

- The concentration of the test atmosphere must be reasonably uniform throughout the chamber and should increase or decrease at a rate close to theoretical at the start or end of the exposure. Silver (1946) showed that the time taken for a chamber to reach a point of equilibrium was proportional to the flow rate of atmosphere passing through the chamber and the chamber volume. From this, the concentration–time relationship during the "run-up" or "run-down" phase could be expressed by the equation

$$t_x = k\frac{V}{F}$$

  where t_x = time required to reach x% of the equilibrium concentration, k = a constant whose value is determined by the value of x, V = chamber volume, and F = chamber flow rate. The t_{99} value is frequently quoted for exposure chambers, representing the time required to reach 99% of the equilibrium concentration and providing an estimate of chamber efficiency. Thus, at maximum efficiency, the theoretical value of k at t_{99} is 4.605; the closer to this value the results of evaluation of actual chamber performance fall, the greater is the efficiency and the better the design of the chamber.
- Flow rates must be controlled in such a way that they are not excessive, lest they cause streaming effects within the chamber, but must be adequate to maintain normal oxygen levels, temperature, and humidity in relation to the number of animals exposed. A minimum of ten air changes per hour is frequently advocated and is appropriate in most cases. However, the chamber design and housing density also need to be taken into account, and some designs, such as that of Doe and Tinston (1981), function effectively at lower air-change rates.
- The chamber or exposure manifold materials should not affect the chemical or physical nature of the test atmosphere.

Table 7.13 Advantages, Disadvantages, and Considerations Associated with Modes of Respiratory Exposure

Mode of Exposure	*Advantages*	*Disadvantages*	*Design Considerations*
Whole-body	Easiest and only practical way of achieving longer term (more than 4 h at a time) exposure	Uses large amounts of test substance Use with powder and liquid aerosols leads to mixed routes of exposure	Chamber mixing Animal heat loads
Head-only	Good for repeated exposure Limited routes of entry into animal More efficient dose delivery	Stress to animal Losses can be large Seal around neck Labor in loading/unloading	Even distribution Pressure fluctuations Sampling and losses Air temperature, humidity Animal comfort Animal restraint
Nose-/mouth-only	Exposure limited to mouth and respiratory tract Uses less material (efficient) Containment of material Can pulse the exposure	Stress to animal Seal about face Effort to expose large number of animals	Pressure fluctuations Body temperature Sampling Airlocking Animals' comfort Losses in plumbing/masks
Lung-only (Tracheal administration)	Precision of dose One route of exposure Uses less material (efficient) Can pulse the exposure	Technically difficult Anesthesia or tracheostomy Limited to small numbers Bypasses nose Artifacts in deposition and response Technically more difficult	Air humidity/ temperature Stress to the animal Physiologic support

Table 7.13 Advantages, Disadvantages, and Considerations Associated with Modes of Respiratory Exposure (Continued)

Mode of Exposure	*Advantages*	*Disadvantages*	*Design Considerations*
Partial-lung	Precision of total dose Localization of dose Can achieve very high local doses Unexposed control tissue from same animal	Anesthesia Placement of dose Difficulty in interpretation of results Technically difficult Possible redistribution of material within lung	Stress to animal Physiologic support

Source: From Gad and Chengelis 1998.

For critical laboratory studies on inhaled drugs, a monodisperse aerosol of a specified range of MMADs should be used to increase the probability of the aerosol reaching the specified target area of the lungs. The Dautrebande aerosol generators (Dautrebande 1962c) and the DeVilbiss nebulizer (Drew and Lippmann 1978) are the classic single-reservoir generators for short-duration inhalation studies. For long-duration inhalation studies, the multiple-reservoir nebulizers (Miller et al. 1981) or the continuous-syringe-metering and elutriating atomizers (Leong et al. 1981) are frequently used. The nebulizers generate a polydisperse droplet aerosol either by the shearing force of a jet of air over a fine stream of liquid or by ultrasonic disintegration of the surface liquid in a reservoir (Drew and Lippmann 1978). The aerosols emerging from a jet nebulizer generally have MMADs ranging between 1.2 and 6.9 μm with GSDs of 1.7 to 2.2, and aerosols from an ultrasonic nebulizer have MMADs ranging between 3.7 and 10.5 μm with GSDs of 1.4 to 2.0 (Mercer 1981).

For testing therapeutic formulations, the liquid aerosols are usually generated by the pressurized metered-dose inhaler (MDI) (Newman, S. P. 1984; Newton 2000; Gad and Chengelis 1998). The pressurized MDI generates a bolus of aerosols by atomizing a well-defined quantity of a drug that is solubilized in a propellant. Of concern in such formulations are the propellants (though these are generally inert gases) and excipients such as stabilizers (see Table 7.14). The aerosols, thus, consist of the drug particles with a coating of the propellant. As the aerosols emerge from the orifice, the mean particle size may be as large as 30 μm (Moren 1981). After traveling through a tubular or cone-shaped spacer, the propellant

Table 7.14 Some Examples of Excipients Used for Dry Powder Aerosols

Active Ingredient	*Excipient Carrier*
Salbutamol sulfate	Lactose (63–90 μm): regular, spray dried, and recrystallized
Budesonide	Lactose (α-monohydrate [<32 μm, 63–90 μm, 125–180 μm])
rhDNase	Lactose (50 wt% < 42 and 115 mm)
	Mannitol (50 wt% < 43 mm)
	Sodium chloride (50 wt% < 87 mm)
Bovine serum	Lactose (a-monohydrate [63–90 mm])
albumin-	Fine-particle lactose (76 wt% < 10 mm)
maltodextrin (50-50)	Micronized polyethylene glycol 6000 (97.5 wt% < 10 mm)
Recombinant human granulocyte-colony stimulating factor-mannitol	Polyethylene glycol 8000 (38–75 mm, 90–125 mm)

may evaporate, reducing the MMADs to a range of 2.8 to 5.5 μm with GSDs of 1.5 to 2.2 (Hiller et al. 1978; Sackner et al. 1981; Newman, S.P., 1984) and making the aerosols more stable for inhalation studies. In a prolonged animal exposure study, multiple MDIs have to be actuated sequentially with an electromechanical gadget (Ulrich et al. 1984) to maintain a slightly pulsatile but relatively consistent chamber concentration.

For generating an aerosol from dry powders, various dust generators, such as the Wright dust feed, air elutriator, or fluidized-bed dust generator, and air impact pulverizer, have been developed for acute and chronic animal inhalation studies and are described in many articles (Hinds 1980; Leong et al. 1981; Phalen 1984; Gad 1998; Valentine and Kennedy 2001; Hext 2000; Gardner and Kennedy 1993). For generating powdery therapeutic agents, a metered-dose dry powder inhaler, spinhaler, or a Rotahaler is used (Newman 1984). The particle size of the drug powder is micronized to a specific size range during manufacture, and the spinhaler or Rotohaler only disperses the powders.

More recently, in a new approach, administering dry powders to both humans and test animals has started. Dry powders, although less frequently used in nasal drug delivery, are becoming more popular. Several devices are available for administering powders, the most common being the insufflator. Many insufflators work with predosed powder in gelatin capsules. To improve patient compliance, a multidose powder inhaler has been developed, which has been used to deliver budesonide. These devices can also be used for administration to animals to test delivery, both in terms of amounts and aerodynamic size of the particles. Early DPIs such as the Rotohaler® used individual capsules of micronized drug, which were difficult to handle; however, modern devices use blister packs (e.g., Diskus®) or reservoirs (e.g., Turbuhaler®). The DPIs rely on inspiration to withdraw drugs from the inhaler to the lung; hence, the effect of inhalation flow rate through various devices has been extensively studied. The major problem to be overcome with these devices is in ensuring that the finely micronized drug is thoroughly dispersed in the airstream. It has been recommended that patients inhale as rapidly as possible from these devices to provide the maximum force to disperse the powder. The quantity of drug and deposition patterns vary enormously depending on the device; for example, the Turbuhaler® produces significantly greater lung delivery of salbutamol than the Diskus®. Vidgren and coworkers (1987) demonstrated by gamma scintigraphy that a typical dry powder formulation of SCG suffers losses of 44% in the mouth and 40% in the actuator nozzle itself.

It must also be emphasized that the major mass of a heterodispersed aerosol may be contained in a few relatively large particles, because the mass of a particle is proportional to the cube of its diameter. Therefore, the particle-size distribution and the concentration of the drug particles in the exposure atmosphere should be sampled using a cascade impactor or membrane filter sampling technique, monitored using an optical or laser particle-size analyzer, and analyzed using optical or electron microscopy techniques.

In summary, many techniques have been developed for generating gas, vapor, and aerosol atmospheres for inhalation toxicology studies. By proper regulation of the operating conditions of the nebulizers and the formulation of MDIs, together with the use of spacer or reservoir attachments to MDIs, more particles within the respirable range can be generated for inhalation. An accurately controlled exposure concentration is essential to accurately determine the dose–response relationship in the safety assessment of an inhalant drug.

Finally, comparisons of various techniques of animal exposures indicate that the whole-body-exposure technique is the most suitable for safety assessment of gases and vapors. It permits simultaneous exposure of a large number of animals to the same concentration of a drug; however, this technique is not suitable for aerosol and powder exposures because the exposure condition represents the resultant effects from inhalation, ingestion, and dermal absorption of the drug (Phalen 1984, 1998).

7.13 REGULATORY GUIDELINES

There is very limited regulatory guidance for the safety evaluation of inhalation drugs. CDER (2002) has issued one piece of guidance, but this speaks more of CMC issues than safety evaluation. The operative guidance is limited to the following:

1. Water-soluble inhalation drugs shall be sterile.
2. Exposure of test animals will be in a manner and by a regimen as similar as possible be employed clinically.

These can lead to unexpected issues, such as conflicts with animal welfare guidance or technical limitations. For example, some therapeutic gases are administered 24 h a day to patients. This is not at all possible by nose-only techniques in animals (too much stress and no access to food and water for lab animals), and not strictly possible even by whole-body regimens due to the requirements of animal husbandry.

7.14 UTILITY OF TOXICITY DATA

Regardless of the type of test and the parameters to be monitored, the ultimate goal is to interpolate or extrapolate from the dose–response data to find a no-observable-adverse-effect level (NOAEL) or a no-observable-effect level (NOEL). By applying a safety factor of 1 to 10 to the NOAEL, a safe single-exposure dose for a phase I clinical trial may be obtained. By applying a more stringent safety factor, a multiple-exposure dose for a clinical trial may also be obtained. After the drug candidate has successfully passed all the drug safety evaluations and entered the production stage, more toxicity tests may be needed for the establishment of a threshold limit value-time-weighted average (TLV-TWA). A TLV-TWA is defined as "the time-weighted average concentration for a normal 8-h workday and a 40-h workweek, to which nearly all workers may be repeatedly exposed, day after day, without adverse effect" (ACGIH 1991). Using TLVs as guides, long-term safe occupational exposures during production and industrial handling of a drug may be achieved. Appropriate safety assessments of pharmaceutical chemicals and drugs will ensure the creation and production of a safe drug for the benefit of humans and animals. Furthermore, inhalation toxicity data are needed for compliance with many regulatory requirements of the Food and Drug Administration, the Occupational Health and Safety Administration, and the Environmental Protection Agency (Gad and Chengelis 1998).

More comprehensive descriptions and discussions on inhalation toxicology and technology may be found in several monographs, reviews, and textbooks (Willeke 1980; Leong 1981; Witschi and Nettesheim 1982; Clarke and Pavia 1984; Phalen 1984; Witschi and Brain 1985; Barrow 1986; McFadden 1986; Salem 1986; Gardner, Crapo, and McClellan 1993; Gad and Chengelis 1998; Valentine and Kennedy 2001; McClellan and Henderson 1989; Hext 2000; Pauluhn 2002).

GLOSSARY OF TERMS

Acceptance criteria Numerical limits, ranges, or other criteria for the test described.

Batch A specific quantity of a drug or other material that is intended to have uniform character and quality, within specified limits, and is produced according to a single manufacturing order during the same cycle of manufacture (21 CFR 210.3(b)(2)).

Container closure system The sum of packaging components that together contain, protect, and deliver the dosage form. This includes primary packaging components and secondary packaging components if the latter are intended to provide additional protection to the drug product (e.g., foil overwrap). The container closure system also includes the pump for nasal and inhalation sprays. For nasal spray and inhalation solution, suspension, and spray drug products, the critical components of the container closure system are those that contact either the patient or the formulation, components that affect the mechanics of the overall performance of the device, or any protective packaging.

CRF Case report form.

CTM Clinical trials material.

CTU Clinical trials unit.

CYP P450 isoenzymes.

Drug product The finished dosage form and the container closure system.

Drug substance An active ingredient that is intended to furnish pharmacological activity or other direct effect in the diagnosis, cure, mitigation, treatment, or prevention of disease or to affect the structure or any function of the human body (21 CFR 314.3(b)).

Excipients Any intended formulation component other than the drug substance.

Extractables Compounds that can be extracted from elastomeric or plastic components of the container closure system when in the presence of a solvent.

GLP Good laboratory practice.

GMP Good manufacturing practice.

Inhalation solutions, suspensions, and sprays Drug products that contain active ingredients dissolved or suspended in a formulation, typically aqueous based, which can contain other excipients and are intended for use by oral inhalation. Aqueous-based drug products for oral inhalation must be sterile (21 CFR 200.51). Inhalation solutions and suspensions are intended to be used with a specified nebulizer. Inhalation sprays are combination products in which the components responsible for metering, atomization, and delivery of the formulation to the patient are a part of the container closure system.

Insufflator Dry powder nasal inhaler used with Rynacrom cartridges. Each cartridge contains one dose; the inhaler opens the cartridge, allowing the powder to be blown into the nose by squeezing the bulb.

Leachables Compounds that leach into the formulation from elastomeric or plastic components of the drug product container closure system.

MDI Metered-dose inhaler, consisting of an aerosol unit and plastic mouthpiece. This is currently the most common type of inhaler and is widely available.

Nasal sprays Drug products that contain active ingredients dissolved or suspended in a formulation, typically aqueous based, which can contain other excipients and are intended for use by nasal inhalation. Container closure systems for nasal sprays include the container and all components that are responsible for metering, atomization, and delivery of the formulation to the patient.

Nociception Perception of pain in the nose.

Placebo A dosage form that is identical to the drug product except that the drug substance is absent or replaced by an inert ingredient.

Pump All components of the container closure system that are responsible for metering, atomization, and delivery of the formulation to the patient.

Specification The quality standard (i.e., test, analytical procedures, and acceptance criteria) provided in the approved application to confirm the quality of drug substances, drug products, intermediates, raw material reagents, components, in-process materials, container closure systems, and other materials used in the production of drug substances or drug products.

Specified impurity An identified or unidentified impurity that is selected for inclusion in the drug substance or drug product specification and is individually listed and limited to ensure reproducibility of the quality of the drug substance and/or drug product.

Spinhaler A dry powder inhaler used with Intal capsules specifically designed for the spinhaler. Each capsule contains one dose; the inhaler opens the capsule such that the powder may be inhaled through the mouthpiece.

Syncroner MDI with elongated mouthpiece, used as training device to see if medication is being inhaled properly.

Turbuhaler A dry powder inhaler. The drug is in the form of a pellet; when the body of the inhaler is rotated, a prescribed amount of the drug is ground off this pellet. The powder is then inhaled through a fluted aperture on top.

REFERENCES

ACGIH. (1991). *Documentation of the Threshold Limit Values and Biological Exposure Indices*. 6th ed. American Conference of Governmental Industrial Hygienists, Cincinnati, OH.

Adams, E.M., Spencer, H.C., Rowe, V.K., and Irish, D.D. (1950). Vapor toxicity of 1,1, 1,-trichloroethane (methylchloroform) determined by experiments on laboratory animals. *Arch Ind Hyg Occup Med* 1: 225–236.

Adams, E.M., Spencer, H.C., Rowe, V.K., McCollister, D.D., and Irish, D.D. (1952). Vapor toxicity of carbon tetrachloride determined by experiments on laboratory animals. *Arch Ind Hyg Occup Med* 6: 50–66.

Agnew, J.E. (1984). Physical properties and mechanisms of deposition of aerosols. In *Aerosols and the Lung, Clinical Aspects* (S.W. Clarke and D. Pavia, Eds.). Butterworth, London, pp. 49–68.

Agu, R.U., Jorissen, M., Kinget, R., Verbeke, N., and Augustigns, P. (2002). Alternatives to in vivo nasal toxicological screening for nasally administered drugs, *STP Pharma Sci* 12: 13–22.

Akoun, G.M. (1989). Natural history of drug-induced pneumonitis. In *Drug Induced Disorders Volume 3: Treatment Induced Respiratory Disorders* (Akoun, G.M., White, J.P., Eds.). New York. Elsevier, pp. 3–9.

Alarie, Y. (1966). Irritating properties of airborne material to the upper respiratory tract. *Arch Environ Health* 13: 433–449.

Alarie, Y. (1973). Sensory irritation by airborne chemicals. *CRC Crit Rev Toxicol* 2: 299–363.

Alarie, Y. (1981a). Toxicological evaluation of airborne chemical irritants and allergens using respiratory reflex reactions. In *Inhalation Toxicology and Technology* (B.K.J. Leong, Ed.). Ann Arbor Science, Ann Arbor, MI, pp. 207–231.

Alarie, Y. (1981b). Bioassay for evaluating the potency of airborne sensory irritants and predicting acceptable levels of exposure in man. *Food Cosmet Toxicol* 19: 623–626.

Alarie, Y. and Luo, J.E. (1984). Sensory irritation by airborne chemicals: a basis to establish acceptable levels of exposure. In *Toxicology of the Nasal Passages* (C.S. Barrow, Ed.). Hemisphere, New York, pp. 91–100.

Alarie, Y., Kane, L., and Barrow, C. (1980). Sensory irritation: the use of an animal model to establish acceptable exposure to airborne chemical irritants. In *Toxicology: Principles and Practice 1* (A.L. Reeves, Ed.). John Wiley & Sons, New York.

Aldridge, S. (2003). Inhaled antibodies work better for chronic sinusitis. *My Health and Age*. March 4, 2003.

Altman, P.L. and Dittmer, D.S. (1974). *Biological Data Book*, Vol. III. Federation of American Societies for Experimental Biology, Bethesda, MD.

Amdur, M.O. and Mead, J. (1958). Mechanics of respiration in unanesthetized guinea pigs. *Am J Physiol* 192: 364–368.

American Thoracic Society (1991). Lung function testing: selection of reference values and interpretative strategies. *Am Rev Respir Dis* 144: 1202–1218.

Andersen, M.E., French, J.E., Gargas, M.L., Jones, R.A., and Jenkins, L.J., Jr. (1979). Saturable metabolism and the acute toxicity of 1,1-dichloroethylene. *Toxicol Appl Pharmacol* 47: 385–393.

Aso, Y., Yoneda, K., and Kikkawa, Y. (1976). Morphologic and biochemical study of pulmonary changes induced by bleomycin in mice. *Lab Invest* 35: 558–568.

Aviado, D.M. (1981). Comparative cardiotoxicity of fluorocarbons. In *Cardiac Toxicology*. Vol. II (T. Balazs, Ed.). CRC Press, Boca Raton, FL, pp. 213–222.

Aviado, D.M. and Micozzi, M.S. (1981). Fluorine-containing organic compounds. In *Patty's Industrial Hygiene and Toxicology*. Vol. 2B (G.D. Clayton and F.E. Clayton, Eds.). John Wiley & Sons, New York, pp. 3071–3115.

Balazs, T. (1981). Cardiotoxicity of adrenergic bronchodilator and vasodilating antihypertensive drugs. In *Cardiac Toxicology*, Vol. II (T. Balazs, Ed.). CRC Press, Boca Raton, FL, pp. 61–73.

Barrow, C.S. (1986). *Toxicology of the Nasal Passages*. Hemisphere, New York.

Bell, K.A. (1978). Local particle deposition in respiratory airway models. In *Recent Developments in Aerosol Science* (D.T. Shaw, Ed.). John Wiley & Sons, New York, pp. 97–134.

Bice, D.E. (1985). Methods and approaches to assessing Immunotoxicology of the lower respiratory tract. In *Immunotoxicology and Immunopharmacology* (J.H. Dean, M.I. Luster, A.E. Munson, and H.A. Amos, Eds.). Raven Press, New York, pp. 145–157.

Boggs, D.F. (1992). Comparative control of respiration. *Comparative Biology of the Normal Lung*, Vol. I. (Parent, R.A., Ed.). CRC Press, Boca Raton, FL, pp. 309–350.

Borison, H.L. (1977). Central nervous system depressants: control-systems approach to respiratory depression, *Pharmacol Ther B* 3: 211–226.

Brain, J.D., Knudson, D.E., Sorokin, S.P., and Davis, M.A. (1976). Pulmonary distribution of particles given by intratracheal instillation or aerosol inhalation. *Environ Res* 11: 13–33.

Brown, L.K. and Miller, A. (1987). Full lung volumes: functional residual capacity, residual volume and total lung capacity. In *Pulmonary Function Tests: A Guide for the Student and House Officer* (Miller, A., Ed.). New York, Grune and Stratton, pp. 53–58.

Budavari, S. (Ed.) (1989). *Merck Index*, 11th ed. Merck and Co., Rahway, NJ, p. 82.

Burham, S.R. (2002). The ideal nasal carticosteroid: balancing efficacy, safety and patient preference. *Clin Exp Allergy Rev* 2: 32–37.

CDER (2002). *Guidance for Industry: Nasal Spray and Inhalation Solution, Suspension, and Spray Drug Products—Chemistry, Manufacturing and Control Documentation*. Food and Drug Administration, Washington, D.C.

Chan, T.L. and Lippmann, M. (1980). Experimental measurements and empirical modelling of the regional deposition of inhaled particles in humans. *Am Ind Hyg Assoc J* 41: 399–409.

Chenoweth, M.B., Leong, B.K.J., Sparschu, G.L., and Torkelson, T.R. (1972). Toxicities of methoxyflurane, halothane and diethyl ether in laboratory animals on repeated inhalation at subanesthetic concentrations. In *Cellular Biology and Toxicity of Anesthetics* (B.R. Fink, Ed.). Williams and Wilkins, Baltimore, pp. 275–284.

Cherniack, N.S. (1988). Disorders in the control of breathing: hyperventilation syndromes. *Textbook of Respiratory Medicine* (Murray, J.F. and Nadal, J.A., Eds.). W.B. Saunders, Philadelphia, pp. 1861–1866.

Clarke, S.W. and Pavia, D., (Eds.) (1984). *Aerosol and the Lung: Clinical and Experimental Aspects.* Butterworth, London.

Cox, J.S.G., J.E. Beach, Blair, A.M.J.N., Clarke, A.J., King, J., Lee, T.B., Loveday, D.E.E., Moss, G.F., Orr, T.S.C., Ritchie, J.T., and Sheard, P. (1970). Disodium cromoglycate. *Adv Drug Res* 5: 115–196.

Curry, S.H., Taylor, A.J., Evans, S., Godfrey, S., and Zeidifard, E. (1975). Disposition of disodium cromoglycate administered in three particle sizes. *Br J Clin Pharmacol* 2: 267–270.

Dahl, A.R. (1990). Dose concepts for inhaled vapors and gases. *Toxicol Appl Pharmacol* 103: 185–197.

Dallas, C.E., Bruckner, J.V., Maedgen, J.L., and Weir, F.W. (1986). A method for direct measurement of systemic uptake and elimination of volatile organics in small animals. *J. Pharmacol. Methods* 16: 239–250.

Dallas, C.E., Ramanathan, R., Muralidhara, S., Gallo, J.M., and Bruckner, J.V. (1989). The uptake and elimination of 1,1,1-trichloroethane during and following inhalation exposure in rats. *Toxicol Appl Pharmacol* 98: 385–397.

Dautrebande, L. (1962a). Importance of particle size for therapeutic aerosol efficiency. In *Microaerosols.* Academic Press, New York, pp. 37–57.

Dautrebande, L. (1962b). Practical recommendation for administering pharmacological aerosols. In *Microaerosols.* Academic Press, New York, pp. 86–92.

Dautrebande, L. (1962c). Production of liquid and solid micromicellar aerosols. In *Microaerosols.* Academic Press, New York, pp. 1–22.

Davies, R.J. and Pepys, J. (1975). Asthma due to inhaled chemical agents — the macrolide antibiotic spiramycin. *Clin Allergy* 5: 99–107.

Davis, B., Marin, M.G., Fischer, S., Graf, P., Widdicombe, J.G., and Nadel, J.A. (1976). New method for study of canine mucus gland secretion *in vivo*: cholinergic regulation. *Am Rev Respir Dis* 113: 257 (abstract).

Davis, C.N., Heyder, J., and Subba Ramu, M.C. (1972). Breathing of half micron aerosols. I. Experimental. *J Appl Physiol* 32: 591–600.

Dennis, W.L. (1961). The discussion of a paper by C.N. Davis: a formalized anatomy of the human respiratory tract. In *Inhaled Particles and Vapours* (C.N. Davis, Ed.). Pergamon Press, London, p. 88.

Diamond, L. and O'Donnell, M. (1977). Pulmonary mechanics in normal rats, *J Appl Physiol: Respir Environ Exercise Physiol*. 43: 942–948.

Doe, J.E. and Tinston, D.J. (1981). Novel chamber for long-term inhalation studies. In *Inhalation Toxicology and Technology*, (Leong, K.J., Ed.). Ann Arbor Science: Ann Arbor, MI.

Dorato, M.A. (1994), Toxicological evaluation of intranasal peptide and protein drugs, *Drugs Pharm Sci* 62: 345–381.

Drazen, J.M. (1984), Physiological basis and interpretation of indices of pulmonary mechanics, *Environ Health Perspect* 56: 3–9.

Drew, R.T. and Lippmann, M. (1978). Calibration of air sampling instruments. In *Air Sampling Instruments for Evaluation of Atmospheric Contaminants*, 5th ed., Sec. I, American Conference of Governmental Industrial Hygienists, Cincinnati, OH, pp. 1–32.

Drew, R.T. and Laskin, S. (1973). Environmental inhalation chambers. In *Methods of Animal Experimentation* Vol. IV. Academic Press, New York, pp. 1–41.

Durham, S.R. (2002). The ideal nasal corticosteroid: balancing efficacy, safety and patient preference, *Clin Exp Allergy Rev* 2: 32–37.

Eldridge, F.L. and Millhorn, D.E. (1981). Central regulation of respiration by endogenous neurotransmitters and neuromodulators, *Annu Rev Physiol* 3: 121–135.

Elliot, G.A. and DeYoung, E.N. (1970). Intranasal toxicity testing of antiviral agents. *Ann N Y Acad Sci* 173: 169–175.

Evans, M.J. (1982). Cell death and cell renewal in small airways and alveoli. In *Mechanisms in Respiratory Toxicology*, Vol. 1. (H, Witschi and P. Nettesheim, Eds.). CRC Press, Boca Raton, FL, pp. 189–218.

Fariba, A., Kellie, M., Stephen, M., Oune, O., Kafi, A., and Mary, H. (2002). Repeated doses of antenatal corticosteroids in animals: a systematic review, *Am J Objectives Gynecol* 186: 843–849.

Ferin, J. (1977). Effect of particle content of lung on clearance pathways. In *Pulmonary Macrophage and Epithelial Cells* (C.L. Sanders, R.P. Schneider, G.E. Dagle, and H.A. Ragan, Eds.). Technical Information Center, Energy Research and Development Administration, Springfield, Virginia, pp. 414–423.

Fieldner, A.C., Kazt, S.H., and Kinney. S.P. (1921). Gas Masks for Gases Met in Fighting Fires. U.S. Bureau of Mines, Technical paper No. 248.

Gad, S.C. (2002). *Drug Safety Evaluation*, Wiley, New York.

Gad, S.C. (2003). *Safety Pharmacology*, CRC Press, Boca Raton, FL.

Gad, S.C. and Chengelis, C.P. (1998). *Acute Toxicology Testing: Perspectives and Horizons*, 2nd ed., Academic Press, San Diego, CA, pp. 404–466.

Gamsu, G., Singer, M.M., Vincent, H.H., Berry, S., and Nadel, J.A. (1976). Postoperative impairment of mucous transport in the lung. *Am Rev Respir Dis* 114: 673–679.

Gardner, D.E., Crapo, J.D., and McClellan, R.O. (1993). *Toxicology of the Lung*. 2nd ed. Raven Press, New York.

Gardner, D.E. and Kennedy, G.L., Jr. (1993). Methodologies and technology for animal inhalation toxicology studies. In *Toxicology of the Lung*, 2nd ed. (Gardner, D.E., Crapo, J.D., and McClellan, R.O., Eds.). Raven Press, New York.

Godfrey, S., Zeidifard, E., Brown, K., and Bell, J.H. (1974). The possible site of action of sodium cromoglycate assessed by exercise challenge. *Clin Sci Mol Med* 46: 265–272.

Haber, F.R. (1924). Funf Vortage aus den jahren 1920–23, No. 3, *Die Chemie im Kriege*. Julius Springer, Berlin.

Hafner, R.E., Jr., Watanabe, P.G., and Gehring, P.J. (1975). Preliminary studies on the fate of inhaled vinyl chloride monomer in rats. *Ann N Y Acad Sci* 246: 135–148.

Hatch, T.F. and Gross, P. (1964). *Pulmonary Deposition and Retention of Inhaled Aerosols*. Academic Press, New York, pp. 16–17, 51–52, 147–168.

Henderson, R. (1988). Use of bronchoalveolar lavage to detect lung damage. In *Toxicology of the Lung* (D.E. Gardner, J.D. Crapo, and E.J. Massaro, Eds.). Raven Press, New York, pp. 239–268.

Henderson, R. (1989). Bronchoalveolar lavage: a tool for assessing the health status of the lung. In *Concepts in Inhalation Toxicology* (R.O. McClellan and R.F. Henderson, Eds.). Hemisphere, Washington D.C., pp. 414–442.

Henderson, R.F. (1984). Use of bronchoalveolar lavage to detect lung damage. *Environ. Health Perspect.* 56: 115–129.

Henderson, R.F. and Loery, J.S. (1983). Effect of anesthetic agents on lavage fluid parameters used as indicators of pulmonary injury. *Lab Anim Sci* 33: 60–62.

Henderson, R.F., Barr, E.B. and Hotchkiss, J.A. (1991). Effect of exposure rate on response of the lung to inhaled particles (abstr.). *Toxicologists* 11: 234.

Henderson, R.F., Mauderly, J.L., Pickrell, J.A., Hahn, F.F., Muhle, F.F., and Rebar, A.H. (1987). Comparative study of bronchoalveolar lavage fluid: effect of species, age, method of lavage. *Exp Lung Res* 1: 329–342.

Henderson, R.F., Benson, J.M., Hahn, F.F., Hobbs, C.H., Jones, R.K., Mauderly, J.L., McClellan, R.O., and Pickrell, J.A. (1985). New approaches for the evaluation of pulmonary toxicity: bronchoalveolar lavage fluid analysis. *Fundam Appl Toxicol* 5: 451–458.

Hensley, M.J., O'Cain, C.F., McFadden, E.R., Jr., and Ingram, R.H., Jr. (1978). Distribution of bronchodilation in normal subjects: Beta agonist versus atropine. *J Appl Physicol* 45: 778–782.

Hext, P.M. (2000). Inhalation toxicology. In *General and Applied Toxicology* (B. Ballantyne, T. Marrs and T. Syversen, Eds.). Macmillan, London, pp. 587–601.

Heyder, J., Gebhart, J., and Stahlhofen, W. (1980). Inhalation of aerosols: particle deposition and retention. In *Generation of Aerosols and Facilities for Exposure Experiments* (K. Willeke, Ed.). Ann Arbor Science, Ann Arbor, MI, pp. 80–99.

Heymans, C. (1955). Action of drugs on carotid body and sinus, *Pharmacol Rev* 7: 119–142.

Heymans, C. and Niel, E. (1958). The effects of drugs on chemoreceptors, In *Reflexogenic Areas of the Cardiovascular Systems* (Heymans, C. and Neil, E., Eds.). Churchill, London, pp. 192–199.

Hiller, F.C., Mazunder, M.K., Wilson, J.D., and Bone, R.C. (1978). Aerodynamic size distribution of metered dose bronchodilator aerosols. *Am Rev Respir Dis* 118: 311–317.

Hinds, W.C. (1980). Dry-dispersion aerosol generators. In *Generation of Aerosols and Facilities for Exposure Experiments* (K. Willeke, Ed.). Ann Arbor Science, Ann Arbor, MI, pp. 171–187.

Hirai, S., Yashiki, T., and Mima, H. (1981). Mechanisms for the enhancement of the nasal absorption of insulin by surfactants. *Int J Pharm* 9: 173–184.

Hocking, W.G. and Golde, D.W. (1979). The pulmonary alveolar macrophage. *New Engl J Med* 310: 580–587, 639–645.

Hodson, M.E., Penketh, A.R., and Batten, J.C. (1981). Aerosol carbenicillin and gentamicin treatment of *Pseudomonas aeruginosa* infection in patients with cystic fibrosis. *Lancet* 2: 1137–1139.

Holma, B. (1967). Lung clearance of mono- and didisperse aerosols determined by profile scanning and whole body counting: a study on normal and SO_2 exposed rabbits. *Acta Med Scand* (Suppl.) 473: 1–102.

Illum, L. and Davis, S.S. (1992). Intranasal insulin—clinical pharmacokinetics. *Clin Pharmacokin* 23: 30–41.

Indans, I. (2002). Nonlethal end-points in inhalation toxicology. *Toxicol Lett* 135(1): 53.

Ingram, R.H., Wellman, J.J., McFadden, E.R., Jr., and Mead, J. (1977). Relative contributions of large and small airways to flow limitation in normal subjects before and after atropine and isoproterenol. *J Clin Invest* 59: 696–703.

Irish, D.D. and Adams, E.M. (1940). Apparatus and methods for testing the toxicity of vapors. *Ind Med Surg* 1: 1–4.

Johnson, H.G., McNee, M.L., Johnson, M.A., and Miller, M.D. (1983). Leukotriene C_4 and dimethylphenylpiperazinium-induced responses in canine airway tracheal muscle contraction and fluid secretion. *Int Arch Allergy Appl Immunol* 71: 214–218.

Johnson, H.G., McNee, M.L., and Braughler, J.M. (1987). Inhibitors of metal catalyzed lipid peroxidation reactions inhibit mucus secretion and 15 HETE levels in canine trachea. *Prostaglandins Leukotrienes Med* 30: 123–132.

Kannisto, S., Voutilainen, R., Remes, K., and Korppi, M. (2002). Efficacy and safety of inhaled steroid and cromane treatment in school-age children: a randomized pragmatic pilot study. *Pediatr Allergy Immunol* 13: 24–34.

Karol, M.H. (1988). Immunologic responses of the lung to inhaled toxicants. In *Concepts in Inhalation Toxicology* (R.O. McClellan and R. Henderson, Eds.). Hemisphere Publishing, Washington, D.C., pp. 403–413.

Karol, M.H., Hillebrand, J.A., and Thorne, P.S. (1989). Characteristics of weekly pulmonary hypersensitivity responses elicited in the guinea pig by inhalation of ovalbumin aerosols. *Toxicology of the Lung* (D.E. Gardner, J.D. Crapo, and E.J. Massaro, Eds.). Raven Press, New York, pp. 427–448.

Kavet, R.I., Brain, J.D., and Levens, D.J. (1978). Characteristics of weekly pulmonary macrophages lavaged from hamsters exposed to iron oxide aerosols. *Lab Invest* 38: 312–319.

Keats, A.S. (1985). The effects of drugs on respiration in man. *Ann Rev Pharmacol Toxicol* 25: 41–65.

King, T.K.C. (1966). Measurement of functional residual capacity in the rat. *J Appl Physiol* 21: 233–236.

Lalej-Bennis, D. (2001). Six month administration of gelified intranasal insulin in 16 type 1 diabetic patients under multiple injections: Efficacy vs. subcutaneous injections and local tolerance. *Diabetes Metab* 27(3): 372–377.

Landy, T.D., Ramsey, J.C., and McKenna, M.J. (1983). Pulmonary physiology and inhalation dosimetry in rats: development of a method and two examples. *Toxicol Appl Pharmacol* 71: 72–83.

Last, J.A. (1982). Mucus production and ciliary escalator. In *Mechanisms in Respiratory Toxicology*, Vol. 1. (H. Witschi and P. Nettesheim, Eds.). CRC Press, Boca Raton, FL, pp. 247–268.

Lauweryns, J.M. and Baert, J.H. (1977). Alveolar clearance and the role of the pulmonary lymphatics. *Am Rev Respir Dis* 115: 625–683.

Lawrence, S. (2002). Intranasal delivery could be used to administer drugs directly to the brain. *Lancet* 359: 1674.

Lee, W.C. and Wang, C.S. (1977). Particle deposition in systems of repeated bifurcating tubes. In *Inhaled Particles IV* (W.H. Walton, Ed.). Oxford, Pergamon, pp. 49–60.

Leigh, D.E. and Mead, J. (1974). *Principles of Body Plethysmography*. National Heart and Lung Institute. NIH, Bethesda, MD.

Leong, B.K.J. and Rop, D.A. (1989). The combined effects of an inhalation anesthetic and an analgesic on the electrocardiograms of beagle dogs. Abstract-475. *International Congress of Toxicology*. Taylor and Francis, London, p. 159.

Leong, B.K.J. and MacFarland, H.N. (1965). Pulmonary dynamics and retention of toxic gases. *Arch Environ Health* 11: 555–563.

Leong, B.K.J., Powell, D.J., and Pochyla, G.L. (1981). A new dust generator for inhalation toxicological studies. In *Inhalation Toxicology and Technology* (B.K.J. Leong, Ed.). Ann Arbor Science, Ann Arbor, MI, pp. 157–168.

Leong, B.K.J., Coombs, J.K., Petzold, E.N., Hanchar, A.H., and McNee, M.L. (1985). Endotracheal nebulization of drugs into the lungs of anesthetized animals (abstr.). *Toxicologist* 5: 31.

Leong, B.K.J., Coombs, J.K., Petzold, E.N., and Hanchar, A.J. (1988). A dosimetric endotracheal nebulization technique for pulmonary metabolic disposition studies in laboratory animals. *Inhal Toxicol* (premier issue): 37–51.

Lichtiger, M., Landa, J.F., and Hirsch, J.A. (1975). Velocity of tracheal mucus in anesthetized women undergoing gynaecologic surgery. *Anesthesiology* 42: 753–756.

Lippmann, M. (1977). “Respirable” dust sampling. *Am Ind Hyg Assoc J* 31: 138–159.

Lippmann, M., Yeates, D.B., and Albert, R.E. (1980). Deposition, retention and clearance of inhaled particles. *Br J Ind Med* 37: 337–362.

Lourenco, R.V. and Cotromanes, E. (1982). Clinical aerosols. II. Therapeutic aerosols. *Arch Intern Med* 142: 2299–2308.

MacFarland, H.N. (1976). Respiratory toxicology. In *Essays in Toxicology*, Vol. 7 (W.J. Hayes, Ed.). Academic Press, New York, pp. 121–154.

Marple, V.A. and Rubow, K.L. (1980). Aerosol generation concepts and parameters. In *Generation of Aerosols and Facilities for Exposure Experiments* (K. Willeke, Ed.). Ann Arbor Science, Ann Arbor, MI, p. 6.

Mauderly, J.L. (1974). The influence of sex and age on the pulmonary function of the beagle dog. *J Gerontol* 29: 282–289.

Mauderly, J.L. (1989). Effects of inhaled toxicants on pulmonary function, In *Concepts in Inhalation Toxicology* (McClellan, R.O., Henderson, R.F., Eds.). New York, Hemisphere Publishing, pp. 347–401.

McClellan, R.O. and Henderson, R.F. (1989). *Concepts in Inhalation Toxicology*. Hemisphere Publishing, Washington, D.C.

McFadden, E.R., Jr. (1986). *Inhaled Aerosol Bronchodilators.* Williams and Wilkins, Baltimore, pp. 40–41.

McNeill, R.S. (1964). Effect of a β-adrenergic blocking agent, propranol, on asthmatics. *Lancet.* November 21: 1101–1102.

Menon, M.P.S. and Das, A.K. (1977). Tetracycline asthma—a case report. *Clin Allergy* 7: 285–290.

Menzel, D.B. and Amdur, M.O. (1986). Toxic responses of the respiratory system. In *Casarett and Doull's Toxicology,* 3rd ed. (C.D. Klaassen, M.O. Amdur, and J. Doull, Eds.). Macmillan, New York, pp. 330–358.

Mercer, T.T. (1981). Production of therapeutic aerosols; principles and techniques. *Chest* 80 (Suppl. 6): 813–818.

Merin, R.G. (1981). Cardiac toxicity of inhalation anesthetics. In *Cardiac Toxicology,* Vol. II (T. Balazs, Ed.). CRC Press, Boca Raton, FL, pp. 4–10.

Milic-Emili, J. (1982). Recent advances in clinical assessment of control of breathing. *Lung* 160: 1–17.

Miller, F.J., Gardner, D.E., Graham, J.A., Lee, R.E., Jr., Wilson, W.E., and Bachmann, J.D. (1979). Size considerations for establishing a standard for inhalable particles. *J Air Pollut Cont Assoc* 29: 610–615.

Miller, J.L., Stuart, B.O., Deford, H.S., and Moss, O.R. (1981). Liquid aerosol generation for inhalation toxicology studies. In *Inhalation Toxicology and Technology* (B.K.J. Leong, Ed.). Ann Arbor Science, Ann Arbor, MI, pp. 121–207.

Moren, F. (1981). Pressurized aerosols for oral inhalation. *Int J Pharm* 8: 1–10.

Moren, F. (1993). Aerosol dosage forms and formulations. In *Aerosols in Medicine.* 2nd ed. (S. Moren, M.B. Dolovich, M.T. Newhouse, S.P. Newman, Eds.). Elsevier, Amsterdam, pp. 329–336.

Mueller, R.A. (1982). The neuropharmacology of respiratory control. *Pharmacol Rev* 34: 255–285.

Murphy, D.J. (1994). Safety pharmacology of the respiratory system: techniques and study design. *Drug Dev Res* 32: 237–246.

Murphy, D.J., Joran, M.E., and Grando, J.C. (1994). Microcapnometry: A noninvasive method for monitoring arterial CO_2 tension during ventilatory measurements in conscious rats. *Toxicol Methods* 4: 177–187.

Mussatto, D.J., Garrad, C.S., and Lourenco, R.V. (1988). The effect of inhaled histamine on human tracheal mucus velocity and bronchial mucociliary clearance. *Am Rev Respir Dis* 138: 775–779.

National Academy of Sciences (NAS). (1958). *Handbook of Respiration.* NAS National Research Council. W.B. Saunders, Philadelphia, p. 41.

Newman, S.P. (1984). Therapeutic aerosols. In *Aerosols and the Lung, Clinical and Experimental Aspects* (S.W. Clarke and D. Pavia, Eds.). Butterworth, London, pp. 197–224.

Newton, P.E. (2000). Techniques for evaluating hazards of inhaled products, in *Product Safety Evaluation Handbook,* 2nd ed., Marcel Dekker: New York, pp. 243–298.

Nuzzo, P.F. and Anton, W.R. (1986). Practical applications of capnography. *Respir Ther* 16: 12–17.

O'Neil, J.J. and Raub, J.A. (1984). Pulmonary function testing in small laboratory mammals. *Environ Health Perspect* 53: 11–22.

Oberst, F.W. (1961). Factors affecting inhalation and retention of toxic vapours. In *Inhaled Particles and Vapours* (C.N. Davis, Ed.). Pergamon Press, New York, pp. 249–266.

Palecek, F. (1969). Measurement of ventilatory mechanics in the rat. *J Appl Physiol* 27: 149–156.

Parent, R.A. (1991). *Comparative Biology of the Normal Lung*. CRC Press: Boca Raton, FL.

Paterson, J.W. (1977). Bronchodilators. In *Asthma* (T.J.H. Clark, and S. Godfrey, Eds.). Chapman and Hall, London, pp. 251–271.

Paterson, J.W., Woocock, A.J., and Shenfield, G.M. (1979). Bronchodilator drugs. *Am Rev Respir Dis* 120: 1149–1188.

Patterson, R. and Kelly. J.F. (1974). Animal models of the asthmatic state. *Annu Rev Med* 25: 53–68.

Patty, F.A. (1958). *Industrial Hygiene and Toxicology*. 2nd ed., Interscience, New York.

Pauluhn, J. (2002). Overview of testing methods used in inhalation toxicology: from facts to artifacts, *Eurotox 2002,* Supp. pp. 5–7.

Paustenbach, D.J., Carlson, G.P., Christian, J.E., Born, G.S., and Rausch, J.E. (1983). A dynamic closed-loop recirculating inhalation chamber for conducting pharmacokinetic and short-term toxicity studies. *Fundam Appl Toxicol* 3: 528–532.

Pavia, D. (1984). Lung mucociliary clearance. In *Aerosols and the Lung, Clinical and Experimental Aspects* (S.W. Clarke and D. Pavia, Eds.). Butterworth, London, pp. 127–155.

Pavia, D., Bateman, J.R.M., Sheahan, N.F., Agnew, J.E., Newman, S.P., and Clarke, S.W. (1980). Techniques for measuring lung mucociliary clearance. *Eur J Respir Dis* 67(Suppl. 110): 157–177.

Pavia, D., Sutton, P.P., Agnew, J.E., Lopez-Vidriero, M.T., Newman, S.P., and Clarke, S.W. (1983). Measurement of bronchial mucociliary clearance. *Eur J Respir Dis* 64(Suppl. 127): 41–56.

Pavia, D., Sutton, P.P., Lopez-Vidriero, M.T., Agnew, J.E., and Clarke, S.W. (1983). Drug effects on mucociliary function. *Eur J Respir Dis* 64(Suppl. 128): 304–317.

Phalen, R.F. (1984). *Inhalation Studies: Foundations and Techniques*. CRC Press, Boca Raton, FL, pp. 35–46, 51–57.

Phalen, R.F. (1997). *Methods in Inhalation Toxicology*, CRC Press, Boca Raton, FL.

Rinehart, W.E. and Hatch, T. (1964). Concentration-time product (CT) as an expression of dose in sublethal exposures to phosgene. *Am Ind Hyg Assoc J* 25: 545–553.

Rosnow, E.C. (1992). Drug-induced pulmonary disease: an update. *Chest* 102: 239–250.

Rylander, R. (1966). Current techniques to measure alterations in the ciliary activity of intact respiratory epithelium. *Am Rev Respir Dis* 93(Suppl): 67–85.

Sackner, M.A., Brown, L.K., and Kim, C.S. (1981). Basis of an improved metered aerosol delivery system. *Chest* 80(Suppl. 6): 915–918.

Salem, H. (1986). Principles of inhalation of inhalation toxicology. In *Inhalation Toxicology: Research Methods, Applications, and Evaluation.* Marcel Dekker, New York, pp. 1–33.

Schalper, M., et al. (1989). Alteration of respiratory Cycle timing by propenanol, *Toxicol. Appl. Pharmacol.* 97: 538–547.

Schreck, R.W., Sekuterski, J.J., and Gross, K.B. (1986). Synchronized intratracheal aerosol generation for rodent studies. In *Aerosols, Formation and Reactivity.* Second International Aerosol Conference, Berlin. Pergamon Press, New York, pp. 37–51.

Schwettmann, R.S. and Casterline, C.L. (1976). Delayed asthmatic response following occupational exposure to enflurane. *Anesthesiology* 44: 166–169.

Shao, Z. and Mitra, A.K. (1992). Nasal membrane and intracellular protein and enzyme release by bile salts and bile salt fatty-acid mixed micelles-correlation with facilitated drug transport, *Pharm Res* 9: 1184–1189.

Shao, Z., Krishnamoorthy, R., and Mitra, A.K. (1992). Cyclodextrins as nasal absorption promoters of insulin-mechanistic evaluations, *Pharm Res* 9(9): 1157–1163.

Sharma, S., White, G., Imondi, A.R., Plaelse, M.E., Vail, D.M., and Dris, M.G. (2001). Development of inhalation agents for oncological use. *J Clin Oncol* 19: 1839–1847.

Sidorenko, G.I. and Pinigin, M.A. (1976). Concentration-time relationship for various regimens of inhalation of organic compounds. *Environ Health Perspect* 13: 17–21.

Silver, S.D. (1946). Constant flow gassing chambers: principles influencing design and operation. *J Lab Clin Med* 31: 1153–1161.

Sinnet, E.E. (1981). Fast integrated flow plethysmograph for small mammals, *J Appl Physiol* 50: 1104–1110.

Smaldone, G.C. (1997). Determinants of dose and response to inhaled therapeutic agents in asthma, In *Inhaled Glucocorticoids in Asthma: Mechanisms and Clinical Actions*. (R.P. Schleimer, W.W. Busse, and P.M. O'Byrne, Eds.). Marcel Dekker, New York, pp. 447–477.

Stahlhofen, W., Gebhart, J., and Heyder, J. (1980). Experimental determination of the regional deposition of aerosol particles in the human respiratory tract. *Am Ind Hyg Assoc J* 41: 385–398.

Stahlhofen, W., Gebhart, J., and Heyder, J. (1981). Biological variability of regional deposition of aerosol particles in the human respiratory tract. *Am Ind Hyg Assoc J* 42: 348–352.

Stott, W.T. and McKenna, M.J. (1984). The comparative absorption and excretion of chemical vapors by the upper, lower and intact respiratory tract of rats. *Fundam Appl Toxicol* 4: 594–602.

Swift, D.B. and Proctor, D.F. (1982). Human respiratory deposition of particles during oronasal breathing. *Atmos Environ* 16: 2279–2282.

Tamulinas, C.B. and Leach, C.L. (2000). Routes of exposure: inhalation and intranasal. In *Excipient Toxicology and Safety* (Weiner, M.L. and Kotkoskie, L.A., Eds.). Marcel Dekker, New York, pp. 185–205.

Task Group on Lung Dynamics. (1966). Deposition and retention models for internal dosimetry of the human respiratory tract. *Health Phys* 12: 173–207.

Tattersfield, A.E. (1986). Beta adrenoreceptor antagonists and respiratory disease, *J Cardiovasc Pharmacol* 8 (Suppl. 4): 535–539.

Taulbee, D.B., Yu, C.P., and Heyder, J. (1987). Aerosol transport in the human lung from analysis of single breaths. *J Appl Physiol* 44: 803–812.

Thorne, P.S. and Karol, M.H. (1989). Association of fever with late-onset pulmonary hypersensitivity responses in the guinea pig. *Toxicol Appl Pharmacol* 100: 247–258.

Touvay, C. and Le Mosquet, B. (2000). Systeme respirataire et pharmacolgie de securite. *Therapie* 55: 71–83.

Ts'o, T.O.T., Leong, B.K.J., and Chenoweth, M.B. (1975). Utilities and limitations of behavioral techniques in industrial toxicology. In *Behavioral Toxicology* (B. Weiss and V.G. Laties, Eds.). Plenum Press, New York. pp. 265–291.

Uemitsu, N., Minobe, Y., and Nakayoshi, H. (1985). Concentration-time-response relationship under conditions of single inhalation of carbon tetrachloride. *Toxicol Appl Pharmacol* 77: 260–266.

Ugwoke, M.I., Verbeke, N., and Kinget, R. (2001). The biopharmaceutical aspects of nasal mucoadhesive drug delivery, *J Pharm Pharmacol* 53: 3–21.

Ulrich, C.E., Klonne, D.R., and Church, S.V. (1984). Automated exposure system for metered-dose aerosol pharmaceuticals. *Toxicologist* 4: 48.

Valentine, R. and Kennedy, G.L., Jr. (2001). Inhalation toxicology. In *Principles and Methods of Toxicology.* 4th ed. (W. Hayes, Ed.). Raven Press, New York, pp. 1085–1143.

Vidgren, M.T., Karkkainen, A., Paronen, T.P., and Karjalainen, P. (1987). Respiratory tract deposition of ^{99}Tc-labeled drug particles administered via a dry powder inhaler, *Int J Pharm* 39: 101–105.

Wanner, A. (1979). The role of mucociliary dysfunction in bronchial asthma. *Am J Med* 67: 477–485.

Wanner, A. (1981). Alteration of tracheal mucociliary transport in airway disease. Effect of pharmacologic agents. *Chest* 80(Suppl. 6): 867–870.

Webber, S.E. and Widdicombe, J.G. (1987). The actions of methacholine, phenylephrine, salbutamol and histamine on mucus secretion from the ferret in-vitro trachea. *Agents and Actions* 22: 82–85.

Weibel, E.R. (1963). *Morphometry of the Human Lung.* Springer-Verlag, Heidelberg.

Weiner, N. (1984). Norepinephrine, epinephrine, and the sympathomimetic amines. In *Goodman and Gilman's, The Pharmacological Basis of Therapeutics.* 7th ed. (A.G. Gilman, L.S. Goodman, T.W. Rall, and F. Murad, Eds.). Macmillan, New York, pp. 145–180.

Weston, R. and Karel, L. (1946). An application of the dosimetric method for biologically assaying inhaled substances. *J Pharmacol Exp Ther* 88:195–207.

Widdicombe, J.G. (1974). Reflex control of breathing. In *Respiratory Physiology.* Vol. 2 (J.G. Widdicombe, Ed.). University Park Press, Baltimore, MD.

Willeke, K. (1980). *Generation of Aerosols and Facilities for Exposure Experiments.* Ann Arbor Science, Ann Arbor, MI.

Williams, M.H. (1974). Steroids and antibiotic aerosols. *Am Rev Respir Dis* 110: 122–127.

Wills, P. and Greenstone, M. (2001). Inhaled hyperosmolar agents for bronchiectasis. *Cochrane Database Syst Rev.* CD002996.

Wilson, A.G.E. (1982). Toxicokinetics of uptake, accumulation, and metabolism of chemicals by the lung. In *Mechanisms in Respiratory Toxicology.* Vol. 1. (H. Witschi and P. Nettesheim, Eds.). CRC Press, Boca Raton, FL, pp. 162–178.

Witschi, H.P. and Brain, J.D. (1985). *Toxicology of Inhaled Materials.* Springer-Verlag, New York.

Witschi, H.P. and Nettesheim, P. (1982). *Mechanisms in Respiratory Toxicology,* Vol. 1. CRC Press, Boca Raton, FL.

Zhang, Y. and Jiang, X. (2001). Detoxification of nasal toxicity of nasal drug delivery system. *Zhongguo Yiyao Gongze Zazhi,* 32: 323–327.

8

VANILLOID RECEPTORS IN THE RESPIRATORY TRACT

Christopher A. Reilly, John M. Veranth, Bellina Veronesi, and Garold S. Yost

CONTENTS

DISCLAIMER

This document has been reviewed by the National Health and Environmental Effects Research Laboratory and approved for publication. Approval does not signify that the contents reflect the views of the Agency, or does mention of trade names or commercial products constitute the endorsement or recommendation for use.

8.1 INTRODUCTION

8.1.1 Vanilloid Receptors and Respiratory Toxicology

Environmental exposure to respiratory toxicants, including industrial and chemical pollutants, ultrafine geological dusts, combustion by-products such as smoke, and many other respirable substances, poses a severe threat to human health (Utell et al., 1988; Amdur, 1989; Pope, 1989; Xu et al., 1991; Dockery, 1993; Dockery et al., 1993; Pope, 1998; Churg and Brauer, 2000; Lighty et al., 2000; Dockery, 2001; Veronesi and Oortgiesen, 2001; Kagawa, 2002; Mortimer et al., 2002). Epidemiology has associated air pollution with adverse health effects ranging from school absences to cardiovascular deaths, and this has motivated many toxicology studies. Recent research has documented a plethora of diverse pathologies associated with inhalation exposure to environmental pollutants. From these studies, a number of hypotheses have been proposed for the mechanisms by which the toxicity of these substances is manifested. One hypothesis, which relates to the toxicological effects resulting from exposure to airborne ambient particulate materials and aerosolized pepper spray products (capsaicinoids), focuses on the activation of the vanilloid receptors, members of the transient receptor potential (TRP) gene family. Specifically, TRPV1 (previously called VR1 or the capsaicin receptor) is believed to be the key molecular determinant and mediator of the inflammatory and cytotoxic effects that are associated with these substances (Veronesi, Oortgiesen et al., 1999; Oortgiesen et al., 2000; Veronesi et al., 2000; Veronesi, de Haar, Roy et al., 2002; Veronesi, de Haar, Lee et al., 2002; Agopyan, Li et al., 2003; Reilly, Taylor et al., 2003; Veronesi et al., 2003). TRPV1 is a member of a continually expanding family of related receptors known as TRP ion channels (Clapham et al., 2001; Montell et al., 2002). TRPV1 and other TRP/vanilloid receptors are polymodal receptors that are activated by a diverse array of chemicals and irritants, pollutants, and environmental conditions such as pH, osmolarity, and temperature (Clapham et al., 2001; Montell et al., 2002; Benham et al., 2003). This chapter will document the research that has contributed to our current understanding of the role that TRPV1 and other TRP/vanilloid receptors play in the etiology of pain perception, respiratory inflammation, and pathologies, including those involving ambient particles and the capsaicinoids present in pepper spray products. Major topics include a detailed review of the cloning and biochemical characterization of TRPV1 and a variety of other related TRP/vanilloid receptors and a discussion of the function and activation of these receptors by diverse ligands and their role in chemical and thermal nociception. A pragmatic discussion of recent progress in research designed to ascertain the importance of these receptors in mediating various respiratory toxicities and pathologies associated with inhalation exposure to respiratory irritants and toxicants is also presented. Throughout the chapter, the major focus will be on describing the effects of capsaicin, the model agonist for TRPV1, which was the first mammalian TRP/vanilloid receptor to be cloned and characterized, and which is the most frequently investigated receptor of its class.

8.1.2 Historical Perspective on Capsaicinoids

Capsaicinoids are a group of structurally related chemicals that are responsible for the pungent properties of hot peppers (Govindarajan, 1985; Govindarajan and Sathyanarayana, 1991). These compounds are present in high concentrations in the fruits of *Capsicum annum* and *C. frutescens*. Six naturally occurring compounds have been identified and their chemical and biological properties characterized: capsaicin, dihydrocapsaicin, nordihydrocapsaicin, homocapsaicin, homodihydrocapsaicin, and nonivamide (Govindarajan, 1985; Govindarajan and Sathyanarayana, 1991; Reilly et al., 2001). All of the capsaicinoid analogs consist of a vanilloid ring moiety

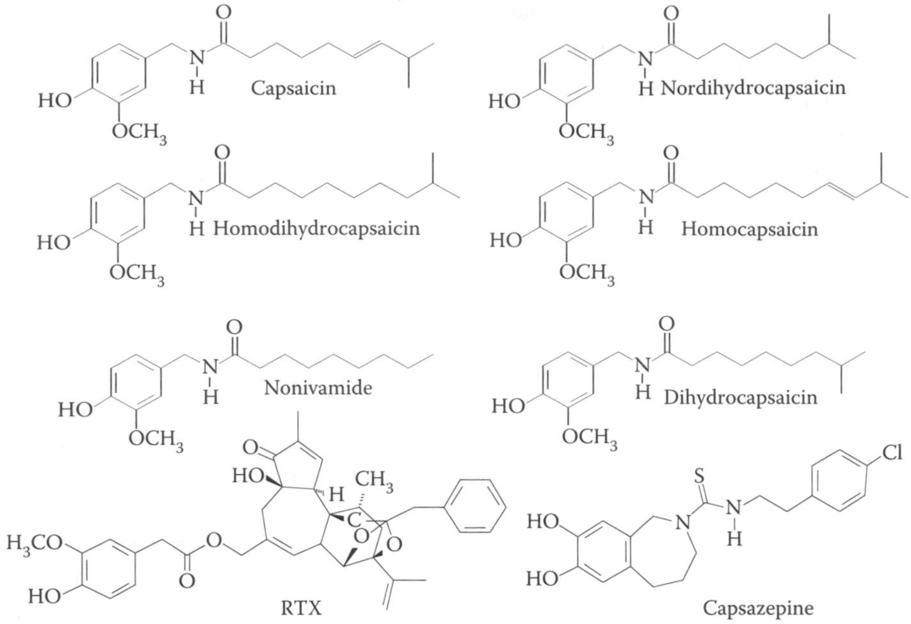

Figure 8.1 Chemical structures of prototypical vanilloid receptor ligands. Structures for the capsaicinoids, resiniferatoxin (Rtx, an agonist), and capsazepine (an antagonist) are shown.

(4-hydroxy-3-methoxybenzylamide) linked to a branched aliphatic chain moiety that is between eight and ten carbons in length and has either zero or one degree of unsaturation (Figure 8.1) (Nelson, 1919). Nonivamide, commonly referred to as synthetic capsaicin because of its facile laboratory synthesis and difficulty of detection in natural mixtures of capsaicinoids, has recently been identified as a natural product (Reilly et al., 2001). Nonivamide differs from capsaicin in that it has a saturated, unbranched aliphatic chain. Capsaicin is the most abundant of the capsaicinoids, constituting approximately 40 to 60% of the total capsaicinoid content in peppers and products derived from pepper extracts. Dihydrocapsaicin constitutes 20 to 40% of the total capsaicinoids in pepper extracts, and the other capsaicinoid analogs constitute the remaining pungent constituents (Govindarajan, 1985; Govindarajan and Sathyanarayana, 1991; Reilly et al., 2001). The absolute and relative capsaicinoid content of peppers can vary, depending on the species and variety of pepper, growing conditions, and time of harvest (Govindarajan, 1985; Govindarajan and Sathyanarayana, 1991; Reilly et al., 2001).

For nearly 7000 yr, humans have cultivated hot peppers for a variety of purposes: to prepare foods, to provide traditional homeopathic medicines, and for self-defense. The first purported use of capsaicinoids as a means of self-defense dates back to the 1500s in Peru when the Inca allegedly burned capsicum plants to ward off invasions by Pizarro and the Spaniards. A more recent example of the use of capsaicin as a chemical deterrent is exemplified by the extensive popularity of pepper sprays as a type of less-than-lethal self-defense weaponry. (Olajos and Salem, 2001; NIJ, March 1994). Use of capsaicinoids in pepper spray products stems from the ability of these compounds to deter an attacker through the production of intense pain, temporary blindness, and uncontrollable coughing. In recent years, capsaicinoids have also been used for a variety of other purposes, e.g., a tool for neurobiological research (Szallasi and Blumberg, 1999), an aid for weight loss (Henry and Emery, 1986; Lim et al., 1997), an agent for local or topical analgesia (particularly in dentistry) (McMahon et al., 1991; Lazzeri et al., 1996; Szallasi and Blumberg, 1999), and an antimicrobial agent (Jones et al., 1997).

8.1.3 Identification and Cloning of TRPV1

For centuries, it has been known that capsaicin and its analogs possess the capacity to cause unpleasant sensations of severe irritation, burning, and itching at exposed sites. In the 1960s, scientists demonstrated that capsaicinoids selectively activated peripheral sensory neurons and produced temporary increases in sensitivity to painful stimuli such as those caused by elevated temperatures, acids, and mustard oil, followed by a period of decreased sensitivity known as *analgesia* (Jancso et al., 1967; Jancso-Gabor et al., 1970). Additional studies were able to demonstrate that the pharmacological properties of capsaicinoids were produced by the activation of a specific receptor that regulated cellular cation fluxes (i.e., calcium and sodium) and that, ultimately, resulted in neuronal depolarization and neuropeptide release (Holzer, 1988; Holzer, 1991). The apparent selectivity for sensory nerves in these tissues correlated with the extensive innervation of these tissues by capsaicin-sensitive and substance-P-containing neurons. The ability of capsaicinoids to promote extended periods of analgesia has recently been used as the basis for topical treatment of severe and chronic neuropathic pain associated with arthritis, HIV-related and diabetic neuropathy, muscle injury, etc., as well as for intravesicular treatment of bladder infection and hyperreactivity disorders that are associated with spinal cord injury (hyperflexia) (Paice et al., 2000; Jensen and Larson, 2001; Szallasi, 2002; Lopez-Rodriguez et al., 2003; Pappagallo and Haldey, 2003).

Early neurobiological research demonstrated that capsaicin acted selectively on a subset of thin-to-medium diameter, unmyelinated to slightly myelinated sensory neurons (Aδ type II and C-fibers) originating from the dorsal root ganglia (DRG) and terminating at or near epithelial surfaces of sensory organs such as the eyes, nose, mouth, and skin, as well as innervating a plethora of visceral organs including the liver, lung, heart, and urinary tract (Holzer, 1991; Szallasi and Blumberg, 1999). These studies showed that capsaicin-induced calcium and sodium influx in the sensory neurons that was correlated to tachykinin release (substance P, neurokinin A, and calcitonin-gene-related peptide, CGRP), localized inflammation, analgesia, and peripheral neuropathy (Szallasi and Blumberg, 1999). The inflammatory effects of capsaicinoids were ultimately attributed to plasma extravasation and infiltration of inflammatory cells, owing to localized, elevated cytokine release at the sites of administration.

In 1997, Dr. David Julius and colleagues at the University of California at San Francisco initiated a search for the identity of the molecular determinant of these highly varied capsaicin-induced responses (Caterina et al., 1997; Tominaga et al., 1998). Using a rat cDNA library and a technique termed *expression cloning*, this group isolated the full-length cDNA (2514 bp) of a protein that imparted sensitivity, in the form of capsaicin-dependent calcium influx into cells, to transfected cells treated with capsaicin and various extracts of peppers. This receptor was called the *capsaicin receptor* (vanilloid receptor, VR1, and subsequently named TRPV1). The sequence of this cDNA coded for a protein product of 838 amino acids (~95 kDa) with a predicted hydropathy profile consistent with a membranc-embedded protein consisting of 6 putative transmembrane (TM) segments, a short pore loop domain between TM5 and TM6, and a long intracellular N-terminal domain and an intracellular C-terminal tail containing a conserved TRP domain and various potential phosphorylation sites; the TRP domain is a highly conserved 25 amino acid segment of unknown function near the C-terminus of most TRP-type proteins (Minke and Cook, 2002). The N-terminal domain contains three ankyrin repeats that anchor TRPV1 to the cytoskeleton, and multiple potential phosphorylation sites that may be related to regulation of receptor activation and function. A graphic representation of TRPV1 is presented in Figure 8.2. The striking resemblance of TRPV1 to the TRP channel first discovered in the visual organs of blind *Drosophila melanogaster* mutants lacking the TRP channel protein placed TRPV1 in the superfamily of nonselective, store-operated cation channels known as TRPs (Minke and Cook, 2002; Montell et al., 2002). Cloning and characterization of the human ortholog of rat TRPV1 soon followed and showed very high similarity in sequence and functional characteristics to the rat gene (Hayes et al., 2000).

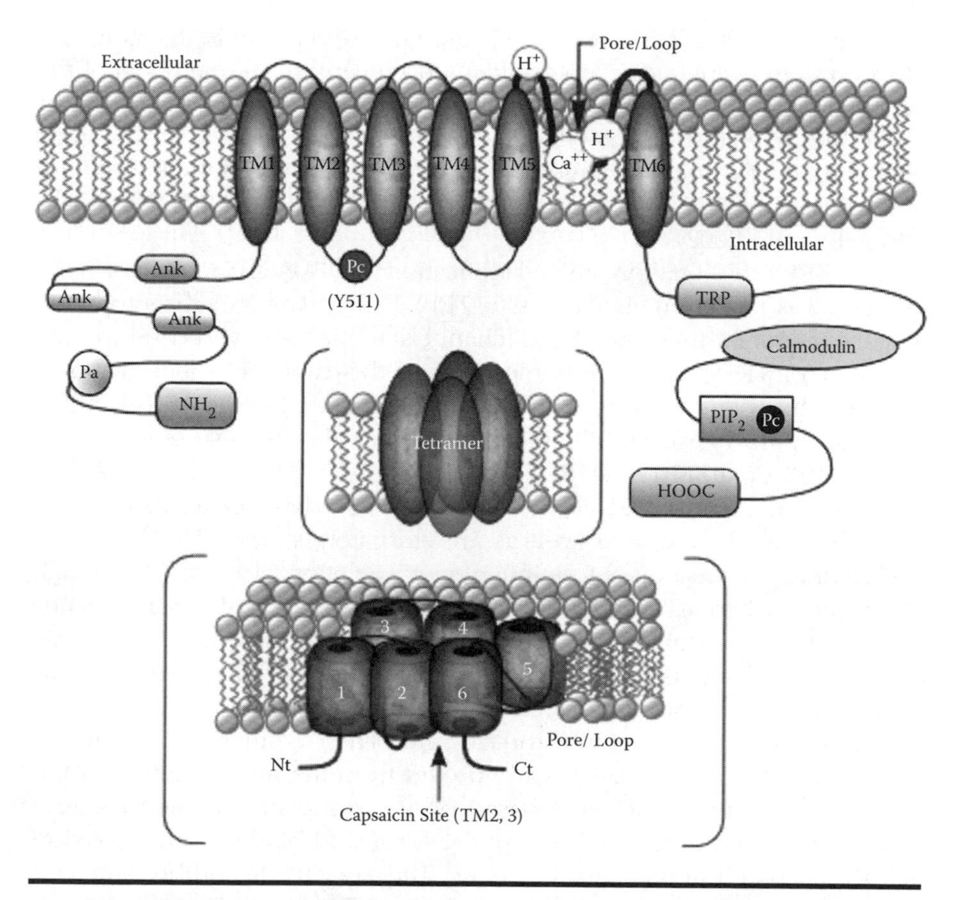

Figure 8.2 Schematic representation of the human TRPV1 receptor. Specific structural features that contribute to the pharmacological and toxicological properties of TRPV1 are shown for the intracellular N- and C-terminal portions of TRPV1. The assembly of four TRPV1 subunits to form an active channel is illustrated in the middle, and a hypothetical model for the structure of a TRPV1 within the cell membrane is shown at the bottom. Ank = ankyrin repeats; TRP = transient receptor potential domain; Pa = phosphorylation site for protein kinase A; Pc = phosphorylation site for protein kinase C; Y511 = tyrosine 511 in which vanilloid aromatic moiety binds; TM1-6 = transmembrane domains; and Pip$_2$ = phosphatidylinositol diphosphate binding site.

8.1.4 Newly Discovered Members of the TRP Superfamily of Ion Channels

Since the discovery of TRPV1, a variety of other TRPV1-like proteins have been identified based on sequence homology, functional similarities, and overlaps in their substrate or ligand-binding and activation profiles

(Montell et al., 2002). There are six known TRPV proteins in mammals. TRPV1 and the vanilloid-receptor-like (VRL) receptors are members of the superfamily of TRP/vanilloid receptor ion channels. Recently, the members of the TRP/vanilloid receptor family of ion channels have been renamed on the basis of a unified nomenclature as TRPV1 to TRPV6, as there are other classes of TRP receptors that include the TRPM and TRPC classes (Montell et al., 2002).

The TRPV designation and classification of TRP channels (formerly OTRPC) was based on similarities to TRPV1 (the capsaicin receptor), the first member of this class of ion channels to be discovered. Since the discovery of TRPV1, numerous members of this cation channel group of receptors have been discovered and characterized: TRPV2 (VRL-1, GRC) (Caterina et al., 1999), TRPV3 (VRL-3) (Smith et al., 2002; Xu et al., 2002), TRPV4 (VRL-2, mTRP12, OTRPC4, and VR-OAC) (Delany et al., 2001), TRPV5 (ECaC1, CaT2), and TRPV6 (ECaC2, CaT1) (den Dekker et al., 2003; Hoenderop et al., 2003), as well as an alternately spliced TRPV1 variant that exhibits a truncated N-terminus region, deletion of exon 7, and lack of function when exposed to traditional TRPV1 agonists (Schumacher et al., 2000; Xue et al., 2001). Each of these members of the TRPV class of TRP ion channels exhibit related, but unique, ligand-binding and activation profiles with the common feature that they all serve as cation channels with high selectivity for calcium. The "V" in this designation represents "vanilloid" and was carried over from the initial naming of the capsaicin receptor as VR1; it is based on the stringent structural requirement of a vanilloid ring pharmacophore for ligand binding and activation of the receptor (Montell et al., 2002). The specific ligand-binding and activation profiles for the TRPV channels, TRPV1 to TRPV4, will be described in the following text.

Designation or classification of TRP channels into the TRPM subfamily is based on the relationship of the receptors to TRPM1 ("M" for melastatin-related proteins), formerly Mlsn1, the first member of this subfamily of receptors (the long TRPs) (Montell et al., 2002). Melastatin, an underexpressed gene product in melanoma cells, was shown to exhibit marked similarity to drosophila TRP and thus it was placed in the TRP superfamily of cation channels. This subfamily contains eight mammalian subtypes and is typically associated with calcium regulation resulting from a variety of events, including reactive oxygen species (ROS) formation and increased concentrations of ADP-ribose (TRPM2) (Birnbaumer et al., 2003), tasting (TRPM5) (Birnbaumer et al., 2003; Perez et al., 2003; Perraud et al., 2003) as well as decrease of temperature (TRPM8) (McKemy et al., 2002; Nealen et al., 2003). Some additional discussion in this chapter will focus on TRPM8 because of its functional similarity to TRPV1 as a temperature-sensitive ion channel (TRPM8 is activated by cold temperatures) and its potential impact on human respiratory health. Recently, a new "cold receptor," ANKTM1, the first member of the newly designated TRPN subfamily of TRP/vanilloid receptors, has been discovered and shown to have properties that are similar to TRPM8 (Story et al., 2003).

Mammalian TRPC ion channels, formerly classified as short TRPs, are the TRP-"canonical" or "classical" channels that are most closely related to the drosophila TRP (TRPC1) (Montell et al., 2002; Beech et al., 2003; Montell, 2003b). TRPC1, formerly known as TRP, was the first TRP channel discovered in drosophila. TRPC1 functions as a mediator of calcium and sodium homeostasis in photoreceptor cells involved in sustaining elevated intracellular calcium in response to light, via a classical G-protein- and phospholipase-C-dependent activation mechanism (Montell, 2003b). The characteristic transient nature of increased cellular calcium in response to light in blind flies (a TRP) ultimately led to the discovery that the loss of function of TRPC1 imparted blindness. There are five functional TRPC channels known to be expressed in humans and seven in mice; they may be involved in diverse, yet undefined, calcium-dependent processes in various cell types derived from glands, lung, heart, skeletal muscle, brain, reproductive organs, and kidney (Beech et al., 2003; Montell, 2003b). One hypothesis is that the multiple TRPC subtypes expressed in smooth muscle are involved in the muscarinic responses to acetylcholine (Beech et al., 2003). In general, however, the physiological functions of these receptors are unknown.

8.1.5 Functional Characteristics of TRPV1 and Other Mammalian TRP Channels

As described in the preceding text, TRPV1, or the capsaicin receptor, was first discovered using capsaicin as a ligand. Since the initial cloning of TRPV1, it has also been characterized for its ability to bind ligands (RTX-type responses) and become activated (calcium-type responses) by a variety of known chemical irritants, endogenous inflammatory mediators, and pain-producing stimuli. Some of the ligands and activators of TRPV1 include capsaicin, resiniferatoxin (RTX), olvanil, phorbol-12-myristate-13-acetate (PMA), lipoxygenase products such as 12(*S*) and 15(*S*)-hydroperoxyeicosatetraenoic acid (15(*S*)- HPETE) and 5(*S*)-hydroxyeicosatetraenoic acid (5(*S*)-HETE), leukotriene B4, phorbol-12-phenylacetate-13-acetate-20-homovanillate (PPAHV), anandamide, elevated temperature, acidic pH, and many others (Appendino et al., 1996; Szallasi et al., 1998; Szallasi and Blumberg, 1999; Hwang et al., 2000; Smart et al., 2000; Caterina and Julius, 2001; Veronesi and Oortgiesen, 2001; Dedov et al., 2002).

The capsaicinoids, and many other chemical ligands, activate TRPV1 and ultimately produce nociception and inflammation through their interaction with, and stimulation of, TRPV1 on peripheral sensory nerves. Capsaicin binding to TRPV1 exhibits stringent structural requirements with respect to the vanilloid ring and alkyl portions of the molecule (Walpole et al., 1993; Szallasi and Blumberg, 1999; Caterina and Julius, 2001). A requirement for the vanilloid ring moiety, or a structurally related conformational analog (e.g., 12(*S*)-HPETE) (Hwang et al., 2000), persists with essentially all of the known TRPV1 ligands including RTX, olvanil, phorbol esters, lipoxygenase products, anandamide, and the prototypical inhibitor capsazepine (which contains the vanilloid ring motif). Capsaicin and nonivamide have been shown to be the most potent of the capsaicinoid analogs (Pyman, 1925). Decreased potency of the other analogs of capsaicin is presumably caused by differences in their acyl chain structures (Pyman, 1925; Walpole et al., 1993b). The ability of capsaicinoids, as well as other TRPV1 agonists, to induce ion fluxes is effectively inhibited by the TRPV1-selective inhibitor capsazepine (Bevan et al., 1992; Walpole et al., 1994), iodo-RTX (Wahl et al., 2001), and the nonselective calcium channel blocker ruthenium red (Maggi et al., 1993). Similarly, the endogenous peptide dynorphin A blocks TRPV1 activation (Planells-Cases et al., 2000). The synergistic activation of TRPV1 is observed with inflammatory mediators such as bradykinin and nerve growth factor (NGF) (Chuang et al., 2001; Prescott and Julius, 2003), possibly histamine (Richardson and Vasko, 2002), a slightly acidic pH (pH 6.0) (Caterina et al., 1997; Jordt et al., 2000), and slight increases in temperature (Caterina et al., 1997; Tominaga et al., 1998). TRPV1 can be activated by pH < 5.3 and temperatures $> 43°C$ through a process that is not dependent on capsaicin. In fact, one hypothesis for the activation of TRPV1 by capsaicinoids proposes that capsaicin binding alters the tertiary confirmation and thermodynamic stability of the protein such that the threshold for thermal activation is decreased to values well below 43°C, the threshold observed in the absence of capsaicin.

Another prototypical TRPV1 agonist that has been the focus of considerable research is RTX. RTX was first isolated from the desert cactus *Euphorbia poisonii* and was found to possess the characteristic vanilloid ring structure and the ability to cause irritation and produce analgesia (Szallasi and Blumberg, 1989; Szallasi and Blumberg, 1999; Caterina and Julius, 2001). RTX has been referred to as an "ultrapotent" TRPV1 agonist, having an EC_{50} for activation of TRPV1 that is estimated to be ~100-fold lower than that of capsaicin. Because of the very high selectivity of this vanilloid for TRPV1, RTX and its tritiated analog, [^{3}H]-RTX, have become valuable tools for understanding the biochemical and pharmacological properties of TRPV1 profiling the expression of TRPV1 and providing insights for the discovery of many new TRPV1 ligands and potential therapeutic agents (Szallasi and Blumberg, 1999).

In contrast to the chemical- or structural-based TRPV1 ligands described earlier, TRPV1 can also be activated by nonselective environmental stimuli such as increased temperature (> 43°C), acidic pH (< 5.3), and even certain types of ambient particulate material (Tominaga et al., 1998; Jordt et al., 2000; Veronesi, de Haar, Lee et al., 2002; Agopyan, Li et al., 2003; Veronesi et al., 2003). Although the precise mechanisms of activation by heat and acidic conditions have not been fully established, specific amino acid residues in the TRV1 protein have been shown to be essential for the activation of TRPV1 by these stimuli (Jordt et al., 2000). Heat or acid appear to alter protein conformation and stability through specific residues of TRPV1, to changing ion flux by disrupting key structural gating mechanisms.

One hypothesis for the physiological necessity to activate TRPV1 by these substances is related to the inflammatory response due to injury or infection. Activation of TRPV1 by inflammatory mediators (e.g., bradykinin) produces painful sensations, which may serve as a protective mechanism against additional tissue injury that may be induced by harmful environmental conditions or substances. Examples could include scenarios such as when an individual accidentally places his or her hand on a hot surface, or when harmful chemical vapors are inhaled, inducing the cough response as a protective measure. Similarly, activation of TRPV1 by endogenous increases in temperature and decreases in pH may also be important in initiating the inflammatory response due to tissue injury.

A very interesting facet related to activation of TRPV1 by capsaicin, and possibly other TRPV1 ligands, is that the binding site for capsaicin on TRPV1 appears to be located on the intracellular surface of the receptor, requiring capsaicin to passively enter the cell prior to interacting with and activating TRPV1. One study has demonstrated that DA-5018-HCl, a water-soluble capsaicinoid analog, was capable of activating TRPV1 ion flux only when injected directly into the cytosol of TRPV1-expressing cells (Jung et al., 1999). This was contrary to the mechanism of capsaicin-induced response, in which capsaicin activated TRPV1 when applied either extracellularly or injected into the cells. Similarly, production of chimeras between rat TRPV1 and the nonresponsive avian TRPV1 ortholog (birds are not sensitive to capsaicin, but are sensitive to temperature and acid) provided supporting evidence that the binding site for capsaicin resided on the intracellular side between TM helices 2 and 3 (Jordt and Julius, 2002). A model for the binding of capsaicin with TRPV1 proposed that the aliphatic moiety of capsaicin extended into the hydrophobic, membrane-embedded core of the assembled protein between TM2 and TM3, whereas the vanilloid ring moiety was associated via aromatic π-stacking interactions with a key tyrosine residue (Y511) that was located on the intracellular surface of the loop between TM2 and TM3 (Jordt and Julius, 2002)

(Figure 8.2). Although the binding sites for a variety of other TRPV1 ligands have not been established, there appears to be significant similarity between them. Essentially, all TRPV1 ligands must possess a vanilloid ring moiety or exist in a conformation that mimics molecules such as capsaicin and RTX in order to modulate TRPV1 activity and displace [^{3}H]-RTX binding (Walpole et al., 1993; Szallasi and Blumberg, 1999; Caterina and Julius, 2001). Recent studies have confirmed this idea by demonstrating that site-directed mutagenesis of the amino acids N676, M677, and L678 diminished [^{3}H]-RTX binding and ameliorated TRPV1 activation by capsaicin and RTX (Kuzhikandathil et al., 2001). Hypothetical three-dimensional models of the TRPV1 subunit configuration suggest that this sequence (on TM6) may potentially be part of the binding site identified for capsaicin on TM2 and TM3, as described earlier (Figure 8.2).

A variety of other TRP receptors have also been identified and characterized. TRPV1 has been described as a molecular integrator of pain-producing chemical and thermal stimuli. Other TRP and vanilloid-receptor-like proteins have been identified and classified based on their similarities to TRPV1 and other TRP ion channels. In general, only a small overlap exists for the types of ligands that activate these TRPV1 receptor analogs. For example, none of the TRPV1 receptor analogs are activated by capsaicinoids, RTX, or acidic pH. However, many of the TRPV, TRPM, and TRPN receptors are activated by abnormal concentrations of diverse physiological ligands, ions, ROS, etc. (Clapham et al., 2001; Montell et al., 2002; Beech et al., 2003; Birnbaumer et al., 2003; den Dekker et al., 2003; Hoenderop et al., 2003; Montell, 2003b; Perraud et al., 2003). Only the TRPV1 to TRPV4, TRPM8, and ANKTM1 receptors are activated by temperature to promote ion flux (Caterina et al., 1997; Caterina et al., 1999; McKemy et al., 2002; Montell et al., 2002; Smith et al., 2002; Xu et al., 2002; Benham et al., 2003; Story et al., 2003). Thus, a general hypothesis has been proposed that certain mammalian TRP/vanilloid receptors serve primarily as physiological mediators of thermal sensation, pain, and inflammatory responses, with additional functions and events resulting from unintended, idiopathic molecular interactions between the receptor and the activating toxin or noxious stimulus. Support for this general hypothesis has been provided, in many instances, by studies on knockout mice (Caterina et al., 2000) and studies of the "painless" mutant drosophila larvae (Goodman, 2003; Montell, 2003a).

From a plethora of *in vitro* studies designed to characterize the function of these receptors, a variety of unique chemical agonists have been identified. In addition, a fascinating picture has begun to emerge that shows the thermal activation of each receptor with temperatures that span specific ranges that may or may not overlap with other members of the TRP superfamily (Caterina et al., 1997; Caterina et al., 1999; Guler et al., 2002;

McKemy et al., 2002; Smith et al., 2002; Watanabe, Vriens et al., 2002; Xu et al., 2002; Benham et al., 2003; Story et al., 2003). Initial characterization of TRPV1 demonstrated that it was activated by temperatures greater than 43°C (Caterina et al., 1997; Tominaga et al., 1998). Interestingly, TRPV2 (VRL-1), a receptor similar in structure to TRPV1, was shown to possess an elevated threshold for thermal activation, approximately 52°C (Caterina et al., 1999). TRPV3, another vanilloid receptor that is expressed in high levels in skin, tongue, and various neurons, is activated by thermal stimuli within the range of 22 to 48°C, with the most significant activation occurring at temperatures slightly below the physiological range (~33°C) (Smith et al., 2002; Xu et al., 2002). TRPM8, a recently cloned receptor that possessed a unique affinity for menthol, was activated by cold temperatures < 22°C (McKemy et al., 2002), whereas yet another "cold" receptor, ANKTM1, appeared to be expressed in a subset of sensory neurons and was activated by temperatures < 20°C, without activation by menthol (Story et al., 2003). Surprisingly, TRPV4 (VRL-2) was activated by phorbol 12,13-dinonanoate 20-homovanillate (4α-PDD), a phorbol-ester, and 5,6-epoxyeicosatetrienoic acids (products of cytochrome-P450-dependent epoxygenase reactions), which are both similar to known agonists of TRPV1 (Hwang et al., 2000; Watanabe, Davis et al., 2002; Watanabe, Vriens, Prenen et al., 2003). However, TRPV4 was not responsive to other traditional vanilloid agonists (Delany et al., 2001). TRPV4 may also become activated by cell swelling induced by hypotonicity, as well as temperatures similar to those that activate TRPV3 (30 to 42°C) (Watanabe, Vriens et al., 2002; Alessandri-Haber et al., 2003; Mizuno et al., 2003; Suzuki et al., 2003; Vriens et al., 2004). Interestingly, the activation of TRPV4 by both hypotonic solutions and temperature appears to be closely linked, because the activation of TRPV4 by heat can be modulated by changes in extracellular osmolarity and *vice versa* (Watanabe, Vriens, Janssens et al., 2003). TRPV5 and TRPV6 are not activated by abnormal temperatures or acidic conditions; rather, they appear to be constitutively active, with tight control of ion flux via membrane potential and a voltage-dependent open pore magnesium blockade that is regulated via a calcium-dependent feedback mechanism (den Dekker et al., 2003; Voets et al., 2003). They appear to be key participants in the regulation of cell volume that occurs in response to changes in extracellular osmolarity at a variety of physiological temperatures (den Dekker et al., 2003; Voets et al., 2003).

8.1.6 Homogenous and Heterogeneous Multisubunit TRP Complexes

A common characteristic for many ion channel receptors is the formation of functional complexes comprising multiple subunits that may or may not be the same. Prime examples of these complexes are the catecholamine neurotransmitter receptors that form functional hetero- or homomeric complexes, each with unique pharmacological and biochemical properties. The TRP receptors appear to mimic such heterogeneity. Studies have shown that TRPC4 and TRPC5, TRPC1-like calcium channels that are activated by inositol triphosphate (IP_3) and are potentially involved in vasorelaxation, could coimmunoprecipitate and associate to form heteromeric receptor complexes in cells (Amiri et al., 2003; Plant and Schaefer, 2003). However, evidence supporting an altered function of the mixed complexes has not been confirmed. Similarly, early studies with TRPV1 demonstrated that the function of the wild-type receptor could be diminished by cotransfecting cells with a nonfunctional dominant negative TRPV1 mutant (Kuzhikandathil et al., 2001). The stoichiometry for complete amelioration of function by the dominant negative form TRPV1 was 4:1 (mutant:normal cDNA), suggesting that the functional form of TRPV1 existed as a tetramer that formed in response to agonist exposure (Kedei et al., 2001; Olah et al., 2001). Formation of tetrameric TRPV1 complexes was also confirmed by the coimmunoprecipitation of TRPV1 that was tagged with the marker proteins EGFP and PKCε (Kedei et al., 2001; Olah et al., 2001), as well as by additional verification of the immunoprecipitation and stoichiometric results using perfluoro-octanoate polyacrylamide gel electrophoresis (PFO-PAGE) and cross-linking agents (Kedei et al., 2001).

Because of the similarities in the sequence and function of the other vanilloid receptors in the TRP/vanilloid receptor family, additional studies on the ability of the different receptor subtypes to form homomeric and mixed complexes with TRPV1 have been performed. Receptors composed of multiple mixed vanilloid receptor proteins would be expected to have unique biochemical properties when compared to homologous receptor complexes. For example, mixed complexes of TRPV1 and TRPV2 or TRPM8 might be expected to exhibit variable activation thresholds to temperature, depending upon their subunit ratios, compared to homotetrameric complexes of each receptor type. Such variations in thermal activation could bridge the current gaps between the ranges of temperatures that specific receptor subtypes respond to and more accurately mimic all of the temperatures that cells respond to. Similarly, differences in ligand binding and activation would also be expected with such mixed receptor complexes. Such differences in the spectrum of properties of these receptor complexes could have a significant impact on responses to temperature and xenobiotics, particularly with respect to temperature sensitivity in respiratory disease.

Receptor activation by elevated temperatures were used to examine if mixed complexes between TRPV1 and TRPV2 could form and function (Caterina et al., 1999). The results of these studies clearly demonstrated two independent activation thresholds for exposure to elevated temperature, ~43°C and ~52°C; identical activation temperatures were observed in cells that were transfected only with the individual TRPV1 and TRPV2 receptor cDNAs. Therefore, the possibility that TRPV1 and TRPV2 could form mixed tetrameric complexes, with altered thermal activation properties, was dismissed. Similar studies with several of the other TRP/vanilloid receptors have been inconclusive. Evidence for the coimmunoprecipitation of TRPV1 with TRPV3 has been presented; however, little functional evidence for the formation of mixed complexes was obtained (Smith et al., 2002). It is intriguing to speculate that heteromeric receptors may control unique responses to diverse ligands, temperatures, and as yet undetermined stimuli. Therefore, additional research to demonstrate the presence and functional significance of mixed TRP receptors is an important goal for scientists in this fascinating field.

8.1.7 Chemical and Thermal Nociception by TRP Channels

The pain-producing properties of capsaicinoids have been known for centuries. However, the molecular mechanisms governing the ultimate sensation of pain following exposure to capsaicin have only recently been established. Identification and characterization of TRPV1 provided a distinct molecular sensor for the responses associated with capsaicin exposure in sensory neurons. Stimulation of TRPV1 by capsaicin results in the sudden influx of calcium and sodium into neurons and has been associated with sodium-dependent neuronal depolarization and calcium-dependent degranulation of the neuron, releasing various neuropeptides such as CGRP and substance P that are critically involved in pain perception and the initiation and progression of inflammatory processes (Szallasi and Blumberg, 1999; Richardson and Vasko, 2002). The critical role of TRPV1 in mediating the perception of pain following exposure to capsaicin was eloquently confirmed using TRPV1 knockout mice (Caterina et al., 2000). Mice lacking TRPV1 lost the typical responses associated with TRPV1 activation by lowered pH, elevated temperature, capsaicin, RTX, or other TRPV1 stimuli. Characteristic behavioral changes included the lack of avoidance of the consumption of capsaicin-fortified water, the loss of capsaicin-induced thermal hyperalgesia, and diminished responses to acid treatment. Responses to mechanical stimuli were not affected in TRPV1 knockouts, suggesting a role for the other TRP receptors that can be activated by mechanical stimuli. Additional studies performed with the TRPV1 knockout mice demonstrated that neurons isolated from the DRG of these mice were not sensitive to temperatures of ~43°C or to capsaicin or other TRPV1 chemical agonists. Surprisingly, these mice remained responsive to temperatures that were greater than 46 to 48°C, suggesting a potential role for TRPV2 or other TRP receptors in the thermal nociception produced by temperatures in the range of 43 to 52°C. Nevertheless, these studies provided strong support for the hypothesis that TRPV1 and the other TRP receptors are vital elements in the detection of painful and potentially harmful environmental (temperature and acid) and chemical stimuli. Experiments utilizing these mice as a means of assessing the function of TRPV1 in responses to inhaled irritants such as acetic acid, capsaicin, and acrolein are underway; however, preliminary data are inconclusive (John B. Morris, University of Connecticut, personal communication).

Extended exposure to capsaicin (and RTX) has also been shown to produce a refractory period (ranging from minutes to an indefinite time) for TRPV1 activation *in vitro* and *in vivo*, resulting in temporary to irreversible loss of pain perception (Caterina et al., 1997; Szallasi and Blumberg, 1999). Although the complete mechanism of this process is not known, evidence suggests that it is multifaceted. For example, TRPV1 can be rendered inactive to further stimulation by exposure to agonists such as capsaicin and RTX (Tominaga et al., 1998; Szallasi and Blumberg, 1999; Caterina and Julius, 2001). Similarly, chronic low-dose or acute high-dose exposures to capsaicin can deplete CGRP and substance P from sensory neurons, inhibit the biosynthesis of substance P, and promote excitotoxic neuronal cell death (peripheral neuropathy) through calcium-dependent processes (Gamse et al., 1980; Cuello et al., 1981; Holzer, 1991; Szallasi and Blumberg, 1999; Caterina and Julius, 2001). In one study, rats were treated systemically throughout development with subchronic and subtoxic doses of capsaicin; these young rats exhibited a marked, selective, and irreversible loss of their expression of thin peripheral sensory (C-fiber and some A-fiber) neurons that contained substance P (Gamse et al., 1980; Cuello et al., 1981). Interestingly, the exposed rats were resistant to the pain-producing and inflammatory effects of capsaicinoids and other noxious chemical stimuli that normally activate TRPV1 in adults. In addition, studies have shown that treatment of capsaicin-sensitive neurons with nontoxic doses of capsaicin and other TRPV1 ligands (RTX and olvanil) promoted the downregulation of TRPV1 expression in these neurons (Wood et al., 1988; Olah et al., 2001). For example, cells treated with vanilloids for short periods of time expressed much less TRPV1, as assessed by functional calcium uptake assays, [^{3}H]-RTX binding assays, and immunoblotting. The ability of capsaicinoids and other TRPV1 agonists to promote transient or irreversible decreases in pain perception is the pharmacological basis for the use of capsaicin and other vanilloids as a topical treatment for chronic neuropathic pain.

8.2 EXPRESSION OF VANILLOID RECEPTORS IN THE RESPIRATORY TRACT

8.2.1 Identification of TRPV1 Expression in Sensory Neurons and Epithelial Cells by Radiolabeled Ligand-Binding and Hybridization Assays

Many TRP receptors have been shown to be expressed at varied levels throughout the human body, including in the nerves and the endothelium and epithelium of respiratory tissues. Prior to the cloning of TRPV1, methods to detect transcripts of this gene, or other TRP genes, were not available. Thus, early studies to define the tissue distribution of TRPV1 utilized [^{3}H]-RTX as a highly selective and sensitive probe for determining TRPV1 expression (Szallasi et al., 1992; Szallasi, 1995). Although the binding properties of RTX had not been established using the cloned TRPV1 receptor, *per se*, it was known to bind and activate the (as of that time undescribed) vanilloid (capsaicin) receptor (now known as TRPV1) that was responsive to capsaicin in DRG neurons. Modifications to the traditional [^{3}H]-RTX binding assays to include α_1-acid glycoprotein permitted identification of TRPV1 in nonneuronal tissues by increasing the sensitivity and selectivity of the assay. The validity of the RTX binding assay for profiling the tissue distribution of TRPV1 in peripheral tissues was confirmed by the displacement of the radioactive probe by capsaicin and the finding that the probe did not bind specifically to the airway tissue of hamsters, a species that is not susceptible to the typical respiratory toxicities of the capsaicinoids (Szallasi et al., 1995). The use of the [^{3}H]-RTX binding assay permitted the identification of TRPV1 expression in various peripheral tissues such as the trachea and bronchi of several species, including humans and guinea pigs (Szallasi, 1995; Szallasi et al., 1995), as well as specific binding to membranes of multiple cell types derived from various neuronal and nonneuronal tissues. Unfortunately, a key limitation of this assay, and the results for TRPV1 expression profiling in respiratory tissues, was that the subcellular location and the cell type that expresses TRPV1 could not be discerned, because tissue/cell homogenates and membrane pellets were used for these assays. As such, in most nonneuronal tissues, particularly in respiratory tissues, specific RTX binding was presumed to be due to the presence of sensory nerve termini that innervate the peripheral tissues. In the case of the results for [^{3}H]-RTX binding in guinea pig and human airways, the presence of TRPV1 was attributed to the presence of peripheral nerve termini at or near the epithelial border, the nerves presumably responsible for mediating the respiratory effects of capsaicinoids (i.e., cough and neurogenic inflammation).

Despite the fact that binding of [^{3}H]-RTX was a suitable assay for establishing TRPV1 expression in peripheral tissue homogenates, its ability to provide cell-type specific data on the expression of TRPV1 was limited. Recently, immunohistochemical, radiolabeled nucleotide hybridization, and PCR-based assays that are based on the cloned capsaicin receptors have been shown to be more sensitive and specific for profiling TRP gene expression in specific cell types and in intact tissues. For example, use of PCR to amplify specific nucleotide sequences of TRPV1 has permitted the identification of TRPV1 in a variety of peripheral nonneuronal tissues of rats and humans; this includes the kidney, lung, testis, pancreas, spleen, liver, stomach, brain (i.e., neuroblastoma cells, spinal cord, cerebellum, hippocampus, and frontal cortex), skin, vascular smooth muscle, perivascular nerve fibers, placenta, cornea, uterus, and bladder (in subepithelial nerve fibers) (Caterina et al., 1997; Hayes et al., 2000; Cortright et al., 2001; Sanchez et al., 2001). Some of the more detailed studies utilizing fluorescent staining techniques were also able to establish the precise subcellular locations of TRPV1 expression on the cell membranes, smooth endoplasmic reticulum, and Golgi complexes (Olah et al., 2001; Liu et al., 2003). The localization of TRPV1 to intracellular compartments has also been shown using functional assays for the induction of calcium release from intracellular stores (Liu et al., 2003; Toth et al., 2004).

Because many of the newly discovered TRP/vanilloid receptors (e.g., TRPV2 to TRPV4, TRPM8, etc.) have not been fully characterized and lack specific radiolabeled ligand probes, PCR, radionucleotide hybridization, and immunohistological assays are the primary mechanisms of assessing their tissuc distributions. These techniques have been instrumental in establishing the tissue distribution of their gene expression and protein products. TRPV2 was initially characterized independently of TRPV1 in a subset of DRG neurons and was also found in various peripheral tissues including the brain and spinal cord, spleen, large and small intestines, vas deferens, lung, and to a lesser extent in heart, kidney, and bladder (Caterina et al., 1999). TRPV3 was expressed in the CNS, pituitary, skeletal muscle, stomach, intestine, adipose tissue, bone, placenta, and to a lesser extent in the heart and lung (Smith et al., 2002; Xu et al., 2002). TRPV4 expression was identified in the trachea, particularly the epithelial cells, skeletal muscle, liver, salivary glands, kidney, placenta, bone marrow, prostate, and to a lesser extent, in the lung (Delany et al., 2001). The nonfunctional TRPV1.5 splice variant has been shown to be expressed in the lung as well as in the brain (hippocampus and cerebellum), kidney, liver, and spleen (Schumacher et al., 2000; Sanchez et al., 2001). The cold receptor TRPM8 was expressed in DRG, trigeminal, and spinal neurons (McKemy et al., 2002), including those of the airway (Wright et al., 1998). Very recent data obtained in our laboratory provide evidence for the expression of TRPM8 in bronchiolar (BEAS-2B) and alveolar (A549) epithelial cells as well as kidney (HEK293) and liver (HepG2) cells (unpublished observations). A second cold receptor, ANKTM1, was recently localized to DRG neurons (Story et al., 2003). Several other TRP receptors, including various other TRPMs and TRPCs, have also been shown to be present in specific cell types originating from respiratory and other peripheral tissues. In general, however, the normal physiological or adverse functions of these receptors in most tissues are still unknown.

8.3 ACTIVATION OF TRP/VANILLOID RECEPTORS IN RESPIRATORY TISSUES

One of the most interesting facets of the study of the TRP/vanilloid receptors is their peripheral tissue distribution, especially in nonsensory tissues such as the lung, liver, kidney, vasculature, etc. A definitive role for some of the TRP/vanilloid receptors in mediating painful thermal and chemical sensations has been well established by recent research, although many issues remain enigmatic regarding the true functions of these receptors in regulating various other sensory functions and nonneuronal actions. Similarly, the significance and specific physiological functions of these receptors in the mediation of responses to endogenous ligands and xenobiotics have not been established. Recent research has provided evidence for a potential role for TRPV1 in the regulation of vascular smooth muscle tone by anandamide, as well as a hypothesis that TRPV1 may have a function in modulating the responses initiated through the cannabainoid (CB_1) receptor (Zygmunt et al., 1999). In general, however, very little is known about TRPV1 or other TRP/vanilloid receptors in tissues such as the lung, liver, and kidney. The following subsections will focus on recent research related to the potential roles of TRPV1 and other TRP/vanilloid receptors in respiratory tissue functions and disorders, including its involvement in the cough reflex, environmentally induced and neurogenic airway inflammation, and in the toxicities related to exposure to capsaicinoids (pepper sprays) and ambient particulate materials.

8.3.1 The Cough Reflex

The cough reflex can be defined as a protective mechanism for the rapid elimination of inhaled toxicants and pollutants, as well as a mechanism to remove accumulated substances from the respiratory tract. Cough is characterized by a deep inspiration followed by a rapid and forced expiratory blast. Animal and human models that utilize inhalation of aqueous citric acid or capsaicin solutions as cough-inducing agents have been instrumental in clarifying the mechanisms of the cough reflex and in developing antitussive agents (Morice et al., 2001). Cough induced by citrate or capsaicin (as well as RTX and anandamide) is a reflex action that is mediated by activation of vagal nerve termini in the respiratory tissues (Laude et al., 1993; Morice, 1996; Jia et al., 2002). Specifically, capsaicin and other vanilloids act by stimulating cellular influx of calcium and sodium in afferent, unmyelinated C-fibers that originate in the cough center of the lower pons/medulla; this is followed by neuropeptide and tachykinin (substance P and CGRP) release and the subsequent medullary activation of spinal and phrenic efferent neurons that innervate the respiratory musculature to ultimately produce cough. The induction of the cough reflex by capsaicin, RTX, anandamide, and even aqueous acid solutions has been shown to be blocked by capsazepine (but not cannabinoid antagonists) and the calcium channel blocker ruthenium red, suggesting a key role for TRPV1 in this process (Jia et al., 2002). In addition to stimulating the cough reflex, inhalation and systemic administration of capsaicin has also been shown to induce the Bezold–Jarisch reflex, a group of responses mediated by TRPV1 that includes cough, bronchospasm, pulmonary hypertension, apnea, bradycardia, hypotension, and coronary vasospasm (Aviado and Guevara Aviado, 2001).

8.3.2 Neurogenic Inflammation

Neurogenic inflammation is the process whereby bioactive substances such as substance P, CGRP, neurokinin A, and possibly glutamate, histamine, catecholamines, and prostaglandins are released from stimulated sensory neurons and act on surrounding target cells, particularly resident immune cells (e.g., macrophages), to promote proinflammatory processes in tissues at, or near, the site of exposure or injury (Richardson and Vasko, 2002). Neurogenic inflammation is characterized by redness and warmth (from vasodilation), swelling (from plasma extravasation), and hypersensitivity or hyperalgesia (due to alterations in neuronal activation thresholds by various endogenous proinflammatory and second-messenger substances) (Jancso et al., 1967; Veronesi and Oortgiesen, 2001; Richardson and Vasko, 2002). This process is depicted in Figure 8.3. One of the principal outcomes of the neurogenic inflammatory process is the stimulation of nonneuronal cells to produce proinflammatory cytokines and substances that ultimately promote and maintain the inflammatory cascade. Capsaicin has been instrumental in defining the mechanisms of neurogenic inflammation. Capsaicin acts directly on the peripheral nerve termini to promote the TRPV1-regulated and calcium-dependent release of substance P and CGRP. Various cell types, including mast cells, epithelial cells, and immune cells, have been shown to release proinflammatory cytokines (e.g., IL-1β, IL-6, IL-8, and TNFα) in response to substance P, CGRP, and

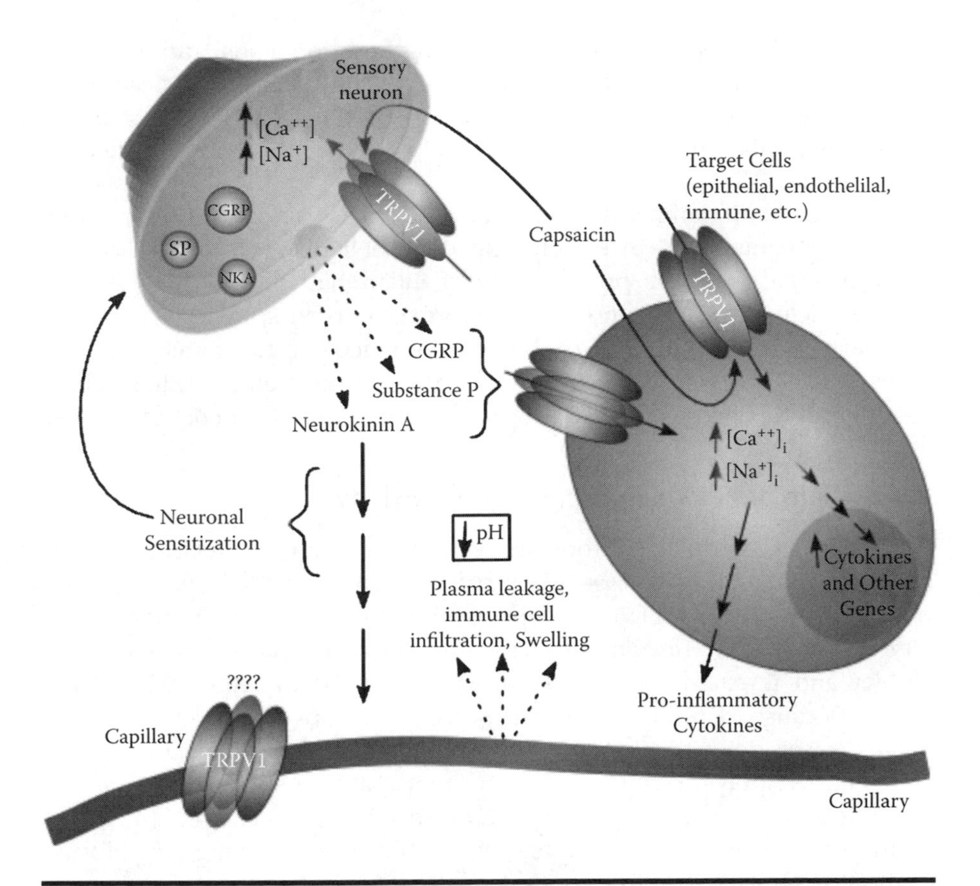

Figure 8.3 Neurogenic inflammatory mechanisms. Interactions between neuronal TRPV1, neighboring target cells, and a number of cellular signaling molecules that participate in neurogenic inflammation are illustrated. Activation of TRPV1 produces Ca^{++} flux and results in the release of a number of neuropeptides and signaling molecules that ultimately produce pain, irritation, and inflammation. CGRP = calcitonin-gene-related peptide.

capsaicin (Richardson and Vasko, 2002). The mixture of neuropeptides and chemokines, in turn, increases the stimulation of capsaicin-sensitive neurons and other cells (Richardson and Vasko, 2002). Additional factors that may exacerbate neurogenic inflammation and cytokine release through TRPV1 activation include direct activators of protein kinase C, ATP, neurotransmitters (5-HT and acetylcholine), histamine, NGF, bradykinin, cyclooxygenase products, decreased pH (frequently, pH is less than 6 at inflamed locations), elevated temperature, and various eicosinoids. Prior exposures to capsaicin or RTX can greatly decrease the capsaicin-induced neurogenic inflammatory process (as discussed earlier), presumably by depleting substance P in these neurons and probably through the destruction and downregulation of TRPV1 in specific capsaicin-sensitive nerve fibers (Gamse et al., 1980). Consonant with this mechanism, neurogenic inflammation that is induced by capsaicin can be blocked by capsazepine.

Neurogenic inflammation has been a crucial focus for respiratory toxicologists interested in establishing the mechanisms by which certain environmental toxicants produce airway inflammation. A variety of substances, including capsaicinoids in the form of pepper sprays, and ambient particulate materials, have been shown to produce overt respiratory pathologies ranging from mild to severe inflammation, as well as cellular death resulting in damage to critical airway architectural and functional entities.

8.3.3 Role of TRPV1 in Pepper Spray Toxicity

Pepper spray products, historically formulated using dilute oily extracts of hot peppers (known as *oleoresin capsicum* or OC), have gained popularity and near uniform acceptance by the law enforcement community as a safe and effective alternative to physical force in incapacitating violent and threatening criminals (Olajos and Salem, 2001; NIJ, March 1994). Because pepper sprays can serve as a convenient and effective means of self-defense, civilians and outdoorsmen also utilize these products. Pepper sprays have been shown to incapacitate subjects through the rapid induction of a painful burning sensation in the skin and eyes, ultimately causing uncontrolled tearing, swelling, and, involuntary, closure of the eyes. Inhalation of pepper spray also contributes to its effectiveness as an incapacitant through induction of uncontrollable cough, dyspnea, inflammation of the mucous membranes throughout the respiratory tract and, potentially, production of apnea. Although pepper sprays have been shown to be generally safe under most routine exposure scenarios, the potential for severe toxicities, particularly cardiovascular and respiratory toxicities, have become abundantly evident (Steffee et al., 1995; Billmire et al., 1996; Busker and van Helden, 1998; Smith and Stopford, 1999; Granfield, March 1994). The potential for respiratory injury and complications from pepper spray exposures may be exacerbated by concomitant illicit drug use and preexisting respiratory disease such as asthma or other airway hypersensitivity disorders.

Although clinical studies have shown that inhalation of very low doses of pepper sprays for short durations produce no measurable adverse effects on respiratory function (Chan et al., 2002), life-threatening and fatal exposures to aerosolized pepper sprays have been documented. In most cases of inhalation exposure to capsaicinoids in pepper sprays, patients complained of general irritation to the eyes and nasal passages, cough, and other general respiratory tract afflictions. However, in some instances, severe inflammation and mucosal and epithelial damage were observed, requiring oxygen therapy and extended hospitalization (Billmire et al., 1996). In one instance, a hospital worker who was inadvertently exposed to pepper spray exhibited severe exacerbation of asthma (Heck, 1995). Several fatalities have also been reported following pepper spray exposure, although the role of the pepper spray in these fatalities was generally dismissed because of the presence of other compounding factors such as concomitant illicit drug use (e.g., methamphetamine, cocaine, and alcohol), physical injury (from other weapons), positional effects on respiration efficiency (e.g., positional asphyxiation due to hog-tying), and preexisting health conditions (e.g., asthma) (Granfield, March 1994).

Recent toxicological evaluation of capsaicinoids and pepper spray products have documented the adverse health effects of capsaicinoids on respiratory tissues (Debarre et al., 1999; Reilly, Taylor et al., 2003) and revealed the molecular mechanisms by which the observed pathologies develop (Reilly, Taylor et al., 2003). In this research, rats (Sprague-Dawley) were exposed to aerosols of capsaicinoids and pepper spray products. In general, moderate-to-marked inflammation of the respiratory tissues, necrotic lesions, hemorrhage, and damage to the epithelial cell lining of the trachea and bronchi and the alveolar epithelium were observed. Representative examples of these toxicities are shown in Figure 8.4. Although the exposures were conducted with sublethal amounts (an LD_{50} for acute intratracheal exposure to capsaicin in mice has previously been reported as 1.8 mg/kg) (Glinsukon et al., 1980), death was occasionally observed in exposed animals, presumably due to cardiovascular or pulmonary dysfunction from the pepper spray exposure (Reilly, Taylor et al., 2003). Although the causes of death were not ascertained, mortality was associated with pepper spray exposures.

To further investigate the mechanisms responsible for the pathologies associated with inhalation exposure to pepper sprays, *in vitro* toxicological studies were performed using cultured human bronchiolar (BEAS-2B) and human alveolar (A549) epithelial cell lines (Reilly, Ehlhardt et al., 2003). Initial studies focused on the hypothesis that capsaicinoids were metabolized by pulmonary cytochrome P450 enzymes (Reilly, Ehlhardt et al., 2003) to produce toxic electrophilic reactive metabolites that may have caused lung cell death. The inflammatory effects of the capsaicinoids were presumed to be caused by neurogenic processes. However, these initial studies demonstrated that rather than playing a critical role in the bioactivation of capsaicinoids as previously proposed by other researchers (Surh and Lee, 1995), pulmonary cytochrome P450 enzymes served a protective role for cells by biotransforming capsaicin and its analogs to metabolites (i.e., O-demethylated, and various aromatic and aliphatic oxygenated and dehydrogenated products) that were less toxic than the parent compounds. In fact, cotreatment of these cells with a P450 inhibitor or with solutions

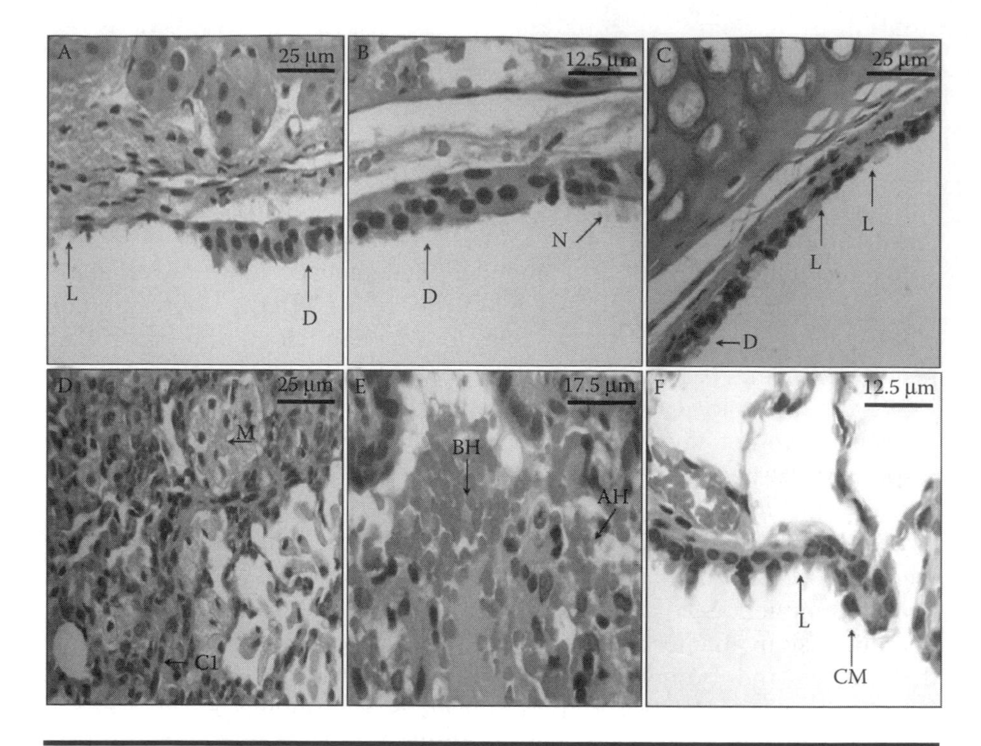

Figure 8.4 Pathologies of lung exposed to pepper spray. Micrographs of lung tissue sections stained with eosin and hematoxylin from rats exposed to capsaicinoids in the form of pepper spray via nose-only inhalation (exposure estimated to be ~1 mg/kg). A and B) Capsaicinoids produced epithelial cell displasia (D), loss of cells (C), and occasional necrosis in nasal turbinates. (C) Similar lesions were observed in tracheal sections. (D through E) Capsaicinoids induced macrophage proliferation (M), accumulation of chronic inflammatory cells (CI), bronchiolar (BH) and alveolar (AH) hemorrhage, cuboidal metaplasia (CM), and loss of alveolar cells (L) in bronchiolar and alveolar portions of the lung. Nasal turbinate sections were obtained by sectioning between the first and second palatal ridges of decalcified heads. Tracheal sections were taken from tissue ~2 mm before the bronchi; the bronchial and alveolar sections were obtained from the right lung at the entry point of the bronchial airway and from the distal tip of the lung, respectively.

containing metabolites of capsaicin drastically decreased cell killing. Therefore, a hypothesis was developed that the metabolism of capsaicin by P450 enzymes decreased the cytotoxicity of capsaicin in lung cells by decreasing the effective concentration of capsaicin or its analogs that interact with and activate TRPV1 to ultimately promote cell death. This hypothesis was also based on previous work that demonstrated a strict structural requirement for capsaicinoid, with respect to both the aromatic vanilloid ring and the aliphatic portions of the molecule in order to bind and activate TRPV1 (Walpole et al., 1993b; Walpole et al., 1993a; Walpole et al., 1993), as well as on the finding that the ability of capsaicin and other vanilloids to promote neuronal cell death was dependent upon the activation of TRPV1 and accumulation of calcium in cells (Wood et al., 1988).

Studies designed to test the validity of this hypothesis were instrumental in providing critical knowledge of the role of TRPV1 in the pulmonary inflammation and cell death caused by capsaicinoids and other TRPV1 agonists (Reilly, Taylor et al., 2003). Using BEAS-2B cells as a surrogate model to investigate the mechanisms of pepper spray (capsaicinoid) toxicity on bronchiolar epithelial cells, it was demonstrated that TRPV1 expression was a requisite factor in initiating both capsaicin-dependent proinflammatory and cell death processes in lung cells. Initial studies focused on the role of TRPV1 in mediating the cytotoxic effects of the capsaicinoids in lung cells. These studies provided support for the hypothesis that TRPV1-mediated cell death by capsaicinoids by demonstrating that the level of TRPV1 mRNA expression in BEAS-2B and A549 cells was greater than in the hepatocarcinoma (HepG2) liver cell line, and that the relative level of expression of TRPV1 mRNA was a predictor of cytotoxicity. In general, TRPV1 expression was observed in all three cell types with the rank order of BEAS-2B > A549 > HepG2, corresponding to the rank order for TRPV1 expression (Figure 8.5). These data suggested that the level of TRPV1 expression in these cells was an important factor in determining cellular susceptibility to the toxic effects of capsaicin; the LC_{50} values were ~100, 110, and >200 μ*M* for BEAS-2B, A549, and HepG2 cells, respectively.

The idea that the level of TRPV1 expression dictated cellular sensitivity to the toxic effects of capsaicin was further supported by the development of a TRPV1-overexpressing cell line that was derived from BEAS-2B cells. Semiquantitative comparison of TRPV1 mRNA expression in both the control BEAS-2B and the overexpressing cells using RT-PCR (Figure 8.6) demonstrated that TRPV1 was stably overexpressed in these cells. Functional consequences of TRPV1 overexpression were verified by dramatic increases in calcium flux into the cells when treated with capsaicin. The most striking finding in these studies was that the overexpression of TRPV1 in BEAS-2B cells imparted a dramatic ~100-fold increase in sensitivity to capsaicin; the LC_{50} concentrations in control and TRPV1-overexpressing cells were 100 μ*M* and ~1 μ*M*, respectively (Figure 8.6). Similar increases in sensitivity were also observed for other traditional TRPV1 agonists such as RTX (~75,000-fold) and olvanil (~15-fold), whereas the toxicities of the newly discovered lower-affinity TRPV1 agonists such as anandamide and

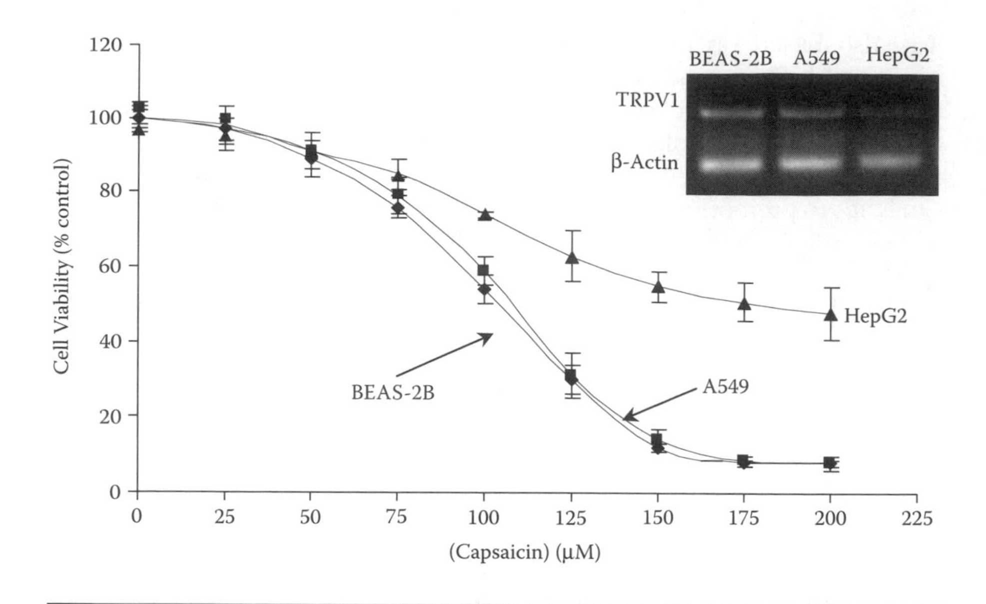

Figure 8.5 Toxicity of capsaicinoids to lung and liver cells. The toxicity of capsaicinoids in lung epithelial and liver cell lines was related to the relative level of TRPV1 expression in the three cell types. BEAS-2B and A549 human bronchial and alveolar (adenocarcinoma) epithelial cells expressed significantly higher levels of TRPV1 than the hepatocarcinoma (HepG2) cell line, as determined by RT-PCR. BEAS-2B (diamonds) and A549 cells (squares) also exhibited higher sensitivity to capsaicinoids than HepG2 cells (triangles). Cells were incubated in the presence of increasing concentrations of capsaicin for 24 h. Viability was assessed using the Dojindo cell counting Kit-8.

scutigeral were increased by about a mere twofold. Contrary to neuronal cell death processes, the ability of capsaicin to induce cell death in BEAS-2B and TRPV1-overexpressing cells was independent of calcium flux from extracellular sources, because cotreatment with capsazepine, ruthenium red, EGTA, or treatment in calcium-free media had no effect on the LD_{50} (Reilly, Taylor et al., 2003). Interestingly, the mechanism of cell death was necrosis for BEAS-2B cells, whereas apoptotic cell death was observed for the TRPV1-overexpressing cells. This difference in the mechanism of cell death may attest to the specificity of the cell death processes that appear to be operative for each cell type. Capsaicin produced a putative TRPV1-mediated cell death process in TRPV1-overexpressing cells, but normal BEAS-2B cells appeared to be killed via composite mechanisms involving TRPV1-mediated and other nonselective processes such as membrane disruption. The collective data provided by these studies provided crucial evidence for the role of TRPV1 in mediating the cytotoxic

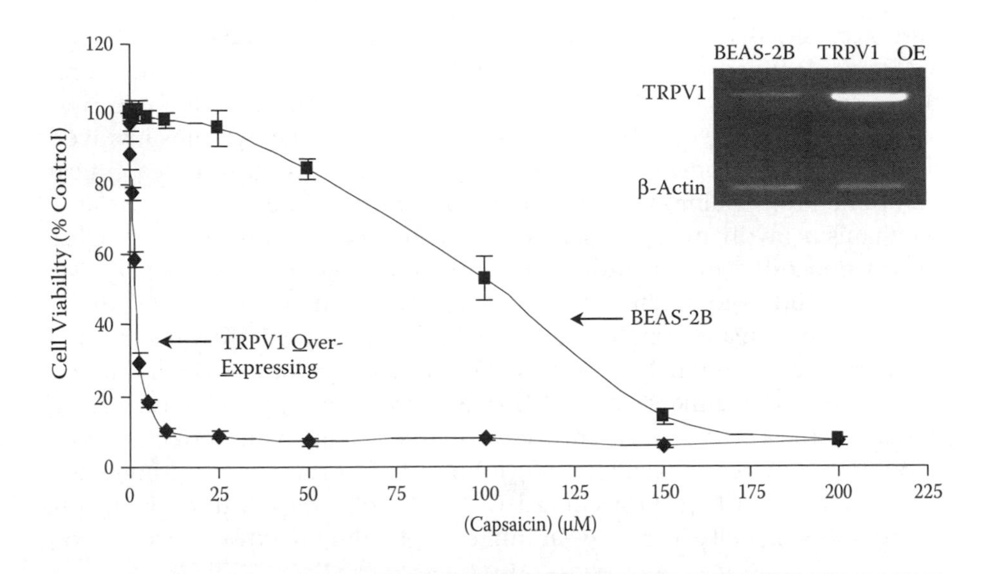

Figure 8.6 Toxicity of capsaicinoids to lung epithelial and TRPV1-overexpressing cells. Overexpression of TRPV1 in human bronchial epithelial (BEAS-2B) cells significantly increased the toxicity of capsaicinoids. Overexpression of TRPV1 (TRPV1-OE; diamonds) resulted in an approximate 100-fold increase in capsaicinoid toxicity relative to control BEAS-2B cells (squares). Cells were incubated in the presence of increasing concentrations of capsaicin for 24 h. Viability was assessed using the Dojindo cell counting Kit-8.

effects of capsaicin, pepper spray products, and possibly other vanilloids in lung cells.

Additional studies designed to assess the role of TRPV1 in the initiation of proinflammatory processes in respiratory tissues were also performed. Previous knowledge regarding the production of pulmonary inflammation by capsaicinoids had focused on neurogenic inflammatory processes, as outlined earlier; these processes were believed to be the chief mechanism of capsaicin-induced pulmonary inflammation. Important studies provided new insight into this process by demonstrating that homogeneous cultures of BEAS-2B cells treated with capsaicin or selected samples of concentrated airborne ambient particulate materials produced the proinflammatory cytokines IL-6, IL-8, and TNFα in a dose- and calcium-dependent manner (Veronesi, Oortgiesen et al., 1999; Veronesi and Oortgiesen, 2001). In addition, it was shown that the elevated production of cytokines could be markedly increased by coexposure to the neuropeptides substance P and CGRP, and inhibited by removal of extracellular calcium or by cotreatment with the TRPV1 antagonist capsazepine (Veronesi, Carter et al., 1999). Additional studies using ambient particulate pollutants demonstrated that the induction of cytokine release proceeded via NF-κB-dependent pathways (Quay et al., 1998; Schins et al., 2000). Therefore, direct stimulation of TRPV1 expressed by lung epithelial cells by capsaicin in the presence (or absence) of the endogenous tachykinins involved in traditional neurogenic inflammatory mechanisms represented a critical component of pulmonary inflammation, proceeding through a new mechanism involving epithelial-cell-mediated proinflammatory cytokine production and acting in addition to known neurogenic processes *in vivo.*

Other studies to confirm the role of TRPV1 in mediating proinflammatory cytokine release from lung epithelial cells using TRPV1-overexpressing cells provided additional support for the proposed processes (Reilly, Taylor et al., 2003). Treatment of TRPV1-overexpressing cells with increasing concentrations of capsaicin showed a marked shift in the effective concentration required for maximum cytokine (IL-6) production vs. untreated controls: from ~100 μM for control BEAS-2B cells to ~0.5 μM for TRPV1-overexpressing cells (Figure 8.7). Interestingly, the effective concentration

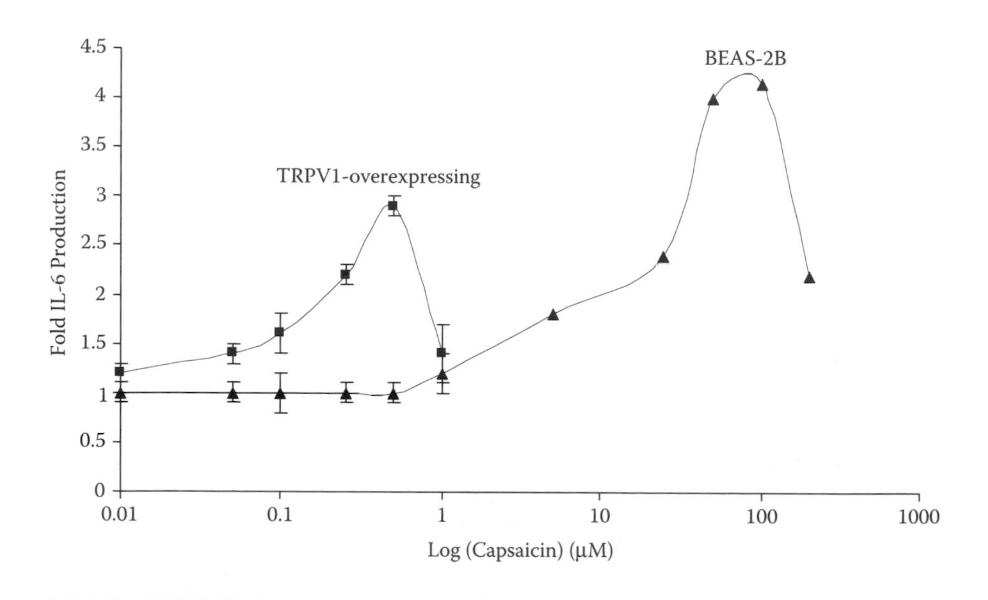

Figure 8.7 Inflammatory responses of lung epithelial and TRPV1-overexpressing cells to capsaicin. Increased expression of TRPV1 in human bronchial epithelial cells (BEAS-2B) significantly decreased the dose of capsaicin required to induce IL-6 expression. A dose–response comparison for the induction of IL-6 release by human bronchial epithelial cells demonstrated a key role for TRPV1 in mediating cytokine release from cells exposed to capsaicinoids. Cells were treated for 24 h with increasing concentrations of capsaicin. After 24 h the culture medium was collected and the amount of IL-6 determined by ELISA.

of capsaicin required to induce maximum cytokine release was roughly equivalent to the LD_{50} value, suggesting that cytokine production by these cells may occur in response to cellular damage. In both cell lines, the production of IL-6 was ameliorated by capsazepine and EGTA, and slightly enhanced by the addition of excess calcium to the treatment solutions. Production of TNFα was not observed in these studies, despite previous research suggesting that it was produced by BEAS-2B cells treated with capsaicin (Veronesi, Carter et al., 1999). Collectively, the data suggested that TRPV1 played a critical role in the initiation of inflammatory processes in respiratory epithelial cells exposed to pepper sprays (capsaicinoids) and that the relative level of TRPV1 expression determined the threshold for both cytotoxicity and pulmonary inflammation.

In summary, these *in vitro* studies on the toxicity of pepper sprays have provided vital mechanistic information regarding the effects of TRPV1 on respiratory cellular responses. It appears that the level of TRPV1 expression is an important determinant of cellular sensitivity to cytotoxic and inflammatory stimuli; elevated levels of TRPV1 sensitized the cells. In addition, the induction of cell death by capsaicin, although mediated by TRPV1, occurs via a distinct pathway that is different from the calcium-dependent induction of cytokine release, because cytotoxicity could not be prevented by capsazepine or by conditions that modified calcium concentrations or flux: EGTA, ruthenium red, or removal of calcium from the treatment solutions (Reilly, Taylor et al., 2003). Furthermore, the proinflammatory effects of pepper sprays in respiratory tissues likely occur via two processes: (1) traditional neurogenic inflammatory processes that act directly on respiratory C-fibers innervating the tissues, stimulating resident immune cells, recruiting additional acute-phase inflammatory cells, and stimulating surrounding epithelial cells to promote cytokine release; and (2) by directly activating airway epithelial cells that express TRPV1 to actively produce proinflammatory cytokines involved in the recruitment and activation of immune cells. Hence, a definitive function for TRPV1 has been established in respiratory tissues: mediation of the cytotoxic and inflammatory effects of pepper sprays and possibly other substances.

8.3.4 TRPV1-Mediated Responses to Ambient Particulate Material

Inhalation exposure to ambient air poses a constant threat to respiratory health because of elevated concentrations of particulate materials and other pollutants, particularly in highly industrialized and populated areas. Recent epidemiological studies have provided compelling evidence that human health is significantly affected by ambient air quality (Utell et al., 1988; Amdur, 1989; Pope, 1989; Xu et al., 1991; Dockery, 1993; Dockery et al., 1993; Pope, 1998; Churg and Brauer, 2000; Lighty et al., 2000; Dockery, 2001;

Veronesi and Oortgiesen, 2001; Kagawa, 2002; Mortimer et al., 2002). Recent studies have demonstrated a strong correlation between increases in complex mixtures of ambient particulate materials and hospital admissions, morbidity and mortality resulting from respiratory and cardiovascular illnesses (Pope, 1989; Pope, 1991; Dockery et al., 1993; Pope, 1996; Dockery, 2001). Epidemiological studies have demonstrated that the composition of the air pollutants can adversely affect health and provided justification for fundamental studies focused on the mechanisms by which airborne pollutants induce respiratory and cardiovascular toxicities (National Research Council, 1998).

The airway epithelium represents the initial barrier to the absorption of potentially toxic inhaled substances. Inhalation exposure to environmental, occupational, and lifestyle pollutants and irritants such as geological dusts, anthropogenic particulate materials, bacteria, and chemical irritants such as pepper sprays or other volatile organics and tobacco smoke, can produce or directly influence the development of a variety of responses, including acute and chronic airway inflammation, cough, bronchitis, chronic obstructive pulmonary disease (COPD), asthma, pulmonary edema, various systemic responses, cell damage, and cancer. In fact, some respiratory pathologies associated with chemical exposure are attributable, in part, to responses initiated by the interaction of the xenobiotics with the airway epithelium. Because lung epithelial cells protect the body from the adverse effects of potentially harmful inhaled substances and are an essential component of normal respiratory physiology, it is important to understand the mechanisms by which these cells are affected or damaged by inhaled toxicants. The previous section of this chapter discussed the role of TRPV1 in initiating and regulating the lung pathologies associated with pepper spray exposure (i.e., inflammation and cell death) and provided a brief introduction to the effects of ambient particulate materials. The discussion will now turn to the role of TRPV1 in regulating respiratory toxicities that are induced by exposure to inhaled ambient particulate materials.

Ambient air comprises a diverse and complex mixture of gas, liquid, and solid substances (Seinfeld and Pandis, 1998). Gaseous components of the air include oxygen, carbon dioxide, nitrogen and nitrogen dioxides, ozone, sulfur dioxides, and a variety of other volatile organic and inorganic substances. The solid particulate matter of ambient air comprises substances originating from diverse sources; they include coarse and fine particles of geological origin, solid combustion by-products such as residual ash and soot from incomplete combustion of organic fuels, secondary aerosols formed by gas-to-particle conversion in the atmosphere, and naturally occurring substances such as mold and pollen. In general, ambient particulate materials are a major source for inhalation exposure to a diverse array of organic and inorganic chemical components, such as adsorbed xenobiotics (e.g., polycyclic aromatic hydrocarbons [PAHs], phenols, pesticides, and quinones), condensed liquids and acids such as sulfuric acid, and various metals and metal oxides, including potentially redox-active iron and copper compounds. A variety of hypotheses have been proposed to explain the role of individual particle components or characteristics in the bioactivity of inhaled particles. (Mauderly et al., 1998). A substantial amount of information has been generated by recent research. However, the principles governing bioactivity *in vivo* remain unclear. Many factors including particle size, shape and surface characteristics, the type and bioavailability of transition metals and other metals, the ability of the particle to catalyze the production of ROS, adsorbed acids, the amount and type of adsorbed organics, and the bioavailability and persistence of the particle in tissues, contribute to significant differences in the short- and long-term effects of these substances on respiratory tissue and physiology (Carter et al., 1997; Smith and Aust, 1997; Lighty et al., 2000; Smith et al., 2000; Aust et al., 2002; Schins et al., 2002; Veronesi, de Haar, Lee et al., 2002; Pagan et al., 2003; Willis et al., 2003). The complexity and heterogeneity of ambient particulate materials have made the elucidation of the culpable chemical and physical entities very difficult, especially because the presence or lack of bioactivity of a substance may be due to complex interactions between different components, rather than an effect caused by the individual substance acting alone.

TRPV1 plays a pivotal role in regulating the cytotoxicity and release of cytokines from epithelial cells induced by both capsaicin and laboratory-generated particle treatment (Veronesi, Oortgiesen et al., 1999; Agopyan, Bhatti et al., 2003; Reilly, Taylor et al., 2003; Agopyan et al., 2004). These observations led to the hypothesis that perhaps airway inflammation induced by ambient particulate material may also occur via a mechanism that is similar to that of capsaicin. Thus, specific characteristics of individual particle types could be responsible for activation of TRPV1-dependent proinflammatory pathways in lung epithelial cells. Initial studies with particulate material obtained from a variety of environmental sources, including combustion by-products such as residual oil and coal fly ash (ROFA and CFA), demonstrated that particles induced an immediate influx of calcium from extracellular sources in BEAS-2B cells with a concomitant increase in the synthesis and secretion of various cytokines, including IL-6, IL-8, and TNFα (Veronesi, Oortgiesen et al., 1999). The cellular responses to the particle treatments were reduced to control levels by cotreatment with the selective TRPV1 antagonist capsazepine, by removal of calcium from the treatment solutions, and by amiloride, an inhibitor of both the acid-sensitive ion channels (ASICs) and TRPV1.

Although these data presented compelling evidence supporting a role for TRPV1 in mediating the proinflammatory responses of lung epithelial cells to ambient particulate materials, the precise physicochemical properties of the particles responsible for activation of TRPV1 and proinflammatory processes remained unclear. Particulate materials are extremely heterogeneous substances that can vary significantly in their composition, depending upon the source. However, morbidity and mortality levels associated with exposures to elevated concentrations of particulate materials are quite uniform (Lai et al., 1996; Wong and Lai, 2004), suggesting that the adverse health effects may be related to the general physical characteristics of particles rather than a specific or unique property. Thus, a hypothesis consistent with this observation was generated: the net electrostatic charge of particulate materials activates TRPV1 to release inflammatory cytokines.

This hypothesis proposed that the chemical composition and physical properties of colloidal particles imparted a unique net electrostatic charge to a particle, that in solution would dictate the amount and type of counter-ions associated with the particle surface (Veronesi, de Haar, Lee et al., 2002; Agopyan, Li et al., 2003; Veronesi et al., 2003). For example, a particle type with a strong net negative surface charge would be surrounded by a large layer of protons or other counter-ions to neutralize the negative electrostatic charge of the particle. The overall dimensions of this counter-ion layer can be measured as a *zeta* potential, the electro kinetic potential of the particles in solution. Initial studies to investigate this hypothesis provided key evidence that the presence of a net negative *zeta* potential (i.e., a large acidic layer of protons surrounding the particle core) was essential for activation of TRPV1 in BEAS-2B cells, as measured by intracellular calcium influx, membrane depolarization, and cytokine production by environmental particulate samples (1 to 10 μm). The validity of this hypothesis was confirmed when the same endpoints were produced by negatively charged synthetic polymeric microspheres (inert polystyrene micelle particles), but not by neutral or positively charged ones. (Veronesi, de Haar, Lee et al., 2002; Agopyan, Li et al., 2003; Reilly, Taylor et al., 2003; Veronesi et al., 2003). In fact, a direct relationship was found to exist between cytokine production and *zeta* potential for a series of environmental particulate samples, as well as a requirement for negative charge on the microsphere surface. In all instances, the responses induced by these particles were inhibited by capsazepine and amiloride, suggesting the simultaneous participation of both TRPV1 and ASICs in these processes (Veronesi, de Haar, Lee et al., 2002; Agopyan, Li et al., 2003; Veronesi et al., 2003).

The data presented above demonstrate a clear role for TRPV1 in mediating airway inflammation and cell death through activation of respiratory epithelial cells via a TRPV1-dependent mechanism similar to that of capsaicin (Reilly, Taylor et al., 2003). Identification of TRPV1 as a critical component of airway cell responsiveness to multiple toxicants provides valuable insight into the mechanisms by which diverse pollutants and inhaled substances promote respiratory dysfunction. Thus, future research about the participation of TRP/vanilloid receptors in the initiation and progression of various airway disorders may supply key information regarding both the mechanisms of respiratory diseases and the exacerbation of these diseases by exposures to environmental contaminants. These studies may ultimately identify new therapeutic targets to treat disorders such as asthma and other airway hyperreactivity disorders.

8.4 ROLE OF VANILLOID RECEPTORS IN HUMAN RESPIRATORY DISEASES

8.4.1 Are Vanilloid Receptors Important in Airway Hypersensitivity Disorders?

With industrialization and technological advances, there has been a rise in the incidence of respiratory diseases and disorders such as environmentally induced asthma, COPD, and other acute and chronic respiratory disorders (Pope, 2000; Strachan, 2000; D'Amato et al., 2002; Mannino, 2002; Weisel, 2002; Costa and Kodavanti, 2003). Because TRP/vanilloid receptors have been shown to be sensitive or responsive to a wide variety of substances, including diverse environmental pollutants and pepper sprays, as well as capable of inducing both neurogenic and cell-initiated inflammatory processes and cell damage, the possibility that these receptors are involved in the development and manifestation of various airway hypersensitivity disorders that can be induced or exacerbated by exposure to environmental factors (i.e., temperature, ambient particles of air pollution, cigarette smoke, xenobiotics, etc.) is intriguing. For example, if an individual's genetic composition or prior exposure to specific xenobiotics results in the upregulation of specific TRP/vanilloid receptors in respiratory tissues, then perhaps this individual would exhibit hypersensitivity to subsequent exposures to these substances (e.g., pepper sprays or particles) that activate the specific receptor and produce airway inflammation or respiratory distress.

An example that appears to fit this hypothesis has recently been demonstrated by respiratory scientists using fine ($PM_{2.5}$) particulate materials obtained from the World Trade Center site (Gavett et al., 2003). In these studies, single exposures to the particulates produced only mild inflammatory responses. However, prior exposure to these particulates induced marked hyperresponsiveness to subsequent methacholine challenges.

Alternatively, downregulation of TRPV1 by substances may prevent protective responses such as inflammation and cough, thus suppressing normal protective physiological functions. Although repeated exposure to capsaicinoids produces desensitization to painful stimuli through a mechanism of decreased substance P content or neuronal cell death (Gamse et al., 1980), there are many reports of unpredicted cellular responses in nonneuronal tissues upon overexpression of TRP receptors. Thus, it would not be surprising to find that overexpression of TRPV1 in asthmatics could be responsible for the hypersensitivity of these people to environmental pollution.

As discussed earlier in this chapter, the overexpression of TRPV1 in human bronchiolar epithelial cells increased the relative cytotoxicity of capsaicin and other TRPV1 agonists by a factor of 10 to 75,000 (Reilly, Taylor et al., 2003). Concomitantly, a marked decrease in the concentration of capsaicin required to induce proinflammatory cytokine production (IL-6) was observed. Therefore, if an individual expressed much higher levels of TRPV1 as a result of altered rates of gene transcription, this person could be expected to exhibit more severe respiratory responses (e.g., airway inflammation and epithelial cell damage) at much lower concentrations of TRPV1 agonists (e.g., pepper sprays, particulate materials, acidic aerosols, elevated air temperatures, etc.). This hypersensitivity may increase the frequency of adverse respiratory symptoms upon exposure to these substances or to conditions similar to that which aggravate asthma and COPD, as well as increase the likelihood of developing respiratory infection because of increased damage to the respiratory epithelium. Recently, it has been shown that treatment of BEAS-2B cells with negatively charged particles promoted a transient induction of TRPV1 mRNA expression, suggesting that exposure to these and related substances may increase the sensitivity of individuals (or cells) to subsequent ambient particle or xenobiotic exposures (Veronesi et al., 2003). Exposures to these substances may also synergistically contribute to the development of the specific respiratory hypersensitivity caused by chronic, nontoxic, low-dose exposures to similar compounds. Although specific data are not available to support the direct participation of TRPV1 or other TRP/vanilloid receptors in the development and manifestation of airway hyperresponsiveness to various environmental stimuli, some indirect reports support this idea. For example, NGF, bradykinin, substance P, and various xenobiotics have been shown to increase TRPV1 expression or activity in various cells (Jordt et al., 2000; Chuang et al., 2001; Ji et al., 2002; Bhave et al., 2003; Prescott and Julius, 2003), as well as to enhance the acute early-phase reaction of bronchial asthma and to sensitize individuals to capsaicin-induced cough (Path et al., 2002). In addition, asthmatics in general tend to exhibit greater sensitivity to capsaicin-induced cough (Doherty et al., 2000; Millqvist et al., 2000). Conversely, chronic occupational inhalation exposure to capsaicinoids in chili pepper plant employees has been shown to decrease the number of capsaicin-sensitive nerves as well as increase the cough threshold for capsaicin, presumably via decreased levels of TRPV1 expression (Blanc et al., 1991). Similarly, it has been shown that the nasal hyperreactivity component of idiopathic rhinitis (rhinorrhea and nasal blockage) could be exacerbated by acute exposure to capsaicinoids (Philip et al., 1996; Sanico et al., 1997), but effectively diminished for extended periods of time by subchronic intranasal capsaicin treatment, with additional blockage of responses to cold and dry air stimuli (Van Rijswijk et al., 2003). Thus, the hypothesis that the levels of expression of specific receptors (e.g., TRPV1) can dictate the response thresholds for toxicities induced by xenobiotics such as pepper sprays or ambient particles and other factors, including heat, moisture, etc., is intriguing, and is likely an important mechanistic determinant for a multitude of adverse respiratory toxicities and disease states related to diverse chemical or pollutant exposures.

8.4.2 Do Other Vanilloid Receptors Modulate Cellular Responses to Respiratory Irritants and Toxicants?

Because of the apparent close relationship between TRPV1 and many of the other TRP/vanilloid receptors, as well as the finding that these receptors respond to specific temperature ranges (although mammals detect all temperatures) and occasionally respond to similar chemical ligands in a functional tetrameric complex, scientists have proposed that perhaps heterogeneous mixed-function receptor complexes composed of multiple TRP/vanilloid receptor subtypes exist and are essential for the detection of temperatures that are between the optimum response ranges for the homologous receptor proteins characterized *in vitro*. For example, varying combinations of TRPV1 and TRPM8 subunits that respond to higher or to moderately cold temperatures, respectively, may be important detectors of thermal sensation between the two temperature ranges. In this case, a receptor complex primarily comprising TRPM8 would respond to cooler temperatures most closely to native homologous TRPM8 complexes, whereas complexes comprising equal numbers of subunits may respond to intermediate temperatures between those observed for either TRPM8 or TRPV1 alone. In addition to altered temperature activation profiles, these mixed complexes would likely exhibit marked differences in their responsiveness to a variety of other potential stimuli including xenobiotics such as capsaicin and other environmental factors such as acidic conditions, diverse ambient particulate materials, and other pollutants that activate TRPV1 and possibly other TRP/vanilloid receptors. Thus, mixed receptor complexes may serve as physiological sensors to control

TRP/vanilloid receptor function in various tissue types, or in various environmental conditions, and as biological mediators of toxicities induced by the diverse variety of environmental toxicants that affect respiratory physiology.

As discussed earlier, initial research to investigate these hypotheses using TRPV1 proved very fruitful, demonstrating that functional tetrameric TRPV1 receptor complexes assemble with agonist treatment and that activation was inhibited by dominant negative mutations of TRPV1 (Kuzhikandathil et al., 2001). Similar studies also investigated the modulation of receptor function (Caterina et al., 1999; Smith et al., 2002). Unfortunately, evidence for the formation and existence of these mixed complexes *in vivo* has not been as easily obtained. Furthermore, more detailed studies on individual receptor subtypes suggest that the formation of mixed receptor complexes may be restricted owing to differences in cellular expression and localization. Regardless, investigating the potential involvement of such molecules in regulating airway responses to inhaled irritants, toxicants, and pollutants may prove to be an interesting topic for future research.

8.4.3 Can Vanilloid Receptor Ligands Be Used Therapeutically?

The idea of using capsaicinoids (and other vanilloids and TRPV1 ligands) for therapeutic purposes, particularly in the treatment of pain, has been known for centuries. The clinical use of capsaicin (and other TRPV1 ligands) engages the three main physiological effects of capsaicinoids as the fundamental basis for treatment: (1) stimulation of sensory fibers to produce counterirritation, (2) hyperstimulation to produce neuronal desensitization and refractory periods of variable duration, and (3) selective downregulation of TRPV1 expression with deletion of TRPV1-expressing sensory neurons. Some of the earliest documented uses of capsaicin in the treatment of illness, including pain, date back several centuries to when people purportedly applied extracts of peppers to treat toothaches (Szallasi and Blumberg, 1999). Similarly, the analgesic properties of capsaicinoids were exploited to lessen the pain associated with castration of the eunuchs serving Chinese emperors (Anderson, 1990; Szallasi and Blumberg, 1999). The first documented modern use of capsaicinoids to treat pain was in the mid-1800s, in which tinctures of hot pepper were the recommended treatment for relief from dental pain (Szallasi and Blumberg, 1999). Since then, capsaicin has been used in a plethora of medicinal products including over-the-counter analgesic topical creams (0.025 to 0.075%) for treatment of itching, general and neuropathic pain (e.g., sports injuries, pain associated with diabetic neuropathy, posttherapeutic neuralgia, and osteoarthritis or rheumatoid arthritis), swelling, and muscle soreness. Some popular over-the-counter products include TheraPatch®, Zostrix®, Axsain®, Capsaicin-P®, Penecine®, Dolorac®, and Arthogesic®. A variety of clinical trials have investigated the potential use of capsaicin (and other vanilloids) as therapeutic agents in many disorders, including acute and chronic pain (arthritic, neuropathic, dental, etc.) (Szabo et al., 1999; Szallasi and Blumberg, 1999; Kissin et al., 2002; Szallasi, 2002; Anand, 2003; Rashid et al., 2003), dermatological conditions (psoriasis, pruritis, etc.) (Ellis et al., 1993; Weisshaar et al., 1998; Weisshaar et al., 2003), headache (Marks et al., 1993; Saper et al., 2002), rhinitis and nasal congestion (Van Rijswijk et al., 2003), as well as bladder hypersensitivity and hyperreactivity disorders (Lazzeri et al., 1996; Kim and Chancellor, 2000; Szallasi and Fowler, 2002). The clinical applications of capsaicin have been reviewed by Robbins (2000).

Despite the apparent popularity and applicability of capsaicin for treatment of pain, conclusive and supportive clinical-trial data are not abundant. Szallasi and Blumberg cite several factors that may explain the paucity of significant clinical-trial data supporting the efficacy of capsaicin for treating pain:

1. Placebo-controlled and blinded studies are difficult to perform due to the characteristic burning sensations induced by capsaicinoids.
2. A high frequency of placebo effects are observed.
3. Limited uptake, extensive metabolism, and little or no systemic distribution of capsaicinoids occur following dermal application.
4. In many instances, treatment with capsaicinoids was not tolerated due to the frequent painful side effects (localized burning and itching, coughing and sneezing, and irritation to the eyes) inherent to the high-dose capsaicinoid exposures required to achieve the desired effect (Szallasi and Blumberg, 1999; Szallasi, 2002).

A potential solution to the inherent pain of capsaicinoid exposure that is currently under investigation is the coadministration of a local anesthetic, such as lidocaine, to counteract the painful sensation associated with capsaicin. This approach has shown some promise for the treatment of pain with high doses of capsaicinoids, effectively abolishing the pain associated with capsaicin treatment while not affecting the benefits of the treatment. However, masking the painful properties of capsaicinoid exposure does not decrease the adverse respiratory and cardiovascular effects that can be associated with the high doses required to achieve the desired effects, despite removal of excess capsaicinoids by extensive washing of the treated areas.

The use of RTX and other more selective and potent vanilloids for treatment of various maladies has, however, gained popularity (Szallasi and Blumberg, 1999; Szallasi, 2002). For example, RTX (Szallasi and Blumberg, 1999; Szallasi, 2002; Szallasi and Fowler, 2002), olvanil (Brand et al., 1987; Brand et al., 1990), civamide (Hua et al., 1997), and the newly discovered SDZ 249-665 (Urban et al., 2000) exhibit greater potency than capsaicin, with markedly decreased side effects. RTX is the most selective TRPV1 agonist known and exhibits potencies that are ~100-fold greater than capsaicin at TRPV1. Olvanil, a more lipophilic synthetic capsaicinoid analog, exhibits oral activity and potencies similar to, or greater than, capsaicin in some pain models. Civimide was also found to be orally active in preventing chemically induced nociception while not producing bronchoconstriction and the traditional release of substance P and CGRP, as observed for capsaicin. SDZ 249-665 also exhibits greater potency, almost identical efficacy, and a much lower potential for toxicity than capsaicin. Furthermore, olvanil and SDZ 249-665 lack the characteristic pain-producing and irritating effects of capsaicin, while achieving their beneficial effects on pain at lower concentrations. As such, clinical use of these substances for more potent analgesics is currently under investigation.

Significant research related to the potential beneficial effects of TRPV1 inhibitors has also been performed. In various cough and pain models, selective and competitive (capsazepine and other chemical derivatives), and nonselective (ruthenium red) TRPV1 antagonists block the effects of capsaicin (the model irritant or stimulant), other vanilloids, and chemical irritants, and the endogenous pain that appears to involve TRPV1-mediated signaling processes (Urban and Dray, 1991; Santos and Calixto, 1997; Szallasi and Blumberg, 1999; Szallasi, 2002; Lopez-Rodriguez et al., 2003; Walker et al., 2003). Unfortunately, clinical use of these compounds has been limited by their lack of specificity and moderate potencies. Currently, there are several new derivatives of capsazepine under investigation for the treatment of pain and other TRPV1-mediated processes. These chemicals exhibit a wide range of properties that include both agonist and antagonist activity, increased binding affinities, greater selectivity for TRPV1, lower effective concentrations, and increased stability. A very recent study has characterized the effectiveness of a newly discovered TRPV1 antagonist with analgesic properties, *N*-(4-tertiarybutylphenyl)-4-(3-chlorophyridin-2-yl)tetrahydropyrazine-1(2*H*)-carboxamide (BCTC), which has potential for treatment of pain by TRPV1 stimulation (Pomonis et al., 2003; Valenzano et al., 2003).

Although considerable knowledge exists regarding the treatment of pain and other disorders with vanilloids (agonists and antagonists), very little information is available that describes the treatment of respiratory disorders with capsaicinoids or other TRPV1 ligands. As mentioned earlier, studies have been published that describe the beneficial effect of treatment of nonallergic, noninfectious rhinitis and rhinorrhea with intranasal capsaicin spray; evidence is also available to show that prior exposure to various substances (e.g., capsaicin or particles) can influence sensitivity to subsequent respiratory challenges or exposures. Similarly, the beneficial effects of TRPV1 antagonists in the prevention of cough and neurogenic inflammation has also been well established using various models (Lou et al., 1991; Lou and Lundberg, 1992; Lalloo et al., 1995; Morice et al., 2001). Despite the fact that vanilloids may have therapeutic potential in respiratory disorders, their use may be limited simply by issues related to safety, organ selective toxicity (capsaicinoids are potent cardiovascular and pulmonary toxicants), adverse side effects, and lack of efficient delivery to respiratory tissues.

Recent cloning of the TRPM8 (menthol) receptor has also provided new insight into the relief of pain with menthol (or peppermint oil) and other botanical phenolic substances such as eugenol, guaiacol, camphor, etc. (Ohkubo and Shibata, 1997; Davies et al., 2002; Anand, 2003). Initially, the relief of oral pain and nasal congestion with the use of these substances was thought to be due to overlapping pharmacologic interaction with the capsaicin receptor because of the ability of capsazepine to block menthol-induced nociceptive responses in animals and the ability of capsaicin to displace [^{3}H]-menthol from receptor-binding sites (Ohkubo and Shibata, 1997; Wright et al., 1998). However, the cloning and characterization of TRPV1 and TRPM8 have demonstrated that menthol and capsaicin interact separately with unique receptors on a specific subset of sensory neurons (McKemy et al., 2002; Nealen et al., 2003). Although the precise molecular mechanisms by which menthol produces analgesia through TRPM8 or other pain receptors have not been established, one could hypothesize that the mechanism is similar to that in the sensation of pain and heat by TRPV1. This mechanism includes receptor activation, alterations in intracellular ion concentrations, nerve cell depolarization, and ultimately, the release of specific second-messenger molecules and their interaction with adjacent cells to produce temporary pain followed by periods of analgesia.

The treatment of respiratory disorders, such as pain, sore throat, cough, and rhinorrhea and upper airway inflammation that is caused by infections and allergies, with throat sprays, syrups, lozenges, topical creams, and vapor rubs containing menthol and other botanical phenols is commonplace and has its roots in traditional homeopathic medicine. Menthol has been used in a variety of over-the-counter products including cough syrups and lozenges (Halls®, Vicks®, Cepacol®, Robitussin®, etc.) and even pain-relief ointments and muscle balms (BenGay®, Theragesic®, Eucalyptamint®, etc.). The main use of menthol is to induce a soothing and cooling sensation when applied on the skin or other areas.

Since the discovery of TRPM8, molecular pharmacologists have begun to investigate the role of this receptor in the treatment of painful symptoms with menthol and other substances. Results suggest that menthol may ameliorate pain (Anand, 2003) and prevent cough induced by citric acid and possibly other cough stimuli (Morice et al., 1994). Menthol has also been used to treat congestion associated with infections such as the common cold (Burrow et al., 1983; Eccles et al., 1990; Eccles, 2000; Eccles, 2003). In these studies, marked increases in the sensation of nasal airflow were observed, providing psychological symptomatic relief for congestion, despite the fact that no measurable increases in airflow through the nasal passages were produced by menthol treatment. Menthol also decreases the breathing rate, suggesting that the beneficial effects on cough induction may be related to a decreased drive to inhale and forcibly exhale air, both of which are components of the cough reflex. However, the principal toxicological effects of menthol are related to respiratory depression (Orani et al., 1991; Eccles, 2003). On the whole, menthol, similar to capsaicin, appears to be a very interesting molecule in regard to its potential therapeutic properties, its use as a tool for determining the functions of TRPM8 and other TRP/vanilloid receptors *in vivo*, and its diverse effects on normal respiratory physiology.

8.5 SUMMARY AND CONCLUSIONS

The TRP/vanilloid receptors are a vital part of the multiple respiratory responses to environmental stimuli. The contributions of these diverse receptors in regulating normal physiological functions and mediating adverse functions are only beginning to be understood. Although TRP/vanilloid receptors may be potential targets in the treatment of specific symptoms and diseases, such as acute and chronic pain and bladder hyperresponsiveness, various detrimental effects elicited by exposure to noxious chemicals and environmental conditions appear to be mediated by these receptors as well. TRPV1 appears to be a key molecular factor governing respiratory cell sensitivity and responsiveness to many inhaled toxicants and pollutants such as pepper sprays and ambient particulate materials. As research continues on the mechanisms by which respiratory diseases and disorders (e.g., asthma, COPD, chronic cough, acute respiratory distress syndrome, etc.) become apparent, a clearer understanding of the participation of the TRP/vanilloid receptors in these processes will emerge. For example, chronic bronchitis may arise through selective activation or upregulation of TRPV1, either by genetic differences in individuals or by chronic low-dose exposure to certain types of pollutants. Similarly, an acute exposure to pollutants or toxicants such as pepper sprays may result in TRPV1-mediated damage to the respiratory epithelium, the protective layer of cells, causing inflammation and increasing the potential for bacterial infection. Furthermore, as the functions of TRP/vanilloid receptors other than TRPV1 are characterized, a multitude of new hypotheses about the contribution of each receptor subtype in regulating respiratory function will undoubtedly emerge. Thus, TRP/vanilloid receptors are a new and intriguing class of proteins that are expressed in respiratory tissues and may play important roles in various respiratory disorders. TRP/vanilloid receptors may also be important molecular targets for pharmacological therapy of many respiratory disorders and for the undesirable consequences associated with exposures to noxious environmental stimuli.

8.6 ACKNOWLEDGMENTS

The authors would like to thank Ms. Diane Lanza for her help with the cell culture manipulations, Ms. Erin Kaser for her help with the figure designs, and Dr. Brian Carr for his assistance with the determination of capsaicinoid-induced cellular toxicities. Funding that supported some of the studies presented here was provided by the National Institutes of Health grants HL69813 and ES11281.

REFERENCES

Agopyan, N., Bhatti, T., Yu, S., and Simon, S.A. (2003). Vanilloid receptor activation by 2- and 10-μm particles induces responses leading to apoptosis in human airway epithelial cells. *Toxicol Appl Pharmacol* 192, 21–35.

Agopyan, N., Head, J., Yu, S., and Simon, S.A. (2004). TRPV1 receptors mediate particulate matter-induced apoptosis. *Am J Physiol Lung Cell Mol Physiol* 286, L563–572.

Agopyan, N., Li, L., Yu, S., and Simon, S.A. (2003). Negatively charged 2- and 10-μm particles activate vanilloid receptors, increase cAMP, and induce cytokine release. *Toxicol Appl Pharmacol* 186, 63–76.

Alessandri-Haber, N., Yeh, J.J., Boyd, A.E., Parada, C.A., Chen, X., Reichling, D.B., and Levine, J.D. (2003). Hypotonicity induces TRPV4-mediated nociception in rat. *Neuron* 39, 497–511.

Amdur, M.O. (1989). Health effects of air pollutants: sulfuric acid, the old and the new. *Environ Health Perspect* 81, 109–113; discussion 121–122.

Amiri, H., Schultz, G., and Schaefer, M. (2003). FRET-based analysis of TRPC subunit stoichiometry. *Cell Calcium* 33, 463–470.

Anand, P. (2003). Capsaicin and menthol in the treatment of itch and pain: recently cloned receptors provide the key. *Gut* 52, 1233–1235.

Anderson, M. (1990). *Hidden Power: The Palace of Eunuchs of Imperial China*. Prometheus Books, Buffalo, NY.

Appendino, G., Cravotto, G., Palmisano, G., Annunziata, R., and Szallasi, A. (1996). Synthesis and evaluation of phorboid 20-homovanillates: discovery of a class of ligands binding to the vanilloid (capsaicin) receptor with different degrees of cooperativity. *J Med Chem* 39, 3123–3131.

Aust, A.E., Ball, J.C., Hu, A.A., Lighty, J.S., Smith, K.R., Straccia, A.M., Veranth, J.M., and Young, W.C. (2002). Particle characteristics responsible for effects on human lung epithelial cells. *Res Rep Health Eff Inst*, 1–65; discussion 67–76.

Aviado, D.M. and Guevara Aviado, D. (2001). The Bezold-Jarisch reflex. A historical perspective of cardiopulmonary reflexes. *Ann N. Y. Acad Sci* 940, 48–58.

Beech, D.J., Xu, S.Z., McHugh, D., and Flemming, R. (2003). TRPC1 store-operated cationic channel subunit. *Cell Calcium* 33, 433–440.

Benham, C.D., Gunthorpe, M.J., and Davis, J.B. (2003). TRPV channels as temperature sensors. *Cell Calcium* 33, 479–487.

Bevan, S., Hothi, S., Hughes, G., James, I.F., Rang, H.P., Shah, K., Walpole, C.S., and Yeats, J.C. (1992). Capsazepine: a competitive antagonist of the sensory neurone excitant capsaicin. *Br J Pharmacol* 107, 544–552.

Bhave, G., Hu, H.J., Glauner, K.S., Zhu, W., Wang, H., Brasier, D.J., Oxford, G.S., and Gereau, R.W.T. (2003). Protein kinase C phosphorylation sensitizes but does not activate the capsaicin receptor transient receptor potential vanilloid 1 (TRPV1). *Proc Natl Acad Sci USA* 100, 12480–12485.

Billmire, D.F., Vinocur, C., Ginda, M., Robinson, N.B., Panitch, H., Friss, H., Rubenstein, D., and Wiley, J.F. (1996). Pepper spray-induced respiratory failure treated with extracorporeal membrane oxygenation. *Pediatrics* 98, 961–963.

Birnbaumer, L., Yidirim, E., and Abramowitz, J. (2003). A comparison of the genes coding for canonical TRP channels and their M, V, and P relatives. *Cell Calcium* 33, 419–432.

Blanc, P., Liu, D., Juarez, C., and Boushey, H.A. (1991). Cough in hot pepper workers. *Chest* 99, 27–32.

Brand, L., Berman, E., Schwen, R., Loomans, M., Janusz, J., Bohne, R., Maddin, C., Gardner, J., Lahann, T., Farmer, R. et al. (1987). NE-19550: a novel, orally active anti-inflammatory analgesic. *Drugs Exp Clin Res* 13, 259–265.

Brand, L.M., Skare, K.L., Loomans, M.E., Reller, H.H., Schwen, R.J., Lade, D.A., Bohne, R.L., Maddin, C.S., Moorehead, D.P., Fanelli, R. et al. (1990). Anti-inflammatory pharmacology and mechanism of the orally active capsaicin analogs, NE-19550 and NE-28345. *Agents Actions* 31, 329–340.

Burrow, A., Eccles, R., and Jones, A. S. (1983). The effects of camphor, eucalyptus, and menthol vapour on nasal resistance to airflow and nasal sensation. *Acta Otolaryngol* 96, 157–161.

Busker, R.W. and van Helden, H.P. (1998). Toxicologic evaluation of pepper spray as a possible weapon for the Dutch police force: risk assessment and efficacy. *Am J Forensic Med Pathol* 19, 309–316.

Carter, J.D., Ghio, A.J., Samet, J.M., and Devlin, R.B. (1997). Cytokine production by human airway epithelial cells after exposure to an air pollution particle is metal dependent. *Toxicol Appl Pharmacol* 146, 180–188.

Caterina, M.J. and Julius, D. (2001). The vanilloid receptor: a molecular gateway to the pain pathway. *Annu Rev Neurosci* 24, 487–517.

Caterina, M.J., Leffler, A., Malmberg, A.B., Martin, W.J., Trafton, J., Petersen-Zeitz, K.R., Koltzenburg, M., Basbaum, A.I., and Julius, D. (2000). Impaired nociception and pain sensation in mice lacking the capsaicin receptor. *Science* 288, 306–313.

Caterina, M.J., Rosen, T.A., Tominaga, M., Brake, A.J., and Julius, D. (1999). A capsaicin-receptor homologue with a high threshold for noxious heat. *Nature* 398, 436–441.

Caterina, M.J., Schumacher, M.A., Tominaga, M., Rosen, T.A., Levine, J.D., and Julius, D. (1997). The capsaicin receptor: a heat-activated ion-channel in the pain pathway. *Nature* 389, 816–824.

Chan, T.C., Vilke, G.M., Clausen, J., Clark, R.F., Schmidt, P., Snowden, T., and Neuman, T. (2002). The effect of oleoresin capsicum "pepper" spray inhalation on respiratory function. *J Forensic Sci* 47, 299–304.

Chuang, H.H., Prescott, E.D., Kong, H., Shields, S., Jordt, S.E., Basbaum, A.I., Chao, M.V., and Julius, D. (2001). Bradykinin and nerve growth factor release the capsaicin receptor from PtdIns(4,5)P2-mediated inhibition. *Nature* 411, 957–962.

Churg, A. and Brauer, M. (2000). Ambient atmospheric particles in the airways of human lungs. *Ultrastruct Pathol* 24, 353–361.

Clapham, D.E., Runnels, L.W., and Strubing, C. (2001). The TRP ion-channel family. *Nat Rev Neurosci* 2, 387–396.

Cortright, D.N., Crandall, M., Sanchez, J.F., Zou, T., Krause, J.E., and White, G. (2001). The tissue distribution and functional characterization of human VR1. *Biochem Biophys Res Commn* 281, 1183–1189.

Costa, D.L. and Kodavanti, U.P. (2003). Toxic responses of the lung to inhaled pollutants: benefits and limitations of lung-disease models. *Toxicol Lett* 140–141, 195–203.

Cuello, A.C., Gamse, R., Holzer, P., and Lembeck, F. (1981). Substance P immunoreactive neurons following neonatal administration of capsaicin. *Naunyn Schmiedebergs Arch Pharmacol* 315, 185–194.

D'Amato, G., Liccardi, G., D'Amato, M., and Cazzola, M. (2002). Respiratory allergic diseases induced by outdoor air pollution in urban areas. *Monaldi Arch Chest Dis* 57, 161–163.

Davies, S.J., Harding, L.M., and Baranowski, A.P. (2002). A novel treatment of postherpetic neuralgia using peppermint oil. *Clin J Pain* 18, 200–202.

Debarre, S., Karinthi, L., Delamanche, S., Fuche, C., Desforges, P., and Calvet, J. H. (1999). Comparative acute toxicity of o-chlorobenzylidene malononitrile (CS) and oleoresin capsicum (OC) in awake rats. *Hum Exp Toxicol* 18, 724–730.

Dedov, V.N., Tran, V.H., Duke, C.C., Connor, M., Christie, M.J., Mandadi, S., and Roufogalis, B.D. (2002). Gingerols: a novel class of vanilloid receptor (VR1) agonists. *Br J Pharmacol* 137, 793–798.

Delany, N.S., Hurle, M., Facer, P., Alnadaf, T., Plumpton, C., Kinghorn, I., See, C.G., Costigan, M., Anand, P., Woolf, C.J., Crowther, D., Sanseau, P., and Tate, S.N. (2001). Identification and characterization of a novel human vanilloid receptor-like protein, VRL-2. *Physiol Genomics* 4, 165–174.

den Dekker, E., Hoenderop, J.G., Nilius, B., and Bindels, R.J. (2003). The epithelial calcium channels, TRPV5 and TRPV6: from identification toward regulation. *Cell Calcium* 33, 497–507.

Dockery, D.W. (1993). Epidemiologic study design for investigating respiratory health effects of complex air pollution mixtures. *Environ Health Perspect* 101(Suppl. 4), 187–191.

Dockery, D.W. (2001). Epidemiologic evidence of cardiovascular effects of particulate air pollution. *Environ Health Perspect* 109(Suppl. 4), 483–486.

Dockery, D.W., Pope, C.A., 3rd, Xu, X., Spengler, J.D., Ware, J.H., Fay, M.E., Ferris, B.G., Jr., and Speizer, F.E. (1993). An association between air pollution and mortality in six U.S. cities. *N Engl J Med* 329, 1753–1759.

Doherty, M.J., Mister, R., Pearson, M.G., and Calverley, P.M. (2000). Capsaicin responsiveness and cough in asthma and chronic obstructive pulmonary disease. *Thorax* 55, 643–649.

Eccles, R. (2000). Role of cold receptors and menthol in thirst, the drive to breathe and arousal. *Appetite* 34, 29–35.

Eccles, R. (2003). Menthol: effects on nasal sensation of airflow and the drive to breathe. *Curr Allergy Asthma Rep* 3, 210–214.

Eccles, R., Jawad, M.S., and Morris, S. (1990). The effects of oral administration of ()-menthol on nasal resistance to airflow and nasal sensation of airflow in subjects suffering from nasal congestion associated with the common cold. *J Pharm Pharmacol* 42, 652–654.

Ellis, C.N., Berberian, B., Sulica, V.I., Dodd, W.A., Jarratt, M.T., Katz, H.I., Prawer, S., Krueger, G., Rex, I.H., Jr., and Wolf, J.E. (1993). A double-blind evaluation of topical capsaicin in pruritic psoriasis. *J Am Acad Dermatol* 29, 438–442.

Gamse, R., Holzer, P., and Lembeck, F. (1980). Decrease of substance P in primary afferent neurones and impairment of neurogenic plasma extravasation by capsaicin. *Br J Pharmacol* 68, 207–213.

Gavett, S.H., Haykal-Coates, N., Highfill, J.W., Ledbetter, A.D., Chen, L.C., Cohen, M.D., Harkema, J.R., Wagner, J.G., and Costa, D.L. (2003). World Trade Center fine particulate matter causes respiratory tract hyperresponsiveness in mice. *Environ Health Perspect* 111, 981–991.

Glinsukon, T., Stitmunnaithum, V., Toskulkao, C., Buranawuti, T., and Tangkrisanavinont, V. (1980). Acute toxicity of capsaicin in several animal species. *Toxicon* 18, 215–220.

Goodman, M.B. (2003). Sensation is painless. *Trends Neurosci* 26, 643–645.

Govindarajan, V.S. (1985). Capsicum production, technology, chemistry, and quality. Part 1: history, botany, cultivation, and primary processing. *Crit Rev Food Sci Nutr* 22, 109–176.

Govindarajan, V.S. and Sathyanarayana, M.N. (1991). Capsicum-production, technology, chemistry, and quality. Part V. Impact on physiology, pharmacology, nutrition, and metabolism; structure, pungency, pain, and desensitization sequences. *Crit Rev Food Sci Nutr* 29, 435–474.

Granfield, J., Onnen, J., and Petty, C.S. (March 1994). Pepper spray in in-custody deaths. In *NIJ/Science and Technology*.

Guler, A.D., Lee, H., Iida, T., Shimizu, I., Tominaga, M., and Caterina, M. (2002). Heat-evoked activation of the ion channel, TRPV4. *J Neurosci* 22, 6408–6414.

Hayes, P., Meadows, H.J., Gunthorpe, M.J., Harries, M.H., Duckworth, D.M., Cairns, W., Harrison, D.C., Clarke, C.E., Ellington, K., Prinjha, R.K., Barton, A.J., Medhurst, A.D., Smith, G.D., Topp, S., Murdock, P., Sanger, G.J., Terrett, J., Jenkins, O., Benham, C.D., Randall, A.D., Gloger, I.S., and Davis, J.B. (2000). Cloning and functional expression of a human orthologue of rat vanilloid receptor-1. *Pain* 88, 205–215.

Heck, A. (1995). Accidental pepper spray discharge in an emergency department. *J Emerg Nurs* 21.

Henry, C.J. and Emery, B. (1986). Effect of spiced food on metabolic rate. *Hum Nutr Clin Nutr* 40, 165–168.

Hoenderop, J.G., Nilius, B., and Bindels, R.J. (2003). Epithelial calcium channels: from identification to function and regulation. *Pflugers Arch* 446, 304–308.

Holzer, P. (1988). Local effector functions of capsaicin-sensitive sensory nerve endings: involvement of tachykinins, calcitonin gene-related peptide, and other neuropeptides. *Neuroscience* 24, 739–768.

Holzer, P. (1991). Capsaicin: cellular targets, mechanisms of action, and selectivity for thin sensory neurons. *Pharmacol Rev* 43, 143–201.

Hua, X.Y., Chen, P., Hwang, J., and Yaksh, T.L. (1997). Antinociception induced by civamide, an orally active capsaicin analogue. *Pain* 71, 313–322.

Hwang, S.W., Cho, H., Kwak, J., Lee, S.Y., Kang, C.J., Jung, J., Cho, S., Min, K.H., Suh, Y.G., Kim, D., and Oh, U. (2000). Direct activation of capsaicin receptors by products of lipoxygenases: endogenous capsaicin-like substances. *Proc Natl Acad Sci USA* 97, 6155–6160.

Jancso, N., Jancso-Gabor, A., and Szolcsanyi, J. (1967). Direct evidence for neurogenic inflammation and its prevention by denervation and by pretreatment with capsaicin. *Br J Pharmacol* 31, 138–151.

Jancso-Gabor, A., Szolcsanyi, J., and Jancso, N. (1970). Irreversible impairment of thermoregulation induced by capsaicin and similar pungent substances in rats and guinea pigs. *J Physiol* 206, 495–507.

Jensen, P.G. and Larson, J.R. (2001). Management of painful diabetic neuropathy. *Drugs Aging* 18, 737–749.

Ji, R.R., Samad, T.A., Jin, S.X., Schmoll, R., and Woolf, C.J. (2002). p38 MAPK activation by NGF in primary sensory neurons after inflammation increases TRPV1 levels and maintains heat hyperalgesia. *Neuron* 36, 57–68.

Jia, Y., McLeod, R.L., Wang, X., Parra, L.E., Egan, R.W., and Hey, J.A. (2002). Anandamide induces cough in conscious guinea pigs through VR1 receptors. *Br J Pharmacol* 137, 831–836.

Jones, N.L., Shabib, S., and Sherman, P.M. (1997). Capsaicin as an inhibitor of the growth of the gastric pathogen *Helicobacter pylori*. *FEMS Microbiol Lett* 146, 223–227.

Jordt, S.E. and Julius, D. (2002). Molecular basis for species-specific sensitivity to "hot" chili peppers. *Cell* 108, 421–430.

Jordt, S.E., Tominaga, M., and Julius, D. (2000). Acid potentiation of the capsaicin receptor determined by a key extracellular site. *Proc Natl Acad Sci USA* 97, 8134–8139.

Jung, J., Hwang, S.W., Kwak, J., Lee, S.Y., Kang, C.J., Kim, W.B., Kim, D., and Oh, U. (1999). Capsaicin binds to the intracellular domain of the capsaicin-activated ion channel. *J Neurosci* 19, 529–538.

Kagawa, J. (2002). Health effects of diesel exhaust emissions — a mixture of air pollutants of worldwide concern. *Toxicology* 181, 182, 349–353.

Kedei, N., Szabo, T., Lile, J.D., Treanor, J.J., Olah, Z., Iadarola, M.J., and Blumberg, P.M. (2001). Analysis of the native quaternary structure of vanilloid receptor 1. *J Biol Chem* 276, 28613–28619.

Kim, D.Y. and Chancellor, M.B. (2000). Intravesical neuromodulatory drugs: capsaicin and resiniferatoxin to treat the overactive bladder. *J Endourol* 14, 97–103.

Kissin, I., Bright, C.A., and Bradley, E.L., Jr. (2002). Selective and long-lasting neural blockade with resiniferatoxin prevents inflammatory pain hypersensitivity. *Anesth Analg* 94, 1253–1258, table of contents.

Kuzhikandathil, E.V., Wang, H., Szabo, T., Morozova, N., Blumberg, P.M., and Oxford, G.S. (2001). Functional analysis of capsaicin receptor (vanilloid receptor subtype 1) multimerization and agonist responsiveness using a dominant negative mutation. *J Neurosci* 21, 8697–8706.

Lai, C.K., Douglass, C., Ho, S.S., Chan, J., Lau, J., Wong, G., and Leung, R. (1996). Asthma epidemiology in the Far East. *Clin Exp Allergy* 26, 5–12.

Lalloo, U.G., Fox, A.J., Belvisi, M.G., Chung, K.F., and Barnes, P.J. (1995). Capsazepine inhibits cough induced by capsaicin and citric acid but not by hypertonic saline in guinea pigs. *J Appl Physiol* 79, 1082–1087.

Laude, E.A., Higgins, K.S., and Morice, A.H. (1993). A comparative study of the effects of citric acid, capsaicin, and resiniferatoxin on the cough challenge in guinea pig and man. *Pulm Pharmacol* 6, 171–175.

Lazzeri, M., Beneforti, P., Benaim, G., Maggi, C. A., Lecci, A., and Turini, D. (1996). Intravesical capsaicin for treatment of severe bladder pain: a randomized placebo controlled study. *J Urol* 156, 947–952.

Lighty, J.S., Veranth, J.M., and Sarofim, A.F. (2000). Combustion aerosols: factors governing their size and composition and implications to human health. *J Air Waste Manag Assoc* 50, 1565–1618; discussion 1619–1622.

Lim, K., Yoshioka, M., Kikuzato, S., Kiyonaga, A., Tanaka, H., Shindo, M., and Suzuki, M. (1997). Dietary red pepper ingestion increases carbohydrate oxidation at rest and during exercise in runners. *Med Sci Sports Exerc* 29, 355–361.

Liu, M., Liu, M.C., Magoulas, C., Priestley, J.V., and Willmott, N.J. (2003). Versatile regulation of cytosolic Ca^{2+} by vanilloid receptor I in rat dorsal root ganglion neurons. *J Biol Chem* 278, 5462–5472.

Lopez-Rodriguez, M.L., Viso, A., and Ortega-Gutierrez, S. (2003). VR1 receptor modulators as potential drugs for neuropathic pain. *Mini Rev Med Chem* 3, 729–748.

Lou, Y.P., Karlsson, J.A., Franco-Cereceda, A., and Lundberg, J.M. (1991). Selectivity of ruthenium red in inhibiting bronchoconstriction and CGRP release induced by afferent C-fibre activation in the guinea pig lung. *Acta Physiol Scand* 142, 191–199.

Lou, Y.P. and Lundberg, J.M. (1992). Inhibition of low pH evoked activation of airway sensory nerves by capsazepine, a novel capsaicin-receptor antagonist. *Biochem Biophys Res Commn* 189, 537–544.

Maggi, C.A., Bevan, S., Walpole, C.S., Rang, H.P., and Giuliani, S. (1993). A comparison of capsazepine and ruthenium red as capsaicin antagonists in the rat isolated urinary bladder and vas deferens. *Br J Pharmacol* 108, 801–805.

Mannino, D.M. (2002). COPD: epidemiology, prevalence, morbidity and mortality, and disease heterogeneity. *Chest* 121, 121S–126S.

Marks, D.R., Rapoport, A., Padla, D., Weeks, R., Rosum, R., Sheftell, F., and Arrowsmith, F. (1993). A double-blind placebo-controlled trial of intranasal capsaicin for cluster headache. *Cephalalgia* 13, 114–116.

Mauderly, J., Neas, L., and Schlesinger, R. (1998). PM Monitoring Needs Related to Health Effects. Atmospheric Observations: Helping Build the Scientific Basis for Decisions Related to Airborne Particulate Matter; PM Measurements Research Workshop, pp. 9–14.

McKemy, D.D., Neuhausser, W.M., and Julius, D. (2002). Identification of a cold receptor reveals a general role for TRP channels in thermosensation. *Nature* 416, 52–58.

McMahon, S.B., Lewin, G., and Bloom, S.R. (1991). The consequences of long-term topical capsaicin application in the rat. *Pain* 44, 301–310.

Millqvist, E., Lowhagen, O., and Bende, M. (2000). Quality of life and capsaicin sensitivity in patients with sensory airway hyperreactivity. *Allergy* 55, 540–545.

Minke, B. and Cook, B. (2002). TRP channel proteins and signal transduction. *Physiol Rev* 82, 429–472.

Mizuno, A., Matsumoto, N., Imai, M., and Suzuki, M. (2003). Impaired osmotic sensation in mice lacking TRPV4. *Am J Physiol Cell Physiol* 285, C96–101.

Montell, C. (2003a). Thermosensation: hot findings make TRPNs very cool. *Curr Biol* 13, R476–478.

Montell, C. (2003b). The venerable inveterate invertebrate TRP channels. *Cell Calcium* 33, 409–417.

Montell, C., Birnbaumer, L., Flockerzi, V., Bindels, R.J., Bruford, E.A., Caterina, M.J., Clapham, D.E., Harteneck, C., Heller, S., Julius, D., Kojima, I., Mori, Y., Penner, R., Prawitt, D., Scharenberg, A.M., Schultz, G., Shimizu, N., and Zhu, M.X. (2002). A unified nomenclature for the superfamily of TRP cation channels. *Mol Cell* 9, 229–231.

Morice, A.H. (1996). Inhalation cough challenge in the investigation of the cough reflex and antitussives. *Pulm Pharmacol* 9, 281–284.

Morice, A.H., Kastelik, J.A., and Thompson, R. (2001). Cough challenge in the assessment of cough reflex. *Br J Clin Pharmacol* 52, 365–375.

Morice, A.H., Marshall, A.E., Higgins, K.S., and Grattan, T.J. (1994). Effect of inhaled menthol on citric-acid-induced cough in normal subjects. *Thorax* 49, 1024–1026.

Mortimer, K.M., Neas, L.M., Dockery, D.W., Redline, S., and Tager, I.B. (2002). The effect of air pollution on inner-city children with asthma. *Eur Respir J* 19, 699–705.

National Research Council (1998). *Research Priorities for Airborne Particulate Matter: I. Immediate Priorities and a Long-Range Research Portfolio.* National Academy Press, Washington, D.C.

Nealen, M.L., Gold, M.S., Thut, P.D., and Caterina, M.J. (2003). TRPM8 mRNA is expressed in a subset of cold-responsive trigeminal neurons from rat. *J Neurophysiol* 90, 515–520.

Nelson, E. (1919). Vanillyl-acyl amides. *J Am Chem Soc* 41, 2121–2130.

NIJ (March 1994). Oleoresin Capsicum: Pepper Spray as a Force Alternative. In National Institute of Justice Technology Assessment Program.

Ohkubo, T. and Shibata, M. (1997). The selective capsaicin antagonist capsazepine abolishes the antinociceptive action of eugenol and guaiacol. *J Dent Res* 76, 848–851.

Olah, Z., Szabo, T., Karai, L., Hough, C., Fields, R.D., Caudle, R.M., Blumberg, P.M., and Iadarola, M.J. (2001). Ligand-induced dynamic membrane changes and cell deletion conferred by vanilloid receptor 1. *J Biol Chem* 276, 11021–11030.

Olajos, E.J. and Salem, H. (2001). Riot control agents: pharmacology, toxicology, biochemistry, and chemistry. *J Appl Toxicol* 21, 355–391.

Oortgiesen, M., Veronesi, B., Eichenbaum, G., Kiser, P.F., and Simon, S.A. (2000). Residual oil fly-ash and charged polymers activate epithelial cells and nociceptive sensory neurons. *Am J Physiol Lung Cell Mol Physiol* 278, L683–695.

Orani, G.P., Anderson, J.W., Sant'Ambrogio, G., and Sant'Ambrogio, F.B. (1991). Upper airway cooling and l-menthol reduce ventilation in the guinea pig. *J Appl Physiol* 70, 2080–2086.

Pagan, I., Costa, D.L., McGee, J.K., Richards, J.H., and Dye, J.A. (2003). Metals mimic airway epithelial injury induced by *in vitro* exposure to Utah Valley ambient particulate matter extracts. *J Toxicol Environ Health* A 66, 1087–1112.

Paice, J.A., Ferrans, C.E., Lashley, F.R., Shott, S., Vizgirda, V., and Pitrak, D. (2000). Topical capsaicin in the management of HIV-associated peripheral neuropathy. *J Pain Symptom Manage* 19, 45–52.

Pappagallo, M. and Haldey, E.J. (2003). Pharmacological management of postherpetic neuralgia. *CNS Drugs* 17, 771–780.

Path, G., Braun, A., Meents, N., Kerzel, S., Quarcoo, D., Raap, U., Hoyle, G.W., Nockher, W.A., and Renz, H. (2002). Augmentation of allergic early-phase reaction by nerve growth factor. *Am J Respir Crit Care Med* 166, 818–826.

Perez, C.A., Margolskee, R.F., Kinnamon, S.C., and Ogura, T. (2003). Making sense with TRP channels: store-operated calcium entry and the ion-channel Trpm5 in taste receptor cells. *Cell Calcium* 33, 541–549.

Perraud, A.L., Schmitz, C., and Scharenberg, A.M. (2003). TRPM2 Ca2+ permeable cation channels: from gene to biological function. *Cell Calcium* 33, 519–531.

Philip, G., Sanico, A.M., and Togias, A. (1996). Inflammatory cellular influx follows capsaicin nasal challenge. *Am J Respir Crit Care Med* 153, 1222–1229.

Planells-Cases, R., Aracil, A., Merino, J.M., Gallar, J., Perez-Paya, E., Belmonte, C., Gonzalez-Ros, J.M., and Ferrer-Montiel, A.V. (2000). Arginine-rich peptides are blockers of VR-1 channels with analgesic activity. *FEBS Lett* 481, 131—136.

Plant, T.D. and Schaefer, M. (2003). TRPC4 and TRPC5: receptor-operated Ca^{2+}-permeable nonselective cation channels. *Cell Calcium* 33, 441–450.

Pomonis, J.D., Harrison, J.E., Mark, L., Bristol, D.R., Valenzano, K.J., and Walker, K. (2003). N-(4-Tertiarybutylphenyl)-4-(3-chlorophyridin-2-yl)tetrahydropyrazine - 1(2H)-carbox-amide (BCTC), a novel, orally effective vanilloid receptor 1 antagonist with analgesic properties: II. *In vivo* characterization in rat models of inflammatory and neuropathic pain. *J Pharmacol Exp Ther* 306, 387–393.

Pope, A. (1989). Respiratory disease associated with community air pollution and a steel mill, Utah Valley. *Am J Pub Health* 89, 623–628.

Pope, A. (1998). Epidemiology investigations of the health effects of particulate air pollution: strengths and limitations. *Appl Occup Environ Hyg* 13, 356–363.

Pope, C.A., 3rd (1991). Respiratory hospital admissions associated with PM10 pollution in Utah, Salt Lake, and Cache valleys. *Arch Environ Health* 46, 90–97.

Pope, C.A., 3rd (1996). Particulate pollution and health: a review of the Utah valley experience. *J Expo Anal Environ Epidemiol* 6, 23–34.

Pope, C.A., 3rd (2000). Epidemiology of fine particulate air pollution and human health: biologic mechanisms and who's at risk? *Environ Health Perspect* 108(Suppl. 4), 713–723.

Prescott, E.D. and Julius, D. (2003). A modular PIP2 binding site as a determinant of capsaicin receptor sensitivity. *Science* 300, 1284–1288.

Pyman, E. J. and Jones, E.C.S. (1925). Relation between chemical composition and pungency in acid amides. *J Am Chem Soc* 127, 2588–2599.

Quay, J.L., Reed, W., Samet, J., and Devlin, R.B. (1998). Air pollution particles induce IL-6 gene expression in human airway epithelial cells via NF-kappaB activation. *Am J Respir Cell Mol Biol* 19, 98–106.

Rashid, M.H., Inoue, M., Kondo, S., Kawashima, T., Bakoshi, S., and Ueda, H. (2003). Novel expression of vanilloid receptor 1 on capsaicin-insensitive fibers accounts for the analgesic effect of capsaicin cream in neuropathic pain. *J Pharmacol Exp Ther* 304, 940–948.

Reilly, C.A., Crouch, D.J., and Yost, G.S. (2001). Quantitative analysis of capsaicinoids in fresh peppers, oleoresin capsicum and pepper spray products. *J Forensic Sci* 46, 502–509.

Reilly, C.A., Ehlhardt, W.J., Jackson, D.A., Kulanthaivel, P., Mutlib, A.E., Espina, R.J., Moody, D.E., Crouch, D.J., and Yost, G.S. (2003). Metabolism of capsaicin by cytochrome P450 produces novel dehydrogenated metabolites and decreases cytotoxicity to lung and liver cells. *Chem Res Toxicol* 16, 336–349.

Reilly, C.A., Taylor, J.L., Lanza, D.L., Carr, B.A., Crouch, D.J., and Yost, G.S. (2003). Capsaicinoids cause inflammation and epithelial cell death through activation of vanilloid receptors. *Toxicol Sci* 73, 170–181.

Richardson, J.D. and Vasko, M.R. (2002). Cellular mechanisms of neurogenic inflammation. *J Pharmacol Exp Ther* 302, 839–845.

Robbins, W. (2000). Clinical applications of capsaicinoids. *Clin J Pain* 16, S86–89.

Sanchez, J.F., Krause, J.E., and Cortright, D.N. (2001). The distribution and regulation of vanilloid receptor VR1 and VR1 5 splice variant RNA expression in rat. *Neuroscience* 107, 373–381.

Sanico, A.M., Atsuta, S., Proud, D., and Togias, A. (1997). Dose-dependent effects of capsaicin nasal challenge: *in vivo* evidence of human airway neurogenic inflammation. *J Allergy Clin Immunol* 100, 632–641.

Santos, A.R. and Calixto, J.B. (1997). Ruthenium red and capsazepine antinociceptive effect in formalin and capsaicin models of pain in mice. *Neurosci Lett* 235, 73–76.

Saper, J.R., Klapper, J., Mathew, N.T., Rapoport, A., Phillips, S.B., and Bernstein, J.E. (2002). Intranasal civamide for the treatment of episodic cluster headaches. *Arch Neurol* 59, 990–994.

Schins, R.P., Duffin, R., Hohr, D., Knaapen, A.M., Shi, T., Weishaupt, C., Stone, V., Donaldson, K., and Borm, P. J. (2002). Surface modification of quartz inhibits toxicity, particle uptake, and oxidative DNA damage in human lung epithelial cells. *Chem Res Toxicol* 15, 1166–1173.

Schins, R.P., McAlinden, A., MacNee, W., Jimenez, L.A., Ross, J.A., Guy, K., Faux, S.P., and Donaldson, K. (2000). Persistent depletion of I kappa B alpha and interleukin-8 expression in human pulmonary epithelial cells exposed to quartz particles. *Toxicol Appl Pharmacol* 167, 107–117.

Schumacher, M.A., Moff, I., Sudanagunta, S.P., and Levine, J.D. (2000). Molecular cloning of an N-terminal splice variant of the capsaicin receptor. Loss of N-terminal domain suggests functional divergence among capsaicin receptor subtypes. *J Biol Chem* 275, 2756–2762.

Seinfeld, J.H. and Pandis, S.N. (1998). *Atmospheric Chemistry and Physics — From Air Pollution to Climate Change.* Wiley, New York.

Smart, D., Gunthorpe, M.J., Jerman, J.C., Nasir, S., Gray, J., Muir, A.I., Chambers, J.K., Randall, A.D., and Davis, J.B. (2000). The endogenous lipid anandamide is a full agonist at the human vanilloid receptor (hVR1). *Br J Pharmacol* 129, 227–230.

Smith, C.G. and Stopford, W. (1999). Health hazards of pepper spray. *N C Med J* 60, 268–274.

Smith, G.D., Gunthorpe, M.J., Kelsell, R.E., Hayes, P.D., Reilly, P., Facer, P., Wright, J.E., Jerman, J.C., Walhin, J.P., Ooi, L., Egerton, J., Charles, K.J., Smart, D., Randall, A.D., Anand, P., and Davis, J.B. (2002). TRPV3 is a temperature-sensitive vanilloid receptor-like protein. *Nature* 418, 186–190.

Smith, K.R. and Aust, A.E. (1997). Mobilization of iron from urban particulates leads to generation of reactive oxygen species *in vitro* and induction of ferritin synthesis in human lung epithelial cells. *Chem Res Toxicol* 10, 828–834.

Smith, K.R., Veranth, J.M., Hu, A.A., Lighty, J.S., and Aust, A.E. (2000). Interleukin-8 levels in human lung epithelial cells are increased in response to coal fly ash and vary with the bioavailability of iron, as a function of particle size and source of coal. *Chem Res Toxicol* 13, 118–125.

Steffee, C.H., Lantz, P.E., Flannagan, L.M., Thompson, R.L., and Jason, D.R. (1995). Oleoresin capsicum (pepper) spray and "in-custody deaths." *Am J Forensic Med Pathol* 16, 185–192.

Story, G.M., Peier, A.M., Reeve, A.J., Eid, S.R., Mosbacher, J., Hricik, T.R., Earley, T.J., Hergarden, A.C., Andersson, D.A., Hwang, S.W., McIntyre, P., Jegla, T., Bevan, S., and Patapoutian, A. (2003). ANKTM1, a TRP-like channel expressed in nociceptive neurons, is activated by cold temperatures. *Cell* 112, 819–829.

Strachan, D.P. (2000). The role of environmental factors in asthma. *Br Med Bull* 56, 865–882.

Surh, Y.J. and Lee, S.S. (1995). Capsaicin, a double-edged sword: toxicity, metabolism, and chemopreventive potential. *Life Sci* 56, 1845–1855.

Suzuki, M., Mizuno, A., Kodaira, K., and Imai, M. (2003). Impaired pressure sensation in mice lacking TRPV4. *J Biol Chem* 278, 22664–22668.

Szabo, T., Olah, Z., Iadarola, M.J., and Blumberg, P.M. (1999). Epidural resiniferatoxin induced prolonged regional analgesia to pain. *Brain Res* 840, 92–98.

Szallasi, A. (1995). Autoradiographic visualization and pharmacological characterization of vanilloid (capsaicin) receptors in several species, including man. *Acta Physiol Scand Suppl* 629, 1–68.

Szallasi, A. (2002). Vanilloid (capsaicin) receptors in health and disease. *Am J Clin Pathol* 118, 110–121.

Szallasi, A., Biro, T., Modarres, S., Garlaschelli, L., Petersen, M., Klusch, A., Vidari, G., Jonassohn, M., De Rosa, S., Sterner, O., Blumberg, P.M., and Krause, J.E. (1998). Dialdehyde sesquiterpenes and other terpenoids as vanilloids. *Eur J Pharmacol* 356, 81–89.

Szallasi, A. and Blumberg, P.M. (1989). Resiniferatoxin, a phorbol-related diterpene, acts as an ultrapotent analog of capsaicin, the irritant constituent in red pepper. *Neuroscience* 30, 515–520.

Szallasi, A. and Blumberg, P.M. (1999). Vanilloid (capsaicin) receptors and mechanisms. *Pharmacol Rev* 51, 159–212.

Szallasi, A. and Fowler, C.J. (2002). After a decade of intravesical vanilloid therapy: still more questions than answers. *Lancet Neurol* 1, 167–172.

Szallasi, A., Goso, C., and Manzini, S. (1995). Resiniferatoxin binding to vanilloid receptors in guinea pig and human airways. *Am J Respir Crit Care Med* 152, 59–63.

Szallasi, A., Lewin, N.E., and Blumberg, P.M. (1992). Identification of alpha-1-acid glycoprotein (orosomucoid) as a major vanilloid binding protein in serum. *J Pharmacol Exp Ther* 262, 883–888.

Tominaga, M., Caterina, M.J., Malmberg, A.B., Rosen, T.A., Gilbert, H., Skinner, K., Raumann, B.E., Basbaum, A.I., and Julius, D. (1998). The cloned capsaicin receptor integrates multiple pain-producing stimuli. *Neuron* 21, 531–543.

Toth, A., Blumberg, P. M., Chen, Z., and Kozikowski, A.P. (2004). Design of a high-affinity competitive antagonist of the vanilloid receptor selective for the calcium entry-linked receptor population. *Mol Pharmacol* 65, 282–291.

Urban, L., Campbell, E. A., Panesar, M., Patel, S., Chaudhry, N., Kane, S., Buchheit, K., Sandells, B., and James, I. F. (2000). *In vivo* pharmacology of SDZ 249-665, a novel, non-pungent capsaicin analogue. *Pain* 89, 65–74.

Urban, L. and Dray, A. (1991). Capsazepine, a novel capsaicin antagonist, selectively antagonises the effects of capsaicin in the mouse spinal cord *in vitro. Neurosci Lett* 134, 9–11.

Utell, M.J., Samet, J.M., Bates, D.V., Becklake, M.R., Dockery, D.W., Leaderer, B.P., Mauderly, J.L., and Speizer, F.E. (1988). Air pollution and health. *Am Rev Respir Dis* 138, 1065–1068.

Valenzano, K.J., Grant, E.R., Wu, G., Hachicha, M., Schmid, L., Tafesse, L., Sun, Q., Rotshteyn, Y., Francis, J., Limberis, J., Malik, S., Whittemore, E.R., and Hodges, D. (2003). N-(4-tertiarybutylphenyl)-4-(3-chloropyridin-2-yl)tetrahydropyrazine -1(2H)-carbox-amide (BCTC), a novel, orally effective vanilloid receptor 1 antagonist with analgesic properties: I. *In vitro* characterization and pharmacokinetic properties. *J Pharmacol Exp Ther* 306, 377–386.

Van Rijswijk, J.B., Boeke, E.L., Keizer, J.M., Mulder, P.G., Blom, H.M., and Fokkens, W.J. (2003). Intranasal capsaicin reduces nasal hyperreactivity in idiopathic rhinitis: a double-blind randomized application regimen study. *Allergy* 58, 754–761.

Veronesi, B., Carter, J.D., Devlin, R.B., Simon, S.A., and Oortgiesen, M. (1999). Neuropeptides and capsaicin stimulate the release of inflammatory cytokines in a human bronchial epithelial cell line. *Neuropeptides* 33, 447–456.

Veronesi, B., de Haar, C., Roy, J., and Oortgiesen, M. (2002). Particulate matter inflammation and receptor sensitivity are target cell specific. *Inhal Toxicol* 14, 159–183.

Veronesi, B., de Haar, C., Lee, L., and Oortgiesen, M. (2002). The surface charge of visible particulate matter predicts biological activation in human bronchial epithelial cells. *Toxicol Appl Pharmacol* 178, 144–154.

Veronesi, B. and Oortgiesen, M. (2001). Neurogenic inflammation and particulate matter (PM) air pollutants. *Neurotoxicology* 22, 795–810.

Veronesi, B., Oortgiesen, M., Carter, J.D., and Devlin, R.B. (1999). Particulate matter initiates inflammatory cytokine release by activation of capsaicin and acid receptors in a human bronchial epithelial cell line. *Toxicol Appl Pharmacol* 154, 106 115.

Veronesi, B., Oortgiesen, M., Roy, J., Carter, J.D., Simon, S.A., and Gavett, S.H. (2000). Vanilloid (capsaicin) receptors influence inflammatory sensitivity in response to particulate matter. *Toxicol Appl Pharmacol* 169, 66–76.

Veronesi, B., Wei, G., Zeng, J. Q., and Oortgiesen, M. (2003). Electrostatic charge activates inflammatory vanilloid (VR1) receptors. *Neurotoxicology* 24, 463–473.

Voets, T., Janssens, A., Prenen, J., Droogmans, G., and Nilius, B. (2003). Mg^{2+}-dependent gating and strong inward rectification of the cation channel TRPV6. *J Gen Physiol* 121, 245–260.

Vriens, J., Watanabe, H., Janssens, A., Droogmans, G., Voets, T., and Nilius, B. (2004). Cell swelling, heat, and chemical agonists use distinct pathways for the activation of the cation channel TRPV4. *Proc Natl Acad Sci USA* 101, 396–401.

Wahl, P., Foged, C., Tullin, S., and Thomsen, C. (2001). Iodo-resiniferatoxin, a new potent vanilloid receptor antagonist. *Mol Pharmacol* 59, 9–15.

Walker, K.M., Urban, L., Medhurst, S.J., Patel, S., Panesar, M., Fox, A.J., and McIntyre, P. (2003). The VR1 antagonist capsazepine reverses mechanical hyperalgesia in models of inflammatory and neuropathic pain. *J Pharmacol Exp Ther* 304, 56–62.

Walpole, C.S., Bevan, S., Bovermann, G., Boelsterli, J.J., Breckenridge, R., Davies, J.W., Hughes, G.A., James, I., Oberer, L., Winter, J. et al. (1994). The discovery of capsazepine, the first competitive antagonist of the sensory neuron excitants capsaicin and resiniferatoxin. *J Med Chem* 37, 1942–1954.

Walpole, C.S., Wrigglesworth, R., Bevan, S., Campbell, E.A., Dray, A., James, I.F., Masdin, K.J., Perkins, M.N., and Winter, J. (1993a). Analogues of capsaicin with agonist activity as novel analgesic agents; structure-activity studies. 2. The amide bond "B-region." *J Med Chem* 36, 2373–2380.

Walpole, C.S., Wrigglesworth, R., Bevan, S., Campbell, E.A., Dray, A., James, I.F., Masdin, K.J., Perkins, M.N., and Winter, J. (1993b). Analogues of capsaicin with agonist activity as novel analgesic agents; structure-activity studies. 3. The hydrophobic side-chain "C-region." *J Med Chem* 36, 2381–2389.

Walpole, C.S., Wrigglesworth, R., Bevan, S., Campbell, E.A., Dray, A., James, I.F., Perkins, M.N., Reid, D.J., and Winter, J. (1993). Analogues of capsaicin with agonist activity as novel analgesic agents; structure-activity studies. 1. The aromatic "A-region." *J Med Chem* 36, 2362–2372.

Watanabe, H., Davis, J.B., Smart, D., Jerman, J.C., Smith, G.D., Hayes, P., Vriens, J., Cairns, W., Wissenbach, U., Prenen, J., Flockerzi, V., Droogmans, G., Benham, C.D., and Nilius, B. (2002). Activation of TRPV4 channels (hVRL-2/mTRP12) by phorbol derivatives. *J Biol Chem* 277, 13569–13577.

Watanabe, H., Vriens, J., Janssens, A., Wondergem, R., Droogmans, G., and Nilius, B. (2003). Modulation of TRPV4 gating by intra- and extracellular Ca^{2+}. *Cell Calcium* 33, 489–495.

Watanabe, H., Vriens, J., Prenen, J., Droogmans, G., Voets, T., and Nilius, B. (2003). Anandamide and arachidonic acid use epoxyeicosatrienoic acids to activate TRPV4 channels. *Nature* 424, 434–438.

Watanabe, H., Vriens, J., Suh, S.H., Benham, C.D., Droogmans, G., and Nilius, B. (2002). Heat-evoked activation of TRPV4 channels in a HEK293 cell expression system and in native mouse aorta endothelial cells. *J Biol Chem* 277, 47044–47051.

Weisel, C.P. (2002). Assessing exposure to air toxics relative to asthma. *Environ Health Perspect* 110(Suppl. 4), 527–537.

Weisshaar, E., Dunker, N., and Gollnick, H. (2003). Topical capsaicin therapy in humans with hemodialysis-related pruritus. *Neurosci Lett* 345, 192–194.

Weisshaar, E., Heyer, G., Forster, C., and Handwerker, H.O. (1998). Effect of topical capsaicin on the cutaneous reactions and itching to histamine in atopic eczema compared to healthy skin. *Arch Dermatol Res* 290, 306–311.

Willis, A., Jerrett, M., Burnett, R.T., and Krewski, D. (2003). The association between sulfate air pollution and mortality at the county scale: an exploration of the impact of scale on a long-term exposure study. *J Toxicol Environ Health* A 66, 1605–1624.

Wong, G.W. and Lai, C.K. (2004). Outdoor air pollution and asthma. *Curr Opin Pulm Med* 10, 62–66.

Wood, J.N., Winter, J., James, I.F., Rang, H.P., Yeats, J., and Bevan, S. (1988). Capsaicin-induced ion fluxes in dorsal root ganglion cells in culture. *J Neurosci* 8, 3208–3220.

Wright, C.E., Bowen, W.P., Grattan, T.J., and Morice, A.H. (1998). Identification of the L-menthol binding site in guinea pig lung membranes. *Br J Pharmacol* 123, 481–486.

Xu, H., Ramsey, I.S., Kotecha, S.A., Moran, M.M., Chong, J.A., Lawson, D., Ge, P., Lilly, J., Silos-Santiago, I., Xie, Y., DiStefano, P.S., Curtis, R., and Clapham, D.E. (2002). TRPV3 is a calcium-permeable temperature-sensitive cation channel. *Nature* 418, 181–186.

Xu, X.P., Dockery, D.W., and Wang, L.H. (1991). Effects of air pollution on adult pulmonary function. *Arch Environ Health* 46, 198–206.

Xue, Q., Yu, Y., Trilk, S.L., Jong, B.E., and Schumacher, M.A. (2001). The genomic organization of the gene encoding the vanilloid receptor: evidence for multiple splice variants. *Genomics* 76, 14–20.

Zygmunt, P.M., Petersson, J., Andersson, D.A., Chuang, H., Sorgard, M., Di Marzo, V., Julius, D., and Hogestatt, E.D. (1999). Vanilloid receptors on sensory nerves mediate the vasodilator action of anandamide. *Nature* 400, 452–457.

9

PULMONARY IMMUNOTOXICOLOGY

Mitchell D. Cohen

CONTENTS

9.1 INTRODUCTION

Research in pulmonary immunotoxicology has been assuming increasing importance in elucidating how workplace or environmental agents cause changes in immune function in the lungs that allow for indirect alterations in the overall health of exposed individuals. This chapter reports how our understanding of the pulmonary immunotoxicology of select agents, i.e., biologics, pollutant gases, cigarette smoke, and metals, has evolved.

9.2 IMMUNOLOGY OF THE RESPIRATORY TRACT

The respiratory system is composed of specialized tissues and cells with important functions apart from those related to respiration, including defense against inhaled infectious and noninfectious agents. The lung immune system is composed of nonspecific innate (natural immunity) and acquired components. Innate immunity (e.g., physical barriers, cells, blood-borne macromolecules) is present before exposure to a xenobiotic and does not discriminate between agents. Acquired (specific) immunity consists of cellular and soluble elements induced by exposure to nonself substances and is characterized by increases in number and defensive capability of distinct macromolecules that recognize and respond to these substances, with the recognition increasing with repeated exposures. The innate and acquired systems are linked; cellular and soluble components of the former influence antigen processing, selection, and presentation, as well as the magnitude of response of the latter.

In the nasal region, large-diameter material is removed by filtration and impaction to prevent entry into deeper regions of the respiratory tract. Further protection is provided by nonspecific mucous, ciliary activity, secreted antimicrobials, and resident phagocytes. Associated with many mucosal surfaces are lymphoid nodules coated with specialized epithelial cells; these constitute the associated lymphoid tissues of the nose, larynx, and bronchus and play a central role in the induced responses against particles and microbial antigens. The tonsils and adenoids comprise the laryngeal-associated lymphoid tissue (LALT) in humans; the upper-tract equivalent structure in other animals is the nasal-associated lymphoid tissue (NALT). Resident immune cells here include populations of T lymphocytes (T cells) of defined phenotypes (e.g., helper $CD3^+CD4^+$, suppressor $CD3^+CD8^+$) and B lymphocytes critical to generating specific antibodies (B cells); macrophages and monocytes are more sparsely distributed. Bronchus-associated lymphoid tissue (BALT), accumulated lymphoid cells situated (as aggregates and clusters) near branch points in the bronchial tree, is covered with a layer of flattened epithelial M cells that act as sentinels at entry ports and transport inhaled antigens to adjoining lymphoid nodules. Mucosa-associated lymphoid tissue (MALT), connected with the nasopharynx and airway mucosal layer, serves as the immunological defense at secretory surfaces.

9.3 IMMUNOLOGIC CELLS IN THE LUNGS

9.3.1 Lymphocytes

Apart from the large numbers of lymphocytes in tracheobronchial lymph nodes and BALT, individual cells or aggregates are also present in the airways and alveolar parenchyma. Lung lymphocytes are mostly T cells that express specific mucosal and memory phenotypes enabling them to recall antigens and respond with quick strong proliferation. Lymphocytes continually recirculate: from blood into tissue, into lymph, and then back into the blood; granulocytes and monocytes leave the blood but do not recirculate. This continuous recirculation increases the likelihood that specific immunocompetent cells will encounter antigen. Upon a first encounter with antigen, lymphocytes acquire a predilection to home in on the encounter site. This highly regulated trafficking is responsible for the integration and control of the local immune response.

Lung lymphocytes arise by migration into the affected site and the secondary lymphoid tissues (peripheral, mesenteric nodes) by extravasation through postcapillary high endothelial venules (HEV). Tissue-specific homing is regulated by receptor interactions with endothelial ligands. The cell class that will predominate is determined by the emigration stimulus; specificity is controlled by endothelial wall changes and local chemokines. Although many cytokines and growth factors are implicated in the regulation of lymphocyte trafficking, chemokines are critical because of their specificity for distinct cell subsets. Postextravasation, B cells are regulated by local antigen-presenting cells (APC) and T cells. Their differentiation into memory or effector cells is then driven by the antigen presentation that occurs in lung-associated lymph nodes.

To initiate a response in the respiratory tract, antigen is presented to naive T cells by dendritic cells (DC) that act as the APC; an intramucosal DC turnover less than 72 h allows for rapid sampling and reaction to antigen. Antigen processing begins with its uptake, followed by its degradation, binding of a fragment to major histocompatability complex (MHC) molecules (one of two forms—Class I or II), and delivery of the complex to the APC or DC surface. Endogenous antigens processed by APC are usually presented to $CD3^+CD8^+$ T cells and exogenous ones to $CD3^+CD4^+$ cells. Antigen solubility influences the manner in which DC are involved in the response. Inhaled solubles are trapped by airway DC, transported to lymph nodes, and presented to naive T cells; response to inhaled particulates is generated only when the lung's protective phagocytic capacity is exceeded.

9.3.2 Alveolar Macrophages

Alveolar macrophages (AM), mobile mononuclear phagocytes that reside in the lung parenchyma, play crucial roles in host defense and the removal of inhaled particles from the alveolar lumen. Their innate activities of phagocytosis and digestion are nonspecific, but these cells become part of the specific immune response when they process ingested material for presentation to lymphocytes. These professional phagocytes are submerged below the surfactant layer in the alveolar lining fluid hypophase and move across epithelial surfaces via pseudopodia. AM keep alveolar surfaces sterile by engulfing foreign materials that land on the surfactant lining. Removal of phagocytized material from the alveolar space is aided by AM migrating up the bronchial tree (along the mucous lining) to the oropharynx, where they are swallowed. Particle-laden macrophages may also reenter interstitial spaces at the entrance to the acini around conducting airways from where they can enter lymphatic vessels and be transported to regional lymph nodes.

Lung macrophage numbers vary, depending on the species and environmental and health conditions. Numbers may increase after exposure to pollutants or infectious agents and decrease in response to some inhaled agents. Numbers also increase through an influx of peripheral monocytes that differentiate into phagocytes. Besides alveolar air spaces, these cells are also present in conducting airways; however, 99% of the total air space macrophages are found in alveoli. Interstitial macrophages reside in the connective tissue around blood vessels and the conducting airways. AM secrete more than 100 substances (e.g., proteases, nitric oxide, cytokines, growth factors, etc.) important in mediating inflammatory responses in the lung. Upon stimulation by microorganisms or their related agents (i.e., endotoxins) or particulates, AM secrete early-response interleukin-1 (IL-1) and tumor necrosis factor-α (TNFα), cytokines that act in autocrine/paracrine fashions to stimulate distal mediator production by nonimmune cells (e.g., interstitial fibroblasts and airway epithelial cells). Compared to blood monocytes and macrophages in other tissues, AM function poorly as accessory cells for fostering immune B or T cell responses.

9.3.3 Mononuclear DC

Because AM are not equipped to perform as APC, professional antigen-presenting DC are abundant in the airways and have high antigen-presenting capacity for immune T cells and naive cells as they traverse the lungs. Lung DC arise from monocytes that differentiate under local microenvironmental influences (i.e., presence of granulocyte-macrophage colony stimulating factor [GM-CSF]). Generally, DC are mainly located in the conducting airways with a lesser number in alveolar parenchyma and visceral pleura. DC are normally present in the epithelial and subepithelial regions of the airways, as well as in BALT. The airway population of DC increases in response to inhaled antigen and decreases after glucocorticoid treatment (possibly through decreased emigration of the cells to lungs or due to apoptosis). In contrast, antigen-stimulated increases in DC populations are thought to be due to local cell proliferation, increased emigration to the airway site from blood and bone marrow, and enhanced residence time in the airway tissue.

9.3.4 Neutrophils (PMN), Eosinophils, Basophils, and Mast Cells

A large marginated pool of PMN resides in the pulmonary capillary bed. PMN, similar to AM, are avid phagocytes but are short lived and less versatile. Normally, these leukocytes remain in lung microvessels, with only a few entering the interstitium or alveolar airspaces. Following lung injury or infection, PMN quickly adhere to endothelial cells and migrate into interstitial or air spaces to phagocytize and kill microorganisms or help repair tissue damage. Phagocytosis stimulates increased oxygen consumption in PMN that results in the generation of oxygen-based radicals important in microbial killing—though it may also injure surrounding lung tissue. PMN also have secretory granules containing numerous enzymes and other proteins designed to kill engulfed organisms.

Basophil and eosinophil polymorphonuclear leukocytes, infrequent in normal lungs, invade the respiratory tract in response to various infections and allergic reactions and may be involved in sensitizing responses to inhaled toxins. Tissue recruitment of eosinophils is primarily controlled by T helper (T_H) lymphocytes and enhanced by cytokines (e.g., IL-5); infiltration is a feature of airway inflammation during bronchial asthma. Although eosinophils can phagocytize, their primary function is to produce basic granule proteins (e.g., major basic protein) and oxidants. Mononuclear mast cells are found in or beneath airway epithelia, around blood vessels in airway submucosa, associated with smooth muscle bundles, and in the interalveolar septa of normal lung. Mast cells have important roles in allergic airway reactions. One specific immunoglobulin isotype (IgE) binds to mast cell receptors via its F_c portion and initiates a series of membrane events that result in discharge of mediators (e.g., histamine) associated with allergic reactions.

9.4 IMMUNOTOXIC EFFECTS OF SPECIFIC CLASSES OF INHALED AGENTS

9.4.1 Biologic Agents

Microbes can be classified as pathogens, opportunists, or nonpathogens. Pathogens (i.e., Mycobacteria, rhinoviruses) circumvent normal pulmonary defenses, multiply, and cause disease. Opportunists (i.e., *Pseudomonas*) cause no disease in immunocompetent hosts but do so in those with impaired immune defenses or if present at levels that overwhelm defenses. Nonpathogens (the majority of microbes) are unable to cause disease via replication or infection. Exposure to microbes is unavoidable due to their common occurrence in dirt and airborne particles. Use of biologic pesticides and recombinant organisms during bioremediation/biodegradation increases the chances of microbe dispersion. The use of bioreactors, contaminated cutting oils, humidifiers, and ice nucleation sprays, as well as air conditioning, composting, and recycling, are also known to increase the risk of host exposures.

Biopesticides are microbes that have adverse effects on bacteria, fungi, insects, or plants (Table 9.1). After inhalation, some cause primary immunologic responses (immediate inflammatory reactions without extensive cellular processing and antigen recognition) in the lungs. Intratracheal instillation of *Bacillus thuringiensis* var. *kurstaki* (Btk) spores in mice caused mortality in less than 24 h, though not from infection (Btk spores cannot replicate in mammals). That the lung weights in these hosts were increased indicated that extensive inflammation occurred. Similar effects were seen after intranasal (IN) instillation of *P. aeruginosa*. Here, associated lipopolysaccharide (LPS) caused overwhelming inflammation that led to rapid mortality. Apart from interspecies differences in effect, intraspecies effects may also vary with strain. Mice given vegetative or spore preparations of Btk, a crystal-minus mutant, *B. cereus*, or *B. subtilis* showed differing responses. Vegetative and spore forms of *B. cereus* and the mutant Btk killed mice; heat inactivated *B. cereus* but not mutant Btk. Similarly, vegetative Btk, but not the spore form, killed mice, whereas neither form of *B. subtilis* had an effect. This suggests that specific properties of microbes cause adverse effects and that these are not possessed by all members of a genus or species or even by all strains in a species.

Opportunists vary in their ability to colonize or cause disease. Colonizing capacity is not common to all species in a genus and also varies among strains.

Table 9.1 Some Microbes Used in Environmental Applications

Microbe	*Use*	*Reference*
Bacillus cereus	Fungicide	Silo Suh et al., 1994
Bacillus thuringiensis	Insecticide	Damgaard et al., 1996
Beauveria bassiana	Insecticide	Wraight et al., 1998
Burkholderia cepacia	Bioremediation, Fungicide	Govan et al., 1996
Comamonas testosteroni	Bioremediation	Bae et al., 1996
Entomophaga grylli	Insecticide	Bidochka et al., 1996
Metarrhizium anisopliae	Insecticide	de Garcia et al., 1997
Paecilomyces sp.	Insecticide	Wraight et al., 1998
Pseudomonas antimicrobica	Fungicide	Walker et al., 1996
Pseudomonas putida	Bioremediation	Ronchel et al., 1995;
Rhizobia	Bioremediation	Damaj and Ahmad, 1996
Trichoderma harzianum	Fungicide	Grondona et al., 1997

Clearance itself is dependent on both the host and the microbe. Mice challenged with *P. aeruginosa* and *B. cepacia* differed in clearance rates. Studies with *Streptococcus sanguis, S. salivarius,* and *Neisseria catarrhalis* show that *S. sanguis* is cleared more easily than *S. salivarius*, and *N. catarrhalis* takes the longest time; AM were responsible for clearing *S. sanguis,* whereas *S. salivarius* and *N. catarrhalis* needed PMN. *Staphylococcus aureus* causes little inflammation at low doses and is cleared by AM; higher doses overwhelm AM and cause PMN infiltration. *B. thuringiensis* strains act as particulates and are cleared biphasically; although most nonpathogens are killed in less than 7 d, these stay viable in the lung for more than 30 d. *Streptococcus zooepidemicus* is rapidly cleared initially, but small numbers escape and create foci of infection. Studies have also shown that a response can change if the infecting dose is varied. Low numbers of *S. aureus* are rapidly cleared without PMN influx, whereas high numbers take longer to clear and require the presence of PMN.

Although most microbes are nonpathogenic, exposure to them is not always innocuous. Daily low-level ($\approx 10^3$ bacteria) challenge with nonpathogens was found to enhance susceptibility to concurrent infection with *S. zooepidemicus*; however, daily challenge with higher levels ($\approx 10^6$ bacteria) was protective even against higher infecting doses. Similar effects occurred with *Hartmanella vermiformis,* enhancing the disease course of concurrent *Legionella pneumophila.* Daily low-level exposure to nonpathogens caused no changes in the numbers or types of lung

Table 9.2 Microorganisms Implicated in Causation of Pulmonary Type I Hypersensitivity Disease

Microbe	*Disease*	*Reference*
Aspergillus fumigatus	Asthma (Allergic bronchopulmonary aspergillosis)	Zhaoming and Lockey, 1996
Fusarium vasinfectum	Asthma	Saini et al., 1998
Gram-negative bacteria	Organic dust toxic syndrome	Zhiping et al., 1996
Gram-negative bacteria	Asthma	Michel et al., 1992
Molds (Cladosporium, Alternaria, Aspergillus)	Asthma	Cross, 1997

immune cells; however, exposure to higher levels caused significant increases in PMN. This effect does not appear to be consistent for all infectious microbes or even for all routes of exposure. The data suggest that repetitive challenge with levels of microbes that fail to elicit an immune response may disturb normal functioning of the local immune system and render it more susceptible to infectious disease.

Although an infection in itself can be deleterious to the lungs, many microbes also cause pulmonary disease through induction of adverse immunologic reactions (e.g., Types I, III, and IV hypersensitivity). Type I (or immediate) hypersensitivity manifests as allergic rhinitis, asthma, or urticaria based on response location and severity and can be induced by several microbes (Table 9.2). Type III hypersensitivity, characterized by antibody–antigen complex formation, complex deposition, complex-bound complement causing vascular permeability, edema, PMN influx, and ultimately, release of tissue damaging cellular enzymes, is induced by several bacteria, actinomycetes, and fungi (Table 9.3). Type IV hypersensitivity (delayed-type hypersensitivity [DTH] or cellular hypersensitivity) caused by microbes associated with the mycobacteria spp. (Table 9.4) is characterized by activation of T cells that recognize specific proteins or protein–hapten conjugates. T cell factors are then released that induce intense cellular infiltrates consisting of activated mononuclear cells whose by-products cause swelling, pain, and tissue destruction.

9.4.2 Ozone

Ozone (O_3) has varied effects on humoral immunity, depending on its concentration, when the exposure occurs relative to antigen stimulation,

Table 9.3 Microorganisms Implicated in Causation of Pulmonary Type III Hypersensitivity Disease

Microbe	*Disease*	*Reference*
Aspergillus fumigatus	Hypersensitivity pneumonitis	Hinojosa et al., 1996
Aspergillus niger	Machine operator's lung	Bernstein et al., 1995
Bacillus pumilus	Machine operator's lung	Bernstein et al., 1995
Epicoccum nigrum	Hypersensitivity pneumonitis	Hogan et al., 1996
Micropolyspora faeni	Hypersensitivity pneumonitis	Hinojosa et al., 1996
Penicillium brevicompactum	Farmer's lung	Nakagawa-Yoshida et al., 1997
Penicillium olivicolor		
Pseudomonas fluorescens	Machine operator's lung	Bernstein et al., 1995
Rhodococcus sp		
Saccharopolyspora rectivirgula	Farmer's lung	Mundt et al., 1996 Reijula, 1993
Staphylococcus capitis	Machine operator's lung	Bernstein et al., 1995
Thermoactinomyces vulgaris	Farmer's lung	Reijula, 1993
Thermoactinomyces vulgaris	Hypersensitivity pneumonitis	Hinojosa et al., 1996

the antigen nature, whether the antigen causes a T-dependent response, and whether a T helper type 1 (Th1) or type 2 (Th2) response is initiated (Table 9.5). Ozone suppresses nonallergic Th1-cell-dependent antibody production in response to antigenic stimulation. B cell IgG formation

Table 9.4 Microorganisms Implicated in Causation of Pulmonary Type IV Hypersensitivity

Microbe	*Disease*	*Reference*
Mycobacterium avium complex	Tuberculosis	Martinez Moragon et al., 1996
Mycobacterium chelonei	Tuberculosis	Martinez Moragon et al., 1996
Mycobacterium flavescens	Tuberculosis	Martinez Moragon et al., 1996
Mycobacterium kansasii	Tuberculosis	Martinez Moragon et al., 1996
Mycobacterium tuberculosis	Tuberculosis	Dannenberg, 1989

Table 9.5 Immunomodulating Effects of Ozone

Immune Parameter Analyzed — Microbiologic	*Regimen*	*Effects*	*Reference*
	Mouse		
In situ bactericidal activity vs. *Staphylococcus aureus*	0.40–4.25 ppm, 17 h	↓	Goldstein et al., 1971
Mortality due to *Streptococcus* sp.	0.1–1.0 ppm, 3 h/d, 1 or 2 d	↑	Gardner and Graham, 1977
Mycobacterium tuberculosis titers in lungs	1.0 ppm, 3 h/d, 5 d/week, 8 week	↑	Thomas et al., 1981
Mortality due to Group C *Streptococcus* sp.	0.1 ppm, 5 d/week, 103 d	↑	Aranyi et al., 1983
In situ bactericidal activity vs. Klebsiella pneumoniae		↑	
Mortality due to *Streptococcus zooepidemicus*	0.1, 0.3, 0.5 ppm, 24 h/d, 5 d/week, 3 week	↑	Graham et al., 1987
Viral titers of influenza A/PR8/34	0.5 ppm, 24 h/d, up to 15 d	N.E.	Jakab and Hmieleski, 1988
Lung lesions from Influenza A/PR8/34		↓	
Mortality, viral titers, lung lesions from influenza A/Hong Kong/68 (H_3N_2)	0.25, 0.5, 1.0 ppm, 3 h/d, 5 d	↑	Selgrade et al., 1988
In situ bactericidal activity vs. *Proteus mirabilis*	0.5 ppm, 24 h/d, 1, 3, 7, 14 d	↑	Gilmour et al., 1991
In situ bactericidal activity vs. *S. aureus*		↑	
Mortality due to *S. zooepidemicus*	0.4, 0.8 ppm, 3 h	↑	Gilmour and Selgrade, 1993
	Rat		
Severity of *Pseudomonas aeruginos* infection	0.64 ppm, 23 h/d, 4 weeks	N.E.	Sherwood et al., 1986
Pulmonary clearance of *S. zooepidemicus*	0.8 ppm, 3 h	↑/↓	Dong et al., 1988

Mortality due to *Listeria monocytogenes*	0.12–1.0 ppm, 24 h/d, 7 d	↑	van Loveren et al., 1988
Pulmonary clearance of *Listeria monocytogenes*		↓	
Mortality due to *S. zooepidemicus*	0.4, 0.8 ppm, 3 h	N.E.	Gilmour & Selgrade, 1993
Mortality due to *Listeria monocytogenes*	0.1 or 0.3 ppm, 4 h/d,	N.E.	Cohen et al., 2001,2002
Pulmonary clearance of *Listeria monocytogenes*	5 d/week, 1 or 3 weeks	N.E./↓	
Immune Parameter Analyzed — Immune Cells	*Regimen*	*Effects*	*Reference*
	Mouse		
AM IFN production	24 h/d, 14 or 21 d	N.E.	Ibrahim et al., 1976
AM superoxide anion (O_2^-) production	0.03–1.2 ppm, 3 h	↓	Ryer-Powder et al., 1988
AM phagocytic activity		N.E.	
AM F_cR-mediated phagocytosis	0.5 ppm, 24 h/d, to 14 d	↓	Canning et al., 1991
Levels of AM and PMN in lungs	0.4 ppm, 12 h/d, 1, 3, 7 d	↑	Oosting et al., 1991
AM F_cR-mediated phagocytosis and O_2^- production		↓	
Levels of AM and PMN in lungs	0.4 ppm, 3, 6, or 12 h	N.E	
AM F_cR-mediated phagocytosis, O_2^- production		↓, N.E	
AM phagocytic activity	0.4 or 0.8 ppm, 3 h	↓	Gilmour and Selgrade, 1993
Lung delayed-type hypersensitivity reactions	0.4, 0.8, or 1.6 ppm, 12 h	↓	Garssen et al., 1997
Lung MIP-2 and MCP-1 mRNA expression	0.6 or 2.0 ppm, 3 h	↑	Zhao et al., 1998
Lung NF-κB activation		↑	
Lung MIP-2, MCP-1, IL-6, and eotaxin mRNA	0.4 ppm, 24 or 96 h	↑	Johnston et al., 1999
IL-1α, IL-1β, IL-1R, IL-10, IL-12, IFNγ mRNA		N.E.	

(continued)

Table 9.5 Immunomodulating Effects of Ozone (Continued)

Immune Parameter Analyzed— Immune Cells	*Regimen*	*Effects*	*Reference*
MIP-2, MCP-1, IL-6, and eotaxin mRNA	1.0 ppm, 1 or 2 h	N.E.	
IL-1α, IL-1β, IL-1R, IL-10, IL-12, IFNγ mRNA		N.E.	
MIP-2, MCP-1, IL-6, and eotaxin mRNA	1.0 ppm, 4 h	↑	
IL-1α, IL-1β, IL-1R, IL-10, IL-12, IFNγ mRNA		N.E.	
MIP-2, MCP-1, IL-6, and eotaxin mRNA	2.5 ppm, 2, 4 or 24 h	↑	
IL-1α, IL-1β, IL-1R, IL-10, IL-12, IFNγ mRNA		N.E.	
Lung MIP-2, MIP-1α, IL-6, iNOS, and eotaxin mRNA expression	1.0 or 2.5 ppm, 4 or 24 h	↑	Johnston et al., 2000
AM LPS + IFNγ-induced NOS expression, NO production, PGE_2 release – wild-type host	0.8 ppm, 3 h	↑	Fakhrzadeh et al., 2002
– knockout ($NOS^{-/-}$) host		N.E.	
Lung MIP-2 and MCP-1 RNA expression	1.0 ppm, 24 h	↑	Johnston et al., 2002
Lung IL-1α, IL-1β, IL-1Ra, IL-6, MIF, MIP-1α, MIP-1 β, RANTES mRNA expression		N.E.	
Lung MIP-2 in knockout ($NOS^{-/-}$) hosts	1.0 ppm, 8 h/d, 3 d	↑	Kenyon et al., 2002
Lung MIP-2 in wild-type hosts		↑	
AM spontaneous and LPS + IFNγ-induced NOS expression and NO production	0.8 ppm, 3 h	↑	Laskin et al., 2002
AM NF-κB activation, STAT-1 activity and expression, protein kinase B, and phosphoinositide-3-kinase activity		↑	

	Rabbit		
AM ability to produce IFN	1.0, 3.0, or 5.0 ppm, 3 h	↓	Shingu et al., 1980
Levels of AM and PMN in lungs, AM size	0.1 ppm, 2 h/d, 1 or 13 d	↑	Driscoll et al., 1987
AM phagocytic activity		↓	
AM release of PGE_2, PGF_2	0.1 or 1.2 ppm, 2 h	↑	Driscoll et al., 1988
AM release of LTB_4, LTC_4		N.E.	
AM O_2^- production	0.1, 0.3, or 0.6 ppm, 3 h	N.E.	Schlesinger et al., 1992
AM phagocytic activity		↓	
AM *ex vivo* LPS-induced TNFα production		N.E.	
	Rat		
AM phagocytic capacity and index	0.8 ppm, 24 h/d, 3, 7, 20 d	↑	Christman et al., 1982
AM rosetting capacity, F_cR expression/activity	0.8 ppm, 4 h	↓	Prasad et al., 1988
Lung AM levels, AM size, and morphologic changes	0.12, 0.8, 1.5 ppm, 6 h	N.E./↑	Hotchkiss et al., 1989
Levels of AM in lungs, AM size	0.2 ppm, 24 h/d, 14 d	↑	Mochitate and Miura, 1989
AM glucose-6-phosphate dehydrogenase, lactate dehydrogenase, glutathione peroxidase, and hexokinase activity		↑	
AM size, morphological changes	0.12–0.75 ppm, 24 h/d, 7 d	↑	Dormans et al., 1990
NK cell activity	0.2–0.8 ppm, 24 h/d, 7 d	↑/↓	van Loveren et al., 1990
Levels of AM in lungs	0.4 ppm, 12 h/d, 1, 3, or 7 d	↑	Oosting et al., 1991
AM F_cR-mediated phagocytosis		↑	

(*continued*)

Table 9.5 Immunomodulating Effects of Ozone (Continued)

Immune Parameter Analyzed— Immune Cells	*Regimen*	*Effects*	*Reference*
AM O_2^- production		↓	
Levels of AM in lungs	0.4 ppm, 3, 6, or 12 h	N.E.	
AM F_cR-mediated phagocytosis/O_2^- production		↓	
Type-II cell IL-1β± IFNγ- and TNFα+ IFNγ-induced NOS mRNA levels & NO production	2.0 ppm, 3 h	↑	Punjabi et al., 1994
Type II cell IL-1α ± IFNγ-, IL-6 ± IFNγ-, and TNFα-induced NOS mRNA levels and NO production		N.E.	
AM adherence to epithelial cultures	0.8 ppm, 3 h	↑	Bhalla, 1996
AM CD11b marker and ICAM-1 expression		↓ and N.E.	
AM motility in response to chemotaxins		↑	
AM spontaneous, LPS, and IFNγ-induced NO production	2.0 ppm, 3 h	↑	Pendino et al., 1996
AM spontaneous and LPS-induced NOS expression		↑	
AM MIP-1α, CINC, TNFα, IL-1β mRNA	1.0 ppm, 6 h	↑	Ishii et al., 1997
AM adherence to epithelial cultures	0.8 ppm, 3 h	↑	Pearson and Bhalla, 1997
AM H_2O_2 and O_2^- production	0.3 ppm, 5 h/d, 5d/week, 4 weeks	↑/↓	Cohen et al., 1998
AM *ex vivo* LPS-induced IL-6, IL-1α, or TNFα production		N.E.	
AM *ex vivo* LPS- or IFNγ induced NO production		N.E.	
AM phagocytic activity	0.8 ppm, 3 h	↓	Dong et al., 1998
AM spontaneous, LPS-, IFNγ-, and LPS + IFN-induced NO production	2.0 ppm, 3 h	↑	Laskin et al., 1998

AM spontaneous, LPS, LPS + IFNγ-induced NOS expression		↑	
AM NF-κB activation		↑	
AM expression of CD18 integrin	1.0 ppm, 2 h	↓	Hoffer et al., 1999
PMN CD11b marker and CD62L selectin expression		↓ and N.E.	
AM O_2^- production	0.1 or 0.3 ppm, 4 h/d,	↑	Cohen et al., 2001, 2002
AM H_2O_2 production	5 d/week, 1 or 3 weeks	↓	
Lung lymphocyte CD3 expression/CD25 expression on $CD3^+$ cells		↓	
IL-2-induced proliferation of lung lymphocytes		N.E.	
BAL cell expression of antigen presentation markers MHC II, B7.1, B7.2, and CD11b/c	1.0 ppm, 24 h/d, 3 d	↑	Koike et al., 2001

in response to pokeweed mitogen (PWM; T-dependent antigen) decreased after *in vitro* exposure to O_3, and IgG production increased in cells provided *S. aureus* Cowan I (T-independent antigen). IgG production decreased after either of these stimulants were added to O_3-exposed T cells (which were then mixed with control B cells). If T and B cells were both exposed, only PWM decreased IgG formation. Thus, T-dependent responses seem to be more sensitive to O_3 than T-independent ones.

Ozone potentiates lung IgE production and anaphylaxis following antigen inhalation. Besides increasing sensitization to antigens, O_3 enhances susceptibility to their challenge. In guinea pigs sensitized to ovalbumin (OA) and exposed to O_3, challenge with OA at a dose sufficient to induce slight anaphylaxis induced severe dyspneic attacks in many. Epidemiologic studies also report findings of altered sensitization. Emergency room visits by asthmatic children were higher on days when O_3 levels were high, and individuals who lived close to highways (with associated high O_3 levels) suffered more frequent and severe allergic reactions to pollen than controls who lived far away. Ozone affects antibody production even in the absence of antigenic stimulation. In mice exposed to O_3, BALT numbers of IgA (but not IgG or IgM)-containing cells were increased. Increases in bronchoalveslar lavage (BAL) IgA, IgG_1, and IgG_2 were also seen, but these ceased to be elevated when the mice were placed back in air.

Whether O_3 affects T-cell-dependent cellular immunity depends on when the exposure occurs relative to infection initiation. In mice infected with *M. tuberculosis* after exposure to O_3, resistance was unaltered. If *M. tuberculosis* was given before exposure, resistance was reduced. Cellular immunity suppression was also dramatic if exposure occurred during a *Listeria monocytogenes* infection. In rats exposed before or during infection, mortality, lung bacterial levels, and lung lesion severity were all enhanced, and the ability of their AM to phagocytize and kill Listeria, Listeria antigen-induced lymphoproliferation, and T/B cell ratios were reduced. These findings, along with data showing that PHA-stimulated lymphocytes from humans exposed to O_3 had depressed mitogenic responses, lend support to the premise that T cells are sensitive targets for O_3 immunotoxicity.

Immunosuppressive effects of O_3 during infection are most evident where host recovery depends on phagocyte ability to engulf and kill pathogens. Receptor-mediated uptake of opsonized *Candida albicans* was suppressed in exposed humans; no decreased F_c or nonspecific receptor-mediated phagocytosis was noted 1 d later. In human AM exposed *in vitro*, phagocytosis of opsonized *Cryptococcus neoformans* was unaffected and membrane opsonin receptor levels were unchanged, but uptake of antibody-coated RBC decreased. Regarding bactericidal functions, studies with rabbit AM showed that *in vivo* or *in vitro* O_3 exposure reduced β-glucuronidase (BG), acid phosphatase, and lysozyme activities. Superoxide anion (O_2^-) formation was inhibited in AM of mice exposed to as little as 0.11 ppm O_3. In a model for respiratory syncytial virus (RSV) infection in human AM, O_3 neither altered susceptibility to infection nor affected AM production of IL-1, IL-6, or TNFα. However, at low multiplicities of infection, O_3 reduced IL-1 and IL-6 production.

Ozone also affects PMN; PMN from O_3-exposed rats adhered to more epithelial cells and had increased motility, spontaneous actin filament redistribution, and surface modifications. The role of these effects on PMN function is unclear. PMN still enter airways after exposures, but their phagocytic and antimicrobial functions are impaired. Exposure of humans to O_3 suppressed PMN capacity to phagocytize and kill *Staphylococcus epidermidis ex vivo* 3 d postexposure; this function returned to normal only after 2 weeks. In mice, PMN were observed in lungs for more than 2 d postinfection, but the infection course was not altered; in rats, microbial inactivation was impaired for the first 48 h, and the disappearance of bacteria corresponded with PMN influx.

Ozone induces immunotoxicity via direct effects on immune or epithelial cells and also by inducing the cells to release cytokines/inflammatory mediators. As an oxidant, O_3 reacts with molecules bearing thiols, amines, or unsaturated double bonds. The resulting alterations of enzymes, glutathione cycling, and lipids lead to cell injury by creating and supporting the presence of free radicals and toxic intermediates (e.g., H_2O_2 and arachidonic acid [AA] metabolites). Direct effects of O_3 upon immune cells have also been suggested by assays that rely on membrane integrity (i.e., cell surface interactions); exposure resulted in alterations in T and B cell resetting, AM adherence to nylon, and PMN adherence to epithelia. Exposure to O_3 also caused tracheal epithelial cells to release PGE_2, PGF_2, 6-keto-PGF_1, and LTB_4; exposed BEAS-S6 bronchial epithelial cells released TBX_2, PGE_2, LTC_4, LTD_4, and LTE_4. Increased levels of these cyclooxygenase and lipoxygenase products indicated augmented eicosanoid metabolism after O_3 exposure. Prostanoids likely have a role in the inhibition of AM phagocytosis after O_3 exposure as indomethacin mitigates the suppression.

An increase in inflammatory cytokines and mediators is also evident in response to O_3. In exposed humans, IL-6 appeared in BAL within 1 h; GM-CSF, IL-8, and fibronectin were also induced. In smokers and non-smokers exposed to O_3, BAL IL-6 and IL-8 levels increased immediately. Inflammatory parameters, including the levels of cytokines released, were also increased in asthmatics as compared to normals. As an aside, these studies show that O_3-induced airway inflammation and cytokine release is independent of smoking status or airway responsiveness. Epithelial cells govern inflammatory responses to O_3 by releasing mediators to control inflammatory cell interactions and shepherd T cells toward Th2 responses. Although epithelia show increased release of macrophage inflammatory protein-2 (MIP-2), RANTES, PGE_2, IL-8, IL-6, and nitric oxide (NO) after exposure, IL-2 and IFN production is depressed. *In vitro* exposures of Type II epithelial-like cells induced DNA-binding activity of transcription factors (NF-κB, NF-IL-6, and AP-1) associated with increased IL-8 gene expression and protein release. This suggests that possible cellular cascades are induced by O_3 that result in inflammatory cell recruitment to the airways and subsequent immunotoxicity.

Although O_3 decreases AM phagocytic and bactericidal abilities, it activates them to produce TNFα, IL-1, IL-6, and IL-8. It was noted with AM from exposed rats that TNFα was released early on (2 to 4 h postexposure), whereas IL-1 formation occurred later (2 to 24 h postexposure); as release of both occurred in conjunction with enhanced hepatic protein synthesis, this suggests an acute phase response. Further *in vitro* research revealed that AM also produced increased amounts of NO and inducible nitric oxide synthase (iNOS) in response to LPS or IFNγ after O_3 exposure. Increased iNOS in AM (as well as in exposed type II pneumocytes) following O_3 inhalation is associated with activation of NF-κB, suggesting that the NF-κB signaling activity may be partly responsible for the increased response of these cells to cytokines after O_3 inhalation.

9.4.3 Sulfur Oxides

Sulfur oxides (SO_x) comprise both gaseous and particulate species. There are four of the former, namely sulfur monoxide, sulfur dioxide (SO_2), sulfur trioxide, and disulfur monoxide. Particulate phase SO_x consist of acidic sulfates, namely sulfuric acid (H_2SO_4) and its products of neutralization with ammonia: letovicite, $(NH_4)_3H(SO_4)_2$; ammonium bisulfate, NH_4HSO_4; and ammonium sulfate, $(NH_4)_2SO_4$.

Much of the database on the effects of SO_x on lung defense is concerned with nonspecific nonimmune mechanisms. Limited evidence on pulmonary humoral- or cell-mediated immunity provides indications of the ability of SO_x to enhance sensitization to antigens or modulate activity of the cells involved in allergic responses. Exposure to SO_2 can facilitate local allergic sensitization. In guinea pigs exposed to 0.1, 4.3, and 16.6 ppm of SO_2 (8 h/d for 5 d) and then to aerosolized OA for each of the last 3 d, measurements of airway obstruction after specific bronchial provocation with inhaled OA yielded positive bronchial reactions compared to controls. Bronchial obstruction was significantly higher in exposed groups for all SO_2 levels, and OA-specific antibodies in serum and lavage fluid were also increased. In guinea pigs exposed to H_2SO_4 at 0.3, 1.0, and 3.2 mg/m^3 for 2 and 4 weeks, antigen-induced histamine release from isolated mast cells was enhanced by exposure to the two highest levels for 2 weeks, but not from the mast cells isolated from the hosts exposed for 4 weeks. These alterations were not accompanied by changes in mast cell numbers in the lungs.

Host resistance studies of effects of the SO_2 on antibacterial and antiviral pathways in rodents exposed to SO_2 alone (100 to 170 μg SO_2/m^3) either 17 h prior to, or 4 h after, infection with aerosols of *S. aureus* or *Streptococci* (Group C) showed no changes in staphylococcal clearance or AM ingestion of the bacteria. Other studies noted that continuous exposure of mice to ~10 ppm SO_2 for up to 3 weeks reduced resistance to *Klebsiella pneumoniae.* Inasmuch as an early study observed no effect with SO_2 on host resistance against Group C *S. pyogenes,* it seems likely that SO_2 may selectively alter lung immune defense mechanisms that are specific for combating Gram-negative microbes. Studies examining the effects of inhaled SO_2 on viral infection have yielded mixed results. One study showed that exposure of mice to SO_2 (as high as 6 ppm) had no effect on influenza virus replication, whereas others have demonstrated either additive or synergistic effects.

Regarding AM, exposure of rats to 1-20 ppm SO_2 for 24 h showed increased *in vitro* phagocytosis for exposures greater than or equal to 5 ppm SO_2. On the other hand, AM from hamsters exposed to 50 ppm SO_2 for 4 h had decreased motility and phagocytic function. Examination of lavage fluid of humans exposed to 8 ppm SO_2 for 20 min showed increased lysozyme-positive AM, indicating cell activation. *In vitro* exposure of human AM to 2.5-12.5 ppm SO_2 for 10 min resulted in dose-dependent increased spontaneous production of reactive oxygen species (ROS); a 30-min exposure of cells to 12.5 ppm SO_2 was cytotoxic.

Effects of H_2SO_4 on pulmonary defense mechanisms have also been examined. AM recovered from rabbits exposed for 3 h to 1 mg/m^3 H_2SO_4 had reduced ability *ex vivo* to phagocytose and kill *S. aureus.* Although the effects did not appear to be due to altered AM F_cR expression, H_2SO_4-induced reductions in O_2^- production may have played a role in the effects on intracellular killing. At levels lower than or equal to 1 mg/m^3, H_2SO_4 affected certain functions of AM from various species following single or repeated inhalations. These included phagocytic activity, adherence, random mobility, intracellular pH, and the release or production of certain cytokines (e.g., TNFα and IL-1) and ROS.

Modulation of pulmonary pharmacological receptors may underlie some SO_2-induced responses. β-Adrenergic stimulation that downregulates AM function was influenced by short-term, repeated exposures to H_2SO_4. Any enhanced downregulation, by affecting ROS production, may create an environment conducive to secondary pulmonary insult, such as bacterial infections.

9.4.4 Nitrogen Oxides

Nitrogen oxides (NO_x) include nitrogen dioxide (NO_2), nitrous oxide (N_2O), nitric oxide (NO), nitrogen trioxide (NO_3), dinitrogen trioxide (N_2O_3), dinitrogen tetroxide (N_2O_4), and dinitrogen pentoxide (N_2O_5). Except for N_2O, the various NO_x are interconvertible and many coexist in the atmosphere.

NO_2 impairs resistance to infectious agents in animals exposed to levels as low as 0.5 ppm NO_2 for 3 months. Mortality due to infection is usually greater if a spike regimen is utilized as compared to continual baseline exposures. In general, effects on bacterial infectivity increase with the exposure concentration and duration, though the former has more influence. Differences between intermittent and continuous exposure also seem to disappear as the number of exposures increases. Studies suggest that as the concentration increases, a shorter exposure time is needed for intermittent and continuous exposures to induce similar effects. Increased mortality due to infection is also proportional to the exposure duration if bacteria are given immediately after exposure, but may not be so when the challenge is much later, suggesting that a critical time interval between NO_2 exposure and infection is needed. Mechanism(s) responsible for the increased microbial infectivity are unclear. However, because NO_2 levels that alter resistance do not generally affect particle clearance processes, these responses may be due to impaired intracellular killing of microbes via AM dysfunction. NO_2 also affects respiratory tract susceptibility to viral infection. Exposure to 5 ppm NO_2 (6 h/d for 2 d before exposure to virus or 4 d after) resulted in MCMV (murine cytomegalovirus) proliferation and bronchopneumonia in mice inoculated with a dose too low to produce replication or histologic abnormalities in the normal course. In addition, the amount of virus required to infect NO_2-exposed mice was 100-fold lower than that needed in controls.

Regarding the effects of NO_2 on humoral and cellular immunity, suppressed responses have clearly been shown to follow exposure to levels higher than 5-ppm NO_2; there are few reports of responses to lower concentrations. Interestingly, enhanced immune function may also follow exposure, and this can be just as detrimental. The direction of change in responses appears to be NO_2 concentration dependent. For example, humoral responses in monkeys chronically exposed to NO_2 were enhanced at 1 ppm and suppressed at 5 ppm. The AM obtained by lavage 3.5 h after exposure to continuous NO_2 tended to inactivate influenza virus *in itro* less effectively than cells collected after air exposure.

AM have been found to suffer functional deficits in response to NO_2. In rats exposed to NO_2 for 0.5 to 10 d, there were complex effects on AM arachidonate metabolism. AM production of TBX_2, LTB_4, and 5-HETE in response to A23187 ionophore was acutely depressed in less than 1 d; their generation recovered to control levels with longer exposures.

In contrast, AM LTB_4 production in response to activated serum was not depressed until after 5 d of exposure (although unstimulated synthesis decreased in less than 1 d). BAL levels of TBX_2, PGE_2 PGF_2, and LTB_4 were depressed within 4 h of exposure, suggesting acute effects on arachidonate metabolism *in situ*. AM obtained from rats exposed for 4 h to NO_2 were more phagocytic to opsonized Sheep red blood cells (SRBC), exhibited increased cytotoxicity to syngeneic adenocarcinoma cells, and were more sensitive to activation by LPS, muramyl dipeptide, and macrophage-activating factor compared to controls. However, repeated exposures over 7- or 14-d periods resulted in AM activity similar to that for controls. *In vitro*, with rat AM exposed for various periods to 10 to 40 ppm NO_2, a dose-related time-dependent enhanced cytotoxic response was observed.

NO_2 had no effect on human lymphocyte subtypes. Among nonsmokers exposed to NO_2 for 2 h/d on four separate days, there was no change in circulating lymphocytes. In their BAL, total lymphocyte levels and the percentages of T cells, B cells, $CD3^+CD8^+$ and $CD3^+CD4^+$ cells, and large granular lymphocytes were not altered by exposure; only a slightly greater proportion of NK cells was seen. At a high level (i.e., 4 ppm), repeated exposures (20 min/d, every second day, six exposures) caused decreases in the numbers of lavaged AM, B cells, and NK cells, and altered T_H/T_S cell ratios; lymphocyte numbers in peripheral blood were also reduced. In mice exposed to 10 ppm NO for 2 h/d up to 30 weeks, leukocytosis was evident by 5 weeks. The ability of spleen cells to mount graft vs. host reaction was stimulated by 20 weeks, but suppressed by 26 weeks. When the ability to reject virus-induced tumors was assessed, fewer exposed animals survived tumor challenge compared to controls; this suggests that NO - at high levels - may affect immunocompetence. In mice exposed continuously to 2 ppm NO ($\approx$4 weeks), there were indications that only female mice had significant increases in mortality from infections.

9.4.5 Wood Smoke, Kerosene Heater Emissions, and Diesel Exhaust

Increased use of wood-burning devices, due to rising energy costs, has caused greater public exposure to pollutants generated by combustion and increased concern about the health effects of wood smoke (WS). Although wood-burning stoves and fireplaces are vented directly to the outside, circumstances (e.g., improper installation, negative indoor air pressure, downdrafts, etc.) may facilitate reentry of the products of incomplete combustion. Health concerns focus on the fact that wood-burning stoves, fireplaces, and furnaces emit significant amounts of toxic agents, including respirable particulates, carbon monoxide (CO), NO_x, SO_x, aldehydes, PAHs, and free radicals. In humans, prolonged inhalation of WS contributes to chronic bronchitis, causes chronic interstitial pneumonitis, fibrosis, and pulmonary arterial hypertension, and alters lung immune defense mechanisms by increasing the levels of pulmonary immunoglobulins (Ig) and immune cell population densities and functions.

Many agents associated with WS alter pulmonary immune defense mechanisms. For example, urban air particles instilled into mice reduced host resistance to bacterial infection. Mice coexposed to carbon black and acrolein had suppressed intrapulmonary killing of *S. aureus*, impaired elimination of *L. monocytogenes* and influenza A virus, and altered intrapulmonary killing of *Proteus mirabilis*. WS-associated agents have also been shown to alter AM activity, i.e., rats exposed nose-only to iron oxide aerosol for 2 h had delayed increases in F_cRI-mediated phagocytosis. Effects on host defense and immune cell function similar to those induced by individual WS components have been seen with intact WS. Single exposure of rabbits to smoke from Douglas fir wood pyrolysis produced increased total AM levels, transitory decreases in AM adherence, altered morphology, and decreased *P. aeruginosa* uptake. AM from WS-exposed humans displayed decreased migration in response to chemoattractants. Studies also showed that short-term exposure of rats to a relevant indoor WS concentration progressively inhibited bacterial clearance even in the absence of histopathological change, lung cell damage, or inflammation; however, similar effects were not seen in rats exposed to particle-free effluents.

Results from *ex vivo* AM functional assays showed that exposure to WS suppressed the production of O_2^-, which is critical for intracellular killing. Inasmuch as AM recovered from WS-exposed rats had an impaired ability to kill *S. aureus in vitro* (compared to control), it appears that short-term WS exposure acts via direct effects on AM function rather than through indirect (i.e., neuroimmune) mechanisms.

Although use in home heating is a primary purpose of kerosene, it is also heavily relied on for lighting and cooking in developing nations. Moreover, many industries in these nations also use kerosene oil (bearing hydrocarbons and trace amounts of *S*- and *N*-derived impurities) as a solvent. Incomplete combustion of kerosene oil used in cooking and illumination generates large volumes of soot, incomplete combustion by-products (i.e., PAHs), and aliphatics. Depending on the operating conditions, heater type, and fuel composition, kerosene heaters can release into small indoor settings unsafe levels of several toxic products (e.g., CO, CO_2, NO_x, SO_x, respirable particles, and hydrocarbons [HC]). Very high indoor pollutant levels also arise from the use of poor-quality fuel, operation in closed rooms, and failure to provide dilution ventilation. Another common source of exposure to kerosene is from the burning of jet fuels (i.e., JP-8) that contain mixtures of kerosene together with aliphatic and aromatic hydrocarbons.

Few studies have investigated the toxic effects of kerosene on the respiratory tract; most have focused on the effects of accidental ingestion, which includes pneumonitis, atelectasis, edema, chronic cough, recurrent bronchitis, and asthma. Unlike the weak epidemiologic data, a number of animal studies of the effects of inhaled kerosene on airway hyperresponsiveness have provided evidence relating exposure to bronchoconstriction. Inhalation of kerosene by guinea pigs 1 h before exposure to acetylcholine potentiated responses of tracheal strips to the agonist and induced decreases in the lethal dose. Inhibition of tracheal acetylcholinesterase activity is thought to explain, in part, the aforementioned symptoms of kerosene intoxication. Other studies investigating possible underlying mechanisms of how kerosene aerosol might induce bronchoconstriction showed that bronchoconstrictive effects in guinea pigs were not modified by administered histamine H_1 antagonist (mepyramine) or steroidal anti-inflammatory (triamcinolone) agents. This suggests that kerosene-induced bronchoconstriction is not mediated by stimulation of histamine H_1 receptors or by release of chemical mediators.

In studies of the effects of kerosene soot on AM, a single instillation of soot was cytotoxic in rats; this effect seemed to be mediated by production of ROS. Oxidative stress was suggested by the increased production of H_2O_2 and thiobarbituric acid (TBA)-reactive substances from AM recovered from soot-exposed rats; decreased antioxidant activity was also noted. The relationship between kerosene, its soot, and asbestos-mediated toxicity was investigated in rats instilled once with chrysotile asbestos, kerosene oil, or kerosene soot, or asbestos with or without oil or soot. The coexposures resulted in increased numbers of AM, elevated levels of H_2O_2 and TBA, and depletion of AM glutathione levels (compared to rats exposed to each agent alone).

Diesel exhaust (DE) is a mixture composed of particulate and gaseous phases. Fine DE particles (DEP) consist of a carbon core with a large surface area to which various HC are adsorbed, including PAHs and nitro-PAHs. The gaseous phase contains a variety of combustion products and HC, including some of the same PAHs found in the particle phase. Respiratory pathologies have been associated with inhaled DE; studies have noted changes in pulmonary function following chronic DE exposure, both restrictive (suggested by decreased lung volumes and compliance) and obstructive (suggested by decreased flow rates and increased airway resistance). Pulmonary inflammation in response to DE is often detected earlier than altered pulmonary function.

A well-studied area of DE toxicity concerns its role in inducing or exacerbating allergic inflammation and airway hyperresponsiveness. Animal studies indicate that DEP (or extracts) enhances IgE production by effects on cytokine or chemokine production as well as on activation of AM and other mucosal cells. Other studies showed that nasal airway resistance and secretions in response to histamine were augmented. Studies also noted that DEP increases the production of OA-specific IgE after repeated instillation in OA-sensitized and challenged mice or after repeated injection with DEP and OA or pollen.

A number of *in vitro* studies also demonstrated the immunomodulating effects of DEP. Human peripheral monocytes and AM exposed to DEP had reduced phagocytosis and increased TNFα release. Exposed human airway epithelial cells had increased releases of GM-CSF. Organic extracts from DEP have been shown to enhance production of IgE from B cells and from transformed B cell lines. In a study of the ability of DEP to influence the number and activity of IgE-secreting cells, peripheral blood mononuclear cells isolated from atopic eczema patients and from healthy nonallergic individuals were exposed to DEP. The number of IgE secretors was increased in some, but not all, DEP-exposed isolates from the eczema patients but in none of the DEP-exposed cells from controls; IgE production by each cell was not modified.

Effects of inhaled DE on aspects of pulmonary immunity other than inflammation and humoral immunity have been noted. Acute and subacute exposure of mice caused increased host mortality in response to *S. pyogenes* infection, but not to challenge with the influenza virus. A second study reported no increase in mortality or other measures of viral infection after 1 month of exposure to DE; after longer exposures (i.e., 3 and 6 months), pulmonary consolidation and virus growth were greater in the DE-exposed animals. Overall, studies of the effects of DE on host susceptibility to respiratory tract infections have been mostly inconclusive.

9.4.6 Tobacco Smoke

The effects of tobacco smoke (TS) on the immune system have become evident from human clinical and epidemiological studies, animal toxicology studies, and *in vitro* models. TS consists of a gaseous and a particulate (organic and inorganic) phase; mainstream, sidestream, and environmental smokes differ chemically and physically, presumably due to chemical transformations that occur as the mix is diluted and aged.

There seems to be substantial, though conflicting, evidence that TS affects Ig levels. Exposed rats had reductions in IgA levels that were persistent, suggesting a selective defect in immune responses initiated at the lung mucosal level. Inhaled TS also caused reduced serum IgG, IgA, and IgM levels, though other reports have indicated somewhat different responses (i.e., effect on IgG only). Smokers also showed biphasic responses in serum IgD: levels increased in light-to-moderate smokers but decreased in heavy ones. Whether the altered Ig levels are biologically significant is questionable, but if these occur along with other immune anomalies, then host resistance to infection may be compromised.

Regarding cellular immunity, PHA-induced lymphoproliferation or tumor-specific T_{CTL} formation is initially increased, but reduced on continued TS exposure. Here, TS containing high levels of tar and nicotine induced changes to a greater degree than TS with low levels. Other studies noted that the relation between smoking and lymphocyte responses to PHA, ConA, or alloantigen (in MLR) was not affected. A T cell functional defect in lung-associated lymphoid tissue, but not in distant lymph nodes, has been reported, suggesting that TS has variable effects in different lymphoid organs. In heavy smokers, total T cells were increased, but the percentage of $CD3^+CD4^+$ subtypes was decreased and that of $CD3^+CD8^+$ increased (resulting in decreased $CD4^+/CD8^+$ ratios). Effects on B cells seem to be more consistent, with several studies reporting increases in total numbers of B cells after TS exposure.

Nonspecific defense mechanisms in the respiratory tract that protect against inhaled airborne chemicals and infectious agents are also affected by TS. TS has been shown to adversely affect mucociliary escalator function in humans, rats, rabbits, and donkeys. Exposure damages tracheal ciliated cells and lung epithelial cells, resulting in a denuded/ulcerated epithelium. TS also stimulates goblet cells to produce excessive mucus, inhibits epithelial ion movement, and initiates changes in mucus composition. Inhaled TS causes hypertrophy of submucosal glands and hyperplasia of mucin-secreting goblet cells in the airway. This, in turn, can alter the rheological properties of airway secretions and result in inhibited mucociliary clearance.

Cells recovered from smokers' lungs differ from those of nonsmokers in the numbers recovered and their composition. TS increases epithelial permeability and causes inflammatory cell migration into the airway lumen. There is evidence that TS activates PMN, producing a shift in favor of proteolysis (elastase release) and enzymatic generation of oxidants. The AM that account for most of the increase in lavage cells have changes in morphology, function, and biochemical properties, including altered abilities to release superoxide, elevated lysosomal enzyme levels, impaired protein/RNA synthesis, and increased migration/chemotactic responses. TS also suppresses the capacity of AM to release select cytokines (e.g., IL-1 and IL-6). AM from exposed rats were unresponsive to LPS *in vitro*, resulting in reduced IL-1β mRNA and protein expression. Lymphocytes from smokers also had reduced responses to PHA or LPS, indicating a lessened ability to release cytokines. In regard to AM phagocytic and bactericidal activity after exposure, some studies noted few differences in the abilities of AM from smokers and nonsmokers to ingest and kill microorganisms; others reported a significant depression in the smokers' AM.

One study noted that AM from smokers were able to phagocytize microbes normally, but showed a deficiency in their ability to kill the engulfed microorganism.

9.4.7 Metals

Toxicities associated with inhaled metals are well known, though the effects from long-term subclinical levels of exposure are less clear. Effects depend on variables such as species, dose, and duration; most studies conclude that inhaled metals are immunosuppressants, with the most consistent finding being a decreased resistance to infection. Metal immunotoxicity occurs via direct effects on specific immune system components or via inhibited immunoregulation. Effects of some environmentally and occupationally important metals (following inhalation or instillation) are reviewed here.

9.4.7.1 Arsenic

Arsenic (As) in the air arises from natural and anthropogenic sources (primarily the burning of coals, oils, or wood, and municipal waste incineration). Airborne As is mostly the trivalent arsenic trioxide (As_2O_3), with smaller contributions from volatile organics (mostly arsine); both trivalent arsenic, As(III), and arsines can undergo oxidation to have pentavalent, As(V). Another major source of inhalable As is TS; As absorbed daily from a pack of cigarettes is ≈6 μg—nonsmokers inhale only ≈0.4 to 0.6 μg As/d. Occupationally, the primary means of exposure is inhalation of As-bearing agents or contaminated dusts used in pesticides, herbicides, fungicides, smelters, mining, glass production, pharmaceuticals, microelectronics, and chemical warfare.

Exposure to As agents is known to produce significant immunotoxicity in the lung (Table 9.6). Instillation of As_2O_3 or gallium arsenide (GaAs) caused lung inflammation (AM and PMN influx) and hyperplasia in rats and hamsters. Hamsters exposed to indium arsenide (InAs) developed local histopathologies, including proteinosis-like lesions, alveolar/bronchiolar hyperplasia, metaplasia, and pneumonia. Studies of effects on humoral immunity are limited. GaAs or sodium arsenite ($NaAsO_2$) suppressed systemic responses to T-dependent SRBC in mice. In workers in a plant burning As-bearing coal, serum IgG, IgA, or IgM levels were no different from those in workers using coal with tenfold less As. Among the workers, blood mononuclear cells had inhibited proliferative responses to PHA. Analyses indicated no change in IL-2 receptor expression, but a diminution in IL-2 secretion; mRNA and intracellular protein assays showed that the inhibition was transcriptional. Electron microscopy revealed that Golgi bodies, mitochondria, and perinuclear membrane ultrastructures were all altered.

Table 9.6 Immunomodulating Effects of Inhaled/Instilled Arsenic Compounds

Compound	*Species*	*Immune Parameter Analyzed*	*Effects*	*Reference*
Arsenic trioxide	Mouse	Mortality due to *Streptococcus pneumoniae*	↑	Aranyi et al., 1985
		Bactericidal activity against inhaled *Klebsiella pneumoniae*	↓	
Arsine	Mouse	Splenic lymphocytes (total), T- and B-lymphocyte percentages	↓	Rosenthal et al., 1989
		NK cell activity (YAC-1 target)	↓	
		Splenic T- and B-lymphocyte proliferation	N.E.	
		Resistance to PYB6 cells, B16F10 melanoma, or influenza virus	N.E.	
		Resistance to *Listeria monocytogenes* and *Plasmodium yoelii*	↓	
		RBC levels, hematocrit, hemoglobin levels	↓	Hong et al., 1989
		White blood cell counts	↑	
		Bone marrow cellularity	N.E.	
Gallium arsenide	Mouse	Spleen cellularity	↑	Sikorski et al., 1989
		PFC formation	↓	
		ConA-/PHA-induced splenic T-lymphocyte proliferation	N.E.	
		LPS-induced splenic B-lymphocyte proliferation	↑	
		Splenic MLR activity	↑	

(continued)

Table 9.6 Immunomodulating Effects of Inhaled/Instilled Arsenic Compounds (Continued)

Compound	*Species*	*Immune Parameter Analyzed*	*Effects*	*Reference*
Gallium arsenide	Mouse	NK cell activity (YAC-1 target)	↑	Sikorski et al., 1989
		Resistance to *P. yoelii* and *S. pneumoniae*	N.E	
		Resistance to *L. monocytogenes*	↑	
		Resistance to viable B16F10 tumor cells	↓	
		Primary IgM antibody response to SRBC	↓	Sikorski et al., 1991
		Macrophage Ia^+ percentage/expression	N.E./ ↓	
		Splenic cell phagocytosis, IL-1 production	N.E.	
		$CD8^+$ and $CD4^+$ cell levels	N.E./ ↓	Burns and Munson, 1993
		LPS-induced B-lymphocyte proliferation	N.E.	
		CD25, LFA-1, and ICAM-1 expression	↓	
		ConA/PHA-induced lymphocyte proliferation	↓	
Sodium arsenate	Rat	BAL PGE_2/TNFα levels	N.E.	Lantz et al., 1994
		PAM induced TNFα, $\cdot O_2^-$, and PGE_2 production	↓/↑/N.E	
Sodium arsenite	Rat	BAL PGE_2/TNFα levels	N.E./N.E.	Lantz et al., 1994
		PAM induced TNFα, $\cdot O_2^-$, and PGE_2 production	↓/N.E./N.E.	
Arsenic-associated fly ash	Human	Serum IgA, IgM, and IgG levels	N.E	Bencko et al., 1988
		Blood α-1-antitrypsin and α-2-macroglobulin levels	N.E	

Immunomodulating Effects of Arsenic Compounds in *In Vitro* Exposures

Compound	*Host Cell*	*Immune Parameter Analyzed*	*Effects*	*Reference*
Arsenic trioxide	Human	PHA-stimulated lymphocyte DNA synthesis	↑/↓	Meng, 1993
Gallium arsenide	Mouse	PFC formation	↓	Burns et al., 1991

Sodium arsenate	Mouse	Vesicular stomatitis virus-induced IFN production	↓	Gainer, 1972
	Rat	PAM O_2^-, TNFα, and PGE_2 production	↓	Lantz et al., 1994
	Human	PBL spontaneous and PHA-induced proliferation	↓	Gonsebatt et al., 1992
		PHA-stimulated lymphocyte DNA synthesis	↑/↓	Meng, 1993
Sodium arsenite	Mouse	Vesicular stomatitis virus-induced IFN production	↓	Gainer, 1972
		PFC formation	↑/↓	Yoshida et al., 1987
		B- and T-lymphocyte levels	↓	
	Rabbit	Vesicular stomatitis virus-induced IFN production	↓	Gainer, 1972
		Zymosan-induced PAM O_2^- and H_2O_2 production	N.E./ ↓	Labedzka et al., 1989
		Zymosan-induced PAM chemiluminescence	↓	Labedzka et al., 1989
	Rat	PAM O_2^- or TNFα production (induced)	↓	Lantz et al., 1994
		PAM PGE_2 production (induced)	N.E.	
	Human	PBL spontaneous or induced proliferation	↓	Gonsebatt et al., 1992
		PHA-stimulated lymphocyte DNA synthesis	↑/↓	Meng, 1993

No studies have specifically examined pulmonary cellular immunity after As exposure. Because data suggests that As interferes with general cellular immunity, it is likely that As is a lung immunomodulant too. Most studies of pulmonary immunomodulation by As focus on innate immunity. Systemically, arsenical exposures caused decreased complement protein levels in mice and altered acute phase proteins in humans. Assessments of the toxicity of instilled trivalent ($NaAsO_2$) and pentavalent (sodium arsenate; Na_2HAsO_4) As on rat AM (using $\cdot O_2^-$, PGE_2, and TNFα production as the end points) showed that only AM from the Na_2HAsO_4-exposed rats had increased O_2^- formation. Exposure to both agents led to decreased AM TNFα production. In studies of trivalent As trisulfide (As_2S_3) and pentavalent calcium arsenate, $Ca_3(AsO_4)_2$, rat AM exposed to $Ca_3(AsO_4)_2$ had increased O_2^- and basal TNFα production, whereas exposure to As_2S_3 had no effect. These alterations in AM may affect host defenses; this is supported by observations that inhaled As_2O_3 increased the susceptibility of mice to infection with *S. pneumonia* and decreased their lung bactericidal activity against *K. pneumoniae*.

In vitro exposure of AM to As(III) suppressed phagocytic ability and O_2^- production. Assays with As_2S_3 and $Ca_3(AsO_4)_2$ indicated that both forms suppressed O_2^- production and LPS-induced TNFα release but had no effect on PGE_2 formation. Using $NaAsO_2$ or Na_2HAsO_4, it was shown that As(III) was tenfold more potent than As(V) in inhibiting O_2^- and TNFα release; curiously, only As(V) blocked LPS-induced PGE_2 release. Other studies have been unable to demonstrate As(V) effects on O_2^- production. In contrast to *in vivo* studies, *in vitro* treatment of rat lung phagocytes with As_2O_3 had no effect on TNFα or IL-1 release, although suppressed inducible O_2^- production was noted.

Possible mechanisms of As toxicity in the lung include: (1) As(III) may interact with enzyme thiol groups and so inhibit function, (2) As(V) may competitively inhibit endogenous substrate binding, (3) As may interact with molecules/substrates to inhibit metabolic reactions, and (4) As may disrupt cell energy systems/ion balances. In a broader sense, the mechanism of the pulmonary immunotoxicity of As is likely related to functional deficits in lung phagocytes—the result being a decreased phagocytic capacity leading to a decrease in *in situ* bactericidal activity and, ultimately, in host resistance to pathogens.

9.4.7.2 Cadmium

Cadmium (Cd), derived from natural and anthropogenic sources, is found in air mostly as relatively insoluble cadmium oxide (CdO). Most Cd exposure is occupational and occurs via inhalation of Cd-bearing fumes and dusts associated with processes or products that use large amounts of Cd, e.g., Ni-Cd batteries, electroplating, alloy production, brazing solders, stabilizers for plastics, pigments, etc.

In studies of effects of Cd on humoral immunity (Table 9.7), acute (2 h) inhalation of cadmium chloride ($CdCl_2$) by mice resulted in suppressed primary (IgM) antibody-forming-cell (AFC) responses, splenocyte viability, and lymphoproliferative responses to LPS even up to 1 week later. Mice inhaling $CdCl_2$ for 10 weeks had suppressed rosette formation; this suggested impaired complement binding to B cells and an inhibited antibody efficacy in eliminating or inactivating bacteria. *In vitro* exposure to $CdCl_2$ has also been shown to inhibit B cell RNA and DNA synthesis, as well as IgG secretion. *In vivo* and *in vitro* studies suggest that Cd suppresses cell-mediated immunity. Mice exposed (1 h) to $CdCl_2$ had reduced T cell proliferative responses to mitogens and allogeneic antigens. Similarly, altered lymphoproliferation by human lymphocytes exposed *in vitro* has been reported. Lymphocyte population shifts, neutropenia, and lymphopenia occurred in rodents exposed for 4 weeks to CdO. Inhibited RNA formation in lymphocytes exposed *in vitro* has been noted, suggesting that, similar to decreases in antibody synthesis and secretion by exposed B cells, decrements in cytokine formation/release may occur with T cells.

Regarding inflammatory responses, rats instilled with cadmium telluride (CdTe) had extensive lymphocyte, AM, and PMN influx into their lungs over a 1-month period; whether these results were affected by the copresent Te ions is unclear. Observations at 1 d differed from those in $CdCl_2$-instilled guinea pigs, in which there were increases in PMN but no other cell type. Rabbit AM production of PGE_2 or LTB_4 to modulate inflammation was differentially modified by Cd. Species- and strain-related differential responses to Cd have been noted in C57Bl6 and DBA mice, and WF rats, exposed to CdO for 3 h. C57Bl6 mice had a faster and greater PMN influx into their lungs than DBAs; rats had more transient inflammatory responses but a higher degree of acute inflammation.

Most studies of the effects of Cd on the lung's innate immunity have been *in vitro*. Inducible O_2^- production by rabbit AM was suppressed by $CdCl_2$, a finding similar to that seen in rat AM cultured with $CdCl_2$ or cadmium acetate; effects on guinea pig AM were absent. Interestingly, other studies indicated that Cd inhibited guinea pig AM phagocytic and microbicidal activity. Syrian hamster AM treated with CdO had altered motion of their phagosomes, a process critical to phagolysosome formation in which intracellular killing occurs; AM motility and lymphokine responsivity were also reduced. Effects on mononuclear cell cytokine production are variable. $CdCl_2$ reduced IL-1, IL-6, and TNFα mRNA levels and production by human peripheral blood mononuclear cells; in contrast, IL-8 production was enhanced.

Several possible mechanisms exist for the immunomodulatory effects of Cd; exposure can alter the AMs' phagocytic and cytotoxic function; surface receptors responsible for the binding of, and toxicity against, pathogens; and O_2^-, IL-1, and TNFα production. The upshot of these

Table 9.7 Immunomodulating Effects of Inhaled/Instilled Cadmium Compounds Exposures

Compound	*Species*	*Immune Parameter Analyzed*	*Effects*	*Reference*
Cadmium chloride	Mouse	Resistance to *Streptococcus pneumoniae*	↓	Gardner et al., 1977
		Resistance to murine cytomegalovirus	N.E	Daniels et al., 1987
		PHA-/allogeneic antigen-induced splenic lymphoproliferation	↓	Krzystyniak et al., 1987
		Spleen cell viability	↓	
		Splenic B-lymphocyte LPS-induced proliferation	↑	
		Splenic PFC (IgM) formation	↓	
		Serum antibody titer (vs. influenza virus)	N.E	Chaumard et al., 1991
Cadmium oxide	Mouse	Thymic atrophy and splenomegaly	↑	Ohsawa and Kawai, 1981
		Anemia, lymphopenia, and neutropenia	↑	
		Numbers of large lymphocytes/ small lymphocytes	↑/↓	
		Resistance to *Pasturella multocida*	↓	Chaumard et al., 1983
		Resistance to Orthomyxovirus influenza A	↑	
	Rat	Thymic atrophy and splenomegaly	↑	Ohsawa and Kawai, 1981
		Anemia, lymphopenia, and neutropenia	↑	
		Numbers of large lymphocytes/small lymphocytes	↑/↓	
Cadmium acetate	Mouse	PAM phagocytic and microbicidal activity, viability	↓	Loose et al., 1978
	Rat	PAM O_2 consumption	↓	Castranova et al., 1980
		PAM glucose metabolism	↓	

Immunomodulating Effects of Cadmium Compounds in *In Vitro* Exposures

Compound	*Host Cell*	*Immune Parameter Analyzed*	*Effects*	*Reference*
Cadmium chloride	Mouse	PAM/PEM phagocytic and microbicidal activity/viability	↓	Loose et al., 1978

	Splenocyte PFC (IgM) formation	↑/↓	Fujimaki et al., 1982
	Lymphocyte RNA synthesis	↑	Gallagher and Gray, 1982
	Splenic B-lymphocyte DNA/RNA synthesis	↓	Daum et al., 1993
	ConA-induced IL-2 production/release, IL-2R expression	↓	Payette et al., 1995
	ConA-induced lymphocyte cell cycling	↓	
Rabbit	PAM EAC rosette formation	↓	Hadley et al., 1977
	PAM phagocytic activity	↓	Graham et al., 1975
Rat	PAM O_2 consumption and glucose metabolism	↓	Castranova et al., 1980
	PAM ROI production	↓	
Human	PHA-/PMA-induced IL-2 production/release, IL-2R expression	↓	Cifone et al., 1989
	PHA-induced lymphocyte blastogenesis	↓	Borella et al., 1990
	Lymphocyte MLR activity	↓	
	Peripheral blood NK cell activity (YAC target)	↓	Cifone et al., 1990
	PBM spontaneous IL-8 production/release/mRNA level	↑	Horiguchi et al., 1993
	Monocytic cell line IL-6 production/release/mRNA level	↓	Funkhouser et al., 1994
	PHA-induced PBM IL-1β and TNFα mRNA and production	↓	Theocharis et al., 1994

effects has been that, in general, hosts treated with Cd (any form) and then challenged with bacteria display decreased resistance. Specifically, mice exposed to $CdCl_2$ or CdO, and then challenged with aerosolized streptococci spp. or *Pasturella multocida* (respectively) had increased mortality rates. In contrast, mice that inhaled CdO had a decreased susceptibility to orthomyxovirus influenza A.

9.4.7.3 Chromium

Inhalation of ambient chromium (Cr; released into the air from both natural and anthropogenic sources) has been estimated to range from less than 0.2 to 0.6 μg/d. Occupational exposure to Cr in the production of stainless steel (SS), chrome alloys, pigments, chrome plating, and welding (of SS) is most often dermal; however, inhalation of particulates and fumes bearing Cr is a primary risk to health.

There have been many studies of the effects of Cr on humoral immunity (Table 9.8). In one, humoral immunity to SRBC was increased in rats exposed 25 or 90 d to sodium dichromate ($Na_2Cr_2O_7$); over a 1-yr period, decreased serum Ig and elevated WBC and RBC levels were noted. Both B cell proliferation and Ig production were inhibited in animals, and this was also noted *in vitro* with human cells. In contrast, no difference in circulating serum IgM, IgG, or IgA was seen in dye workers despite reduced blood lymphocyte levels. With respect to cell-mediated immune responses, most studies examined the ability of Cr to produce DTH responses. *In vitro* studies indicated that Cr affected proliferation of T cells. With human T cells, Cr(VI) enhanced proliferation at low concentrations and suppressed it at the higher ones; Cr(III) was uniformly inactive. Many of these effects may be related to the clastogenic/cytogenetic effects of Cr in lymphocytes. Recent studies have shown that exposure to Cr(VI) caused an increase in lymphocytes with DNA strand breaks or micronuclei, but not in other clastogenicity markers, i.e., sister chromatid exchanges. These studies suggest that prolonged exposure to Cr or high Cr levels has a potential to produce immunosuppression and that the oxidative state/speciation of the Cr is critical to the postexposure outcomes.

In studies of the immunotoxic effects of Cr in the lung, rats exposed for 4 weeks to $Na_2Cr_2O_7$ had enhanced levels of AM with increased phagocytic activity; if exposure was extended up to 8 weeks, AM numbers decreased. In rabbits exposed for 6 weeks to sodium chromate (Na_2CrO_4), there was an influx of AM with unaltered function but having morphological changes (primarily enlarged lysosomes). In contrast, rabbits getting chromic (III) nitrate had no increase in AM but had more cells with decreased functions (metabolic, phagocytic) that correlated with an increase in enlarged lysosomes and Cr-bearing inclusions.

Table 9.8 Immunomodulating Effects of Inhaled/Instilled Chromium Compounds

Compound	*Host Cell*	*Immune Parameter Analyzed*	*Effects*	*Reference*
Calcium chromate	Rat	Zymosan-induced PAM chemiluminescence	N.E.	Galvin and Oberg, 1984
		PAM O_2 consumption	N.E.	
Chromic nitrate	Rabbit	PAM levels in BAL/PAM size	↑	Johansson et al., 1986a
		PAM morphological changes (lamellar inclusions, larger Golgi)	↑	
		PAM ROI production	↓	
		PAM phagocytic activity	N.E.	
Chromium trioxide	Rat	Zymosan-induced PAM chemiluminescence	N.E.	Galvin and Oberg, 1984
		PAM O_2 consumption	N.E.	
Sodium chromate	Rabbit	PAM levels in BAL/PAM size	N.E.	Johansson et al., 1986
		PAM ROI production and phagocytic activity	N.E.	
		PAM morphological changes	↑	
Sodium dichromate	Rat	Serum Ig levels	N.E./↑/↓	Glaser et al., 1985
		PAM levels in BAL	N.E./↓	
		Lymphocyte and neutrophil levels in BAL	↑/↓	
		PAM phagocytic activity	↑/↓	

Immunomodulating Effects of Chromium Compounds in *In Vitro* Exposures

Compound	*Host Cell*	*Immune Parameter Analyzed*	*Effects*	*Reference*
Calcium chromate	Rat	PAM zymosan-induced chemiluminescence	↓	Galvin and Oberg, 1984
		PAM O_2 consumption	↓	

(continued)

Table 9.8 Immunomodulating Effects of Chromium Compounds in *In Vitro* Exposures (Continued)

Compound	*Host Cell*	*Immune Parameter Analyzed*	*Effects*	*Reference*
Chromic chloride	Rat	PAM O_2 consumption (stimulated), ROI production, viability	↓	Castranova et al., 1980
		Thymocyte ATP/GTP production	N.E.	Lazzarini et al., 1985
		Thymocyte O_2 consumption	N.E.	
	Rabbit	PAM phagocytosis	↓	Graham et al., 1975
	Human	PBL mitogen responsiveness	N.E	Borella et al., 1990
Chromium trioxide	Rat	Zymosan-induced PAM chemiluminescence	↓	Galvin and Oberg, 1984
		PAM O_2 consumption	↓	
Potassium chromate	Rat	PAM viability	↓	Pasanen et al., 1986
		PAM LDH release	↑	
Potassium dichromate	Rat	Thymocyte ATP/GTP production	↓	Lazzarini et al., 1985
		Thymocyte O_2 consumption	↓	
	Human	PBL mitogen responsiveness	↑/↓	Borella et al., 1990
Sodium chromate	Cow	PAM viability/phagocytosis	↓	Hooftman et al., 1988
	Mouse	PEM random migration	↓	Christensen et al., 1992
		PEM phagocytic activity and IFNα/β production	↓	
Sodium dichromate	Human	T-lymphocyte mitogen responsiveness	N.E./ ↓	Kucharz and Sierakowski, 1987
		T-lymphocyte IL-2 production	↓	

Exposure to Cr results in lung inflammation. Rats exposed (13 weeks) to chromic (III) oxide (Cr_2O_3) or chromic sulfate $Cr_2(SO_4)_3$, had changes in bronchial and mediastinal lymphatic tissues consisting of increases in Cr-laden AM, lymphoid hyperplasia, and interstitial (with Cr_2O_3) or granulomatous [with $Cr_2(SO_4)_3$] inflammation. Inflammatory effects were also noted after only one instillation; exposure caused granuloma formation in the entire airway and increased the number of alveoli with progressive fibrotic changes. In a study of the role of Cr solubility on inflammatory effects, exposure (2 or 4 weeks) of rats to soluble potassium chromate (K_2CrO_4) or insoluble barium chromate ($BaCrO_4$) induced differing levels of PMN and monocyte infiltration. Although both agents caused significant increases in total cell numbers, the soluble form resulted in increases in lavageable PMN and monocytes after only 2 weeks; after an added 2 weeks of exposure, levels of both cell types were still elevated but trending lower; insoluble Cr had no effect on PMN or monocytes.

This latter study also evaluated the effects on the formation of inflammatory cytokines, ROS, and basal and inducible NO in AM. Inducible IL-1 and IL-6 production by $BaCrO_4$-exposed rat AM was unaffected (but that of TNFα was reduced). Surprisingly, inducible IL-1 and TNFα (but not IL-6) production by K_2CrO_4-exposed rat AM was reduced. This contrasts with *in vitro* findings in which Cr-exposed cells had enhanced IL-1 and TNFα release. Solubility-related differences were also apparent with respect to the degree of O_2^- and H_2O_2 formation as only the AM from $BaCrO_4$-exposed rats had reduced levels. Effects on this parameter were worsened if the AM were pretreated with IFNγ in an attempt to boost cell activity. This $BaCrO_4$-specific effect on IFNγ responsivity was also apparent with NO production; however, these AM already had higher spontaneous NO levels compared to the control AM.

9.4.7.4 Nickel

Ambient Ni, normally released during natural processes, is also introduced into the air via fossil fuel combustion, Ni mining and refining, alloy production, and waste incineration. Overall, daily exposures to Ni are ≈0.1 to 1.0 mg/d; daily intake of Ni is higher in smokers as tobacco can contain 1 to 3 mg/cigarette. In occupational settings in which Ni and its salts are used in the production of stainless steel, alloys, cast iron, and batteries, and as a catalyst and pigment, inhalation and skin contact are the major exposure routes.

Nickel exposure appears to suppress humoral immune responses (Table 9.9). Acute inhalation of nickel chloride ($NiCl_2$) reduced murine splenic responses to T-dependent SRBC. Rats exposed for 4 months to nickel oxide (NiO) particles had suppressed serum anti-SRBC levels. In Ni-exposed workers, serum IgM, IgG, and IgA levels were all decreased compared to controls. Although not common, cases of occupational

Table 9.9 Immunomodulating Effects of Inhaled/Instilled Nickel Compounds

Compound	*Species*	*Immune Parameter Analyzed*	*Effects*	*Reference*
Nickel chloride	Mouse	Mortality due to *Streptococcus pyogenes*	N.E ./↑	Adkins et al., 1979
		Clearance of viable *S. pyogenes*	↓	
		Mortality due to murine cytomegalovirus	N.E	Daniels et al., 1987
	Rabbit	PAM size and phagocytic activity, ultrastructural changes	↑	Wiernik et al., 1983
		Ex vivo PAM NBT reduction (basal and particle-induced)	↑	
		Ex vivo PAM bactericidal activity	N.E./↓	
		PAM containing laminated structures	↑	Johansson et al., 1987
		Recovered PAM lysozyme level	↓	
	Rat	PAM levels and viability in BAL	N.E./↓	Adkins et al., 1979
		Ex vivo PAM phagocytic activity	↑/↓	
		BAL total protein, lactate dehydrogenase, β-glucuronidase	N.E./↑	Benson et al., 1986
		BAL total nucleated cells, PAM and PMN levels	N.E./↑	
		BAL lymphocyte and eosinophil levels	N.E	
Nickel (metallic)	Rabbit	*Ex vivo* PAM phagocytic activity	↑	Camner et al., 1978
		PAM size, ultrastructural changes	↑	
		PAM size, ultrastructural changes, surface smoothing	↑	Johansson et al., 1980
		Ex vivo PAM basal and induced NBT reduction	↑	

		Ex vivo PAM phagocytic activity, induced NBT reduction	N.E.	Johansson et al., 1983
		Ex vivo PAM basal NBT reduction	↑	
Nickel oxide	Hamster	Mortality due to influenza A/PR/8 virus:		Port et al., 1975
		Ni exposure postinfection	N.E./↑	
		Ni exposure prior to infection	↓	
		Neutrophil and PAM infiltration into infected lungs	↑	
Nickel oxide	Mouse	PAM levels in BAL	N.E./↓	Spiegelberg et al., 1984
		Granulocyte, lymphocyte levels, PAM size in BAL	↑	
		Ex vivo PAM phagocytic activity	↑	
		BAL PMN, PAM, lymphocyte, and eosinophil levels	N.E.	Benson et al., 1986
		BAL total protein, lactate dehydrogenase, β-glucuronidase	N.E.	
		BAL total protein, lactatedehydrogenase, β-glucuronidase	↑	Benson et al., 1989
		BAL percentage of neutrophils	↑	
		BAL percentage of PAM	↓	
Nickel subsulfide	Monkey	Lung PFC formation	N.E.	Haley et al., 1987
		Lung NK cell activity (K562 target)	↑	
		Ex vivo PAM phagocytic activity	↓	

(*continued*)

Table 9.9 Immunomodulating Effects of Inhaled/Instilled Nickel Compounds (Continued)

Compound	*Species*	*Immune Parameter Analyzed*	*Effects*	*Reference*
	Mouse	Bronchial lymph node, spleen, thymus lymphoid depletion	N.E./↑	Benson et al., 1987
		BAL total protein, lactate dehydrogenase, β-glucuronidase	↑	Benson et al., 1989
		BAL percentage of neutrophils	↑	
		BAL percentage of PAM	↓/N.E	
	Rat	BAL total protein, lactate dehydrogenase, β-glucuronidase	N.E./↑	Benson et al., 1986
		BAL total nucleated cells, PMN and PAM levels	N.E./↑	
		BAL lymphocyte and eosinophil levels	N.E.	
		Bronchial lymph node, spleen, thymus lymphoid depletion	N.E./↑	Benson et al., 1987
		BAL total protein, lactate dehydrogenase, β-glucuronidase	↑	Benson et al., 1989
		BAL total nucleated cells and percentage of PMN	↑	
		BAL percentage of PAM	↓	
Nickel sulfate	Mouse	Mortality due to *S. pyogenes*	↑	Adkins et al., 1979
		Bronchial lymph node, spleen, thymus lymphoid depletion	N.E./↑	Benson, Henderson et al., 1988
		BAL total protein, lactate dehydrogenase, β-glucuronidase	↑	Benson et al., 1989
		BAL percentage of PAM	↓/N.E	

	Rat	BAL total protein, lactate dehydrogenase, β-glucuronidase	N.E./↑	Benson et al., 1986
		BAL total nucleated cells, PMN, and PAM levels	N.E./↑	
		BAL lymphocyte and eosinophil levels	N.E.	
		Bronchial lymph node lymphoid depletion	N.E./↑	Benson, Henderson et al., 1988
		Mediastinal lymph node lymphoid depletion	N.E./↑	
		BAL total protein, lactate dehydrogenase, β-glucuronidase	↑	Benson et al., 1989
		BAL total nucleated cells, percentage of neutrophils	↑	
		BAL percentage of PAM	↓	
Nickel (undefined; during occupational/environmental exposure)				
		Using Nickel-Sensitive Individuals		
		$NiSO_4$-induced PBL proliferation	↑	Everness et al., 1990
		ConA-induced PBL proliferation	N.E.	
		Ni-specific T-lymphocyte		Kapsenberg et al., 1991, 1992
		TNFα, GM-CSF, IL-2, and IFNγ production	↑	
		IL-4 and IL-5 production	N.E.	
		HLA-A, -B, -C, or -DR antigen expression	N.E.	Karvonen et al., 1984
		Asthma events ($NiSO_4$-induced), serum Ni-specific IgE	↑	Novey et al., 1983

(continued)

Table 9.9 Immunomodulating Effects of Inhaled/Instilled Nickel Compounds (Continued)

Compound	*Species*	*Immune Parameter Analyzed*	*Effects*	*Reference*
		Ni-specific T cell CD3, CD4 expression, IL-2, IFNγ production	↑	Sinigaglia et al., 1985
		In Hard-Metal Asthmatic Patients		
		ConA-induced PBL proliferation	N.E.	Kusaka et al., 1991
		$NiSO_4$-induced PBL proliferation	N.E./ ↑	

Immunomodulating Effects of Nickel Compounds in *In Vitro* Exposures

Compound	*Host Cell*	*Immune Parameter Analyzed*	*Effects*	*Reference*
Nickel chloride	Dog	PAM viability	↓	Benson, Henderson et al., 1986
	Mouse	PEM (nonelicited) adherence to glass	N.E./ ↑	Hernandez et al., 1991
		Splenic lymphocyte mitogen responsiveness: to LPS (↑/N.E.); to ConA (↓/↑); to PHA (N.E./↓).		Lawrence, 1981
		Splenic lymphocyte MLR reactivity	N.E./↑	
		IL-2-induced HT-2 lymphocyte cell line proliferation	N.E./↓	Warner and Lawrence, 1988
		ConA-induced splenocyte IL-2 production	N.E.	
		Ni-induced splenocyte IL-2 formation, receptor expression	↑	
		Ni-induced splenocyte proliferation	↑	

	Rabbit	Zymosan-induced PAM O_2^- production	N.E.	Graham et al., 1975
		PAM adherencer	N.E.	Lundborg et al., 1987
		PAM lysozyme release (cultured postharvest)	↓	
		Zymosan-induced PAM O_2^- and H_2O_2 production	N.E.	Labedzka et al., 1989
		Zymosan-induced PAM chemiluminescence	N.E.	
	Rat	Resting PAM O_2 consumption and glucose metabolism	↓	Castranova et al., 1980
		Phagocytozing PAM O_2 consumption, ROI formation	↓	
		PAM membrane integrity	N.E.	
		PAM viability	↓	Benson, Henderson et al., 1986
	Human	Vascular endothelium ICAM-1, VCAM-1, E-selectin	↑	Goebeler et al., 1993
		Vascul Resting PAM O_2 consumption and glucose metabolism ar endothelial cell NF-κB activity, IL-6 protein & mRNA	↑	Goebeler et al., 1995
Nickel-copper oxide	Dog	PAM viability	↓	Benson, Burt et al., 1988
	Mouse	PAM viability	↓	
	Rat	PAM viability	↓	
Nickel oxide	Dog	PAM viability	↓	Benson, Burt et al., 1988
	Mouse	PAM viability	N.E.	
	Rabbit	Zymosan-induced PAM O_2^- production	N.E./↑	Labedzka et al., 1989

(continued)

Table 9.9 Immunomodulating Effects of Nickel Compounds in *In Vitro* Exposures (Continued)

Compound	*Host Cell*	*Immune Parameter Analyzed*	*Effects*	*Reference*
		Zymosan-induced PAM H_2O_2 production	N.E./↓	
		Zymosan-induced PAM chemiluminescence	↓/N.E.	
	Rat	PAM viability	N.E.	Benson, Burt et al., 1988
		PEM viability	N.E.	Kuehn et al., 1982
Nickel oxide (black)	Rat	PAM viability	↓	Takahashi et al., 1992
		PAM lactate dehydrogenase release	↑	
Nickel oxide (green)	Dog	PAM viability	↓	Benson, Henderson et al., 1986
	Rat	PAM viability	N.E.	
		PAM viability	N.E./↓	Takahashi et al., 1992
		PAM lactate dehydrogenase release	N.E./↑	
Nickel subsulfide	Cow	PAM viability	↓	Finch et al., 1987
		PAM degranulation and membrane tearing	↑	
	Rat	PEM viability	N.E.	Kuehn et al., 1982
Nickel subsulfide (crystalline)	Dog	PAM viability	↓	Benson, Henderson et al., 1986
	Rat	PAM viability	↓	
Nickel sulfate	Dog	PAM viability	↓	Benson, Henderson et al., 1986
	Rat	PAM viability	↓	
		Vascular endothelium ICAM-1, VCAM-1, E-selectin	↑	Wildner et al., 1992
Nickel sulfide (amorphous)	Mouse	Embryo fibroblast polyI: C-induced IFNα, β production	N.E./↓	Sonnenfeld et al., 1983

	Rat	Poly I:C-induced L-929 IFNα and IFNβ production	N.E.	Jaramillo and Sonnenfeld, 1989
		ConA-induced lymphocyte IL-2 and IFNγ production	N.E.	
Nickel sulfide (crystalline)	Mouse	Embryo fibroblast polyI: C-induced IFNα/IFNβ production	N.E./↓	Sonnenfeld et al., 1983
		Poly I:C-induced L-929 IFNα and IFNγ production	N.E./↓	Jaramillo and Sonnenfeld, 1989,1992b
		IFNα/β-induced L-929 antiviral activity	N.E.	
	Rat	ConA-induced splenic lymphocyte IL-2, IFNγ production	N.E.	Jaramillo and Sonnenfeld, 1989, 1992a
		NiS-, ConA- or LPS-induced spleen cell proliferation	↑	
		NiS-induced splenic $CD4^+$ T-lymphocyte proliferation	↑	
		NiS-induced splenic $CD8^+$ T-lymphocyte proliferation	N.E.	
		NiS-induced splenic B-lymphocyte proliferation	N.E.	
		NiS-induced spleen cell IL-1/IL-2 production	↑	
Nickel Sulfide (crystalline)	Rat	ConA-induced PEM IL-1 production	N.E.	Jaramillo and Sonnenfeld, 1993
		PEM phagocytic activity: — IgG-opsonized SRBC/PEM	N.E.	

(*continued*)

Table 9.9 Immunomodulating Effects of Nickel Compounds in *In Vitro* Exposures (Continued)

Compound	*Host Cell*	*Immune Parameter Analyzed*	*Effects*	*Reference*
		PEM phagocytic activity:—Phagocytic index	N.E./↓	
		PEM phagocytic activity:—Binding index	N.E./↑	
Nickel	Rat			Kuehn et al., 1982
Antimonide, Disulfide, Monoarsenide, Monoselenide, Subselenide, Sulfarsenide		PEM viability	N.E	
Ferro Alloy, Ferrosulfide, Subarsenide, Telluride, Titanate		PEM viability	↓	

asthma have been reported in Ni-sensitive workers; some are Type I hypersensitivity reactions and Ni-specific IgE antibodies have been noted. With cell-mediated immunity, exposure to Ni has resulted in type IV DTH responses. Although Ni is described as a strong sensitizer in humans, significant animal data have been difficult to generate.

Regarding inflammatory responses to Ni, a general one with AM hyperplasia was noted in lungs of rodents exposed for 13 weeks to Ni_3S_2, NiO, or nickel carbonyl. Exposure (4 months) of rats to NiO resulted in fewer AM. Rats exposed for up to 22 d to Ni_3S_2 had increases in BAL lavageable cells after 2 d of exposure. Rats receiving a high dose had increased numbers of lung PMN after just one exposure. Only low Ni_3S_2 levels caused continually elevated numbers of AM; higher doses caused spikings after 2 or 12 d. In rats inhaling $NiCl_2$ for 5 d, lavageable cell numbers increased during exposure but returned to control levels later; during exposure and recovery, AM levels were depressed and PMN levels increased. Similar effects from $NiSO_4$ were seen in mice exposed for 24 h: whereas PMN levels remained elevated up to 3 d, AM levels were unaffected. Interestingly, in mice and rats exposed for 2 to 6 months to $NiSO_4$ or NiO, persistent states of AM hyperplasia became evident in rats exposed to either, whereas only NiO induced changes in mice. A species-dependent difference in the inflammatory responses was not unexpected. Studies using several strains of mice have shown that there are also significant strain-dependent variations in lung responsivity to Ni agents (i.e., to $NiSO_4$). The basis for all of these variations is still unknown.

In studies to assess the effects of Ni on the release of inflammatory cytokines/chemokines, some researchers reported increased CINC levels in exposed host BAL. Unfortunately, induction of these types of mediators might be agent specific; rats instilled with ultrafine Ni did not manifest changes in MIP-2 levels. Oddly, *in vitro*, ultrafine Ni caused AM to release TNFα. Compound-specific effects are borne out by studies that indicated that exposure of mice only to $NiSO_4$ caused increases in lung mRNA levels of MIP-2, IFNγ, MCP-1, IL-6, IL-1β, and TNFα.

It seems that AM are targets for the toxic effects of Ni. Parenteral $NiCl_2$ exposure caused AM activation followed by suppression of phagocytosis. Similarly, lungs of rabbits exposed to Ni dust for 1 to 6 months contained AM with activated appearance but reduced phagocytic capacities. In rabbits exposed for 4 months to $NiCl_2$, the percentages of AM with inclusion bodies and surface smoothing increased 40- and 8-fold, respectively. In rats exposed once to ultrafine Ni, levels of foamy AM and degenerated AM within alveoli were increased within 1 to 4 weeks. Decreased AM phagocytosis was also seen in mice exposed for ≈9 weeks to NiO or nickel subsulfide (Ni_3S_2); effects from nickel sulfate ($NiSO_4$) were inconsistent. Rabbit and rat AM exposed to Ni each exhibited reduced phagocytic ability and metabolic capacity. *In vitro* exposure of rabbit AM to $NiCl_2$ produced dose-related decreases in lysozyme activity (confirming a direct effect of Ni ions on AM) as well as in $\cdot O_2^-$ production. In a study comparing the effects of six NiO compounds on AM from dogs, mice, and rats, species sensitivity of AM to Ni was clear, with cells from dogs being most sensitive, AM of rats and mice were equal in their sensitivity. Inhibited AM activity appears to be at least partially responsible for the changes in host resistance after Ni exposure. Mice exposed for 2 h to $NiCl_2$ or $NiSO_4$ had decreased resistance to *S. pyogenes* aerosol challenge; this increased susceptibility correlated with decreases in pathogen clearance and AM phagocytic ability. Similarly, hamsters instilled with NiO had an increased mortality on subsequent challenge with influenza.

9.4.7.5 Zinc

Outside of highly industrialized regions, ambient zinc (Zn) levels are relatively low. The Zn in ambient air is derived from automobile exhaust, soil erosion, and local commercial, industrial, or construction activity. Individuals are exposed to higher levels via TS or occupation; zinc metal may be used in alloys and chemical processes, and its salts are used in wood preservatives, fertilizers, pesticides, textiles, ceramics, and rubber. Immunotoxic effects of zinc oxide (ZnO) in the lung have been extensively studied due to its role in *metal fume fever*; far less is known about the effects of most other Zn-bearing agents.

Exposure to ZnO induces strong inflammatory responses, Type II cell hyperplasia, and increased fibrosis. Mice exposed for 1 to 5 d to ZnO developed a tolerance, with PMN infiltration decreasing as exposure frequency increased; PMN-related resistance was not apparent in mice reexposed after a 5-d reprieve. Contradictory results were seen in naive subjects and sheet-metal workers; the former had reduced PMN and IL-6 in their BAL on repeat exposure to ZnO, whereas the latter had mild PMN recruitment and higher IL-6 after one exposure. Although zinc chloride ($ZnCl_2$) instillation induced similar effects, it also caused lymphocytic infiltration and foamy AM aggregates. Except for the latter, the immunohistologic changes in response to $ZnCl_2$ are similar to those in the lungs of rats instilled with zinc hydroxide ($Zn(OH)_2$) colloid. In rats instilled with ROFA having high Zn levels (or its leachate), these "soluble" forms were very inflammatory. Mice instilled once with $ZnCl_2$ had dose-dependent increases in PMN. However, a recent study noted that although instillation of rats with $ZnSO_4$ or Zn-bearing PM induced PMN influx, inhalation of intact PM did not — rather, there were dose- and time-dependent increases in particle-loaded AM levels instead.

In studies to examine effects of Zn on the ability of the lungs to resist bacteria, exposure of mice (3 h) to increased levels of $ZnSO_4$ resulted in greater mortality from subsequently inhaled *S. pyogenes* compared to controls or to mice exposed to equal or greater amounts of zinc ammonium sulfate, i.e., $Zn_2(NH_4)_2(SO_4)_2$. Interestingly, if Zn were first instilled and the hosts then infected, mice that received $Zn_2(NH_4)_2(SO_4)_2$ did not have greater mortality rates. No differences in mortality (between agents or when compared with sham/air-exposed infected controls) were evident when mice were permitted to inhale either compound rather than have them instilled.

The effects of Zn on AM have been well studied. AM from rats instilled with ZnO had changes in size and ultrastructure. A single 4-h exposure of hamsters to increasing amounts of $ZnSO_4$ or $Zn_2(NH_4)_2(SO_4)_2$ resulted in decreased AM phagocytic activity. In the context of exposure level, the results parallel those seen in the mouse bacterial resistance studies in that effects from $ZnSO_4$ were greater than those due to $Zn_2(NH_4)_2(SO_4)_2$. A species-dependent effect on phagocytic activity was noted in AM recovered 24 h after guinea pigs were exposed for 3 h to ZnO; no effects were found in cells from exposed rabbits. With AM recovered from hosts instilled with $Zn(OH)_2$, it was found that the cells displayed increased levels of proliferating cell nuclear antigen. Conversely, treatment of the hosts with $ZnSO_4$ failed to induce any effect on the nuclear antigen levels.

Most information about lung immunotoxicologic effects of Zn derive from studies of ZnO and metal fume fever. In guinea pigs exposed 1, 2, or 3 d (3 h/d), there were consistent dose- and number of exposure-dependent increases in BAL total protein, LDH, BG, and AP activities. It was unclear if the effects seen after the second exposure were due to acute-onset responses or latent effects of the first exposure. Studies have since clarified that it was the latter; guinea pigs and rats that had a single 3-h ZnO exposure did not display changes in these parameters until 4 to 24 h later. Interestingly, this same study indicated that interspecies variations in response to ZnO occur; rabbits exposed once for 2 h to ZnO only had a change in BG activity (a decrement rather than an increase).

One expects pyrogenic cytokine levels in the lungs to increase after ZnO inhalation. Studies of humans exposed for 15 to 30 min to welding fumes indicate that several pyrogenic, chemotactic, or anti-inflammatory cytokines are released postexposure. There was a rapid (<3 h) increase in IL-1, TNFα, and IL-8 levels in exposed host BAL; with increased time, post-exposure these levels decreased but that of IL-6 rose; IL-4 or IL-10 anti-inflammatory cytokine analysis was inconclusive. In studies with humans exposed to ZnO particles for 2 h, plasma IL-6 levels underwent continual increases in the period after exposure. Similar studies reported elevations in IL-8 and TNFα in BAL even 24 h after exposure.

9.5 CONCLUSION

Pulmonary toxicology has been recognized as a specialized field of study for more than 50 yr, and immunotoxicology has been expanding since its inception in the early 1970s as a separate area of research. With increasing efforts in both areas, it was not unexpected that in the quest to understand how inhaled toxins and toxicants can affect human health, there would be a convergence; this has given rise to the discipline of pulmonary immunotoxicology. The goal of this chapter was to provide both immunotoxicologists and pulmonary toxicologists up-to-date information concerning the effects of various classes of potentially inhalable materials on the immune function (as well as individual components therein) of the respiratory tract.

The data presented here clearly show that:

1. The pulmonary immune system is a sensitive target for toxicity from a large number of natural and anthropogenic agents.
2. There is a wide spectrum of effects that can manifest as immunomodulation—hypersensitivity, asthma, fibrosis, inflammation, autoimmunity, and/or immunosuppression—following an environmental or occupational encounter.
3. The type of effect that develops is often a function of the type, dose, and physicochemical characteristics of the inhaled toxicant/ mixture.
4. The effects of a given agent are not universal—these can vary across species, between sexes, as a function of age, depending on the health status of the host at the time of the encounter. Although much has been done to determine the overall immunomodulatory potentials of many agents in the various classes outlined here, current research is increasingly focused on elucidating the mechanisms by which these workplace and environmental agents produce their changes in immunologic function in the respiratory tract that may subsequently alter host health status.

For the reader, recommended references for each of the sections presented in this chapter, as well as those cited in the various tables throughout, are presented below.

REFERENCES/BIOLOGIC AGENTS

Bae, H.S. et al., Biodegradation of the mixtures of 4-chlorophenol and phenol by *Comamonas testosteroni* CPW301, *Biodegradation*, 7, 463, 1996.

Bernstein, D.I. et al., Machine operator's lung. A hypersensitivity pneumonitis disorder associated with exposure to metalworking fluid aerosols, *Chest*, 108, 636, 1995.

Bidochka, M.J. et al., Fate of biological control introductions: monitoring an Australian fungal pathogen of grasshoppers in North America, *Proc. Natl. Acad. Sci. USA*, 93, 918, 1996.

Brieland, J. et al., Coinoculation with *Hartmannella vermiformis* enhances replicative *Legionella pneumophila* lung infection in a murine model of Legionnaire's disease, *Infect. Immun.*, 64, 2449, 1996.

Clapp, W.D. et al., The effects of inhalation of grain dust extract and endotoxin on upper and lower airways, *Chest*, 104, 825, 1993.

Cross, S., Mould spores: unusual suspects in hay fever, *Commun. Nurs.*, 3, 25, 1997.

Damaj, M. and Ahmad, D., Biodegradation of polychlorinated biphenyls by rhizobia: a novel finding, *Biochem. Biophys. Res. Commn.*, 218, 908, 1996.

Damgaard, P.H. et al., Enterotoxin-producing strains of *Bacillus thuringiensis* isolated from food, *Lett. Appl. Microbiol.*, 23, 146, 1996.

Dannenberg, A.M., Jr., Immune mechanisms in the pathogenesis of pulmonary tuberculosis, *Rev. Infect. Dis.*, Suppl. 2, S369, 1989.

de Garcia, M.C. et al., Fungal keratitis caused by *Metarhizium anisopliae* var. *anisopliae, J. Med. Vet. Mycol.,* 35, 361, 1997.

George, S.E. et al., Distribution, clearance, and mortality of environmental pseudomonads in mice upon intranasal exposure, *Appl. Environ. Microbiol.*, 57, 2420, 1991.

Govan, J.R., Hughes, J.E., and Vandamme, P., *Burkholderia cepacia*: medical, taxonomic and ecological issues, *J. Med. Microbiol.*, 45, 395, 1996.

Grondona, I. et al., Physiological and biochemical characterization of *Trichoderma harzianum,* a biological control agent against soilborne fungal plant pathogens, *Appl. Environ. Microbiol.*, 63, 3189, 1997.

Hinojosa, M. et al., Hypersensitivity pneumonitis in workers exposed to esparto grass (*Stipa tenacissima*) fibers, *J. Allergy Clin. Immunol.*, 98, 985, 1996.

Hogan, M.B. et al., Basement shower hypersensitivity pneumonitis secondary to *Epicoccum nigrum*, *Chest*, 110, 854, 1996.

Martinez Moragon, E. et al., Lung diseases due to opportunistic environmental *Mycobacteria* in patients uninfected with human immunodeficiency virus. Risk factors, clinical and diagnostic aspects and course, *Arch. Bronchoneumol.,* 32, 170, 1996.

Mayes, M.E. et al., Characterization of mammalian toxicity of the crystal polypeptides of *Bacillus thuringiensis* subsp. *Israelensis*, *Fundam. Appl. Toxicol.,* 13, 310, 1989.

Michel, O. et al., Inflammatory response to acute inhalation of endotoxin in asthmatic patients, *Am. Rev. Respir. Dis.*, 146, 352, 1992.

Mundt, C., Becker, W.M., and Schlaak, M., Farmer's lung: patients IgG_2 antibodies specifically recognize *Saccharopolyspora rectivirgula* protein and carbohydrate structures, *J. Allergy Clin. Immunol.*, 98, 441, 1996.

Nakagawa-Yoshida, K. et al., Fatal cases of farmer's lung in a Canadian family. Probable new antigens, *Penicillium brevicompactum* and *P. olivicolor. Chest*, 111, 245, 1997.

Onofrio, J.M. et al., Granulocyte-alveolar-macrophage interaction in the pulmonary clearance of *Staphylococcus aureus, Am. Rev. Respir. Dis.,* 127, 335, 1983.

Reijula, K.E., Two bacteria causing farmer's lung: fine structure of *Thermoactinomyces vulgaris* and *Saccharopolyspora rectivirgula, Mycopathologia,* 121, 143, 1993.

Ronchel, M.C. et al., Construction and behavior of biologically contained bacteria for environmental application in bioremediation, *Appl. Environ. Microbiol.*, 61, 2990, 1995.

Saini, S.K. et al., Allergic bronchopulmonary mycosis to *Fusarium vasinfectum* in a child, *Ann. Allergy Asthma Immunol.*, 80, 377, 1998.

Sherwood, R.L., Biological agents, in *Pulmonary Immunotoxicology*. Cohen, M.D., Zelikoff, J.T., and Schlesinger, R.B., Eds., Kluwer Academic, Boston, 2000, p. 181.

Sherwood, R.L. et al., Comparison of *Streptococcus zooepidemicus* and influenza virus pathogenicity in mice by three pulmonary exposure routes, *Appl. Environ. Microbiol.*, 54, 1744, 1988.

Siegel, J.P., Shadduck, J.A., Szabo, J., Safety of the entomopathogen *Bacillus thuringiensis* var. *israelensis* for mammals, *J. Econ. Entomol.*, 80, 717, 1987.

Silo Suh, L.A. et al., Biological activities of two fungistatic antibiotics produced by *Bacillus cereus* UW85, *Appl. Environ. Microbiol.*, 60, 2023, 1994.

Walker, R., Emslie, K.A., and Allan, E.J., Bioassay methods for the detection of antifungal activity by *Pseudomonas antimicrobica* against the grey mould pathogen *Botrytis cinerea*, *J. Appl. Bacteriol.*, 81, 531, 1996.

Wraight, S.P. et al., Pathogenicity of the entomopathogenic fungi *Paecilomyces* sp. and *Beauveria bassiana* against the silverleaf whitefly, *Bemisia argentifolii*, *J. Invert. Pathol.*, 71, 217, 1998.

Zhaoming, W. and Lockey, R.F., A review of allergic bronchopulmonary aspergillosis, *J. Invest. Allergy Clin. Immunol.*, 6, 144, 1996.

Zhiping, W. et al., Exposure to bacteria in swine-house dust and acute inflammatory reactions in humans, *Am. J. Respir. Crit. Care Med.*, 154, 1261, 1996.

REFERENCES/OZONE

Aranyi, C. et al., Effects of subchronic exposure to a mixture of O_3, SO_2, and $(NH_4)_2SO_4$ on host defenses of mice, *J. Toxicol. Environ. Health*, 12, 55, 1983.

Bhalla, D.K., Alteration of alveolar macrophage chemotaxis, cell adhesion, and cell adhesion molecules following ozone exposure of rats, *J. Cell. Physiol.*, 169, 429, 1996.

Canning, B.J. et al., Ozone reduces murine alveolar and peritoneal macrophage phagocytosis: the role of prostanoids, *Am. J. Physiol.*, 261, L277, 1991.

Christman, C.A. et al., Enhanced phagocytosis by alveolar macrophages induced by short-term ozone insult, *Environ. Res.*, 28, 241, 1982.

Cohen, M.D. et al., Effects of inhaled ozone on pulmonary immune cells critical to antibacterial responses *in situ*, *Inhal. Toxicol.*, 14, 599, 2002.

Cohen, M.D. et al., Ozone-induced modulation of cell-mediated immune responses in the lungs, *Toxicol. Appl. Pharmacol.*, 171, 71, 2001.

Cohen, M.D. et al., Immunotoxicologic effects of inhaled chromium: role of particle solubility and co-exposure to ozone, *Toxicol. Appl. Pharmacol.*, 152, 30, 1998.

Dong, W. et al., Altered alveolar macrophage function in calorie-restricted rats, *Am. J. Respir. Cell Mol. Biol.*, 19, 462, 1998.

Dormans, J.A. et al., Surface morphology and morphometry of rat alveolar macrophages after ozone exposure, *J. Toxicol. Environ. Health*, 31, 53, 1990.

Driscoll, K.E., Vollmuth, T.A., and Schlesinger, R.B., Acute and subchronic ozone inhalation in the rabbit: response of alveolar macrophages, *J. Toxicol. Environ. Health*, 21, 27, 1987.

Driscoll, K.E. et al., Effects of *in vitro* and *in vivo* ozone exposure on eicosanoid production by rabbit alveolar macrophages, *Inhal. Toxicol.*, 1, 109, 1988.

Fakhrzadeh, L., Laskin, J.D., and Laskin, D.L., Deficiency in iNOS protects mice from O_3-induced lung inflammation and tissue injury, *Am. J. Respir. Cell Mol. Biol.*, 26, 413, 2002.

Gardner, D.E. and Graham, J.A., Increased pulmonary disease mediated through altered bacterial defenses, in *Pulmonary Macrophage and Epithelial Cells: Proceedings of 16th Annual Hanford Biology Symposium*, Sanders, C.L., Schneider, R.P., Dagle, G.E., and Ragan, H.A., Eds., Energy Research and Development Administration, Washington, D.C. Available from: NTIS, Springfield, VA, CONF-760927, 1977, 1.

Garssen, J. et al., Ozone-induced impairment of pulmonary Type IV hypersensitivity and airway hyperresponsiveness in mice, *Inhal. Toxicol.*, 9, 581, 1997.

Gilmour, M.I. et al., Suppression and recovery of the alveolar macrophage phagocytic system during continuous exposure to 0.5 ppm ozone, *Exp. Lung Res.*, 17, 547, 1991.

Gilmour, M.I. and Selgrade, M.K., A comparison of the pulmonary defenses against Streptococcal infection in rats and mice following O_3 exposure: differences in disease susceptibility and neutrophil recruitment, *Toxicol Appl. Pharmacol.*, 123, 211, 1993.

Goldstein, E. et al., Ozone and the antibacterial defense mechanisms of the murine lung, *Arch. Intern. Med.*, 127, 1099, 1971.

Graham, J.A. et al., Influence of exposure patterns of NO_2 and modifications by O_3 on susceptibility to bacterial infections in mice, *J. Toxicol. Environ. Health*, 21, 113, 1987.

Hoffer, E. et al., Adhesion molecules of blood polymorphonuclear leukocytes and alveolar macrophages in rats: modulation by exposure to ozone, *Hum. Exp. Toxicol.*, 18, 547, 1999.

Ibrahim, A.L. et al., Effects of O_3 on respiratory epithelium and alveolar macrophages of mice. I. Interferon production, *Proc. Soc. Exp. Biol. Med.*, 152, 483, 1976.

Ishii, Y. et al., Rat alveolar macrophage cytokine production and regulation of PMN recruitment following acute O_3 exposure, *Toxicol. Appl. Pharmacol.*, 147, 214, 1997.

Jakab, G.J. and Hmieleski, R.R., Reduction of influenza virus pathogenesis by exposure to 0.5 ppm ozone, *J. Toxicol. Environ. Health*, 23, 455, 1988.

Johnston, C.J. et al., Endotoxin potentiates ozone-induced pulmonary chemokine and inflammatory responses, *Exp. Lung Res.*, 28, 419, 2002.

Johnston, C.J. et al., Antioxidant and inflammatory response after acute nitrogen dioxide and ozone exposures in C57Bl/6 mice, *Inhal. Toxicol.*, 12, 187, 2000.

Johnston, C.J. et al., Inflammatory and antioxidant gene expression in C57BL/6J mice after lethal and sublethal ozone exposures, *Exp. Lung Res.*, 25, 81, 1999.

Kenyon, N.J. et al., Susceptibility to ozone-induced acute lung injury in iNOS-deficient mice, *Am. J. Physiol.—Lung Cell. Mol. Physiol.*, 282, L540, 2002.

Koike, E., Kobayashi, T., and Shimojo, N., Ozone exposure enhances expression of cell-surface molecules associated with antigen-presenting activity on bronchoalveolar lavage cells in rats, *Toxicol. Sci.*, 63, 115, 2001.

Laskin, D.L. et al., Upregulation of phosphoinositide 3-kinase and protein kinase B in alveolar macrophages following ozone inhalation. Role of NF-κB and STAT-1 in O_3-induced NO production and toxicity, *Mol. Cell. Biochem.*, 234, 91, 2002.

Laskin, D.L. et al., Increased nitric oxide synthase in the lung after ozone inhalation is associated with activation of NF-κB, *Environ. Health Perspect.*, 106(Suppl. 5), 1175, 1998.
Oosting, R.S., van Golde, L.M., and van Bree, L., Species differences in impairment and recovery of alveolar macrophage function following single and repeated ozone exposures, *Toxicol. Appl. Pharmacol.*, 110, 170, 1991.
Prasad, S.B. et al., Effects of pollutant atmospheres on surface receptors of pulmonary macrophages, *J. Toxicol. Environ. Health*, 24, 385, 1988.
Punjabi, C.J. et al., Production of nitric acid by rat type II pneumocytes: increased expression of inducible nitric oxide synthase following inhalation of a pulmonary irritant, *Am. J. Respir. Cell Mol. Biol.*, 11, 165, 1994.
Ryan, L.K., Ozone, in *Pulmonary Immunotoxicology,* Cohen, M.D., Zelikoff, J.T., and Schlesinger, R.B., Eds., Kluwer Academic, Boston, 2000, p. 301.
Ryer-Powder, J.E. et al., Inhalation of ozone produces a decrease in superoxide anion production in mouse alveolar macrophage, *Am. Rev. Respir. Dis.*, 138, 1129, 1988.
Schlesinger, R.B. et al., Assessment of toxicologic interactions resulting from acute inhalation exposure to H_2SO_4 and O_3 mixtures, *Toxicol. Appl. Pharmacol.*, 115, 183, 1992.
Selgrade, M.K et al., Evaluation of effects of ozone exposure on influenza infection in mice using several indicators of susceptibility, *Fundam. Appl. Toxicol.*, 11, 169, 1988.
Sherwood, R.L. et al., Effect of 0.64 ppm O_3 on alveolar macrophage lysozyme levels with rats with chronic pulmonary bacterial infection, *Environ. Res.*, 41, 378, 1986.
Shingu, H. et al., Effects of ozone and photochemical oxidants on interferon production by rabbit alveolar macrophages, *Bull. Environ. Contam. Toxicol.*, 24, 433, 1980.
Thomas, G.B. et al., Effects of exposure to ozone on susceptibility to experimental tuberculosis, *Toxicol. Lett.*, 9, 11, 1981.
van Loveren, H. et al., Effects of O_3 on respiratory *Listeria monocytogenes* infection in the rat. Suppression of macrophage function and cellular immunity and aggravation of histopathology in lung and liver during infection, *Toxicol. Appl. Pharmacol.*, 94, 374, 1988.
van Loveren H. et al., Effects of ozone, hexachlorobenzene, and bis(tri-*n*-butyltin)oxide on NK activity in the rat lung, *Toxicol. Appl. Pharmacol.*, 102, 21, 1990.
Zhao, Q. et al., Chemokine regulation of ozone-induced neutrophil and monocyte inflammation, *Am. J. Physiol.*, 274(1 Pt. 1), L39, 1998.

REFERENCES/SO_X AND NO_X

Amoruso, M.A., Witz, G., and Goldstein, B.D., Decreased superoxide anion radical production by rat alveolar macrophages following inhalation of ozone or nitrogen dioxide, *Life Sci.*, 28, 2215, 1981.
Azoulay, E., Bouley, G., and Blayo, M.C., Effects of nitric oxide on resistance to bacterial infection in mice, *J. Toxicol. Environ. Health,* 7, 873, 1981.
Azoulay-Dupuis, E., Bouley, G., and Blayo, M.C., Effects of sulfur dioxide on resistance to bacterial infection in mice, *Environ. Res.*, 29, 312, 1982.

Clarke, R.W. et al., Inhaled particle-bound sulfate: effects on pulmonary inflammatory responses and alveolar macrophage function, *Inhal. Toxicol.*, 12, 169, 2000.

D'Amato, G. et al., Respiratory allergic diseases induced by outdoor air pollution in urban areas, *Monaldi Arch. Chest Dis.*, 57, 161, 2002.

Devlin, R.B. et al., Inflammatory response in humans exposed to 2.0 ppm nitrogen dioxide, *Inhal. Toxicol.*, 11, 89, 1999.

Ehrlich, R., Findlay, J., and Gardner, D.E., Effects of repeated exposures to peak concentrations of NO_2 and ozone on resistance to streptococcal pneumonia, *J. Toxicol. Environ. Health*, 5, 631, 1979.

Fairchild, G.A., Effects of ozone and sulfur dioxide on virus growth in mice, *Arch. Environ. Health*, 32, 28, 1977.

Frampton, M.W. et al., Nitrogen dioxide exposure *in vivo* and human alveolar macrophage inactivation of influenza virus *in vitro, Environ. Res.*, 48, 179, 1989.

Fujimaki, H., Katayama, N., and Wakamori, K., Enhanced histamine release from lung mast cells of guinea pigs exposed to sulfuric acid aerosols, *Environ. Res.*, 58, 117, 1992.

Fujimaki, H., Shimizu, F., and Kubota, K., Effect of subacute exposure to NO_2 on lymphocytes required for antibody response, *Environ. Res.*, 29, 280, 1982.

Gardner, D.E. et al., Nonrespiratory function of the lungs: host defenses against infection, in *Air Pollution by Nitrogen Oxides,* Schneider, T. and Grant, L, Eds., Elsevier, New York, 1982, p. 401.

Goldstein, E. et al., Effect of near-ambient exposures to SO_2 and ferrous sulfate particles on murine pulmonary defense mechanisms, *Arch. Environ. Health,* 34, 424, 1979.

Graham, J.A. et al., Influence of exposure patterns of NO_2 and modifications by ozone on susceptibility to bacterial infectious disease in mice, *J. Toxicol. Environ. Health,* 21, 113, 1987.

Holt, P.G. et al., Immunological function in mice chronically exposed to nitrogen oxides (NO_x), *Environ. Res.*, 19, 154, 1979.

Jakab, G.J., Modulation of pulmonary defense mechanisms against viral and bacterial infections by acute exposures to nitrogen dioxide, *HEI Research reports,* 20, 1, 1988.

Kienast, K. et al., Realistic *in vitro* study of oxygen radical liberation by alveolar macrophages and mononuclear cells of the peripheral blood after short-term exposure to SO_2, *Pneumologie,* 47, 60, 1993.

Knorst, M.M. et al., Effect of sulfur dioxide on cytokine production of human alveolar macrophages *in vitro, Arch. Environ. Health,* 51, 150, 1996.

Kodavanti, U.P. et al., Variable pulmonary responses from exposure to concentrated ambient air particles in a rat model of bronchitis, *Toxicol. Sci.*, 54, 441, 2000.

Koike, E. et al., Effect of exposure to NO_2 on alveolar macrophage-mediated immunosuppressive activity in rats, *Toxicol. Lett.*, 121, 135, 2001.

Maigetter, R.Z. et al., Effect of exposure to nitrogen dioxide on T- and B-cells in mouse spleens, *Toxicol. Lett.*, 2, 157, 1978.

Mautz, W.J. et al., Respiratory tract responses to repeated inhalation of an oxidant and acid gas-particle air pollutant mixture, *Toxicol. Sci.*, 61, 331, 2001.

McGovern, T.J. et al., Ozone-induced alteration in β-adrenergic pharmacological modulation of pulmonary macrophages, *Toxicol. Appl. Pharmacol.*, 137, 51, 1996.

Miller, F.J. et al., Evaluating the toxicity of urban patterns of oxidant gases. II. Effects in mice from chronic exposure to NO_2, *J. Toxicol. Environ. Health,* 21, 99, 1987.

Osebold, J.W., Gershwin, L.J., and Zee, Y.C., Studies on enhancement of allergic lung sensitization by inhalation of ozone and sulfuric acid aerosol, *J. Environ. Pathol. Toxicol.*, 3, 221, 1980.

Parker, R.F. et al., Short-term NO_2 exposure enhances susceptibility to murine respiratory *Mycoplasmosis* and decreases intrapulmonary killing of *Mycoplasma pulmonis, Am. Rev. Respir. Dis.*, 140, 502, 1990.

Persinger, R.L. et al., Molecular mechanisms of NO_2-induced epithelial injury in the lung, *Mol. Cell. Biochem.*, 234, 71, 2002.

Richters, A. and Damji, K.S., Changes in T-lymphocyte subpopulations and NK cells following exposure to ambient levels of NO_2, *J. Toxicol. Environ. Health,* 25, 247, 1988.

Riedel, F. et al., Effects of SO_2 exposure on allergic sensitization in the guinea pig, *J. Allergy Clin. Immunol.*, 82, 527, 1988.

Robison, T.W. et al., Depression of stimulated arachidonate metabolism and superoxide production in rat alveolar macrophages following *in vivo* exposure to 0.5 ppm NO_2, *J. Toxicol. Environ. Health,* 38, 273, 1993.

Rose, R.M. et al., Pathophysiology of enhanced susceptibility to murine cytomegalovirus respiratory infection during short-term exposure to 5 ppm NO_2, *Am. Rev. Respir. Dis.*, 137, 912, 1988.

Rubinstein, I. et al., Effects of 0.6 ppm nitrogen dioxide on circulating and bronchoalveolar lavage lymphocyte phenotypes in healthy subjects, *Environ. Res.*, 55, 18, 1991.

Sandström, T. et al., Effects of repeated exposure to 4 ppm NO_2 on bronchoalveolar lymphocyte subsets and macrophages in healthy men, *Eur. Respir. J.*, 5, 1092, 1992.

Schelseinger, R.B., Chen, L.C., and Zelikoff, J.T., Sulfur and nitrogen oxides, in *Pulmonary Immunotoxicology* Cohen, M.D., Zelikoff, J.T., and Schlesinger, R.B., Eds., Kluwer Academic, Boston, 2000, p. 337.

Sandström, T. et al., Cell response in bronchoalveolar lavage fluid after exposure to sulfur dioxide: a time-response study, *Am. Rev. Respir. Dis.*, 140, 1828, 1989.

Skornik, W.A. and Brain, J.D., Effect of sulfur dioxide on pulmonary macrophage endocytosis at rest and during exercise, *Am. Rev. Respir. Dis.*, 142, 655, 1990.

Sone, S., Brennan, L.M., and Creasia, D.A., *In vivo* and *in vitro* NO_2 exposures enhance phagocytic and tumoricidal activities of rat alveolar macrophages, *J. Toxicol. Environ. Health,* 11, 151, 1983.

Suzuki, T. et al., Decreased phagocytosis and superoxide anion production in alveolar macrophages of rats exposed to NO_2, *Arch. Environ. Contam. Toxicol.*, 15, 733, 1986.

van Bree, L. et al., Biochemical and morphological changes in lung tissue and isolated cells of rats induced by short-term NO_2 exposure, *Hum. Exp. Toxicol.*, 19, 392, 2000.

Wegmann, M., Renz, H., and Herz, U., Long-term NO_2 exposure induces pulmonary inflammation and progressive development of airflow obstruction in C57BL/6 mice: a mouse model for COPD?, *Pathobiology,* 70, 284, 2002.

Zelikoff, J.T. et al., Effects of inhaled sulfuric acid aerosols on pulmonary immunocompetence: a comparative study in humans and animal, *Inhal. Toxicol.*, 9, 731, 1997.

Zelikoff, J.T. et al., Immunotoxicity of sulfuric acid aerosol: effects on pulmonary macrophage effector and functional activities critical for maintaining host resistance against infectious diseases, *Toxicology,* 92, 269, 1994.

REFERENCES/WOOD SMOKE, KEROSENE, AND DIESEL EXHAUST

Amakawa, K. et al., Suppressive effects of diesel exhaust particles on cytokine release from human and murine alveolar macrophages, *Exp. Lung Res.*, 29, 149, 2003.

Arashidani, K. et al., Indoor pollution from heating, *Ind. Health,* 34, 205, 1996.

Arif, J.M. et al., Effect of kerosene and its soot on the chrysotile-mediated toxicity to the rat alveolar macrophages, *Environ. Res.*, 72, 151, 1997.

Arif, J.M. et al., Modulation of macrophage-mediated cytotoxicity by kerosene soot: possible role of reactive oxygen species, *Environ. Res.*, 61, 232, 1993.

Boman, B.C., Forsberg, A.B., and Jarvholm, B.G., Adverse health effects from ambient air pollution in relation to residential wood combustion in modern society, *Scand. J. Work Environ. Health,* 29, 251, 2003.

Brunekreef, B. et al., Air pollution from truck traffic and lung function in children living near motorways, *Epidemiology,* 8, 293, 1997.

Butterfield, P. et al., Woodstoves and indoor air, *J. Environ. Health,* 59, 172, 1989.

Casaco, A. et al., Effects of kerosene on airway sensitization to egg albumin in guinea pig, *Allergol. et Immunopathol.*, 13, 235, 1985a.

Casaco, A. et al., Induction of acetylcholinesterase inhibition in the guinea pig trachea by kerosene, *Respiration,* 48, 46, 1985b.

Casillas, A.M. et al., Enhancement of allergic inflammation by DEPs: permissive role of reactive oxygen species, *Ann. Allergy Asthma Immunol.*, 83, 624, 1999.

Cass, G.R. and Gray, H.A., Regional emissions and atmospheric concentrations of diesel engine particulate matter: Los Angeles as a case study, in *Diesel Exhaust: A Critical Analysis of Emissions, Exposure, and Health Effects,* Health Effects Institute, Cambridge, MA, 1995, p. 125.

Castranova, V. et al., Effect of exposure to DEPs on the susceptibility of the lung to infection, *Environ. Health Perspect.*, 109(Suppl. 4), 609, 2001.

Cohen A.J. and Nikula, K., Health effects of diesel exhaust: laboratory and epidemiologic studies, in *Air Pollution and Health,* Cohen A.J. and Nikula, K., Eds., Academic Press, San Diego, 1999, 707.

Diaz-Sanchez, D. et al., *In vivo* nasal challenge with DEPs enhances expression of CC chemokines rantes, MIP-1α, and MCP-3 in humans, *Clin. Immunol.*, 97, 140, 2000.

Fick, R.B. et al., Alterations in the antibacterial properties of rabbit pulmonary macrophage exposed to woodsmoke, *Am. Rev. Respir. Dis.*, 129, 76, 1984.

Fujimaki, H. et al., Intranasal instillation of DEPs and antigen in mice modulate cytokine production in cervical lymph node cells, *Int. Arch. Allergy Immunol.*, 108, 268, 1995.

Fujimaki, H. et al., Inhalation of diesel exhaust enhances antigen-specific IgE antibody production in mice, *Toxicology,* 116, 227, 1997.

Fujimaki, H. et al., Roles of $CD4^+$ and $CD8^+$ T-cells in adjuvant activity of diesel exhaust particles in mice, *Int. Arch. Allergy Immunol.*, 124, 485, 2001.

Goodwin, S.R. and Berman, L.S., Kerosene aspiration: immediate and early pulmonary and cardiovascular effects, *Vet. Hum. Toxicol.*, 30, 521, 1988.

Harris, D.T. et al., Effects of short-term JP-8 jet fuel exposure on cell-mediated immunity, *Toxicol. Ind. Health,* 16, 78, 2000.

Harris, D.T. et al., Immunotoxicological effects of JP-8 jet fuel exposure, *Toxicol. Ind. Health,* 13, 43, 1997.

Harris, D.T. et al., JP-8 jet fuel exposure results in immediate immunotoxicity, which is cumulative over time, *Toxicol. Ind. Health,* 18, 77, 2002.

Hiramatsu, K. et al., Inhalation of diesel exhaust for three months affects major cytokine expression and induces BALT formation in murine lung, *Exp. Lung Res.*, 29, 607, 2003.

Holgate, S.T. et al., Health effects of acute exposure to air pollution. Part I: Healthy and asthmatic subjects exposed to diesel exhaust, *Res. Rep. Health Eff. Inst.*, 112, 1, 2003.

Honicky, R.E. and Osborne, J.S., Respiratory effects of wood heat: clinical observations and epidemiologic assessment, *Environ. Health Perspect.*, 95, 105, 1991.

Hsu, T.H. and Kou, Y.R., Airway hyperresponsiveness to bronchoconstrictor challenge after woodsmoke exposure in guinea pigs, *Life Sci.*, 68, 2945, 2001.

Leonard, S.S. et al., Woodsmoke particles generate free radicals and cause lipid peroxidation, DNA damage, NF-κB activation and TNFα release in macrophages, *Toxicology,* 150, 147, 2000.

Lovik, M. et al., DEPs and carbon black have adjuvant activity on the local lymph node response and systemic IgE production to ovalbumin, *Toxicology,* 121, 165, 1997.

Ma, J.Y. and Ma, J.K., Dual effect of the particulate and organic components of DEPs on the alteration of pulmonary immune/inflammatory responses and metabolic enzymes, *J. Environ. Sci. Health Part C Environ. Carc. Ecotoxicol. Rev.*, 20, 117, 2002.

Mauderly, J.L., Diesel exhaust, in *Environmental Toxicants: Human Exposures and Their Health Effects,* Lippmann, M., Ed., Van Nostrand Reinhold, New York, 1992, p. 119.

Mesa, M.G., Alvarez, R.G., and Parada, A.C., Biochemical mechanisms in the effects of kerosene on airway of experimental animals, *Allergol. Et Immunopathol.*, 16, 363, 1988.

Nel, A.E. et al., Enhancement of allergic inflammation by the interaction between diesel exhaust particles and the immune system, *J. Allergy Clin. Immunol.*, 102, 539, 1998.

Pacheco, K.A. et al., Influence of diesel exhaust particles on mononuclear phagocytic cell-derived cytokines: IL-10, TGFα and IL-1β, *Clin. Exp. Immunol.*, 126, 374, 2001.

Rhodes, A.G. et al., The effects of jet fuel on immune cells of fuel system maintenance workers, *J. Occup. Environ. Med.*, 45, 79, 2003.

Robledo, R.F. et al., Short-term pulmonary response to inhaled JP-8 jet fuel aerosol in mice, *Toxicol. Pathol.*, 28, 656, 2000.

Saito, Y. et al., Long-term inhalation of diesel exhaust affects cytokine expression in murine lung tissues: comparison between low- and high-dose diesel exhaust exposure, *Exp. Lung Res.*, 28, 493, 2002.

Saito, Y. et al., Effects of diesel exhaust on murine alveolar macrophages and a macrophage cell line, *Exp. Lung Res.*, 28, 201, 2002.

Saxena, R.K. et al., Effect of diesel exhaust particulate on bacillus *Calmette-Guerin* lung infection in mice and attendant changes in lung interstitial lymphoid subpopulations and IFNγ response, *Toxicol. Sci.*, 73, 66, 2003.

Siegel, P.D. et al., Effect of DEP on immune responses: contributions of particulate versus organic soluble components, *J. Toxicol. Environ. Health,* 67, 221, 2004.

Suzuki, T., Kanoh, T., and Kanbayashi, M., The adjuvant activity of pyrene in diesel exhaust on IgE antibody production in mice, *Japan. J. Allergology,* 42, 963, 1993.

Takano, H. et al., Diesel exhaust particles enhance antigen-induced airway inflammation and local cytokine expression in mice, *Am. J. Respir. Crit. Care Med.*, 156, 36, 1997.

Takano, H. et al., DEPs enhance lung injury related to bacterial endotoxin through expression of proinflammatory cytokines, chemokines, and intercellular adhesion molecule-1, *Am. J. Respir. Crit. Care Med.*, 165, 1329, 2002.

Tesfaigzi, Y. et al., Health effects of subchronic exposure to low levels of wood smoke in rats, *Toxicol. Sci.*, 65, 115, 2002.

Thomas, P.T. and Zelikoff, J.T., Air pollutants: Modulators of pulmonary host resistance against infection, in *Air Pollution and Health,* Holgate, S.T., Samet, J.M., Koren, H.S., and Maynard, R.I., Eds., Academic Press, San Diego, 1999, p. 357.

Thorning, D.R. et al., Pulmonary responses to smoke inhalation: morphologic changes in rabbits exposed to pine woodsmoke, *Hum. Pathol.*, 13, 355, 1982.

Tuthill, R.W., Woodstoves, formaldehyde, and respiratory disease, *Am. J. Epidemiol.*, 120, 952, 1984.

Wang, S., Young, R.S., and Witten M., Age-related differences in pulmonary inflammatory responses to JP-8 jet fuel aerosol inhalation, *Toxicol. Ind. Health,* 17, 23, 2001.

Wong, S.S. et al., Inflammatory responses in mice sequentially exposed to JP-8 jet fuel and influenza virus, *Toxicology,* 197, 139, 2004.

Yang, H.M. et al., Respiratory exposure to DEPs decreases the spleen IgM response to a T-cell-dependent antigen in female $B_6C_3F_1$ mice, *Toxicol. Sci.*, 71, 207, 2003.

Yang, H.M. et al., Effects of DEP, carbon black, and silica on macrophage responses to lipopolysaccharide: evidence of DEP suppression of macrophage activity, *J. Toxicol. Environ. Health,* 58, 261, 1999.

Yin, X.J. et al., Alteration of pulmonary immunity to *Listeria monocytogenes* by DEPs. I. Effects of DEPs on early pulmonary responses, *Environ. Health Perspect.*, 110, 1105, 2002.

Yin, X.J. et al., Alteration of pulmonary immunity to Listeria monocytogenes by DEPs. II. Effects of DEPs on T-cell-mediated immune responses in rats, *Environ. Health Perspect.*, 111, 524, 2003.

Zeikoff, J.T., Woodsmoke, kerosene heater emissions, and diesel exhaust, in *Pulmonary Immunotoxicology,* Cohen, M.D., Zelikoff, J.T., and Schlesinger, R.B., Eds., Kluwer Academic, Boston, 2000, p. 369.

Zelikoff, J.T. et al., The toxicology of inhaled woodsmoke, *J. Toxicol. Environ. Health,* 5, 269, 2002.

REFERENCES/TOBACCO SMOKE

Ando, M., Sugimoto, M., Nishi, R., Surface morphology and function of pulmonary alveolar macrophages from smokers and nonsmokers, *Thorax,* 39, 850, 1984.

Bahna, S.L., Heiner, D.C., and Myhre, B.A., Changes in serum IgD in cigarette smokers, *Clin. Exp. Immunol.*, 51, 624, 1983.

Benner, C.L. et al., Chemical composition of environmental tobacco smoke: particulate-phase compounds, *Environ. Sci. Technol.*, 23, 688, 1989.

Burton, R.C. et al., Effects of age, gender, and cigarette smoking on human immunoregulatory T-cells subsets, *Diagn. Immunol.*, 1, 216, 1983.

Chang, J.C., Distler, S.G., and Kaplan, A.M., Tobacco smoke suppresses T-cell but not APCs in the lung-associated lymph nodes, *Toxicol. Appl. Pharmacol.*, 102, 514, 1990.

Chong, I.W. et al., Expression and regulation of the MIP-1α gene by nicotine in rat alveolar macrophages, *Eur. Cytokine Net.*, 13, 242, 2002.

Churg, A. et al., Macrophage metalloelastase mediates acute cigarette-smoke-induced inflammation via TNFα release, *Am. J. Respir. Crit. Care Med.*, 167, 1083, 2003.

Coggins, C.R., OSHA review of animal inhalation studies with environmental tobacco smoke, *Inhal. Toxicol.*, 8, 819, 1996.

Coggins, C.R., A review of chronic inhalation studies with mainstream cigarette smoke in rats and mice, *Toxicol. Pathol.*, 26, 307, 1998.

Cosio, M.G., Majo, J., and Cosio, M.G., Inflammation of the airways and lung parenchyma in COPD: role of T-cells, *Chest*, 121(Suppl. 5), 160S, 2002.

Dean, J. and Muray R., Toxic response of the immune system, in *The Health Consequences of Smoking: Toxicology,* United States Department of Health and Human Services, Washington, D.C., 1991.

Edwards, K. et al., Mainstream and sidestream cigarette smoke condensates suppress macrophage responsiveness to interferon gamma, *Hum. Exp. Toxicol.*, 18, 233, 1999.

EPA: Environmental Protection Agency. *Indoor Air Facts: Environmental Tobacco Smoke June 1989.* U.S. Government Printing Office, Washington, D.C., 1989.

Finch, G.L. et al., Surface morphology and functional studies of alveolar macrophages from cigarette smokers and nonsmokers, *J. Reticuloendothel. Soc.*, 32, 1, 1982.

Gairola, C.G., Free lung cells response of mice and rats to mainstream cigarette smoke exposure, *Toxicol. Appl. Pharmacol.*, 84, 567, 1986.

Gardner, D.E., Direct and indirect injury to respiratory tract, in *Air Pollution and Lung Disease in Adults,* Witorsch, P. and Spagnolo, S.V., Eds., CRC, Boca Raton, FL 1994, p. 19.

Gardner, D.E., Tobacco smoke, in *Pulmonary Immunotoxicology,* Cohen, M.D., Zelikoff, J.T., and Schlesinger, R.B., Eds., Kluwer Academic, Boston, 2000, p. 387.

Grashoff, W.F. et al., Chronic obstructive pulmonary disease: role of bronchiolar mast cells and macrophages, *Am. J. Pathol.*, 151, 1785, 1997.

Green, G.M., Mechanism of tobacco smoke toxicity on pulmonary macrophage cells, *Eur. J. Respir. Dis.*, 66, 82, 1985.

Hoidal, J.R. and Niewoehner, D.E., Lung phagocyte recruitment and metabolic alterations induced by cigarette smoke in humans and in hamsters, *Am. Rev. Respir. Dis.*, 126, 548, 1982.

Holt, P.G. et al., Low-tar high-tar cigarettes: comparison of effects in mice, *Arch. Environ. Health,* 31, 258, 1976.

Holt, P.G., Immune and inflammatory function in smokers, *Thorax,* 42, 241, 1987.

Hughes, D.A. et al., Numerical and functional alteration in circulatory lymphocytes in cigarette smokers, *Clin. Exp. Immunol.*, 61, 459, 1985.

Hunninghake, C.W. and Crystal, R.G., Cigarette smoking and lung destruction: accumulation of neutrophils in lungs of smokers, *Am. Rev. Respir. Dis.*, 128, 833, 1983.

Hwang, D. et al., Activation and inactivation of cyclooxygenase in rat alveolar macrophages by aqueous cigarette tar extracts, *Free Radic. Biol. Med.*, 27, 673, 1999.

IARC, *Monograph on the Evaluation of Carcinogenic Risk of Chemicals to Humans — Tobacco Smoking,* Vol. 38. World Health Organization, Lyon, France, 1986.

Jeffery, P.K., Lymphocytes, chronic bronchitis and chronic obstructive pulmonary disease, Novartis Foundation Symposium, 234, 149, 2001.

Johnson, J.D. et al., Effects of mainstream and environmental tobacco smoke on the immune system in animals and humans, *CRC Crit. Rev. Toxicol.*, 20, 369, 1990.

King, T.E. et al., Phagocytosis and killing of *Listeria monocytogenes* by alveolar macrophages: smokers versus nonsmokers, *Am. Rev. Respir. Dis.*, 158, 1309, 1988.

Miller, K. et al., Effects of cigarette smoke exposure on biomarkers of lung injury in the rat, *Inhal. Toxicol.*, 89, 803, 1996.

Miller, L. et al., Reversible alterations in immunoregulatory T-cells in smoking: analysis by monoclonal antibodies and flow cytometry, *Am. Rev. Respir. Dis.*, 126, 265, 1982.

Mochida-Nishimura, K. et al., Differential activation of MAP kinase pathways and NF-κB in bronchoalveolar cells of smokers and nonsmokers, *Mol. Med.*, 7, 177, 2001.

Ozan, E. et al., Histological and biochemical effects of cigarette smoke on lungs, *Acta Physiol. Hung.*, 88, 301, 2001.

Rodgman, A., Environmental tobacco smoke, *Reg. Toxicol. Pharmacol.*, 16, 223, 1992.

Seagrave, J. et al., Effects of cigarette smoke exposure and cessation on inflammatory cells and matrix metalloproteinase activity in mice, *Exp. Lung Res.*, 30, 1, 2004.

Soliman, D.M. and Twigg, H., Cigarette smoking decreases bioactive IL-6 secretion by macrophages, *Am. J. Physiol.*, 263, 471, 1992.

Sopori, M.L., Goud, N.S., and Kaplan, A.M., Effects of tobacco smoke on the immune system, in *Immunotoxicology and Immunopharmacology,* 2nd ed., Dean, J.H., Luster, M.J., Munson, A.E., and Kimber, I., Eds., Raven Press, New York, 1994, 412.

Sterling, T.B. and Kobayashi, D., Indoor by-product levels of tobacco smoke: a critical review of the literature, *J. Air Pollut. Control. Assoc.*, 32, 250, 1982.

Tager, M. et al., Evidence of a defective thiol status of alveolar macrophages from COPD patients and smokers, *Free Radic. Biol. Med.,* 29, 1160, 2000.

Takeuchi, M. et al., Inhibition of lung natural killer cell activity by smoking: the role of alveolar macrophages, *Respiration,* 68, 262, 2001.

Tetley, T.D., Macrophages and the pathogenesis of COPD, *Chest,* 121(Suppl. 5), 156S, 2002.

Tollerud, D.J., Clark, J.W., and Brown, L.M., Association of cigarette smoking with decreased number of circulating NK cells, *Am. Rev. Respir. Dis.*, 139, 194, 1989.

van der Vaart, H. et al., Acute effects of cigarette smoke on inflammation and oxidative stress: a review, *Thorax,* 59, 713, 2004.

Wang, S. et al., Functional alterations of alveolar macrophages subjected to smoke exposure and antioxidant lazaroids, *Toxicol. Ind. Health,* 15, 464, 1999.

Wu, Z.X. et al., Airway hyperresponsiveness to cigarette smoke in ovalbumin-sensitized guinea pigs, *Am. J. Respir. Crit. Care Med.*, 161, 73, 2000.

Yamaguchi, E. et al., Interleukin-1 production of macrophages is decreased in smokers, *Am. Rev. Respir. Dis.*, 14, 397, 1989.

Yu, M., Pinkerton, K.E., and Witschi, H., Short-term exposure to aged and diluted sidestream cigarette smoke enhances ozone-induced lung injury in $B_6C_3F_1$ mice, *Toxicol. Sci.*, 65, 99, 2002.

REFERENCES/METALS

Adamson, I.Y. et al., Zinc is the toxic factor in the lung response to an atmospheric particulate sample, *Toxicol. Appl. Pharmacol.*, 166, 111, 2000.

Adkins, B., Jr., Richards, J.H., and Gardner, D.E., Enhancement of experimental respiratory infection following nickel inhalation, *Environ. Res.*, 20, 33, 1979.

Aranyi, C, Bradof, J.N., and O'Shea, W.J., Effects of arsenic trioxide inhalation exposure on pulmonary antibacterial defenses in mice, *J. Toxicol. Environ. Health,* 15, 163, 1985.

Arfsten, D.P., Aylward, L.L., and Karch, N.J., Chromium, in *Immunotoxicology of Environmental and Occupational Metals,* Zelikoff, J.T. and Thomas, P.T., Eds., Taylor and Francis, London, 1998, p. 63.

Arsalane, K. et al., Effects of nickel hydroxycarbonate on alveolar macrophage functions, *J. Appl. Toxicol.*, 12, 285, 1992.

Bencko, V. et al., Immunological profiles in workers of a power plant burning coal rich in arsenic content, *J. Hyg. Epidemiol. Microbiol. Immunol.*, 32, 137, 1988.

Benson, J.M. et al., Biochemical responses of rat and mouse lung to inhaled nickel compounds, *Toxicology,* 57, 255, 1989.

Benson, J.M. et al., Comparative acute toxicity of four nickel compounds to F344 rat lung, *Fundam. Appl. Toxicol.*, 7, 340, 1986.

Benson, J.M. et al., Comparative inhalation toxicity of nickel subsulfide to F344/N rats and $B_6C_3F_1$ mice exposed for twelve days, *Fundam. Appl. Toxicol.*, 9, 251, 1987.

Benson, J.M., Burt, D.G., and Carpenter, R.L., Comparative inhalation toxicity of nickel sulfate to F344/N rats and $B_6C_3F_1$ mice exposed for twelve days, *Fundam. Appl. Toxicol.*, 10, 164, 1988.

Benson, J.M., Henderson, R.F., and McClellan, R.O., Comparative cytotoxicity of four nickel compounds to canine and rodent alveolar macrophages *in vitro, J. Toxicol. Environ. Health,* 19, 105, 1986.

Benson, J.M., Henderson, R.F., and Pickrell, J.A., Comparative *in vitro* cytotoxicity of nickel oxides and nickel-copper oxides to rat, mouse, and dog pulmonary alveolar macrophages, *J. Toxicol. Environ. Health,* 24, 373, 1988.

Berche, P. et al., Susceptibility of mice to bacterial infections after chronic exposure to cadmium, *Ann. Microbiol.*, 131B, 145, 1980.

Blakley, B.R., The effect of cadmium chloride on the immune response in mice, *Can. J. Comp. Med.*, 49, 104, 1985.

Blanc, P.D., Boushey, H.A., Wong, H., Wintermeyer, S.F., and Bernstein, M.S., Cytokines in metal fume fever, *Am. Rev. Respir. Dis.*, 147, 134, 1993.

Borella, P., Manni, S., and Giardino, A., Cadmium, nickel, chromium, and lead accumulate in human lymphocytes and interfere with PHA-induced proliferation, *J. Trace Elem. Electrolytes Health Dis.*, 4, 87, 1990.

Brown, R.F. et al., Histopathology of rat lung following exposure to zinc oxide/hexachloroethane smoke or instillation with zinc chloride followed by treatment with 70% oxygen, *Environ. Health Perspect.*, 85, 81, 1990.

Burns, L.A. and Munson, A.E., Gallium arsenide selectively inhibits T cell proliferation and alters expression of CD15 (IL-2R/p55), *J. Pharmacol. Exp. Ther.*, 265, 178, 1993.

Burns, L.A. et al., Evidence for arsenic as the primary immunosuppressive component of gallium arsenide, *Toxicol. Appl. Pharmacol.*, 110, 157, 1991.

Burns-Naas, L.A., Arsenic, cadmium, chromium, and nickel, in *Pulmonary Immunotoxicology,* Cohen, M.D., Zelikoff, J.T., and Schlesinger, R.B., Eds., Kluwer, Boston, 2000, 241.

Camner, P., Johansson, A., and Lundborg, M., Alveolar macrophages in rabbits exposed to nickel dust. Ultrastructural changes and effect on phagocytosis, *Environ. Res.*, 16, 226, 1978.

Castranova, V. et al., Effects of heavy metal ions on selected oxidative metabolic processes in rat alveolar macrophages, *Toxicol. Appl. Pharmacol.*, 53, 14, 1980.

Chaumard, C., Forestier, F., and Quero, A.M., Influence of inhaled cadmium on the immune response to influenza virus, *Arch. Environ. Health,* 46, 50, 1991.

Chen, L.C. and Gordon, T., Development of pulmonary tolerance in mice exposed to zinc oxide fumes, *Toxicol. Sci.*, 60, 144, 2001.

Christensen, M.M., Enrst, E., and Ellermann-Ericksen, S., Cytotoxic effects of hexavalent chromium in cultured murine macrophages, *Arch. Toxicol.*, 66, 347, 1992.

Cifone, M.G. et al., Cadmium inhibits spontaneous (NK), antibody-mediated (ADCC) and IL-2-stimulated cytotoxic functions of NK cells, *Immunopharmacology,* 20, 73, 1990.

Cifone, M.G. et al., Effects of cadmium on lymphocyte activation, *Biochim. Biophys. Acta,* 1011, 25, 1989.

Cohen, M.D. and Costa, M., Chromium, in *Environmental and Occupational Medicine,* 3rd ed., Rom, W.N., Ed., Lippincott-Raven, Philadelphia, 1998, p. 1045.

Cohen, M.D., Other metals: aluminum, copper, manganese, selenium, vanadium, and zinc, in *Pulmonary Immunotoxicology,* Cohen, M.D., Zelikoff, J.T., and Schlesinger, R.B., Eds., Kluwer, Boston, 2000, p. 267.

Cohen, M.D., Pulmonary immunotoxicology of select metals: aluminum, arsenic, cadmium, chromium, copper, manganese, nickel, vanadium, and zinc, *J. Immunotoxicol.*, 1, 39, 2004.

Cohen, M.D., Zelikoff, J.T., Chen, L.C., and Schlesinger, R.B., Immunotoxicologic effects of inhaled chromium: role of particle solubility and co-exposure to ozone, *Toxicol. Appl. Pharmacol.*, 152, 30, 1998.

Conner, M.W. et al., Lung injury in guinea pigs caused by multiple exposure to ultrafine ZnO: changes in pulmonary lavage fluid, *J. Toxicol. Environ. Health,* 25, 57, 1988.

Daniels, M.J. et al., Effects of $NiCl_2$ and $CdCl_2$ on susceptibility to murine cytomegalovirus and virus-augmented NK cell and interferon responses, *Fundam. Appl. Toxicol.*, 8, 443, 1987.

Daum, J.R., Shepherd, D.M., and Noelle, R.J., Immunotoxicology of cadmium and mercury on B-lymphocyte: effects on function, *Int. J. Immunopharmacol.*, 15, 383, 1993.

Everness, K.M. et al., Discrimination between nickel-sensitive and non-nickel-sensitive subjects by an *in vitro* lymphocyte transformation test, *Br. J. Dermatol.*, 122, 293, 1990.

Finch, G.L. et al., *In vitro* interactions between pulmonary macrophages and respirable particles, *Environ. Res.*, 44, 241, 1987.

Fine, J.M. et al., Metal fume fever: characterization of clinical and plasma IL-6 responses in controlled human exposures to zinc oxide fume at and below the threshold limit value, *J. Occup. Environ. Med.*, 39, 722, 1997.

Fujimaki, H., Murakami, M., and Kubota, K., *In vitro* evaluation of cadmium-induced alteration of the antibody response, *Toxicol. Appl. Pharmacol.*, 62, 288, 1982.

Funkhouser, S.W. et al., Cadmium inhibits IL-6 production and IL-6 mRNA expression in a human monocytic cell line, THP-1, *Environ. Res.*, 66, 77, 1994.

Gainer, J.H., Effects of arsenicals on interferon formation and action, *Am. J. Vet. Res.*, 33, 2579, 1972.

Gallagher, K.E. and Gray, I., Cadmium inhibition of RNA metabolism in murine lymphocytes, *J. Immunopharmacol.* 3, 339, 1982.

Galvin, J.B. and Oberg, S.G., Toxicity of hexavalent chromium to the alveolar macrophage *in vivo* and *in vitro*, *Environ. Res.*, 33, 7, 1984.

Gardner, D.E. et al., Alterations in bacterial defense mechanisms of the lung induced by inhalation of cadmium, *Bull. Eur. Physiopathol. Respir.*, 13, 157, 1977.

Gavett, S.H., Madison, S.L., Dreher, K.L., Winsett, D.W., McGee, J.K., and Costa, D.L., Metal and sulfate composition of residual oil fly ash determines airway hyperreactivity and lung injury in rats, *Environ. Res.*, 72, 162, 1997.

Glaser, U. et al., Low-level chromium(VI) inhalation effects on alveolar macrophages and immune functions in Wistar rats, *Arch. Toxicol.*, 57, 250, 1985.

Goebeler, M. et al., Activation of NF-κB and gene expression in human endothelial cells by the common haptens nickel and cobalt, *J. Immunol.*, 155, 2459, 1995.

Goebeler, M. et al., Nickel chloride and cobalt chloride, two common contact sensitizers, directly induce expression of ICAM-1, VCAM-1, and ELAM-1 by endothelial cells, *J. Invest. Dermatol.*, 100, 759, 1993.

Gonsebatt, M.E. et al., Inorganic arsenic effects on human lymphocyte stimulation and proliferation, *Mutat. Res.*, 283, 91, 1992.

Gordon, T. et al., Pulmonary effects of inhaled zinc oxide in human subjects, guinea pigs, rats, and rabbits, *Am. Ind. Hyg. Assoc. J.*, 53, 503, 1992.

Graham, J.A. et al., Effect of trace metals on phagocytosis by alveolar macrophages, *Infect. Immun.*, 11, 1278, 1975.

Hadley, J.G. et al., Inhibition of antibody-mediated rosette formation by alveolar macrophages: sensitive assay for metal toxicity, *J. Reticuloendothel. Soc.*, 22, 417, 1977.

Haley, P.J. et al., Immunopathologic effects of nickel subsulfide on primate pulmonary immune system, *Toxicol. Appl. Pharmacol.*, 88, 1, 1987.

Hassoun, E.A. and Stohs, S.J., Chromium-induced production of reactive oxygen species, DNA single-strand breaks, NO production, and lactate dehydrogenase leakage in J774A.1 cell cultures, *J. Biochem. Toxicol.*, 10, 315, 1995.

Hatch, G.E. et al., Correlation of effects of inhaled versus intratracheally injected metals on susceptibility to respiratory infection in mice, *Am. Rev. Respir. Dis.*, 124, 167, 1981.

Hernandez, M. et al., Cadmium and nickel modulation of adherence capacity of murine macrophages and lymphocytes. *Int. J. Biochem.*, 23, 541, 1991.

Hong, H.L., Fowler, B.A., and Boorman, G.A., Hematopoietic effects in mice exposed to arsine gas, *Toxicol. Appl. Pharmacol.*, 97, 173, 1989.

Hooftman, R.N., Arkesteyn, C.W., and Roza, P., Cytotoxicity of some types of welding fume particles to bovine alveolar macrophages, *Am. Occup. Hyg.*, 32, 95, 1988.

Horiguchi, H. et al., Cadmium induces interleukin-8 production in human peripheral blood mononuclear cells with the concomitant generation of superoxide radicals, *Lymphokine Cytokine Res.*, 12, 421, 1993.

Ishiyama, H. et al., Histopathological changes induced by zinc hydroxide in rat lungs, *Exp. Toxicol. Pathol.*, 49, 261, 1997.

Jaramillo, A. and Sonnenfeld, G., Potentiation of lymphocyte proliferative responses by nickel sulfide, *Oncology,* 49, 396, 1992a.

Jaramillo, A. and Sonnenfeld, G., Characteristics of inhibition of interferon production after crystalline nickel sulfide exposure, *Environ. Res.*, 57, 88, 1992b.

Jaramillo, A. and Sonnenfeld, G., Effects of amorphous and crystalline nickel sulfide on induction of interferons and IL-2, *Environ. Res.*, 48, 275, 1989.

Jaramillo, A. and Sonnenfeld, G., Effects of nickel sulfide on induction of LL-1 and phagocytic activity, *Environ. Res.*, 63, 16, 1993.
Johansson, A. et al., Lysozyme activity in ultrastructurally defined fractions of alveolar macrophages after inhalation exposure to nickel, *Br. J. Ind. Med.*, 44, 47, 1987.
Johansson, A. et al., Morphology and function of alveolar macrophages after long-term nickel exposure, *Environ. Res.*, 23, 170, 1980.
Johansson, A. et al., Rabbit alveolar macrophages after inhalation of hexa- and trivalent chromium, *Environ. Res.*, 39, 372, 1986.
Johansson, A. et al., Rabbit lungs after long-term exposure to low nickel dust concentrations. II. Effects on morphology and function, *Environ. Res.*, 30, 142, 1983.
Kapsenberg, M.L. et al., Functional subsets of allergen-reactive human $CD4^+$ T-cells, *Immunol. Today,* 12, 392, 1991.
Kapsenberg, M.L. et al., T_H1 lymphokine production profiles of nickel-specific $CD4^+$ T-lymphocyte clones from nickel contact allergic and non-allergic individuals, *Invest. Dermatol.*, 98, 59, 1992.
Karvonen, J. et al., HLA antigens in nickel allergy, *Ann. Clin. Res.*, 16, 211, 1984.
Kodovanti, U.P. et al., Pulmonary and systemic effects of zinc-containing emission particles in three rat strains: Multiple exposure scenarios, *Toxicol. Sci.*, 70, 73, 2002.
Koller, L.D. and Roan, J.G., Effects of lead, cadmium and methylmercury on immunological memory, *J. Environ. Pathol. Toxicol.*, 4, 47, 1980.
Krzystyniak, K. et al., Immunosuppression in mice after inhalation of cadmium aerosol, *Toxicol. Lett.*, 38, 1, 1987.
Kucharz, E.J. and Sierakowski, S.J., Immunotoxicity of chromium compounds: effect of sodium dichromate on T-cell activation *in vitro*, *Arh. Hig. Rada. Toksikol.*, 38, 239, 1987.
Kuehn, K., Fraser, C.B., and Sunderman, F.W., Jr., Phagocytosis of particulate nickel compounds by rat peritoneal macrophages *in vitro*, *Carcinogenesis,* 3, 321, 1982.
Kusaka, Y. et al., Lymphocyte transformation test with nickel in hard metal asthma: another sensitizing component of hard metal, *Ind. Health,* 29, 153, 1991.
Kuschner, W.G. et al., Early pulmonary cytokine responses to zinc oxide fume inhalation, *Environ. Res.*, 75, 7, 1997.
Kuschner, W.G. et al., Pulmonary responses to purified zinc oxide fume, *J. Invest. Med.*, 43, 371, 1995.
Labedzka, M. et al., Toxicity of metallic ions and oxides to rabbit alveolar macrophages, *Environ. Res.*, 48, 255, 1989.
Lantz, R.C. et al., Effect of arsenic exposure on alveolar macrophage function I. Effect of soluble As(III) and As(V), *Environ. Res.*, 67, 183, 1994.
Lawrence, D.A., Heavy metal modulation of lymphocyte activities. I. *In vitro* effects of heavy metals on primary humoral immune response, *Toxicol. Appl. Pharmacol.*, 57, 439, 1981.
Lazzarini, A. et al., Effects of chromium(VI) and chromium(III) on energy charge and oxygen consumption in rat thymocytes, *Chem. Biol. Interact.*, 53, 273, 1985.
Loose, S.D., Silkworth, J.B., and Simpson, D.W., Influence of cadmium on the phagocytic and microbial activity of murine peritoneal macrophages, pulmonary alveolar macrophages, and polymorphonuclear neutrophils, *Infect. Immun.*, 22, 378, 1978.

Lundborg, M., Johansson, A., and Camner, P., Morphology and release of lysozyme following exposure of rabbit lung macrophages to nickel or cadmium *in vitro*, *Toxicology*, 46, 191, 1987.

Meng, Z., Effects of arsenic on DNA synthesis in human lymphocytes, *Arch. Environ. Contam. Toxicol.*, 25, 525, 1993.

Novey, H.S., Habib, M., and Wells, I.D., Asthma and IgE antibodies induced by chromium and nickel salts, *J. Allergy Clin. Immunol.*, 72, 407, 1983.

Ohsawa, M. and Kawai, K., Cytological shift in lymphocytes induced by cadmium in mice and rats, *Environ. Res.*, 24, 192, 1981.

Ostrosky-Wegman, P. et al., Lymphocyte proliferation kinetics and genotoxic findings in a pilot study on individuals chronically exposed to arsenic, *Mutat. Res.*, 250, 477, 1991.

Otsuka, F. and Ohsawa, M., Differential susceptibility of T- and B-lymphocyte proliferation to cadmium: relevance to zinc requirement in T-lymphocyte proliferation, *Chem. Biol Interact.*, 78, 193, 1991.

Pasanen, J.T. et al., Cytotoxic effects of four types of welding fumes on macrophages *in vitro*: a comparative study, *J. Toxicol. Environ. Health*, 18, 143, 1986.

Payette, Y. et al., Decreased IL-2 receptor and cell cycle changes in murine lymphocytes exposed *in vitro* to low doses of cadmium chloride, *Int. J. Immunopharmacol.*, 17, 235, 1995.

Port, C.D. et al., Interaction of nickel oxide and influenza infection in the hamster, *Environ. Health Perspect.*, 10, 268, 1975.

Prieditis, H. and Adamson, I., Comparative pulmonary toxicity of various soluble metals found in urban particulate dusts, *Exp. Lung Res.*, 28, 563, 2002.

Rosenthal, G.J. et al., Effect of subchronic arsine inhalation on immune function and host resistance, *Inhal. Toxicol.*, 1, 113, 1989.

Sikorski, E.E. et al., Immunotoxicity of the semiconductor gallium arsenide in female $B_6C_3F_1$ mice, *Fundam. Appl. Toxicol.*, 13, 843, 1989.

Sikorski, E.E. et al., Suppression of splenic accessory cell function in mice exposed to gallium arsenide, *Toxicol. Appl. Pharmacol.*, 110, 143, 1991.

Sinigaglia, F. et al., Isolation and characterization of Ni-specific T-cell clones from patients with Ni-contact dermatitis, *J. Immunol.*, 135, 3929, 1985.

Skornik, W.A. and Brain, J.D., Relative toxicity of inhaled metal sulfate salts for pulmonary macrophages, *Am. Rev. Respir. Dis.*, 128, 297, 1983.

Sonnenfeld, G., Streips, U.N., and Costa, M., Differential effects of amorphous and crystalline NiS on murine interferon production, *Environ. Res.*, 32, 474, 1983.

Spiegelberg, T., Kordel, W., and Hichrainer, D., Effects of NiO inhalation on alveolar macrophages and humoral immune system of rats, *Ecotoxicol. Environ. Safety*, 8, 516, 1984.

Takahashi, S. et al., Cytotoxicity of nickel oxide particles in rat alveolar macrophages cultured *in vitro*, *J. Toxicol. Sci.*, 17, 243, 1992.

Theocharis, S.E., Souliotis, T., and Panayiotisid, P., Suppression of IL-1β and TNFα bio-synthesis by cadmium in *in vitro* activated human peripheral blood mononuclear cells, *Arch. Toxicol.*, 69, 132, 1994.

Thomas, P.T. et al., Evaluation of host resistance and immune function in cadmium-exposed mice, *Toxicol. Appl. Pharmacol.*, 80, 446, 1985.

Warner, G.L. and Lawrence, D.A., The effects of metals on IL-2-related lymphocyte proliferation, *Int. J. Immunopharmacol.*, 10, 629, 1988.

Wiernik, A. et al., Rabbit lung after inhalation of soluble nickel. I. Effects on alveolar macrophages, *Environ. Res.*, 30, 129, 1983.

Wildner, O., Lipkow, T., and Knop, J., Increased expression of ICAM-1, E-selectin, and VCAM-1 by cultured human endothelial cells upon exposure to haptens, *Exp. Dermatol.*, 1, 191, 1992.

Yoshida, T., Shimamura, T., and Shigeta, S., Enhancement of the immune response *in vitro* by arsenic, *Int. J. Immunopharmacol.*, 9, 411, 1987.

10

ANIMAL MODELS OF RESPIRATORY ALLERGY

Jürgen Pauluhn

CONTENTS

10.1 INTRODUCTION

Allergic asthma is a complex chronic inflammatory disease of the airways, and its etiology is multifactorial. It involves the recruitment and activation of many inflammatory and structural cells, all of which release inflammatory mediators that result in pathological changes typical of asthma (Barnes et al., 1998; O'Byrne and Postma, 1999). Several features of asthma can be suitably investigated in animal models. These features include cellular infiltrations in the lung, antigen-specific IgE production, and a predominant Th2-type immune response characterized by elevations in the levels of typical cytokines seen upon allergen (hapten) sensitization and challenge. The number of mediators involved in the process of sensitization to an allergen and/or the development of a chronic inflammatory process in the mucosa of the lower airways, including airway remodeling, indicates that these processes are extremely complex. *Airway hyperreactivity* is defined as an exaggerated acute obstructive response of the airways to one or more nonspecific stimuli. Increased hyperresponsiveness may also be associated with direct exposure of sensory nerve endings (Barnes, 1986) or with the loss of enzymes that metabolize sensory neuropeptides (Frossard et al., 1989; Barnes et al., 1991). Neurotrophins are important mediators between the (systemic) immune system and the local nervous system (Carr et al., 2001). They are produced locally during allergic reactions and serve as amplifiers of Th2 effector functions and, thus, play an important role in the development of inflammation and airway hyperresponsiveness (AHR) (Braun et al., 2000). This demonstrates that the microenvironment adjacent to the site of injury and/or sensitization and challenge may be important for the progression of disease. Commonly, animals are sensitized by two successive encounters for the test agent, followed by challenge with the antigen (or hapten). This primary allergen challenge results in an asthmatic phenotype. However, secondary allergen challenges, after prolonged gaps, are often used so that the phenotype more closely resembles the human disease. The route, method, and dose of allergen exposure also determine the phenotype of the allergen response (Kannan and Deshpande, 2003). The response to short-term, high-level exposures causes airway lesions different from those caused by low-level, chronic antigen challenge. Also, the methods used to assess immediate or delayed bronchoconstriction, the kind and extent of airway inflammation, and tissue remodeling have an impact on the outcome of studies involving allergen (hapten)-sensitized animal models of asthma.

The emphasis of this chapter is to review methodologies suitable for identification of respiratory allergens for the purpose of hazard identification in animal models currently used in toxicology. This chapter includes an analysis of how the test design, i.e, route, dose, frequency of dosing, timing of challenge exposures with the hapten or antigen, and the end point selected, can affect the outcome of a study. The validity and limitations of the various approaches are also discussed.

10.2 MECHANISMS LEADING TO AN ASTHMATIC PHENOTYPE

Asthma is characterized by allergic chronic airway inflammation and related inflammatory mediator cell products that orchestrate the disease. Much of our understanding of the response to allergens comes from the study of animal models, especially guinea pigs, rats, and mice. These investigations have not only shed light on the immunologic mechanisms underlying allergy, but have also allowed evaluation of potential therapeutic agents. Mediator interaction may also occur by "priming" of inflammatory cells, leading to augmented release of secondary mediators. This implies some uncertainty as to whether exogenously administered agents exert similar effects when the exposure is direct (by inhalation) or via systemic routes. To date, more than 50 different mediators have been identified in asthma, although, at the same time, current gene technology identifies an ever-increasing range of molecules that could be involved in the process of sensitization to an allergen and/or the development of a chronic inflammatory process in the mucosa of the lower airways. The intricate interaction of structural and mobile cell populations and the mediators that orchestrate their influx and activation, including their predilection to induce a somewhat self-perpetuating and self-amplifying type of inflammation, is not yet fully understood. Mediators may act synergistically to enhance each other's effect, or one mediator may modify the release or action of another.

This disease is clinically characterized by the recruitment and activation of specific inflammatory cells, chemotaxis, bronchoconstriction, increased airway secretion (and mucous cell hyperplasia), plasma exudation, neural effects, hyperplasia and hypertrophy of airway smooth muscle cells, and increased AHR defined as an exaggerated acute obstructive response of the airways to one or more nonspecific stimuli and is often associated with airway epithelial damage and disruption, a common feature of even mild asthma (Laitinen et al., 1985). How this increased bronchial responsiveness is triggered, amplified, and sustained and how it relates to inflammatory events remain, to a certain extent, incompletely elucidated (Kimber and Dearman, 1997). Many inflammatory cells are recruited to asthmatic airways or are activated *in situ*. These include mast cells, macrophages, eosinophilic and neutrophilic granulocytes, T lymphocytes, dendritic cells, basophils, and platelets. Although much attention has been paid in the past to acute inflammatory responses, such as bronchoconstriction, plasma exudation, and mucus hypersecretion, in asthma, it is increasingly recognized that chronic inflammation and airway remodeling are the most important aspects (Roche et al., 1989). Some of these changes may be irreversible, leading to fixed narrowing of the airways. The activation and coordination of allergic inflammatory or immune responses depend on a complex interaction of multiple cell populations and diverse mechanisms. Numerous experimental and clinical studies have established that $CD4^+$ T cell-mediated inflammation of airways is central to the pathogenesis of asthma (Foster et al., 2002). According to dogma, allergic airway responses are dependent on Th2-type cytokines, including local production of IL-4, IL-5, and IL-13, and signal transduction proteins (Wils-Karp, 1999). Several studies have suggested that Th1 cytokines actually contribute to the overall allergic microenvironment and play a significant role in detrimental responses. Thus, the Th2-type cytokines produced during asthmatic or allergic responses may not be the only cytokines needed for inflammatory responses in the development of severe asthma (Raman et al., 2003) and may explain IgE-independent mechanisms of allergic inflammation (Wilder et al., 1999; Scheerens et al., 1999). The components of the airway wall that are critical in determining the response to inhaled substances are organized in the epithelial mesenchymal trophic unit (Evans et al., 1999). The basic assumption is that the epithelial, interstitial, nervous (neural), and immunological components have an active interaction with each other. Neural control of airways, including axon reflexes, is complex, and the extent of contribution of neurogenic mechanisms to the pathogenesis of airway disease is still uncertain. Perturbation of one compartment creates imbalance in all parts of the airway wall; rather, a metabolic response in one compartment will produce alterations in the other compartments. This demonstrates that the microenvironment adjacent to the site of injury and/or sensitization and challenge may be important for the progression of disease.

For high-molecular-weight agents, the mechanisms of sensitization by which these incriminated agents cause asthma are believed to be IgE mediated, whereas for low-molecular-weight agents, the immunologic and nonimmunologic factors that incite and sustain the inflammatory cascade, which ultimately results in clinical asthma, are still not well understood. The reasons for the induction of immediate hypersensitivity to the antigen mediated by a predominant IgE and IgG_1 response may be in the properties of the antigen or allergen itself, the time frame investigated, or the route of sensitization.

10.3 AGENTS THAT CAUSE RESPIRATORY ALLERGY

Agents used in animal models have broadly been categorized into low-molecular-weight chemical haptens (molecular weight < 1000 Da) and high-molecular-weight agents that include proteins from diverse sources such as plants, bacteria, athropods, or animals. Most low-molecular-weight agents are reactive chemicals and act as allergens when bound to appropriate carrier molecules, such as autologous proteins, forming a complete hapten-carrier antigen. The antigenic determinant may be the hapten or it may be a new antigenic determinant formed by a chemical reaction between the hapten and the carrier protein. All low-molecular-weight agents believed to be etiologic share a common toxicological characteristic of being irritants. In some cases, these agents are present as a gas; in others, the inciting agent is an aerosol or mixtures of a volatile hapten partitioned with the aerosol phase. For these types of irritant chemicals, it is not yet clear, for instance, whether induced AHR is a dose-dependent phenomenon or whether a brief high-level exposure plays a more important role. The most prevalent class of low-molecular-weight asthmagens includes platinum salts, acid anhydrides (e.g., trimellitic anhydride [TMA] and phthalic anhydride), diisocyanates (such as diphenylmethane diisocyanate [MDI], toluene diisocyanate [TDI], and hexamethylene diisocyanate [HDI]), reactive dyes, and derivatives of cyanuric chloride, to mention a few. For a more comprehensive discussion of agents, see Selgrade et al. (1994).

The influence of coexposure of particles (e.g., alum and airborne particles) and/or irritants (e.g., ozone) along with allergens has provided direct evidence of adjuvant activity or allergen-induced atopy-exacerbating airway dysfunction upon exposures to these irritants or particles (Steerenberg et al., 2003). Coexposures of particles or irritants along with allergens have shown a profound increase in immunoglobulins (IgE and IgG) and cytokines typifying a predominant Th2 response and an influx of eosinophils; however, the timing of coexposures (concurrent vs. sequential) appears to be critical to the outcome of a test. Last et al. (2004) showed that ozone appears to antagonize the specific inflammatory effects of ovalbumin exposure, especially when given before or during exposure to ovalbumin.

10.4 ADVANTAGES AND DISADVANTAGES OF CURRENT ANIMAL MODELS

Studies of allergic sensitization employed single and multiple sensitization dosage regimens with concentrations ranging from nonirritant (low dose) to irritant (high dose). Routes may be injection (intra- or subcutaneous and intraperitoneal with or without adjuvant), epicutaneous exposure, bolus instillation (intratracheal or intranasal), or steady-state inhalation exposures over short or long periods of time (Sarlo and Karol, 1994). Little attention has been directed toward the identification and quantification of induction-related local irritant or inflammatory responses and the ensuing systemic effects. Thus, it is difficult to judge the impact of a specific protocol design on specific findings common factors among the protocols. With regard to antibody production (e.g., specific IgG_1 and total IgE), a relationship between concentration and frequency of booster, or challenge exposure, has been noted in many studies. For example, Kumar et al. (2000) found that although IgG-synthesizing cells were undetectable in normal airways, the overwhelming majority of immunoglobulin-producing cells in mice chronically challenged by inhalation with aerosolized antigen exhibited cytoplasmic immunoreactivity for IgG. The local immunoglobulin production does not contribute significantly to the induction of an IgE-mediated response. Animal models can be tailored for the production of IgE; even in guinea pigs, large doses of antigen tend to support earlier and more intense bronchopulmonary hypersensitivity and favor IgG, whereas smaller doses favor mixed IgE and IgG production (Pretolani and Vargaftig, 1993). The exposure concentration of the free chemical to elicit pulmonary responses in short-term animal models is usually much higher than that recognized to elicit responses in humans. The available experimental evidence appears to suggest that a threshold concentration needs to be exceeded for responses to occur. The high-level exposure commonly used in animal models triggers acute inflammation in the lung parenchyma, especially perivascular and peribronchiolar inflammation, which is markedly different from the acute-on-chronic inflammation of the airway wall observed in human asthma (Cohn, 2001; Kumar and Foster, 2002). Moreover, the pattern of acute inflammation differs in several respects from that observed in individuals with asthma.

Perhaps the most significant deficiency of the commonly used models of allergic bronchopulmonary inflammation is that, by their very nature, they involve short-term experiments. Thus, they do not exhibit many of the lesions that typify chronic human asthma, such as the intraepithelial accumulation of eosinophils in the intrapulmonary airways, chronic inflammation of the airway wall, and its remodeling, which may have important consequences for both airway hyperreactivity and development of fixed airflow obstruction (Kumar and Foster, 2002). Whether or not these divergent and conflicting results from animal models and humans are related to the extreme variety of sensitization and challenge protocols employed, including the different readouts to define disease or response to challenge, remains obscure. Differences in susceptibilities related to the dosing of critical structures within the respiratory tract of obligate-nasal-breathing laboratory animals and oronasal-breathing humans inflicted with airway disease contributes to the difficulty of defining "most appropriate" animal model. Moreover, for highly reactive low-molecular-weight chemicals, there is need for the synthesis of a chemical-carrier protein conjugate to elicit pulmonary hypersensitivity when the challenge by inhalation is to adequately dose the critical target within the lung. However, when instilled intranasally, positive responses were shown to occur with the nonconjugated hapten also (Vanoirbeek et al., 2003; Scheerens et al., 1996).

Despite these variations, animal models are indispensable in studies requiring an intact immune system, especially for studying the pathogenetic mechanisms in atopic diseases, modulator regulation, and related biologic effects. When designing an *in vivo* model of respiratory allergy, it is important to clearly decide which of the aspects of this complex disease are the focus. In addition to expedience and practicability, the species and strain selection, sensitization and challenge protocols, as well as the end points chosen are vital to the outcome of the test. The relative abundance of biomarkers to probe disease may change from one compartment to another. For example, it may be different in serum, lung parenchyma, and draining lymph nodes, or in bronchoalveolar lavage fluid (BALF) (Schuster et al., 2000). End points that independently integrate a series of complex events might be most practical for probing positive response in animal models to detect lung-sensitizing agents. Suffice it to say, the limitations of most of the currently applied bioassays are that they model the acute, rather than the chronic, manifestation of asthma. Apart from the difficulties in extrapolating rodent data to human data, it has to be borne in mind that each animal model is for a particular trait associated with asthma, rather than for the entire asthma phenotype (Kips et al., 2003).

10.5 SPECIES SELECTION

Several rodent and nonrodent species have been used for respiratory allergy studies. The choice of species is made primarily with a view to extrapolate the experimental results to man. However, when selecting a particular species for a study, additional factors must be considered, such as the comparative morphology of the respiratory tract, the presence or absence of lung disease or susceptible states, and the similarity of biochemical and physiological responses to those in man. The choice of animal models may also be based on practical considerations rather than their validity for use in human beings. An animal species must be small enough to allow handling and exposure in sufficient numbers in relatively small inhalation chambers, and large enough to allow measurement of all end points relevant to identifying the inherent toxicity of the substance under investigation. Most of the animal models are associated with allergic alveolitis or hypersensitivity pneumonitis, which may overshadow the inflammatory lesions of the airways. However, use of controlled exposure to low mass concentrations of aerosolized antigen succeeded in eliciting acute-on-chronic allergic inflammation of the airways without an accompanying alveolitis. Diisocyanate-induced extrinsic alveolitis has been reported to occur in humans also (Mapp et al., 1985; Vandenplas et al., 1993; Baur, 1995). Notwithstanding the wealth of valuable data that have been obtained in these models, their outcome *per se* is highly dependent on protocol design variables.

An obvious advantage of the current mouse models is the availability of genetically characterized inbred strains and genetically manipulated strains at reasonably low costs. This species allows for the *in vivo* application of an extremely wide diversity of immunological tools, including gene deletion technology. However, in this species, because of the poorly developed bronchial musculature, it is more difficult to detect bronchoconstriction. Mice subjected to repeated challenge protocols exhibit abnormalities of airway epithelium similar to those observed in human asthma, as well as evidence of airway hyperreactivity to methacholine. Remarkable strain-specific differences were found in mice (Shinagawa and Kojima, 2003; Whitehead et al., 2003; Takeda et al., 2001). However, it could be speculated that at least some of the differences among different strains reflect the differences in the method and, especially, the associated dose of airway antigen exposure. Excellent papers are available on the advantages and disadvantages of murine systems in the study of asthma pathogenesis (Tu et al., 1995; Gelfand, 2002; Dearman et al., 1992; Dearman, Betts et al., 2003; Dearman, Spence et al., 2003).

Many features of respiratory allergy can be aptly modeled in the high-IgE-responding Brown-Norway (BN) rat (Arts et al., 1997, 1998; Schuster et al., 2000; Cui et al., 1997). BN rats have displayed both immediate- and delayed-phase responses following aeroallergen challenge of sensitized animals, airway hyperreactivity to methacholine, acetylcholine or serotonin, and the accumulation of neutrophils, lymphocytes, and, particularly, activated eosinophils in lung tissue and BALF. However, the rat is only a weak bronchoconstrictor, and higher levels of agonist are required to induce the same level of response as compared to guinea pigs. Thus, this animal model focuses on the induction of airway inflammation and comprises most of the characteristic features of asthma, depending on the protocol employed.

The guinea pig model for assessment of pulmonary allergenicity of low-molecular-weight chemicals and proteins, including pharmacological intervention studies, has been used extensively. (Briatico-Vangosa et al., 1994; Blaikie et al., 1995; Sun and Chung, 1997; Griffiths-Johnson and Karol, 1991; Hayes and Newman-Taylor, 1995; Hayes et al., 1992a,b; Karol and Thorne, 1988; Karol, 1983, 1994; Pauluhn and Eben, 1991; Pauluhn, 1994a; Pauluhn and Mohr, 1998; Sarlo and Clark, 1992; Sarlo and Karol, 1994; Sugawara et al., 1993; Welinder et al., 1995). The particular advantages and disadvantages of this animal model have been reviewed extensively (Bratico-Vangosa et al., 1994; Sarlo and Karol, 1994; Karol, 1994; Pretolani and Vargaftig, 1993).

Conventionally, the guinea pig has been the species of choice for toxicological evaluation of chemical-related respiratory allergy, primarily because it is possible to elicit and measure with relative ease, in this species, the challenge-induced pulmonary reactions that in some ways resemble the acute clinical manifestations of human allergic asthma. The guinea pig is known to respond vigorously to inhaled irritants by developing an asthmatic-like bronchial spasm. This species possesses a developed bronchial smooth muscle, which contracts intensively and rapidly in response to *in vivo* or in vitro exposure to antigen. This is an anatomical prerequisite for both the expression of bronchoconstriction of the immediate hypersensitivity reaction, which evolves in minutes, and for its late component, which evolves in hours. This anatomical feature renders this species especially susceptible to a nonspecific airway hyperreactivity bronchoconstrictive as well as a specific hypersensitivity response. Therefore, this species has been used for decades for the study of protein-evoked anaphylactic shock and pulmonary hypersensitivity, and can experience both immediate-onset and late-onset responses. However, they are quite prone to respiratory tract infections, which limit their use for chronic inhalation studies (Chengelis, 1990). Guinea pig anaphylactic responses usually involve IgG_1 antibodies even though the model can be tailored for the production of IgE. Large doses of antigen tend to support earlier and more intense bronchopulmonary hypersensitivity and favor IgG, whereas smaller doses favor mixed IgE and IgG production (Pretolani and Vargaftig, 1993). As with other animal models, the inflammatory response varies according to the protocol used.

10.6 MANIFESTATION OF RESPIRATORY ALLERGY

Bronchial hyperreactivity and symptoms of airway obstruction clinically characterize respiratory allergy (Temelkovski et al., 1998). Underlying the clinical manifestations is acute-on-chronic inflammation of the airway mucosa, with degranulation of mast cells, recruitment of eosinophils and neutrophils, as well as accumulation of activated T lymphocytes and other chronic inflammatory cells. Pulmonary parenchymal inflammation is not a significant feature of asthma, although limited numbers of eosinophils may be present in the alveoli immediately adjacent to the airways involved. Abnormalities of the airway epithelium are prominent, including shedding and regeneration of ciliated cells, goblet cell hyperplasia and/or metaplasia, as well as a distinctive pattern of subepithelial fibrosis, which, in humans, reverses upon pharmacological intervention (Fabbri et al., 1998).

To elicit changes comparable to those of human asthma, it is necessary to allow for repeated inhalation exposure to the agent under consideration. Of considerable interest is the evidence that airway hyperreactivity does not correlate with IgE titers or with the magnitude of the inflammatory response in the airway wall (Temelkovski et al., 1998). Indeed, this model is relatively labor intensive and requires specialized equipment but does succeed in achieving a new level of fidelity of asthmatic lesions. For highly reactive low-molecular-weight volatile chemicals, acute alveolitis or hypersensitivity pneumonitis often confound the inflammatory response localized in airways. Although the inflammatory response within the airways has an atopic or allergic basis in at least two thirds of asthmatic patients (McFadden and Gilbert, 1992), the mechanisms involved in the development of airway obstruction remain poorly understood. Accumulation of inflammatory cells and exudate, increased airway smooth muscle mass, deposition of connective tissue, and epithelial hyperplasia may all contribute to thickening of the airway wall, which appears to be the basis for the excessive diminution of airflow that accompanies bronchoconstriction in asthmatics. The respiratory epithelium has long been recognized to be the primary barrier protecting the underlying tissue from inhaled irritants and noxious stimuli. The loss of this physical barrier function could be envisaged as playing a role in hyperreactivity by permitting greater access of bronchoconstrictor stimuli to the underlying smooth muscle. Also, the metabolic activity of this barrier is increasingly appreciated.

Investigation of the pathophysiological mechanisms of chronic asthma has been limited by the lack of a satisfactory animal experimental model. Attempts have been made to model the disease in a number of species (Karol, 1994; Fügner, 1985; Bice et al., 2000). The relevance of experimental systems employing antigens that bear little relationship to triggers of human asthma or using pathophysiologically inappropriate methods for the delivery of these antigens is debatable. However, for highly reactive agents that are preferentially deposited in the upper airways of obligate-nasal-breathing laboratory rodents, adequate doses of antigenic determinants can only be delivered when the irritant properties of the chemical are concealed.

Accordingly, for such agents, hapten inhalation challenge exposures are biased to be (false) negative, whereas immediate-onset respiratory responses are commonly observed upon hapten-conjugate challenges. Models based upon sensitization by systemic administration of protein antigens, such as ovalbumin, and subsequent (repeated) inhalation challenges have gained widespread acceptance when high-molecular-weight antigens are used.

10.7 END POINTS OF RESPIRATORY ALLERGY

As detailed earlier, a common pathologic accompaniment or cause of increased AHR is prolonged airway inflammation and remodeling after inhalation of specific haptens or antigens. It is suggested that inflammation and/or remodeling are responsible for the change in increased agonist responsiveness. In contrast to the assays relying on an induction of a specific set of characteristic immunologic biomarkers, some models do not depend on a preconceived mechanism of sensitization and response. Rather, they function by reproducing the characteristics, which typify the hypersensitivity reactions, i.e., the immediate-onset physiologic response of the airways, which is bronchoconstriction. The clinical manifestation of respiratory allergy is not characterized by a unique end point, except anaphylaxis that occurs in guinea pigs. Depending on the type of plethysmography applied, an analysis of additional characteristics typifying a change in breathing patterns is useful. In spontaneously breathing animals, in addition to measurement of respiratory rate, flow-derived end points, such as peak inspiratory and expiratory flows, and respiratory-pattern-related end points, such as inspiratory, expiratory, and relaxation times and enhanced pause (Penh), have been shown to improve the sensitivity of analysis (Pauluhn, 2003, 2004). As detailed earlier, the magnitude of respiratory changes is markedly less pronounced in mice and rats as compared to guinea pigs.

The delayed-onset physiologic response is more difficult to quantify by physiological methods because of the spontaneous activities in unrestrained animals in whole-body plethysmographs (see the following text); however, the more sustained inflammatory component of this response can be suitably probed by bronchoalveolar lavage (BAL) and histopathology. However, the characteristic features of the acute and chronic types of allergic inflammation of airways also need to be considered. These end points appear to depend on the intensity of the priming response at the site of sensitization. For a better appreciation of dose- and concentration-related effects observed following challenge, the specific responses occurring at the site of induction, including draining lymph nodes, should be observed.

10.8 ANTIBODY DETERMINATIONS AND BIOMARKERS OF EXPOSURE

Recent studies have focused on the initial reactivity of diisocyanates with target airway macromolecules and peptides, including glutathione and lung epithelial proteins. Specific airway epithelial cell proteins that are selectively adducted by diisocyanates have been identified (Redlich and Karol, 2002). Moreover, HDI bound either to keratin-18 or keratin-10 was identified in human endobronchial and skin biopsy samples as the major protein adducted with HDI following inhalation and dermal exposure, respectively. In BALF albumin was found to be the major HDI-adducted protein (Wisnewski et al., 2000).

Common serological measurements assume that the end points addressed in the systemic circulation are reflective of changes at the sites of allergic inflammation. Before testing for specific immunologic responses to chemicals, appropriate antigens must be prepared and fully characterized to determine the number and kind of chemical ligands bound to each protein carrier molecule. Especially for multifunctional chemicals, the selection of the most "antigenic" hapten–protein conjugate is critical because different conjugates have to be tested against *in vivo* antisera, assuming that the synthesized conjugate that is recognized best by the antiserum is likely to be most similar to the conjugate formed *in vivo* (Pauluhn et al., 2002a). Accordingly, the quality and degree of conjugation affects recognition in the ELISA assay. Subtle differences in laboratory procedures may result in conjugates of different antigenic activity, which renders the comparison among different laboratories difficult.

Measurements of total serum IgE in the mouse and BN rat are considered end points for the identification and assessment of chemicals used to induce respiratory allergy. Especially in mice, this test is further extended using a cytokine profiling (Th1 vs. Th2), in which cytokine levels produced by local draining lymph node cells are measured after topical induction (Dearman, Betts et al., 2003; Dearman, Spence et al., 2003; Kimber and Dearman, 2002). In guinea pigs, successful sensitization to a low-molecular-weight chemical (hapten) or allergen has been demonstrated by the assessment of anti-IgG_1 antibody. Andersson (1980) demonstrated that low-dose antigen causes production of both IgE and IgG in the guinea pig, whereas only IgG_1 was produced when guinea pigs were sensitized with larger amounts of antigen. Repeated exposure to ovalbumin has been reported to produce immunological tolerance (Schramm et al., 2004) and enhance IgE production in the BN rat (Siegel et al., 1997).

Dose–response relationships are observed between the induction dose (in terms of exposure concentration and number of exposures) and antibody titers (Siegel et al., 2000; Pauluhn and Eben, 1991; Potter and Wederbrand, 1995). In studies using various concentrations and fractional doses to obtain similar cumulative doses, it was shown that the apparent threshold for antibody production significantly increases with the number of applications of TDI (Potter and Wederbrand, 1995). A synopsis of diverse repeated-inhalation-exposure studies with HDI homopolymers in Wistar rats and guinea pigs demonstrates an apparent relationship between acute lung irritation and (sub)chronic lung inflammation in rats and an induction of IgG_1 antibodies in guinea pigs (Pauluhn, 2000). Hence, it appears that increased titers of IgG_1 antibodies could only be detected at exposure concentrations causing membranolytic or cytotoxic effects of airway epithelial cells. In guinea pigs sensitized to the HDI monomer by three intradermal injections ($3 \times 0.3\%$) or to the HDI homopolymer by six intradermal injections ($6 \times 30\%$), almost equal IgG_1 antibody titers were observed, showing that the monomer is a more potent inducer of specific IgG_1 antibodies than the corresponding homopolymer (Pauluhn et al., 2002a). Thus, for complex mixtures containing chemically related monomers and homopolymers, antibody cross-reactivity might preclude a robust analysis of the causal relationship of mixed monomer or homopolymer exposure and antibody response.

Dose–response relationships were observed between the cumulative induction doses and antibody titers, respiratory hyperresponsiveness, or morphological changes. However, the animals experiencing marked respiratory reactions upon challenge were not always associated with an increased magnitude of IgG or IgG_1 antibody response. This means that dose–response curves based on specific IgG_1 and severity of pulmonary response (anaphylaxis) do not necessarily parallel those in guinea pigs sensitized to and challenged with ovalbumin (Figure 10.1).

It appears to be generally accepted that a specific IgG_1 antibody response provides the potential for the elicitation of a pulmonary response; however, there is no clear relationship between antibody titer and pulmonary responsiveness (Blaikie et al., 1995; Pauluhn and Eben, 1991; Lushniak et al., 1998). However, the exact role of locally synthesized IgG in the pathogenesis of asthma is unknown. In mice, IgG may play a role in type I hypersensitivity responses (Kumar et al., 2000). Recently, an alternative to the guinea pig anaphylactic antibody model has been proposed, which assesses the formation of specific IgG_1 antibodies in a mouse intranasal test (MINT), the assumption being that a specific IgG_1 antibody is a surrogate for anaphylactic antibody in this species. However, it was concluded that this model requires substantial further validation before it can be adopted as an alternative model (Blaikie and Basketter, 1999).

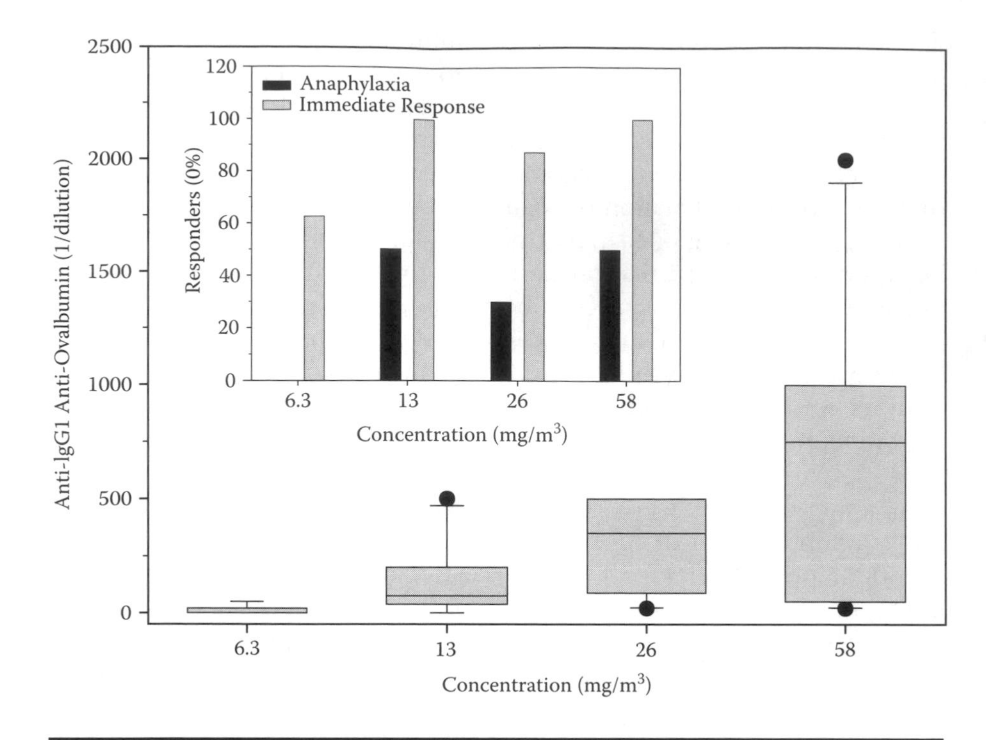

Figure 10.1 Guinea pigs were sensitized to ovalbumin aerosol on four consecutive days (days 0–3, duration: 15-min/d, concentrations: 6 ± 1.9, 13 ± 2.2, 27 ± 5.9, and 58 ± 14.4 mg/m², mass median aerodynamic diameter [MMAD] ≈1.5 µm). A tukey box plot of IgG_1 antibodies (ELISA) from blood collected prior to challenge with aerosolized ovalbumin is shown. The boundaries of the box represent the 10th and 90th percentiles; the medians are displayed as solid lines. The insert shows the percentage of guinea pigs (eight per group) displaying either a change in respiratory rate or anaphylaxis.

10.9 IMMEDIATE- OR DELAYED-ONSET PHYSIOLOGICAL RESPONSE

The analysis of responses associated with the clinical manifestation of respiratory allergy is based on the analysis of immediate- and/or delayed-onset pulmonary sensitivity using pulmonary function tests. There are integrative tools available to inhalation toxicologists to assess the function of the lung. Measurements can be made during or after exposure in spontaneously breathing conscious animals by using various types of plethysmographs (e.g., whole-body, bias-flow or barometric and nose-only, and volume displacement). Investigators also used more in-depth measurements in anesthetized, cannulated animals for the determination of lung resistance (R_L) and dynamic compliance (C_{dyn}) (Hamelmann et al., 1997) or for forced maneuvers (Pauluhn et al., 1995). The definition of positive responses and their categorization varies from one method to another and from one laboratory to another (Karol, 1983; Sarlo and Clark, 1992; Botham et al. 1988; Botham et al., 1989; Blaikie et al., 1995; Pauluhn, 1997; Pauluhn et al., 2002b). Changes in respiratory rates and patterns are commonly recorded by either plethysmographic fluctuations of pressure or by signs of respiratory distress. Most protocols employ single and/or repeated challenge periods of 10–30 min to elicit respiratory hyperresponsiveness. Sometimes, measurements are followed by extended postchallenge measurements for the detection of delayed-onset reactions (Alarie et al., 1990; Karol et al., 1985; Pauluhn and Eben, 1991; Zhang et al., 2004; Thorne and Karol, 1988).

Ideally, respiratory patterns are measured during the challenge exposure in volume displacement plethysmographs (Figure 10.2). For this purpose nose-only animal restrainers are equipped with a wire mesh pneumotachograph and a differential pressure transducer fitted directly onto the exposure restrainer (plethysmograph). The head and body compartments are separated using a double-layer latex neck seal separated by a stabilizer. Precautions have to be taken to avoid artifacts due to restraint and tight-fitting seals around the neck. This type of plethysmograph provides a means

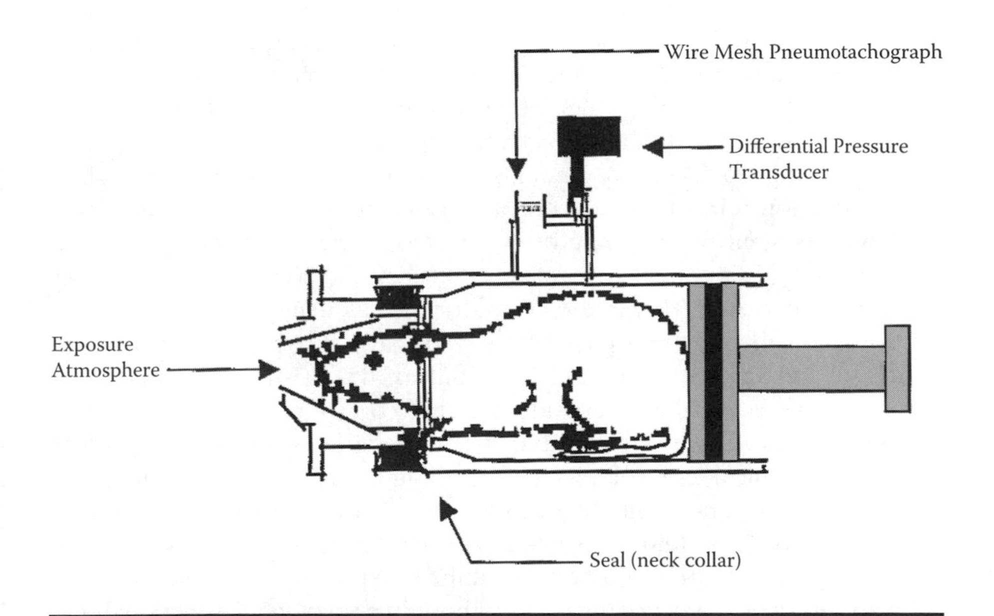

Figure 10.2 A rat in a volume displacement plethysmograph for measurement of respiratory patterns during challenge.

to challenge animals to well-characterized atmospheres with regard to actual breathing-zone concentrations and particle size (Pauluhn and Eben, 1991; Pauluhn, 1994b; Pauluhn and Mohr, 2000) as well as to analyze reflexively induced changes in breathing patterns. Confounding effects due to habituation and grooming usually as observed in whole-body plethysmographs do not occur. The tidal volumes calculated based on box pressure changes usually differ from those obtained by volume displacement plethysmography because of the difficulty of precise volume measurements in whole-body barometric plethysmographs (Epstein and Epstein, 1978). In contrast, measurements in barometric, whole-body chambers allow for a continuous measurement of the box pressure–time wave, do not require restraint, and, therefore, are suitable for the analysis of postchallenge respiratory reactions delayed in onset, as suggested by Zhang et al. (2004).

Common parameters derived from the breathing pattern and breath structure include the breathing frequency (or respiratory rate, RR), tidal volume (TV), inspiratory and expiratory times (T_i and T_e), and peak flow rates during inspiration and expiration (PIF and PEF). Analysis of additional parameters may be useful to better understand the quality of changes observed, i.e., whether they originate from the stimulation of receptors on the upper or lower respiratory tract or if they are related to lung resistance. These include the relaxation time, RT, which is commonly defined as the time from the start of expiration to the return to 35% of TV, apneic pause (pause = Te/RT − 1), end inspiratory pause (EIP), which is a measure of the pause between inspiration and expiration and enhanced pause (Penh = pause × PEF/PIF). Further details are given elsewhere (Chong et al., 1998; Hamelmann et al., 1997; Kimmel et al., 2002; Epstein and Epstein, 1978; Drorbough and Fenn, 1955).

Depending on the agent under study, concentrations high enough to cause irritation-related stimulation of sensory nerve endings in the upper or lower respiratory tract may be associated with a decrease or increase in respiratory frequency, respectively (Arts et al., 2001; Pauluhn, 1994a; Vijayaraghavan et al., 1993). However, for the calculation of Penh, a series of flow-derived breathing parameters, especially those dominated by the shape of the expiratory phase, are utilized. The pattern of this phase appears to be particularly dependent on the control of respiration, including pulmonary sensory or inflation/deflation reflexes and those involved in the mechanical control of the breathing apparatus (Jacquez, 1979). Changes in baseline Penh have been shown to be associated with an increase in BAL protein associated with changes in elastic tissue recoil rather than bronchoconstriction (Pauluhn, 2004). Hence, Penh may be viewed as a sensitive noninvasive, although nonspecific, functional end point suitable for probing changes in respiration in nonrestrained conscious rats (Figure 10.3). However, identifying and quantitating (especially)

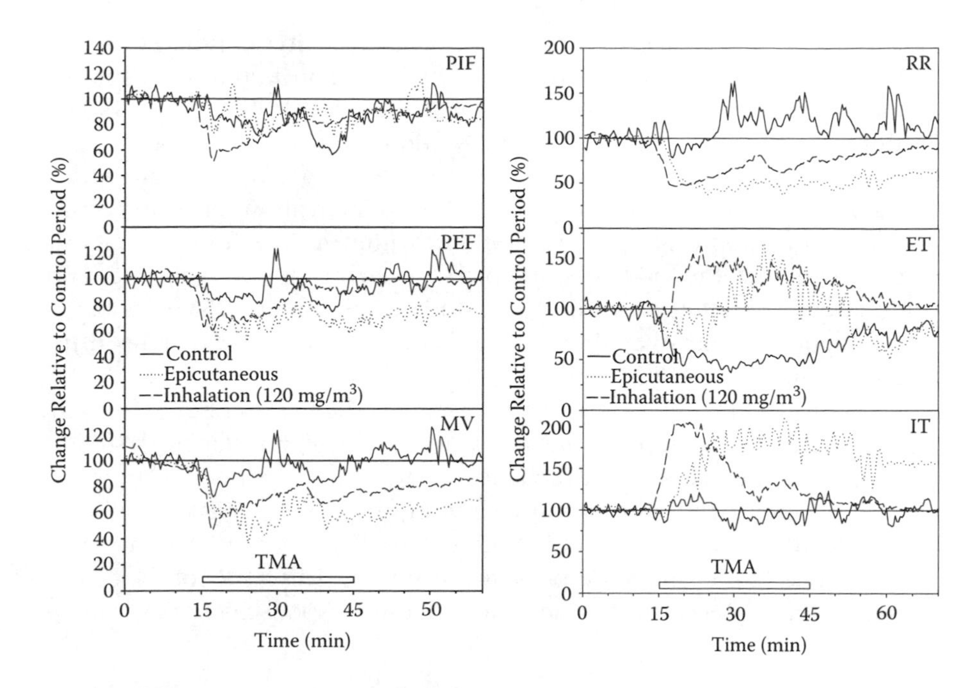

Figure 10.3 Change of respiratory patterns during a challenge with ≈23-mg TMA/m³ air (duration of challenge: 30 min). BN rats were sensitized either by epicutaneous administration of TMA in acetone:olive oil or by 5 × 3-h/d inhalation exposures to 120 mg TMA/m³. Respiratory response data were normalized to the mean of a 15-min prechallenge exposure period (=100%). Before and after the challenge, the rats were exposed to conditioned air. The animals selected represent those displaying maximum responsiveness upon challenge.

Note: PIF: Peak Inspiratory Flow During Tidal Breathing, PEF: Peak Expiratory Flow During Tidal Breathing, MV: Respiratory Minute Volume, RR: Respiratory Rate, ET: Expiratory Time, and IT: Inspiratory Time.

airway bronchoconstriction remains an elusive end point. Caution is advised in regard to the quantitative interpretation of Penh, because the associated changes may be related to alterations in breathing patterns secondary to changes in tissue mechanics. Thus, Penh appears to integrate several physiological end points in a wholly noninvasive and nondisturbing manner so that nonspecific functional changes can readily be identified in studies in which incremental, rather than absolute, changes are the primary focus.

Similarly, methodological differences in the sensitivity in detecting changes related to upper or lower respiratory tract irritation (Pauluhn and Mohr, 2000) or mild-to-moderate hypersensitivity responses appear to be among the reasons for the wide range of concentrations used for challenge exposures, e.g., as used for TMA or MDI (Pauluhn, 1994a; Pauluhn et al., 1999). Each method of determination, end point, and procedure to define positive changes in breathing patterns may affect the sensitivity both in identification and qualification of changes in breathing patterns (Figure 10.3). For both guinea pigs and BN rats, it was shown that the particular respiratory change chosen as the end point affects the interpretation of the study (Pauluhn et al. 2002b, 2003). These methodological variables are considered to be a major reason for interlaboratory variability.

Standardized procedures for evaluation have been described (Pauluhn, 1997; Pauluhn and Mohr, 1998; Pauluhn, 2003). For instance, baseline data are collected during a prechallenge adaptation period of approximately 15 min, and during or following the subsequent inhalation challenge, any response exceeding the mean ±3 standard deviations (SD) of this period is taken as a positive response. This type of objective, quantitative analysis calculates the area under the response curve exceeding

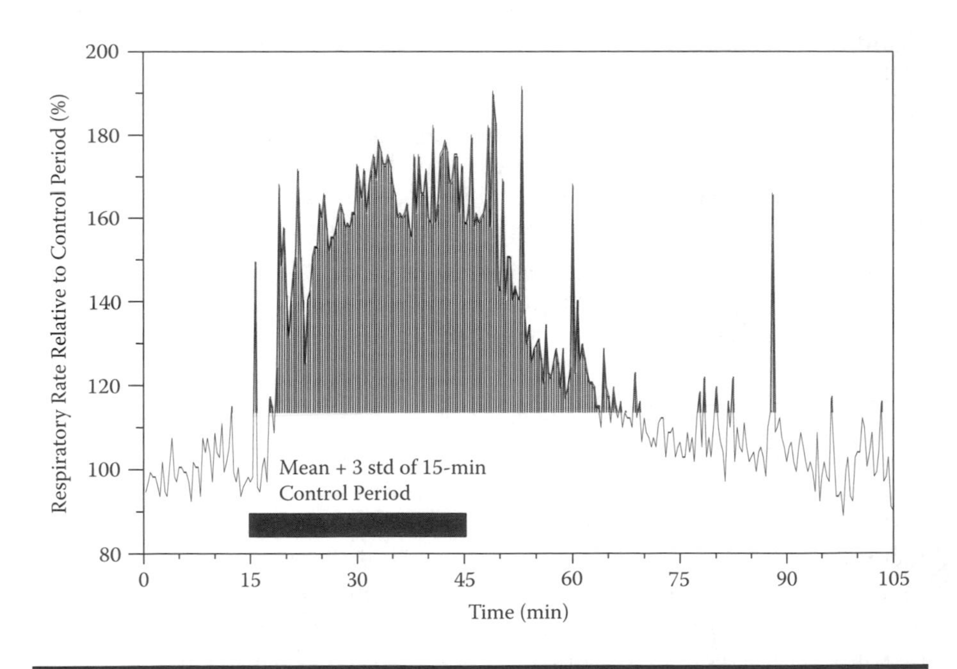

Figure 10.4 Integration of respiratory response prior to, during, and following challenge in guinea pigs (TDI conjugate).

the mean ±3 SD of the prechallenge period and can be used to express objectively the "intensity" of individual animal responses (Figure 10.4). As illustrated in Figure 10.5, the changes in respiratory rate in guinea pigs sensitized to and challenged with ovalbumin vary markedly from one animal to another with regard to the onset of anaphylaxis (animal 1 and animal 2), the time-related rapid change (animal 2 and animal 3), a more sustained immediate decrease (animal 5), or delayed increase (animal 4) in breathing frequency. The corresponding changes observed in BN rats using whole-body barometric plethysmography were markedly less pronounced, as shown in Figure 10.6, for TMA. Thus, unique changes in breathing patterns, as a result of respiratory allergy evoked by ovalbumin or TMA aerosols, have not been observed either in guinea pigs or in BN rats.

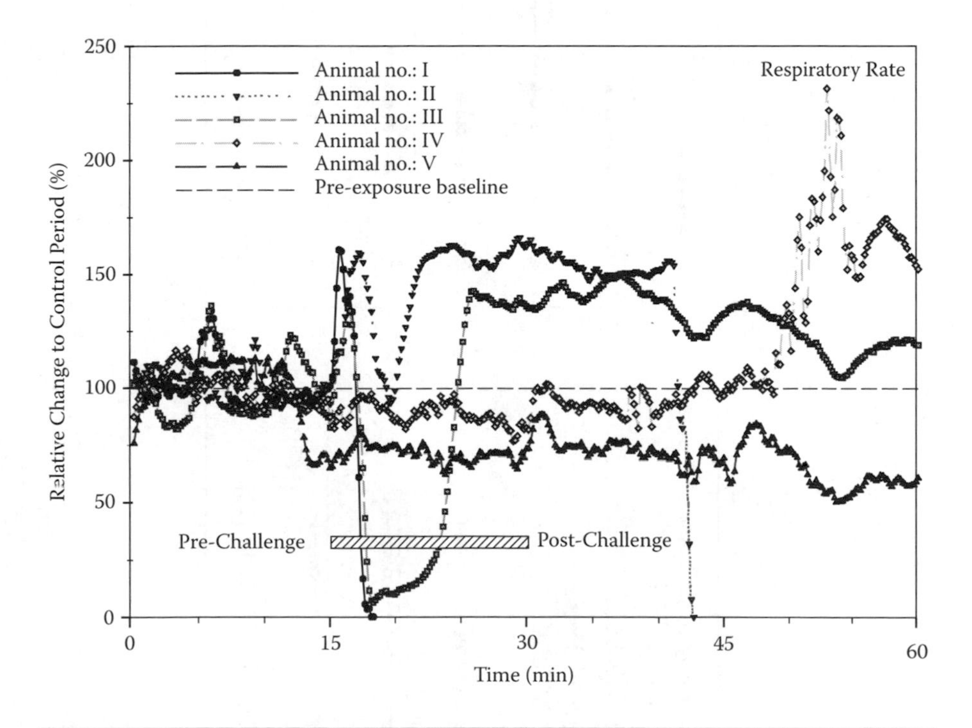

Figure 10.5 Guinea pigs were sensitized to ovalbumin aerosol on four consecutive days (days 0–3, duration: 15 min/d, concentration: 27 mg/m³) and challenged with ovalbumin aerosol 11.5 ± 3.6 mg/m³ for 15 min (mass median aerodynamic diameter, MMAD, ≈1.5 μm). The individual animals' respiratory rate was normalized to the 15-min preexposure period.

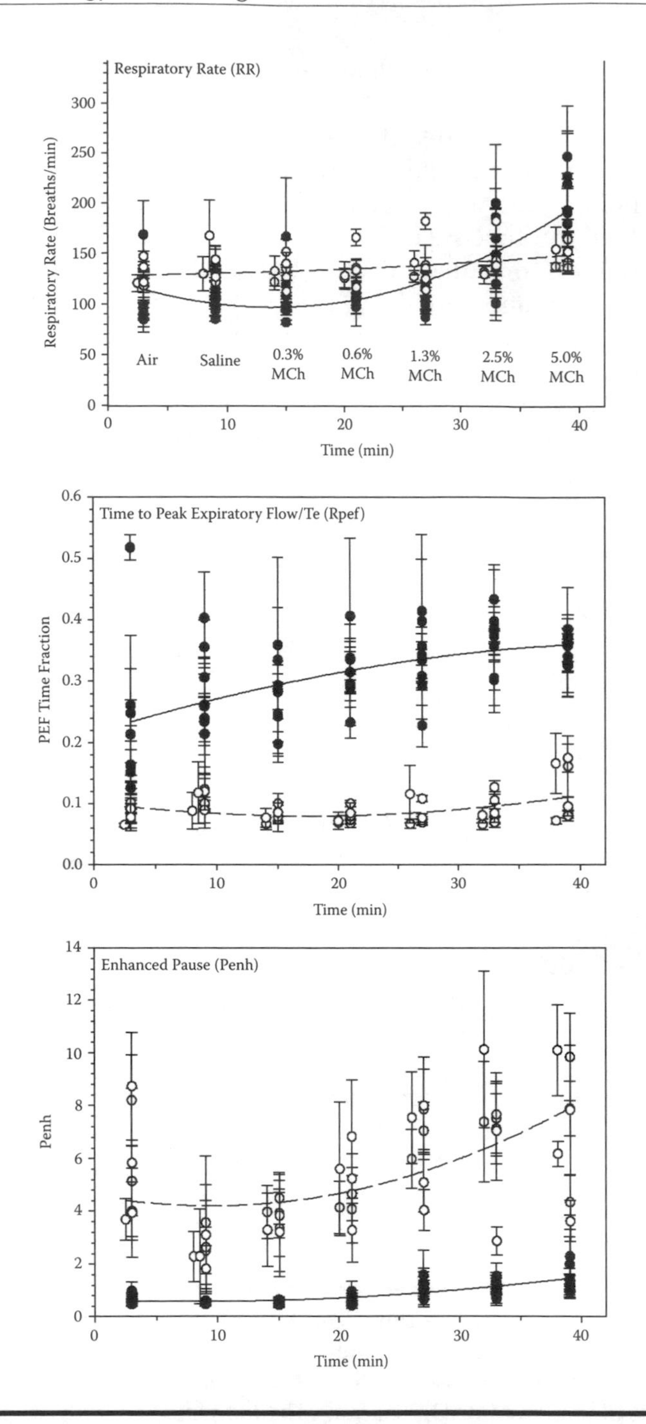

Figure 10.6 Change in breathing patterns of sham control BN rats (filled symbols, solid lines) and rats exposed for 1 × 6 h the previous day to 103 mg MDI/m³ (open symbols, dashed lines). The rats were exposed sequentially to air,

10.10 AHR TO PHARMACOLOGICAL AGONISTS

AHR is a central feature of asthma (Busse and Lemanske, 2001) and is characterized by exaggerated airway narrowing after exposure to nonspecific stimuli such as acetylcholine, methacholine, histamine, or exercise (Cockroft, 1997). The dysfunction underlying AHR includes hypersensitivity (a shift to the left of the bronchoconstrictor dose–response curves), hyperreactivity (increased slopes of these curves), and a greater maximum degree of induced bronchoconstriction. However, the mechanisms underlying these pathophysiologic abnormalities remain unclear.

In small laboratory animals, nonspecific airway responsiveness can be measured in the manner described in the preceding text. The common measurements utilize either volume displacement or whole-body (barometric) plethysmography. In either mode, animals are exposed to concentrations of the aerosolized agonist increased stepwise. AHR to acetylcholine has been observed in guinea pigs, for instance, sensitized to TDI using a single, brief, high-level exposure regimen. Eight guinea pigs were sensitized by inhalation to target TDI concentrations of 150 mg/m^3 (vapor) and 300 mg/m^3 (mixture of vapor and aerosol) for a duration of 15 min. After 3 weeks, the hyperresponsiveness elicited by acetylcholine aerosol was determined by volume displacement plethysmography one day after the TDI challenge (0.6 mg/m^3 for 30 min). EC_{50} was 32 and 7.2 mg acetylcholine/m^3 (breathing-zone concentrations of acetylcholine aerosol) in the TDI induction and control groups, respectively (Figure 10.7). Upon challenge with the TDI conjugate three out of eight guinea pigs of the aerosol-vapor induction group displayed a typical increase in respiratory rate, whereas guinea pigs sensitized to the vapor phase alone were indistinguishable from the control group. In comparison with both the naive control and the vapor-only induction groups, a significantly increased recruitment of eosinophils was present in the trachea, bronchi, and lung-associated lymph nodes of the aerosol-vapor group. In comparison with repeated-exposure protocols using lower concentrations but higher cumulative exposure dosages of TDI, the brief, high-level exposure regimen produced more pronounced responses to acetylcholine aerosol (Pauluhn and Mohr, 1998).

Figure 10.6 (Continued): saline, and aerosolized methacholine (MCh) in increasing concentrations (aqueous spray solutions of 0.3%, 0.6%, 1.3%, 2.5%, and 5%) in barometric plethysmographs. The time axis (abscissa) contains the cumulative time required for each step, *viz.*, a 1-min challenge period followed by a data collection period of 5 min. The 5-min mean values (±SD) of each step were calculated for each end point and rat.

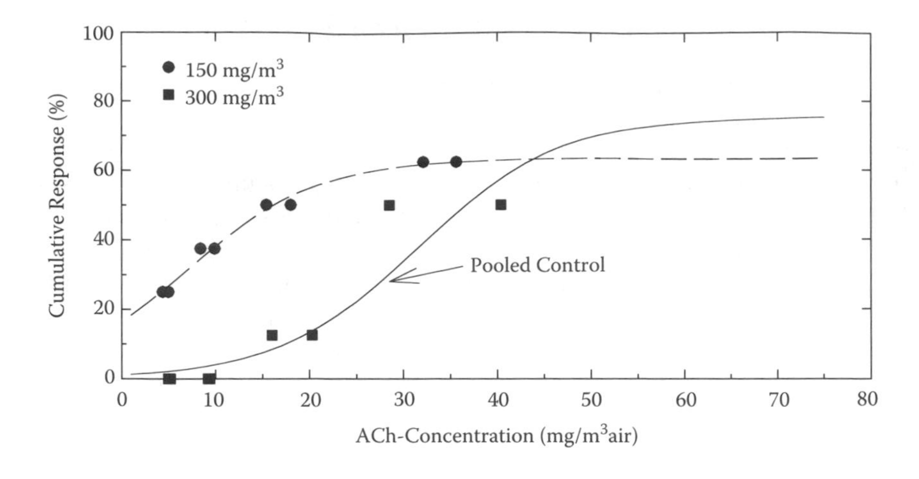

Figure 10.7 Acetylcholine bronchoprovocation assay in guinea pigs exposed for 1 × 15 min to TDI. Stepped exposure of ACh: 0.1%–0.2%–0.4%–0.8% (nebulized solution).

Methacholine-evoked changes in Penh using barometric plethysmography have frequently been used. Several factors can affect the various parameters involved in the calculation of Penh, and its indiscriminate use is subject to misleading interpretations of response. As almost all respiratory mechanics variables show qualitative correlations (Mitzner and Tankersley, 1998; Mitzner et al., 2003), it appears to be difficult to causally relate changes in Penh to any specific pathophysiological event, such as airway constriction. It is also important to recall that nasal airway resistance is the largest component of the total airway resistance in obligate-nasal-breathing rodents (Lung, 1987).

10.11 STRUCTURAL CHANGES OF THE LUNG: AIRWAY REMODELING

As mentioned earlier, chronic inflammation and airway remodeling are among the hallmarks of asthma. In humans, pulmonary function measurements and sometimes histopathology (lung biopsies) address these changes. Although changes in pulmonary function are not pathognomonic for specific lesions, much about the lung structure can be inferred from functional changes. In such cases, a battery of tests of different facets of lung function are usually applied, and results are expressed as classes of function disorders (e.g., obstructive or restrictive) that are consistent with classes of morphological changes (e.g., emphysema, bronchitis, and fibrosis). In animal models, the principle of pulmonary function measurements in anesthetized rats can be described as follows: Transpulmonary pressure is commonly measured using a water-filled catheter placed in the esophagus of the animal and referenced to the breathing port of the plethysmograph. The focus is on the measurements of the divisions of lung volume, mechanical properties of lung, maximal expiratory flow volume, and pressure–volume curves in forced maneuvers. Spontaneous-breathing parameters obtained during tidal breathing, such as R_L and C_{dyn}, are also addressed (Diamond and O'Donnell, 1977).

In Wistar rats exposed to the volatile phenyl isocyanate (PhI), a steep concentration–response relationship was observed following a 5-d/week, 6-h/d inhalation exposure regimen over either two or four weeks. The most prominent histopathologic changes were reflective of the anterior–posterior gradient of injury of airways, typical of reactive volatile agents. Concentrations of PhI of 3 mg/m^3 and lower were apparently scrubbed in the nasal passages, whereas at higher concentration, a precipitously increased penetration of the vapor into the lower respiratory tract occurred. Accordingly, at 3 mg/m^3 no effects were observed at all, whereas at 4 mg/m^3, in the main bronchi, evidence existed of goblet cell hyperplasia and peribronchial influx of inflammatory cells, including eosinophilic granulocytes. Exposure to slightly higher exposure concentrations (7 and 10 mg/m^3) not only caused mortality (32% and 66%, respectively) but also additional characteristic features of asthma in airways, such as bronchial muscular hypertrophy, eosinophilic bronchitis, goblet cell hyperplasia, increased intraluminar mucus, increased mitogenic activity in the epithelium, and subepithelial collagen deposition. These findings were accompanied by pneumonitis, septal thickening, and edema. Functional changes were indicated by increased total lung capacity, left shift of the pressure–volume curve, and decreased maximal forced expiratory maneuvers (Figure 10.8). Such changes are typical of obstructive lung disease. After a 2-month recovery period, in the absence of booster challenges, the airway inflammation was still apparent but appeared to be superimposed by collagen deposition and fibrogenic responses, i.e., the functional changes observed reflect sustained, slowly reversible, mixed obstructive or restrictive changes suggestive of lung remodeling. Karol and Kramarik (1996) verified the allergic etiology of PhI-induced airway inflammation.

In fact, this study shows that for agents exhibiting a concentration-dependent depth of penetration into the lung, demonstration of asthmagenic effects in rodent species are experimentally demanding because of the extremely steep concentration–response curve. Similarly, airway plugging

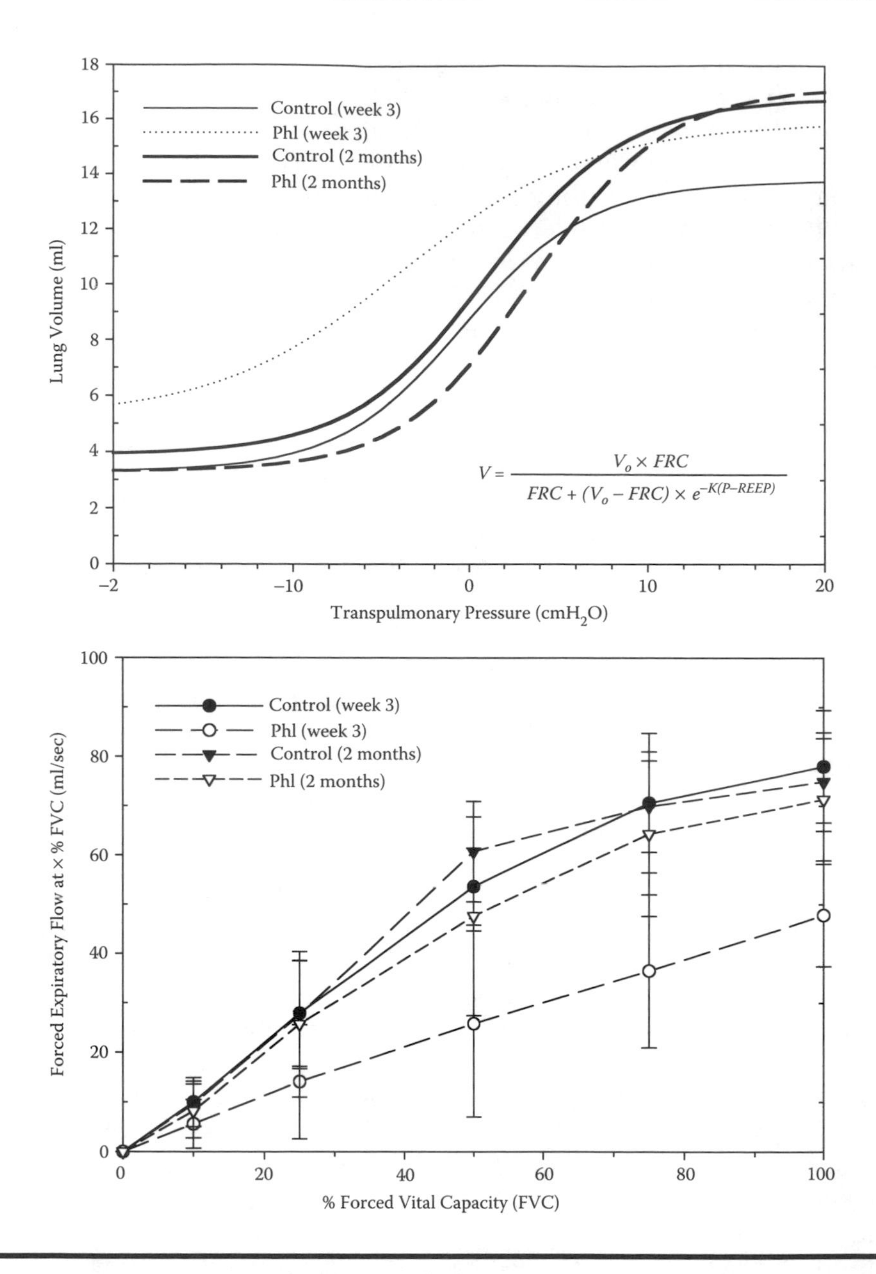

Figure 10.8 Exposure of Wistar rats for 2 weeks (5 × 6 h/d/week) to 7-mg/m³ phenyl isocyanate and assessment of lung function following an observation period of 1 week and 2 months. Upper panel: analysis of quasistatic deflation pressure–volume loops using the Sigmoid model of Paiva et al. (1975).

Note: Lower Panel: maximum expiratory flow volume curves (Means ± SD). Vo: Maximum inflation volume, FRC: functional residual capacity, and REEP: resting end-expiratory pressure.

and *bronchiolitis obliterans* associated with the high breathing frequency of small laboratory rodents may dramatically increase the dead-space ventilation and venous admixture so that rats inflicted with asthma-like airway lesions may succumb (Pauluhn et al., 1995). This appears to support the common view held, *viz.*, the extent of epithelial disruption and desquamation is most critical for the initiation and progression of this disease. The disadvantage of this animal model is that it is biased to be false negative because of the all-or-nothing type of inflammatory response.

10.12 ASSESSMENT OF INTRALUMINAL AIRWAY INFLAMMATION

Acute, irritant- or challenge-related injury to the respiratory tract results in an intraluminal exudate composed of cellular and glandular secretions, substances derived from degenerating or dead epithelial cells, and a large variety of cellular and humoral mediators and markers of acute inflammation (Henderson, 1989; Pauluhn, 2004; Folkesson et al., 1998). The precise composition of the exudate depends on the cause of the injury and the site at which it occurs. However, so far it appears to be difficult to unequivocally distinguish the relative contribution of biomarkers produced locally and those due to increased plasma exudation. Two important concepts must be borne in mind concerning the use of BAL in the assessment of allergic inflammation and injury of the respiratory tract. One is the possibility of sampling errors because the most damaged region may not necessarily be sufficiently accessible to the lavage fluid. The other is that the relative proportion of the various components of interstitial inflammation is not necessarily the same as that found in blood and BALF. Similarly, it is important to remember that BAL, as do most pulmonary function tests, evaluates the integrated function of the entire organ; focal lesions can exist without measurably affecting total organ function and, accordingly, remain undetected by these more-integrated measurements.

It is clear from the preceding discussion that neither pulmonary function tests nor BAL are a substitute for histopathological evaluations. It should also be clear that the question of the relative sensitivity of pulmonary function tests and histopathology is irrelevant in a general sense. Either assay could be the more sensitive one under a given circumstance. The two approaches are complementary and are used to the best advantage in concert. As a general principle, qualitative changes in lung structure and quantitative changes in lung function are detectable using light microscopy and appropriate tests, respectively, at about the same time in chronic, progressive lung diseases (Mauderly, 1986).

10.13 EXPERIMENTAL PROTOCOLS

Despite the wealth of valuable data that has been obtained from studies in short-term animal models of asthmatic inflammation or responses, the limitations of each model must be recognized. Some of the divergent and conflicting results from animal models of respiratory allergy may be related to spontaneously ocurring pulmonary diseases of specific strains used for study. Also, the variety of sensitization and challenge protocols employed, including the different readouts used to assess responses and responsiveness to specific stimuli, as well as the timing of these assessments, adds to the variables of these models. Importantly, although short-term exposure to very-high-mass concentrations of aerosolized allergen or hapten is experimentally convenient and leads to successful priming of animals, this is quite unlike the recurrent long-term exposure to low-mass concentrations of allergen experienced by humans with asthma. High-level exposure triggers acute inflammation in the lung parenchyma, which is markedly different from the acute-on-chronic inflammation of the airway wall observed in human asthma. Moreover, the pattern of acute inflammation differs in several respects from that observed in individuals with asthma. Perhaps the most significant deficiency of the commonly used models of allergic bronchopulmonary inflammation is that, by their very nature, they involve short-term experiments in regard to induction or single exposures for the elicitation of respiratory allergy. These models do not exhibit many of the lesions that typify chronic human asthma as detailed earlier.

The approaches for developing models of chronic asthma have involved factors ranging from pulmonary overexpression of cytokine patterns to the processes of lung remodeling. However, some of these models do not involve hapten or allergen challenge. There have been relatively few descriptions of satisfactory rodent models of chronic hapten or antigen challenges associated with airway inflammation and remodeling resembling human asthma. Experimental protocols have involved repeated inhalation exposure (Pauluhn, 2004; Palmans et al., 2000; Haczku, 1994) or repeated bolus delivery of antigen intratracheally or intranasally (Blyth et al., 1996). The benefits and shortcomings associated with each approach are addressed in the following subsections.

10.13.1 Protocol Determinants: Route, Dose, and Timing

Most of the currently applied assays utilize two phases, viz., an induction phase, which includes single to multiple exposures to the test compound (sensitization) either via the respiratory tract, dermal contact, or intraperitoneal or cutaneous injection, followed by a challenge or elicitation phase. As detailed in the following text, challenge exposures are by inhalation exposures or technically less demanding procedures, such as intranasal or intratracheal instillations. The challenge can either be with the chemical (hapten), homologous protein conjugate of the hapten, or antigen. The choice depends both on the irritant potency and the physical form (vapor or aerosol) of the hapten. End points to characterize positive response range from the induction of immunoglobulins (e.g., total IgE and specific IgG), cytokines, or lymphokines in serum to (patho-)physiological reaction occurring in the lung (e.g., bronchoconstriction and influx of inflammatory cells). For the identification of chemical irritants, concentrations causing minimal irritation must be selected for successful challenge exposure, as changes in breathing patterns caused by marked irritation may be clinically indistinguishable from allergic response.

10.13.1.1 Route

Among the issues that may contribute to some of the inconsistencies among animal studies is the fact that the doses and dosing regimens vary substantially from one animal model or laboratory to another. Also, the mode of administration of the agent for sensitization and elicitation of respiratory allergy and the techniques chosen to analyze response differ appreciably. Especially in mechanistic studies, animals receive various amounts of alum-precipitated antigen (e.g., ovalbumin) by intraperitoneal injection, once or repeatedly. For example, almost unlimited amounts of haptens or allergens can be applied to the skin for any time period. Different skin areas may be dosed at specified time intervals. Associated local irritant responses may not be test limiting because the sites of induction and elicitation (respiratory tract) differ. Apart from the analysis of activation of draining lymph nodes, little attention has been paid to the characterization of the induction-related irritant inflammatory response occurring in the skin or respiratory tract. Also, the metrics of dosing, i.e., dose per surface area vs. total dose per animal, concentration, frequency of dosing, anatomical site of exposure, and impact of the vehicle chosen, have received little attention but are viewed to be important for the outcome of a test (Boukhman and Maibach, 2001). Also, dry TMA dust applied to the skin of BN rats caused a dose-dependent production of specific IgE and IgG (Zhang et al., 2002). In contrast, in inhalation studies, the exposure intensity is reasonably well defined by the duration of exposure, and concentration, and particle size of the agent present in the inhalation chamber. Especially for irritant chemicals, steep concentration–response curves are commonly observed, and sufficiently high excursions of the concentrations used during the sensitization phase might render the interpretation of findings following challenge exposures difficult because of preexisting, irritation-related lung inflammation. Challenge exposures have utilized intratracheal, intranasal, or inhalation routes. The first two are bolus techniques and may be suitable for nonirritant allergens, and methodological variables exist that are decisive for the site of dosing within the respiratory tract and the residence time of the bolus administered. The inhalation route of exposure represents the most elaborate technique; however, it allows the highest degree of standardization, precision, and reproducibility when appropriately conducted, and exposure conditions might easily be translated to workplace exposure regimens.

Intranasal instillation techniques have gained popularity because they are relatively inexpensive, can be used repeatedly, are less labor intensive, and are technically less demanding than controlled inhalation exposures. Detailed recommendations for this technique have been published (Gizurarson, 1990). However, the volume of the intranasally instilled bolus, the breathing pattern, and the position of the animal during and after dosing might determine the extent to which it penetrates the thoracic airways, as shown by Ebino et al. (1999). Anesthesia may increase aspiration into lower airways. Marked differences may occur in the delivered dose because of differences in the intranasal instillation techniques, e.g., the dosing orifice may be placed into the nostril or positioned at the external nares. Ebino et al. (1999) conclude that TDI dissolved in a nonaqueous vehicle (olive oil:ethyl acetate) does not reach the trachea and/or lower airways of 11-week-old mice when using a volume of 20 μl/instillation divided equally between the two nares. In contrast, when using an indicator dye and water as the vehicle, evidence of lower respiratory tract dosing existed at this instillation volume. Analysis of published data appears to suggest that in younger mice, instillation volumes of aqueous solutions up to 50 μl/instillation have been used. Other authors have shown that a delivery volume of 5 μl of a radioactive substance did apparently not gain access to the lower respiratory tract of mice; this increased to a maximum of 56% when 50 μl was instilled (Southam et al., 2002). These authors conclude that for optimal delivery to the lower respiratory tract, a volume of ≥35 μl should be delivered to anesthetized mice. Hence, the technical designs of intranasal studies need to be carefully observed, and an understanding of the physicochemical characteristics of the agent and its diluent is necessary to achieve appropriate pulmonary distribution patterns of the chemical following this alternative route of dosing.

10.13.1.2 Dose and Concentration Used for Induction and Challenge

Human and animal studies have each provided evidence that perhaps the most important factor for the development of respiratory allergy is the concentration of the agent (Karol and Thorne, 1988). Usually, more episodes of brief, high-concentration exposures are required. High priming doses followed by repeated booster or subclinical challenge exposures appear to be more critical than long-term exposures to subthreshold concentrations. Animal studies have indicated more precisely the relationship between exposure and production of respiratory allergy. Dose-related effects have been investigated systematically in guinea pigs, because this species had been found to be capable of reproducing the most pronounced acute immediate-onset pulmonary hypersensitivity reaction to inhaled agents.

The importance of exposure concentration in effective respiratory sensitization of experimental animals has been demonstrated. For example, Karol (1983) examined in guinea pigs the influence of increasing concentrations of inhaled TDI on the stimulation of specific antibody responses and on changes in respiratory rate following inhalation challenge with TDI–protein conjugate. Exposure of animals for 3 h/d for 5 consecutive days to TDI concentrations equal to or greater than 0.36 ppm caused significant challenge-induced respiratory reactions and the appearance of specific antibodies in some of the treated guinea pigs. Similar exposure to lower concentrations of TDI (0.12 ppm) failed to cause either antihapten antibody production or respiratory hypersensitivity. It is important to emphasize here that although there is a clear requirement for a critical minimum concentration of chemical allergen, other factors influence what this minimum concentration actually is in different circumstances. Thus, the frequency of exposure in relation to dose is important (as with contact sensitization) insofar as the local exposure concentration at any one time appears to be of greater importance than the cumulative delivered dose over a more protracted period. In addition, it is likely that effective sensitization of the respiratory tract can be induced by routes of exposure other than inhalation (for example, via skin contact). The critical dose or concentration required for sensitization varies with the route of exposure. More recent studies with TDI, utilizing either very short duration of exposure or inhalation exposures along with intradermal injections, suggest that the concentration at the site of initial deposition in the respiratory tract appears to be among the most important variables. TDI aerosol is more likely to penetrate the airways of the lung of guinea pigs than TDI vapor and, accordingly, will produce a more distinct allergic reaction (Pauluhn and Mohr, 1998). The relative potency of “large” and “small” aerosols of MDI was studied in guinea pigs. There was an apparent greater response in animals sensitized by intradermal injections or by inhalation exposure with the large aerosol and then challenged with the small aerosol, i.e., the size of an aerosol reaching its airway target site in sufficiently high dosages (Pauluhn et al., 2000).

Published evidence suggests that for TMA, low-dose intradermal induction regimens appear to evoke a more vigorous anaphylactic respiratory response upon challenge with the free or TMA–protein conjugate when compared with high-dose regimens (Hayes et al., 1995; Pauluhn et al, 1999). Also, for HDI, the sensitizing dose was shown to be an important determinant of airway inflammation and HDI-specific antibody response. BALB/c mice that were epicutaneously sensitized to HDI and then challenged intranasally with the HDI–protein conjugate displayed lung inflammation when sensitized with 0.1% HDI but not at a concentration ten times higher, the optimal dose for inducing anti-HDI IgG_1 antibody production (Herrick et al., 2002). Similar inverse concentration-dependent responses have also been shown by Scheerens et al. (1996), who reported that dermal induction with lower concentrations of TDI (1% vs. 0.1%) was more efficient in causing an increased total cell count and neutrophilic inflammation 1 d after intranasal challenge. A comparable inverse dose–response relationship has not been reported for the inhalation route of induction. However, it was shown for TMA that repeated low-dose inhalation challenge exposures were more effective in eliciting respiratory allergy-like responses than high priming dosages (Pauluhn, 2003). Conversely, in BN rats, a clear concentration dependence of respiratory responses, including lung inflammation, and the topical concentration of TMA in a mixture of acetone and olive oil was demonstrated (Pauluhn et al., 2002; Pauluhn, 2003). Whether or not this paradoxical phenomenon is related to immunological tolerance (Kurts et al., 1999), shifts in the respective dose–response curves, or modifications of antigen retention and/or its presentation in an noninflamed or inflamed microenvironment of the skin, remains unresolved.

For irritant volatile chemical haptens, sufficiently high bronchial bolus doses can be delivered by either intranasal instillations of the chemical in a vehicle or inhalation exposures using the protein conjugate of the hapten. To maximize the magnitude of response, the critical location in the lung must be dosed sufficiently. Accordingly, in obligate-nasal-breathing species, for reactive volatile chemicals, inhalation challenge exposure with the hapten conjugate appears to be indispensable for state-of-the-art hazard identification.

Several studies in sensitized BN rats and guinea pigs have elaborated on the relationship between the provocation dose of the chemical, e.g., TMA or MDI, and airway response. In both species, a dose-dependent change in functional breathing parameters was observed after inhalation challenge. Zhang et al. (2004) have shown that in BN rats the late airway response in terms of intensity and duration was more pronounced at higher (mildly irritant) concentrations of TMA than at lower concentrations.

A study in guinea pigs showed that the relative contributions of immediate and/or delayed responses also depend on the allergen concentration and exposure duration (Santing et al., 1994). Currently, the relative contribution of early and late responses to the overall response is difficult to quantify in animal models because of the varying sensitivity of methods and end points to probe for such changes.

10.13.2 Time

Most of the current animal models of allergic airway inflammation have been restricted to acute inflammatory changes following relatively short periods of hapten or allergen exposure. Repeated challenge exposure may exacerbate preceding events, allowing more conclusive assessments. Thus, studies investigating the sequence of inflammatory events after allergen challenge and the temporal association with specific end points appear to be critical to the detection of response and to further our understanding of the factors involved in pathogenesis. Only few studies have evaluated the sequence of inflammatory events taking place after allergen challenge. It was demonstrated that increases in maximal bronchoconstriction, associated increase in airway contractile tissue, and subepithelial fibrosis require a chronic, but not brief, allergen challenge protocol (Tomkinson et al., 2001; Leigh et al., 2002).

The known human asthmagen MDI has been evaluated in diverse animal models. In most studies, one challenge with the MDI aerosol, acute respiratory responses were equivocal because changes in breathing patterns caused by stimulation of pulmonary irritant receptors and allergic response are phenotypically indistinguishable. An increased peribronchial or perivascular recruitment of eosinophilic granulocytes was taken as indirect evidence that the response is typified by a delayed allergic inflammation rather than immediate-onset responses observed in the case of TMA (Pauluhn et al., 1999). The impact of protocol determinants with regard to the timing and frequency of challenge exposures can be instrumental to the outcome of the study. When BN rats were sensitized to and sequentially challenged with a minimally irritant concentration of MDI on days 21, 35, 50, and 64, a time-related increased response (BAL neutrophils) indicative of an allergic inflammation (Figure 10.9) was observed. With regard to the recruitment of eosinophilic granulocytes, remarkable time-related diminution was observed in the control group, suggesting that breeder environments might trigger preexisting pulmonary disease in this susceptible rat strain. Accordingly, an apparent influx of eosinophilic granulocytes existed; however, the background levels present in the control render the interpretation difficult (Figure 10.9). Despite clear evidence

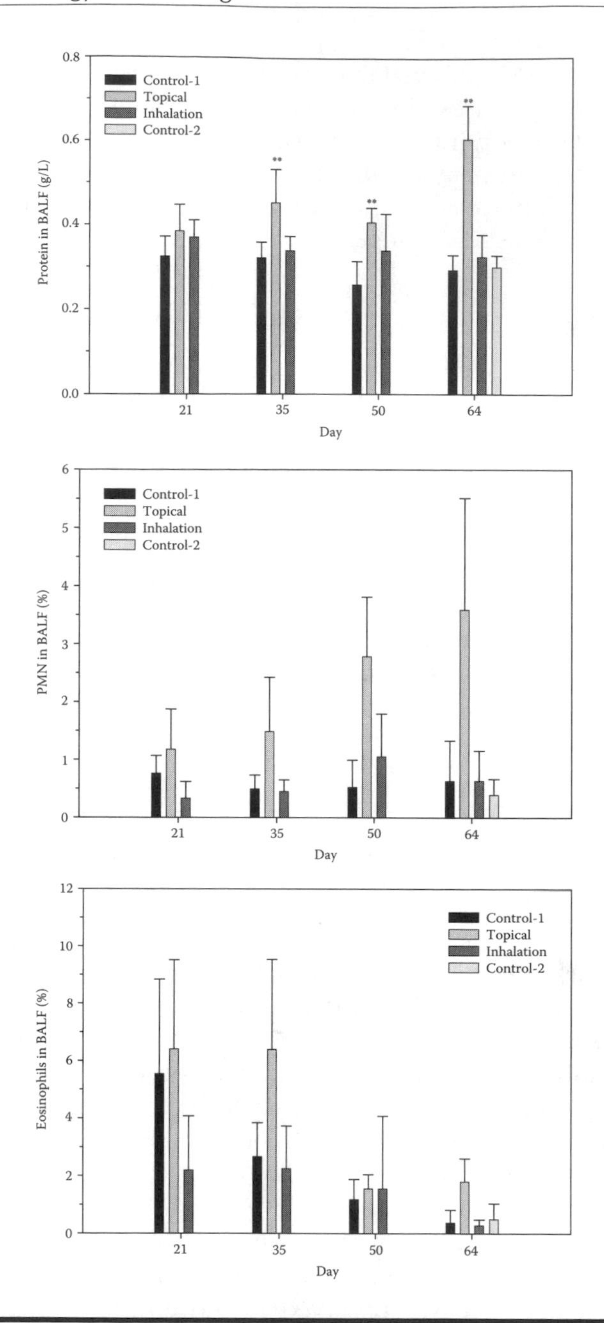

Figure 10.9 BN rats Sensitized to MDI by high-dose topical induction or by repeated inhalation exposure (5 × 3-h, 28 mg MDI/m³). All rats were challenged with 15 mg MDI/m³ for 30 min on days 21, 35, 50, and 64 2 days prior to sacrifice (days shown on the abzissa). Naive rats of control-1 were repeatedly challenged whereas those of control-2 were challenged on day 64 only. Protein and differential cell counts from bronchoalveolar lavage (mean±SD, asterisks denote statistical significant differences to control-1; * $P \leq 0.05$, ** $P \leq 0.01$).

of a neutrophilic type of allergic inflammation probed by BAL, histology findings were characterized by a significantly increased peribronchial or perivascular influx of eosinophils rather than neutrophils. This finding corroborates the notion that each method chosen to probe generalized or focal responses within the heterogeneous lung has its particular advantages and disadvantages.

10.14 CONCLUSIONS

In conclusion, the outcome of each animal model currently used for studying asthma is highly contingent upon protocol variables. Acute models have the advantage of being cost- and time-effective; however, the major disadvantage of most of the acute models is that only some selected features of human asthma are modeled. One of the issues of highest concern in studies addressing respiratory allergy is that the dose and/or concentration as well as the frequency of dosing are decisive for the outcome of the test. The available experimental evidence also underscores the necessity to standardize allergen sensitization and challenge protocols and the inclusion of multiple posthapten or postallergen challenge time points. The analysis of changes in allergen-induced airway or lung inflammation should include physiological, biochemical or immunological, and histological end points, as responses may be either localized and compartmentalized or generalized. Because of the complex interplay of epithelial, mesenchymal, and neurohumoral modulating factors that orchestrate this type of chronic airway inflammation, the approaches chosen for the characterization of cells and cytokines in the responding lung may range from those associated with the Th2/Th1 system to the specific responses believed to be involved in the phenotypic development of human asthma. So far, none of the currently applied animal models duplicate all the features of human asthma. Accordingly, the results from each animal model must be interpreted cautiously with regard to the protocol variables and, especially, dosing regimens, animal species, and strain selected before a meaningful extrapolation to humans can be arrived at.

REFERENCES

Alarie, Y., Iwasaki, M., and Schaper, M. (1990). Whole-body plethysmography in sedentary or exercise conditions to determine pulmonary toxicity, including hypersensitivity induced by airborne toxicants. *J Am Coll Toxicol* 9: 407–439.

Andersson, P. (1980). Antigen-induced bronchial anaphylaxis in actively sensitized guinea pigs. *Allergy* 35: 61–71.

Arts, J.H., De Koning, M.W., Bloksma, N., and Kuper, C.F. (2001). Respiratory irritation by trimellitic anhydride in Brown Norway and Wistar rats. *Inhal Toxicol* 13: 719–728.

Arts, J.H.E., Dröge, S.C.M., Spanhaak, S., Bloksma, N., and Kuper, C.F. (1997). Local lymph node activation and IgE responses in Brown Norway and Wistar rats after dermal application of sensitizing and non-sensitizing chemicals. *Toxicology* 117: 229–237.

Arts, J.H.E., Kuper, C.F., Spoor, S.M., and Bloksma, N. (1998). Airway morphology and function of rats following dermal sensitization and respiratory challenge with low molecular weight chemicals. *Toxicol Appl Pharmacol* 152: 66–76.

Barnes, P.J. (1986). Asthma as an axon reflex. *Lancet* 1: 242–245.

Barnes, P.J., Chung, K.F., and Page, C.P. (1991). Inflammatory mediators. In *Pharmacology of Asthma (Handbook of Experimental Pharmacology)*, Page, C.P. and Barnes P.J. (Eds.), Springer-Verlag, Berlin, Vol. 98, pp. 54–106.

Barnes, P.J., Chung, K.F., and Page, C.P. (1998). Inflammatory mediators of asthma: an update. *Pharmacol Rev* 50: 515–596.

Baur, X. (1995). Hypersensitivity pneumonitis (extrinsic allergic alveolitis) induced by isocyanates. *J Allergy Clin Immunol* 95: 1004–1010.

Bice, D.E., Seagrave, J.C., and Green, F.H.Y. (2000). Animal models of asthma: potential usefulness for studying health effects of inhaled particles. *Inhal Toxicol* 12: 829–862.

Blaikie, L. and Basketter, D.A. (1999). Experience with a mouse intranasal test for the predictive identification of respiratory sensitization potential of proteins. *Food Chem Toxicol* 37: 889–896.

Blaikie, L., Morrow, T., Wilson, A.P., Hext, P., Hartop, P.J., Rattray, N.J., Woodcock, D., and Botham, P.A. (1995). A two-centre study for the evaluation and validation of an animal model for the assessment of the potential of small molecular weight chemicals to cause respiratory allergy. *Toxicology* 96: 37–50.

Blyth, D.I., Pedrick, M.S., Savage, T.J., Hessel, E.M., and Fattah, D. (1996). Lung inflammation and epithelial changes in a murine model of atopic asthma. *Am J Respir Cell Mol Biol* 14: 425–428.

Botham, P.A., Hext, P.M., Rattray, N.J., Walsh, S.T., and Woodcock, D.R. (1988). Sensitization of guinea-pigs by inhalation exposure to low-molecular-weight chemicals. *Toxicol Lett* 41: 159–173.

Botham, P.A., Rattray, N.J., Woodcock, D.R., Walsh, S.T., and Hext, P. (1989). The induction of respiratory allergy in guinea pigs following i.d. injection of trimellitic anhydride. A comparison of response to 2,4-dinitrochlorobenzene. *Toxicol Lett* 47: 25–39.

Boukhman, M.P., and Maibach, H.I. (2001). Thresholds in contact sensitization: immunologic mechanisms and experimental evidence in humans—an overview. *Food Chem Toxicol* 39: 1125–1134.

Braun, A., Lommatzsch, M., and Renz, H. (2000). The role of neutrophins in allergic bronchial asthma. *Clin Exp Allergy* 30: 178–186.

Briatico-Vangosa, C., Braun, C.J.L., Cookman, G., Hofmann, T., Kimber, I., Loveless, S.E., Morrow, T., Pauluhn, J., Sørensen, T., and Niessen, H.J. (1994). Review. Respiratory allergy: hazard identification and risk assessment. *Fundam Appl Toxicol* 23: 145–158.

Busse, W.W. and Lemanske, R.F. (2001). Asthma. *N Engl J Med* 344: 350–362.

Carr, M.J., Hunter, D.D., and Undem, B.J. (2001). Neutrophins and asthma. *Curr Opin Pulm Med* 7: 1–7.

Chengelis, C.P. (1990). Examples of alternative use in toxicology for common species. *J Am Coll Toxicol* 9: 319–342.

Chong, B.T.Y., Agrawal, D.K., Romero, F.A., and Townley, R.G. (1998). Measurement of bronchoconstriction using whole-body plethysmographs: comparison of freely moving versus restrained guinea pigs. *J Pharmacol Toxicol Methods* 39: 163–168.

Cockroft, D.W. (1997). Airway responsiveness. In Barnes, P.J., Grunstein, M.M., Leff, A.R., Woolcock, A.J. (Eds.). *Asthma.* Philadelphia: Lippincott-Raven, pp. 1253–1266.

Cohn, L. (2001). Food for thought: can immunological tolerance be induced to treat asthma? *Am J Respir Cell Mol Biol* 24: 509–512.

Cui, Z.-H., Sjöstrand, M., Pullerits, T., Andius, P., Skoogh, B.-E., and Lötvall, J. (1997). Bronchial hyperresponsiveness, epithelial damage, and airway eosinophilia after single and repeated allergen exposure in a rat-model of anhydride-induced asthma. *Allergy* 52: 739–746.

Dearman, R.J., Betts, C.J., Humphreys, N., Flanagan, B.F., Gilmour, N.J., Basketter, D.A., and Kimber, I. (2003). Chemical allergy: considerations for the practical application of cytokine profiling. *Toxicol Sci* 71: 137–145.

Dearman, R.J., Skinner, R.A., Humphreys, N.E., and Kimber, I. (2003). Methods for the identification of chemical respiratory allergens in rodents: comparison of cytokine profiling with induced changes in IgE. *J Appl Toxicol* 23: 199–207.

Dearman, R.J., Spence, L.M., and Kimber, I. (1992). Characterization of murine immune response to allergenic diisocyanates. *Toxicol Appl Pharmacol* 112: 190–197.

Diamond, L. and O'Donnell, M. (1977). Pulmonary mechanics in normal rats. *J Appl Physiol Respir Environ Exercise Physiol* 43: 942–948.

Drorbough, J.E. and Fenn, W.O. (1955). A barometric method for measuring ventilation in newborn infants. *Pediatrics* 16: 81–87.

Ebino, K. Lemus, R., and Karol, M.H. (1999). The importance of the diluent for airway transport of toluene diisocyanate frollowing intranasal dosing of mice. *Inhal Toxicol* 11: 171–185.

Epstein, M.A. and Epstein, R.A. (1978). A theoretical analysis of the barometric method for measurement of tidal volume. *Respir Physiol* 32: 105–120.

Evans, M.J., van Winkle, L.S., Fanucchi, M.V., and Plopper, C.G. (1999). The attenuated fibroblast sheath of the respiratory tract epithelial-mesenchymal trophic unit. *Am J Respir Cell Mol Biol* 21: 655–657.

Fabbri, L.M., Caramori, G., Beghé, B., Papi, A., and Ciaccia, A. (1998). Physiological consequences of long-term inflammation. *Am J Respir Crit Care Med* 157: S195–S198.

Folkesson, H.G., Weström, B.R., and Karlsson, B.W. (1998). Effects of systemic and local immunization on alveolar epithelial permeability to protein in the rat. *Am J Respir Crit Care Med* 157: 324–327.

Foster, P.S., Yang, M., Herbert C., and Kumar R.K. (2002). CD4(+) T-lymphocytes regulate airway remodeling and hyper-reactivity in a mouse model of chronic asthma. *Lab Invest* 82: 455–462.

Frossard, N., Rhoden, K.J., and Barnes, P.J. (1989). Influence of the epithelium on guinea pig airway responses to tachykinins: role of endopeptidase and cyclooxygenase. *J Pharmacol Exp Ther* 248: 292–299.

Fügner, A. (1985). Pharmacological aspects of immediate hypersensitivity reactions *in vivo*. In *Pulmonary and Antiallergic Drugs*, P. Devlin (Ed.). John Wiley & Sons, New York, pp. 123–189.

Gelfand, E.W. (2002). Mice are good models of human airway disease. *Am J Respir Crit Care Med* 166: 5–8.

Gizurarson, S. (1990). Animal models for intranasal drug delivery. *Acta Pharm Nord* 2: 105–122.

Griffiths-Johnson, D.A. and Karol, M.H. (1991). Validation of a non-invasive technique to assess development of airway hyperreactivity in an animal model of immunologic respiratory hyper-sensitivity. *Toxicology* 65: 283–294.

Haczku, A., Moqbel, R., Elwood, W., Sun, J., Bay, A.B., Barnes, P.J., and Chung, K.F. (1994). Effect of prolonged repeated exposure to ovalbumin in sensitized Brown Norway rats. *Am J Respir Crit Care Med* 150: 23–27.

Hamelmann, E., Schwarze, J., Takeda, K., Oshib, A., Larsen, G.L., Irvin, C.G., and Gelfand, E.W. (1997). Noninvasive measurement of airway responsiveness in allergic mice using barometric plethysmography. *Am J Respir Crit Care Med* 156: 766–775.

Hayes, J.P. and Newman-Taylor, A.J. (1995). *In vivo* models of occupational asthma due to low molecular weight chemicals. *Occup Environ Med* 52: 539–543.

Hayes, J.P., Daniel, R., Tee, R.D., Barnes, P.J., Chung, K.F., and Newman Taylor, A.J. (1992a). Specific immunological and bronchopulmonary responses following intradermal sensitization to free trimellitic anhydride in guinea-pigs. *Clin. Exper. Allergy* 22: 694–700.

Hayes, J.P., Daniel, R., Tee R.D., Barnes, P.J., Taylor, A.J., and Chung, K.F. (1992b). Bronchial hyperreactivity after inhalation of trimellitic anhydride dust in guinea pigs after intradermal sensitization to the free hapten. *Am. Rev. Respir. Dis.* 146: 1311–1314.

Hayes, J.P. and Newman Taylor A.J. (1995). *In vivo* models of occupational asthma due to low molecular weight chemicals. *Occup. Environ. Med.* 52: 539–543.

Henderson, R.F. (1989). Bronchoalveolar lavage: a tool for assessing the health status of the lung. In *Concepts in Inhalation Toxicology*, R.O. McClellan and R.F. Henderson (Eds.). Hemisphere Publishing, New York, pp. 415–444.

Herrick, C.A., Xu, L., Wisnewski, A.V., Das, J., Redlich, C.A., and Bottomly, K. (2002). A novel mouse model of diisocyanate-induced asthma showing allergic-type inflammation in the lung after inhaled antigen challenge. *J Allergy Clin Immunol* 109: 873–878.

Jacquez, J.A. (1979). *Respiratory Physiology—Reflexes Affecting Respiration.* Hemisphere Publishing, Washington, pp. 331–350.

Kannan, M.S. and Deshpande, D.A. (2003). Allergic asthma in mice: what determines the phenotype? *Am J Physiol Lung Cell Mol Physiol* 285: L29–L31.

Karol, M.H. (1983). Concentration-dependent immunologic response to toluene diisocyanate (TDI) following inhalation exposure. *Toxicol Appl Pharmacol* 68: 229–241.

Karol, M.H., Stadler, J., and Magreni, C. (1985). Immunotoxicologic evaluation of the respiratory system: animal models for immediate- and delayed-onset respiratory hypersensitivity. *Fundam Appl Toxicol* 5: 459–472.

Karol, M.H. and Thorne, P.S. (1988). Pulmonary hypersensitivity and hyperreactivity: implications for assessing allergic responses. In *Toxicology of the Lung*, Eds., D.E. Gardner, J.D. Crapo, and E.J. Massaro, Raven Press, New York, pp. 427–448.

Karol, M.H. and Kramarik, J.A. (1996). Phenyl isocyanate is a potent chemical sensitizer. *Toxicol Lett* 89: 139–146.

Karol, M.H. (1994). Animal models of occupational asthma. *Eur Respir J* 7: 555–568.

Kimber, I. and Dearman, R.J. (1997). *Toxicology of Chemical Respiratory Hypersensitivity*. Taylor and Francis, London.

Kimber, I., and Dearman, R.J. (2002). Chemical respiratory allergy: role of IgE antibody and relevance of route of exposure. *Toxicology* 181–182: 311–315.

Kimmel, E.C., Whitehead, G.S., Reboulet, J.E., and Carpenter, R.L. (2002). Carbon dioxide accumulation during small animal, whole body plethysmography: effects on ventilation, indices of airway function, and aerosol deposition. *J Aerosol Med* 15: 37–49.

Kips, J.C., Anderson, G.P., Fredberg, J.J., Herz, U., Inman, M.D., Jordana, M., Kemeny, D.M., Lötvall, J., Pauwels, R.A., Plopper, C.G., Schmidt, D., Sterk, P.J., van Oosterhout, A.J.M., Vargaftig, B.B., and Chung, K.F. (2003). Murine models of asthma. *Eur Respir J* 22: 374–382.

Kumar, R.K., Temelkovski, J., McNeil, H.P., and Hunter, N. (2000). *Clin Exp Allergy* 30: 1486–1492.

Kumar, R.K. and Foster, P.S. (2002). Translational review—modeling allergic asthma in mice—pitfalls and opportunities. *Am J Respir Cell Mol Biol* 27: 267–272.

Kurts, C., Sutherland, R.M., Davey, G., Li, M., Lew, A.M., Blanas, E., Carbone, F.R., Miller, J.F., and Heath, W.R. (1999). CD8 cell ignorance or tolerance to islet antigens depends on antigen dose. *Proc Natl Acad Sci USA*. 96: 12703–12707.

Laitinen, L.A., Heino, M., Laitinen, A., Kava, T., and Haahtela, T. (1985). Damage of the airway epithelium and bronchial reactivity in patients with asthma. *Am Rev Respir Dis* 131: 599–606.

Last, J.A., Ward, R., Temple, L., and Kenyon, N.J. (2004). Ovalbumin-induced airway inflammation and fibrosis in mice exposed to ozone. *Inhal Toxicol* 33–43.

Leigh, R., Ellis, R., Wattie, J., Southam, D.S., deHoogh, M., Gauldie, J., O'Byrne, M., and Imman, M.D. (2002). *Am J Respir Cell Mol Biol* 27: 526–535.

Lung, M.A. (1987). Effects of lung inflation on nasal airway resistance in the anesthetized rat. *J Appl Physiol* 63: 1339–1343.

Lushniak, B.D., Reh, C.M., Bernstein, D.I., and Gallagher, J.S. (1998). Indirect assessment of 4,4-diphenylmethane diisocyanate (MDI) exposure by evaluation of specific humoral immune responses to MDI conjugated to human serum albumin. *Am J Ind Med* 33: 471–477.

Mapp, C.E., DalVecchio, L., Boschetto, P., and Fabbri, L.M. (1985). Combined asthma and alveolitis due to diphenylmethane diisocyanate (MD) with demonstration of no crossed respiratory reactivity to toluene diisocyanate (TDI). *Ann Allergy* 54: 424–429.

Mauderly, J.L. (1986). Respiration of F344 rats in nose-only inhalation exposure tubes. *J. Appl. Toxicol.* 6: 25–30.

McFadden, E.R. and Gilbert, I.A. (1992). Asthma. *J Engl J Med* 327: 1928–1937.

Mitzner, W., Tankersley, C., Lundblad, L.K., Adler, A., Irvin, C.G., and Bates, J. (2003). Interpreting Penh in mice. *J Appl Physiol* 94: 828–832.

Mitzner, W. and Tankersley, C. (1998). Noninvasive measurement of airway responsiveness in allergic mice using barometric plethysmography. *Am J Respir Crit Care Med* 158: 340–341.

O'Byrne, P.M. and Postma, D.S. (1999). The many faces of airway inflammation. *Am J Respir Crit Care Med* 159: S41–S66.

Paiva, M., Yernault, J.C., van Eer de Weghe, P., and Englert, M. (1975). A sigmoid model of the static volume-pressure curve of human lung. *Respir Physiol* 23: 317–323.

Palmans, E., Kips, J.C., and Pauwels, R.A. (2000). Prolonged allergen exposure induces structural airway changes in sensitized rats. *Am J Respir Crit Care Med* 161: 627–635.

Pauluhn, J. (1994a). Assessment of chemicals for their potential to induce respiratory allergy in guinea pigs: a comparison of different routes of induction and confounding effects due to pulmonary hyperreactivity. *Toxicol in Vitro* 8: 981–985.

Pauluhn, J. (1994b). Validation of an improved nose-only exposure system for rodents. *J Appl Toxicol* 14: 5–62.

Pauluhn, J. (1997). Assessment of respiratory hypersensitivity in guinea pigs sensitized to toluene diisocyanate: improvements on analysis of response. *Fundam Appl Toxicol* 40: 211–219.

Pauluhn, J. (2000). Inhalation toxicity of 1,6-hexamethylene diisocyanate-homopolymer (HDI-IC) aerosol: results of single inhalation exposure studies. *Toxicol Sci* 58: 173–181.

Pauluhn, J. (2003). Respiratory hypersensitivity to trimellitic anhydride in brown Norway rats: analysis of dose-response following topical induction and time course following repeated inhalation challenge. *Toxicology* 194: 1–17.

Pauluhn, J. (2004). Comparative analysis of pulmonary irritation by measurements of Penh in bronchoalveolar lavage fluid in Brown Norway rats and Wistar rats exposed to irritant aerosols. *Inhal Toxicol* 16: 159–175.

Pauluhn, J. and Eben, A. (1991). Validation of a non-invasive technique to assess immediate or delayed onset of airway hypersensitivity in guinea pigs. *J Appl Toxicol* 11: 423–431.

Pauluhn, J., Rüngeler, W., and Mohr, U. (1995). Phenylisocyanate induced asthma in rats following a two-week exposure period. *Fundam Appl Toxicol* 24: 217–228.

Pauluhn, J. and Mohr, U. (1998). Assessment of respiratory hypersensitivity in guinea pigs sensitized to toluene diisocyanate: a comparison of sensitization protocols. *Inhal Toxicol* 10: 131–154.

Pauluhn, J., Dearman, R., Doe, J., Hext, P., and Landry, T.D. (1999). Respiratory hypersensitivity to diphenylmethane-4,4-diisocyanate in guinea pigs: comparison with trimellitic anhydride. *Inhal Toxicol* 11: 187–214.

Pauluhn, J. and Mohr, U. (2000). Inhalation studies in laboratory animals—current concepts and alternatives. Review. *Toxicol Pathol* 28: 734–753.

Pauluhn, J., Thiel, A., Emura, M., and Mohr, U. (2000). Respiratory sensitization to diphenyl-methane-4,4-diisocyanate (MDI) in guinea pigs: impact of particle, size on induction and elicitation of response. *Toxicol Sci.* 56: 105–113.

Pauluhn, J., Eidmann, P., and Mohr, U. (2002a). Respiratory hypersensitivity in guinea pigs sensitized to 1,6-hexamethylene diisocyanate (HDI): comparison of results obtained with the monomer and homopolymer of HDI. *Toxicology* 171: 147–160.

Pauluhn, J., Eidmann, P., Freyberger, A., Wasinska-Kempka, G., Vohr, H.-W. (2002). Respiratory hypersensitivity to trimellitic anhydride in brown Norway rats: a comparison of endpoints. *J Appl Toxicol* 22: 89–97.

Potter, D.W. and Wederbrand, K.S. (1995). Total IgE antibody production in BALB/c mice after dermal exposure to chemicals. *Fundam Appl Toxicol* 26: 127–135.

Pretolani, M. and Vargaftig, B.B. (1993). Commentary: from lung hypersensitivity to bronchial hyperreactivity. *Biochem Pharmacol* 45: 791–800.

Raman, K., Kaplan, M.H., Hogaboam, C.M., Berlin, A., and Lukacs, N.W. (2003). STAT4 signal pathways regulate inflammation and airway physiology changes in allergic airway inflammation locally via alteration of chemokines. *J Immunol* 170: 3859–3865.

Redlich, C.A. and Karol, M.H. (2002). Diisocyanate asthma: clinical aspects and immunopathogenesis. *Int Immunopharmacol* 2: 213–224.

Roche, W.R., Beasely, R., Williams, J.H., and Holgate, S.T. (1989). Subepithelial fibrosis in the bronchi of asthmatics. *Lancet* 1: 520–524.

Santing, R.E., Olymulder, C.G., Zaagsma, J., and Meurs, H. (1994). Relationship among allergen-induced early and late phase airway obstructions, bronchial hyperreactivity, and inflammation in conscious, unrestrained guinea pigs. *J Allergy Clin Immunol* 93: 1021–1030.

Sarlo, K. and Clark, E.D. (1992). A tier approach for evaluating the respiratory allergenicity of low-molecular-weight chemicals. *Fundam Appl Toxicol* 18: 107–114

Sarlo, K. and Karol, M.H. (1994). Guinea pig predictive tests for respiratory allergy. *Immunotoxicology and Immunopharmacology*, 2nd ed., Eds., J.H. Dean, M.I.Luster, A.E. Munson, and I. Kimber. Raven Press, New York, pp. 703–720.

Scheerens, H., Buckley, T.L., Davidse, E.M., Garssen, J., Nijkamp, F.P., and van Loveren, H. (1996). Toluene diisocyanate-induced in vitro tracheal hyperactivity in the mouse. *Am J Respir Crit Care Med* 154: 858–865.

Scheerens, H., Buckley, T.L., Muis, T.L., Garssen, J., Dormans, J., Nijkamp, F.P., and van Loveren, H. (1999). Long-term topical exposure to toluene diisocyanate in mice leads to antibody production and in vitro airway hyperresponsiveness three hours after intranasal challenge. *Am J Respir Crit Care Med* 159: 1074–1080.

Schramm, C.M., Puddington, L., Wu, C., Guernsey, L., Gharaee-Kermani, M., Phan, S.H., and Thrall, R.S. (2004). Chronic inhaled ovalbumin exposure induces antigen-dependent but not antigen-specific inhalational tolerance in a murine model of allergic airway disease. *Am J Pathol* 164: 295–304.

Schuster, M., Tschernig, Th., Krug, N., and Pabst, R. (2000). Lymphocytes migrate from the blood into the bronchoalveolar lavage and lung parenchyma in the asthma model of the brown Norway rat. *Am J Respir Crit Care Med* 161: 558–566.

Selgrade, M.K., Zeiss, C.R., Karol, M.H., Sarlo, K., Kimber, I., Tepper, J.S., and Henry, M.C. (1994). *Inhal Toxicol* 6: 303–319.

Shinagawa, K., and Kojima, M. (2003). Mouse model of airway remodeling—strain differences. *Am J Respir Crit Care Med* 168: 959–967.

Siegel, P.D., Al-Humadi, N.H., Millecchia, L.L., Robinson, V.A., Hubbs, A.F., Nelson, E.R., and Fedan, J.S. (2000). Ovalbumin aeroallergen exposure-response in brown Norway rats. *Inhal Toxicol* 12: 245–261.

Siegel, P.D., Al-Humadi, N.H., Nelson, E.R., Lewis, D.M., and Hubbs, A.F. (1997). Adjuvant effect of respiratory irritation on pulmonary sensitization: time and site dependency. *Toxicol Appl Pharmacol* 144: 356–362.

Southam, D.S., Dolovich, M., O'Byrne, P.M., and Inman, M.D. (2002). Distribution of intranasal instillations in mice: effect of volume, time, body position, and anesthesia. *Am J Physiol Lung Cell Mol Physiol* 282: L833–L839.

Steerenberg, P.A., van Dalen, W.J., Withagen, C.E.T., Dormans, J.A.M.A., and van Loveren, H. (2003). Optimization of route of administration for coexposure to ovalbumin and particle matter to induce adjuvant activity in respiratory allergy in the mouse. *Inhal Toxicol* 15: 1309–1325.

Sugawara, Y., Okamoto, Y., Sawahata, T., and Tanaka, K. (1993). An asthma model developed in the guinea pig by intranasal application of 2,4-toluene diisocyanate. *Ant Arch Allergy Immunol* 101: 95–101.

Sun, J. and Chung, K.F. (1997). Interaction of ozone exposure with airway hyperresponsiveness and inflammation induced by trimellitic anhydride in sensitized guinea pigs. *J. Toxicol Environ Health* 51: 77–87.

Takeda, K., Haczku, A., Lee, J.J., Irvin, C.G., and Gelfand, E.W. (2001). Strain dependence of airway hyperresponsiveness reflects differences in eosinophil localization in the lung. *Am J Physiol Lung Cell Mol Physiol* 281: L394–L402.

Temelkovski, J., Hogan, S.P., Shepherd, D.P., Foster, P.S., and Kumar, R.K. (1998). An improved murine model of asthma: selective airway inflammation, epithelial lesions and increased methacholine responsiveness following chronic exposure to aerosolised antigen. *Thorax* 53: 849–856.

Thorne, P.S. and Karol, M.H. (1988). Assessment of airway reactivity in guinea-pigs: comparison of methods employing whole body plethysmography. *Toxicology* 52: 141–163.

Tomkinson, A., Cieslewicz, G., Duez, C., Larson, C.D., Lee, J.J., and Gelfand, E.W. (2001). Temporal association between airway hyperresponsiveness and airway eosinophilia in ovalöbumin-sensitized mice. *Am J Respir Crit Care Med* 163: 721–730.

Tu, Y.-P., Larsen, G.L., and Irvin, C.G. (1995). Utility of murine systems to study asthma pathogenesis. *Eur Respir Rev* 5: 224. 230.

Vandenplas, O., Malo, J.-L., Saetta, M., Mapp, C.E., and Fabbri, L.M. (1993). Occupational asthma and extrinsic alveolitis due to isocyanates: current status and perspectives. *Br J Ind Med* 50: 213–228.

Vanoirbeek, J.A., Mandervelt, C., Cunningham, A.R., Hoet, P.H., Xu, H., Vanhooren, H.M., Nemery, B. (2003). Validity of methods to predict the respiratory sensitizing potential of chemicals: a study with a piperidinyl chlorotriazine derivative that caused an outbreak of occupational asthma. *Toxicol Sci* 76: 338–346.

Vijayaraghavan, R., Schaper, M., Thompson, R., Stock, M.F., and Alarie, Y. (1993). Characteristic modifications of the breathing pattern of mice to evaluate the effects of airborne chemicals on the respiratory tract. *Arch Toxicol* 67: 478–490.

Welinder, H., Zhang, X., Gustavsson, C., Björk, B., and Skerfving, S. (1995). Structure-activity relationship of organic anhydrides as antigens in an animal model. *Toxicology* 103: 127–136.

Whitehead, G.S., Walker, J.K.L., Berman, K.G., Foster, W.M., and Schwartz, D.A. (2003). Allergen-induced airway disease is mouse strain dependent. *Am J Physiol Lung Cell Mol Physiol* 285: L32–L42.

Wilder, J.A., Collie, D.D., Wilson, B.S., Bice, D.E., Lyons, C.R., and Lipscomb, M.F. (1999). Dissociation of airway hyperresponsiveness from immunoglobulin E and airway eosinophilia in a murine model of allergic asthma. *Am J Respir Cell Mol Biol* 20: 1326–1334.

Wils-Karp, M. (1999). Immunological basis of antigen-induced airway hyperresponsiveness. *Annu Rev Immunol* 17: 255.

Wisnewski, A.V., Srivastava, R., Herrick, C., Xu, L., Lemus, R., Cain, H., Magoski, N.M., Karol, M.H., Bottomly, K., and Redlich, C.A. (2000). Identification of human lung and skin proteins conjugated with hexamethylene diisocyanate in vitro and *in vivo*. *Am J Respir Crit Care Med* 162: 2330–2336.

Zhang, X.D., Murray, D.K., Lewis, D.M., and Siegel, P.D. (2002). Dose-response and time course of specific IgE and IgG after single and repeated topical skin exposure to dry trimellitic anhydride powder in a brown Norway rat model. *Allergy* 57: 620–626.

Zhang, X.D., Fedan, J.S., Lewis, D.M., and Siegel, P.D. (2004). Asthma-like biphasic airway responses in brown Norway rats sensitized by dermal exposure to dry trimellitic anhydride powder. *J Allergy Clin Immunol* 113: 320–326.

11

FIBER TOXICOLOGY

David M. Bernstein

CONTENTS

Fibers, due to their unique characteristics and effects, have prompted extensive scientific research over the years. The emphasis was initially on asbestos after the discovery of the association of asbestos exposure with mesothelioma and lung cancer. Subsequently, as interest in synthetic mineral fibers (SMFs) developed, these fibers were investigated using more advanced techniques that incorporated not only our understanding of the physical and chemical properties of fibers but also the limitations imposed by the test system used for these toxicological investigations. It is largely through these studies that a more complete understanding of the factors influencing fiber toxicology has developed.

A fiber is unique because of its shape. This uniqueness is enhanced because of the aerodynamic properties of fibers that allow them to align with the airflow and thus pass through openings more in proportion to their diameter rather than their length. This uniqueness becomes important biologically for that subset of fibers that in the above fashion can reach the alveolar region of the lung. If the fiber is longer than the size that a macrophage can fully engulf, the fiber cannot be cleared by the macrophage and will remain in the alveolus unless it dissolves.

Fibers that have been used commercially span a large range of chemical and physical characteristics. These fibers include amphibole and serpentine asbestos, synthetic vitreous fibers (SVFs), refractory ceramic fibers, organic fibers, carbon fibers, and the new nanofibers.

Historically, a fiber has been defined as a particle that has a length >5 μm, a width <3 μm, and an aspect ratio > 3:1 (WHO, 1985 and NIOSH, 1994). This definition was not based upon any criteria relating to health but, rather, was related to facilitating the counting of fiber using light microscopy. In addition, no consideration was given to the composition of the fibers, which has since been found to be important in determining pathogenicity of fibers.

There have been numerous articles that have reviewed fiber toxicology (Elmore, 2003; Britton, 2002; Warheit, Hart, et al. 2001; Warheit, Reed, and Webb, 2001; Hesterberg, 2001; Moolgavkar, 2001; Nicholson, 2001; Mast, 2000; Vu, 1997; Maxim, 1999; McClellan, 1992; Ilgren, 1991; Churg, 1988; Huncharek, 1986; Davis, 1986). More recently, IARC re-reviewed the evaluation of the carcinogenicity in humans of man-made vitreous fibers (MMVF) as part of their IARC Monograph series (IARC, 2002). In addition, the USEPA commissioned a review, by an expert panel, of short-term assays and testing strategies for assessment of fiber toxicology (Olin et al., 2005).

11.1 FACTORS INFLUENCING FIBER TOXICOLOGY

Mineral fiber toxicology has been associated with three key factors: dose, dimension, and durability. The dose is determined by the fibers' physical characteristics and dimensions, how the fibrous material is used, and the control procedures that are implemented. In addition, the thinner and shorter fibers will weigh less and thus can remain suspended in air longer than thicker and longer fibers. Most asbestos fibers are thinner than commercial insulation fibers; however, they are thicker than the new nanofibers that are currently being developed.

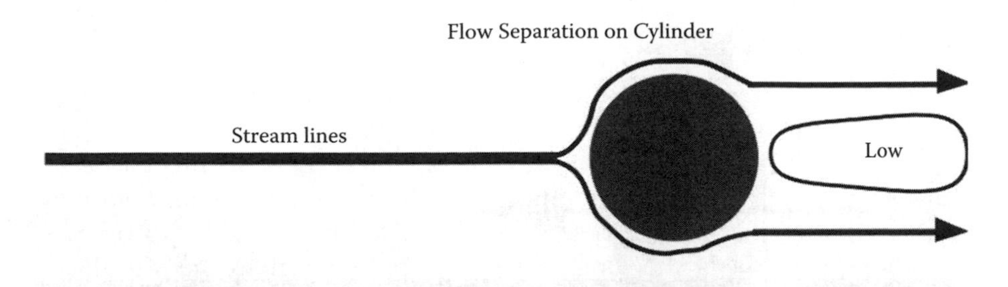

Figure 11.1 A fiber aligned perpendicular to the airflow. when the air accelerates around the front of the fiber, the pressure decreases. A low-pressure area forms above and below the fiber. When the air attempts to flow around the back of the fiber, the adverse pressure gradient soon causes the boundary layer to separate and the drag to increase dramatically.

The fiber dimensions govern two factors: first, whether the fiber is respirable and, second, if it is respirable, their response in the lung milieu.

The aerodynamic properties of fibers allow them to align with the airflow and thus pass through openings more in proportion to their diameter rather than their length. In an airstream, such as is created during inhalation, the fiber will tend to align with the airflow. This happens effectively because of the effect of streamlining, which reduces the pressure drag caused by the fiber. If the fiber is aligned perpendicular to the airflow (Figure 11.1), then as the air accelerates around the front of the fiber, the pressure decreases. A low-pressure area forms above and below the fiber. When the air attempts to flow around the back of the fiber, the adverse pressure gradient soon causes the boundary layer to separate and the drag to increase dramatically.

However, in the presence of air gradients such as may develop during inhalation, the fiber will experience a torque until it is oriented parallel to the direction of the airflow (Figure 11.2). In this orientation, the fiber is much more streamlined and will produce less pressure drag (Baron, 1993).

If a fiber does penetrate into the lung, then the fiber dimensions will also determine the response to the fiber in the lung milieu. Shorter fibers that can be fully engulfed by the macrophage are likely to undergo successful phagocytosis and clearance by the macrophage system. It is only the longer fibers that the macrophage cannot fully engulf that, if they are persistent, will lead to disease. This is illustrated in Figure 11.3 (from Bernstein et al., 1984), which shows a scanning electron micrograph of a macrophage that has engulfed a sized fiber that was instilled in the rat lung. In this study, the fibers manufactured were all of a uniform diameter of 1.5 μm.

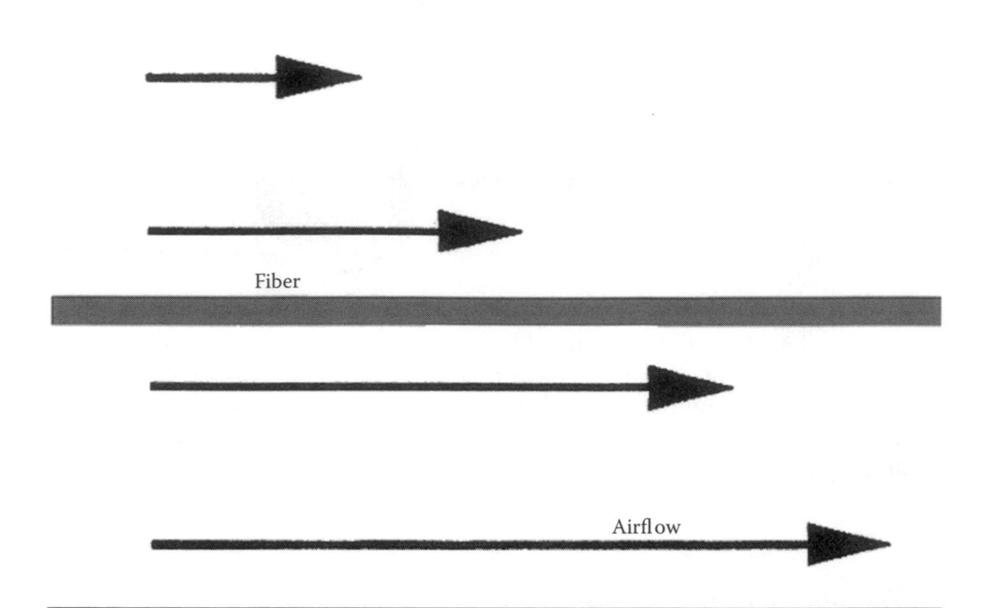

Figure 11.2 A fiber will tend to align with the airstream. In the presence of air gradients such as during inhalation, the fiber will experience a torque until the fiber is oriented parallel to the direction of the airflow. In this orientation, the fiber is much more streamlined and will produce less pressure drag. (From Baron, P.A. 1993. Measurement of Asbestos and Other Fibers. In *Aerosol Measurement*, Eds. K. Willeke and P.A. Baron. New York: Van Nostrand Reinhold.)

The fiber engulfed by the macrophage has been dissolved and has a thinner diameter.

This leads to the third factor, that of durability. Those fibers whose chemical structure renders them wholly or partially soluble, once deposited in the lung, are likely to either dissolve completely or dissolve until they are sufficiently weakened focally to undergo breakage into shorter fibers. The resulting short fibers are then likely to undergo successful phagocytosis and clearance by the macrophage system.

These factors have been shown to be important determinants of potential toxicity for SMFs (Donaldson and Tran, 2004; Bernstein et al., 2001a and 2001b; Oberdoester, 2000; Miller et al., 1999; Hesterberg, Chase et al., 1998; and Hesterberg, Hart et al., 1998).

With other types of fibers, such as organic fibers, not enough is known to fully evaluate the importance of these and other factors in the causation of disease. Most likely, if an organic fiber is rigid (similar to mineral fibers), then length and biopersistence are important. However, many organic fibers are curly and less rigid than mineral fibers and may not frustrate the macrophages' ability to clear them. With organic fibers, as with organic

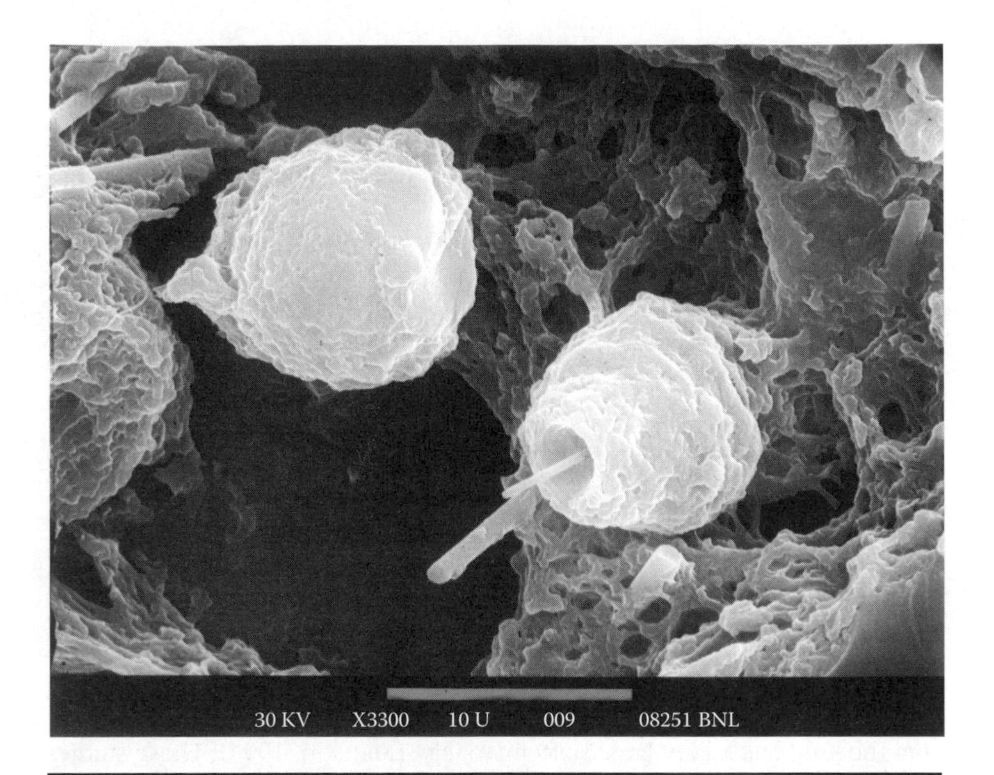

Figure 11.3 **Scanning electron micrograph of a macrophage that has engulfed a sized fiber that was instilled in the rat lung. In this study, the fibers were manufactured to be of uniform diameter of 1.5 μm. These size fibers are seen adjacent to the macrophages. The fiber engulfed by the macrophage has been dissolved and is of a thinner diameter. (From Bernstein, D.M., Drew, R. Schidlovsky, G. and Kuschner. M. 1984. Pathogenicity of MMMF and the Contrasts with Natural Fibres. In *Biological Effects of Man-Made Mineral Fibres: Proceedings of a WHO/IARC Conference in Association with JEMRB and TIMA*. Copenhagen: WHO. With Permission.)**

particles, consideration should be given to the chemical interactions with surrounding cells and the effect this may have on pathogenicity.

11.2 *IN VITRO* TOXICOLOGY

In vitro toxicology studies are often very helpful in elucidating the possible mechanisms involved in pathogenesis. However, as used in the assessment of fiber toxicology, they are difficult to interpret. This stems from several factors. The *in vitro* test system is a static system and thus is not sensitive to differences in fiber solubility. High doses of fibers are used to obtain a positive response, and it is difficult to extrapolate from these short-term cellular exposures to lower-dose chronic exposures *in vivo*. In addition, the number of fibers and the size distribution are usually often not quantified. Most important, however, these end points have not been validated as screening assays that are predictive of long-term pathological effects *in vivo*. Although *in vitro* tests may be useful tools for the identification and evaluation of possible mechanisms, these test systems are of limited use in differentiating between the fiber types.

11.3 BIOPERSISTENCE

Within the lung, fibers that can be fully engulfed by the macrophage can be removed similar to any other particle. However, those fibers that are too long to be fully engulfed by the macrophage cannot be cleared by this route. It has been estimated that the upper limit for fiber length that the macrophage can engulf is between 10 and 15 μm in the rat and between 15 and 20 μm in humans (Miller, 2000a). Fibers longer than this will remain in the lung and can be cleared from the lung only if they can dissolve.

In the lung, extensive work on modeling the dissolution of SMFs using *in vitro* dissolution techniques and inhalation biopersistence has shown that the lung has a very large buffer capacity (Mattson, 1994). These studies have shown that an equivalent *in vitro* flow rate of up to 1 ml/min is required to provide the same dissolution rate of SMF as that which occurs in the lung. This large fluid flow within the lung results in the dissolution of the more soluble fibers.

The association that long fibers (20 to 50 μm) have with both lung and peritoneal disease, as opposed to shorter ball-milled fibers (3 μm or less), was reported as early as 1951 (Vorwald et al., 1951). In one of the earlier studies investigating the biopersistence of SMFs, Hammad et al. (1988) found that fibers <5 μm in length had the longest retention following short-term inhalation, with longer fibers clearing more rapidly and fibers >30 μm in length clearing very rapidly. He proposed that clearance of mineral wools is a result of biological clearance and the elimination of fibers by dissolution and subsequent breakage. However, there was no relationship between these phenomena and long-term toxicological effects.

Early chronic inhalation studies of fibers were often performed without considering the respirability of the fibers in the rat and without preserving the length distribution of the fibers. As commercial insulation SMFs are usually thick and long, investigators would often grind the fibers to produce a more respirable fraction. This process frequently pulverized the rat-respirable long-fiber fraction. In 1988, a series of well-designed chronic inhalation studies on SMFs was initiated at the Research and Consulting

Company Ltd. (Itingen, Switzerland) by the Thermal Insulation Manufacturers Association, TIMA (Hesterberg et al., 1993, 1995; Mast et al., 1995a, 1995b; McConnell et al., 1994, 1995). These studies were the first of their kind to take into account the respirability of mineral fibers in rats and the importance of fiber length. The results of the studies indicated that the more soluble fibers tested showed little or no pathogenic response, whereas the less soluble fibers showed some response. To further investigate this, a protocol was developed for the evaluation of the biopersistence of SMF (Bernstein et al., 1994; Musselman et al., 1994), and all the fibers from TIMA's chronic inhalation studies were evaluated for biopersistence. Subsequently, additional studies were performed on some of the SMFs included in this analysis (Bernstein et al., 1996, Hesterberg, Chase et al., 1998). In these early studies, there was no standardization of the statistical methods for determining the clearance half-times of the fibers tested; this resulted in values that were not always comparable between studies. To resolve this difficulty and permit the determination of the relationship of biopersistence to toxicity, standardized biopersistence protocols ("Ispra Protocols") were developed by the European Commission at the European Chemicals Bureau (ECB) and were subsequently published (Bernstein and Riego-Sintes, 1999).

For mineral fibers, the clearance half-time of fibers longer than 20 μm ranges from a few days to less than 100 d. This is illustrated in Table 11.1. Also shown in this table are the results from biopersistence studies performed on chrysotile and amphiboles using the same protocol. For SVFs, the European Commission has established a directive that states that if the inhalation biopersistence clearance half-time of a fiber is less than 10 d, then it is not classified as a carcinogen.

Recent studies on the serpentine asbestos chrysotile have shown that it is not very biopersistent in the lung. Serpentine asbestos is a naturally occurring mined fiber, and there appear to be some differences in biopersistence depending on where it was mined. However, chrysotile lies at the soluble end of this scale and ranges from being the least biopersistent fiber to a fiber with a biopersistence similar to that of glass and stone wools. It is less biopersistent than ceramic and special-purpose glasses and more than an order of magnitude less biopersistent than amphiboles. In contrast to SVF, the rapid clearance of chrysotile appears to result from the loss of stability of its chemical structure in the lung resulting in the disintegration of the longer fibers into amorphous particles.

11.3.1 The Macrophage as a Mediator of Biopersistence

Kamstrup (2002) described possible mechanisms that could account for the rapid clearance half-time of the long HT (high temperature) fibers.

Table 11.1 Clearance Half-Times of a Range of Mineral Fibers as Determined by the Fiber Inhalation Biopersistence Protocol

Fiber	*Type*	*Weighted $T_{1/2}$ Fibers $L > 20$ μm (days)*	*Reference*
Calidria chrysotile	Serpentine asbestos	0.3	Bernstein et al., 2005
Brazilian chrysotile	Serpentine asbestos	2.3	Bernstein et al., 2004a
Fiber B	B01.9	2.4	Bernstein et al., 1996
Fiber A	Glass wool	3.5	Bernstein et al., 1996
Fiber C	Glass wool	4.1	Bernstein et al., 1996
Fiber G	Stone wool	5.4	Bernstein et al., 1996
MMVF34	HT stone wool	6	Hesterberg et al, 1998b
MMVF22	Slag wool	8	Bernstein et al., 1996b
Fiber F	Stone wool	8.5	Bernstein et al., 1996
MMVF11	Glass wool	9	Bernstein et al., 1996
Fiber J	X607	9.8	Bernstein et al., 1996
MMVF 11	Glass wool	13	Bernstein et al., 1996
Fiber H	Stone wool	13	Bernstein et al., 1996
Canadian chrysotile	Serpentine asbestos	11.4	Bernstein et al., 2004b
MMVF10	Glass wool	39	Bernstein et al., 1996
Fiber L	Stone wool	45	Bernstein et al., 1996
MMVF33	Special-purpose glass	49	Hesterberg et al., 1998b
RCF1a	Refractory ceramic	55	Hesterberg et al., 1998b

(*continued*)

Table 11.1 Clearance Half-Times of a Range of Mineral Fibers as Determined by the Fiber Inhalation Biopersistence Protocol (Continued)

Fiber	*Type*	*Weighted $T_{1/2}$ Fibers L > 20 μm (days)*	*Reference*
MMVF21	Stone wool	67	Hesterberg et al., 1998b
MMVF32	Special-purpose glass	79	Hesterberg et al., 1998b
MMVF21	Stone wool	85	Bernstein et al., 1996
Amosite	Amphibole asbestos	418	Hesterberg et al., 1998
Crocidolite	Amphibole asbestos	536	Bernstein et al., 1996

He stated that the HT fiber is characterized by relatively low silica and high alumina content, with a high dissolution rate at pH 4.5 and relatively low rate at pH 7.4 (Knudsen et al., 1996). Apart from possible exposure to the acidic environment of the phagolysosomes within the macrophages (Oberdörster, 1991), measurements have shown that the microenvironment at the surface of activated macrophages is acidic, with a pH < 5 between attached macrophages and a nonporous glass surface (Etherington et al., 1981). It is, therefore, probable that long HT fibers, highly soluble at pH 4.5, are subject to extracellular dissolution and consequent breakage when exposed to the acidic environment of attached macrophages without being engulfed completely.

As mentioned earlier, at acidic pH chrysotile also becomes less stable, and a similar mechanism may help accelerate the clearance or disintegration of the long chrysotile fibers.

11.3.2 Short Fiber Clearance

For all fiber exposures, there are many shorter fibers of less than 20 μm in length and even more that are less than 5 μm in length. The clearance of the shorter fibers has in these studies been shown to be either similar to or faster than the clearance of insoluble nuisance dusts (Muhle et al., 1987; Stoeber et al., 1970). In a recent report issued by the Agency for Toxic Substances and Disease Registry entitled "Expert Panel on Health Effects of Asbestos and Synthetic Vitreous Fibers: The Influence of Fiber

Length," the experts stated: "Given findings from epidemiologic studies, laboratory animal studies, and *in vitro* genotoxicity studies, combined with the lung's ability to clear short fibers, the panelists agreed that there is a strong weight of evidence that asbestos and SVFs (synthetic vitreous fibers) shorter than 5 μm are unlikely to cause cancer in humans" (ATSDR, 2003; USEPA, 2003). In addition, Berman and Crump (2003), in their technical support document to the EPA on asbestos-related risk, also found that shorter fibers do not appear to contribute to disease.

11.4 CHRONIC INHALATION TOXICOLOGY STUDIES

11.4.1 The Difficulty of Designing Chronic Inhalation Toxicology Studies with Fibers

Although many chronic inhalation toxicology studies of fibers ranging from amphibole asbestos to soluble glass fibers and organic fibers have been performed, their design and subsequent interpretation are often confounded by the fiber size distribution and the ratio of longer fibers to shorter fibers and nonfibrous particles. In some of these studies, the exposures often approach and exceed the levels that have been shown to produce what is now termed *lung overload* in the rat. Thus, it can become very difficult to compare the effects of such a study with those of another.

High concentrations of insoluble nuisance dusts have been shown to compromise the clearance mechanisms of the lung, causing inflammation and a tumorigenic response in the rat — a phenomenon often referred to as lung overload (Bolton et al., 1983; Muhle et al., 1988; Morrow, 1988; Oberdorster, 2002; Miller 2000b).

This is illustrated in Table 11.2, which shows for a series of studies performed at the RCC laboratories, the exposure concentrations and lung burdens for a number of SVFs and serpentine and amphibole asbestos. The asbestos exposures in these studies were included as positive controls; however, as shown in Table 11.3, it would be very difficult to compare the chrysotile exposure, and even the crocidolite exposure, to those of the SVF on a comparative fiber number basis. The exposure concentration and resulting lung burden of chrysotile was so large that it is very likely that a lung overload effect did occur. It would have been much more useful and interesting in these studies if the exposure concentrations and fiber size distributions were comparable between the positive controls and the SVFs.

In a recent study, Bellmann et al. (2003) reported on a calibration study to evaluate the end points in a 90-d subchronic inhalation toxicity study of MMVFs (with a range of biopersistence) and amosite. One of the fibers was a calcium-magnesium-silicate (CMS) fiber for which the stock preparation had a large concentration of particulate material in addition to the fibers. After chronic inhalation of the fiber X607, which

Table 11.2 Fiber Exposure and Lung Burden Data for a Series of Fiber Inhalation Toxicology Studies Performed Using Similar Protocols

Fiber Type	*Aerosol Total (fibers/cm³)*	*Aerosol[a] Number (fibers/cm³, WHO Standard)*	*Aerosol Number (fibers/cm³, length >20 μm)*	*Lung Total Fiber Number (in lung after 24 months of exposure)*	*Lung No. of Fibers of length > 20 μm (In lung after 24 months of exposure)*	*Reference*
MMVF11[C]	NA	41	14	93,000,000	5,580,000	Hesterberg et al., 1993
MMVF11[C]	NA	153	50	692,000,000	24,912,000	Hesterberg et al., 1993
MMVF11[C]	273	246	84	1,284,000,000	30,816,000	Hesterberg et al., 1993
MMVF21[D]	44	34	13	112,142,857	14,130,000	McConnell et al., 1994
MMVF21[D]	185	150	74	548,173,913	50,432,000	McConnell et al., 1994
MMVF21[D]	264	243	114	622,884,615	80,975,000	McConnell et al., 1994
MMVF22[D]	33	30	10	21,984,733	2,880,000	McConnell et al., 1994
MMVF22[D]	158	131	50	320,625,000	7,695,000	McConnell et al., 1994
MMVF22[D]	245	213	99	596,750,000	23,870,000	McConnell et al., 1994
RCF1[C]	36	26	13	99,900,000	12,787,200	Mast et al., 1995b
RCF1[C]	91	75	35	233,120,000	31,238,080	Mast et al., 1995b
RCF1[C]	162	120	58	578,240,000	58,402,240	Mast et al., 1995b

(continued)

Table 11.2 Fiber Exposure and Lung Burden Data for a Series of Fiber Inhalation Toxicology Studies Performed Using Similar Protocols (Continued)

Fiber Type	*Aerosol Total (fibers/cm³)*	*Aerosol[a] Number (fibers/ cm³, WHO Standard)*	*Aerosol Number (fibers/cm³, length >20 μm)*	*Lung Total Fiber Number (in lung after 24 months of exposure)*	*Lung No. of Fibers of length > 20 μm (In lung after 24 months of exposure)*	*Reference*
RCF1[C]	234	187	101	1,017,500,000	132,275,000	Mast et al., 1995a
Chrysotile[b]	102,000	10,600	NA	54,810,000,000	NA	Hesterberg et al., 1993
Crocidolite (26w)	4,214	1,610	236	2,025,000,000	88,452,000	McConnell et al., 1994

Note: NA = Data not available.

[a] WHO fibers: Fibers with length >5 μm, width <3 μm, and aspect ratio > 3:1 (WHO. 1985. *Reference Methods for Measuring Airborne Man-Made Mineral Fibres (MMMF), WHO/EURO MMMF Reference Scheme*. Vol. EH-4. Copenhagen: World Health Organization and NIOSH. 1994 (2003). *NIOSH Manual of Analytical Methods*. Eds. P.C. Schlecht and P.F. O'Conner. 4th ed., Vol. 1st Supplement Publication 96-135, 2nd Supplement Publication 98-119, 3rd Supplement 2003-154. http://www.cdc.gov/niosh/nmam/: DHHS (NIOSH) Publication 94-113.).

[b] Calculated based upon the lung dry weight as 0.1 of the wet weight.

[c] The number of fibers L > 20 μm per lung was not reported in the publications and was calculated based upon counting data provided by Owens-Corning.

[d] The total number of fibers per lung was not reported in the publications and was calculated based upon counting data provided by Owens-Corning.

Table 11.3 Chronic Inhalation Toxicology Studies of Insulation Glass Wools and Special-Purpose Glass Fibers in Rodents (Rats and Hamsters)

Test Substance	*Aerosol Fibers (Numbers and Dimensions)*	*Positive Control*	*Test System (No. at Risk); Observation Time*	*Exposure*	*Lung Dose*	*No. of Thoracic Tumors/ No. of Animals*	*Comments (Positive Control Tumor Incidence)*	*Reference*
				Insulation glass wools				
Rat								
Glass fiber	~168 WHO f/cm^3	Amosite, ~1178 WHO f/cm^3	Sprague-Dawley rats, male; 24 mo	Whole-body, 5 h/d, 5 d/wk, for 3 mo		2111 (adenomas)	Short exposure period, small number (11) of animals at risk (amosite: 3/12 tumors)	Lee et al. (1981)
Glass fiber + resin	240 WHO f/cm^3 (10 fL > 20 μm, 10 mg/m^3 respirable dust; D, 52% a 1 μm; L, 72% 5–20 μm	Chrysotile, 10 mg/m^3, 3800 WHO f/cm^3	56 (48) SPF Fischer rats, male and female; 24 mo; lifetime	Whole-body. 7 h/d, 5 d/wk, for 12 mo	mg f/lung: 1.9 at 12 mo 0.5 at 21 mo	1/148 (adeno-carcinoma); fibrosis, 0	Type of glass fiber not specified (chrysotile: 12/48 tumors; fibrosis)	Wagner et al. (1984)

(continued)

Table 11.3 Chronic Inhalation Toxicology Studies of Insulation Glass Wools and Special-Purpose Glass Fibers in Rodents (Rats and Hamsters) (Continued)

Test Substance	*Aerosol Fibers (Numbers and Dimensions)*	*Positive Control*	*Test System (No. at Risk); Observation Time*	*Exposure*	*Lung Dose*	*No. of Thoracic Tumors/ No. of Animals*	*Comments (Positive Control Tumor Incidence)*	*Reference*
Glass fiber –resin	323 WHO f/cm^3 (19 f L > 20 μm, 10 mg/m^3 respirable dust; D, 47% < 1 μm; L, 58% 5–20 μm	Chrysotile, 10 mg/m^3, 3800 WHO f/cm^3	56 (47) SPF Fischer rats, male and female; 24 mo; lifetime	Whole-body. 7 h/d, 5 d/wk, for 12 mo	mg f/lung: 0.9 at 12 mo 0.2 at 21 mo	1/47 (adenoma); fibrosis, 0	Type of glass fiber not specified (chrysotile: 12/48 tumors; fibrosis)	Wagner et al. (1984)
Glass fiber; French (Saint Gobain)	48 WHO f/cm^3, 5 mg/m^3 respirable dust; D, 69% < 1 μm; L, 42% > 10 μm	Chrysotile, 5 mg/m^3; L, 6%> 5 μm	45 Wistar rats, male/female; 28 mo	Whole-body for 24 mo	ND	1/45	Type of glass fiber not specified (chrysotile: 9/47 tumors; fibrosis)	Le Bouffant et al. (1984)
Insulsafe II building insulation	30 f/cm^3; L, > 10 μm; D, < 1 μm; 100 total f/cm^3; 10 mg/m^3; D, 1.4 μm mean; GMD, 1.2 μm; L, 37 μm mean; GML, 24 μm	Crocidolite, 3000 total f/cm^3, 90 f/cm^3 (L > 10 μm)	52 Osborne-Mendel rats, female; lifetime	Nose-only, 6 h/d, 5 d/wk, for 24 mo	3×10^4 f/mg dry lung at 3 mo	Tumors: 0/52; fibrosis, 0	Aerosolized fibers were short. Asbestos: low tumor incidence (3/157; ~5%) + fibrosis (some)	Smith et al. (1987)

Manville 901 building insulation	232 WHO f/cm^3 (73 fL > 20 μm), 30 mg/m^3; D, 1.4 μm mean; GMD, 1.3 μm; L, 16.8 μm mean; GML, 13.1 μm	Chrysotile, 10 mg/m^3, 10 600 WHO f/cm^3	140 (119) Fischer rats, male; lifetime	Nose-only, 6 h/d, 5 d/wk, for 24 mo	fL > 5 μm/lung: 42 × 10^6 at 12 mo; 82 × 10^6 at 24 mo f L > 20 μm/lung: 3 × 10^6 at 12 mo; 5 × 10^6 at 24 mo	Lung tumors: 7/119 (1 carcinoma); fibrosis, 0	Chrysotile: 13/69 lung tumors; 1/69 mesothelioma	Hesterberg et al. (1993); Hesterberg & Hart (2001)
Insulsafe II building insulation	246 WHO f/cm^3 (90 f L > 20μm), 30 mg/m^3; D, 0.9 μm mean; GMD, 0.7 μm; L, 183 μm mean; GML, 13.7μm	Chrysotile, 10 mg/m^3, 10 600 WHO f/cm^3	140 (112) Fischer rats, male; lifetime	Nose-only, 6 h/d, 5 d/wk, for 24 mo	fL> 5 μm/lung: 69 × 10^6 at 12 mo; 182 × 10^6 at 24 mo fL > 20 μm/lung: 7 × 10^6 at 12 mo; 6 × 10^6 at 24 mo	Lung tumors: 3/112; fibrosis, 0	Chrysotile: 13/69 lung tumors; 1/69 mesothelioma	Hesterberg et al. (1993); Hesterberg & Hart (2001)
Manville building insulation	25 f/cm^3; L, > 10 μm D, < 1 μm;100 total f/cm^3; 12 mg/m^3; D, 1.4 μm mean; GMD, 1.1 μm; L, 31 μm mean; GML, 20 μm	Crocidolite, 3000 f/cm^3, total f/cm^3, 90 f/cm^3; L > 10 μm	57 Osborne-Mendel rats, female; lifetime	Nose-only, 6 h/d, 5 d/wk, for 24 mo	2000 f/mg dry lung at 3 mo	Tumors: 0/57; fibrosis, 0	Aerosolized fiber concentration was very low. Asbestos: few tumors (5%) + fibrosis (some)	Smith et al. (1987)
Owens Corning building	5 f/cm^3; L > 10 μm; D, < 1 μm; 25 total f/cm^3;	Crocidolite, 3000 f/cm^3, total f/cm^3,	58 Osborne-Mendel rats, female, lifetime	Nose-only, 6 h/d, 5 d/wk, for	600 f/mg dry lung at 3 mo	Tumors: 0/58; fibrosis, 0	Aerosolized fiber concentration was very	Smith et al. (1987)

(continued)

Table 11.3 Chronic Inhalation Toxicology Studies of Insulation Glass Wools and Special-Purpose Glass Fibers in Rodents (Rats and Hamsters) (Continued)

Test Substance	*Aerosol Fibers (Numbers and Dimensions)*	*Positive Control*	*Test System (No. at Risk); Observation Time*	*Exposure*	*Lung Dose*	*No. of Thoracic Tumors/ No. of Animals*	*Comments (Positive Control Tumor Incidence)*	*Reference*
insulation	9 mg/m³; D, 3 μm mean; GMD, 3 μm; L, 114 μm mean; GML, 83 μm	90 f/cm³; L > 10 μm		24 mo			low and most fibers were very coarse and thick. Asbestos: few tumors (5%) + fibrosis (some)	
Owens Corning building insulation	5 and 15 mg/m³, no fiber dimensions; no aerosol concentrations specified	None	500 Fischer 344 rats; lifetime	Whole-body, 7 h/d, 5 d/wk, for 86 wks	ND	Tumor, 0/500; fibrosis, 0	No data on concentrations of fibers in aerosol or in the lung. No positive asbestos control	Moorman et al. (1988)
Hamster								
Insulsafe II building insulation	30 f/cm³; L, > 10 μm; D, < 1 μm; 100 total f/cm³, 10 mg/m³; D, 1.4 μm mean; GMD, 1.2 μm; L, 37 μm mean; GML, 24 μm	Crocidolite, 3000 total f/cm³, 90 f/cm³ (L > 10 μm)	60 Syrian golden hamsters, male; lifetime	Nose-only, 6 h/d, 5 d/wk, for 24 mo	1 × 10⁴ f/mg dry lung at 3 mo	Tumors: 0160; fibrosis, 0	Aerosolized fibers were short. Asbestos-exposed animals had no tumors (some did have fibrosis)	Smith et al. (1987)

Glass fiber	~168 WHO f/cm^3	Amosite ~1178 WHO f/cm^3	Hamsters; 24 months	Whole-body, 5 h/d, 5 d/wk, for 3 mo	ND	0/9	Short exposure period, small number (9) of animals at risk (amosite: 0/5 tumors)	Lee et al. (1981)
Manville 901 building insulation	339 WHO f/cm^3 (134 f L > 20 μm), 30 mg/m^3; (GMD. 0.84 μm; GMI, 12.4 μm	Amosite (mid), 165 WHO f/cm^3 (38 f L> 20 μm); amosite (high), 263 WHO f/cm^3 (69 f L > 20 μm)	125 (81) Syrian golden hamsters, male; lifetime	Nose-only, 6 Wd, 5 d/wk, for 18 mo	fL > 5 μm/lung: 32 × 10^6 at 12 mo; 77 × 10^6 at 18 mo f L > 20 pmllung: 1 × 10^6 at 12 mo; 5 × 10^6 at 18 mo	Lung tumors or mesotheliomas, 0/81; fibrosis, 0	Amosite (mid): 22/85 mesotheliomas, no lung tumors, fibrosis Amosite (high) 17/87 mesotheliomas, no lung tumors, fibrosis	Hesterberg et al. (1999); McConnell et al. (1999)
Manville building insulation	25 f/cm^3; L, >10 μm D, < 1 μm; 100 total f/cm^3; 12 mg/m^3; D, 1.4 μm mean; GMD, 1.1 μm; L, 31 μm mean; GML, 20 μm	Crocidolite, 3000 f/cm^3, total f/cm^3, 90 f/cm^3; L > 10 μm	66 Syrian golden hamsters, male; lifetime	Nose-only, 6 h/d, 5 d/wk, for 24 mo	1000 f/mg dry lung at 3 mo	Tumors: 0/66; fibrosis, 0	Aerosolized fiber concentration was very low. Asbestos-exposed animals had no tumors (some did have fibrosis)	Smith et al. (1987)

(continued)

Table 11.3 Chronic Inhalation Toxicology Studies of Insulation Glass Wools and Special-Purpose Glass Fibers in Rodents (Rats and Hamsters) (Continued)

Test Substance	*Aerosol Fibers (Numbers and Dimensions)*	*Positive Control*	*Test System (No. at Risk); Observation Time*	*Exposure*	*Lung Dose*	*No. of Thoracic Tumors/No. of Animals*	*Comments (Positive Control Tumor Incidence)*	*Reference*
Owens Corning building insulation	5 /cm^3; L, > 10 μm; D, < 1 μm; 25 total f/cm^3; 9 mg/m^3; D, 3 μm mean; GMD, 3 μm; L, 114 μm mean; GML, 83 μm	Crocidolite, 3000 f/cm^3, total f/cm^3, 90 f/cm^3; L > 10 μm	61 Syrian golden hamsters, male; lifetime	Nose-only, 6 h/d, 5 d/wk, for 24 mo	500 f/mg dry lung at 3 mo	Tumors: 0/61; Fibrosis, 0	Aerosolized fiber concentration was very low and most fiber were very coarse and thick. Asbestos-exposed animals had no tumors (some did have fibrosis)	Smith et al. (1987)
Special-purpose glass fibers								
Rat								
JM 100	10 mg/m^3 respirable dust; D, 0.3 μm; L, 71% < 10 μm	Chrysotile, 10 mg/m^3	100 Fischer rats, 50 male, 50 female; lifetime	Whole-body, 7 h/d, 5 d/wk. for 12 mo	ND	0/55; fibrosis, 0	JM 100 fibers were short (chrysotile: 11/56 tumors; fibrosis)	McConnell et al. (1984)

JM 100	1436 WHO f/cm^3 (108 f L > 20 μm), 10 mg/m^3 respirable dust; D, 97% < 1 μm; L, 93% 5–20 μm	Chrysotile, 10 mg/m^3 3800 WHO f/cm^3	56 (48) SPF Fischer rats, male and female; 24 mo; li Mime	Whole-body, 7 h/d, 5 d/wk, for 12 mo	mg f/lung: 4.5 at 12 mo 2.1 at 21 mo	1/48 (adeno-carcinoma); fibrosis, 0	JM 100 fibers were short (chrysotile: 12/48 tumors; fibrosis)	Wagner et al. (1984)
JM 100	332 WHO f/cm^3, 5 mg/m^3 respirable dust; D, 95% 1 μm; L, 60% > 10 μm, 25% > 20 μm	Chrysotile, 5 mg/m^3, 6000 WHO f/cm^3	48 Wistar rats, male and female; 28 mo	Whole-body, 5 h/d, 5 d/wk, for 24 mo	ND	0/48	JM 100 fibers were short (chrysotile: 9/47 tumors; fibrosis)	Le Bouffant et al. (1984; 1987)
JM 104/475	252 WHO f/cm^3, 3 mg/m^3; D, 0.42 μm median; L, 4.8 μm median	Crocidolite, 162 WHO f/cm^3; chrysotile, 131 WHO f/cm^3	108 Wistar rats, female; lifetime	Nose-only, for 12 mo	WHO f × 10^6: 70 at 12 mo 25 at 24 mo	1/107	JM 104/475 fibers were short (crocidolite:1/50 tumors; chrysotile: 0/50 tumors)	Muhle et al. (1987)
Manville Code 100	530 f/cm^3; L, > 10 μm D, ≤ 1 μm; 3000 total f/cm^3; 3 mg/m^3; D, 0.4 μm mean; GMD, 0.4 μm; L, 7.5 μm mean; GML, 4.7 μm	Crocidolite, 3000 f/cm^3, total f/cm^3, 90 f/cm^3; L > 10 μm	57 Osborne-Mendel rats, female; lifetime	Nose-only, 6 h/d, 5 d/wk, for 24 mo	2 × 10^6 f/mg dry lung at 2 mo	Tumors: 0/57; fibrosis, 0	Low survival rates: < 50% of rats survived to 24 mo (including controls). Asbestos: few tumors (3/57; 5%) + fibrosis	Smith et al. (1987)

(continued)

Table 11.3 Chronic Inhalation Toxicology Studies of Insulation Glass Wools and Special-Purpose Glass Fibers in Rodents (Rats and Hamsters) (Continued)

Test Substance	*Aerosol Fibers (Numbers and Dimensions)*	*Positive Control*	*Test System (No. at Risk); Observation Time*	*Exposure*	*Lung Dose*	*No. of Thoracic Tumors/ No. of Animals*	*Comments (Positive Control Tumor Incidence)*	*Reference*
JM 475	5 and 15 mg/m^3; no fiber dimensions; no aerosol concentrations specified	None	500 Fischer 344 rats; 18 mo; lifetime	Whole-body, 7 h/d, 5 d/wk, for 86 wks	ND	Tumor, 0; fibrosis, 0	No data on fiber concentrations in aerosol or in the lung. No positive asbestos control	Moorman et al. (1988)
JM 475	1066 WHO f/cm^3 (38 f > 20 μm/cm^3); 0.2 μm < D (More than 60% of fibers) < 0.4 μm	Amosite, 981 WHO f/cm^3 (89 f> 20 μm/cm^3)	24 mo, 83 (38) Wistar rats	Whole-body, 7 h/d, 5 d/wk, for 12 mo	2241 × 10^6 WHO f/lung; 11 × 10^6 f > 20 μm/lung	4/38 (adenomas); fibrosis: negligible	Authors concluded tumors/fibrosis similar to that of controls (5% tumors in air control).	Davis et al. (1996); Cullen et al. (2000)
104E	975 WHO f/cm^3 (72 f > 20 μm/cm^3); 0.2 μm < D (more than 60% of fibers) < 0.4 μm	Amosite, 981 WHO f/cm^3 (89 f> 20 μm/cm^3)	24 mo, 83 (43) Wistar rats	Whole-body, 7 h/d, 5 d/wk, for 12 mo	2356 × 10^6 WHO f/lung; 83 × 10^6 f > 20 μm/lung	12/43 (7 carcinomas, 3 adenonmas and 2 mesotheliomas); fibrosis: +	Authors concluded that amosite and 104 E were fibrogenic and tumorigenic.	Davis et al. (1996); Cullen et al. (2000)

Hamster								
Manville Code 100	530 f/cm³; L, > 10 μm; D, ≤ 1 μm; 3000 total f/cm³; 3 mg/m³; D, 0.4 μm mean; GMD, 0.4 μm; L, 7.5 μm mean; GML, 4.7 μm	Crocidolite. 3000 f/cm³, total f/cm³, 90 f/cm³; L, > 10 μm	69 Syrian golden hamsters, male; lifetime	Nose-only, 6 h/d, 5 d/wk, for 24 mo	1 × 10⁶ f/mg dry lung at 2 mo	Tumors: 0/69; fibrosis, 0	Low survival rates: < 25% of hamsters survived to 24 mo (including controls). Asbestos-exposed animals had no tumors (some did have fibrosis).	Smith et al. (1987)
JM 475	310 WHO f/cm³ (109 fL > 20 μm), 37 mg/m³; OMD, 0.70 μm, GML 11.8 μm	Amosite (mid), 163 WHO f/cm³ (38 f L > 20μm), 3.5 mg/m³; amosite (high) 263 WHO f/cm³ (69 fL >20 μm), 7 mg/m³	125 (83) Syrian golden hamsters, male; lifetime	Nose-only, 6 h/d, 5 d/wk, for 18 mo	fL> 5 μm/lung: 49 × 10⁶ at 12 mo 234 × 10⁶ at 18 mo f L > 20 μm/lung: 6 × 10⁶ at 12 mo 30 ×10⁶ at 18 mo	Lung tumors: 0/83; mesothelioma, 1/83; fibrosis, 0	Amosite (mid): 22/85 mesotheliomas, no lung tumors, fibrosis Amosite (high) 17/87 masotheliomas, no lung tumors, fibrosis	Hesterberg et al. (1999); McConnell et al. (1999)

Note: f = fiber; L = length; D = diameter, total f, any particle having L:D ≥ 3; f > 20 = fibers with length > 20 μm; f/cm³ = no. of fibers per cm³ of air, (some concentrations were expressed as fibers > 5 μm length, others as total fibers; authors did not always specify); GMD = geometric mean diameter, GML = geometric mean length; lifetime = until survival rate of air controls is ≤ 20%; (typically ~30 months); thoracic tumors = lung tumors, including adenomas and carcinomas; WHO = respirable fibers as defined by World Health Organization, L > 5 μm, D < 3 μm, L:D ≥ 3; h = hour; d = day; wk = week; mo = month; ND = not determined.

Source: Reproduced from IARC. 2002. *IARC Monographs on the Evaluation of Carcinogenic Risks to Humans: Man-Made Vitreous Fibers.* Vol. 81. Lyon: IARC Press. With permission.

is similar to CMS but which had considerably fewer particles present, no lung tumor or fibrosis was detected (Hesterberg, Hart et al., 1998). In the Bellmann et al. 90-d study, due to the method of preparation, the aerosol exposure concentration for the CMS fiber had 286 fibers/cm^3 of length < 5 μm, 990 fibers/cm^3 of length > 5 μm, and 1793 particles/cm^3, a distribution that is not observed in manufacturing. The total CMS exposure concentration was 3069 particles and fibers per cm^3. The authors point out that "the particle fraction of CMS that had the same chemical composition as the fibrous fraction seemed to cause significant effects." For the CMS fiber exposure, the authors reported that the number of polymorphonuclear leukocytes (PMN) in the bronchoalveolar lavage fluid (BALF) was higher and interstitial fibrosis was more pronounced than had been expected on the basis of biopersistence data. In addition, the interstitial fibrosis persisted through 14 weeks after cessation of the 90-d exposure. This effect, attributed to the particles, was observed with an exposure concentration of 3069 particles and fibers per cm^3, 50% of which were particles or short fibers. It follows directly from this, as well as the evidence from the many publications on overload, that a dramatically more pronounced effect could be expected to occur at higher exposure concentrations.

Even within the SVF exposures, as the long fibers are the most important in terms of potential toxicity, in high-dose exposures, the number of fibers longer than 20 μm varies from 84, 114, and 99 to 101 per cubic centimeter. Thus, although the studies are relatively comparable, there may be quantitative differences. However, in consideration of the percentage of long fibers in the total aerosol, and the fact that the rat-respirable fibers were preselected from the commercial bulk fiber and represented less than 2% of the commercial product, these studies are still remarkable for the uniformity of exposures.

These discrepancies in study design put in question the value, in terms of comparative analysis, of the chrysotile study listed in Table 11.3 in particular and perhaps the crocidolite study as well. McConnell et al. (1999) reported on perhaps the only well-designed multiple-dose study on any asbestos, in which amosite particle and fiber number and length were chosen to be comparable to the SVF exposure groups. In this hamster inhalation toxicology study, the amosite aerosol concentration ranged from 10 to 69 fibers/cm^3 (longer than 20 μm); these values were chosen based upon a previous, multidose 90-d subchronic inhalation toxicology study (Hesterberg et al., 1999).

Table 11.3 summarizes the chronic inhalation studies of insulation glass wools and special-purpose glass fibers in rodents (rats and hamsters) (from IARC, 2002). Although the importance of fiber length was proposed as early as 1951 (Vorwarld et al., 1951), many chronic inhalation studies did not specify the number of long fibers in the exposure atmosphere and often ground the fibers either prior to, or as part of, the aerosol generation process. In addition, in many early studies of SVF, there was no adequate consideration of the respirability of fibers in the rodent, and as a result little of the actual aerosol often reached deep into the lungs. In these cases, it should not be surprising that little or no effect occurred as there was hardly any deposition of the longer fibers that have the greatest potential for producing a response.

The studies presented in Table 11.2 indicate that there is a large difference in the biopersistence between amphibole asbestos and the soluble SVFs, as well as the serpentine chrysotile. These differences appear to be related to the differences in chemical structure between the amphiboles and the SVF and serpentines as well as, possibly, the influence of the acidic pH associated with the macrophage on the chrysotile fiber. Yet, when the chronic inhalation studies that have been performed on chrysotile and amphiboles are examined, these differences are not always apparent.

In an analysis by Berman and Crump (2004) of many of the animal dose–response studies that have been performed on asbestos, they concluded that:

- Short fibers (i.e., less than or somewhere between 5 and 10 μm in length) do not appear to contribute to cancer risk.
- Beyond a fixed, minimum length, potency increases with increasing length, at least up to a length of 20 μm (and possibly up to a length of as much as 40 μm).
- The majority of fibers that contribute to cancer risk are thin, with diameters less than 0.5 μm, and the most potent fibers may be even thinner. In fact, it appears that the fibers that are most potent are substantially thinner than the upper limit defined by respirability.
- Identifiable components (fibers and bundles) of complex structures (clusters and matrices) that exhibit the requisite size range may contribute to overall cancer risk because such structures likely disaggregate in the lung. Therefore, such structures should be individually enumerated during analysis to determine the concentration of asbestos.
- For asbestos analyses to adequately represent biological activity, samples need to be prepared by a direct-transfer procedure.
- Based on animal dose–response studies alone, fiber type (i.e., fiber mineralogy) appears to have only a modest effect on cancer risk (at least among the various asbestos types).

Concerning the lack of differentiation seen in the dose–response studies, the authors stated that this may be due at least in part to the limited lifetime of the rat relative to the biodurability of the asbestos fiber types evaluated in these studies.

Perhaps more important in understanding these results, as mentioned earlier, are the differences in study design and the relatively recent understanding of the effect of high concentrations of insoluble particles (and short fibers) in the rat lung overload.

The issue of using equivalent fiber number for exposures was approached in a study reported by Davis et al. (1978) in which chrysotile, crocidolite, and amosite were compared on an equal-mass and equal-number basis; however, the fiber number was determined by phase-contrast optical microscopy (PCOM) and, thus, the actual number, especially of the chrysotile fibers, was probably greatly underestimated. As an example, the 10 mg/m^3 exposure to chrysotile was reported by PCOM as approximately 2,000 fibers/cm^3 of length greater than 5 μm, whereas, when a similar mass concentration of another chrysotile was measured by SEM, 10,000 fibers/cm^3 of length greater than 5 μm was reported with a total fiber count of 100,000 fibers/cm^3 (Mast et al., 1995). There is little quantitative data presented in these publications on the nonfibrous particle concentration in the test substances to which the animals were exposed. Pinkerton et al. (1983) presents summary tables of length measurements of Calidria chrysotile by scanning electron microscopy (SEM) in which the number of nonfibrous particles counted is stated; however, from the data presented, the aerosol exposure concentration of nonfibrous particles cannot be extracted. In all studies, the asbestos was ground prior to aerosolisation, a procedure that would produce a lot of short fibers and dust. Most of the studies prior to Mast et al. (1995) used, for aerosolisation of the fibers, an apparatus that had a rotating steel blade to push or chop the fibers off a compressed plug and into the airstream. As some of the authors state, the steel used in the grinding apparatus and the aerosolisation apparatus often wore, resulting sometimes in considerable exposure to the metal fragments as well. These factors contribute significantly to the difficulty in interpreting the results of the serpentine chrysotile and the amphibole inhalation exposures studies.

11.5 INTRAPERITONEAL INJECTION STUDIES

As these studies on fiber toxicology were being developed in the 1980s and the early and mid-1990s, two schools or camps emerged that differed in their views on which chronic animal rodent study was most appropriate for assessing fiber toxicology. On one side were the scientists who considered the inhalation toxicology study the most appropriate, as it used the route by which humans were exposed and because it gave rise to similar lesions and tumors (McClellan et al., 1992). On the other side were those scientists who considered the chronic inhalation model as not being sensitive enough and adhered to the chronic intraperitoneal (IP) injection study as more sensitive and the most appropriate, because with this study any fiber could be shown to be carcinogenic if sufficient fibers were injected into the peritoneal cavity (Pott et al., 1987, 1995).

Table 11.4 summarizes a range of fibers that have been evaluated using the IP study. The fibers range from ceramic and E-glass to the very soluble experimental fiber B01-09. Of particular importance in interpreting the IP study is that there is no maximum tolerated dose that can be used as an indicator of when a fiber should be of concern. By increasing the dose and frequency of injection, 20 trillion fibers were injected in the B01-09 study to produce a tumorigenic response. The B01-09 fiber was so soluble that the inhalation clearance half-time could not be measured. Comparison was often made with amphibole asbestos as the sole example of a fiber known to be carcinogenic in humans. For fibers with a carcinogenic potential less than that of amphibole asbestos, the only other measure was the response in chronic inhalation studies.

With the detailed evaluation of both the inhalation and IP chronic animal data that was mandated by the European Commission in 1996, it was found that in fact both data sets are coherent when analyzed using appropriate statistical methods, taking into account for the IP studies, not only the dose but the fiber dimensions as well. As is discussed in the text that follows, both the pathogenic response following inhalation and the IP tumor response following injection can be predicted on the basis of the biopersistence of the fibers (Bernstein et al., 2001a and 2001b).

11.6 THE IMPORTANCE OF FIBER LENGTH

Vorwald et al. (1951) first reported on the finding that long fibers (20 to 50 μm) are associated with both lung and peritoneal disease as opposed to shorter ball-milled fibers (3 μm or less) that were not. However, for many years little research was performed to follow up on these results. Hammad et al. (1988) found that fibers < 5 μm in length had the longest retention following short-term inhalation, with longer fibers clearing more rapidly and fibers > 30 μm in length clearing very rapidly. He proposed that clearance of mineral wools is a result of biological clearance, the elimination of fibers being by dissolution and subsequent breakage. However, he found no relationship between these phenomena and long-term toxicological effects.

In the series of SVF chronic inhalation studies performed at RCC in the 1980s, the relationship of the more durable fibers with disease became more apparent and resulted in the design of the inhalation biopersistence study as described earlier. The importance of fiber length for the potential to produce a pathogenic effect has been documented (McClellan et al., 1992; Coin et al., 1992; Goodglick and Kane, 1990; Lippmann, 1990: WHO, 1988).

Table 11.4 The Tumorigenic Response in Rats Following Intraperitoneal Injection of a Range of Synthetic Vitreous Fibers

Fiber	*Median Diameter (μm)*	*Median Length (μm)*	*Number of Fibers Injected × 10³*	*Mass Injected (mg)*	*Tumor Fraction*	*Reference*
B01-09	0.7	9	10,000,000	500	0.21	Roller et al., 1996
B01-09	0.7	9	20,000,000	1000	0.66	Roller et al., 1996
B01-09	0.7	9	2,500,000	125	0.08	Roller et al., 1997
B01-09	0.7	9	5,000,000	250	0.11	Roller et al., 1997
C	0.44	12	10,000	0.7	0.10	Lambré et al., 1998
C	0.44	12	3,000	2.1	0.08	Lambré et al., 1998
C	0.44	12	100,000	7	0.02	Lambré et al., 1998
C	0.44	12	500,000	35	0.02	Lambré et al., 1998
Wollastonite	0.71	5.6	93,000	30	0.00	Muhle and Pott, 1991
Wollastonite	1.1	8.1	430,000	100	0.00	Pott et al., 1989
G	0.52	10.5	10,000	1.1	0.04	Lambré et al., 1998
G	0.52	10.5	70,000	7.7	0.02	Lambré et al., 1998
G	0.52	10.5	500,000	55	0.04	Lambré et al., 1998
Slag wool	0.18	2.7	30	5	0.05	Pott et al., 1984
MMVF22	1.1	20	1,000,000	129.6	0.54	Jones et al., 1997
MMVF11	0.77	14.6	400,000	70	0.40	Roller et al., 1996
MMVF11	0.77	14.6	1,000,000	180	0.70	Roller et al., 1996
MMVF21	1.1	17.7	1,000,000	183.1	0.95	Jones et al., 1997
MMVF-21	1.02	16.9	400,000	60	0.97	Roller et al., 1997
MMVF-21	1.02	16.9	1,000,000	150	0.87	Roller et al., 1997
RCF1	0.88	17.8	1,000,000	110.9	0.88	Jones et al., 1997

(*continued*)

Table 11.4 The Tumorigenic Response in Rats Following Intraperitoneal Injection of a Range of Synthetic Vitreous Fibers (Continued)

Fiber	Median Diameter (μm)	Median Length (μm)	Number of Fibers Injected × 10^3	Mass Injected (mg)	Tumor Fraction	Reference
RCF1	0.86	15.9	1,000,000	188.8	0.72	Jones et al., 1997
F	0.57	9.9	10,000	1.1	0.06	Lambré et al., 1998
F	0.57	9.9	70,000	7.7	0.02	Lambré et al., 1998
F	0.57	9.9	500,000	55	0.06	Lambré et al., 1998
A	0.5	8.8	10,000	0.7	0.06	Lambré et al., 1998
A	0.5	8.8	30,000	2.1	0.02	Lambré et al., 1998
A	0.5	8.8	100,000	7	0.02	Lambré et al., 1998
A	0.5	8.8	500,000	35	0.06	Lambré et al., 1998
H	0.51	13	10,000	1.1	0.06	Lambré et al., 1998
H	0.51	13	70,000	7.7	0.02	Lambré et al., 1998
H	0.51	13	500,000	55	0.18	Lambré et al., 1998
MMVF10	1.1	15	1,000,000	144.4	0.59	Jones et al., 1997
JM 104/475	0.14	2.3	320,000	2	0.17	Pott et al., 1991
JM 104/475	0.15	2.6	680,000	5	0.64	Pott et al., 1989
JM 104/475	0.18	3.2	57,000	0.5	0.17	Pott et al., 1987
JM 104/475	0.18	3.2	228,000	2	0.26	Pott et al., 1987
JM 100475	0.24	1.4	50,000	2	0.05	Pott et al., 1984
JM 100475	0.33	2.4	280,000	2	0.05	Pott et al., 1984
JM 106475	0.47	2.2	120,000	10	0.03	Pott et al., 1976
JM 100475	0.33	2.4	1,300,000	10	0.45	Pott et al., 1987
JM 106475	0.47	2.2	24,000	2	0.00	Pott et al., 1976

(continued)

Table 11.4 The Tumorigenic Response in Rats Following Intraperitoneal Injection of a Range of Synthetic Vitreous Fibers (Continued)

Fiber	Median Diameter (μm)	Median Length (μm)	Number of Fibers Injected × 10³	Mass Injected (mg)	Tumor Fraction	Reference
JM 106475	0.47	2.2	120,000	10	0.11	Pott et al., 1976
JM 106475	0.47	2.2	24,000	2	0.03	Pott et al., 1976
JM 106475	0.47	2.2	1,200,000	100	0.72	Pott et al., 1976
JM 100475	0.4	4.9	400,000	25	0.32	Smith et al., 1987
JM 100475	0.32	4	1,000,000	8.3	0.33	Jones et al., 1997
JM 104E	0.2	10	12,000	2	0.27	Pott et al., 1976
JM 104E	0.2	10	60,000	10	0.53	Pott et al., 1976
JM 104E	0.2	10	300,000	50	0.71	Pott et al., 1976
JM 104E	0.26	3.5	430,000	10	0.49	Pott et al., 1984
JM 104E	0.26	4	290,000	10	0.66	Pott et al., 1984
JM 104E	0.26	4	58,000	2	0.32	Pott et al., 1984
JM 104E	0.29	4.8	505,000	5	0.44	Pott et al., 1987
JM 104E	0.29	4.8	505,000	5	0.81	Pott et al., 1987
JM 104E	0.29	4.8	1,010,000	10	0.73	Pott et al., 1984
JM 104E	0.39	2.7	300,000	10	0.09	Pott et al, 1984
JM 104E	0.3	3.5	4,300	10	0.50	Pott et al., 1987
JM 104E	0.3	3.5	4,300	10	0.55	Pott et al., 1987
JM 100E	0.32	4.4	155,000	2	0.48	Pott et al., 1987

Source: Bernstein, D.M., Riego Sintes, J.M., Ersboell, B.K., and Kunert, J. 2001b. Biopersistence of synthetic mineral fibers as a predictor of chronic intraperitoneal injection tumor response in rats. *Inhal Toxicol* 13(10): 851–875. With permission.

Bernstein et al. (1996), in a comparison of nine different fibers, evaluated the influence of fiber length on clearance using the inhalation biopersistence protocol; he studied the clearance of fibers in three length categories: fibers with length < 5 µm, 5 to 20 µm, and > 20 µm. The classes were chosen as follows: The fibers < 5 µm in length could be completely phagocytized by macrophages and treated by the lung essentially as nonfibrous particles. The length fraction > 20 µm was considered as being greater than the size of the macrophage and represented a length that the macrophage most likely could not fully phagocytize. It was pointed out that these length fractions were not considered as strict cutoffs but rather as representative of mechanistic categories with the transition between categories thought to occur over a range of lengths. The fiber clearance half-times in each of these length fractions were then examined for association with the fiber durability as determined using an acellular *in vitro* flow-through system using a modified Gambles solution. This analysis provided a good indication that the disappearance of fibers longer than 20 µm was mediated by their dissolution at pH 7.4. Although not reported in the publication, the authors chose the 20-µm cutoff for the longer fibers by performing iterative analyses starting at a fiber length of 10 µm and greater, and then increasing the minimum fiber length to determine at what lower cutoff the best correlation of half-time was obtained with the *in vitro* durability data (personal communication). The authors found that the best association was at approximately 18 to 20 µm, and above this the correlation diminished as a result of fewer and fewer fibers being available in the longer size ranges. Based on this iterative analysis, the 20-µm size was chosen as the size best representative of the longer fibers.

Subsequently, in an analysis that provided the basis for the European Commission's directive on SMFs, Bernstein et al. (2001a and 2001b) reported that there exists an excellent correlation between the biopersistence of fibers longer than 20 µm and the pathological effects following either chronic inhalation or chronic IP injection studies. This analysis showed that it was possible, using the clearance half-time of the fibers longer than 20 µm as obtained from the inhalation biopersistence studies, to predict the number of fibers longer than 20 µm remaining following 24 months of chronic inhalation exposure. These studies, which included only SMFs, are discussed in more detail in the following text.

Berman et al. (1995) statistically analyzed nine different asbestos types in 13 separate studies. Due to limitations in the characterization of asbestos structures in the original studies, new exposure measures were developed from samples of the original dusts that were regenerated and analyzed by transmission electron microscopy. The authors reported that although no univariate model was found to provide an adequate description of the lung tumor responses in the inhalation studies, the measure most highly correlated with tumor incidence was the concentration of structures (fibers)

20 μm in length. However, using multivariate techniques, measures of exposure were identified that adequately describe the lung tumor responses.

The potency appears to increase with increasing length, with structures (fibers) longer than 40 μm being about 500 times more potent than structures between 5 and 40 μm in length. Structures < 5 μm in length do not appear to make any contribution to lung tumor risk. This analysis also did not find a difference in the potency of chrysotile and amphibole in the induction of lung tumors. The authors stated that the mineralogy appears to be important in the induction of mesothelioma, with chrysotile being less potent than amphibole; however, no assessment was made of the influence of the number of particles and short fibers on lung overload.

11.7 RELATIONSHIP OF BIOPERSISTENCE TO CHRONIC TOXICITY

As described earlier, the relationship of the biopersistence of the fibers to their potential pathological effect following chronic inhalation and chronic IP injection has been examined in various studies (Bernstein et al., 2001a and 2001b; Hesterberg, Chase et al., 1998; and Hesterberg, Hart et al., 1998).

These studies compared (for SVFs) the relationship of fiber clearance half-time as determined in the short-term biopersistence studies to both the number of fibers remaining and the pathological response in chronic inhalation toxicology studies (24 months of exposure).

Tumorigenic response is usually considered the most important indicator of pathogenicity in fiber studies. However, in these studics, only RCF 1 at the highest dose resulted in a statistically significant number of lung tumors compared to the air controls. As such, the number of tumors could not be used in determining an association with biopersistence. As a good indicator of the continuum of pathogenic response, collagen deposition at the bronchoalveolar junction was used as a measure of the early response at the site where fibrosis can occur following fiber exposure.

As shown in Figure 11.4 (from Bernstein et al., 2001a), an excellent relationship was found between the number of fibers remaining (L > 20 μm) in the lung at the end of 2 yr of exposure and the collagen score in the lung at the same time point. The results indicate that there exists a threshold below which no effect occurs. Crossing this threshold can be avoided by having low exposures or by using biosoluble fibers that do not persist in the lung.

The relationship with biopersistence is seen in Figure 11.5 and Figure 11.6 (from Bernstein et al., 2001a). In Figure 11.5, the clearance half-time of fibers longer than 20 μm from the short-term biopersistence studies is shown to be linearly related to the number of longer fibers remaining in the lung following 2 yr of exposure in the chronic inhalation toxicology studies.

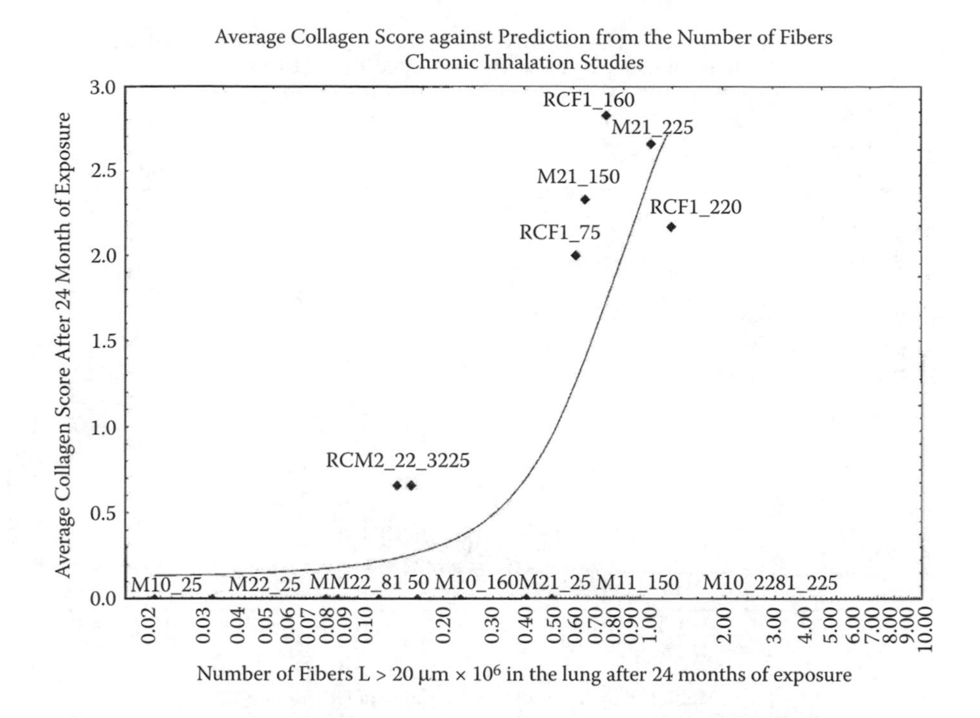

Figure 11.4 Plot of the relationship of the number of fibers with L > 20 μm in the lung following 24 months of exposure in the chronic inhalation studies to the collagen deposition at the bronchoalveolar junction following 24 months of exposure, as determined using the proportional odds model. The collagen deposition was fitted to the number of fibers itself and not the logarithm; however, the plot is in a logarithmic scale for clarity. The high chi square of 64.9 for covariates implies that the influence of the number of fibers on the collagen score is highly significant (a *p* value of .0001). (From Bernstein, D.M., Riego Sintes, J.M., Ersboell, B.K., and Kunert. J. 2001a. Biopersistence of synthetic mineral fibers as a predictor of chronic inhalation toxicity in rats. *Inhal Toxicol* 13(10): p. 832, Figure 2. With permission.)

Figure 11.6 shows that, similarly, the biopersistence of these longer fibers can be used to predict the collagen levels in the chronic inhalation studies.

Similar correlations were determined between the short-term biopersistence clearance half-times and the number of tumors and the fiber dose and dimensions in the IP injection studies (Bernstein et al., 2001b).

The results of these analyses are summarized in Table 11.5. The biopersistence of fibers longer than 20 μm correlates not only with the results of the chronic inhalation studies but also with the results of the chronic IP injection studies, showing the consistency of all data related to chronic exposure to fibers. As there is no maximum dose that can be injected

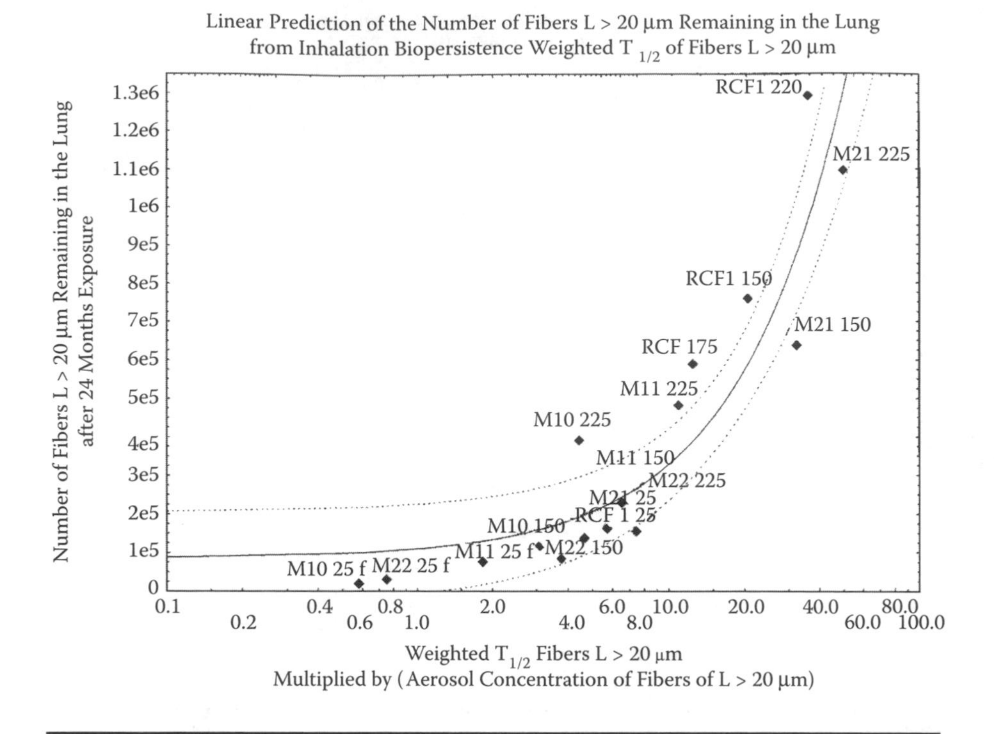

Figure 11.5 Plot of the linear prediction of the number of fibers with L > 20 μm remaining in the lung after 24 months of exposure in the chronic inhalation studies from the weighted $T_{1/2}$ of fibers L > 20 μm, as determined in the inhalation biopersistence studies. The weighted $T_{1/2}$ was found to be a good predictor, explaining 83% of the variance in fiber number (L > 20 μm) in the lung. The relationships shown in these figures are linear; however, for clarity the half-times were plotted on a logarithmic scale. (From Bernstein, D.M., J.M. Riego Sintes, B.K. Ersboell, and J. Kunert. 2001a. Biopersistence of synthetic mineral fibers as a predictor of chronic inhalation toxicity in rats. *Inhal Toxicol* 13(10): p. 834, Figure 3. With permission.)

in the IP studies, they can only be understood in comparison to fibers that have shown a response in humans epidemiologically or that have elicited a positive response in chronic inhalation toxicity studies.

11.8 SUMMARY

Appreciation of the key concepts influencing mineral fiber toxicology is essential in the design and interpretation of fiber toxicology studies. However, this is only part of what is required to understand the complex

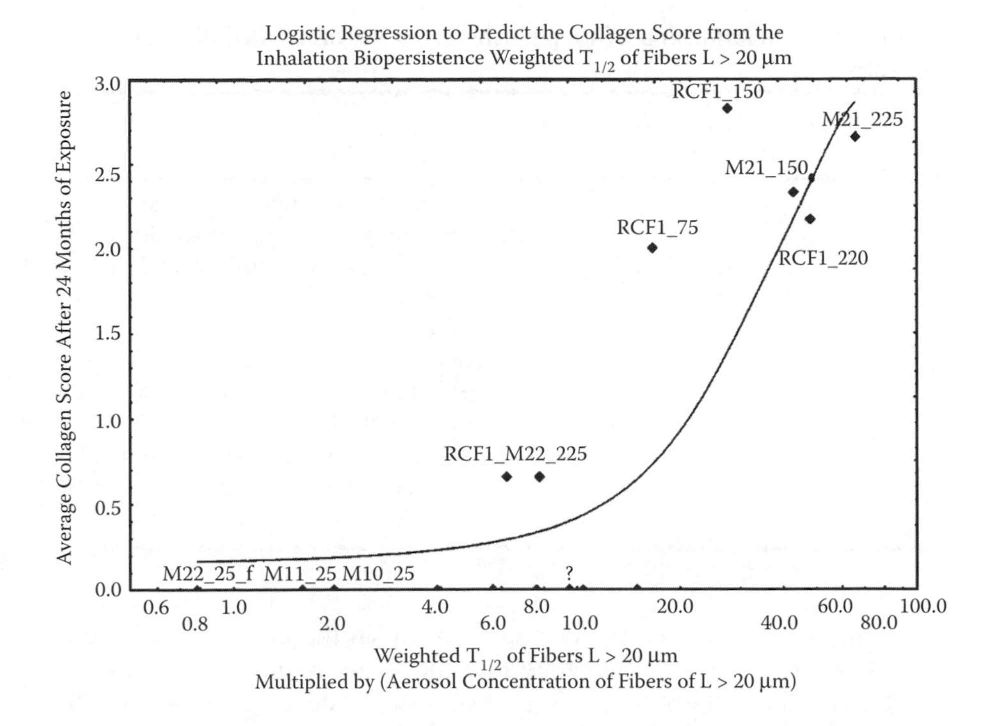

Figure 11.6 Plot of the logistic regression function showing the prediction of the collagen score at the bronchoalveolar junction after 24 months of exposure in the chronic inhalation studies from the weighted $T_{1/2}$ of the fibers with L > 20 μm, as determined in the inhalation biopersistence studies. The chi square of 68.6 is highly significant (a *p* value of .0001). (From Bernstein, D.M., Riego Sintes, J.M., Ersboell, B.K., and Kunert, J. 2001a. Biopersistence of Synthetic Mineral Fibers as a Predictor of Chronic Inhalation Toxicity in Rats. *Inhal Toxicol* 13(10): p. 837, Figure 5. With Permission.)

set of data that exists. As important is understanding the use of the animal model, whether rat or another species, in the design and interpretation of the inhalation toxicology studies. Consideration should be given to the length and diameter distribution of the fibers to ensure that sufficient long fibers reach the alveolar region of the lung for the specific animal model used. However, care should be taken as well to ensure that lung overload is not induced as a result of the presence of large numbers of shorter fibers and particles. With the strong inflammatory response induced in overload conditions, it might be difficult, or even impossible, to determine what, if any, response the fibers may have caused. It is even possible that overload can dampen the response that might have occurred from the

Table 11.5 Relationship of Biopersistence to Chronic Inhalation and IP Results

The Biopersistence of Fibers Longer than 20 μm	*Correlates With*
	The number of fibers $L > 20\ \mu m$ remaining in chronic inhalation toxicology studies following 2 yr of exposure
	The early fibrotic response (collagen deposition) observed after 24 months of exposure in the chronic inhalation toxicology studies
	The number of tumors and the fiber dose and dimensions in the chronic IP injection studies

biologically active fibers, by locking them up in the cellular response so that they are not free to penetrate into the target cells.

With mineral fibers, there is a clear association of toxicity with the biopersistence of the longer fibers (i.e., those fibers that the macrophage cannot fully engulf and clear from the lung as it does particles). The fiber length and biopersistence of the fiber are factors that determine the potential hazard of the fibers. Also important, however, in risk assessment is the dose, which is determined through the respirable fraction of the exposure concentration. Thus, fibers that have shown a tumorigenic response in animal studies may not necessarily produce a similar response in humans; this would depend upon the cumulative exposure that has occurred.

REFERENCES

ATSDR. 2003. Report on the Expert Panel on Health Effects of Asbestos and Synthetic Vitreous Fibers: The Influence of Fiber Length. Atlanta, GA: Prepared for Agency for Toxic Substances and Disease Registry Division of Health Assessment and Consultation.

Baron, P.A. 1993. Measurement of asbestos and other fibers. In *Aerosol Measurement*. Eds. K. Willeke and P.A. Baron. New York: Van Nostrand Reinhold.

Bellmann, B., Muhle, H., Creutzenberg, O., Ernst, H., Muller, M., Bernstein, D.M. and Riego Sintes, J.M. 2003. Calibration study on subchronic inhalation toxicity of man-made vitreous fibers in rats. *Inhal Toxicol* 15(12): 1147–1177.

Berman, D.W. and K.S. Crump. 2003. Technical Support Document for a Protocol to Assess Asbestos-Related Risk. Washington, D.C. Office of Solid Waste and Emergency Response. U.S. Environmental Protection Agency.

Berman, D.W., K.S. Crump, E.J. Chtfield, J.M. Davis, and A.D. Jones. 1995. The sizes, shapes, and Minerology of abestos structures that induce lung tumors or mesothelioma in AF/HAN rats following inhalation. *Risk Anal* 15(a): 181–95.

Bernstein, D.M. and J.M.R. Riego-Sintes. 1999. Methods for the Determination of the Hazardous Properties for Human Health of Man Made Mineral Fibers (MMMF). Vol. EUR 18748 EN, April. 93 pp. http://ecb.ei.jrc.it/DOCUMENTS/Testing-Methods/mmmfweb.pdf: European Commission Joint Research Centre, Institute for Health and Consumer Protection, Unit: Toxicology and Chemical Substances, European Chemicals Bureau.

Bernstein, D.M., C. Morscheidt, H.-G. Grimm, P. Thévenaz, and U. Teichert. 1996. Evaluation of soluble fibers using the inhalation biopersistence model, a nine-fiber comparison. *Inhal Toxicol* 8(4): 345–385.

Bernstein, D.M., J. Chevalier, and P. Smith. 2005. Final Results of the Comparison of Calidria chrysotile asbestos to pure tremolite: inhalation biopersistence and histopathology following short-term exposure. *Inhal Toxicol* 17(9): 427–449.

Bernstein, D., J. Chevalier, and P. Smith. 2005. Comparison of Calidria asbestos to pure tremolite: final results of the inhalation biopersistence and histopathology examination following short-term exposure. *Inhal Toxicol* 17(19): 427–449.

Bernstein, D.M., J.M. Riego Sintes, B.K. Ersboell, and J. Kunert. 2001a. Biopersistence of synthetic mineral fibers as a predictor of chronic inhalation toxicity in rats. *Inhal Toxicol* 13(10): 823–849.

Bernstein, D.M., J.M. Riego Sintes, B.K. Ersboell, and J. Kunert. 2001b. Biopersistence of synthetic mineral fibers as a predictor of chronic intraperitoneal injection tumor response in rats. *Inhal Toxicol* 13(10): 851–875.

Bernstein, D.M., R. Mast, R. Anderson, T.W. Hesterberg, R. Musselman, O. Kamstrup, and J. Hadley. 1994. An experimental approach to the evaluation of the biopersistence of respirable synthetic fibers and minerals. *Environ Health Perspect* 102(Suppl. 5): 15–18.

Bernstein, D.M., R. Rogers, and P. Smith. 2005. The biopersistence of Canadian chrysotile asbestos following inhalation. *Inhal Toxicol* 17(1): 1–14.

Bernstein, D., R. Rogers, P. Smith, 2005. The biopersistence of Canadian Chrysotile asbestos following inhalation: final results through 1 year after cessation of exposure. *Inhal Toxicol* 17(1): 1–14.

Bernstein, D.M., R. Drew, G. Schidlovsky, and M. Kuschner. 1984. Pathogenicity of MMMF and the contrasts with natural fibres. In *Biological Effects of Man-Made Mineral Fibres: Proceedings of a WHO/IARC Conference in Association with JEMRB and TIMA*. Copenhagen: World Health Organization.

Bolton, R.E., J.H. Vincent, A.D. Jones, J. Addison, and S.T. Beckett. 1983. An overload hypothesis for pulmonary clearance of UICC amosite fibres inhaled by rats. *Br J Ind Med* 40: 264–272.

Britton, M. 2002. The epidemiology of mesothelioma. *Semin Oncol* 29(1): 18–25.

Churg, A. 1988. Chrysotile, tremolite, and malignant mesothelioma in man. *Chest* 93(3): 621–628.

Coin, P.G., V.L. Roggli, and A.R. Brody. 1992. Deposition, clearance, and translocation of chrysotile asbestos from peripheral and central regions of the rat lung. *Environ Res* 58 (1): 97–116.

Davis, J.M. 1986. A review of experimental evidence for the carcinogenicity of man-made vitreous fibers. *Scand J Work Environ Health* 12(Suppl. 1): 12–17.

Davis, J.M., S.T. Beckett, R.E. Bolton, P. Collings, and A.P. Middleton. 1978. Mass and number of fibres in the pathogenesis of asbestos-related lung disease in rats. *Br J Cancer* 37(5): 673–688.

Davis, J.M.G., D.M. Brown, R.T. Cullen, K. Donaldson, A.D. Jones, B.G. Miller, C. McIntosh, and A. Searl. 1996. A comparison of methods of determining and predicting the pathogenicity of mineral fibres. *Inhal Toxicol* 8: 747–770.

Donaldson, K. and C.L. Tran. 2004. An introduction to the short-term toxicology of respirable industrial fibres. *Mutat Res* 553(1–2): 5–9.

Elmore, A.R. 2003. Final report on the safety assessment of aluminum silicate, calcium silicate, magnesium aluminum silicate, magnesium silicate, magnesium trisilicate, sodium magnesium silicate, zirconium silicate, attapulgite, bentonite, Fuller's earth, hectorite, kaolin, lithium magnesium silicate, lithium magnesium sodium silicate, montmorillonite, pyrophyllite, and zeolite. *Int J Toxicol* 22(Suppl. 1): 37–102.

Etherington, D.J., D. Pugh, and I.A. Silver. 1981. Collagen degradation in an experimental inflammatory lesion: studies on the role of the macrophage. *Acta Biol Med Ger* 40(10–11): 1625–1636.

Goodglick, L.A. and A.B. Kane. 1990. Cytotoxicity of long and short crocidolite asbestos fibers *in vitro* and *in vivo*. *Cancer Res* 50(16): 5153–5163.

Hammad, Y., W. Simmons, H. Abdel-Kader, C. Reynolds, and H. Weill. 1988. Effect of chemical composition on pulmonary clearance of man-made mineral fibers. *Ann Occup Hyg* 22: 769–779.

Hesterberg, T.W. and G.A. Hart. 2001. Synthetic vitreous fibers: a review of toxicology research and its impact on hazard classification. *Crit Rev Toxicol* 31(1): 1–53.

Hesterberg, T.W., C. Axten, E.E. McConnell, G.A. Hart, W. Miller, J. Chevalier, J. Everitt, P. Thevenaz, and G. Oberdorster. 1999. Studies on the inhalation toxicology of two fiberglasses and amosite asbestos in the Syrian golden hamster. Part I. Results of a subchronic study and dose selection for a chronic study. *Inhal Toxicol* 11(9): 747–784.

Hesterberg, T.W., G.A. Hart, J. Chevalier, W.C. Miller, R.D. Hamilton, J. Bauer, and P. Thevenaz. 1998a. The importance of fiber biopersistence and lung dose in determining the chronic inhalation effects of X607, RCF1, and chrysotile asbestos in rats. *Toxicol Appl Pharmacol* 153(1): 68–82.

Hesterberg, T.W., G. Chase, C. Axten, W.C. Miller, R.P. Musselman, O. Kamstrup, J. Hadley, C. Morscheidt, D.M. Bernstein, and P. Thevenaz. 1998b. Biopersistence of synthetic vitreous fibers and amosite asbestos in the rat lung following inhalation. *Toxicol Appl Pharmacol* 151(2): 262–275.

Hesterberg, T.W., W.C. Miller, E.E. McConnell, J. Chevalier, J.G. Hadley, D.M. Bernstein, P. Thevenaz, and R. Anderson. 1993. Chronic inhalation toxicity of size-separated glass fibers in Fischer 344 rats. *Fundam Appl Toxicol* 20 (4): 464–476.

Hesterberg, T.W., W.C. Miller, P. Thevenaz, and R. Anderson. 1995. Chronic inhalation studies of man-made vitreous fibres: characterization of fibres in the exposure aerosol and lungs. *Ann Occup Hyg* 39(5): 637–653.

Huncharek, M. 1986. The biomedical and epidemiological characteristics of asbestos-related diseases: a review. *Yale J Biol Med* 59(4): 435–451.

IARC. 2002. *IARC Monographs on the Evaluation of Carcinogenic Risks to Humans: Man-Made Vitreous Fibres.* Vol. 81. Lyon: IARC Press.

Ilgren, E.B. and K. Browne. 1991. Asbestos-related mesothelioma: evidence for a threshold in animals and humans. *Regul Toxicol Pharmacol* 13(2): 116–132.

Kamstrup, O., A. Ellehauge, C.G. Collier, and J.M. Davis. 2002. Carcinogenicity studies after intraperitoneal injection of two types of stone wool fibres in rats. *Ann Occup Hyg* 46(2): 135–142.

Knudsen, T., M. Guldberg, V.R. Christensen, and S.L. Jensen. 1996. New type of stonewool (HT fibres) with a high dissolution rate at pH = 4.5. *Glastech Ber Glass Sci Technol* 69(10): 331–337.

Le Bouffant, L., J.P. Henin, J.C. Martin, C. Normand, G. Tichoux, and F. Trolard. 1984. Distribution of inhaled MMMF in the rat lung: long-term effects. In *Biological Effects of Man-Made Mineral Fibres: Proceedings of a WHO/IARC Conference in Association with JEMRB and TIMA*. Copenhagen: World Health Organization.

Lee, K.P., C.E. Barras, F.D. Griffith, R.S. Waritz, and C.A. Lapin. 1981. Comparative pulmonary responses to inhaled inorganic fibers with asbestos and fiberglass. *Environ Res* 24 (1): 167–191.

Lippmann, M. 1990. Effects of fiber characteristics on lung deposition, retention, and disease. *Environ Health Perspect* 88: 311–317.

Mast, R.W., L.D. Maxim, M.J. Utell, and A.M. Walker. 2000. Refractory ceramic fiber: toxicology, epidemiology, and risk analyses — a review. *Inhal Toxicol* 12 (5): 359–399.

Mast, R.W., E.E. McConnell, R. Anderson, J. Chevalier, P. Kotin, D.M. Bernstein, P. Thevenaz, L.R. Glass, W.C. Miller, and T.W. Hesterberg. 1995b. Studies on the chronic toxicity (inhalation) of four types of refractory ceramic fiber in male Fischer 344 rats. *Inhal Toxicol* 7(4): 425–467.

Mast, R.W., E.E. McConnell, T.W. Hesterberg, J. Chevalier, P. Kotin, P. Thevenaz, D.M. Bernstein, L.R. Glass, W. Miiller, and R. Anderson. 1995a. Multiple-dose chronic inhalation toxicity study of size-separated kaolin refractory ceramic fiber in male Fischer 344 rats. *Inhal Toxicol* 7(4): 469–502.

Mattson, S.M. 1994. Glass fibres in simulated lung fluid: dissolution behavior and analytical requirements. *Ann Occup Hyg* 38: 857–877.

Maxim, L.D., R.W. Mast, M.J. Utell, C.P. Yu, P.M. Boymel, B.K. Zoitos, and J.E. Cason. 1999. Hazard assessment and risk analysis of two new synthetic vitreous fibers. *Regul Toxicol Pharmacol* 30(1): 54–74.

McClellan, R.O., F.J. Miller, T.W. Hesterberg, D.B. Warheit, W.B. Bunn, A.B. Kane, M. Lippmann, R.W. Mast, E.E. McConnell, and C.F. Reinhardt. 1992. Approaches to evaluating the toxicity and carcinogenicity of man-made fibers: summary of a workshop held November 11–13, 1991, Durham, NC. *Regul Toxicol Pharmacol* 16(3): 321–364.

McConnell, E.E. 1995. Fibrogenic effect of wollastonite compared with asbestos dust and dusts containing quartz. *Occup Environ Med* 52(9): 621.

McConnell, E.E., C. Axten, T.W. Hesterberg, J. Chevalier, W.C. Miiller, J. Everitt, G. Oberdorster, G.R. Chase, P. Thevenaz, and P. Kotin. 1999. Studies on the inhalation toxicology of two fiberglasses and amosite asbestos in the Syrian golden hamster. Part II. Results of chronic exposure. *Inhal Toxicol* 11(9): 785–835.

McConnell, E.E., J.C. Wagner, J.W. Skidmore, and J.A. Moore. 1984. A comparative study of the fibrogenic and carcinogenic effects of UICC Canadian chrysotile B asbestos and glass microfibre (JM 100). In *Biological Effects of Man-Made Mineral Fibres: Proceedings of a WHO/IARC Conference in Association with JEMRB and TIMA*. Copenhagen: World Health Organization.

McConnell, E.E., O. Kamstrup, R. Musselman, T.W. Hesterberg, J. Chevalier, W.C. Miller, and P. Thievenaz. 1994. Chronic inhalation study of size-separated rock and slag wool insulation fibers in Fischer 344/N rats. *Inhal Toxicol* 6: 571–614.

Miller, B.G., A.D. Jones, A. Searl, D. Buchanan, R.T. Cullen, C.A. Soutar, J.M. Davis, and K. Donaldson. 1999. Influence of characteristics of inhaled fibres on development of tumours in the rat lung. *Ann Occup Hyg* 43: 167–179.

Miller, F.J. 2000b. Dosimetry of particles in laboratory animals and humans in relationship to issues surrounding lung overload and human health-risk assessment: a critical review. *Inhal Toxicol* 12(1–2): 19–57.

Miller, F.J. 2000a. Dosimetry of particles: critical factors having risk assessment implications. *Inhal Toxicol* 12 (Suppl. 3): 389–395.

Moolgavkar, S.H., R.C. Brown, and J. Turim. 2001. Biopersistence, fiber length, and cancer risk assessment for inhaled fibers. *Inhal Toxicol* 13(9): 755–772.

Moorman, W.J., R.T. Mitchell, A.T. Mosberg, and D.J. Donofrio. 1988. Chronic inhalation toxicology of fibrous glass in rats and monkeys. *Ann Occup Hyg* 32(Suppl. 1): 757–767.

Morrow, P.E. 1988. Possible mechanisms to explain dust overloading of the lungs. *Fundam Appl Toxicol* 10(3): 369–384.

Muhle, H., B. Bellman, and U. Heinrich. 1988. Overloading of lung clearance during chronic exposure of experimental animals to particles. *Ann Occup Hyg* 32(Suppl. 1): 141–147.

Muhle, H., F. Pott, B. Bellmann, S. Takenaka, and U. Ziem. 1987. Inhalation and injection experiments in rats to test the carcinogenicity of MMMF. *Ann Occup Hyg* 31(4B): 755–764.

Musselman, R.P., W.C. Miiller, W. Eastes, J.G. Hadley, O. Kamstrup, P. Thevenaz, and T.W. Hesterberg. 1994. Biopersistences of man-made vitreous fibers and crocidolite fibers in rat lungs following short-term exposures. *Environ Health Perspect* 102(Suppl 5): 139–143.

Nicholson, W.J. 2001. The carcinogenicity of chrysotile asbestos — a review. *Ind Health* 39(2): 57–64.

NIOSH. 1994 (2003). *NIOSH Manual of Analytical Methods.* Eds. P. C. Schlecht, P.F. O'Conner. 4th ed., Vol. 1st Supplement Publication 96–135, 2nd Supplement Publication 98–119, 3rd Supplement 2003–154. http://www.cdc.gov/niosh/nmam/: DHHS (NIOSH) Publication 94–113.

Oberdörster, G. 1991. Deposition, elimination and effects of fibers in the respiratory tract of humans and animals. *VDI Ber.* 17–37.

Oberdorster, G. 1995. Lung particle overload: implications for occupational exposures to particles. *Regul Toxicol Pharmacol* 21(1): 123–135.

Oberdorster, G. 2000. Determinants of the pathogenicity of man-made vitreous fibers (MMVF). *Int Arch Occup Environ Health* 73 Suppl: S60–8.

Oberdorster, G. 2002. Toxicokinetics and effects of fibrous and nonfibrous particles. *Inhal Toxicol* 14(1): 29–56.

Olin, S., and ILSI Fiber Toxicity Assay Working Group. 2005. Testing of fibrous particles: short term assays and strategies. *Inhal Toxicol* 17: 497–537.

Pinkerton, K.E., A.R. Brody, D.A. McLaurin, B. Adkins, Jr., R.W. O'Connor, P.C. Pratt, and J.D. Crapo. 1983. Characterization of three types of chrysotile asbestos after aerosolization. *Environ Res* 31(1): 32–53.

Pott, F. 1995. Detection of mineral fibre carcinogenicity with the intraperitoneal test — recent results and their validity. *Ann Occup Hyg* 39(5): 771–779.

Pott, F., K.H. Friedrichs, and F. Huth. 1976. Ergebnisse aus Tierversuchen zur kanzerogenen Wirkung faserformiger Staube und ihre Deutung im Hinblick auf die Tumorentstehung beim Menschen. *Zbl Bakt Hyg I Abt Orig* B162: 467–505.

Pott, F., M. Roller, R.M. Rippe, P.-G. Germann, and B. Bellmann. 1991. Tumours by the intraperitoneal and intrapleural routes and their significance for the classification of fibers. In *Mechanisms of Fiber Carcinogenesis*, Eds. R.C. Brown, J.A. Hoskins, and N.F. Johnson, pp. 547–565. New York: Plenum Press.

Pott, F., M. Roller, U. Ziem, F.J. Reiffer, B. Bellmann, M. Rosenbruch, and F. Huth, 1989. Carcinogenicity studies on natural and man-made fibers with the intraperitoneal test in rats. In *Nonoccupational Exposure to Mineral Fibers*, Eds. J. Bignon, J. Peto, and R. Saracci, Vol. 90, pp. 173–179. Lyon: IARC.

Pott, F., H.W. Schlipköter, U. Ziem, K. Spurny, and F. Huth. 1984. New results from implantation experiments with mineral fibers. In *Biological Effects of Mineral Fibers*, Vol. 2, pp. 286–302. Copenhagen: WHO.

Pott, F., U. Ziem, F.J. Reiffer, F. Huth, H. Ernst, and U. Mohr. 1987. Carcinogenicity studies on fibres, metal compounds, and some other dusts in rats. *Exp Pathol* 32(3): 129–152.

Roller, M., F. Pott, K. Kamino, G.H. Althoff, and B. Bellmann. 1996. Results of current intraperitoneal carcinogenicity studies with mineral and vitreous fibres. *Exp Toxicol Pathol* 48(1): 3–12.

Roller, M., F. Pott, K. Kamino, G.H. Althoff, and B. Bellmann. 1997. Dose-response relationship of fibrous dusts in intraperitoneal studies. *Environ Health Perspect* 105(Suppl. 5): 1253–1256.

Smith, D.M., L.W. Ortiz, R.F. Archuleta, and N.F. Johnson. 1987. Long-term health effects in hamsters and rats exposed chronically to man-made vitreous fibres. *Ann Occup Hyg* 31(4B): 731–754.

Smith, D.M., L.W. Ortiz, R.F. Archuleta, and N.F. Johnson. 1987. Long-term health effects in hamsters and rats exposed chronically to man-made vitreous fibers. *Ann Occup Hyg* 31: 731–750.

Stoeber, W., H. Flachsbart, and D. Hochrainer. 1970. Der Aerodynamische Durchmesser von Latexaggregaten und Asbestfassern. *Staub-Reinh. Luft* 30: 277–285.

Vorwald, A.J., T.M. Durkan, and P.C. Pratt. 1951. Experimental studies of asbestosis. *AMA Arch Ind Hyg Occup Med* 3(1): 1–43.

Vu, V.T. and D.Y. Lai. 1997. Approaches to characterizing human health risks of exposure to fibers. *Environ Health Perspect* 105S(Suppl. 5): 1329–1336.

Wagner, J.C., G.B. Berry, R.J. Hill, D.E. Munday, and J.W. Skidmore. 1984. Animal experiments with MMM(V)F — effects of inhalation and intrapleural inoculation in rats. In *Biological Effects of Man-Made Mineral Fibres: Proceedings of a WHO/IARC Conference in Association with JEMRB and TIMA*. Copenhagen: WHO.

Warheit, D.B., G.A. Hart, T.W. Hesterberg, J.J. Collins, W.M. Dyer, G.M. Swaen, V. Castranova, A.I. Soiefer, and G.L. Kennedy, Jr. 2001. Potential pulmonary effects of man-made organic fiber (MMOF) dusts. *Crit Rev Toxicol* 31(6): 697–736.

Warheit, D.B., K.L. Reed, and T.R. Webb. 2001. Man-made respirable-sized organic fibers: what do we know about their toxicological profiles? *Ind Health* 39(2): 119–125.

WHO. 1985. *Reference Methods for Measuring Airborne Man-Made Mineral Fibres (MMMF), WHO/EURO MMMF Reference Scheme*. Eds. Vol. EH-4. Copenhagen: WHO.

WHO. 1988. Environmental Health Criteria 77 Man-Made Mineral Fibres. Vol. 77. World Health Organigation Geneva.

USEPA. 2003. Report on the Peer Consultation Workshop to discuss a proposed protocol to assess asbestos-related risk. Prepared for U.S. Environmental Protection AGency, Office of Solid Waste and Emergency Response, Washington, DC 20460, EPA Contract No. 68-C-98-148, Work Assignment 2003-05. Prepared by Eastern Research Group, Inc., 110 Hartwell Avenue, Lexington, MA 02421. Final Report May 30, 2003.

12

REPRODUCTIVE TOXICOLOGY TESTING OF INHALED PHARMACEUTICALS AND BIOTECHNOLOGY PRODUCTS

André Viau and Keith Robinson

CONTENTS

12.1 INTRODUCTION

For toxicity testing of pharmaceuticals to be administered clinically by inhalation, it is accepted that much of the testing will be performed using the clinical route, with nose-only or oronasal exposure systems being commonly used. These types of systems have several advantages over whole-body systems, including a lack of external deposition of the test article, resulting in avoidance of dermal absorption or oral ingestion via grooming and the utilization of less test article due to lower airflows and smaller chamber volumes.

Inhalation exposures for reproductive toxicology testing of industrial chemicals and agrochemicals, typically using whole-body exposures, have been performed for over 20 yr (John et al., 1983). In these experiments, rodent (usually rat) fertility and multigeneration reproduction studies and also rat and rabbit developmental toxicity studies were performed. Observations of fertility effects with chemicals such as 1,2-dibromo-3-chloropropane (DBCP) in both rats and rabbits (Rao et al., 1982; Rao et al., 1983) and embryo–fetal effects in rabbits with 2-methoxypropanol-1 (Hellwig et al., 1994) were made.

The use of nose-only exposure systems for inhalation reproductive toxicology testing has become more common in the last 15 yr following the publication of a number of validation studies of the potential effects of tube restraint upon major organogenesis in rodents (i.e., Beyrouty et al., 1990; Tyl et al., 1994). Nose-only exposure allows reduced drug use in comparison to whole-body chambers and provides exposure via the respiratory tract only to mimic human use. In our laboratory, we have conducted a number of method development and validation studies to assess the applicability of these treatment procedures to various reproductive toxicology studies.

Pharmaceutical testing International Conference on Harmonization guidelines (ICH, 1993) have been in broad use for reproductive toxicology studies since the mid-1990s. Three standard studies are described: a rodent fertility and early embryonic development study (ICH-1), embryo–fetal development studies (ICH-3) in rodents and nonrodents, and a pre- and postnatal study (ICH-2), again in rodents.

For juvenile pediatric studies, nose-only exposure/oronasal masks have been used for a number of years. In our laboratory, weanling rats have been treated in nose-only restraint tubes for over 15 yr. With the increased concern for the generation of data for pharmaceuticals used to treat the pediatric population (FDA, 1998) and the consequent need for preclinical studies (FDA, 2003), combined with the frequent use of inhalation as the treatment route for these drugs, there is a need for studies with inhalation exposures in preweaning animals. Consideration of the formulation as well as the active ingredient must be made, especially where it may be different from the adult product. Furthermore, the oronasal mask exposure system in the dog may be adapted to incorporate the clinical actuation device, which may be of particular importance if the design of the device is modified for the pediatric population.

It is important that toxicokinetic data are collected to ensure adequate blood levels are achieved for both reproductive and pediatric nonclinical studies. Data from adult toxicity studies provide a guide to the likely toxicokinetics that will be seen.

12.2 EXPERIMENTAL PROCEDURES, STUDY DESIGNS, AND RESULTS

The following subsections describe standard study designs for ICH reproductive toxicology studies that might be applied to the safety testing of a novel pharmaceutical or biotechnology product by the inhalation route. Also described are the experimental procedures for various method development and validation studies performed in our laboratory.

For the nonclinical pediatric studies, each study is custom-designed so that only the method development and validation studies are described in the following text in detail. A general description of the types of end points that can be added to these pediatric studies to answer the concerns of regulatory agencies regarding the development of a variety of organ systems is given.

For reproductive toxicology, testing outbred strains of animals is usually recommended. In our laboratory, reproductive toxicology studies are typically conducted with CD-IGS® rats and New Zealand White rabbits. For some rat studies described in the following text, CD rats, the substrain that preceded the CD-IGS® animals, were used, whereas for some other studies, Han Wistar animals were employed.

For ICH reproductive toxicology studies, a control and three treated groups are required. Additional groups can be added such as air controls where an unusual vehicle is used. Group sizes to provide at least 15 or 16 pregnancies are required. This typically requires group sizes of 22 to 24 animals. To provide for the selection of dosages, not only preceding toxicity studies are used but typically range-finding studies also precede the definitive embryo–fetal development studies. These often contain more dose groups but with fewer animals, and as they represent scaled down versions of the main studies, they will not be discussed further.

For all studies described, the conditions for the animal room environment and photoperiod were: temperature 22 ± 3°C for rats and 17 ± 3°C for rabbits; humidity 50 ± 20%; and light cycle 12 h light and 12 h dark. The room air was exchanged with fresh air 10 to 15 times per hour.

All animals were fed standard certified pelleted commercial laboratory diets. Rats had free access to PMI 5002 (PMI Feeds, Inc.), and rabbits were fed 180 g per day of PMI 5322 (PMI Feeds, Inc.). Municipal tap water that had been softened, purified by reverse osmosis, and sterilized by ultraviolet light was freely available.

12.2.1 Rodent Inhalation Exposure Systems

Standard nose-only inhalation chambers were utilized in these experiments. Each chamber is of modular design with 20 separate ports in each row into which the conical front section of a polycarbonate restraint tube is inserted. The top section of the inhalation chamber has an opening for inlet air into which the test or control articles are introduced. The bottom section of the chamber has a corresponding air extraction port and a drain valve for cleaning the chamber. To prevent outward leakage of the test atmospheres, the chambers are operated under slight negative pressure, maintained by means of a gate valve located in the exhaust line. This is monitored as a differential pressure across a constriction in the exhaust line using a Magnehelic® gauge. A vacuum pump is used to exhaust the inhalation chamber at the required flow rate and to draw out the contaminated air through a purifying system consisting of a 5-μm coarse filter before expelling the air from the building.

Photo 12.1 Nose-only exposure chamber and powder delivery system.

The tubes, which serve to restrain individual animals, are custom-molded polycarbonate of various sizes. One end of the tube is tapered to approximately fit the shape of the animal's head and the diameter of the cylindrical portion was selected to prevent the animal turning in the tube. The back portion of the tube is sealed. The tube containing the animal is fastened to the inhalation chamber by means of "o" rings with the nose portion of the tube protruding into the port of the chamber. This permits the animal to breathe in the test or control atmospheres within the inhalation chamber without otherwise coming in contact with the atmosphere. As the animal's body weight increases throughout the study, the appropriate larger-size restraint tube is selected for use.

Photo 12.2 Restraint tube for rats and mice of various sizes.

Powder vehicle and test article are generated using an extended-duration powder delivery system (EDPDS) and introduced into the inhalation chamber via a Venturi T-section supplied with predried compressed air. The various test article concentrations are achieved by altering the rate of test article introduction into the feed nozzle of the T-section.

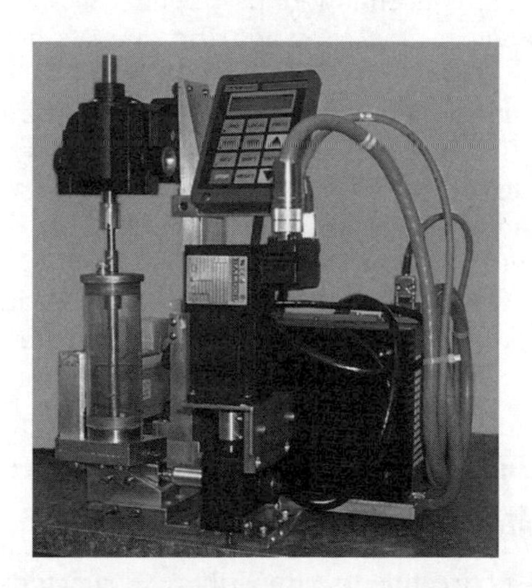

Photo 12.3 Extended-Duration Powder Delivery System (EDPDS).

For liquids, the test and control atmospheres are generated into the chamber air inlet using nebulizers supplied with predried compressed air. The various test article concentrations are achieved by altering the formulation strength or the duration of exposure.

The test and vehicle control atmospheres of metered dose inhalers are generated into the chamber using an actuating device under the control of an electronic timer to discharge aerosol cans at preset intervals into the inlet side of the exposure system. The various concentrations are achieved by altering the rate of canister actuation and the number of canisters being actuated at the same time.

The inhalation chamber and generation system for the test and vehicle control article groups are contained within separate ventilated walk-in fume hoods to prevent possible contamination of the room air with trace amounts of the test article.

During the acclimation period, the animals are conditioned to the laboratory setting and restraint tubes. Acclimation to restraint tubes is performed for increasing periods of time prior to initiation of treatment up to the intended duration of exposure.

12.2.1.1 Exposure Procedures

All animals are subjected to nose-only exposure for up to 4 h every day for several consecutive days depending on the study design.

For each exposure group, animals are placed onto the chamber only after aerosol has been actively generated for the time required to reach 95% of the target concentration (t_{95}) and after confirmation that the appropriate chamber concentration has been established. Time zero is defined as that point in time at which the animals are placed onto the exposure chamber. Animals can be exposed for up to 4 h and the generator turned off. The animals are removed from the exposure chamber and returned to their home cages after exposure. Flow rate through the chamber is calculated to yield a t_{95} of no more than approximately 10% of the exposure duration. The t_{95} and t_{05} (the time to decay to 5% of the established concentration) are calculated.

The vehicle control group is restrained and handled in the same manner as the test article groups and is exposed to the vehicle formulation at the same concentration and duration as the high-dose test group.

12.2.2 Monitoring of the Experimental Atmosphere

12.2.2.1 Prestudy Atmosphere Characterization

Prior to the start of treatment, atmosphere characterization of the test aerosols is performed.

Airflow through the inhalation chamber is set at a level determined during restraint tube acclimation of the animals to be adequate to maintain a chamber environment of 20 to 24°C, 20 to 70% RH, and at least 19% O_2 within each group (prior to test article introduction). This level represents a minimum of 12 air changes per hour. Chamber airflow is continuously monitored in the exhaust line by means of a precalibrated Magnehelic® gauge.

The exposure system operational conditions required to establish each target aerosol concentration are determined gravimetrically from open-face glass fiber filter samples collected at a representative animal breathing point during the prestudy period. Each test and vehicle control atmosphere is continually monitored throughout the exposure period by a real-time Microdust® proaerosol monitoring system (mainly used for powder and metered dose formulations) to provide instantaneous qualitative feedback on temporal atmosphere concentration stability. Concentrations of active ingredient in treated groups are confirmed analytically from the deposit on appropriate gravimetric filters.

Analysis of the aerosol particle size distribution for all groups is performed using a cascade impactor. The method consists of classification into a series of size ranges followed by gravimetric analysis. In addition, chemical analysis of the filters from all groups is performed. The mass median aerodynamic diameter (MMAD) and its geometric standard deviation (GSD) are calculated (MMAD ± GSD) from the gravimetric and analytical data using a computer program.

The homogeneity of chamber atmosphere concentration is determined at low- and high-dose concentrations, by collecting filter samples in duplicate for gravimetric analysis from three equidistantly spaced breathing ports located about the circumference of the chamber (overall gravimetric concentration). Additional samples are collected from a reference port to assess total and within-port variation of test article distribution within the chamber.

12.2.2.2 Animal Exposure Atmosphere Measurements

Inhalation chamber airflow, temperature, and humidity are recorded at least twice during each exposure and oxygen concentration is measured once on the second day of treatment.

Actual chamber concentrations of aerosol are measured at least twice daily for each group, including controls, from a sampling port from the animal breathing zone using a gravimetric method. Each test atmosphere is continually monitored throughout the exposure period by a real-time Microdust® proaerosol monitoring system to provide instantaneous qualitative feedback on temporal atmosphere concentration stability. Chamber concentrations of active ingredient are determined analytically, in all groups, from the deposit on the gravimetric filters collected on day 1 and weekly thereafter during treatment.

Particle size distribution analysis is performed for each group weekly using a cascade impactor. The method consists of classification into a series of size ranges followed by gravimetric analyses and chemical analyses. The MMAD and its GSD are calculated (MMAD ± GSD).

Nominal chamber concentrations are calculated from the airflow through the chamber and the quantities of test article used.

12.2.2.3 *Achieved-Dose Calculation*

The achieved total body dose of active test material (mg/kg/day) for each inhalation treatment level is determined as follows:

Achieved total dose of active test material (mg/kg/d)	$= \frac{\text{RMV} \times \text{active concentration} \times \text{T} \times \text{D}}{\text{BW}}$
Where RMV (l/min)	= respiratory minute volume, calculated[a]
Active concentration (mg/l)	= chamber concentration of active ingredient determined by chemical analysis
T (min)	= treatment time (60 min)
BW (kg)	= mean body weight per sex per group from the regular body weight occasions during treatment
D[b]	= Total deposition fraction based on the particle size

[a] $0.499 \times [\text{body weight (kg)}]^{0.809}$ l/min. (From Bide, R.W., Armour, S.J., and Yee, E. 2000. Allometric respiration/body mass data for animals to be used for estimates of inhalation toxicity to young adult humans, *J. Appl. Toxicol.* 20: 273–290.) It is assumed that this parameter is unaffected by exposure to the test article.

[b] Assuming total body dose deposition fraction is 100%. Based on particle size, reference to the literature is used.

12.2.3 ICH-1 — Fertility and Early Embryonic Development Study

12.2.3.1 *Fertility and Early Embryonic Development Study— Standard Design*

These studies may be conducted as separate male and female studies in which each sex is mated with untreated partners; the treated males are paired with females that have normal estrous cycles and the treated females with proven breeder males. Alternatively, a combined study may be performed in which the treated males and females are paired within the treatment group. There is also some flexibility in the design as to the premating treatment period for the males, which is normally 4 weeks, such that it can be reduced in some circumstances to 2 weeks (ICH, 2000).

At the start of treatment, the males are 12 weeks of age and the females 10 weeks of age. Treatment is typically performed for 4 weeks for males and 2 weeks for females prior to placement for mating, so that the animals are 16 and 12 weeks of age at mating, respectively. Males are usually treated throughout a 3-week mating period and for another 2 to 3 weeks postmating (to provide a total dosing period of 9 to 10 weeks) and allow for evaluation of their female partners reproductive status before necropsy at approximately 21 weeks of age. The dams are dosed during the mating period and following mating up to day 7 of gestation. In-life observations include individual body weights and food consumption measured twice weekly, commencing the day of randomization and extending through the treatment period. Mated females are weighed on days 0, 3, 7, 10, and 13 of gestation. The estrous cycles of the females are determined prior to placement for mating by examination of a vaginal lavage. Following the premating treatment period, one female is placed with one male in the same treatment group for a maximum of 21 d. The females are examined daily for evidence of mating by examination of a vaginal lavage for spermatozoa. The day of positive identification of spermatozoa in the vaginal lavage is termed day 0 of gestation. Females failing to show signs of mating are euthanized at the end of the mating period.

For each male, the epididymides, prostate, seminal vesicles, and testes are dissected free of fat and weighed. The left cauda epididymis is used to provide samples for assessments of motility (with a computer-assisted sperm analyzer), sperm concentration (millions per gram of epididymis), morphology (the percentage of abnormal sperm was assessed). The right testis is prepared for histological examination by embedding in glycol methacrylate, sectioned and stained with PAS hematoxylin, and examined. On day 13 of gestation the reproductive tract of the females is dissected out, the ovaries weighed and the corpora lutea counted. The uterine contents are examined and the number and position of live embryos, dead embryos, and resorptions is recorded. The uterus of any euthanized animal judged to be nonpregnant is stained with 10% aq. ammonium sulfide solution and examined for implantation sites (Salewski, 1964).

12.2.3.2 Fertility and Early Embryonic Development Study—Method Development and Validation Studies

For a validation study to assess the effects of tube restraint on estrous cyclicity prior to mating, groups of female CD-IGS® rats with normal estrous cycles were randomly assigned to either room control or to nose-only exposure in restraint tube (7 weeks of age; 9 weeks of age at the start

Table 12.1 Rat Female Fertility Study

Parameter	*Room Control*	*Tube Restraint 9 Weeks*	*Tube Restraint 11 Weeks*
Pretube acclimation (14 days)			
Number of days in estrous	3.9	3.9	3.7
Number of cycles seen	3.5	3.6	3.5
Average cycle length (days)	4.2	4.2	4.1
During tube acclimation/ restraint (14 days)			
Number of days in estrous	4.1	3.6	4.1
Number of cycles seen	3.2	3.2	3.3
Average cycle length (days)	4.2	4.4	4.4

of cycle assessment). The cycles were assessed for 14 d, then the females in the tube acclimation and restraint groups were acclimated to the tubes for 0.5 or 1 h per day followed by tube restraint for 1 h per day.

12.2.3.3 Fertility and Early Embryonic Development Study—Results of CTBR Studies

The tube restraint assessment of estrous cyclicity showed no adverse effect of tube acclimation or restraint for up to 1 h per day upon average estrous cycle lengths of females at 9 and 11 weeks of age at the start of tube restraint (Table 12.1). These ages equate to placement for mating at 11 and 13 weeks of age, respectively.

Fertility data shown in Table 12.2 and Table 12.3 for males and females are compared to historical control data from studies dosed by other routes (gavage, etc.). As can be seen, there are no effects of the exposure procedure upon any parameters examined.

12.2.4 ICH-2—Pre- and Postnatal Study

12.2.4.1 Pre- and Postnatal Study—Standard Design

Females for these studies start treatment at 12 weeks of age. The animals are treated from days 6 to 20 of gestation and resumed on days 1 or 2 *postpartum* until at least day 21 *postpartum.* A complete detailed examination is performed on the days of body weight assessment. Mated females are weighed on days 0, 3, 6, 9, 12, 15, 18, and 20 of gestation and on days 0, 4, 7, 14, 17, and 21 *postpartum.* Individual food consumption is

Table 12.2 Female Fertility of CD-IGS® Rats

Parameter	*Inhalation Control*	*Historical Control Range*
Mean day to mating	2.4	2.6–4.0
Mating index (%)	96.0	85.0–100.0
Conception rate (%)	90.0	62.5–100.0
Fertility index (%)	91.9	66.7–100.0
Number of corpora lutea/dam	17.4	16.1–20.3
Number of implantations/dam	15.7	15.1–18.6
Preimplantation loss (%)	10.6	3.4–11.9
Number of live embryos/dam	15.0	14.3–17.3
Number of dead embryos/dam	0.0	0.0–0.2
Number of resorptions/dam	0.8	0.8–1.4
Postimplantation loss (%)	8.7	4.3–10.7

Table 12.3 Male Fertility of CD Rats

Parameter	*Air Control*	*Vehicle Control*	*Historical Control Range*
Mean day to mating	3.0	2.8	2.6–4.0
Mating Index (%)	100.0	100.0	85.0–100.0
Conception rate (%)	72.0	100.0	62.5–100.0
Fertility Index (%)	72.0	100.0	66.7–100.0
Male reproductive assessments			
Epididymal sperm count	871	926	643–979
Motility	82.9	84.1	77.2–86.4
Number of corpora lutea/dam	19.1	18.8	17.3–19.4
Number of implantations/dam	16.3	17.4	16.0–17.4
Preimplantation loss (%)	16.3	7.0	7.0–16.3
Number of live fetuses/dam	15.7	16.2	14.6–16.8
Number of dead fetuses/dam	0.1	0.0	0.0–0.1
Number of resorptions/dam	0.6	1.2	0.6–1.7
Postimplantation loss (%)	9.0	3.6	3.6–17.2
Fetal weight (g)	3.9	4.0	3.8–4.1
Sex ratio (% male)	48.1	48.5	44.2–51.7

measured at similar intervals up to day 18 of gestation. Females are observed for signs of parturition and where possible, parturition is observed, the time of onset and completion of parturition is recorded and any sign of dystocia noted. The females' behavior immediately *postpartum* is observed. The day of completion of littering is termed day 0 *postpartum*.

12.2.4.1.1 The F_1 Generation Pups

After parturition (day 0 *postpartum*) the pups are examined for malformations, sexed, and the numbers of live and dead recorded. The live pups are weighed individually. Daily litter observations during lactation include general condition of the pups. In addition to the assessment of body weight at birth, the pups are weighed individually on days 4, 7, 14, and 21 *postpartum*. On day 4 *postpartum* the litter are culled to eight pups, where necessary, to give a litter of four males and four females, where possible. The selected pups are identified by tattooing of the paws. Parameters of physical and reflexological development may be assessed. At weaning, on day 21 *postpartum*, the F_1 generation are separated from their dams and at least one male and one female rat are randomly selected from each litter, where possible, to provide the F_1 adult generation.

12.2.4.1.2 Observations — F_1 Adult Generation

Observations of clinical condition, body weight, and observations at parturition and during lactation are similar to those of the F_0 generation dams and for the F_2 generation pups, observations up to day 4 *postpartum* are similar to the F_1 generation pups. On day 21 *postpartum*, the pupillary closure and visual placing responses are tested for visual function. Vaginal opening for females and preputial separation for males is assessed. For behavioral performance, locomotor activity, auditory startle habituation, and water maze are examined.

12.2.4.1.3 Terminal Procedures for Adult Generations

At termination, on day 21 of lactation for dams with litters the uteri of all females are examined and, if appropriate, the number of implantation site scars recorded.

12.2.4.1.4 Terminal Procedures for Pups (F_0 and F_1 Generations)

Pups found dead or dying on or before day 7 *postpartum* and any pups born malformed are euthanized and placed in Bouin's fluid for subsequent examination using a modified Barrow and Taylor technique (1969). F_1 generation pups dying between days 8 and 20 *postpartum* and unselected weanlings are given a complete gross external examination.

12.2.4.2 Pre- and Postnatal Study—Method Development and Validation Studies

In whole-body reproduction studies (i.e., Hamm et al., 1985; Breslin et al., 1989) exposure of the dams is suspended for up to 4 d around parturition to avoid littering during the exposure period. To study the most appropriate days to suspend treatment around parturition for nose-only inhalation treatment, groups of ten time-mated Han Wistar rats were acclimated to restraint tubes for 15, 30, and 60 min per day on days 13, 14, and 15 of gestation, respectively. They were then tube-restrained daily for 60 min per day until day 20 of gestation. Subsequently, tube restraint was restarted on days 1, 2, 3, or 4 *postpartum.* A room control group, of similar size, was not restrained. The dams were allowed to litter and the pups examined daily and weighed on days 0, 1, 4, 7, and 10 *postpartum.*

12.2.4.3 Pre- and Postnatal Study—Results of CTBR Studies

The validation study comparing the effects of various intervals of interruption of maternal exposure around parturition showed that intervals as short as from day 20 of gestation to day 1 *postpartum* had no deleterious effect upon pup survival or growth when the dams were removed for 1h of tube restraint (Table 12.4).

12.2.5 ICH-3—Embryo–Fetal Development Study

12.2.5.1 Rat Embryo–Fetal Development Study—Standard Design

Time-mated females of 11 to 12 weeks of age at the start of dosing are used. The day of positive identification of spermatozoa in a vaginal lavage is termed *day 0 of gestation.* The animals are treated from day 6 to day 17 of gestation, inclusive. A complete detailed examination is performed on the days of body weight assessment. Individual body weights and food consumption are measured on days 0, 3, 6, 9, 12, 15, 18, and 21 of gestation.

On gestation day 21, the reproductive tract is dissected out, the ovaries removed and the corpora lutea counted. The gravid uterus is weighed, the uterine contents, including the placentas, are examined and the number and position of live fetuses, dead fetuses, and early, middle, and late resorptions is recorded.

For fetal examinations, each fetus is weighed, given a detailed external examination, the external sex recorded and the fetus euthanized by

Table 12.4 Maternal Performance, Pup Viability, and Growth

Parameter	*Room Control*	*Tube Restraint*			
		1PP	*2PP*	*3PP*	*4PP*
Maternal performance					
Number of dams dying/euthanized	0	0	0	0	0
Body weight gain (g)					
Days 0 to 4 PP	7.0	7.2	1.6	2.6	8.3
Days 4 to 10 PP	23.7	26.9	22.0	33.4	18.3
Gestation length (days)	21.7	21.5	21.7	21.2	21.6
Number of live pups/dam	9.9	9.8	8.2	9.8	9.6
Number of dead pups/dam	0.0	0.0	0.1	0.0	0.0
Pup viability					
Viability index	99.2	97.8	96.5	95.3	100.0
Survival index—day 7	100.0	100.0	100.0	98.9	100.0
Survival index—day 10	100.0	100.0	100.0	98.9	100.0
Pup weights (g)					
Day 0 PP	5.9	5.8	5.9	5.5	5.7
Day 1 PP	6.7	6.6	6.9	6.3	6.7
Day 4 PP	10.1	9.8	10.4	9.4	10.3
Day 7 PP	14.6	14.2	15.4	13.7	14.8
Day 10 PP	20.2	19.8	21.6	19.4	20.5

Note: PP = *postpartum*.

subcutaneous injection of euthanasia solution. A detailed internal examination using a dissecting microscope is performed on approximately half of the fetuses in each litter, which are then eviscerated. The heads of these fetuses are removed and placed in Bouin's fluid for examination by the technique of Wilson (1965). The remaining half of the fetuses in each litter are eviscerated and placed in 85% ethanol/15% methanol for subsequent staining with alizarin red S (using a modified Dawson technique (1926) and skeletal examination. Abnormalities are classified as major malformations, minor visceral or skeletal anomalies, or common skeletal variants (Palmer, 1977).

12.2.5.2 Rat Embryo–Fetal Development Study—Method Development and Validation Studies

A validation of nose-only restraint for inhalation rat teratology studies designed to meet the FDA guidelines that preceded the ICH guidelines

(D'Aguanno, 1976) was performed (Beyrouty, 1990). Mated rats were assigned as either room controls or air controls. They were assigned to tube restraint for 1, 2, and 4 h per day on days 3, 4, and 5 of gestation, respectively, prior to 5 h per day restraint on days 6 to 15 of gestation. Evaluations were similar to those described earlier except that the dams and fetuses were evaluated on day 20 of gestation.

12.2.5.3 Rat Embryo–Fetal Development Study—Results of CTBR Studies

In an initial validation study (conducted prior to the ICH guidelines) by Beyrouty et al. (1990), no adverse effects of an exposure of 5 h per day for days 6 to 15 of gestation, inclusive, were noted upon maternal, ovarian, uterine, or fetal parameters (Table 12.5).

A slightly lower body weight gain among the dams was noted between days 0 to 6 of gestation, the period of acclimation to the restraint tubes.

Table 12.5 Rat Teratology Study

Parameter	*Room Control*	*Tube Restraint*
Number of dams dying/euthanized	0	0
Body weight gain (g)		
Gestation days 0 to 6	25.1	20.3
Gestation days 6 to 15	35.0	34.8
Gestation days 15 to 20	59.0	66.2
Pregnancy (%)	92.3	92.3
Number of dams with total resorption	0	0
Number of corpora lutea/dam	16.4	15.3
Number of implantations/dam	15.0	15.1
Preimplantation loss (%)	9.3	1.6
Number of live fetuses/dam	13.1	14.1
Number of dead fetuses/dam	0.0	0.0
Number of resorptions/dam	1.9	1.0
Postimplantation loss (%)	12.0	6.7
Fetal weight (g)	3.3	3.5
Sex ratio (% male)	51.0	47.0
Major malformations—number of litters affected (percentage of fetuses affected)	0 (0.0)	0 (0.0)
Skeletal anomalies—number of litters affected (percentage of fetuses affected)	10 (19.7)	12 (21.0)

Control groups from subsequent studies designed to meet the ICH guidelines have similarly shown no differences from control groups of studies dosed by gavage or other standard routes.

12.2.5.4 Rabbit Embryo–Fetal Development Study—Standard Design

Time-mated or artificially inseminated rabbits of 5 to 6 months of age at the start of dosing are used. The day of mating or insemination is termed day 0 of gestation. The does are treated from day 7 to day 19 of gestation, inclusive of both days. The does are examined twice daily and any evidence of abortion recorded. A complete detailed examination is performed on the days of body weight assessment. Individual body weights are measured on days 0, 4, 7, 10, 13, 16, 19, 22, 25, and 29 of gestation, and food consumption is measured daily.

On gestation day 29, examinations of the does and fetuses are performed in a similar manner to those described in the preceding text for the rats, except that for the does, the uterus is also examined for empty implantation sites (which denote an abortion). For the fetal examinations, all fetuses are given a detailed internal examination and skeletal examination. The heads may be examined by a coronal section between the frontal and parietal bones of all fetuses or one half of the heads may be examined by the technique of Wilson (1965).

12.2.5.5 Rabbit Inhalation Exposure Systems

Rabbits were placed in restraint tubes on a modified nose-only inhalation chamber and exposed to predried conditioned air generated through a nebulizer. One end of the custom-molded polycarbonate restraint tubes used is tapered to approximately fit the shape of the head and the diameter of the cylindrical portion of the tube is such that it is difficult for the animal to turn. The tube containing the animal is fastened to the inhalation chamber, with the nose portion protruding through a gasket into the chamber.

Airflow through the chamber is set to maintain a chamber environment of 18 to 20°C and 30 to 70% RH with the animal load.

12.2.5.6 Rabbit Embryo–Fetal Development Study—Method Development and Validation Studies

A group of inseminated does were restrained in tubes for treatment on days 7 to 19 of gestation following acclimation periods of 1, 2, and 4 h on days 4, 5, and 6 of gestation, respectively. Evaluations were similar to those described in the preceding text (Salame et al., 1997).

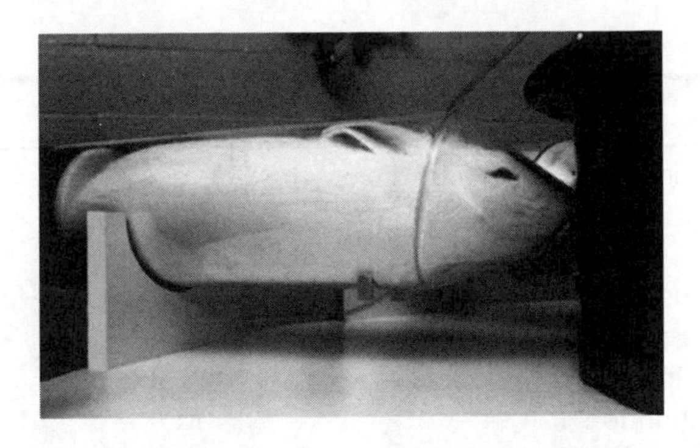

Photo 12.4 Rabbit in restraint tube on nose-only exposure chamber.

12.2.5.7 Rabbit Embryo–Fetal Development Study—Results of CTBR Studies

A validation study (Salame et al., 1997) of 4 h per day tube restraint showed no differences for the maternal, ovarian, uterine, or fetal parameters (Table 12.6).

Body temperatures, measured before and after acclimation, were unaffected by tube restraint with peak values of daily measurements being 39.8°C and 39.7°C. Maternal observations showed no mortality, adverse clinical findings, effects upon body weight, or food consumption.

12.2.6 Toxicokinetics for ICH Reproductive Toxicology Studies

Plasma levels of the test article are usually measured at intervals during the studies. For embryo–fetal development studies, typically, measurements are made at the beginning and end of the dosing period with enough time points to allow for the calculation of standard toxicokinetic parameters, including Cmax, Tmax, and AUC. The fetal plasma levels can be measured in both rats and rabbits (Pinsonneault et al., 1998). In the fertility studies, only limited toxicokinetics may be required if data for males and nonpregnant females are available from preceding toxicity studies at comparable dosages. Measurements for pre- and postnatal studies are more varied but should include maternal plasma levels during late gestation and lactation, with either milk levels or pup plasma levels being possible additions to examine pup exposure postnatally.

12.2.7 Rat Nonclinical Pediatric Studies

12.2.7.1 Rat Nonclinical Pediatric Studies—Material and Methods

Mated female CD-IGS® rats (as described in the preceding text) are supplied time mated on gestation day 18 and housed in solid-bottomed

Table 12.6 Rabbit Teratology Study

Parameter	*Room Control*	*Tube Restraint*
Number of does dying/euthanized	0	0
Pregnancy rate (%)	100	100
Abortion rate (%)	0.0	12.5
Number of does with total resorption	0	0
Number of corpora lutea/doe	9.4	10.0
Number of implantations/doe	8.2	8.7
Preimplantation loss (%)	14.9	12.3
Number of live fetuses/doe	7.7	8.6
Number of dead fetuses/doe	0.0	0.0
Number of resorptions/doe	0.5	0.1
Postimplantation loss (%)	6.1	1.1
Fetal weight (g)	46.3	45.3
Sex ratio (percentage of males)	51.1	51.2
Fetal findings		
Number of litters (percentage of fetuses)		
Major malformations	1 (2.8)	0 (0.0)
External and visceral anomalies	2 (6.5)	2 (3.3)
Skeletal anomalies	9 (7.4)	5 (8.3)
Skeletal variants (percentage of fetuses affected)		
13th ribs	50.7	57.4
Sternebral	37.7	37.6

cages (with similar automatic watering) and housed on corncob bedding (Bed'O-Cob®). They are allowed to litter and on day 4 *postpartum* litters are culled to eight pups, comprising four males and four females, where possible. Litters are weaned on day 21 *postpartum* and the weanlings are housed by litter for several days postweaning prior to being housed individually in stainless-steel mesh-bottomed cages each equipped with an automatic watering valve. All cages are clearly labeled with a color-coded cage indicating project, group and animal numbers, and sex. Each animal is uniquely identified using the AIMS® system. The pups are identified within the litters with tattooing of the paws performed at culling (day 4 *postpartum*). Pups can also be identified using the AIMS® tail tattoo system, which can be performed as early as day 9 *postpartum*, if required.

The conditions for the animal room environment and photoperiod, as well as the diet, are as described before.

12.2.7.2 Rat Nonclinical Pediatric Studies—Inhalation Exposure

Standard stainless-steel cylindrical "flow-through" nose-only inhalation chambers are utilized in these experiments. Each chamber is of modular design with 20 ports in each row into which the conical front section of a polycarbonate restraint tube can be inserted. The chambers are operated under slight negative pressure maintained by means of a gate valve located in the exhaust line.

Photo 12.5 Nose-only exposure of neonates (day 10 to day 12).

12.2.7.3 Rat Nonclinical Pediatric Studies—Inhalation Exposure for Method Development and Validation Studies

In the validation study described by Stoute et al. (2003), the pups were conditioned to the laboratory setting and restraint tubes. Pups (2 pups/sex/litter) were acclimated to the restraint tubes from days 7 to 9 *postpartum* for increasing periods of time (1, 2, and 4 h, respectively). The remaining 2 pups/sex/litter comprised the cage control group.

The pups were then subjected to nose-only exposure to room air for 4 h from days 10 to 20 *postpartum*. During daily exposure, air through the nebulizer was actively generated for 4 h. Time zero was defined as the time at which the animals were placed onto the exposure chamber. The rats were removed after the treatment period and returned to their home cages.

Inhalation chamber airflow, temperature, and humidity were recorded approximately every hour during each exposure. Oxygen levels were determined once during the first week.

One pup/sex/litter was selected to continue from days 21 to 34 *postpartum* in the nose-only restraint tube validation study and 1 pup/sex/litter was selected to continue as the control group.

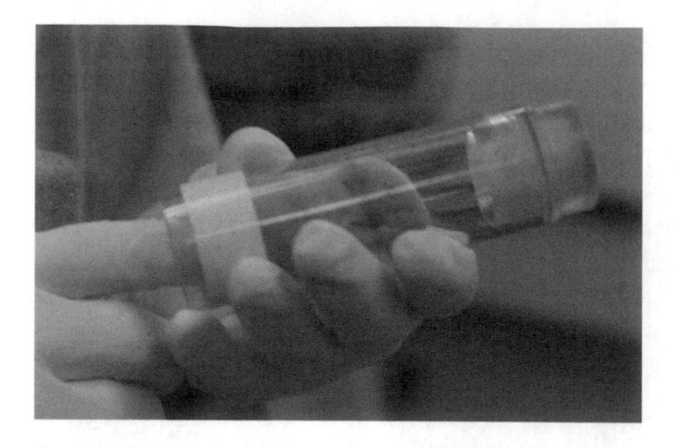

Photo 12.6 Pup acclimation to restraint tube at day 7 *postpartum*.

12.2.7.4 Rat Nonclinical Pediatric Studies—Method Development and Validation Studies

The general condition of the pups was evaluated daily during the lactation period and postweaning. All animals were examined once daily for mortality and signs of ill health and a complete detailed examination was performed at least weekly.

The pups were weighed individually on days 4 (postculling), 7, 10, 14, 17, and 21 *postpartum* prior to weaning and postweaning on days 24, 28, 32, and 35 *postpartum*.

On completion of the acclimation period, animals were euthanized by CO_2 asphyxiation (on day 21 *postpartum*) and intraperitoneal injection of sodium pentobarbital (on day 35 *postpartum*). In order to avoid autolytic change, a complete gross pathology examination of the carcass was conducted immediately on all animals euthanized. All necropsies were conducted under the supervision of a pathologist. Necropsy consisted of an external examination, including identification of all clinically recorded lesions, as well as a detailed internal examination.

On completion of the necropsy of each animal, tissues and organs were retained. Neutral buffered 10% formalin was used for fixation and preservation of the following tissues, bronchi, lungs (all lobes, infused with neutral buffered 10% formalin), lymph nodes (tracheobronchial), nasal cavities with sinuses, trachea, and larynx.

12.2.7.5 Rat Nonclinical Pediatric Studies—Results of Studies at CTBR

In the evaluation of the effect of air treatment on rat pups, no mortality or adverse clinical findings resulted from 4 h of tube restraint from day

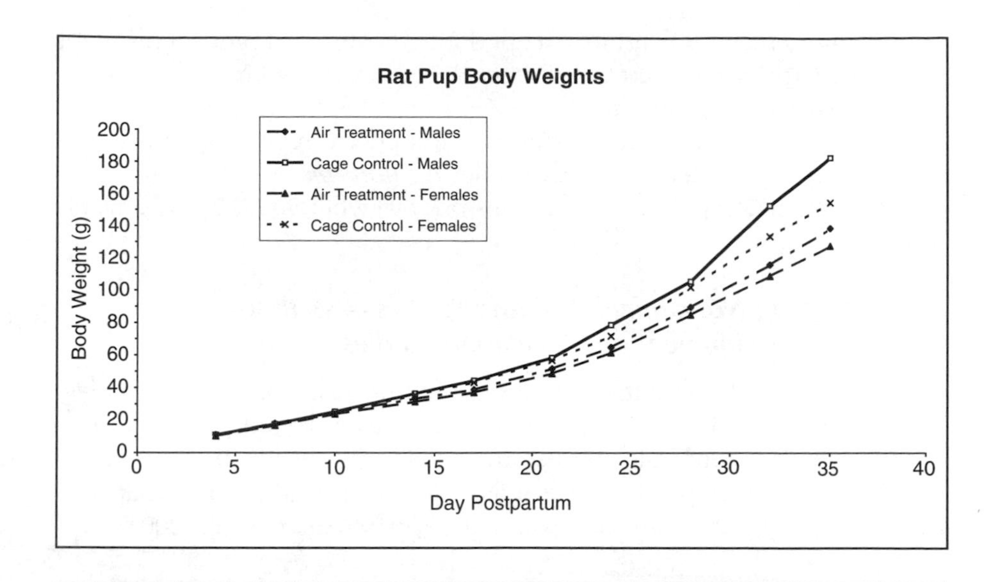

Figure 12.1

10 *postpartum* (Stoute et al., 2003). Body weight gain was slightly reduced during the preweaning period, no gross pathological findings could be related to the restraint procedures, and the findings remained within the historical control range for untreated rat pups. Postweaning body weight differences were more marked; however, the cage control group values were unusually high and the extent of these differences were comparable to those seen when litters of CD rats are not culled; therefore, these differences are not considered adverse (Figure 12.1).

12.2.8 Dog Nonclinical Pediatric Studies

Pups from gravid female dogs (*Canis familiaris*), of the Beagle strain, and approximately 2 to 4 yr of age, are used for these studies.

Each adult female is identified with an individual ear tattoo (ventral aspect of 1 pinna).

Animals are housed individually in stainless-steel cages (approximately 18 ft^2) that are equipped with an automatic watering valve and with modified floors and nesting boards, feed pans and water bottles or bowls when necessary. Each pup is identified with tattoo ink codes on the abdominal surface or by color patches on their fur and sex for the initial study.

The animal room environment and photoperiod are controlled (conditions: temperature 24 ± 3°C and approximately 6 weeks later reduced to 21 ± 3°C, humidity 50 ± 20%, 12 h light, 12 h dark).

Animals had access to a standard pelleted commercial dog food (600 g—Eukanuba premium performance formula) daily prior to whelping and *ad libitum* after whelping, except during designated procedures. Pups are offered the same feed *ad libitum* dry or mixed with water, starting four or five weeks *postpartum*.

Tap water that has been softened, purified by reverse osmosis, and sterilized by ultraviolet light is available, *ad libitum*.

The acclimation period for the dams prior to whelping is approximately 2 to 4 weeks.

12.2.8.1 Dog Nonclinical Pediatric Studies—Method Development and Validation Studies

Following littering, the pups were identified and assigned to the treatment groups such that one pup per litter per sex was assigned to each of two groups. For the initial validation study, a split litter design was employed in which two pups per litter were assigned to an inhalation group and two pups per litter to an oral gavage group (Robinson et al., 2002).

12.2.8.2 Dog Nonclinical Pediatric Studies—Inhalation Exposure

Prior to the start of treatment, pups were acclimated to the exposure mask for gradually increasing periods of time.

The test atmospheres were generated using Pari LC Plus nebulizers supplied with predried compressed air. Pups were dosed from day 10 *postpartum* using an oronasal face mask fitted with inlet and outlet tubes. During dosing, the pups were held by technical personnel in a walk-in inhalation booth inside the husbandry room. Each pup was given one exposure session per day. Individual doses were calculated based on the most recent body weight of each pup.

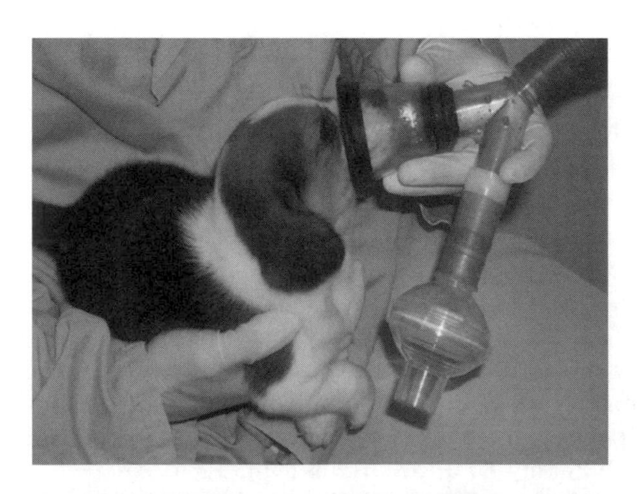

Photo 12.7 Dosing of pup at day 28 *postpartum*.

12.2.8.2.1 Dog Nonclinical Pediatric Studies—Prestudy Atmosphere Characterization

Prior to the start of the treatment, during (postdosing), and at the end of the treatment period, atmosphere characterization of the test article aerosols was performed.

For each nebulizer used, a minimum triplicate evaluation of the drug output at the facemask was performed gravimetrically. Particle size distribution analysis of the test aerosols of at least one nebulizer through the dog mask, and the MMAD ± GSD was calculated from the gravimetric data using a cascade impactor. Chemical analysis of the gravimetric filters for drug amount was performed.

12.2.8.2.2 Dog Nonclinical Pediatric Studies—Results of Studies at CTBR

In the validation study (Robinson et al., 2002), there were neither deaths nor clinical signs associated with oronasal treatment. Body weight gain was unaffected by oronasal treatment as compared to the gavage treated pups (Figure 12.2). Ophthalmology, clinical pathology, and pathology evaluations showed no effects of inhalation treatment (Figure 12.2).

Clinical pathology parameters for both oronasal and gavage showed clear patterns across time (Pinsonneault et al., 2002) as shown in Figure 12.3 to Figure 12.6. The reticulocyte count decreased and the number of red

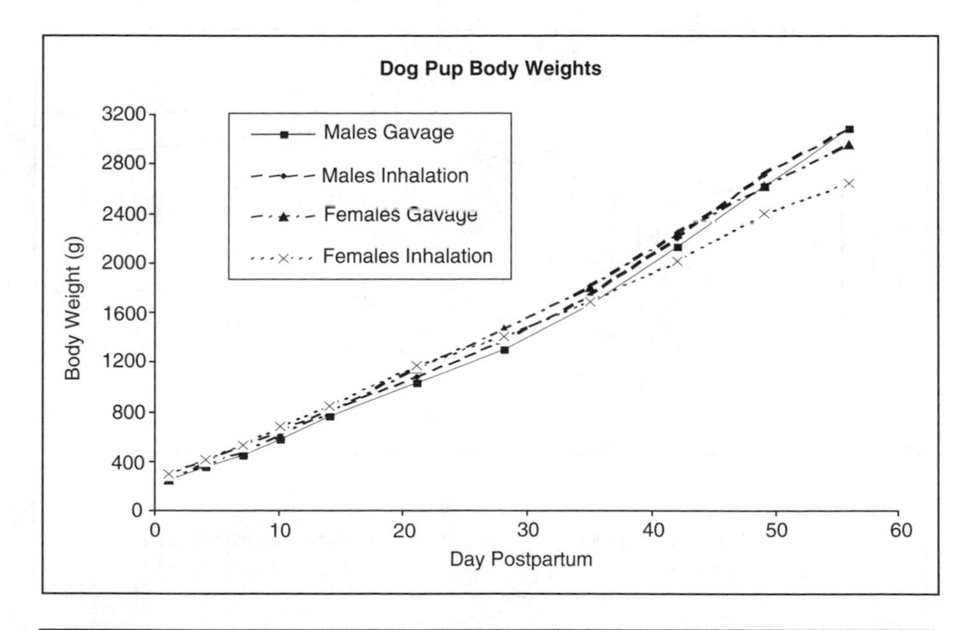

Figure 12.2

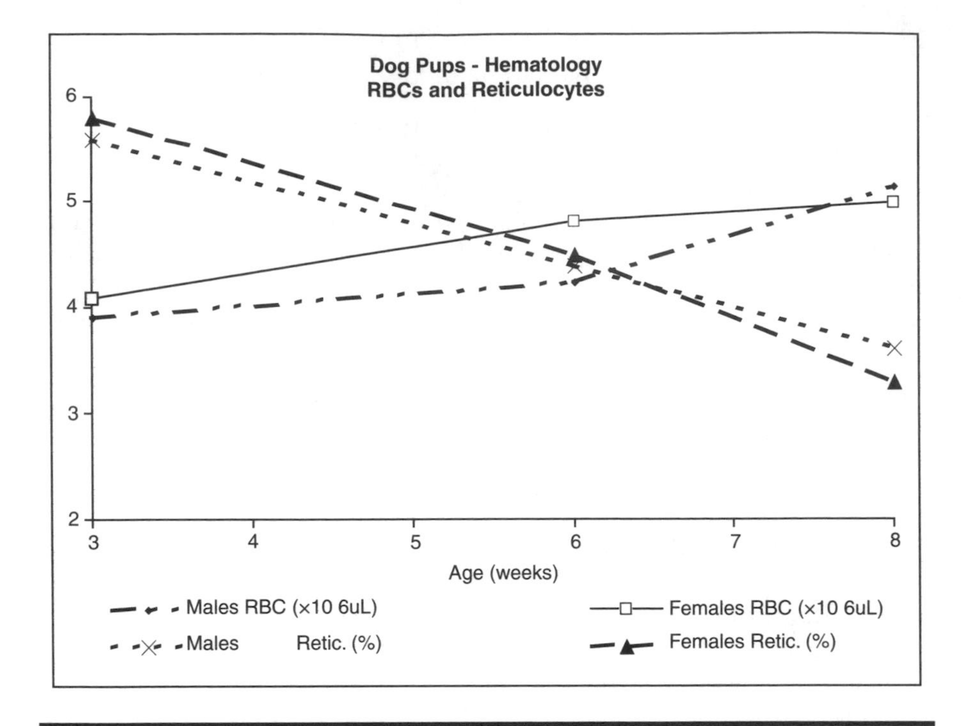

Figure 12.3

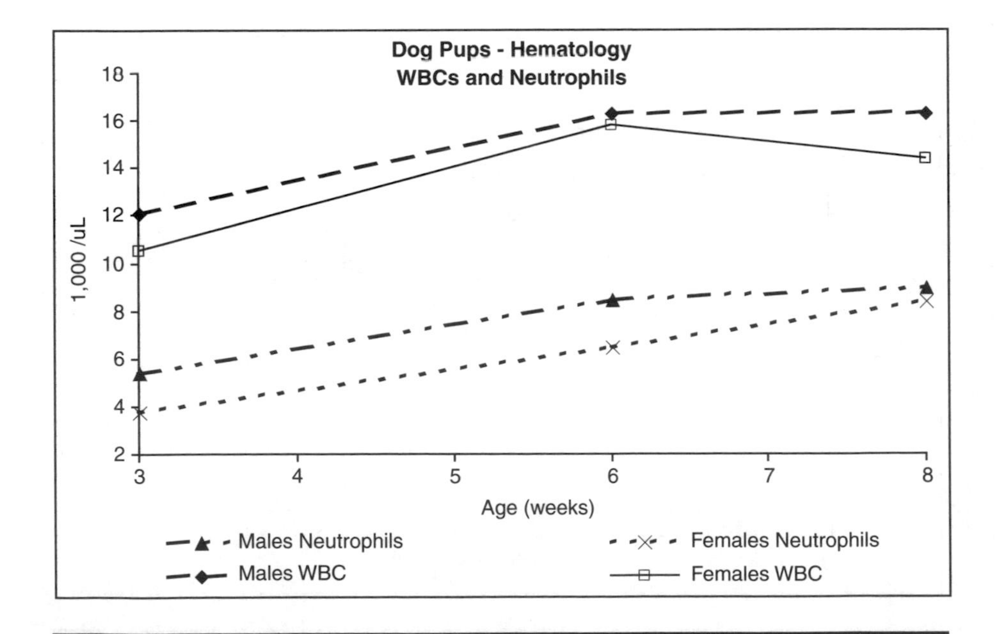

Figure 12.4

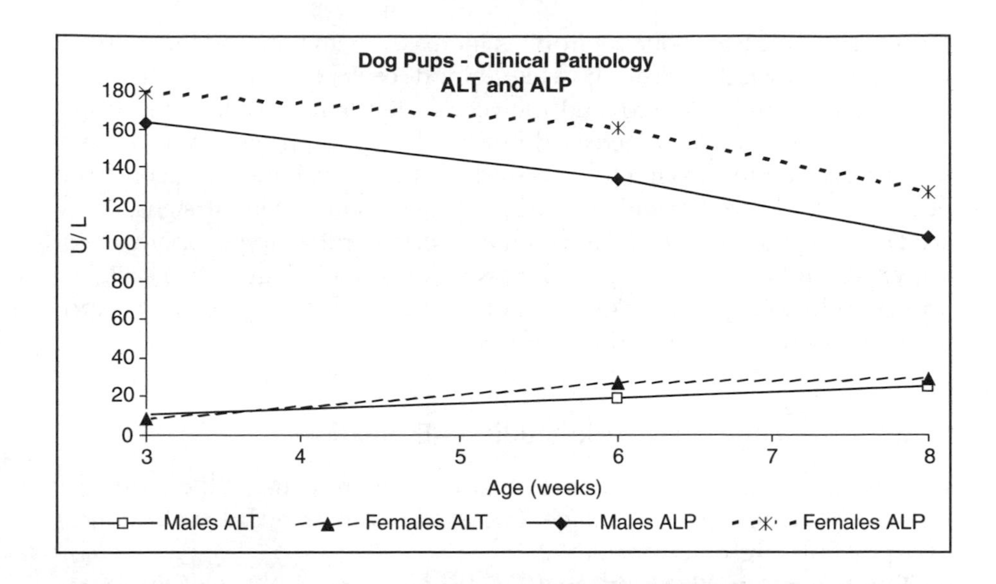

Figure 12.5

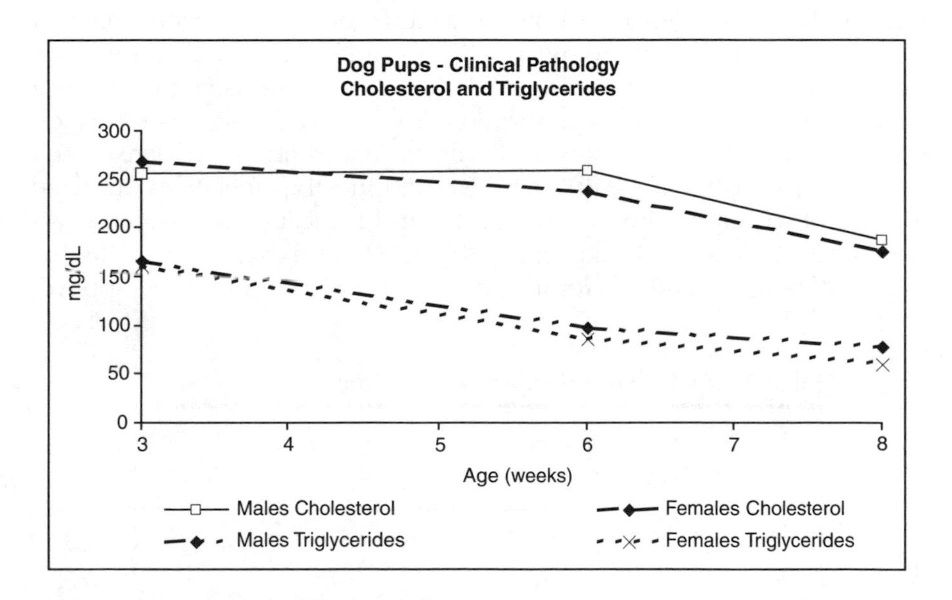

Figure 12.6

blood cells increased. Four neutrophils increased with time and this resulted in a trend toward increasing white blood cells. For clinical chemistry parameters, both cholesterol and triglyceride levels fell, while for levels of alanine transferase (ALT) increased but alkaline phosphatase (ALP) tended to decrease. Results from a 14-day study with a novel antiviral agent (Lee et al., 2003) showed no adverse effects on mortality, clinical signs, body weights, respiratory rate, tidal volume, electrocardiography, ophthalmoscopy, hematology and serum biochemistry, gross pathology, organ weights, and histopathology. Toxicokinetic monitoring confirmed parenteral exposure levels proportional to dose.

12.2.9 Nonclinical Pediatric Studies—Evaluations

Because each study is designed to meet the requirements of the particular test article, a description of a generic study with standard toxicological end points is presented (Table 12.7).

In addition, the FDA guidance (2003) identified specific organ systems for evaluation, including, skeletal growth, neurobehavioral development and immunological, pulmonary, and renal function.

A variety of *in vivo* and *ex vivo* evaluations can be applied to assess skeletal growth ranging from simple measures of crown–rump length in rats and height and length in dogs, to a range of imaging techniques that can be applied to both these species. X-rays can be used for the evaluation of longitudinal bone growth, morphological abnormalities and epiphyseal closure. Dual, energy X-ray absorptiometry (DXA) provides information on whole-body, axial, and appendicular bone area and bone density and concentration (BMD and BMC), whereas peripheral quantitative computed tomography (pQCT) measures trabecular and cortical bone compartments and bone geometry (Robinson et al., 2004). Terminal evaluations including histomorphometry and *ex vivo* imaging (DXA and pQCT) can be applied.

Table 12.7 Pediatric Testing—Evaluations

	(Earliest Day Postpartum)	
Evaluation	*Rat*	*Dog*
In life (clinical signs, body weight)	1	1
Food consumption	22	42
Clinical pathology	1	1
Ophthalmoscopy	21	21
Toxicokinetic plasma sampling	1	1
Gross pathology, histopathology	1	1

To assess the developing immune system of rats, a range of assays are available from a screen using hematology and immunohistopathology to tests of innate and adaptive immunity including lymphocyte phenotyping, NK cell activity, and T-cell-dependent antibody response (TADR) (Desilets et al., 2004).

Neurobehavioral assays are applied to rat inhalation pediatric studies and screens may include a functional observational battery (FOB), a motor activity test, and assessments of learning and memory such as passive avoidance and a water maze test. For dogs neurological examinations can be performed. For both species, the central and peripheral nervous systems can be examined histopathologically.

As of day 30 *postpartum,* rats are placed in "head-out" plethysmographs to monitor lung functions. Animals are allowed to acclimate to environmental conditions for approximately 15 min prior to each data collection period. Immediately following the acclimation period, ventilatory parameters (tidal volume, respiratory rate, and derived minute volume) are measured for an approximate 15-min period. The respiratory waveforms are analyzed and the parameters calculated. The respiratory parameters can be recorded as 20-sec means, and any 20-sec mean value with a respiratory rate above 240 breaths/min is excluded from the data and labeled as "apparent sniffing" artifact.

Photo 12.8 Nose-only inhalation chamber with head-out plethysmographs used to monitor tidal volume and respiratory rate.

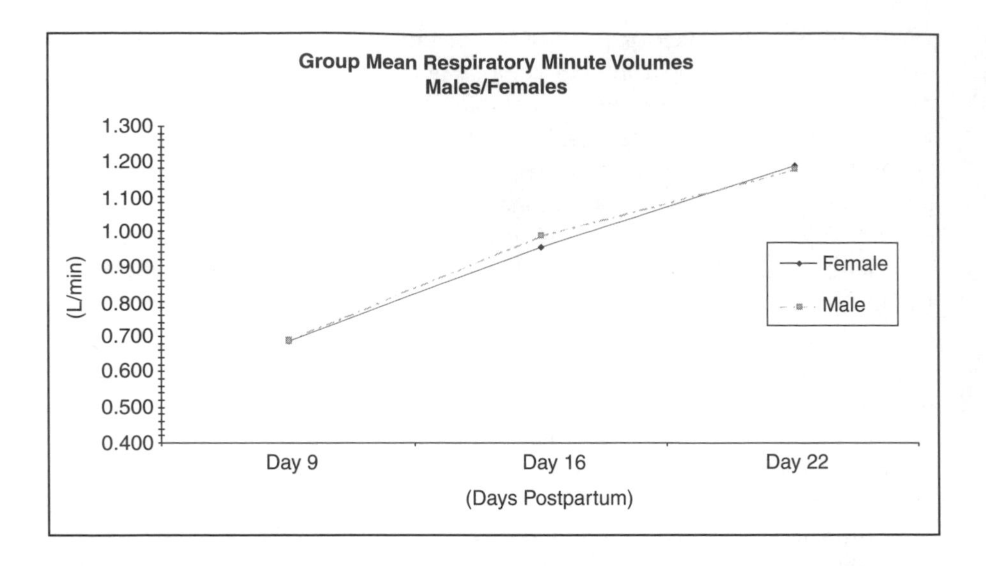

Figure 12.7

For dogs, respiratory minute volume (based upon tidal volume and respiratory rate) as early as day 9 *postpartum* can be monitored. The animal is held with a mask attached to the pneumotach and continuously monitored for 10 to 15 min. Respiratory minute volume increases slightly as the animals get older (Figure 12.7 to Figure 12.9).

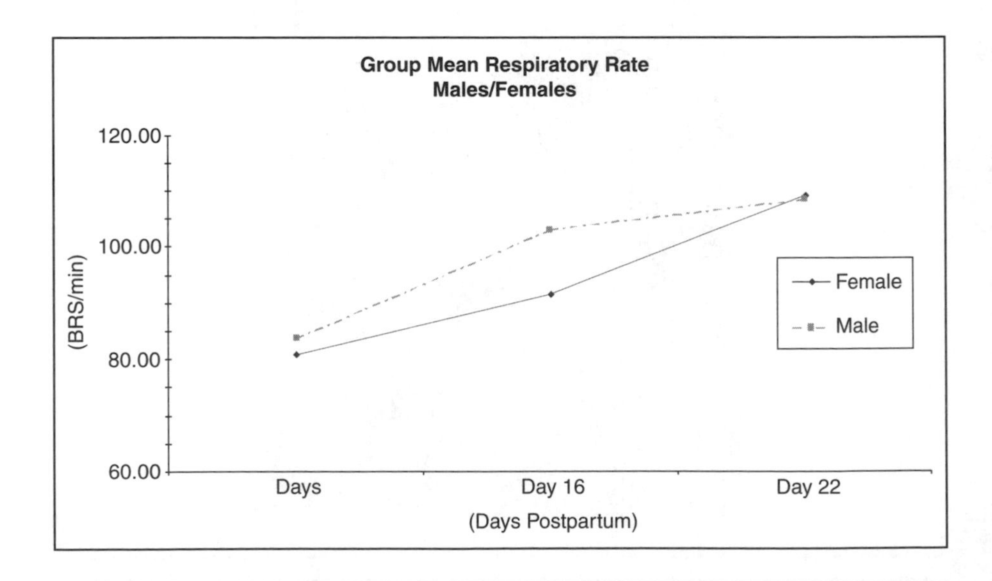

Figure 12.8

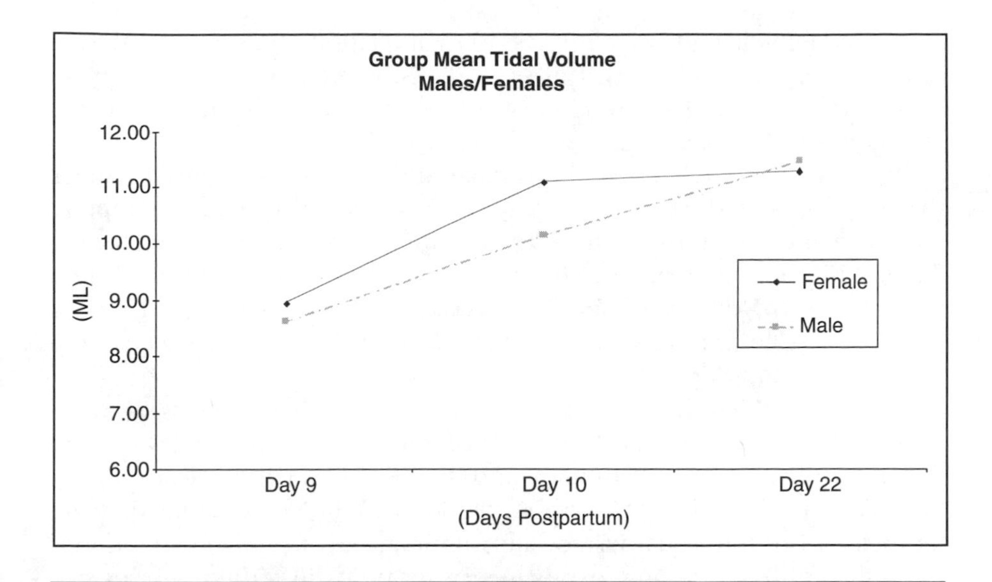

Figure 12.9

For evaluation of renal function, urinalysis can be performed upon rats postweaning.

12.2.10 Nonclinical Pediatric Studies—Toxicokinetics

Studies typically contain assessments of toxicokinetics, as in younger pups, metabolism is often poor and absorption from the neonatal lung may be different from that of the adult. Usually, the measurements at the beginning and end of treatment are obtained and, for longer exposure periods, additional interim occasions may be added. For rats, separate subsets of animals are treated and terminal blood samples obtained from the abdominal aorta in preweaning pups, whereas for older rats samples are obtained from the jugular vein. For dog pups, samples can be from the jugular veins from the start of treatment. To provide sufficient data to calculate standard parameters such as Cmax and AUG, typically, six time points per occasion are required.

12.3 DISCUSSION

12.3.1 ICH Reproductive Toxicology Studies

In our laboratory, potential effects of tube restraint were addressed in a variety of validation studies performed in our laboratory. In reproductive toxicity studies, effects upon both males and females need to be considered.

Preceding general toxicity type studies at CTBR had shown no adverse effects upon the pathology (gross and histopathological) of the male reproductive system. However, considerable literature exists discussing the effects noted by others with nose-only exposures.

Brock et al. (1996) reported testicular and related epididymal changes that were attributed to the nose-only procedures and increased temperature in a 4-week study with HFC-143a. In this study, Sprague-Dawley (Crl:CD BR) rats were exposed 6 h per day, 5 d per week. Similar changes were noted by Dodd et al. (1997) in studies with HFC-143a; in a 13-week study with Fisher, 344 rats were exposed 2 h per day, 5 d per week, and testicular lesions were observed.

Lee et al. (1993) described the testicular lesions associated with nose-only dosing as degeneration, with exfoliated germ cells and low sperm density in the epididymal tubules and suggested that these lesions may be related to the stress of restraint. Dodd et al. (1996) thought that the temperature of the testis may be an important factor. Other laboratories have shown that nose-only exposures of male rats in fertility studies can be performed without deleterious effects on reproductive function. In a series of studies with HFA-134a, Alexander et al. (1996) reported no adverse effects.

Rothenberg et al. (2000) postulated that exposures greater than 4 h per day, irrespective of acclimation, exposure conditions, or strain of the rat, would result in testicular atrophy.

Data from our laboratory shows nose-only inhalation exposure of animals in restraint tubes does not result in these types of testicular changes. Effects on neither reproductive performance nor testicular histopathology were observed. The key factors in the etiology of this lesion are most likely the types of restraint tubes used and the acclimation to the restraint tubes. No effects were seen in animals acclimated to the restraint tubes 3 d prior to exposure for increasing periods of time until reaching the exposure duration, normally 1 to 4 h. Previous assessments in restraint tubes at our laboratory has shown that Sprague-Dawley rats that were restrained in tubes and placed on an inhalation chamber for 1 h at a chamber exhaust flow rate of 12.5 l/min or 50 l/min remained within a body temperature range of 35.8 to 37.6°C.

Stress can have adverse effects upon many aspects of female rodent reproduction. Because restraint is a procedure used to induce stress-related changes in rodents (gastric ulceration), stress from tube restraint must be considered for both rat and rabbits.

Tube restraint in stress experiments is known to increase body temperature. However, this is not seen with our restraint tubes and acclimation systems. Temperature measurements for pregnant rabbits showed that tube restraint for up to 4 h did not increase rectal body temperatures with peak values of 39.8°C before and 39.7°C after tube restraint.

Few effects upon estrous cyclicity from tube restraint have been reported. Increased preimplantation loss following prolonged restraint of 5 h per day has been reported in rats (MacKiven et al., 1992).

Although stress has been shown to increase malformation rates in A/JAX mice (Rosenzweig and Blaustein, 1970), CD-1 mice failed to show an adverse effect for nose-only vs. whole-body exposure (Tyl et al., 1966). Rats have generally been shown to be resistant to the effects of stress upon the malformation rates (Golub et al., 2004). Similarly, in our laboratory, no adverse effects of 5 h per day of tube restraint were seen. Other laboratories have shown comparable results (Christian et al., 1996).

For the postnatal removal of the dams for treatment for 1 h for the pre- and postnatal study, no adverse effect was noted upon survival or growth even when the dams were removed for treatment as early as day 1 *postpartum*. Barlow (1978) showed adverse effects on postnatal growth following prolonged (9 h) restraint during gestation. Other adverse effects of stress upon reproduction of the male offspring have been reported (Ward et al., 1983). Also, effects upon behavior of different levels of stress have been reported (Kofman, 2002).

For rabbit embryo–fetal studies, custom-designed restraint tubes have been shown to be suitable by several investigators (Salame et al., 1997; Alexander et al., 1997) to allow for nose-only exposures. Other systems involving head-only exposures (Rothenberg et al., 1996) result in local dermal and auricular deposition of the test article and the need to fit the rabbits with goggles and apply ointment to the eyes to avoid ocular exposure.

Therefore, we can conclude that all ICH studies can be conducted in rats and embryo–fetal development studies in rabbits by nose-only inhalation exposures with only minor design modification in the pre- and postnatal study and with no adverse effects upon reproduction of the exposure procedures.

12.3.2 Nonclinical Pediatric Studies—Effects on Pups

In reviewing the development of the rat lung, Zoetis and Hurtt (2003) stated that the rat is an acceptable animal model because of similarities to the development of the human lung. The rat is born with lungs less developed than human lungs. So, alveolar development occurs between days 7 and 21 *postpartum* in the rat but from late gestation to childhood (about 2 yr) in the human. The nose-only exposure system for preweaning rats has several advantages over the previous procedures of whole-body exposures of the dam and pups together. Not only is external deposition avoided but also potential maternal transmission of the drug via the milk and potential confounding due to maternal toxicity.

The dog pup lung is comparable to a newborn human lung by day 11 *postpartum* (Zoetis and Hurtt, 2003). Treatment of the dog pups from day 10 *postpartum* allows treatment to cover the equivalent of alveolar development of the dog lung (completed by 16 weeks), with the critical period in humans continuing up to 2 yr. The oronasal masks used in our laboratory allow for the same dosing procedure to be used throughout the dosing period with sizes of masks being increased as the dog pups grow. Some dosing systems provide for head-only exposure at younger ages resulting in the aforementioned problems and also a change in exposure to oronasal at some point during longer studies.

In both rats and dogs, lung development can be monitored by in-life assessments of the respiratory minute volumes by measurements of tidal volume and respiratory rate. Postmortem evaluations of lung development can involve both histopathology and histomorphometry.

In summary, rat and dog pups can be treated with appropriate dosing procedures that allow for dosing using similar systems from neonatal to adult ages. Treatment can begin when these test species lungs are comparable in development to the human neonate, providing a suitable animal model.

REFERENCES

Alexander, D.J., Liberetto, S.E., Adams, M.J., Hughes, E.W., Bannerman, M. 1996. HFA-134a (1,1,1,2-tetrafluoroethane): effects of inhalation exposure upon reproductive performance, development and maturation of rats. *Hum. Exp. Toxicol.* 15: 508–517.

Alexander, D.J., Mather, A., and Dines, G.D. 1997. A snout-only inhalation exposure system for use in rabbit teratology studies. *Inhal. Toxicol.* 9: 477–490.

Barlow, S., Knight, A., and Sullivan, F.M. 1978. Delay in postnatal growth and development of offspring produced by maternal restraint stress during pregnancy in the rat. *Teratology* 18: 211–218.

Barrow, M.V. and Taylor, W.J. 1969. A rapid method for detecting malformations in rat fetuses. *J. Morph.* 127: 291–306.

Beyrouty, P., Gordon, C., Robinson, K., and Pinsonneault, L. 2000. An embryo-fetal development study of the potential adverse effects associated with jugular blood collection in the pregnant rat. *Teratology.* 61: 478, P19.

Beyrouty, P., Viau A., Robinson, K., Goldsmith, G., and Lulham, G. 1990. Suitability of Nose-Only Tube Restraint Procedures for Rat Inhalation Teratology Studies. American Association for Aerosol Research Meeting.

Bide, R.W., Armour, S.J., and Yee, E. 2000. Allometric respiration/body mass data for animals to be used for estimates of inhalation toxicity to young adult humans. *J. Appl. Toxicol.* 20: 273–290.

Breslin, W.J., Kirk, H.D., Streeter, C.M., Quast, J.F., and Szabo, J.R. 1989. 1,3-dichloropropene: two generation inhalation reproduction study in Fischer 344 rats. *Fundam. Appl. Toxicol.* 12: 129–143.

Brock, W.J., Trockimowicz, H.J., Farr, C.H., Millischer, R.-J., and Rusch, G.M. 1996. Acute, subchronic and developmental toxicity and genotoxicity of 1,1,1-trifluroethane (HFC-143a). *Fundam. Appl. Toxicol.* 31: 200–209.

Christian, M.S., Hoberman, A.M., Rothenberg, S., Foss, J.A. and Parker, R.E. 1996. Methods to reduce stress in nose-only and head-only rabbit reproductive toxicity studies. *Teratology* 53(5): 27A.

D'Aguanno, W. 1976. Guidelines for Reproduction Studies for Safety Evaluation of Drugs for Human Use. (Based on FDA for reproduction studies for safety evaluation of drugs for human use, 1966).

Dawson, A.B. 1926. A note on the staining of the skeleton of cleared specimens with alizarin red S. *Stain Technol.* 1: 123.

Desilets, G., Rouleau, N., Pouliot, L., and Lesauteur, L. 2004. Monitoring the primary and secondary antibody resposnse to klh in a developmental immunotoxicity study. *Toxicologist* 78: 1439.

Dodd, D.E., Kinkead, E.R., Wolfe, R.E., Leahy, H.E., English, J.H., and Vinegar, A. (1997). Acute and subchronic inhalation studies on trifluoroiodomethane vapor in Fischer 344 rats. *Fundam Appl. Toxicol.* 35: 64–77.

Golub, M.S., Campbell, M.A., Kaufman, F.L., Iyer, P., Ling-Hong, L., Donald, J.M., and Morgan J.E. 2004. Effects of restraint stress in gestation: implications for rodent developmental toxicology studies. *Birth Defects Res.* (Part B) 71: 26–36.

Hamm, T.E., Raynor, T.M., Phelps, M.C., Auman, C.D., Adams, W.T., Procter, J.E., and Wolkski-Tyl, R. 1985. Reproduction in Fischer-344 rats exposed to methyl chloride by inhalation for two generations. *Fundam. Appl. Toxicol.* 5: 568–577.

Hellwig, J., Klimisch, H.-J., and Jackh, R. 1994. Prenatal toxicity of inhalation exposure to 2-methoxypropanol-1. *Fundam. Appl. Toxicol.* 23: 608–613.

International Conference on Harmonisation. 1993. Detection of Toxicity to Reproduction for Medicinal Products. June 24, 1993.

International Conference on Harmonisation. 2000. Maintenance of the ICH guideline on Toxicity to Male Fertility. An Addendum to the Tripartite Guideline on: Detection of Toxicity to Reproduction for Medicinal Products.

John, J.A., Quast, J.F., Murray, F.J., Calhoun, L.G., and Staples R.E. 1983. Inhalation toxicity of epichlorhydrin: effects on fertility in rats and rabbits. *Toxicol. Appl. Pharmacol.* 68: 415–423.

Kofman, O. 2002. The role of prenatal stress in the etiology of developmental behavioural disorders. *Neurosci. Biobehav. Rev.* 26: 457–470.

Lee, K.-P., Frame, S.R., Sykes, G., Valentine, R. 1993. Testicular degeneration and spermatid retention in young male rats. *Toxicol. Pathol.* 21: 292–302.

Lee, W., Gordon, C., Viau, A., Addjiri-Awere, A., McCartney, J., Nash, J.A., Hincks, J.R., Rhodes, G., and Davies M.H. 2003. VP14637: Two week inhalation toxicity study in neonatal dogs. *Toxicologist* 72: 1413, 290–291.

MacKiven, E., DeCatanzaro, D., and Younglai, E.V. 1992. Chronic stress increases estrogen and other steroids in inseminated rats. *Physiol. Behav.* 52: 152–162.

Palmer, A.K. 1977. Incidence of sporadic malformations, anomalies and variations in random-bred laboratory animals. In *Methods in Prenatal Toxicology*, Eds., Neubert, D., Merker, H.-J., and Kwasigroch, T.E. Georg Thieme, Stuttgart. 52–71.

Pinsonneault, L., Robinson, K., Ducharme, S., Sey, S. and Kam, M. 1998. A novel technique for the collection of rat and rabbit fetal blood. *Teratology* 57: 254.

Pinsonneault, L., Gordon, C., Robinson, K., Martin, A., and Viau, A. 2002. Gavage dosing in neonatal Beagle dogs—growth and clinical pathology. *Teratology* 65(6): 327.

Rao, K.S., Burek, J.D., Murray, F.J., John, J.A., Schwetz, B.A., Bell, T.J., Potts, W.J., and Parker, C.M. 1983. Toxicologic and reproductive effects of inhaled 1,2-dibromo-3-chloropropane in rats. *Fundam. Appl. Toxicol.* 3: 104–110.

Rao, K.S., Burek, J.D., Murray, F.J., John, J.A., Schwetz, B.A., Bell, T.J., Beyer, J.E. and Parker, C.M. 1982. Toxicologic and reproductive effects of inhaled 1,2-dibromo-3-chloropropane in male rabbits. *Fundam. Appl. Toxicol.* 2: 241–251.

Robinson, K., Gordon, C., Salame, R., Viau A., and Pinsonneault, L. 2002. Procedures for inhalation treatment of neonatal dogs. *Teratology* 65(6): 328.

Robinson, K., Stoute, M., Viau, A., and L. Pouliot. 2003. Inhalation nose-only exposure of neonatal and juvenile rats. *Teratology* 67(5): 350.

Robinson, K., Varela, A., Doyle, N., Jolette, J., Chouinard, L., Sabourin, M., Chevrier, C., and Smith, 5. 2004. Feasibility study to assess skeletal development in non-clinical pediatric studies. *Toxicologist* 78: 2099.

Rothenberg, S.J., Parker, R.M., York, R.G., Dearlove, G.E., Martim, M.E., Denny, K.H., Lief, S.D., Hoberman, A.M., Christian, M.S. 2000. Lack of effects of nose-only inhalation exposure on testicular toxicity in male rats. *Toxicol. Sci.* 53: 127–134.

Rozenzweig, S. and Blaustein, F.M. 1970. Cleft palate in A/J mice resulting from restraint and deprivation of food and water. *Teratology* 3: 47–52.

Salame, R., Pinsonneault, L., Robinson, K., Banks, C., and Perkin, C.J. 1997. A validation study of nose-only inhalation exposure of pregnant rabbits. *Toxicologist* 36: 1842, 362.

Stoute, M., Viau, A., Robinson, K., and Banks, C. 2003. Inhalation nose-only exposure of neonatal and juvenile rats. *Toxicologist* 72: 1442, 297.

Stoute, M., Maguire, S., Robinson, K., Viau, A., and Banks, C. 2004. The effects of pup viability and growth during nose-only inhalation of wistar-han rats for pre- and postnatal studies. *Toxicologist* 78: 707.

Salewski, E. 1964. staining methods for the macroscopic identification of implantation sites in rat uteri (färbemethode zum makroskopischen nachweis von implantations—stellen am uterus der ratte. Von E. Salewski[Koln]), *Naunym-Schmiedebergs Arch. Exp. Path. Pharmak.* 247: 367.

Tyl, R.W., Ballantyne, B., Fisher, L.C., Fait, D.L., Savine, T.A., Pritts, I.M., and Dodd, D. E. 1994. Evaluation of exposure to water aerosol or air by nose-only or whole-body inhalation procedures for CD-1 mice in developmental toxicity studies. *Fundam. Appl. Toxicol.* 23: 251–260.

US FDA Draft Guidance for Industry: Non-Clinical Safety Evaluation of Pediatric Drug Products Center for Drug Evaluation and Research. 2003 (http//www.fda.gov/cder/guidance/index.htm).

US FDA Guidance for Industry: Immunotoxicology Evaluation of New Drugs. Center for Drug Evaluation and Research. 2002 (http//www.fda.gov/cder/guidance/index.htm).

Ward, I.L. 1983. Effects of maternal stress on the sexual behaviour of male offspring. *Monogr. Neural. Sci.* 9: 169–175.

Wilson, J.G. 1965. Methods for administering agents and detecting malformations in experimental animals. In *Teratology, Principles and Techniques*, Eds., J.G. Wilson and J. Warkany. Chicago: The University of Chicago Press, 262–277.

Zoetis, T. and Hurrt, M.E. 2003. Species comparison of lung development. *Birth Defects Res.* (Part B) 68: 121–124.

13

EFFECTS OF ENGINEERED NANOSCALE PARTICULATES ON THE LUNG

David B. Warheit

CONTENTS

13.1 INTRODUCTION AND GENERAL BACKGROUND

The development of new products using nanomaterials, frequently referred to as *nanotechnology*, is an emerging multidisciplinary technology that involves the synthesis of molecules in the nanoscale (i.e., 10^{-9} m) size range. The term nanotechnology is derived from the Greek word *nano*, meaning dwarf. Sometimes the nanosize issue is difficult to put into proper perspective, but for relating size comparisons to biological end points, the diameter of an erythrocyte (i.e., red blood cell) is approximately 7 µm or 7000 nm; bacteria generally are in the range of 1 µm or 1000 nm, and some viruses measure in the 60 to 100 nm size range. From a material science and chemistry standpoint, what makes nanotechnology so exciting is the fact that as one decreases the particle size range, i.e., moves down the nanoscale, for a given material, the laws of physics appear to change, often yielding completely new physical properties. For instance, titanium dioxide, a white particle type, loses its color and becomes transparent at decreasing size ranges less than 50 nm. Other particle types utilized for electrical insulating can suddenly become conductive, and insoluble substances can become more soluble below 100 nm. In the aggregate, these changes in physical properties enhance versatility and, thus, are likely to give rise to new industrial and medical applications as well as more eclectic products. These possibilities have generated great interest in this potentially new technology (Colvin, 2003).

This brief review is designed to identify some of the physicochemical factors that are likely to influence and impact pulmonary effects following aerosol exposure to nanoscale particulates. In addition, the review should provide a current perspective on external perceptions of the potential health effects (or lack thereof) following exposure to nanoscale or ultrafine particles (UFP). In this regard, it is widely considered that, owing to surface area or particle number considerations, nanoparticles are significantly more toxic than fine-sized particles of identical chemistry. This perception, however, is based on a paucity of hazard data; moreover, particle size is only one factor that is likely to influence the lung's response to inhaled materials. The remainder of the chapter will describe some recent studies that provide conflicting results regarding the influence of particle size, as evidenced by the findings with nanoscale titania, nanoscale quartz particles, and single-wall carbon nanotubes (SWCNT). Finally, a proposed methodology for safe handling of nanomaterials in the laboratory will be presented.

The production of engineered nanoscale particulates is not new. Some nanoscale particulate types have been produced for decades. For example, nanoscale carbon black particles, utilized in the manufacture of rubber products and pigments, have been in production for more than a century. Similarly, nanoscale fumed amorphous silica particulates and other metal oxide particles such as titania, alumina, and zirconia have been produced as nanomaterials for over 50 yr and used as thixotropic agents in pigments and cosmetics applications, and more recently, as the basis for polishing powders in the microelectronics industry. Many of these nanoscale particle types are produced using gas-phase flame reactions, carried out under very controlled conditions (Borm and Kreyling, 2004).

Newer engineered nanoscale particulates are now being developed for many different applications, using more advanced preparation techniques. Production volumes currently are low, often on a laboratory scale producing less than 10 kg per day. Some nanomaterial-type examples include the following: magnetic materials for high-density data storage, high-current electrode materials for fuel cells, and materials with new surface properties for paints, coatings for self-cleaning windows, and stain-resistant textiles. These engineered nanoscale materials are significantly more costly to produce relative to conventional (bulk scale) materials. As a consequence, they must deliver very high performance or enhanced value to customers in order to justify the enhanced costs. In the future, as production costs are reduced via enhanced material volumes concomitant with the maturation of nanotechnology methods, it seems likely that the competitive advantages of products containing nanomaterials will ultimately result in the displacement of bulk (i.e., micro- and macroscale) materials in many high-value applications (Colvin, 2003).

It is important to note that particle surfaces and interfaces play important roles for engineered nanoscale particulates and confer advantages over larger-size particles. In this regard, as the particle size decreases, the proportion of atoms at the surface is substantially enhanced relative to the volume of the particle. As a consequence, nanoparticle types have greater reactivity and function as more effective catalysts or more efficient filler materials, thus permitting weight reduction of composite materials. The higher surface energy can also facilitate stronger interactions between nanoparticles, causing particles to aggregate. A nanoparticle with a diameter of 5 nm and density of 5 g/cm^3 has a surface area of 240 m^2/g when assuming a sphere-like shape. This would suggest that 20% of the particle atoms are associated with its surface. The physical and chemical properties of particles can be substantially altered by varying its composition, size, or surface composition (Colvin, 2003).

Engineered nanoscale particulates differ in significant ways from the heterogeneous and polydisperse airborne particulates traditionally associated with ambient, combustion-derived UFP inherent in air pollution particulate matter. Although both UFP and engineered nanoscale particles are defined as particles with diameters under 100 nm, their broad size distributions and heterogeneous compositions make comparisons with engineered nanomaterials problematic. In contrast with combustion-derived UFP, engineered nanomaterials are produced in liquid-phase or in closed gas-phase reactors. Because particle–particle interactions increase significantly in the nanoscale-size regime, engineered nanoparticles become strongly associated into bulk aggregates following generation and synthesis via the liquid or gas phase. It is thus difficult to generate respirable nanoparticles from dried powders, although this can be altered by appropriate control over surface coatings. The high uniformity of the engineered nanoscale samples is a common characteristic feature of many nanoparticulate types, in which the control of particle size is necessary to define commercial properties (Borm and Kreyling, 2004).

13.2 WHAT IS KNOWN ABOUT THE PULMONARY RISKS OF NANOPARTICLES

Pulmonary toxicity studies in rats demonstrate that exposures to UFP or nanoparticles result in elevated inflammatory responses when compared with larger-sized particles of identical chemical composition at equivalent mass concentrations (Donaldson et al., 2001; Oberdorster et al., 2000). Particle surface area and particle number determinations have been postulated to play important roles in the development of nanoparticle lung toxicity. Contributing to the impact of nanoparticle-mediated toxicity is their very high size-specific deposition rate when inhaled experimentally as singlet UFP rather than as aggregated particles. Some evidence indicates that inhaled UFP or nanoparticles, following deposition in the alveolar regions of the lung, largely escape alveolar macrophage surveillance and transmigrate to the pulmonary interstitium by translocation processes from alveolar spaces through epithelium (Donaldson et al., 2001; Oberdorster et al., 2000).

It is interesting to note that very few studies have been reported that have assessed the inhalation toxicity of UFP by laboratory animals at very high particle concentrations. Some of these were hazard-based toxicity studies conducted to investigate pulmonary effects caused by lung particle overload, i.e., induction of lung tumors in rats at high retained particulate lung burdens. In this regard, 2-yr inhalation studies with ultrafine (uf) (P-25) and fine-sized TiO_2 particles (average primary particle sizes ~20 nm and ~270 nm, respectively) have demonstrated that less than one-tenth the inhaled mass concentrations of the uf TiO_2 particles, compared with the fine particles, produced equivalent numbers of lung tumors in rats in these 2-yr studies (approximately 16 to 30%) (Lee et al., 1985; Heinrich et al., 1995). In addition to the long-term studies, shorter-term pulmonary toxicity studies with uf and fine carbon black, nickel, as well as TiO_2 particles in rats (Ferin et al., 1992; Oberdorster et al., 1994; Li et al., 1996) have supported the notion of enhanced lung inflammatory potency of the UFP compared to exposures of fine-sized particulates of similar composition. When the instilled doses of the various particulates were expressed in terms of particle surface area, the responses of the uf and fine TiO_2 particles were aligned on the same dose–response curve. This is because a given mass of UFP has a much greater surface area when compared to the mass of fine-sized particles and, thus, is more likely to produce particle overload and consequent inflammation in the lung. Thus, from a hazard and regulatory viewpoint (i.e., from a risk perspective), it is important to delineate the pulmonary effects of UFP in rats at overload vs. nonoverload conditions.

13.2.1 Systematic Comparisons of Fine and Ultrafine Particles

It may be surprising to note that the lung toxicity database for conducting systematic comparisons of the pulmonary effects of UFP or nanoparticles vs. fine-sized particles in rats is extremely limited and consists of studies with only two particle types: titanium dioxide and carbon black particles (Borm and Kreyling, 2004). In addition, most if not all of the nano vs. fine size comparisons have been conducted in studies with rats, a species that is known to be uniquely sensitive in developing adverse lung responses to low-solubility particles, particularly at overload concentrations. Thus, long-term (2 yr) high-dose inhalation toxicity studies with rats using poorly soluble, low-toxicity dusts can ultimately produce pulmonary fibrosis and lung tumors via an overload mechanism. These lung-tumor-related effects are unique to rats and have not been reported in other particle-exposed rodent species such as mice or hamsters under similar chronic conditions (Hext, 1994). For the mechanistic/pathogenetic connection, it has been hypothesized that the particle-overload effects in rats result in the development of uniquely sensitive lung defense responses, characterized by enhanced and sustained levels of pulmonary inflammation, cellular proliferation, fibroproliferative effects, and subsequent inflammation-derived mutagenesis of epithelial cells, and this ultimately results in the development of lung tumors (Warheit, 1999).

13.2.2 Species Differences in Lung Responses to Inhaled Fine and Ultrafine TiO_2 Particles

Data from two recently completed subchronic inhalation toxicity studies have provided important insights into species differences in lung responses to inhaled pigment-grade and uf titanium dioxide particles. The studies provided a systematic comparison of lung responses of female rats, mice, and hamsters exposed for 13 weeks to fine TiO_2 particles at concentrations of 10, 50, or 250 mg/m^3, or to uf TiO_2 (Degussa P-25) particles at concentrations of 0.5, 2, or 10 mg/m^3. Following the end of exposures, animals were held for recovery periods of 4, 13, 26, or 52 weeks (46 or 49 weeks for particle-exposed hamsters) (Bermudez et al., 2002; 2004). The interspecies comparative results from the two studies were rather similar and indicated the following:

1. There existed rodent species differences in lung responses to inhaled fine-sized and uf titanium dioxide particles, with rats being the most sensitive.
2. The lung inflammatory responses of rats to uf TiO_2 particles were approximately five times greater (by mass) than the responses to fine-sized particles, as shown by rat lung responses and potency comparisons between pigment-grade or fine-sized (particle size ~ 300 nm) and uf (particle size = 15 to 40 nm) TiO_2 particles.

As indicated in the preceding text, with the exception of aerosol exposure concentrations, the experimental designs of the two interspecies studies were nearly identical. In the 90-d uf-TiO_2 study, female rats, mice, and hamsters were exposed by inhalation exposure to 0.5, 2.0, or 10 mg/m^3 uf TiO_2 particles (10 to 40 nm in diameter for 13 weeks (6 h/d, 5 d/week). Following the end of the exposure period, lungs of animals were assessed immediately after, as well as 4, 13, 26, or 52 weeks (49 weeks for the uf-TiO_2-exposed hamsters) postexposure. At each time point, uf-TiO_2 burdens in the lung and lymph nodes concomitant with biomarkers of lung injury, including inflammation, cytotoxicity, lung cell proliferation, and histopathological alterations, were assessed. Results demonstrated that mice and rats had similar retained-particle lung burdens at the end of the 13-week exposures, whereas hamsters had retained-lung burdens that were significantly lower. Lung burdens in all three species decreased with increasing transit time following completion of exposure, and 1 yr later, the percentages of the lung particle burden remaining in the high-dose exposure (10 mg/m^3) group were 57, 45, and 3% for rat, mouse, and hamster, respectively. The significant reduction of particle clearance from the lungs of mice and rats of the 10-mg/m^3 exposure groups were evidence of pulmonary particle overload. Significant pulmonary inflammation was evident, particularly in rats, as well as in mice exposed to 10 mg/m^3, concomitant with increased concentrations of soluble biomarkers of cell injury in bronchoalveolar lavage (BAL) fluids. The initial neutrophil response in rats was substantially greater than in mice and, in contrast, the response of hamsters was minimal. At the highest exposure concentration of 10 mg/m^3, BAL fluid biomarkers of lung injury (e.g., lactate dehydrogenase [LDH] and protein) were highest in rats > mice > hamsters and diminished with time postexposure. It was noteworthy, but not surprising, that progressive epithelial and fibroproliferative alterations were observed in the lungs of rats but not mice or hamsters. These fibroproliferative lesions were characterized by foci of alveolar epithelial proliferation of metaplastic epithelial cells or alveolar bronchiolization surrounding areas of heavily particle-laden macrophages. Additional observations in the lungs of rats included interstitial particle accumulation and alveolar septal fibrosis. All of these particle-overload-related lesions in the rat became more pronounced with increasing time postexposure. The investigators concluded that significant species differences existed in the pulmonary responses to inhaled uf TiO_2 particles. Under conditions of equivalent particle lung burdens, rats developed a more severe inflammatory response than mice, which subsequently resulted in progressive epithelial and fibroproliferative changes. These findings were consistent with the results of a companion study using inhaled pigment-grade (fine mode) TiO_2 (Bermudez et al., 2002). These species differences can be explained both by pulmonary responses and by particle dosimetry differences between these rodent species (Bermudez et al., 2004).

13.2.3 Other Factors and Conflicting Results

13.2.3.1 Role of Particle Size—Studies with Different Samples of Nanoscale TiO_2 Particles

In a further complication of the nanoparticle toxicity scenario, the results of recent pulmonary bioassay studies in rats suggest that, on a mass basis, not all nanoparticle types are more toxic or inflammogenic compared to fine-sized particles of similar chemical composition. As mentioned previously, the limited numbers of studies that have been reported suggest that uf (P-25) TiO_2 particles produced greater pulmonary inflammation when compared with fine-sized TiO_2 particles. However, in contrast with the conclusions of the earlier studies with P-25-type uf TiO_2 particles, the results of recent preliminary studies comparing the effects of nano- vs. fine-sized particles have indicated that pulmonary exposures in rats to uncoated TiO_2 nanoscale rods (200-nm lengths × 30-nm diameters) and TiO_2 nanoscale dots (particle size < 30 nm) did not produce enhanced lung inflammation in rats when compared to fine-sized TiO_2 particle exposures (particle size ~ 270 nm) at similar mass doses. These studies are currently being repeated.

As discussed in the section "Introduction and General Background," biological interactions between particle surfaces and cells may be more important in determining toxicity than the diameter or particle size of the particulate *per se*. With regard to surface coatings, it has also been recently reported that, in general, hydrophobic TiO_2 seems to be less inflammatory than naive TiO_2, regardless of particle size (Hohr et al., 2002; Rehn et al., 2003; Warheit et al., 2003a).

13.2.3.2 Studies with Nanoquartz Particles

Using a similar pulmonary bioassay protocol, lung bioassay studies have compared the toxicity effects in rats of intratracheally instilled, uncoated nanoscale quartz particles (50 nm) vs. fine-sized quartz particles (particle size ~ 1600 nm). In the first study, at equivalent mass doses, the nanoquartz particles produced less intense and sustained pulmonary inflammatory and cytotoxic responses when compared to the effects produced by the Min-U-Sil quartz particles (Warheit et al., 2005a). These preliminary findings are intriguing because crystalline quartz silica particles are classified as a category-one human carcinogen by the International Agency for Research on Cancer (IARC). A second study is in progress comparing the pulmonary effects of smaller 10-nm nanoquartz particles, fine-sized quartz particles (400 to 500 nm), and Min-U-Sil quartz particles (1600 nm as described earlier). In this ongoing study, preliminary findings have indicated that the lung effects of nanoquartz were equal to or greater than the Min-U-Sil effects. Thus, it will be interesting to reconcile the conclusions of the two studies with nanoquartz particles.

To summarize this section, a number of factors are likely to influence the pulmonary toxicity of nanoparticles. These include the following:

- 1. Particle number and size
- 2. Surface dose
- 3. Surface coatings on particles, particularly for engineered nanoparticulates (Warheit et al., 2003b)
- 4. The degree to which ambient UFP "age" and become aggregates or engineered nanoparticles aggregate/agglomerate owing, in large part, to surface characteristics
- 5. Surface charges on particles, as well as particle shape and electrostatic attraction potential, as is the case for engineered SWCNT, which readily agglomerate (Warheit et al., 2004; Maynard et al., 2004)
- 6. Method of particle synthesis (i.e., whether formed by gas-phase (fumed) or liquid-phase (colloidal/precipitated) synthesis and postsynthetic modifications

13.3 PULMONARY BIOASSAY STUDY WITH CARBON NANOTUBES

One example of a factor that may influence the toxicity/hazard assessment and corresponding risk of engineered nanoparticulates is the electrostatic attraction/aggregation or agglomeration potential of some nanoscale

Table 13.1 Experimental Groups for SWCNT Pulmonary Bioassay Study

Group 1	Phosphate buffered saline (PBS)
Group 2	PBS + 1% Tween 80
Group 3	Single-wall carbon nanotubes (SWCNT) (A & B)
Group 4	Crystalline silica particles (quartz) (A&B) (positive particle control)
Group 5	Carbonyl iron particles (A&B) (negative particle control)

Note: Intratracheal instillation exposures [1 (A) and 5 (B) mg/kg (+ Tween)] in rats.

Source: From Warheit, D.B., Laurence, B.R., Reed, K.L., Roach, D.H., Reynolds, G.A.M., and Webb, T.R. Lung toxicity bioassay study in rats with single wall carbon nanotubes. In press, *J Am Chem Soc* 2004b. With permission.

materials, and in particular, SWCNT (Single wall carbon nanotubes). Two pulmonary bioassay studies in mice (Lam et al., 2004) and in rats (Warheit et al., 2004) with SWCNT were recently reported. Individual SWCNT have diameters of 1 nm and lengths greater than 1 μm. However, SWCNT rarely, if ever, exist as discrete individual particles, and because of their strong electrostatic attraction, form agglomerates of "nanoropes" or "nanomats" consisting of 10 to 200 individual SWCNT (Warheit et al., 2004).

In a recent study, rats were intratracheally instilled with multiple doses of SWCNT, quartz particles (positive control), or carbonyl iron particles (negative control) (Table 13.1 and Table 13.2). This study was designed as a hazard screen to assess whether SWCNT exposure produces significant toxicity in the lungs of rats and to compare the pulmonary effects with a low- and high-toxicity particulate samples, using a well-developed, short-term lung bioassay (Table 13.3 and Table 13.4). Bridging studies can be important in providing an inexpensive preliminary safety screen when assessing the hazards of a variety of new developmental compounds or when making small modifications to an existing chemical product (Figure 13.1). The strength of the bridging technique depends on the availability of good inhalation toxicity data on one of the compounds. The particulate for which inhalation data exist can then be used as a control material for an intratracheal instillation bridging study (see Figure 13.1). The basic concept of the bridging assessment is that the lung effects of the instilled material, which serve as a control (known) material, are "bridged" on the one hand to the inhalation toxicity data for that material, and on the other hand, to the new materials being tested. The

Table 13.2 Protocol for Carbon Nanotube Bioassay Study

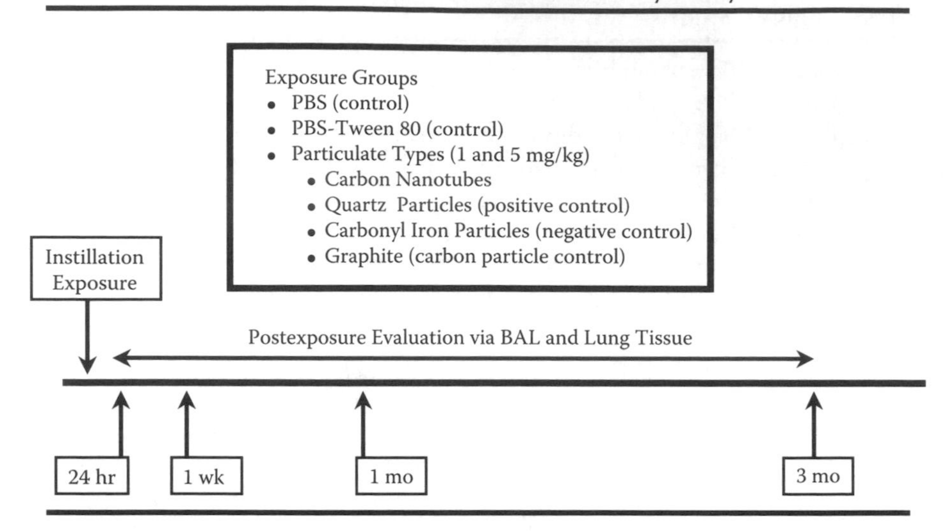

Source: From Warheit, D.B., Laurence, B.R., Reed, K.L., Roach, D.H., Reynolds, G.A.M., and Webb, T.R. Lung toxicity bioassay study in rats with single wall carbon nanotubes. In press, *J Am Chem Soc* 2005b. With permission.

results of bridging studies in rats are then useful as preliminary lung toxicity screening (i.e., hazard) data, because consistency in the response of the inhaled and instilled control material serves to validate the responses with the newly tested dust material.

Exposures to high-dose (5 mg/kg) SWCNT produced mortality in ~15% of the instilled rats within 24 h postinstillation exposure. This mortality resulted from mechanical blockage of the large airways by the instilled aggregate, and not from toxicity *per se* (Figure 13.2).

Table 13.3 Pulmonary Bioassay Studies

Working hypothesis
Four factors influence the development of pulmonary fibrosis
inhaled materials that cause cell/lung injury
inhaled materials that promote ongoing inflammation
inhaled materials that reduce alveolar macrophage function
inhaled materials that persist in the lung

Table 13.4 Pulmonary Bioassay Components

Bronchoalveolar lavage assessments
Lung inflammation and cytotoxicity
Cell differential analysis
BAL fluid lactate dehydrogenase (cytotoxicity)
BAL fluid alkaline phosphatase (epithelial cell toxicity)
BAL fluid protein (lung permeability)
Lung tissue analysis
Lung weights
Lung cell proliferation (BrdU)
Parenchymal
Airway
Lung histopathology

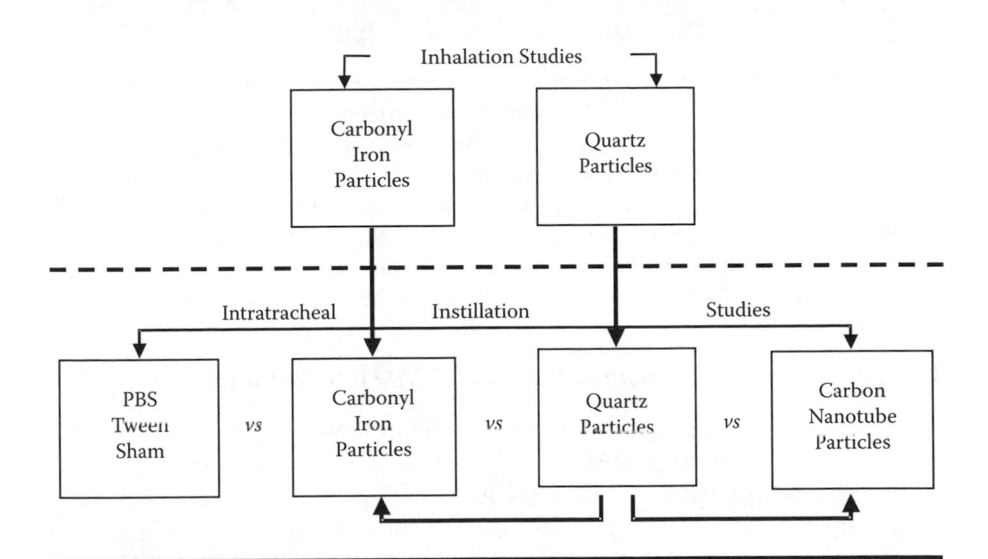

Figure 13.1 Schematic demonstrating the strategy for conducting pulmonary bioassay bridging studies. Bridging studies can have utility in providing an inexpensive preliminary safety screen when evaluating the hazards of new developmental compounds. The basic idea of the bridging concept is that the effects of the instilled material serve as a control (known) material and then are "bridged" on the one hand to the inhalation toxicity data for that material, and on the other hand, to the new materials being tested. (Reproduced from Warheit, D.B., Laurence, B.R., Reed, K.L., Roach, D.H., Reynolds, G.A.M., and Webb, T.R. Lung toxicity bioassay study in rats with single wall carbon nanotubes. In press, *J Am Chem Soc* 2005b. With permission.)

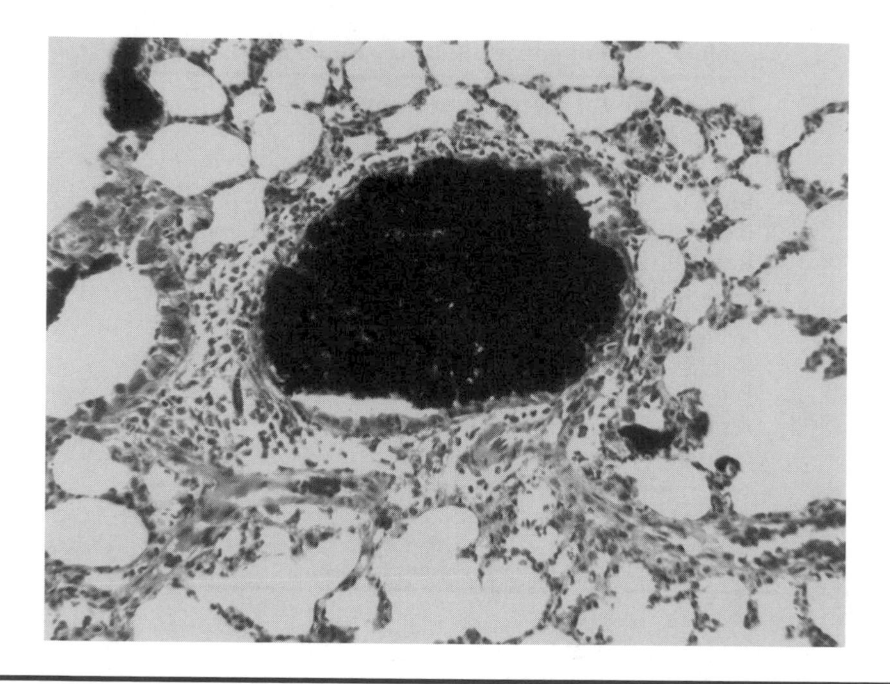

Figure 13.2 Light micrograph of lung tissue from a rat exposed to 5-mg/kg carbon nanotubes (CNT) (a few hours after exposure). The major airways are mechanically blocked by the CNT instillate. This led to suffocation in 15% of the CNT-exposed rats and was not evidence of inherent pulmonary toxicity of CNT. (Reproduced from Warheit, D.B., Laurence, B.R., Reed, K.L., Roach, D.H., Reynolds, G.A.M., and Webb, T.R., Lung toxicity bioassay study in rats with single wall carbon nanotubes. In press, *J Am Chem Soc* 2005b. With permission.)

13.3.1 Pulmonary Inflammation and BAL Fluid Parameters

Results from the BAL fluid biomarker studies demonstrated that the numbers of cells recovered by BAL from the lungs of high-dose quartz-exposed (5 mg/kg) groups were significantly higher than any of the other groups for all postexposure time periods. Intratracheal instillation exposures to quartz particles (1 and 5 mg/kg) produced persistent lung inflammation, as measured through 3 months postexposure; in contrast, instillation of Carbonyl iron (CI) and SWCNT resulted in a transient lung inflammatory response, as demonstrated by enhanced percentages of BAL-recovered neutrophils, measured at 24 h postexposure (Figure 13.3).

Transient increases in BAL fluid LDH values were enhanced in BAL fluids of high-dose (5 mg/kg) SWCNT-exposed rats at 24 h postexposure, but were not sustained. In contrast, exposures to 5-mg/kg quartz particles produced a persistent increase in lung fluid LDH values compared to controls through the 3-month postexposure period (Figure 13.4). Similar

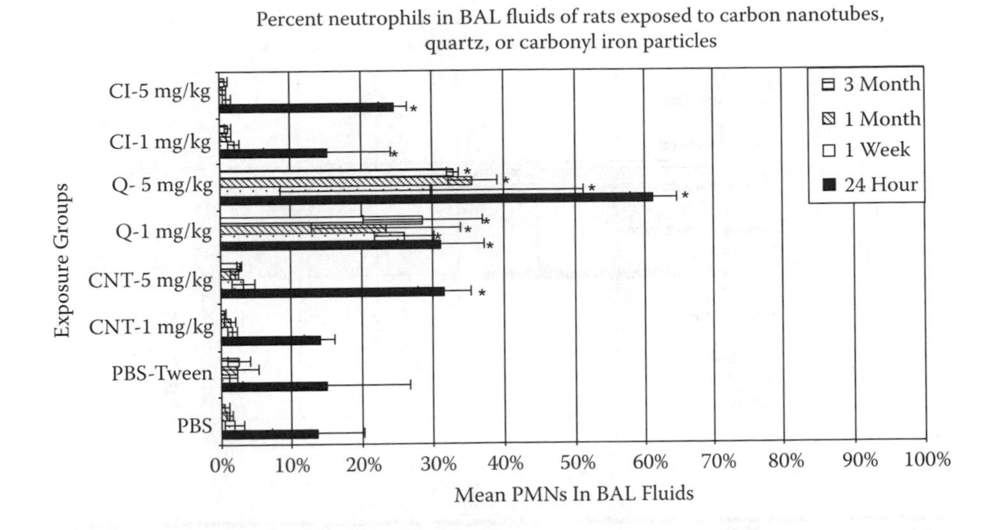

Figure 13.3 Pulmonary inflammation in particulate-exposed rats and controls as evidenced by percentage of neutrophils (PMN) in BAL fluids at 24 h, 1 week, 1 month, and 3 months postexposure (pe). Instillation exposures resulted in transient inflammatory responses for nearly all groups at 24 h pe. However, exposures to quartz particles at 1 and 5 mg/kg produced a sustained lung inflammatory response. $^*p < .05$. (Reproduced from Warheit, D.B., Laurence, B.R., Reed, K.L., Roach, D.H., Reynolds, G.A.M., and Webb, T.R., Lung toxicity bioassay study in rats with single wall carbon nanotubes. In press, *J Am Chem Soc* 2005b. With permission.)

BAL fluid microprotein results were measured in rats exposed to quartz particles, indicating a sustained increase at 24 h, 1 month, and 3 months postexposure, whereas transient increases in BAL fluid microprotein values were measured in the lung fluids recovered from high-dose (5 mg/kg) SWCNT-exposed rats at 24 h postexposure (Figure 13.5).

13.4 CELL PROLIFERATION AND LUNG HISTOPATHOLOGY

The BAL and cell proliferation results demonstrated that lung exposures to quartz particles produced significant increases as compared to controls in pulmonary inflammation, cytotoxicity, and lung parenchymal cell proliferation indices, whereas exposures to SWCNT produced only transient lung inflammation (Figure 13.6). Histopathological analyses revealed that quartz particles produced inflammation, foamy alveolar macrophage accumulation, and tissue thickening. In contrast, pulmonary exposures to

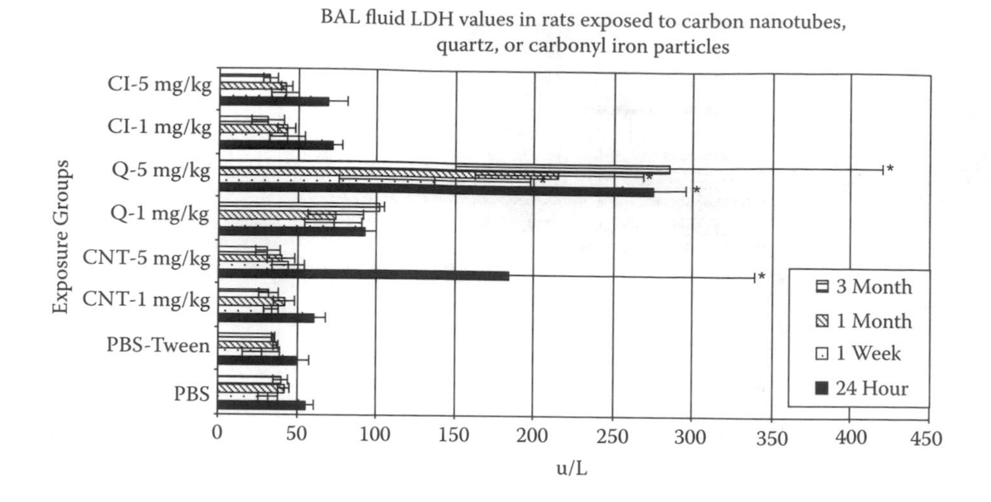

Figure 13.4 BAL fluid LDH values for particulate-exposed rats and corresponding controls at 24 h, 1 week, 1 month, and 3 months postexposure (pe). Significant increases in BALF LDH vs. controls were measured in the CNT 5-mg/kg exposed group at 24 h pe and the 5-mg/kg quartz-exposed animals at all four time periods pe. *$p < .05$. (Reproduced from Warheit, D.B., Laurence, B.R., Reed, K.L., Roach, D.H., Reynolds, G.A.M., and Webb, T.R., Lung toxicity bioassay study in rats with single wall carbon nanotubes. In press, *J Am Chem Soc* 2005b. With permission.)

SWCNT produced a non-dose-dependent series of multifocal granulomas, which was evidence of a foreign tissue body reaction (Figure 13.7). In the center of each of the granulomas were agglomerated carbon nanotubes surrounded by mononuclear cell types (Warheit et al., 2004). Similar findings were reported by Lam et al. (2004) in SWCNT-exposed mice. It is noteworthy that, unlike the results with quartz particles, the finding of unusual pulmonary lesions (i.e., multifocal granulomas) in rats was not consistent with the following: enhanced cell proliferation indices, sustained lung inflammation, a dose–response relationship, and a uniform and progressive distribution of lesions (Warheit et al., 2004). In addition, the results of two recent exposure assessment studies indicate very low respirable (Maynard et al., 2004) aerosol SWCNT concentration exposures at the workplace. Thus, the physiological relevance of these findings remains to be determined by conducting an inhalation toxicity study in rats with aerosols of SWCNT. It is also important to note here that SWCNT, owing to their unique electrostatic characteristics, do not appear to be representative of other nanoscale particulates.

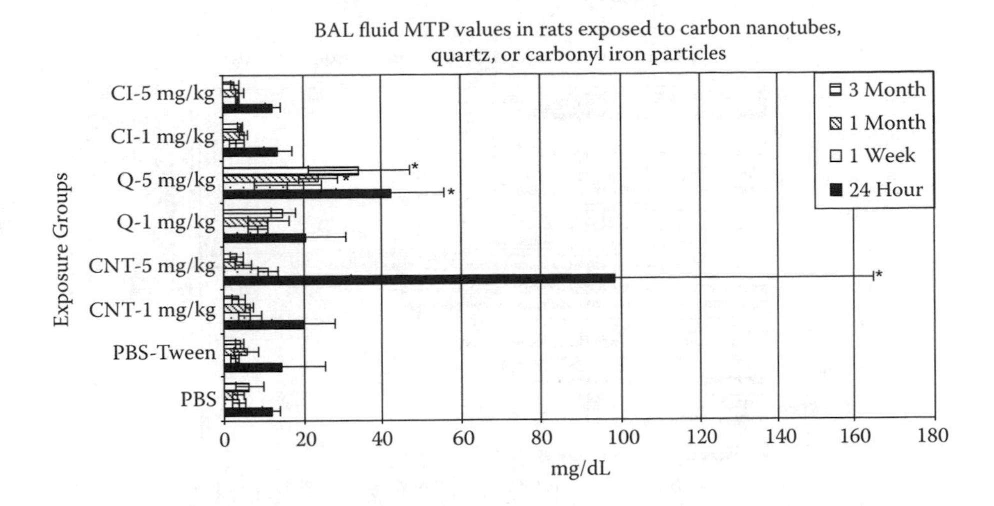

Figure 13.5 BAL fluid protein (MTP) values for particulate-exposed rats and corresponding controls at 24 h, 1 week, 1 month, and 3 months postexposure (pe). Significant increases in BALF MTP vs. controls were measured in the CNT 5-mg/kg exposed group at 24 h pe and the 5-mg/kg quartz-exposed animals at three time periods pe. * $p < .05$. (Reproduced from Warheit, D.B., Laurence, B.R., Reed, K.L., Roach, D.H., Reynolds, G.A.M., and Webb, T.R., Lung toxicity bioassay study in rats with single wall carbon nanotubes. In press, *J Am Chem Soc* 2005b. With permission.)

13.4.1 Safe Handling of Nanomaterials in the Laboratory

As discussed in the section "Introduction and General Background," engineered nanoscale particles generally are defined as particulates in the size range less than 100 nm. Having been designated as "the next big thing" in science and the foundation for the "next industrial revolution," governments and companies around the world are providing billions of dollars for nanotechnology-related research efforts. Because of the emerging and eclectic nature of this exciting technology, the aforementioned significant expenditures on research activities, the popularity of science fiction, and the substantial public relations hype as evidenced by overly robust revenue projections, nanotechnology has attracted the attention of the public and others. This has led to concerns regarding potential adverse human health effects as well as environmental impacts of the development and use of nanomaterials. In this regard, it is of concern that many researchers are handling nanomaterials in the laboratory.

Given the paucity of safety data on health risks associated with exposures to nanoparticles, a prudent product stewardship tool or safe-handling

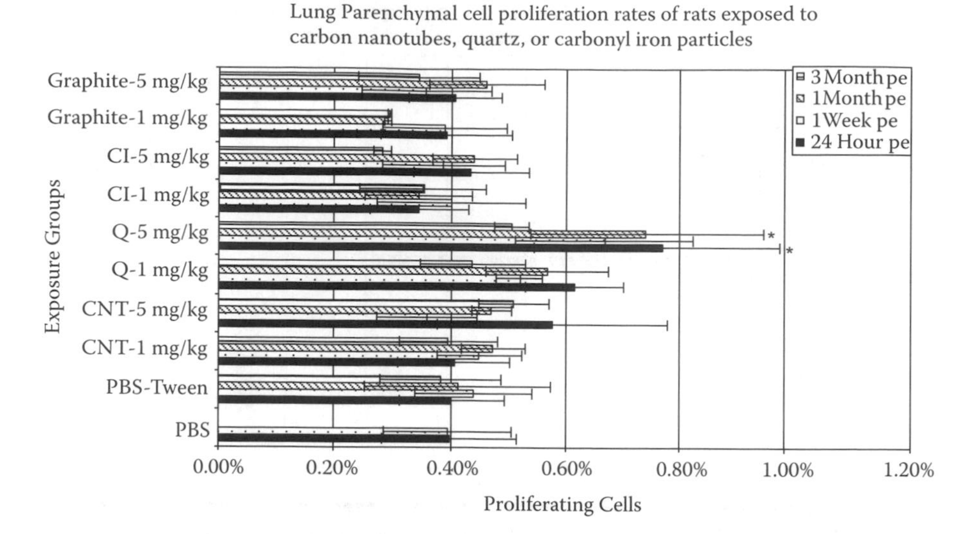

Figure 13.6 Lung parenchymal cell proliferation rates (BrdU) in particulate-exposed rats and corresponding controls at 24 h, 1 week, 1 month, and 3 months postexposure (pe). Significant increases in BrdU immunostained cells were measured in the 5-mg/kg quartz-exposed rats vs. controls at 24 h and 1 month pe. $p < .05$. (Reproduced from Warheit, D.B., Laurence, B.R., Reed, K.L., Roach, D.H., Reynolds, G.A.M., and Webb, T.R., Lung toxicity bioassay study in rats with single wall carbon nanotubes. In press, *J Am Chem Soc* 2005b. With permission.)

methodology for nanotechnology would be based on the general concept that health risk is a function of hazard assessment and exposure assessment (Table 13.5). With regard to the hazard assessment tool, some of the initial questions that should be addressed include the following (Table 13.6):

1. What is the presumed exposure? The four major routes of occupational exposure are the respiratory tract (i.e., inhalation exposure), the skin, eyes, and the gastrointestinal tract (via oral or inhalation exposures).
2. Which route of exposure predominates? This should be the first concern, because personal protective clothing can often prevent dermal and eye exposures in many cases.

In the absence of adequate toxicological data on the nanoparticle type of interest, it is important to ascertain whether any hazard information is available on the micro- or macroscale chemical product. This may provide some initial clues on the potential toxicity of the nanoparticle type, but must be followed up by relevant hazard studies with the nanomaterial.

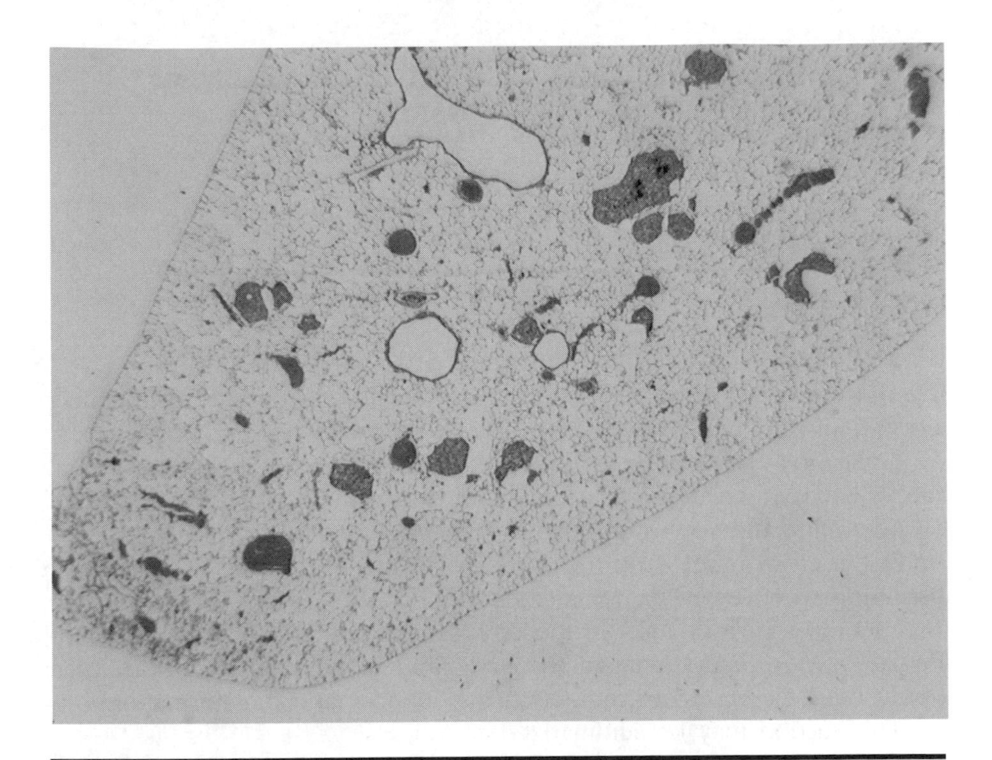

Figure 13.7 Low-magnification micrograph of lung tissue from a rat exposed to single-wall carbon nanotubes (1 mg/kg) at 1 month postinstillation. Note the diffuse pattern of granulomatous lesions (arrows). It was interesting to note that few, if any, lesions existed in some lobes, whereas other lobes contained several granulomatous lesions, likely due to the nonuniform deposition pattern following carbon nanotube instillation. Magnification = × 20. (Reproduced from Warheit, D.B., Laurence, B.R., Reed, K.L., Roach, D.H., Reynolds, G.A.M., and Webb, T.R., Lung toxicity bioassay study in rats with single wall carbon nanotubes. In press, *J Am Chem Soc* 2005b. With permission.)

Adequate pulmonary, dermal, and oral toxicity tests/bioassays are available that can provide important hazard information on the nanoparticle type of interest. Exposure assessment determinations are also an important tool for determining health risks (Table 13.7). However, current methodologies

Table 13.5 Product Stewardship/Safe-Handling Tool for Nanotechnology

General concept
Risk is a function of hazard evaluation and exposure assessment

Table 13.6 Hazard/Risk Assessment Tool

What the presumed exposure is, e.g., powder vs. aqueous suspension
Life cycle issues
What information is available on the macro- or microscale chemical/product

are not designed for quantifying nanoparticulate exposures and thus may not be sufficiently sensitive; these should therefore be validated.

Similar to health-based studies, current information on the environmental fate of nanoparticulates is very limited. In future studies, it will be important to determine the routes through which engineered nanoscale particles enter the environment and the modes of dispersion in the environment, assessing whether the nanomaterials are persistent and bioaccumulative or undergo transformation in the environment. An estimate of nanoparticulate exposure potential in the environment will require information about releases, emissions, transport, distribution, and transformation.

This section may be summarized as follows:

- Evaluating the potential hazards and exposure to engineered nanoscale particulates is an emerging area in toxicology and health risk assessment.
- A related issue involves the extent to which nanoparticle toxicity evaluations can be extrapolated from existing toxicology databases for bulk material particulates of identical chemical composition.
- Generally, it cannot be assumed that nanomaterials have the same toxicity profiles as their bulk counterparts.
- A prudent product stewardship tool for nanotechnology would be based on the general concept that health risk is a function of hazard and exposure.

Table 13.7 Inhalation Hazards

Start with a literature review
1. Pulmonary screen bioassay methodology
2. As product becomes more successful, or if already in widespread use, recommended short-term to subchronic inhalation studies

- There is a paucity of hazard information on engineered nanoscale particulates.
- With regard to the hazard assessment tool, one of the initial questions that should be addressed is: What is the presumed exposure? This information should be utilized to prioritize testing strategies.
- Adequate pulmonary, dermal, and oral toxicity tests/bioassays are available that can provide useful hazard information on nanoparticulates.
- With regard to environmental fate, estimates of nanoparticulate exposure will require information about releases, emissions, transport, distribution, and transformation.

13.5 REGULATORY IMPLICATIONS

Regulatory decisions regarding nanoparticulates have not been finalized. This is because many of the nanoparticle types have been regulated under bulk-material criteria, and up to this point have not been considered to be "new" materials. However, recent reports indicate that the United States Environmental Protection Agency (USEPA) has been considering development of a pilot program in which companies making nanoscale materials with existing bulk chemical would voluntarily submit to the agency information regarding the nanoscale developmental compounds. This program could be developed as a component of broader agency efforts to develop a comprehensive strategy to deal with nanoscale materials (Phibbs, 2004).

REFERENCES

Bermudez, E., Mangum, J.B., Asgharian, B., Wong, B.A., Reverdy, E.E., Janszen, D.B., Hext, P.M., Warheit, D.B., Everitt, J.I. Long-term pulmonary responses of three laboratory rodent species to subchronic inhalation of pigmentary titanium dioxide particles. *Toxicol Sci* November 2002, 70(1): 86–97.

Bermudez, E., Mangum, J.B., Wong, B.A., Asgharian, B., Hext, P.M., Warheit, D.B., Everitt, J.I. Pulmonary responses of mice, rats, and hamsters to subchronic inhalation of ultrafine titanium dioxide particles. *Toxicol Sci* February 2004, 77(2): 347–357.

Borm, P.J.A. and Kreyling, W. Toxicological hazards of inhaled nanoparticles—potential implications for drug delivery. *J Nanosci Nanotechnol* 2004, in press.

Born, P.J.A. and Kreyling, W. Toxicological hazards of inhaled nanoparticles—potential implications for drug delivery. *J Nanosci Nanotechnol* 2004, 4: 521–531.

Colvin, V.L. The potential environmental impact of engineered nanomaterials. *Nat Biotechnol* October 2003, 21(10): 1166–1170.

Donaldson, K., Stone, V., Clouter, A., Renwick, L., MacNee, W. Ultrafine particles. *Occup Environ Med* 2001, 58: 211–216.

Ferin, J., Oberdorster, G., Penney, D.P. Pulmonary retention of ultrafine and fine particles in rats. *Am J Respir Cell Mol Biol* 1992, 6: 535–542.

Heinrich, U., Fuhst, R., Rittinghausen, S., Creutzenberg, O., Bellmann, B., Koch, W., Levsen, K. Chronic inhalation exposure of Wistar rats and two different strains of mice to diesel engine exhaust, carbon black, and titanium dioxide. *Inhal Toxicol* 1995, 7: 533–556.

Hext, P.M. Current perspectives on particulate-induced pulmonary tumours. *Hum Exp Toxicol* October 1994, 13(10): 700–715. Review.

Hohr, D., Steinfartz, Y., Schins, R.P.F., Knaapen, A.M., Martra, G., Fubini, B., and Borm, P.J.A. The surface area rather than the surface coating determines the acute inflammatory response after instillation of fine and ultrafine TiO2 in the rat. *Int J Hygiene Environ Health* 2002, 205: 239–244.

Lam, C.W., James, J.T., McCluskey, R., Hunter, R.L. Pulmonary toxicity of single wall carbon nanotubes in mice 7 and 90 days after intratracheal instillation. *Toxicol Sci* January 2004, 77(1): 126–134.

Lee, K.P., Trochimomicz, H.J., Reinhardt, C.F. Pulmonary response of rats exposed to titanium dioxide (TiO_2) by inhalation for 2 years. *Toxicol Appl Pharmacol* 1985, 79: 179–192.

Li, X.Y., Gilmour, P.S., Donaldson, K., MacNee, W. Free radical activity and pro-inflammatory effects of particulate air pollution (PM10) in vivo and in vitro. *Thorax* 1996, 51: 1216–1222.

Maynard, A.D., Baron, P.A., Foley, M., Shvedova, A.A., Kisin, E.R., Castranova, V. Exposure to carbon nanotube material: aerosol release during the handling of unrefined single-walled carbon nanotube material. *J Toxicol Environ Health* Part A, 2004, 67: 87–107.

Oberdorster, G., Ferin, J., Lehnert, B.E. Correlation between particle size, in vivo particle persistence, and lung injury. *Environ Health Perspect* 1994, 102 (Suppl. 5): 173–179.

Oberdorster, G. Toxicology of ultrafine particles: in vivo studies. *Phil Trans R Soc London A* 2000, 358: 2719–2740.

Phibbs, P. Agency considers nanoscale pilot program involving voluntary submission of information. *BNA Daily*, October 5, 2004.

Rehn, B., Seiler, F., Rehn, S, Bruch, J., and Maier, M. Investigations on the inflammatory and genotoxic lung effects of two types of titanium dioxide: untreated and surface treated. *Toxicol Appl Pharmacol* 2003, 189: 84–95.

Warheit, D.B., Reed, K.L., and Webb, T.R. Development of pulmonary bridging studies: pulmonary toxicity studies in rats with triethoxyoctylsilane (OTES)-coated, pigment-grade titanium dioxide particles. *Exp Lung Res* 29: 593–606, 2003a.

Warheit, D.B. Mechanistic Considerations for Toxicity Assessments: In Vivo. Report to the Medical Research Council at the IEH (Institute for Environment and Health, U.K.). IEH Report on Approaches to Predicting Toxicity from Occupational Exposure to Dusts. Report R11. Page Bros. Norwich, U.K. 68–77, 1999.

Warheit, D.B. Pulmonary toxicity of occupational and environmental exposures to fibers and nano-sized particulate. In *Air Pollutants and the Respiratory Tract.* W.M. Foster and D. Costa, Eds., in press, 2004.

Warheit, D.B., Webb, T.R., and Reed, K.L. Pulmonary toxicity studies with TiO_2 particles containing various commercial coatings. *Toxicologist* 2003b, 72(1), p. 298A.

Warheit, D.B., Laurence, B.R., Reed, K.L., Roach, D.H., Reynolds, G.A., Webb, T.R. Comparative pulmonary toxicity assessment of single wall carbon nanotubes in rats. *Toxicol Sci* 2004, 77: 117–125.

Warheit, D.B., Laurence, B.R., Reed, K.L., Roach, D.H., Reynolds, G.A.M., and Webb, T.R. Lung toxicity bioassay study in rats with single wall carbon nanotubes. In press, *J Am Chem Soc* 2004b.

Warheit, D.B., Laurence, B.R., Reed, K.L., Roach, D.H., Reynolds, G.A.M., and Webb, T.R. Comparative pulmonary toxicity assessment of single wall carbon nanotubes in rats. *Toxicol Sci*, 2004, 77: 117–125.

Warheit, D.B., Pulmonary toxicity of occupational and environmental exposures to fibers and nano-sized particulate. In: *Air Pollutants and the Respiratory Tract*, Foster, W.M. and Costa, D. Eds. second edition, *Lung Biology in Health and Disease*, Vol 204; Taylor and Francis, Atlanta, 2005a, 303–327.

Warheit, D.B., Laurence, B.R., Reed, K.L., Roach, D.H., Reynolds, G.A.M., and Webb, T.R. Lung toxicity bioassay study in rats with single wall carbon nanotubes. In: American Chemical Socieyty Volume: Nanotechnology and the Environment: Applications and Implications. Karn, B., Masciangioli, T., Zhang, W., Colvin, V. and Alivisatos, P. ACS Symposium Series 890, American Chemical Society Washington, D.C. 2005b, 67–90.

14

DIESEL EXHAUST AND VIRAL INFECTIONS

Ilona Jaspers

CONTENTS

14.1 INTRODUCTION

Over the past several years, interest in the adverse effects induced by exposure to diesel exhaust (DE) has grown, at least partially owing to the findings suggesting a link between exposure to DE and enhanced susceptibility to developing asthma as well as exacerbation of existing asthma (Diaz-Sanchez et al., 2003; Nel et al., 2001; Pandya et al., 2002). Many studies have demonstrated that exposure to DE adversely affects the immune system by shifting the immune responses to those of a more allergic phenotype (Diaz-Sanchez, 1997; Wang et al., 1999). Although these findings are extremely important, little work has been done to examine the effects of DE on respiratory virus infections, which can also exacerbate asthma symptoms. This chapter will review the epidemiological findings linking particulate matter (PM) and respiratory virus infections and discuss potential interactions between DE and viral infections that have been shown in controlled experiments. Because, thus far, there have been only a few studies showing a direct link between exposure to DE and respiratory virus infections, some of the potential interactions are hypothetical, based on preliminary evidence or extrapolated from studies using other sources of particles, such as ambient PM or carbon black (CB) particles.

14.2 SIGNIFICANCE OF INTERACTIONS

Acute respiratory infections (ARIs) continue to be among the most important causes of morbidity and mortality in children and the elderly (>65 yr). The viruses primarily associated with respiratory infections include picornaviruses, coronaviruses, adenoviruses, parainfluenza viruses, influenza viruses, and the respiratory syncytial virus (RSV). Despite large-scale vaccination efforts and hygiene awareness, viral infections of the respiratory tract remain an enormous public health problem all over the world. According to the WHO, ARIs are the most common cause of death in children below 5 yr in developing countries. Children suffer from frequent ARI, most of which are self-limiting; but ARI can also be associated with pneumonia and, thus, become life threatening. Most ARIs are caused by RSV, influenza, parainfluenza, or adenovirus. In addition, respiratory virus infections are associated with the development and exacerbation of asthma, thus enhancing the occurrence and severity of other respiratory diseases (Cohen and Castro, 2003; Gern, 2004; Peebles, 2004; Wilson, 2003; Yamaya and Sasaki, 2003). Although many of the respiratory virus infections resolve on their own, there are several risk factors that can enhance the severity and, thus, the morbidity associated with viral infections, making them potentially lethal. Among the intrinsic risk factors that could adversely affect the efficiency with which the host can overcome a viral infection are age and immune competence. Extrinsic factors, such as nutritional status of the host and the influences of exposure to tobacco smoke or air pollutants, such as DE, could also adversely affect respiratory virus infections. Because of the large number of people, especially children and the elderly, exposed to DE and infected with respiratory viruses every year, potential interactions between DE and viral infections that adversely affect the resolution of the infection have a significant public health impact.

14.3 EPIDEMIOLOGICAL EVIDENCE

The vast majority of currently available epidemiological data linking ambient particulate air pollution with respiratory virus infections and the associated morbidity focus on the association between PM levels and respiratory infections. DE particles (DEP) are thought to be a major contributor to ambient PM levels, contributing as much as 70% to the total (Weinhold, 2002). Therefore, we will review studies that have associated exposure to ambient PM levels with respiratory virus infections, assuming that some, if not most, of these effects are caused by DEP.

Epidemiological analyses have demonstrated that exposure to air pollution is significantly associated with increased respiratory morbidity and mortality (Pope, 2000; Zanobetti et al., 2000). Long-term exposure to PM air pollution is also associated with increased risk of cardiovascular mortality (Pope et al., 2004). Interestingly, this study also pointed out that while long-term exposure to PM contributes to the progression of cardiovascular disease, it exacerbates already existing respiratory disease. Furthermore, this epidemiological analysis of data collected by the American Cancer Society, which linked specific causes of mortality with air pollution data for metropolitan areas throughout the U.S., also demonstrated that in people who have never smoked, exposure to PM was significantly associated with increased risk of influenza and pneumonia (Pope et al., 2004).

A study examining the relationship between traffic-related air pollution and respiratory infections in a birth cohort comprising 2-yr-olds in the Netherlands found that ear, nose, and throat infections and flu or serious colds were positively associated with air pollution, although the results were not statistically significant (Brauer et al., 2002). However, the observed associations were robust and consistent with previous studies, which have also demonstrated the association between traffic-related air pollution and incidence of runny nose (van Vliet et al., 1997), which is most likely caused by respiratory virus infections. Specifically, this study found that respiratory symptoms, such as cough, wheeze, and runny nose were reported more often in children living within 100 m of the freeway (van Vliet et al., 1997). Furthermore, a study examining the correlation between emergency hospital admission and air pollution levels in Hong Kong showed that, among other pollutants, PM10 was significantly associated with admissions for influenza (Wong et al., 1999). This study also indicated that the strongest association between PM and influenza infections was seen during the winter months and that the elderly (>65 yr) were particularly at risk.

A study conducted in Finland comparing upper respiratory infections in children living in an area with low levels of air pollution to those of children living in an area of high-level air pollution demonstrated that in the highly polluted city, children had more respiratory infections during a 12-month period than those in two other less-polluted reference cities (Jaakkola et al., 1991). However, the air pollution in the highly polluted city was mainly derived from a cellulose pulp mill, a power plant, and a chemical plant and, though the PM levels were also higher in this area, the main differences with regard to air pollution were in the nitrogen oxide and hydrogen sulfite levels (Jaakkola et al., 1991).

Thus, taken together, there is epidemiological evidence that suggests a correlation between PM air pollution and increased incidence and severity of upper respiratory infections, especially in children and the elderly.

14.4 COMMON RESPIRATORY VIRUSES

14.4.1 RSV

RSV is an enveloped, nonsegmented, single-stranded RNA virus of the family of Paramyxoviridae, which infects about 50% of all infants within the first year of life and is therefore considered to be a major cause of lower respiratory tract infections in children. Outbreaks of RSV infections occur all over the world, and in temperate climate zones usually occur in the winter and in the spring. Data from the CDC suggest that 120,000 hospital admissions of infants and young children in the U.S. are caused annually by RSV. Although most RSV infections in infants resolve on their own, RSV-positive pneumonia or bronchiolitis can be very dangerous in very young children (Shay et al., 2001). In fact, 70% of the especially severe forms of bronchiolitis in infants can be attributed to RSV infections (Anderson and Heilman, 1995). Furthermore, morbidity associated with RSV infections in adults, especially the elderly, is being increasingly recognized (Falsey and Walsh, 2000; Mlinaric-Galinovic et al., 1996). Similar to infants and young children, RSV-associated pneumonia in the elderly has been estimated to result in tens of thousands of hospitalizations in the U.S. each year (Han et al., 1999). In addition, it is increasingly recognized that RSV infections are the major contributor to flu-like symptoms in otherwise healthy adults (Zambon, 2001). Specifically, RSV, together with influenza, is the most prevalent cause of flu-like symptoms in adolescents and adults (Zambon et al., 2001; Irmen and Kelleher 2000; Lina et al., 1996). Thus, RSV, besides being a major contributor to respiratory-related hospitalizations in infants and very young children, can also be a major health concern in adults. In the adult population, RSV infections can aggravate underlying conditions, especially cardiopulmonary diseases (Walsh et al., 1999), further adding to the importance of these in adults.

RSV primarily infects airway epithelial cells, which upon infection release a large repertoire of inflammatory cytokines and chemokines, which, in turn, trigger further inflammatory responses (Harris and Werling, 2003). Thus, epithelial cells, together with alveolar macrophages, play an important role in activating cellular immunity following RSV infections. Cell-mediated immune responses are important in the response to RSV and subsequent recovery from the infection, because subjects with compromised cell-mediated immune responsiveness develop an unusually severe form of RSV infection with prolonged proliferation of the virus and progressive pneumonia (Fishaut et al., 1980). Among the proinflammatory cytokines and chemokines released after RSV infections, which mediate airway inflammation and bronchial hyperresponsiveness, are interleukin (IL)-1, IL-6, IL-8, tumor necrosis factor alpha (TNF-α), macrophage-inflammatory protein (MIP)-1α, monocyte chemoattractant protein (MCP)-1, granulocyte-macrophage colony-stimulating factor (GM-CSF), RANTES (regulated upon activation, normal T cell expressed and secreted), IFN-β, and leukotrienes (Domachowske et al., 2001).

Two proteins found on the surface of RSV mediate attachment and entry into the host cells. The G protein of the RSV mediates attachment of the virus to its target epithelial cell, whereas the F protein facilitates fusion of the virus lipid membrane with the cellular lipid membrane, thus permitting insertion of the viral RNA into the host cell (Harris and Werling, 2003; Welliver, 2003). The formation of syncytia, a typical feature of RSV infection, is also caused by the F protein, which promotes fusion of uninfected host cells.

14.4.2 Influenza

Despite widespread immunization efforts and the introduction of antiviral substances, influenza continues to be a major threat to public health. Similar to RSV, influenza infections occur predominantly in the winter months. Each year, influenza virus infections result in about 20,000 deaths and over 100,000 hospitalizations in the U.S. alone (Brammer et al., 1997; Simonsen et al., 2003). The influenza virus belongs to the family of Orthomyxoviridae, which are classified into three types, A, B, and C, with influenza A being the most pathogenic one. The influenza virus is an enveloped, single-stranded, segmented RNA virus. Its genome consists of eight RNA segments, which encode ten viral proteins: envelope glycoproteins hemagglutinin (HA) and neuraminidase (NA), matrix protein (M1), nucleoprotein (NP), three polymerases (PB1, PB2, and PA), ion channel (M2), and nonstructural proteins (NS1 and NS2). Influenza viruses are categorized based on their two surface antigens, HA (H1–H15) and NA (N1–N9), of which influenza viruses able to infect humans are limited to H1, H2, H3, N1, and N2.

Influenza viruses infect and replicate in epithelial cells of the upper respiratory tract, but can also infect monocytes or macrophages and other leukocytes. Influenza HA attaches to sialic-acid-containing glycoproteins on cell surfaces, which function as receptors for the virus (Couceiro et al., 1993). Upon binding to sialic acid residues, the influenza virus is internalized into the host cell endosome via clathrin-dependent endocytosis. A drop in endosomal pH triggers the fusion of viral and endosomal membranes, which causes the liberation of influenza virus ribonucleoproteins (vRNP) complexes into the host cell cytoplasm. vRNP complexes are transported into the nucleus, where viral RNA polymerases (PB1, PB2, and PA) mediate viral RNA (mRNA and vRNA) synthesis, thus initiating viral RNA replication and transcription, translation of viral mRNA, and ultimately, assembly of *de novo* synthesized viral particles.

Infections with influenza A and B can result in serious diseases, such as pneumonia and encephalitis (Wang et al., 2003). Locally, infection with influenza results in destruction of host cells via cytolytic and apoptotic mechanisms. Apoptosis of virus-replicating cells is actually regarded as a mechanism by which the host cell attempts to contain the viral infection, preventing release of viral particles and thus, infection of neighboring cells (Lowy, 2003). Epithelial cells respond to influenza virus infections by synthesizing and releasing a number of proinflammatory cytokines and chemokines, which recruit and activate immune cells to aid in removing the virus and virus-infected cells, thus ultimately clearing the viral infection. Among the long list of proinflammatory cytokines and chemokines produced by influenza-infected cells are IL-6, IL-8, TNF-α, RANTES, and eotaxin (Adachi et al., 1997; Julkunen et al., 2000).

14.4.3 Rhinovirus

Respiratory infections due to rhinovirus usually cause symptoms associated with the common cold, such as rhinorrhea, nasal congestion, sore and scratchy throat, cough, and headache and, sometimes, fever and malaise

(Hayden, 2004). Respiratory tract complications, such as otitis media, acute sinusitis, and exacerbation of asthma are also associated with rhinovirus infections (Hayden, 2004). Rhinoviruses belong to the family of Picornaviridae, which consist of very small, positive single-stranded RNA viruses made up of a simple viral capsid and the RNA. There are over 100 serotypes of rhinoviruses, of which over 90% attach to the host cell via the intercellular adhesion molecule (ICAM-1). The incidence of rhinovirus infections is very high, with an estimated 1 billion colds occurring annually in the U.S. alone. Although substantial morbidity, medical costs, and drain on productivity, such as days missed from work or school, are associated with rhinovirus infections, these infections are usually not considered to be a serious threat to public health. Rhinovirus infections are most prevalent during early spring and fall in temperate climates, but can also occur during other seasons. The airway epithelial cells are the primary site for rhinovirus infection. The rhinovirus itself does not appear to cause cytotoxicity of infected cells, but rather, it is the host response to the virus that is responsible for the symptoms. Specifically, rhinovirus-infection-associated symptom severity increases and decreases as infection-induced secretion of proinflammatory mediators, such as IL-1β, IL-6, and IL-8, increases and decreases. Other proinflammatory mediators produced by rhinovirus-infected epithelial cells include RANTES and Gro α (van Kempen et al., 1999).

14.5 HOST DEFENSE MECHANISMS

Cells of the respiratory tract have developed a number of antiviral strategies aimed at clearing a viral infection. These antiviral strategies can be subdivided into intracellular and intercellular antiviral strategies. *Intracellular antiviral strategies* are the mechanisms by which the infected host cell by itself attempts to stop viral replication and rid itself of the virus. *Intercellular antiviral strategies* are the mechanisms by which other cells are recruited to the site of infection and help in clearing the infection. Antiviral strategies include apoptosis of infected cells, activation of the interferon system, production of defensins, toll-like receptor-mediated innate immune responses and release of inflammatory cytokines and chemokines, as well as the adaptive immune response.

14.5.1 Apoptosis

Apoptosis of virus-infected and virus-replicating host cells is generally regarded as an efficient mechanism of host defense and viral clearance, because it permits the infected host to dispose of virus-infected cells on a single-cell basis without inducing a large-scale inflammatory response. Virus-infected cells undergoing apoptosis are eventually removed by scavenger cells, such as macrophages and other PMNs, via phagocytosis, without releasing the viral particles into the surrounding environment. Although RSV-infected epithelial cells do express a number of apoptosis-associated genes, such as caspase 3, interferon regulatory factor-1, and CD95 (Fas) (Kotelkin et al., 2003), they do not show the typical features of apoptosis, such as membrane blebbing, chromosomal DNA fragmentation, or changes in nuclear morphology. This could be due to RSV-induced expression of the IEX-1L genes, which is an antiapoptotic gene conferring resistance to TNF-α-induced apoptosis (Domachowske et al., 2000). Influenza-virus-induced apoptosis has been observed *in vitro* (Lowy, 2003) and *in vivo* (Mori et al., 1995) and is thought to be mediated by the upregulation of Fas and Fas ligand (FasL) following infection (Fujimoto et al., 1998). Apoptosis in influenza can also be induced by latent TGF-β on the cell surface, which is cleaved by influenza neuraminidase into its active form (Lowy, 2003).

14.5.2 Interferons

There are two main types of interferons (IFN), type I and type II. Type I IFNs include IFN-α, IFN-β, IFN-ω, and IFN-τ, whereas type II IFN is IFN-γ. Most cell types can produce IFN-α and IFN-β; IFN-γ is only produced by certain immune cells, such as monocytes or macrophages, natural killer (NK) cells, CD4+ T helper 1 (Th1) cells, and CD8+ cytotoxic T cells (Samuel, 2001; Ronni et al., 1995). IFNs mediate their effect by interacting with specific membrane receptors, initiating complex signaling cascades that ultimately result in the transcription of numerous IFN-stimulated genes (ISGs). There may be hundreds of ISGs whose expression is either up- or downregulated in response to IFN. Among the best-studied ISGs are dsRNA-activated serine/threonine protein kinase (PKR), 2′5′-oligoadenylate synthetase (OAS) and RNaseL, MHC molecules, and interferon regulatory factors (IRFs) (Samuel, 2001; Ward and Samuel, 2002; Khabar et al., 2003). Together, the expression of these ISGs is aimed at turning off viral replication and enhancing immune responses to the viral infection, thus ultimately clearing the viral infection.

14.5.3 β-Defensins

Epithelial cells can also protect themselves against pathogens by secreting a number of antimicrobial peptides, such as β-defensins. The human β-defensin (hBD) family includes several members, with hBD-1 being constitutively expressed and hBD-2 being induced in epithelial cells upon stimulation with cytokines (Yang et al., 2002; Yang et al., 2001). The activities of hBD include mast cell degranulation (Niyonsaba et al., 2003), chemotactic activity for monocytes, immature dendritic cells, and memory T cells (Yang et al., 2001; Yang et al., 2002), thus potentially forming a link between innate and specific immunity to viral infections. Recent studies have shown that rhinovirus as well as dsRNA can enhance the levels of hBD-2 and hBD-3 in human bronchial epithelial cells (Duits et al., 2003; Proud et al., 2004).

14.5.4 Toll-Like Receptors

Toll-like receptors (TLRs) are a family of receptors that mediate innate immune responses against invading pathogens (Akira and Sato, 2003; Janssens and Beyaert, 2003; Takeda and Akira, 2004; Akira, 2004). Virus-derived dsRNA or high CpG content, molecular signatures of viral infections, can interact with TLR3 and TLR7, respectively, and trigger signaling cascades that culminate in the production of type I IFN and inflammatory cytokines and chemokines (Matsumoto et al., 2004; Heil et al., 2004; Lund et al., 2004; Diebold et al., 2004). Thus, TLR3 and TLR7 are increasingly recognized as playing an important role in mediating the innate immune response against viral infections.

14.5.5 Cytokines and Chemokines

As mentioned earlier, respiratory epithelial cells respond to virus infections by releasing a number of inflammatory cytokines and chemokines, which recruit and activate immune cells. For example, RSV-infected epithelial cells produce RANTES, MCP-1, IL-8, and MIP-1α, which may lead to the recruitment of monocytes and other inflammatory cells (Anderson et al., 1994; Harrison et al., 1999; Welliver, 2003). Influenza-infected epithelial cells release RANTES, MCP-1, and IL-8, and influenza-infected monocytes or macrophages secrete MIP-1α/β, RANTES, MCP-1, MCP-3, MIP-3α, and IP-10, all of which preferentially recruit monocytes to the site of infection (Adachi et al., 1997; Van Reeth, 2000). Influenza-infected monocytes or macrophages and epithelial cells also produce IL-1β, IL-6, and TNF-α, which do not directly contribute to the antiviral activity of the cells, but are involved in the enhanced expression of other cytokines and chemokines, such as MCP-1 and MCP-3 and, thus, contribute to the maturation of tissue macrophages and dendritic cells. Furthermore, type I IFNs, which by themselves exert antiviral effects, can also, in combination with IL-18, enhance NK and T cell IFN-γ production and the development of a Th1-type immune response, including upregulation of IL-12 receptor expression and enhanced T cell survival (Liu, Mori, et al., 2004). IFN-α/β increase the expression of MCP-1, MCP-3, and IP-10, resulting in further recruitment of monocytes or macrophages and Th1-type cells to the site of infection.

In addition, IFN-α/β upregulate HLA gene expression, thus increasing antigen presentation of macrophages and dendritic cells.

14.5.6 Adaptive Immune Response

In RSV infection, cellular immunity appears to be most important in clearing the infection. However, antibodies against RSV can partially protect against reinfection, and maternal antibodies transferred transplacentally or through breast milk also confer some protection. In contrast, antibodies against the influenza surface protein HA prevent the virus from attaching to the cells and neutralize infectivity, and antibodies against NA inhibit the release of newly synthesized virus from the cells and, thus, prevent spreading of the infection to neighboring cells and other people.

14.6 EFFECTS OF DE

DEP make up a large fraction (up to 70%) (Weinhold, 2002) of respirable PM (aerodynamic diameter <10 μm). Numerous controlled *in vitro* and *in vivo* studies have been conducted to examine the acute and chronic effects of DEP exposures on the respiratory system and immune responses. Many of these studies have been conducted to examine the effects of DE on allergy-related Th2 immune responses, which are reviewed elsewhere (Diaz-Sanchez et al., 2003; Gershwin, 2003). Studies on the effects of DE on immune responses that pertain to viral infections will be addressed here.

14.6.1 Respiratory System

Acute nasal challenges of human volunteers with DEP resulted in increased numbers of neutrophils, lymphocytes, and macrophages 24 h after challenge (Nel et al., 2001). Furthermore, nasal lavage levels of RANTES, MIP-1α, and MCP-3 protein, as well as nasal lavage cell mRNA levels for IL-2, IL-4, IL-5, IL-6, IL-10, IL-13, and IFN-γ, were enhanced in human volunteers 24 h postchallenge with DEP (Diaz-Sanchez et al., 2000; Diaz-Sanchez et al., 1996). Exposure of nonasthmatic human volunteers to DE increased the number of neutrophils and lymphocytes, as well as IL-8 protein in the bronchoalveolar lavage fluid (BALF), and enhanced P-selectin levels in biopsy tissue obtained from these individuals. (Holgate et al., 2003). Intratracheal instillation of DEP into mice enhanced the levels of MIP-1α, IL-1β, and MCP-1 in the lung 24 h postinstillation (Takano et al., 2002). This study also demonstrated that exposure to DEP enhanced the levels of TLR-4 in the lungs of mice (Takano et al., 2002). Taken together, these data show that acute exposures to DE cause a neutrophilic proinflammatory response in the respiratory tract. Chronic exposures of mice to DEP for 3 months resulted in increased mRNA levels for TNF-α, IL-12p40, IL-4, and IL-10 in total lung tissue, whereas IL-1β and iNOS expression were somewhat decreased (Hiramatsu et al., 2003). Interestingly, results obtained from the same group demonstrated that macrophages derived from mice exposed to DE for 1 month had decreased mRNA levels for TNF-α, IL-1β, IL-6, IL-12p40, IFN-γ, and iNOS and increased mRNA levels for IL-10 (Saito et al., 2002). These data demonstrate that exposure to DE can elicit different effects in different cell types. This notion is further supported by *in vitro* experiments, which show that human and murine respiratory epithelial cells respond to DE exposure by increasing the expression of IL-8, GM-CSF, RANTES, soluble ICAM-1, and IL-1β (Bayram et al., 1998; Takizawa et al., 2000b; Hashimoto et al., 2000; Boland et al., 1999), whereas DE exposure of human or murine macrophages or monocytic cell lines increased the basal expression of IL-8 and decreased the expression of TNF-α, IL-6, IL-8, and IL-12p40 in activated macrophages (Li et al., 2002; Nilsen et al., 2003; Amakawa et al., 2003). Thus, different cell types that come in direct contact with DEP release different mediators, and exposure to DEP changes their responsiveness to other inflammatory challenges.

As indicated earlier, IFNs play an important role in the defense against viral infections. Recent evidence demonstrates that exposure to DEP inhibits basal *in vivo* IFN-γ production in mice and LPS-induced IFN-γ expression in splenic NK cells (Finkelman et al., 2004). Furthermore, these studies showed that DEP decreased IL-2, IL-12, and Poly I:C-induced *in vivo* levels of IFN-γ.

14.7 POTENTIAL MECHANISMS OF VIRUS AND DE INTERACTIONS

As described earlier, viral infection of respiratory epithelial cells or monocytes/macrophages require interactions between the virus and the host cell and result in inflammatory and antiviral defense responses, which ultimately clear the infection. Adaptive immune responses are important in clearing viral infections, but the focus of this section will be the potential interactions at the level of the infected host cells, rather than the effects of DE on the immune system. The flowchart in Figure 14.1 illustrates potential steps of the viral infection process that could be affected by exposure to DE; this will be discussed in more detail in the following subsections.

14.7.1 Virus Attachment

As described earlier, respiratory viruses, such as the rhinovirus and influenza virus, attach to the host cell through defined receptors. Rhinoviruses

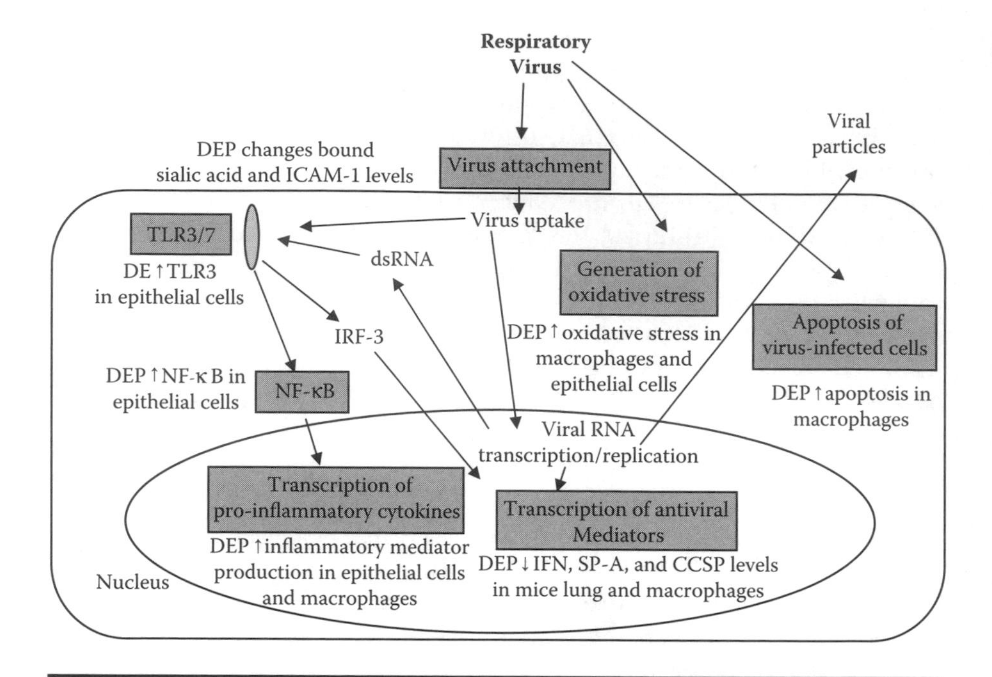

Figure 14.1 Schematic of potential interactions between diesel exhaust and viral infections.

use host cell ICAM-1 as the receptor for attachment. Similarly, influenza viruses attach to sialic acid residues on host cell membrane glycoproteins (Gagneux et al., 2003; Yamaya and Sasaki, 2003). Therefore, upregulation of ICAM-1 or sialic acid residues following exposure to DE could potentially enhance the susceptibility to viral infection. Weekly instillation of DEP caused enhanced levels of bound sialic acid in the BALF of mice (Sagai et al., 1996). Although bound sialic acid measurements were used as an indicator for mucus production, these data suggest that exposure to DE may affect the levels or distribution of sialic acid residues, which would change influenza virus attachment. Furthermore, *in vitro* and *in vivo* exposure to DE enhanced the expression of ICAM-1 on the epithelial cell surface (Salvi et al., 1999; Takizawa et al., 2000a), which could significantly affect subsequent attachment and infection with rhinovirus.

14.7.2 Oxidative Stress

Generation of oxidative stress plays an important role in the pathogenesis of respiratory viral infections. For example, rhinovirus infection of human bronchial epithelial cells increases the levels of reactive oxygen intermediates in the supernatants, and rhinovirus-induced IL-8 production in these cells was found to be mediated by oxidative stress (Biagioli et al., 1999; Kaul et al., 2000). Similarly, RSV infection of human respiratory epithelial cells induces reactive oxygen species synthesis, and RSV-induced expression of cytokines and chemokines, such as RANTES, was mediated by oxidative stress (Liu, Castro, et al., 2004; Casola et al., 2001). Numerous studies applying antioxidant therapies have shown that oxidative stress plays an important role in influenza pathogenesis; for example, influenza virus replication and influenza-induced apoptosis are inhibited by reduced GSH (Cai et al., 2003; Nencioni et al., 2003), treatment with antioxidants inhibits influenza-induced proinflammatory mediator production (Knobil et al., 1998), and overexpression of extracellular superoxide dismutase (EC-SOD) protects against influenza-induced lung injury in mice (Suliman et al., 2001). Furthermore, the levels of xanthine oxidase, an enzyme involved in the generation of superoxide, were enhanced in the BALF of influenza-infected mice, suggesting that the production of reactive oxygen species is enhanced during influenza virus infection (Peterhans, 1997). In addition, studies have demonstrated that influenza-induced pneumonia in mice involves the formation of both superoxide and nitric oxide (and potentially peroxynitrite) (Akaike and Maeda, 2000). Taken together, these and many other studies have demonstrated the importance of oxidative stress as an intermediary step in respiratory virus infections.

Several studies have demonstrated that many of the adverse effects induced by exposure to DE are mediated by DE-induced oxidative stress (Li et al., 2003; Baulig et al., 2003). In particular, polycyclic aromatic hydrocarbons (PAH) adsorbed onto the surface of DEP appear to be responsible for DE-induced oxidative stress (Baulig et al., 2003; Hiura et al., 1999). Specifically, cellular GSH levels decrease with exposure to DE (Matsuo et al., 2003), and antioxidants that enhance cellular GSH levels inhibit adverse effects, such as proinflammatory mediator production, associated with exposure to DE (Whitekus et al., 2002; Hiura et al., 1999; Li et al., 2002). Evidence that exposures to DE and viral infection do in fact add, or even synergize, in their ability to generate oxidative stress comes from our own preliminary studies. A549 cells, a human alveolar type-II-like cell line, were exposed to DEP 2 h prior to infection with influenza A virus. At 24 h postinfection, whole-cell lysates were analyzed for the levels of oxidized protein carbonyl levels, using a commercially available kit. Figure 14.2 shows that infection of human respiratory epithelial cells with the influenza virus enhances oxidized protein carbonyl levels, a marker of oxidative stress, and that pretreatment with DEP augments this effect even further. Thus, exposure to DE and infection with respiratory viruses could synergize in their ability to induce oxidative stress, which, in turn, could adversely affect virus-induced lung injury and viral clearance.

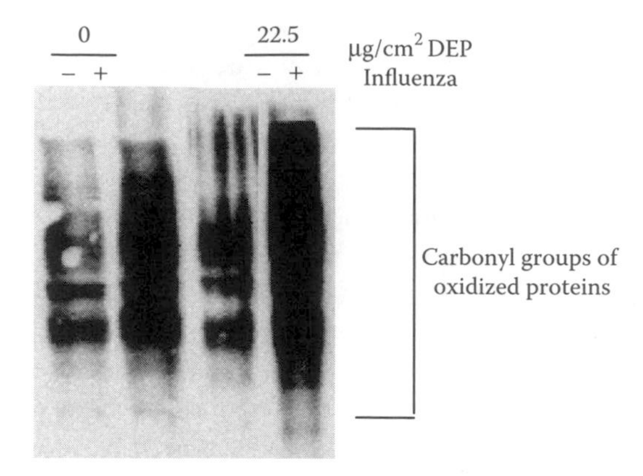

Figure 14.2 Exposure to DEP and treatment with influenza enhance levels of carbonyl groups from oxidized proteins, which are markers of oxidative stress. A549 cells were exposed to DEP for 2 h and subsequently infected with influenza A Bangkok 1/79. Levels of carbonyl groups were analyzed 24 h postinfection using a commercially available kit.

14.7.3 Host Defense

As indicated earlier, there are several antiviral defense strategies used by the host to limit a viral infection and eventually clear the pathogen. Apoptotic cell death of virus-producing host cells and the subsequent removal of apoptotic cells by phagocytosis is a naturally occurring mechanism by which the host eliminates the pathogen (Fujimoto et al., 1998) and prevents viral replication and spreading of the virus to neighboring cells. Thus, apoptosis of virus-producing cells is important in limiting the infection. Studies have shown that exposure to DEP causes necrosis in human bronchial epithelial cells (Matsuo et al., 2003), although it induces apoptosis in respiratory macrophages (Hiura et al., 2000). Thus, influenza infection and exposure to DE can induce cell death in lung epithelial cells by apoptotic and necrotic pathways, respectively. It is conceivable that exposure to DEP could modify apoptotic cell death induced by influenza infection or that influenza infection could alter diesel-induced necrosis. Enhanced necrosis or apoptosis would imply greater injury to the epithelium, whereas decreased apoptosis could mean a more widespread infection because apoptotic cell death of virus-replicating cells is a way to limit the infection.

Type I and type II interferons, and other antiviral mediators whose expression is regulated by interferons, are essential for the ability of host cells to limit viral propagation before other host immune cell responses can eliminate the viral infection. As indicated earlier, several studies have demonstrated that exposure to DE decreases the levels of lung IFN-γ (Saito et al., 2002) and the ability of immune cells to release IFN-γ in response to other stimuli (Yin et al., 2004; Yin et al., 2002). In addition, mice repeatedly exposed to DE for several months showed decreased ability to clear a subsequent influenza infection, which was attributed, at least partially, to the suppressed lung IFN-γ levels (Hahon et al., 1985). Thus, the decreased ability to produce IFNs following exposure to DE could have significant implications for the ability to clear a viral infection. However, other studies have demonstrated either no effect of DE on virus-induced IFN levels or increased IFN levels in animal or cell culture models exposed to DE and subsequently infected with respiratory virus. For example, Harrod et al. (2003) found that mice exposed to DE for several days and infected with RSV showed increased lung IFN-γ levels and our own studies have suggested that exposure to DEP increased IFN-β levels in human respiratory epithelial cells infected with the influenza virus (data not shown).

However, the expression of other mediators of first-line host defense against viral infections is modulated by exposure to DE. For example, Harrod et al. (2003) found that exposure to DE decreased the levels of RSV-induced surfactant protein A (SP-A) and Clara cell secretory protein (CCSP or also called CC-10 or CC-16). SP-A is a member of the collectin family and a "pattern recognition molecule," which interacts with glyco-conjugates on the surface of microorganisms such as the influenza virus or RSV (Li et al., 2002; Griese, 2002; Haagsman, 2002; Harrod et al., 1999). Interaction of virus with SP-A neutralizes its infectivity and enhances the elimination of the pathogen by phagocytosis (Li et al., 2002; Haagsman, 2002). Similarly, absence or decreased levels of CCSP have been shown to enhance the severity of subsequent infections with RSV (Wang et al., 2001; Harrod et al., 1998). Thus, DE-induced suppression of CCSP and SP-A could significantly enhance the susceptibility to respiratory virus infections, as was shown in the model of DE exposure and infection with RSV (Harrod et al., 2003).

14.7.4 Inflammation

Both exposure to DE and infection with the respiratory virus enhance the production and release of numerous proinflammatory mediators. Thus, it is conceivable that exposure to DE modifies virus-induced proinflamma-tory responses. This notion is supported by studies demonstrating that exposure of mice to DE enhances RSV-induced inflammatory cell count, inflammatory histological scores, and proinflammatory cytokine produc-tion, thus increasing overall lung injury associated with RSV infection (Harrod et al., 2003). Similarly, our own preliminary studies have shown that exposure of differentiated human nasal epithelial cells to DEP enhanced influenza-induced proinflammatory mediator production. Specifically, differentiated human nasal epithelial cells were exposed to DEP for 2 h and subsequently infected with influenza A virus. At 24 h postinfection, the total RNA was analyzed for IL-6 mRNA levels using real-time RT-PCR. Figure 14.3 shows that exposure to DEP enhances influenza virus-induced IL-6 mRNA levels. Interestingly, studies by Hahon et al. (1985) have demonstrated that repeated exposures of mice to DE for 3 to 6 months had no effect on the inflammatory response as evaluated by histopathological examination. Inflammatory cytokine and chemokine production is necessary to recruit and activate immune cells and ultimately clear the infection, but mediators released by activated inflammatory cells can also contribute to lung injury and are responsible for some of the symptoms associated with viral infections. Therefore, DE-induced augmentation of virus-induced inflammatory responses could increase lung injury, whereas DE-induced suppression of virus-induced inflammatory responses could impair host defenses and clearance of infection.

TLRs are increasingly recognized as being important receptors mediating innate immune responses against a wide range of pathogens (Akira and Sato 2003; Takeda and Akira, 2004; Barton and Medzhitov, 2003).

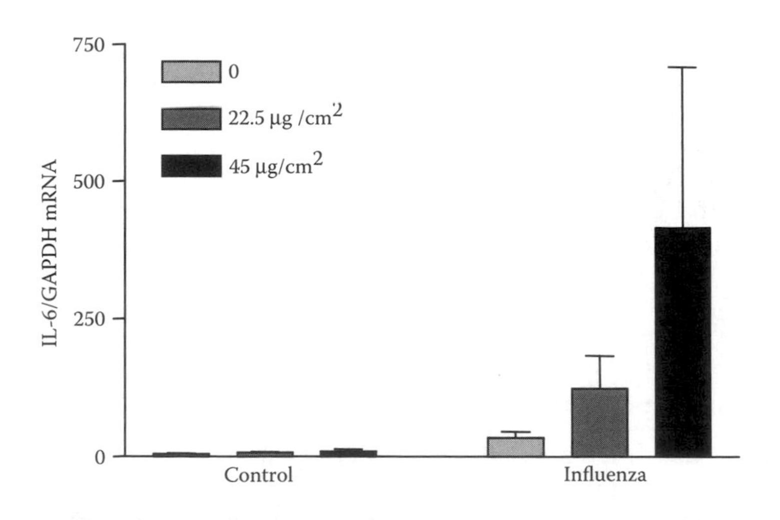

Figure 14.3 Exposure to DEP enhances influenza-induced IL-6 mRNA levels in differentiated human nasal epithelial (HNE) cells. Differentiated HNE cells were exposed to DEP for 2 h and subsequently infected with influenza A Bangkok 1/79. IL-6 mRNA levels were analyzed 24 h postinfected and normalized to GAPDH mRNA levels. Data represent mean ± SEM.

Recent evidence suggests that receptor-mediated signaling pathways are involved in host cell responses to viral infections. Specifically, TLR3 and TLR7 have been shown to mediate cellular responses to dsRNA, a by-product of many viral infections, and single-stranded RNA viruses, respectively (Lund et al., 2004; Alexopoulou et al., 2001). The evidence came from experiments that showed that mice without functional TLR3 were unable to respond to a synthetic form of dsRNA (poly I:C), but were still able to respond to other microbial products, such as LPS, which is derived from Gram-negative bacteria (Alexopoulou et al., 2001). Furthermore, TLR7 recognized the high CpG content of single-stranded RNA viruses such as the influenza virus, and activation of TLR7 by the influenza virus resulted in production of inflammatory cytokines (Lund et al., 2004). Interaction of TLR3 with dsRNA, or of TLR7 with high CpG content of single-stranded RNA viruses, leads to the activation of signaling cascades (for a review see Takeda and Akira, 2004), ultimately resulting in the activation of transcription factors, such as NF-κB and IRF-3, which mediate the transcription of proinflammatory cytokines (NF-κB) and type I IFNs (IRF-3). Thus, DE-induced modifications of TLR3 and TLR7 levels or their activity could significantly affect subsequent viral infections. Preliminary results obtained in our laboratory indicate that exposure of human respiratory epithelial cells to DEP can enhance the activity of TLR3. Figure 14.4 shows that exposure of A549 cells to DEP enhanced IL-8 mRNA production in response to poly I:C, a known ligand for TLR3. These preliminary data thus indicate that exposure to DEP can enhance the activity of TLR3 on human respiratory epithelial cells.

14.8 EXPERIMENTAL CONSIDERATIONS

14.8.1 Susceptibility to Infections vs. Exacerbation of an Existing Infection

Exposure of alveolar macrophages to PM has been shown to impair their ability to phagocytize RSV and mount an inflammatory response against the invading microbe (Becker and Soukup, 1999). Specifically, these studies have demonstrated that exposure of human alveolar macrophages to urban PM impairs phagocytosis of RSV and subsequent generation and release of proinflammatory cytokines (Becker and Soukup, 1999). However, the effect of air pollution on virus infections may also depend on the sequence of exposure and infection. For example, Kaan et al. have shown that although exposure of guinea pig alveolar macrophages to PM prior to infection with RSV decreased the number of macrophages staining positively for RSV, no effect on RSV immunopositivity was observed when the macrophages were infected first and subsequently exposed to PM (Kaan and Hegele, 2003). Interestingly, this study also showed that independent of the sequence by

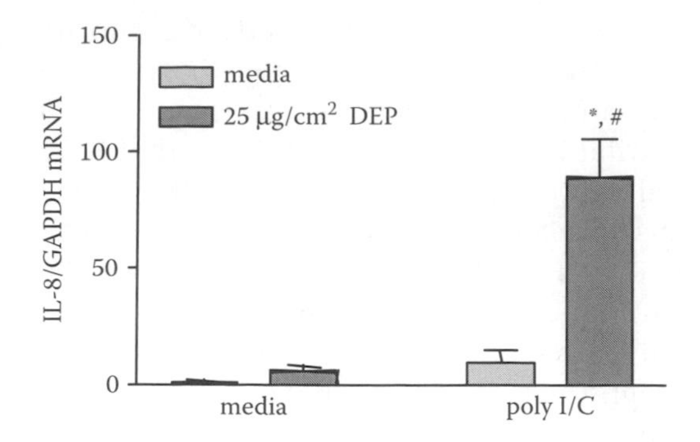

Figure 14.4 Exposure to DEP enhances the activity of TLR3 in human respiratory epithelial cells. A549 cells were exposed to DEP for 2 h and treated with 100 μg/ml poly I/C. Total RNA isolated 24 h after poly I:C treatment was analyzed for IL-8 mRNA levels using real-time RT-PCR. Data are mean + SEM. * Significantly different from poly I:C–treated but unexposed cells; $p > .05$.

which macrophages were exposed to PM and infected with RSV, RSV viral replication and yield were decreased by PM exposure (Kaan and Hegele, 2003). Similarly, RSV-induced production of IL-8 and IL-6 was suppressed by PM exposure, with RSV-induced IL-6 release being most strongly suppressed by PM exposure when macrophages were infected with RSV before exposure to PM (Kaan and Hegele, 2003). The importance of the sequence of exposure to particles and viral infection is further supported by studies from Lambert et al. (Lambert, Trasti et al., 2003; Lambert, Mangum, et al., 2003). These studies demonstrated that exposure to ultrafine CB particles 1 d prior to infection with RSV decreased the number of lymphocytes and IP-10, IL-1Rα, lymphotactin, and IFN-γ mRNA levels in the BALF from mice 4 d postinfection, whereas there was no effect of CB particles on RSV-induced RANTES, MCP-1, MIP-1α, IL-6, and eotaxin mRNA levels (Lambert, Trasti et al., 2003). Furthermore, there was no effect of CB on RSV titers or viral clearance. In contrast, exposure to CB 3 d after infection with RSV increased the number of BALF neutrophils and lymphocytes, RSV-induced airway hyperresponsiveness, and BALF MCP-1 and MIP-1α levels in mice 4 d postinfection (Lambert, Mangum et al., 2003). Again, no effect on RSV titers or viral clearance was observed. Taken together, these studies highlight the importance of the sequence of exposure to particles and viral infections and also suggest that exposure to particles can affect susceptibility and response to subsequent viral infections, as well as modulate the course and severity of an already existing infection.

14.8.2 Acute vs. Chronic Exposure to DE Prior to Infection

The effect of DE on subsequent viral infections could also depend on the exposure period prior to infection. Acute and chronic exposures to DE have different effects on respiratory cells that are targets for viral infections (epithelial cells and monocytes/macrophages). For example, chronic exposures to DE cause metaplastic and dysplastic changes in the human nasal epithelium, resulting in enhanced numbers of mucus-producing goblet cells and metaplastic squamous cells, as well as increased numbers of lymphocytes (Gluck et al., 2003). Similarly, rats exposed to DE for up to 24 months showed increased numbers of inflammatory cells and enhanced mucus production in the lung (Ishihara and Kagawa, 2003). In addition, mice exposed to DE for 1 month had suppressed lung mRNA levels for TNF-α, IL-1β, IL-6, IL-12p40, IFN-γ, and iNOS in their lungs (Saito et al., 2002), which are all mediators important in host defense against invading pathogens. Thus, chronic exposures to DE prior to infection could significantly alter the cell population and cellular environment an invading pathogen would encounter. The suppressed expression of mediators that are important in host defense, together with enhanced mucus production, which promotes attachment of, and colonization by, invading pathogens, could create an environment in favor of viral infection and replication.

Acute exposure studies in human volunteers have demonstrated that a single exposure to DE causes neutrophil influx and decreased phagocytosis in alveolar macrophages within 18 h of exposure; it also enhances the levels of markers of inflammation, such as expression of ICAM-1, IL-8, and Gro-α, which could contribute to the increased recruitment of inflammatory cells (Salvi et al., 1999; Salvi et al., 2000). These findings are supported by *in vitro* exposure studies, which largely demonstrate that single exposures to DE enhance the expression of inflammatory mediators in epithelial cells and decrease phagocytic activity in macrophages (Castranova et al., 1985; Yin et al., 2002). Taken together, these findings suggest that acute exposure to DE would create an inflammatory environment, which could affect subsequent viral infections. Although many of the inflammatory mediators generated after exposure to DE are beneficial in the clearance of the virus, additional production of inflammatory mediators induced by the viral infection may amplify preexisting, that is DE-induced, airway inflammation and, therefore, result in enhanced tissue injury. Such interaction could potentially lead to the exacerbation of existing inflammatory lung diseases, such as asthma.

14.9 CONCLUSIONS

Although epidemiological evidence suggests a link between exposures to PM and increased morbidity due to respiratory infections, there is no direct link between DEP and respiratory virus infections. Because DE is a major contributor to ambient PM levels, one could extrapolate from the existing PM epidemiological data to DEP. However, epidemiological studies that can reveal a direct association between exposure to DE and the incidence and morbidity of respiratory virus infections need to be conducted.

More controlled exposure studies are necessary to determine the potential mechanisms by which exposure to DE could enhance the incidence and severity of respiratory virus infections. This chapter discussed some of the potential steps during viral infections that could be affected by exposure to DE and supported the discussion, when possible, with either published studies or preliminary data. However, more detailed studies are necessary to determine how exposure to DE affects respiratory infections in humans.

REFERENCES

Adachi, M., Matsukura, S., Tokunaga, H., and Kokubu, F. 1997. Expression of cytokines on human bronchial epithelial cells induced by influenza virus A. *Int Arch Allergy Immunol* 113, 307–311.

Akaike, T. and Maeda, H. 2000. Nitric oxide and virus infection. *Immunology* 101, 300–308.

Akira, S. 2004. Toll receptor families: structure and function. *Semin Immunol* 16, 1–2.

Akira, S. and Sato, S. 2003. Toll-like receptors and their signaling mechanisms. *Scand J Infect Dis* 35, 555–562.

Alexopoulou, L., Holt, A.C., Medzhitov, R., and Flavell, R.A. 2001. Recognition of double-stranded RNA and activation of NF-kappaB by Toll-like receptor 3. *Nature* 413, 732–738.

Amakawa, K., Terashima, T., Matsuzaki, T., Matsumaru, A., Sagai, M., and Yamaguchi, K. 2003. Suppressive effects of diesel exhaust particles on cytokine release from human and murine alveolar macrophages. *Exp Lung Res* 29, 149–164.

Anderson, L.J. and Heilman, C.A. 1995. Protective and disease-enhancing immune responses to respiratory syncytial virus. *J Infect Dis* 171, 1–7.

Anderson, L.J., Tsou, C., Potter, C., Keyserling, H.L., Smith, T.F., Ananaba, G., and Bangham, C.R. 1994. Cytokine response to respiratory syncytial virus stimulation of human peripheral blood mononuclear cells. *J Infect Dis* 170, 1201–1208.

Barton, G.M. and Medzhitov, R. 2003. Toll-like receptor signaling pathways. *Science* 300, 1524–1525.

Baulig, A., Sourdeval, M., Meyer, M., Marano, F., and Baeza-Squiban, A. 2003. Biological effects of atmospheric particles on human bronchial epithelial cells. Comparison with diesel exhaust particles. *Toxicol In Vitro* 17, 567–573.

Bayram, H., Devalia, J.L., Sapsford, R.J., Ohtoshi, T., Miyabara, Y., Sagai, M., and Davies, R.J. 1998. The effect of diesel exhaust particles on cell function and release of inflammatory mediators from human bronchial epithelial cells *in vitro*. *Am J Respir Cell Mol Biol* 18, 441–448.

Becker, S. and Soukup, J.M. 1999. Exposure to urban air particulates alters the macrophage-mediated inflammatory response to respiratory viral infection. *J Toxicol Environ Health A* 57, 445–457.

Biagioli, M.C., Kaul, P., Singh, I., and Turner, R.B. 1999. The role of oxidative stress in rhinovirus-induced elaboration of IL-8 by respiratory epithelial cells. *Free Radic Biol Med* 26, 454–462.

Boland, S., Baeza-Squiban, A., Fournier, T., Houcine, O., Gendron, M.C., Chevrier, M., Jouvenot, G., Coste, A., Aubier, M., and Marano, F. 1999. Diesel exhaust particles are taken up by human airway epithelial cells *in vitro* and alter cytokine production. *Am J Physiol* 276, L604–13.

Brammer, L., Fukuda, K., Arden, N., Schmeltz, L.M., Simonsen, L., Khan, A., Regnery, H.L., Schonberger, L.B., and Cox, N.J. 1997. Influenza surveillance—United States, 1992–1993 and 1993–1994. *MMWR CDC Surveill Summ* 46, 1–12.

Brauer, M., Hoek, G., Van Vliet, P., Meliefste, K., Fischer, P.H., Wijga, A., Koopman, L.P., Neijens, H.J., Gerritsen, J., Kerkhof, M., Heinrich, J., Bellander, T., and Brunekreef, B. 2002. Air pollution from traffic and the development of respiratory infections and asthmatic and allergic symptoms in children. *Am J Respir Crit Care Med* 166, 1092–1098.

Cai, J., Chen, Y., Seth, S., Furukawa, S., Compans, R.W., and Jones, D.P. 2003. Inhibition of influenza infection by glutathione. *Free Radic Biol Med* 34, 928–936.

Casola, A., Burger, N., Liu, T., Jamaluddin, M., Brasier, A.R., and Garofalo, R.P. 2001. Oxidant tone regulates RANTES gene expression in airway epithelial cells infected with respiratory syncytial virus. Role in viral-induced interferon regulatory factor activation. *J Biol Chem* 276, 19715–19722.

Castranova, V., Bowman, L., Reasor, M.J., Lewis, T., Tucker, J., and Miles, P.R. 1985. The response of rat alveolar macrophages to chronic inhalation of coal dust and/or diesel exhaust. *Environ Res* 36, 405–419.

Cohen, L. and Castro, M. 2003. The role of viral respiratory infections in the pathogenesis and exacerbation of asthma. *Semin Respir Infect* 18, 3–8.

Couceiro, J.N., Paulson, J.C., and Baum, L.G. 1993. Influenza virus strains selectively recognize sialyloligosaccharides on human respiratory epithelium; the role of the host cell in selection of hemagglutinin receptor specificity. *Virus Res* 29, 155–165.

Diaz-Sanchez, D. 1997. The role of diesel exhaust particles and their associated polyaromatic hydrocarbons in the induction of allergic airway disease. *Allergy* 52, 52–56; discussion 57–58.

Diaz-Sanchez, D., Jyrala, M., Ng, D., Nel, A., and Saxon, A. 2000. *In vivo* nasal challenge with diesel exhaust particles enhances expression of the CC chemokines rantes, MIP-1alpha, and MCP-3 in humans. *Clin Immunol* 97, 140–145.

Diaz-Sanchez, D., Proietti, L., and Polosa, R. 2003. Diesel fumes and the rising prevalence of atopy: an urban legend? *Curr Allergy Asthma Rep* 3, 146–152.

Diaz-Sanchez, D., Tsien, A., Casillas, A., Dotson, A.R., and Saxon, A. 1996. Enhanced nasal cytokine production in human beings after *in vivo* challenge with diesel exhaust particles. *J Allergy Clin Immunol* 98, 114–123.

Diebold, S.S., Kaisho, T., Hemmi, H., Akira, S., and Reis e Sousa, C. 2004. Innate antiviral responses by means of TLR7-mediated recognition of single-stranded RNA. *Science* 303, 1529–1531.

Domachowske, J.B., Bonville, C.A., Mortelliti, A.J., Colella, C.B., Kim, U., and Rosenberg, H.F. 2000. Respiratory syncytial virus infection induces expression of the anti-apoptosis gene IEX-1L in human respiratory epithelial cells. *J Infect Dis* 181, 824–830.

Domachowske, J.B., Bonville, C.A., and Rosenberg, H.F. 2001. Gene expression in epithelial cells in response to pneumovirus infection. *Respir Res* 2, 225–233.

Duits, L.A., Nibbering, P.H., van Strijen, E., Vos, J.B., Mannesse-Lazeroms, S.P., van Sterkenburg, M.A., and Hiemstra, P.S. 2003. Rhinovirus increases human beta-defensin-2 and -3 mRNA expression in cultured bronchial epithelial cells. *FEMS Immunol Med Microbiol* 38, 59–64.

Falsey, A.R. and Walsh, E.E. 2000. Respiratory syncytial virus infection in adults. *Clin Microbiol Rev* 13, 371–384.

Finkelman, F.D., Yang, M., Orekhova, T., Clyne, E., Bernstein, J., Whitekus, M., Diaz-Sanchez, D., and Morris, S.C. 2004. Diesel exhaust particles suppress *in vivo* IFN-gamma production by inhibiting cytokine effects on NK and NKT cells. *J Immunol* 172, 3808–3813.

Fishaut, M., Tubergen, D., and McIntosh, K. 1980. Cellular response to respiratory viruses with particular reference to children with disorders of cell-mediated immunity. *J Pediatr* 96, 179–186.

Fujimoto, I., Takizawa, T., Ohba, Y., and Nakanishi, Y. 1998. Co-expression of Fas and Fas-ligand on the surface of influenza virus-infected cells. *Cell Death Differ* 5, 426–431.

Gagneux, P., Cheriyan, M., Hurtado-Ziola, N., van der Linden, E.C., Anderson, D., McClure, H., Varki, A., and Varki, N.M. 2003. Human-specific regulation of alpha 2-6-linked sialic acids. *J Biol Chem* 278, 48245–48250.

Gern, J.E. 2004. Viral respiratory infection and the link to asthma. *Pediatr Infect Dis J* 23, S78–86.

Gershwin, L.J. 2003. Effects of air pollutants on development of allergic immune responses in the respiratory tract. *Clin Dev Immunol* 10, 119–126.

Gluck, U., Schutz, R., and Gebbers, J.O. 2003. Cytopathology of the nasal mucosa in chronic exposure to diesel engine emission: a five-year survey of Swiss customs officers. *Environ Health Perspect* 111, 925–929.

Griese, M. 2002. Respiratory syncytial virus and pulmonary surfactant. *Viral Immunol* 15, 357–363.

Haagsman, H.P. 2002. Structural and functional aspects of the collectin SP-A. *Immunobiology* 205, 476–489.

Hahon, N., Booth, J.A., Green, F., and Lewis, T.R. 1985. Influenza virus infection in mice after exposure to coal dust and diesel engine emissions. *Environ Res* 37, 44–60.

Han, L.L., Alexander, J.P., and Anderson, L.J. 1999. Respiratory syncytial virus pneumonia among the elderly: an assessment of disease burden. *J Infect Dis* 179, 25–30.

Harris, J. and Werling, D. 2003. Binding and entry of respiratory syncytial virus into host cells and initiation of the innate immune response. *Cell Microbiol* 5, 671–680.

Harrison, A.M., Bonville, C.A., Rosenberg, H.F., and Domachowske, J.B. 1999. Respiratory syncytical virus-induced chemokine expression in the lower airways: eosinophil recruitment and degranulation. *Am J Respir Crit Care Med* 159, 1918–1924.

Harrod, K.S., Jaramillo, R.J., Rosenberger, C.L., Wang, S.Z., Berger, J.A., McDonald, J.D., and Reed, M.D. 2003. Increased susceptibility to RSV infection by exposure to inhaled diesel engine emissions. *Am J Respir Cell Mol Biol* 28, 451–463.

Harrod, K.S., Mounday, A.D., Stripp, B.R., and Whitsett, J.A. 1998. Clara cell secretory protein decreases lung inflammation after acute virus infection. *Am J Physiol* 275, L924–30.

Harrod, K.S., Trapnell, B.C., Otake, K., Korfhagen, T.R., and Whitsett, J.A. 1999. SP-A enhances viral clearance and inhibits inflammation after pulmonary adenoviral infection. *Am J Physiol* 277, L580–8.

Hashimoto, S., Gon, Y., Takeshita, I., Matsumoto, K., Jibiki, I., Takizawa, H., Kudoh, S., and Horie, T. 2000. Diesel exhaust particles activate p38 MAP kinase to produce interleukin 8 and RANTES by human bronchial epithelial cells and N-acetylcysteine attenuates p38 MAP kinase activation. *Am J Respir Crit Care Med* 161, 280–285.

Hayden, F.G. 2004. Rhinovirus and the lower respiratory tract. *Rev Med Virol* 14, 17–31.

Heil, F., Hemmi, H., Hochrein, H., Ampenberger, F., Kirschning, C., Akira, S., Lipford, G., Wagner, H., and Bauer, S. 2004. Species-specific recognition of single-stranded RNA via toll-like receptor 7 and 8. *Science* 303, 1526–1529.

Hiramatsu, K., Azuma, A., Kudoh, S., Desaki, M., Takizawa, H., and Sugawara, I. 2003. Inhalation of diesel exhaust for three months affects major cytokine expression and induces bronchus-associated lymphoid tissue formation in murine lungs. *Exp Lung Res* 29, 607–622.

Hiura, T.S., Kaszubowski, M.P., Li, N., and Nel, A.E. 1999. Chemicals in diesel exhaust particles generate reactive oxygen radicals and induce apoptosis in macrophages. *J Immunol* 163, 5582–5591.

Hiura, T.S., Li, N., Kaplan, R., Horwitz, M., Seagrave, J.C., and Nel, A.E. 2000. The role of a mitochondrial pathway in the induction of apoptosis by chemicals extracted from diesel exhaust particles. *J Immunol* 165, 2703–2711.

Holgate, S.T., Sandstrom, T., Frew, A.J., Stenfors, N., Nordenhall, C., Salvi, S., Blomberg, A., Helleday, R., and Soderberg, M. 2003. Health effects of acute exposure to air pollution. Part I: Healthy and asthmatic subjects exposed to diesel exhaust. *Res Rep Health Eff Inst* 1–30; discussion 51–67.

Irmen, K.E. and Kelleher, J.J. 2000. Use of monoclonal antibodies for rapid diagnosis of respiratory viruses in a community hospital. *Clin Diagn Lab Immunol* 7, 396–403.

Ishihara, Y. and Kagawa, J. 2003. Chronic diesel exhaust exposures of rats demonstrate concentration and time-dependent effects on pulmonary inflammation. *Inhal Toxicol* 15, 473–492.

Jaakkola, J.J., Paunio, M., Virtanen, M., and Heinonen, O.P. 1991. Low-level air pollution and upper respiratory infections in children. *Am J Public Health* 81, 1060–1063.

Janssens, S. and Beyaert, R. 2003. Role of toll-like receptors in pathogen recognition. *Clin Microbiol Rev* 16, 637–646.

Julkunen, I., Melen, K., Nyqvist, M., Pirhonen, J., Sareneva, T., and Matikainen, S. 2000. Inflammatory responses in influenza A virus infection. *Vaccine* 19(Suppl. 1), S32–7.

Kaan, P.M. and Hegele, R.G. 2003. Interaction between respiratory syncytial virus and particulate matter in guinea pig alveolar macrophages. *Am J Respir Cell Mol Biol* 28, 697–704.

Kaul, P., Biagioli, M.C., Singh, I., and Turner, R.B. 2000. Rhinovirus-induced oxidative stress and interleukin-8 elaboration involves p47-phox but is independent of attachment to intercellular adhesion molecule-1 and viral replication. *J Infect Dis* 181, 1885–1890.

Khabar, K.S., Siddiqui, Y.M., al-Zoghaibi, F., al-Haj, L., Dhalla, M., Zhou, A., Dong, B., Whitmore, M., Paranjape, J., Al-Ahdal, M.N., Al-Mohanna, F., Williams, B.R., and Silverman, R.H. 2003. RNase L mediates transient control of the interferon response through modulation of the double-stranded RNA-dependent protein kinase PKR. *J Biol Chem* 278, 20124–20132.

Knobil, K., Choi, A.M., Weigand, G.W., and Jacoby, D.B. 1998. Role of oxidants in influenza virus-induced gene expression. *Am J Physiol* 274, L134–42.

Kotelkin, A., Prikhod'ko, E.A., Cohen, J.I., Collins, P.L., and Bukreyev, A. 2003. Respiratory syncytial virus infection sensitizes cells to apoptosis mediated by tumor necrosis factor-related apoptosis-inducing ligand. *J Virol* 77, 9156–9172.

Lambert, A.L., Mangum, J.B., DeLorme, M.P., and Everitt, J.I. 2003. Ultrafine carbon black particles enhance respiratory syncytial virus-induced airway reactivity, pulmonary inflammation, and chemokine expression. *Toxicol Sci* 72, 339–346.

Lambert, A.L., Trasti, F.S., Mangum, J.B., and Everitt, J.I. 2003. Effect of preexposure to ultrafine carbon black on respiratory syncytial virus infection in mice. *Toxicol Sci* 72, 331–338.

Li, G., Siddiqui, J., Hendry, M., Akiyama, J., Edmondson, J., Brown, C., Allen, L., Levitt, S., Poulain, F., and Hawgood, S. 2002. Surfactant protein-A—deficient mice display an exaggerated early inflammatory response to a beta-resistant strain of influenza A virus. *Am J Respir Cell Mol Biol* 26, 277–282.

Li, N., Hao, M., Phalen, R.F., Hinds, W.C., and Nel, A.E. 2003. Particulate air pollutants and asthma. A paradigm for the role of oxidative stress in PM-induced adverse health effects. *Clin Immunol* 109, 250–265.

Li, N., Wang, M., Oberley, T.D., Sempf, J.M., and Nel, A.E. 2002. Comparison of the pro-oxidative and proinflammatory effects of organic diesel exhaust particle chemicals in bronchial epithelial cells and macrophages. *J Immunol* 169, 4531–4541.

Lina, B., Valette, M., Foray, S., Luciani, J., Stagnara, J., See, D.M., and Aymard, M. 1996. Surveillance of community-acquired viral infections due to respiratory viruses in Rhone-Alpes (France) during winter 1994 to 1995. *J Clin Microbiol* 34, 3007–3011.

Liu, B., Mori, I., Hossain, M.J., Dong, L., Takeda, K., and Kimura, Y. 2004. Interleukin-18 improves the early defence system against influenza virus infection by augmenting natural killer cell-mediated cytotoxicity. *J Gen Virol* 85, 423–428.

Liu, T., Castro, S., Brasier, A.R., Jamaluddin, M., Garofalo, R.P., and Casola, A. 2004. Reactive oxygen species mediate virus-induced STAT activation: role of tyrosine phosphatases. *J Biol Chem* 279, 2461–2469.

Lowy, R.J. 2003. Influenza virus induction of apoptosis by intrinsic and extrinsic mechanisms. *Int Rev Immunol* 22, 425–449.

Lund, J.M., Alexopoulou, L., Sato, A., Karow, M., Adams, N.C., Gale, N.W., Iwasaki, A., and Flavell, R.A. 2004. Recognition of single-stranded RNA viruses by toll-like receptor 7. *Proc Natl Acad Sci USA* 101, 5598–5603.

Matsumoto, M., Funami, K., Oshiumi, H., and Seya, T. 2004. Toll-like receptor 3: a link between toll-like receptor, interferon and viruses. *Microbiol Immunol* 48, 147–154.

Matsuo, M., Shimada, T., Uenishi, R., Sasaki, N., and Sagai, M. 2003. Diesel exhaust particle-induced cell death of cultured normal human bronchial epithelial cells. *Biol Pharm Bull* 26, 438–447.

Mlinaric-Galinovic, G., Falsey, A.R., and Walsh, E.E. 1996. Respiratory syncytial virus infection in the elderly. *Eur J Clin Microbiol Infect Dis* 15, 777–781.

Mori, I., Komatsu, T., Takeuchi, K., Nakakuki, K., Sudo, M., and Kimura, Y. 1995. *In vivo* induction of apoptosis by influenza virus. *J Gen Virol* 76(Pt. 11), 2869–2873.

Nel, A.E., Diaz-Sanchez, D., and Li, N. 2001. The role of particulate pollutants in pulmonary inflammation and asthma: evidence for the involvement of organic chemicals and oxidative stress. *Curr Opin Pulm Med* 7, 20–26.

Nencioni, L., Iuvara, A., Aquilano, K., Ciriolo, M.R., Cozzolino, F., Rotilio, G., Garaci, E., and Palamara, A.T. 2003. Influenza A virus replication is dependent on an antioxidant pathway that involves GSH and Bcl-2. *FASEB J* 17, 758–760.

Nilsen, A.M., Hagemann, R., Eikas, H., Egeberg, K., Norkov, T., and Sundan, A. 2003. Reduction of IL-12 p40 production in activated monocytes after exposure to diesel exhaust particles. *Int Arch Allergy Immunol* 131, 201–208.

Niyonsaba, F., Hirata, M., Ogawa, H., and Nagaoka, I. 2003. Epithelial cell-derived antibacterial peptides human beta-defensins and cathelicidin: multifunctional activities on mast cells. *Curr Drug Targets Inflamm Allergy* 2, 224–231.

Pandya, R.J., Solomon, G., Kinner, A., and Balmes, J.R. 2002. Diesel exhaust and asthma: hypotheses and molecular mechanisms of action. *Environ Health Perspect* 110(Suppl. 1), 103–112.

Peebles, R.S. Jr. 2004. Viral infections, atopy, and asthma: is there a causal relationship? *J Allergy Clin Immunol* 113, S15–8.

Peterhans, E. 1997. Oxidants and antioxidants in viral diseases: disease mechanisms and metabolic regulation. *J Nutr* 127, 962S–965S.

Pope, C.A., 3rd. 2000. Epidemiology of fine particulate air pollution and human health: biologic mechanisms and who's at risk? *Environ Health Perspect* 108(Suppl. 4), 713–723.

Pope, C.A., 3rd, Burnett, R.T., Thurston, G.D., Thun, M.J., Calle, E.E., Krewski, D., and Godleski, J.J. 2004. Cardiovascular mortality and long-term exposure to particulate air pollution: epidemiological evidence of general pathophysiological pathways of disease. *Circulation* 109, 71–77.

Proud, D., Sanders, S.P., and Wiehler, S. 2004. Human rhinovirus infection induces airway epithelial cell production of human beta-defensin 2 both *in vitro* and *in vivo*. *J Immunol* 172, 4637–4645.

Ronni, T., Sareneva, T., Pirhonen, J., and Julkunen, I. 1995. Activation of IFN-alpha, IFN-gamma, MxA, and IFN regulatory factor 1 genes in influenza A virus-infected human peripheral blood mononuclear cells. *J Immunol* 154, 2764–2774.

Sagai, M., Furuyama, A., and Ichinose, T. 1996. Biological effects of diesel exhaust particles (DEP). III. Pathogenesis of asthma-like symptoms in mice. *Free Radic Biol Med* 21, 199–209.

Saito, Y., Azuma, A., Kudo, S., Takizawa, H., and Sugawara, I. 2002. Long-term inhalation of diesel exhaust affects cytokine expression in murine lung tissues: comparison between low- and high-dose diesel exhaust exposure. *Exp Lung Res* 28, 493–506.

Salvi, S., Blomberg, A., Rudell, B., Kelly, F., Sandstrom, T., Holgate, S.T., and Frew, A.J. 1999. Acute inflammatory responses in the airways and peripheral blood after short-term exposure to diesel exhaust in healthy human volunteers. *Am J Respir Crit Care Med* 159, 702–709.

Salvi, S.S., Nordenhall, C., Blomberg, A., Rudell, B., Pourazar, J., Kelly, F.J., Wilson, S., Sandstrom, T., Holgate, S.T., and Frew, A.J. 2000. Acute exposure to diesel exhaust increases IL-8 and GRO-alpha production in healthy human airways. *Am J Respir Crit Care Med* 161, 550–557.

Samuel, C.E. 2001. Antiviral actions of interferons. *Clin Microbiol Rev* 14, 778–809.

Shay, D.K., Holman, R.C., Roosevelt, G.E., Clarke, M.J., and Anderson, L.J. 2001. Bronchiolitis-associated mortality and estimates of respiratory syncytial virus-associated deaths among U.S. children, 1979–1997. *J Infect Dis* 183, 16–22.

Simonsen, L., Blackwelder, W.C., Reichert, T.A., and Miller, M.A. 2003. Estimating deaths due to influenza and respiratory syncytial virus. *JAMA* 289, 2499–2500; author reply 2500–2.

Suliman, H.B., Ryan, L.K., Bishop, L., and Folz, R.J. 2001. Prevention of influenza-induced lung injury in mice overexpressing extracellular superoxide dismutase. *Am J Physiol Lung Cell Mol Physiol* 280, L69–78.

Takano, H., Yanagisawa, R., Ichinose, T., Sadakane, K., Yoshino, S., Yoshikawa, T., and Morita, M. 2002. Diesel exhaust particles enhance lung injury related to bacterial endotoxin through expression of proinflammatory cytokines, chemokines, and intercellular adhesion molecule-1. *Am J Respir Crit Care Med* 165, 1329–1335.

Takeda, K. and Akira, S. 2004. TLR signaling pathways. *Semin Immunol* 16, 3–9.

Takizawa, H., Ohtoshi, T., Kawasaki, S., Abe, S., Sugawara, I., Nakahara, K., Matsushima, K., and Kudoh, S. 2000a. Diesel exhaust particles activate human bronchial epithelial cells to express inflammatory mediators in the airways: a review. *Respirology* 5, 197–203.

van Kempen, M., Bachert, C., and Van Cauwenberge, P. 1999. An update on the pathophysiology of rhinovirus upper respiratory tract infections. *Rhinology* 37, 97–103.

Van Reeth, K. 2000. Cytokines in the pathogenesis of influenza. *Vet Microbiol* 74, 109–116.

van Vliet, P., Knape, M., de Hartog, J., Janssen, N., Harssema, H., and Brunekreef, B. 1997. Motor vehicle exhaust and chronic respiratory symptoms in children living near freeways. *Environ Res* 74, 122–132.

Walsh, E.E., Falsey, A.R., and Hennessey, P.A. 1999. Respiratory syncytial and other virus infections in persons with chronic cardiopulmonary disease. *Am J Respir Crit Care Med* 160, 791–795.

Wang, M., Saxon, A., and Diaz-Sanchez, D. 1999. Early IL-4 production driving Th2 differentiation in a human *in vivo* allergic model is mast cell derived. *Clin Immunol* 90, 47–54.

Wang, S.Z., Rosenberger, C.L., Espindola, T.M., Barrett, E.G., Tesfaigzi, Y., Bice, D.E., and Harrod, K.S. 2001. CCSP modulates airway dysfunction and host responses in an ova-challenged mouse model. *Am J Physiol Lung Cell Mol Physiol* 281, L1303–L1311.

Wang, Y.H., Huang, Y.C., Chang, L.Y., Kao, H.T., Lin, P.Y., Huang, C.G., and Lin, T.Y. 2003. Clinical characteristics of children with influenza A virus infection requiring hospitalization. *J Microbiol Immunol Infect* 36, 111–116.

Ward, S.V. and Samuel, C.E. 2002. Regulation of the interferon-inducible PKR kinase gene: the KCS element is a constitutive promoter element that functions in concert with the interferon-stimulated response element. *Virology* 296, 136–146.

Weinhold, B. 2002. Fuel for the long haul: diesel in America. *Environ Health Perspect* 110, A458–A464.

Welliver, R.C. 2003. Respiratory syncytial virus and other respiratory viruses. *Pediatr Infect Dis J* 22, S6–10; discussion S10–S12.

Whitekus, M.J., Li, N., Zhang, M., Wang, M., Horwitz, M.A., Nelson, S.K., Horwitz, L.D., Brechun, N., Diaz-Sanchez, D., and Nel, A.E. 2002. Thiol antioxidants inhibit the adjuvant effects of aerosolized diesel exhaust particles in a murine model for ovalbumin sensitization. *J Immunol* 168, 2560–2567.

Wilson, N.M. 2003. Virus infections, wheeze and asthma. *Paediatr Respir Rev* 4, 184–192.

Wong, T.W., Lau, T.S., Yu, T.S., Neller, A., Wong, S.L., Tam, W., and Pang, S.W. 1999. Air pollution and hospital admissions for respiratory and cardiovascular diseases in Hong Kong. *Occup Environ Med* 56, 679–683.

Yamaya, M. and Sasaki, H. 2003. Rhinovirus and asthma. *Viral Immunol* 16, 99–109.

Yang, D., Biragyn, A., Kwak, L.W., and Oppenheim, J.J. 2002. Mammalian defensins in immunity: more than just microbicidal. *Trends Immunol* 23, 291–296.

Yang, D., Chertov, O., and Oppenheim, J.J. 2001. The role of mammalian antimicrobial peptides and proteins in awakening of innate host defenses and adaptive immunity. *Cell Mol Life Sci* 58, 978–989.

Yin, X.J., Dong, C.C., Ma, J.Y., Antonini, J.M., Roberts, J.R., Stanley, C.F., Schafer, R., and Ma, J.K. 2004. Suppression of cell-mediated immune responses to listeria infection by repeated exposure to diesel exhaust particles in brown Norway rats. *Toxicol Sci* 77, 263–271.

Yin, X.J., Schafer, R., Ma, J.Y., Antonini, J.M., Weissman, D.D., Siegel, P.D., Barger, M.W., Roberts, J.R., and Ma, J.K. 2002. Alteration of pulmonary immunity to *Listeria monocytogenes* by diesel exhaust particles (DEPs). I. Effects of DEPs on early pulmonary responses. *Environ Health Perspect* 110, 1105–1111.

Zambon, M.C. 2001. The pathogenesis of influenza in humans. *Rev Med Virol* 11, 227–241.

Zambon, M.C., Stockton, J.D., Clewley, J.P., and Fleming, D.M. 2001. Contribution of influenza and respiratory syncytial virus to community cases of influenza-like illness: an observational study. *Lancet* 358, 1410–1416.

Zanobetti, A., Schwartz, J., and Dockery, D.W. 2000. Airborne particles are a risk factor for hospital admissions for heart and lung disease. *Environ Health Perspect* 108, 1071–1077.

15

ADAPTATION TO TOXICANT EXPOSURE DURING RODENT LUNG TUMORIGENESIS

Geoffrey M. Curtin, Ryan J. Potts, Paul H. Ayres, David J. Doolittle, and James E. Swauger

CONTENTS

A considerable amount of experimental data supports the concept that the pathogenesis of human lung cancer is dependent upon the sequential alteration of oncogenes and tumor suppressor genes, which, in turn, impact signaling pathways responsible for the regulation of cellular proliferation, differentiation, and apoptosis. Within this conceptual framework, a number of specific gene and chromosomal changes have been associated with lung tumor formation, although the exact sequence and timing of these changes are yet to be elucidated. During the course of the last several years, our laboratory has engaged in a series of studies intended to expand current understanding of the molecular events contributing to toxicant-induced tumorigenicity. These efforts have benefited significantly through the application of experimental findings from mouse skin studies examining the multistage process of tobacco-smoke-induced tumor formation; specifically, the contribution of specific toxicants to the seemingly discrete yet complementary stages of initiation, promotion (including conversion and propagation), and progression. This chapter summarizes the current understanding of the molecular changes associated with lung tumorigenesis and proposes a model for the specific events contributing to this multistage process. This model is discussed in terms of an emerging theory that secondary (indirect) genotoxicity and eventual adaptation to toxicant-induced damage contribute significantly to lung tumor formation.

15.1 MOLECULAR PATHOGENESIS OF LUNG CANCER

The molecular pathogenesis of lung cancer is most often described in terms of sequentially altered genes with the potential to impact signaling pathways responsible for controlling cellular proliferation, differentiation, and/or apoptosis. Oncogenic changes likely involved in lung tumor development include alterations in autocrine signaling loops, receptor and nonreceptor tyrosine kinases, membrane-associated G proteins, cytoplasmic serine/threonine kinases, and nuclear transcription factors (1). Autocrine signaling systems potentially contributing to lung cell transformation include gastrin-releasing peptide/bombesin-like peptides and their corresponding receptors; neuregulins that interact with the erbB family of transmembrane receptor tyrosine kinases, including epidermal growth factor receptor; and insulin-like growth factors (IGFs) functioning through the type I IGF receptor.

Notable among the signal transduction proteins implicated in the neoplastic process is p21-ras, which provides a proliferative signal to the cell nucleus while in its active configuration, i.e., when bound to guanosine triphosphate (GTP). This protein possesses an intrinsic GTPase activity and assumes an inactive configuration upon hydrolysis of GTP to guanosine diphosphate. Mutation of the *ras* gene leads to an inability of the corresponding protein to hydrolyze GTP and reacquire its inactive configuration, resulting in continued or dysregulated signal transduction to the nucleus (2).

Downstream effectors of p21-ras signaling include nuclear proto-oncogene products such as myc, which modulates cell proliferation and/or differentiation through direct activation of the genes regulating DNA synthesis, RNA metabolism, and cell-cycle progression (3). Equally significant during the neoplastic process is the ability of the tumor cell to escape the constraints of programmed cell death, or apoptosis. This homeostatic process, entailing the apoptotic susceptibility of a cell, is thought to be modulated by the ratio of pro- and antiapoptotic proteins. For instance, Bcl-2 appears to protect cells from apoptotic signaling, whereas Bax functions to promote programmed cell death and appears to act as a tumor suppressor (4).

With regard to tumor suppressor genes that inhibit cell proliferation, the p53 gene encodes for a protein that serves as the primary modulator of transcription, particularly in response to DNA damage. Activation of p53 leads to a rapid increase in corresponding protein levels and contributes to the activation of p53 as a sequence-specific transcription factor for the regulation of downstream genes responsible for modulating proliferation/cell-cycle arrest, restriction point (G_1/S phase) transition, and/or apoptosis (1). As an example, p21(*Waf*1/*Cip*1) represents a p53-responsive gene that inhibits cyclin-dependent kinase (CDK)/cyclin complexes during the first gap (G_1) phase of the cell cycle, whereas the oncoprotein Mdm2 binds to the transcriptional activation domain of p53 and inhibits its ability to regulate target genes. Consequently, loss of p53 function allows for the continued survival of genetically damaged cells, which, in turn, can contribute to neoplastic transformation.

Central to the regulation of cell-cycle progression, i.e., transition from the G_1 to the DNA synthesis (S) phase, is the p16-cyclin D_1-CDK4-Rb pathway (1). The retinoblastoma (*Rb*) gene encodes for a nuclear phosphoprotein that, when hypophosphorylated, binds and modulates the cellular proteins required for G_1/S phase transition; hypophosphorylated Rb effectively inhibits S phase entry. Cyclin D_1/CDK4 functions with other cyclin/CDK complexes to phosphorylate Rb, which, in turn, inhibits Rb binding to transcription factors and allows restriction point (G_1/S phase) transition through the cell cycle. The $p16^{INK4}$ protein represents the third target for change within the p16-cyclin D_1-CDK4-Rb pathway, regulating Rb function by inhibiting cyclin D_1/CDK4 kinase activity. It would appear that most, if not all, lung cancers possess acquired genetic or epigenetic (transcriptional silencing via hypermethylation) abnormalities associated with this critical regulatory pathway.

Morphologically distinct changes such as hyperplasia, metaplasia, dysplasia, and carcinoma *in situ* have been reported in bronchial epithelium prior to the appearance of a clinically overt lung cancer, with preneoplastic cells and adjacent bronchial epithelium exhibiting a number of the genetic abnormalities previously discussed (1). Allelotyping of microdissected preneoplastic foci suggests that 3p allele loss constitutes an early molecular change associated with lung tumorigenesis, followed by additional losses at 9p, 17p (and p53 mutation), and 5p, and eventual *ras* mutation (5–9). A high prevalence of deletion of one copy of the short arm of chromosome 3 in both small-cell (>90%) and non-small-cell (>80%) lung cancer provides strong evidence for the existence of one or more critical genes on this chromosomal arm. Ki-*ras* activation (along with p53 mutation) has been proposed as an early causative event during lung tumorigenesis (10–13). This association is supported by the observation that mutations for this gene are present during atypical alveolar hyperplasia, a potential precursor lesion of adenocarcinomas (14,15). These observations are consistent with the multistep model of carcinogenesis and a "field cancerization" process whereby the entire tissue region repeatedly exposed to potential carcinogenic damage is at risk of developing multiple, separate foci of neoplasia (16).

15.2 SIMILARITIES DEMONSTRATED WITHIN MAMMALIAN SPECIES

The rationale for using animal studies to evaluate human health risk is that mammalian species generally possess similar genetic, biochemical, and physiologic makeups. These similarities extend to toxification and detoxication mechanisms, as well as to the target sites for the adverse effects of toxicants. With advances in cellular and molecular biology supporting the concept that human cancer develops as a multistage process, mouse skin has emerged as a convenient model for investigating the tumorigenic process (17,18). In fact, much of the experimental evidence supporting a multistage process for tumor development, i.e., involving the complementary stages of initiation, promotion, and progression, has been provided by this experimental model.

Use of the mouse skin tumorigenicity model to assess potential human lung cancer risk is supported by broad evidence that epithelial cells from diverse organ systems exhibit significant similarities. These similarities include architectural arrangement, response to mitogenic and/or inhibitory signaling factors, mechanisms of cell-cycle regulation, metabolic activation and/or detoxication of carcinogens, and DNA repair (19). Epithelial cell populations provide the first line of defense against the many exogenous toxicants to which the body is exposed and possess specific defense mechanisms to protect cells important for long-term tissue survival. Cell compartmentalization within the epithelia has important practical implications with regard to neoplastic transformation, including the localization of stem cells in a manner that minimizes both exposure to toxicants and the probability of genetic insult. Additionally, the coexistence of other potential target (nonstem) cells, many of which are approaching the end of their proliferative life span, minimizes the likelihood that mutations will result in tumor formation.

Other protective mechanisms involve cell-cycle control and/or the response of particular cell types to potentially damaging DNA lesions. The crucial stem cells are long-lived, slowly cycling cells that undergo division only when necessary for tissue repopulation, e.g., after removal of cells by wounding or as a function of cell turnover. This presumably allows ample time to repair any DNA damage that might have occurred during the long resting phase of the cell cycle. Stem cell populations are particularly sensitive to certain threshold levels of DNA damage, and undergo self-destruction through apoptosis if the damage cannot be adequately repaired (19). Despite the multiplicity of defense systems guarding against neoplastic transformation, multiple genetic events that target individual facets of cell growth and tissue architecture (growth factor signaling; cell-cycle control and the balance between proliferation and differentiation; cell location, adhesion, and longevity within the tissue; and responses to DNA damage) can lead to the formation of tumors.

Additional support for the use of mouse skin as a model for assessing potential human lung cancer risk is provided by the commonality of molecular events involved in tumor formation. Both skin and lung epithelium possess the metabolic capacity to activate procarcinogens to forms capable of inducing genetic damage. In turn, irreversible changes occur at the genetic level to convert normal epithelium to a population of preneoplastic or dormant tumor cells. These changes, regardless of tissue origin, modulate the activity of oncogenes that function in a positive manner to accelerate tumorigenesis and tumor suppressor genes that fail in their negative modulation of growth and induction of differentiation or apoptosis. Chemical interaction with the stem cell DNA, which induces an irreversible genetic change, frequently results in dysregulation of proliferative control, with altered *ras* signaling being a prominent genetic event associated with chemically initiated mouse skin (20–29). Oncogenic activation of *ras* has similarly been reported to be an early, common event in human lung tumorigenesis (10–13).

Tumor promotion involves clonal expansion of the initiated cell population and leads to the formation of macroscopic (usually benign) tumors. Significant progress in discerning the mechanisms involved in tumor promotion has been made through studies examining DNA methylation, i.e., the 5-methylcytosine content of DNA (30). The concept that altered methylation of DNA, involving an epigenetic mechanism, may be important during tumor promotion is consistent with the reversible nature of this step; specifically, *de novo* methylation provides an opportunity for the reversal of altered methylation. Moreover, altered DNA methylation may result in a changed pattern of gene expression that provides subpopulations of cells with a selective growth advantage, consistent with the clonal expansion of initiated cells. Altered patterns of gene expression would likewise explain the multiple tumor phenotypes observed during tumorigenesis. Studies focusing on inhibitory factors impacting Rb control of cell-cycle progression have identified aberrant methylation of $p16^{INK4a}$, a putative tumor suppressor gene, as functioning in a cooperative manner with the *ras* oncogene to promote tumor formation in both mouse skin and human lung (19,31,32).

15.3 MULTISTAGE NATURE OF TUMORIGENESIS

Separation of the tumorigenic process, at least conceptually, into discrete stages provides a framework in which tumor development can be examined. Nonetheless, it is important to recognize that the stages of tumorigenesis are invariably intertwined. For example, recent studies illustrate that the stages of initiation and promotion often function in a synergistic manner during tumor formation. Studies employing Tg.AC mouse skin support the existence of non-Ha-*ras*, 7,12-dimethylbenz(*a*)anthracene (DMBA)-initiated target genes capable of increasing tumor formation within the context of a classical initiation–promotion protocol (33). Based on transgene expression of oncogenic v-Ha-*ras* within the skin, the Tg.AC mouse model would no longer require chemical initiation for tumor development. Moreover, the available initiated cell population (due to transgene expression) would be expected to far exceed that induced with high concentrations of chemical initiator, the standard practice for mouse skin tumorigenicity studies. Interestingly, limited DMBA initiation and subsequent 12-*O*-tetradecanoyl-phorbol-13-acetate (TPA) promotion of Tg.AC mice induced a tenfold increase in tumor-related end points as compared to acetone-initiated TPA-promoted controls. The implication is that there exist additional target genes that cooperate with mutated Ha-*ras*, an established target of DMBA in mouse skin, to increase tumorigenicity. Cells containing a mutant Ha-*ras* and presumably an additional initiator-induced genetic change were two orders of magnitude more responsive to promoter treatment and exhibited an increased frequency for malignant conversion.

Subsequent studies examined the practical impact of a multihit scenario during initiation and the consequences of such a scenario in terms of tumor formation (34). Results from these studies demonstrated that a D/T/A/T (D = DMBA, T = TPA, and A = acetone) dosing regimen yielded a tumor multiplicity (tumors/mouse) of 2.8, with D/A/D/T and D/T/D/T regimens yielding multiplicities of 5.8 and 12.4, respectively; hence, the D/T/D/T regimen induced a synergistic response. More than 90% of tumors resulting from the D/T/A/T and D/T/D/T treatment regimens contained a mutated Ha-*ras* allele, although a normal Ha-*ras* allele persisted in all cases, indicating that a gene other than the remaining normal allele was the target for the second genotoxic event. It was concluded that tumor promotion (clonal expansion) between the first and second mutation has a profound outcome on tumorigenesis by increasing the probability that the second genotoxic event will occur in a previously initiated cell.

Consequently, the mouse skin carcinogenesis model can arguably be extended to a multihit, multistage scenario. Repetitive exposure to mutagenic agents promotes the accumulation of additional genetic alterations. Moreover, cells with this additional genetic damage may exhibit a selective growth advantage with promoter treatment. By clonally expanding DMBA-initiated cells with TPA promotion, the probability of a second mutation occurring in a cell with a preexisting alteration is significantly increased. Initiation is typically considered to be a relatively infrequent event, with approximately 0.1% of Ha-*ras* genes reportedly sustaining a codon 61 mutation 24 h following chemical exposure; at 9 d postexposure, codon mutations were undetectable (34). The probability that additional mutational events from a second application of carcinogen would occur in the small population of preinitiated cells in the absence of clonal expansion would be minimal. However, the population of mutant Ha-*ras*-containing cells is likely increased by a factor of $>10^4$ during promoter-driven clonal expansion of preneoplastic mouse skin, thereby significantly increasing the target population of initiated cells capable of sustaining additional genetic damage.

Initiation and promotional events can function in a synergistic manner, with tumor promotion effectively maintaining and expanding the initiated phenotype; hence, these seemingly discrete stages are revealed to be invariably intertwined. Adding to this complexity is the view that the promotion stage may be more accurately described in terms of two distinct processes. Although it is generally accepted that promotion by TPA represents a single process related to the strong and persistent hyperplasia induced by its repeated application to chemically initiated skin, this view may be somewhat inadequate in explaining the functionality of tumor promoters during mouse skin tumorigenesis. Early studies by Boutwell (35) revealed that application of the promoter croton oil to initiated skin for 6 weeks resulted in relatively few skin tumors, whereas treatment for 14 weeks completed the promotion process with the appearance of numerous tumors. The active promoting ingredient in croton oil has since been demonstrated to be TPA (36). Substitution of turpentine for croton oil during the final 8 weeks of promotion similarly induced multiple tumors, whereas the exclusive use of turpentine produced none, suggesting a qualitative difference between these two promoter treatments.

The substitution of wound healing for turpentine treatment following limited croton oil application to initiated mouse skin yielded similar results (35). Wound healing alone in initiated mouse skin yielded no tumors, but many of the mice treated with croton oil for 5 weeks post-initiation developed one or more tumors along the line of an isolated wound. These findings provided the basis for a three-stage sequence of tumorigenesis, proposed as: (1) *initiation* to a precancer state, carried out by a single, small dose of carcinogen and involving an irreversible genetic change; (2) *conversion* of an initiated cell into a preneoplastic, or dormant tumor cell (induced by croton oil—first-stage promotion); and (3) *propagation* of the dormant tumor cell population, dependent upon cellular proliferation (induced by turpentine or wound healing—second-stage promotion). The complex nature of the molecular interactions and cellular events involved in this process would likely result in synergy under certain experimental designs, including the tumorigenicity testing of complex mixtures.

Furstenberger et al. (37) examined the stability of conversion (first-stage promotion) using an experimental protocol that employed limited application of TPA to initiated mouse skin, followed by twice-weekly promotion with a second-stage promoter, i.e., the TPA analog 12-*O*-retinoylphorbol-13-acetate (RPA). It was observed that RPA retained some promotional activity even when its repeated application was delayed up to 8 weeks after completion of the brief TPA exposure, raising the additional question of whether conversion could be established prior to initiation. In an effort to avoid the complications associated with short-term TPA effects (hyperplasia and inflammation, which disappear within a few weeks of exposure), initiation by DMBA was conducted 2 to 6 weeks after TPA application. The result of TPA exposure (prior to DMBA initiation) and subsequent long-term application of RPA was the production of tumors to an extent comparable to that obtained with the conventional TPA application after initiation (38).

With regard to the manner in which conversion may be affected by first-stage promoters, several lines of evidence suggest that although promoters generally do not interact directly with DNA and are not necessarily mutagenic, compounds such as TPA possess significant biological activity that may secondarily result in chromosomal or mutational events (39). Birnboim reported that a single application of TPA resulted in DNA strand breaks in human leukocytes, with the observed damage apparently related to a sharp increase in reactive oxygen species (ROS) production occurring shortly after promoter addition (40). This conclusion was based on the observation that enzymes capable of removing ROS demonstrated a pronounced inhibitory effect on the TPA-induced DNA damage. Numerical and structural aberrations of chromosomes were similarly produced during TPA treatment of mouse keratinocyte cultures (41), whereby increases in gaps and chromatid breaks were observed, accompanied by intra- and inter-chromosomal exchanges. These aberrations were visible within 24 h of treatment, increased with longer and multiple exposures, and persisted for several days following removal of TPA from the culture medium.

These findings were in good agreement with studies demonstrating that brief treatment of mouse keratinocyte cultures with TPA produced double minute chromosomes, or the cytogenetic equivalent of gene amplification (42). Treatments that prevented promotion of initiated mouse skin by TPA similarly inhibited the formation of chromosomal aberrations, indicating a possible causal or functional relationship between the aberrations and the process of conversion (first-stage promotion). As a further indication of a primary role for chromosomal aberrations during conversion, the alkylating and clastogenic agent methyl methanesulfonate (MMS) was demonstrated to induce chromosome breaks and gaps in epidermal cells when topically applied to mouse skin, although no initiating activity was observed (43). Moreover, MMS proved to be a powerful agent of conversion; in contrast to TPA, MMS was a weak inducer of DNA synthesis and, hence, had to be accompanied by a second-stage promoter to exert a promoting effect comparable to TPA.

Collectively, these results are consistent with a role for chromosomal aberrations in the conversion stage of tumor promotion. Tumor progression, which represents the transition from a benign to malignant neoplasm, is characterized by a high level of genetic instability and increased numbers of chromosomal alterations (44–49). Conversion (first-stage promotion) and tumor progression both appear to be based on chromosomal effects, suggesting that the two events are closely related. Given the degree of overlap between the two stages of promotion and the synergy exhibited for initiation and promotion, perhaps the concept of tumor progression as an independent, largely stochastic occurrence requires reevaluation. Although describing the events involved in chemically induced tumorigenesis as occurring in discrete stages is useful, it should be recognized that there are a myriad of complex molecular and cellular changes that occur.

15.4 TUMORIGENICITY TESTING OF TOBACCO SMOKE

Although lung cancer is one of the more significant human health effects attributed to tobacco use (50), development of a useful rodent lung tumorigenicity model that provides both reliable and reproducible dose-related results in response to tobacco smoke inhalation exposure has been difficult. Whereas classical rodent inhalation models exposed to tobacco smoke have generally demonstrated limited lung tumor induction (51), some measure of success has been observed with genetically predisposed rodent models.

Witschi and colleagues (52–57) published a series of reports providing evidence for tobacco-smoke-induced lung tumor formation within the A/J mouse model. Whole-body inhalation exposure to a mixture of 89% sidestream and 11% mainstream tobacco smoke (MTS) was reported to yield statistically significant increases for tumor indices relative to air-exposed controls. Collectively, mice were exposed to inhalation chamber concentrations ranging from 78.5 to 137 mg total suspended particulates (TSP) per cubic meter of air (6 h/d, 5 d/week), with *ad libitum* access to feed and water maintained during exposures. Demonstration of increased tumor formation required that the standardized 20-week exposure period be followed by a 16-week recovery period, during which time mice were provided filtered air. Justification for the recovery period was provided by Bogen and Witschi (58), who suggested that tobacco smoke exposure suppressed the growth of preneoplastic foci. Tobacco-smoke-induced lung tumor formation was proposed to occur predominantly via a genotoxic mechanism, with the recovery period allowing for smoke-induced genetic damage to progress to tumors. The conclusion that tobacco smoke exposure was capable of suppressing preneoplastic foci was likewise supported by results from cocarcinogenesis studies employing either urethane or 3-methylcholanthrene (3-MC) followed by inhalation exposure to the tobacco smoke mixture. Twenty weeks exposure to tobacco smoke significantly reduced lung tumor formation induced by either of these rodent lung carcinogens (52).

Although these results regarding the ability of tobacco smoke to induce lung tumors in A/J mice were initially confirmed by D'Agostini et al. (59), subsequent experiments employing a similar testing protocol (whole-body exposure to 89% sidestream and 11% MTS, 113 mg TSP/m^3 of air; 6 h/d for 20 weeks, with 18 weeks recovery) failed to demonstrate increased tumorigenicity (60). In an effort to examine the impact of apoptotic signaling during tobacco-smoke-induced tumor formation, the latter study additionally employed a transgenic (Tg) mouse (UL53-3 X A/J) possessing a dominant-negative p53 mutation (60). Although the sensitivity of the mutant mice to tobacco-smoke-induced tumorigenicity was not striking (e.g., 0.8 tumors/animal for tobacco-smoke-exposed females compared to 0.2 for corresponding sham controls), it was significantly enhanced relative to wild-type A/J mice (e.g., 0.3 tumors/animal for smoke-exposed females compared to 0.3 for sham). It is worth noting that studies employing MTS, in contrast to a mixture of sidestream and mainstream smoke, have been largely unsuccessful in terms of demonstrating increased tumor formation in A/J mice. Finch et al. (61) exposed A/J mice whole-body to MTS at a concentration of 248 mg TSP/m^3 in air for 24 weeks (6 h/d, 5 d/week). Following 5 weeks recovery, MTS-exposed animals were devoid of grossly observable lung tumors, whereas air-exposed shams possessed a background multiplicity of 0.3 (tumors/animal) and an incidence of 26% (tumor-bearing animals, TBA). In addition, D'Agostini et al. (59) reported results from a 20-week whole-body exposure and 16-week recovery regimen, with A/J mice provided MTS at a concentration of 1300 mg TSP/m^3 in air (1 h/d, 5 d/week); neither tumor multiplicity nor incidence was increased relative to sham controls following completion of the recovery period.

Attempts by our laboratory to develop a rodent lung tumorigenicity model for the evaluation of MTS employed both the A/J mouse and rasH2 Tg mouse models. The A/J mouse was selected, in part, based on reports that the increased sensitivity to lung carcinogens exhibited by this model was likely linked to genes designated as *Pas*, for pulmonary adenoma susceptibility (62). Experimental evidence suggests that the Ki-*ras* proto-oncogene is a likely candidate for *Pas-1* (63). A similar sensitivity to lung carcinogens has been demonstrated for the rasH2 Tg mouse, which carries the prototype human c-Ha-*ras* gene (~5–6 copies of human c-Ha-*ras* gene integrated into the genome in tandem array) within its own promoter region (64). Results from our studies demonstrated that whole-body MTS inhalation exposure (0.200 mg of wet total particulate matter per liter [WTPM/l] of air, or 200 mg TSP/m^3; 6 h/d, 5 d/week for 20 weeks) induced statistically significant increases in both tumor multiplicity and incidence in both mouse strains (65). In each case, detection of statistical differences required the inclusion of a 16-week recovery period, as well as a more stringent assessment of the lung (microscopic verification using serially sectioned tissue) than reported for previous studies (gross determination of intact lung). MTS exposure of A/J mice for 20 weeks (assessed prior to recovery) resulted in reductions for lung tumor formation (refer to Figure 15.1), extending published findings regarding the ability of tobacco smoke to potentially suppress the growth of preneoplastic foci, possibly via increased apoptotic signaling.

Whole-body exposure of rasH2 Tg mice for 20 weeks yielded increases for tumor multiplicity (0.71 microscopically confirmed tumors/animal compared to 0.23 for shams) and incidence (57% TBA compared to 23% for shams). Although these changes were not statistically significant, they represented a contrast to the A/J mouse findings of reductions for both indices immediately following cessation of MTS exposure (refer to Figure 15.1). That neither of these models demonstrated significant increases for lung tumor formation in response to tobacco smoke inhalation exposure illustrates the difficulty of trying to identify a useful rodent lung tumorigenicity-testing protocol. Nonetheless, the observed responses

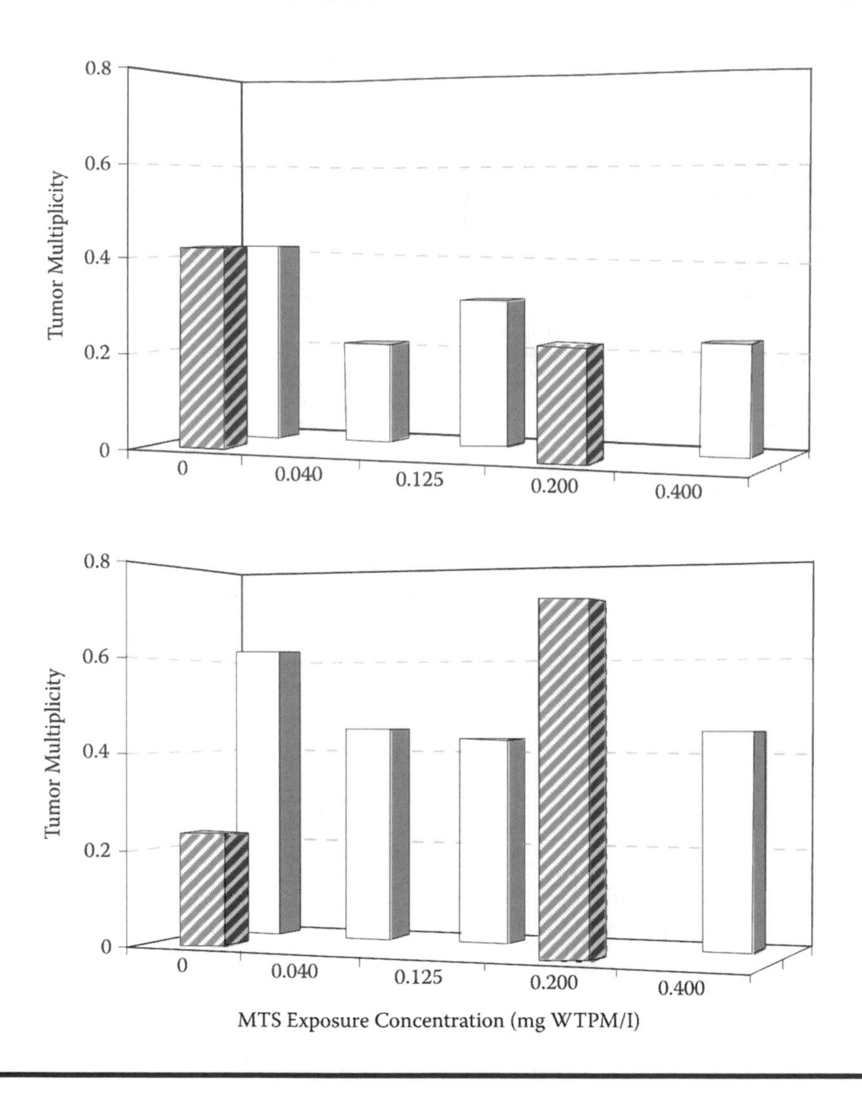

Figure 15.1 Microscopically confirmed lung tumor multiplicity in response to MTS inhalation exposure. A/J (*top panel*) and RasH2 Tg (*bottom panel*) mice exposed either whole-body (*slashed bars*) or nose-only (*solid bars*) for 6 h/d (5 d/week, 20 weeks) or 3 h/d (5 d/week, 28 weeks), respectively.

lead to the speculation that the mutation carried by A/J mice results in increased sensitivity to toxicant-induced apoptosis and, hence, the killing of preneoplastic or neoplastic cells. Conversely, the altered *ras* signaling exhibited by rasH2 Tg mice (gene dosage effect) may be sufficient to either support MTS-induced tumor development or allow additional genetic damage in the absence of apoptosis. This conclusion is complicated by the finding that nose-only exposure of both A/J and rasH2 Tg mice (0.040, 0.125, or 0.400 mg WTPM/l; 3 h/d, 5 d/week for 28 weeks) inhibited lung tumor formation (relative to shams) immediately following cessation of dosing. It is important to note that all tumors were microscopically confirmed from serially sectioned lung (400-μm step sections), and that the whole-body exposure concentration employed for the corresponding studies (200 mg WTPM/l) fell within the range of nose-only concentrations tested, the only differences being the route and duration of exposure. Possible explanations for these findings, as well as the experimental approaches currently being employed to elucidate the underlying reasons for this disparity, are discussed in subsequent sections.

That toxicant-induced apoptosis is effective in removing preneoplastic or neoplastic cells to the extent that it represents a potential confounder during tobacco smoke testing constitutes a significant finding. Additional support for this finding is provided by studies demonstrating that MTS exposure induces statistically significant increases in apoptosis within the lungs of rodents. D'Agostini et al. (66) reported that whole-body MTS exposure induced a twofold increase in apoptotic cells residing within the bronchial/bronchiolar epithelium following 18 d of continuous dosing, with an approximate tenfold increase observed at 100 d dosing; the number of proliferating cells (evidenced by proliferating cell nuclear antigen, or PCNA, staining) was increased threefold at both 18 and 100 d of exposure. In addition, similar exposure to a mixture of sidestream and MTS (66) for 28 consecutive days yielded a tenfold increase in apoptosis and a corresponding fourfold increase in PCNA staining. Similar findings have been generated within our own laboratory (refer to Figure 15.2), with MTS inducing statistically significant increases in apoptosis within the lung following either acute (14–28 d) or subchronic (90 d; data not shown) exposure. MTS exposure failed to induce a proliferative response (as indicated by 5-bromo-2-deoxyuridine, or BrdU, labeling) that was greater than the level of apoptosis during either the 28-d exposure period or the subsequent 21-d recovery, a finding that is inconsistent with increased tumor formation.

In fact, several laboratories have reported that tobacco smoke exposure of A/J mice leads to a reduction in lung tumor formation (relative to sham controls) immediately following cessation of dosing, even inhibiting the responses of the known rodent lung carcinogens urethane and 3-MC (52). It would appear that tobacco smoke is actually providing an antipromotional stimulus during exposure. Accepting that the A/J mouse constitutes a genotoxicity model, as proposed by Bogen and Witschi (58), this widely employed testing protocol fails to explain the means by which preneoplastic cells are provided a selective growth advantage for the development of tumors. If, as supported by results from the early assessments of tobacco smoke constituents (discussed in the following text), the overriding tumorigenic potential

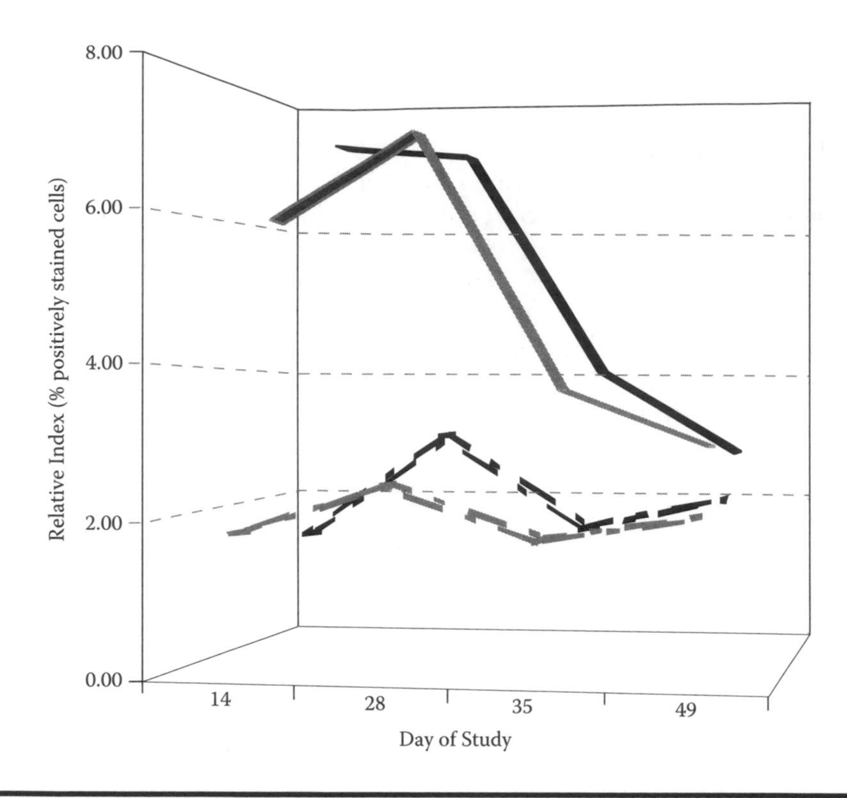

Figure 15.2 Relative lung cell proliferation and apoptosis in A/J mice following MTS inhalation exposure. Mice exposed nose-only to MTS (0.400 mg WTPM/l, 3 h/d, 7 d/week) for 28 d, followed by 21-d recovery; the percentages of BrdU+ (*gray lines*) and apoptotic+ (*black lines*) cells were determined for both sham- (*dashed lines*) and MTS-exposed (*solid lines*) mice.

of this complex mixture resides as promotional activity, then the A/J mouse lung model would appear to be largely incapable of adequately assessing the tumorigenic activity of tobacco smoke. Moreover, the absence of a promotional stimulus, critical for the selection and propagation of initiated cells, during either tobacco smoke exposure or the often-employed recovery period, would likely result in the formation of relatively few tumors.

15.5 TUMORIGENICITY TESTING OF CIGARETTE SMOKE CONDENSATE

Recognizing that neoplasms of the lung, larynx, mouth, and esophagus are not consistently produced in animals as a result of inhalation exposure to tobacco smoke (50,67) and acknowledging that the widely used A/J mouse has yet to be effectively validated as a model for assessing tobacco-smoke-induced lung tumorigenicity (68,69), our laboratory continues to rely primarily on comparative evaluations of cigarette smoke condensates (CSCs) during toxicological testing. This is accomplished primarily through the use of a standardized mouse skin dermal tumor promotion protocol that has provided both reliable and reproducible results in terms of numbers of total tumors and TBA (70,71).

During the 1970s, the National Cancer Institute (NCI) engaged in a series of studies with the expressed objective of reducing health risks associated with tobacco consumption through the development of less hazardous cigarettes (72); these efforts relied principally on the predictive power of the chronic mouse skin tumorigenesis model. Influencing the NCI's selection of this model were reports demonstrating the ability of this assay to assess the tumorigenic potential of CSC. Wynder and Hoffmann (73) had confirmed the existence of tumor-promoting agents in the condensate, whereby repetitive promotion (52 weeks) of DMBA-initiated Ha/ICR mice resulted in approximately 45% of experimental animals developing dermal tumors. A follow-up study examined the initiating and promoting activities of specific CSC subfractions (74) and reported that whole tar and the neutral portion of fractionated CSC exhibited tumor-promoting potentials significantly greater than the activities of these materials in a complete tumorigenesis assay (i.e., tested in the absence of initiation).

Studies by Wynder and colleagues (75–79), examining the tumorigenic capacity of both whole and fractionated CSC, suggested that carcinogenic polycyclic aromatic hydrocarbons (PAHs) are present at a concentration far too low to unilaterally account for the activity exhibited in mouse skin; the combined concentration of known PAH carcinogens was deemed to account for approximately 3% of the detectable carcinogenic potential. Moreover, although CSC contains numerous mutagenic constituents, the overall initiating activity was assessed to be relatively weak (80), with the content of cocarcinogens (promoters) being sufficient to elicit a tumorigenic response in mouse skin previously initiated with either a single dose of PAH or the repeated simultaneous application of low doses of PAH and various condensate constituents (80–84). The notion that PAHs account for only a small percentage of the tumorigenic activity exhibited by CSC, and that some PAHs that are borderline or noncarcinogenic can act as initiators when followed by CSC promoters, remain as accepted views by many in the field of tobacco smoke toxicology (85,86).

Collective review of results from mouse skin studies examining CSC tumorigenic potential underscores the notion that this complex mixture constitutes a weak initiator, yet possesses measurable tumor-promoting activity (87). Only Slaga and coworkers (88) have attributed any initiating potential to CSC, reporting that SENCAR mice treated with CSC and promoted twice weekly with TPA (50 weeks) exhibited a tumor multiplicity slightly greater than controls at study termination. Comparing the initiating potentials of diesel extracts, roofing tar, coke oven emissions, and CSC, it was concluded that the initiating properties of CSC in mouse skin were relatively weak. This finding is in good agreement with mouse skin tumorigenicity studies examining a number of presumed human respiratory carcinogens (89,90); relative initiating potencies for coke oven emissions, roofing tar, and CSC were found to be 1.0, 3.9, and 0.0024, respectively. Finally, Slaga and colleagues (91) recently reported on efforts to correlate sustained epidermal hyperplasia and increased epidermal ornithine decarboxylase (ODC) activity with tumor-promoting potential. During an examination of more than 50 suspected carcinogens, it was concluded that CSC contributed to tumor formation by providing a moderate promoting activity.

Findings from our laboratory, likewise employing SENCAR mouse skin within a standardized dermal tumor promotion protocol, demonstrated that CSC failed to induce statistically significant increases in the frequency of mutated *ras* (codon 61), implicated during chemical initiation of mouse skin (92). Conversely, statistically significant increases for the indices related to sustained hyperplasia (epidermal thickness and proliferative index) and inflammation (myeloperoxidase activity and leukocyte invasion) were observed, consistent with tumor-promoting potential. These changes were evident in both DMBA-initiated and noninitiated mouse skin. CSC-induced hyperplasia and inflammation diminished with continued promotion (evaluated at 4, 8, and 12 weeks of exposure), consistent with an amelioration of grossly observed changes (desquamation and erythema) specific to the skin. Reduced clinical findings within the chemical application site of DMBA-initiated, CSC-promoted skin were temporally associated with the development of tumors, which continued to develop in a time- and dose-dependent manner throughout the 29-week promotion period. Continued promotion of noninitiated mouse skin failed to induce tumors in a manner that was statistically different from that observed for vehicle controls.

These results suggest that CSC testing protocols likely require chemical initiation using subthreshold levels of a carcinogen to adequately assess tumorigenic potential. If tumor promotion encompasses both the conversion of initiated cells and the selective propagation of preneoplastic cells as proposed by Boutwell (35), CSC would appear to possess the potential for both conversion and propagation based on its ability to induce significant numbers of tumors in initiated mouse skin. Consistent with published studies examining conversion and propagation (35,37,38), short-term promotion of initiated mouse skin with CSC yields modest increases in tumors, with continued application required to complete the promotion process. That condensate is capable of inducing an irreversible change that complements prior chemical initiation (i.e., conversion of initiated cells to preneoplastic cells) is supported by recent studies demonstrating that CSC induces ODC expression within the perifollicular epidermis in a dose-dependent and irreversible manner following 9 weeks of treatment (submitted for publication). Moreover, inhibition of ODC expression through the use of a specific inhibitor, difluoromethylornithine provided in drinking water markedly reduces CSC-induced dermal tumor formation (unpublished data). Increased ODC expression has been demonstrated by other laboratories to provide a promotion stimulus that is both necessary and sufficient for tumor development (93–98), with the perifollicular epidermis reported to be the origin of mouse skin tumors (99–101).

15.6 ADAPTATION DURING CADMIUM-INDUCED TUMORIGENICITY

Cadmium (Cd) is a toxic transition metal of significant occupational and environmental concern and has been classified as a human carcinogen (102) based on epidemiological data and experimental findings demonstrating lung tumor formation in animals following long-term, low-level inhalation exposure (103,104). The underlying mechanism responsible for Cd-induced carcinogenicity has yet to be fully elucidated, due in part to the consideration that this carcinogen is poorly mutagenic and probably acts through indirect genotoxic or epigenetic mechanisms, e.g., the aberrant activation of oncogenes and/or suppression of apoptosis (105). Although Cd is capable of eliciting intracellular oxidant production and lipid peroxidation (106), it does not exhibit Fenton chemistry and does not appear to directly induce oxidative DNA damage. Cd has been reported to increase the mutation frequency of known genotoxic agents (107), as well as effectively inhibit base excision and nucleotide excision repair processes (108,109).

In terms of molecular pathogenesis, Cd has been reported to activate several proto-oncogenes associated with cell proliferation (c-*myc*, c-*jun*, c-*fos*), both *in vitro* and *in vivo* (110–115), as well as induce upregulation of signaling pathways contributing to increased mitogenesis (116). More recently, Cd-induced transformation of rodent liver cells was proposed to be associated with errors in DNA methylation (117), possibly due to inhibition of DNA methyltransferase during the early stages of toxicant exposure. With regard to tumor suppressor genes, Cd has been reported to inhibit proteins critical to cell-cycle control by disrupting native (wild-type) p53 conformation, which, in turn, inhibits its binding to DNA and downregulates transcriptional activation (118). Given that this transcription factor serves to allow sufficient time for the cell to repair genetic damage prior to DNA replication and/or cell division, the modulation of p53 function could potentially allow proliferation even after exposure to endogenous or exogenous DNA-damaging agents.

Toxicant-induced apoptosis, a process that functions to eliminate genetically damaged cells, is effectively blocked by Cd (119,120), reportedly through its ability to inhibit a key enzyme of the cell death signaling pathway (119). Human prostate epithelial cells transformed by Cd demonstrate a marked resistance to this carcinogen, with the acquired resistance likely due to a global decrease in caspase expression coupled with decreased production of the proapoptotic regulatory protein, Bax, and overexpression of anti-apoptotic Bcl-2 (121). Cellular redox balance, as determined by a complex system of enzymatic and nonenzymatic antioxidant defenses (122), constitutes an important modulator of toxicant-induced apoptosis. Specifically, diminished antioxidant capacity would be predicted to favor apoptotic signaling, whereas enhanced antioxidant capacity would be expected to inhibit this type of cell death (122–127). Although it remains to be established whether Cd-induced suppression of apoptosis contributes in a significant manner to lung carcinogenesis, repeated inhalation of Cd aerosols does appear to convey an increased resistance to higher doses, as well as cross-tolerance to otherwise-lethal levels of oxygen (128,129).

A pulmonary rat model of Cd adaptation has been developed by periodically exposing animals to an atmosphere containing 1.6 mg Cd/m^3 of air. Initial weeks of the adaptive process were characterized by the release of cellular enzymes (lactic dehydrogenase, acid and alkaline phosphatases, and lysozymes) into the alveolar space and by an intense inflammatory response induced by an influx of activated polymorphonuclear leukocytes and macrophages into the lung. By the third week of exposure, pulmonary cell injury began to resolve; with the exception of hyperplasia and hypertrophy of the alveolar epithelium, Cd-exposed animals resembled sham controls by the fourth week of exposure (129,130). Initial studies employing lung cells adapted to Cd, *in vitro* or *in vivo*, revealed an increased oxidant resistance, which in turn was tentatively ascribed to changes in antioxidant defenses (131,132). The ability of Cd to induce apoptosis in nonadapted cells would potentially have the effect of selecting for apoptotic-resistant variants arising in neoplastic populations (133,134). Subsequent studies (135) demonstrated that alveolar epithelial cells, adapted to Cd by repeated *in vitro* exposure, exhibited a marked resistance to hydrogen peroxide-induced apoptosis compared to similarly challenged nonadapted cells. This suppression was due, in part, to an observed upregulation in the gene expression of several resistance factors (discussed later). It was concluded from these studies, as well as other published reports (122,124,136), that the total cellular balance existing between antioxidants and pro-oxidant intermediates, rather than a single factor, likely determined susceptibility to apoptosis.

Hallmarks of Cd adaptation include hyperplasia and hypertrophy of type II alveolar epithelial stem cells, an inflammatory response involving polymorphonuclear leukocytes, and the increased gene and protein expression of several resistance factors (137). With regard to the latter, upregulation of metallothionein (cysteine-rich, metal-binding protein possessing considerable free-radical-scavenging ability), increased levels of glutathione (GSH), and the induction of enzymes involved with both the synthesis of GSH (γ-glutamylcysteine synthetase regulatory and catalytic subunits) and its metabolism (GSH *S*-transferases) have been implicated as contributing to the pulmonary adaptive response. This enhancement of important cellular defense systems in response to Cd exposure, although initially beneficial, may ultimately contribute to the carcinogenic process. For example, Cd-adapted alveolar epithelial cells have been reported to exhibit a reduced ability to repair DNA damage (108,109), due, in part, to the inhibition of two base excision repair enzymes (8-oxoguanine-DNA glycosylase and endonuclease III). Cells with genetic aberrations resulting from unrepaired DNA lesions normally would be removed from the lung by apoptosis, but, in this case, oxidant-induced apoptotic signaling is significantly attenuated in Cd-adapted cells relative to their nonadapted counterparts. Suppression of apoptotic signaling, particularly within stem cells that can be stimulated to proliferate, could lead to the retention of preneoplastic or neoplastic cells, favor their clonal expansion by providing a selective growth advantage relative to nonadapted cells, and possibly promote tumor development.

15.7 ADAPTATION DURING TOBACCO-SMOKE-INDUCED LUNG TUMORIGENICITY

For carcinogens with low mutagenic potential, the adaptive response detailed for Cd may constitute a universal mechanism for understanding tumorigenicity. As previously discussed, adaptation to Cd generally entails hyperplasia and hypertrophy of the target cell population, a sustained inflammatory response, and the increased gene and protein expression of numerous resistance factors (137). Biochemical changes associated with increased resistance include elevated intracellular levels of GSH concomitant with induction of enzymes responsible for both its synthesis and metabolism, a reduced DNA repair capacity due to inhibition of repair enzymes and/or reduced gene expression, and an attenuated apoptotic response.

Studies conducted by our laboratory effectively demonstrate that repetitive application of CSC induces a statistically significant hyperplastic and inflammatory response in chemically initiated mouse skin (92). In fact, a strong correlation was demonstrated for molecular end points associated with condensate-induced hyperplasia (proliferative index) and inflammation (myeloperoxidase activity) relative to numbers of microscopically confirmed tumors generated as a consequence of 29 weeks of promoter treatment. The magnitude of hyperplasia and inflammation generally subsides with time, as the skin appears to adapt to repeated exposure. This amelioration of induced changes is consistent with the concurrent disappearance of grossly observable clinical signs (desquamation and erythema) and the emergence of identifiable tumors. Results from rodent inhalation studies suggest that tobacco smoke induces lung cell hyperplasia in a statistically significant manner compared to air-exposed controls (66,138). Moreover, experimental evidence suggests that the induced hyperplasia represents a transient response and may not continue with repeated inhalation exposure (138).

Efforts to identify the mechanism responsible for Cd-induced carcinogenicity have been complicated, in part, due to the consideration that this carcinogen is poorly mutagenic and probably acts through indirect or epigenetic mechanisms to alter oncogene expression and suppress apoptosis. A similar situation may exist for tobacco-smoke-induced lung tumorigenicity. Considering that similarly exposed A/J and rasH2 Tg mice exhibited dissimilar MTS-induced tumor responses (65), at least in terms of the whole-body exposure regimen employed, our initial hypothesis regarding the contribution of *ras* signaling during toxicant-induced lung tumorigenesis may have been overly simplistic. Recent studies by Witschi and colleagues (139) suggest that the Swiss-Webster (SW) mouse may constitute a rodent inhalation model that is responsive to tobacco-smoke-induced lung tumorigenicity. In this study, 20-week whole-body exposure followed by a 16-week recovery period yielded a tumor multiplicity and incidence of 0.35 and 19%, respectively, compared to 0.04 and 4% for sham controls. These results were confirmed by De Flora et al. (140), employing a similar testing regimen and demonstrating increases greater than fourfold for multiplicity and incidence when comparing tobacco-smoke-exposed mice to shams. Even more significant were the results from studies that employed a 36-week exposure period devoid of recovery; tobacco-smoke-exposed SW mice exhibited 0.68 tumors/animal and 50% TBA compared to 0.14 and 9.1% for sham controls (140).

Although the underlying basis for the increased responsiveness exhibited by rasH2 Tg and SW mice to tobacco-smoke-induced lung tumorigenesis has yet to be determined, the possibility that overexpression of cyclooxygenase-2 (COX-2) imparts the observed sensitivity has become a focus of our investigative studies. The COX-2 isoform is generally expressed at low or undetectable levels in most tissues, but is transiently induced in response to growth factors, proinflammatory cytokines, and mechanical tissue damage (141,142). Noting that overexpression of COX-2 is evident in a high percentage of human tumors, including lung adenocarcinomas, Wardlaw et al. (143) characterized the expression of this inducible protein in lung tissue from a number of cancer-susceptible and cancer-resistant mouse strains. Results from these studies suggested a possible correlation between COX-2 expression in type II alveolar epithelial cells and lung cancer susceptibility, with A/J and SW mice exhibiting high levels of protein and mRNA expression within the lung prior to carcinogen exposure. Similar results were reported by Bauer et al. (144) during an examination of A/J mouse lung tissue, with spontaneous tumors in aged mice staining positively for COX-2; these findings suggested that elevated protein levels were not dependent upon chemical induction. COX-2 immunostaining was observed in normal bronchiolar and alveolar epithelia, with expression levels being much higher in urethane-induced adenomas and carcinomas as compared to controls.

Interestingly, transgenic COX-2 overexpression has been reported to sensitize mouse skin to carcinogen-induced tumorigenesis (145). Although deemed insufficient for tumor induction, it was reported that overexpression of COX-2 transformed the epidermis into an autopromoted state, whereby the tissue was dramatically sensitized to genotoxic carcinogens. Although little information exists regarding the disposition of COX-2 in lung tissue of rasH2 Tg mice, it is plausible, based on the gene dosage effect of c-Ha-*ras*, that expression may be elevated in a manner consistent with A/J and SW mice. Beyond the increased sensitivity reported for these transgenic mice (146,147), the ras/ERK (extracellular-signal-related kinase) signaling pathway reportedly plays a role in the regulation of COX-2. Human non-small-cell lung cancer cell lines harboring Ki-*ras* mutations exhibit high COX-2 expression levels, with inhibition of ras activity in these cell lines having been shown to decrease expression of this protein (148). Moreover, rat intestinal cells transfected with Ha-*ras* tend to overexpress COX-2, with inhibitors of ERK signaling ameliorating the transfection-induced response (149). Finally, Han et al. (150) provided evidence that COX-2 is induced by p53-mediated activation of the ras/raf/ERK signaling cascade, and that elevation of this protein effectively counteracts p53-mediated apoptosis. Additional downstream signaling associated with COX-2 expression includes the activation of growth-regulated genes, c-*myc* and *ODC* (151). Hence, this protein could effectively modulate both the antiapoptotic and proliferative signaling implicated as contributing to the tumorigenic process.

15.8 ROLE OF SECONDARY GENOTOXICITY DURING TOBACCO-SMOKE-INDUCED ADAPTATION

There is increasing experimental evidence that sustained inflammation and cellular proliferation contribute to the genotoxic and, ultimately, tumorigenic effects of several diverse lung toxicants. An excessive and persistent inflammatory response and the subsequent secretion of cell-derived oxidants can result in genetic damage (152), a phenomenon that has been termed *secondary genotoxicity*. These inflammatory mediators have been implicated in the activation of proto-oncogenes (153), similarly inducing local tissue damage and tissue remodeling (154). Studies have demonstrated that enhanced epithelial cell proliferation increases the likelihood that oxidant-induced DNA damage will become fixed in a dividing cell and that clonal expansion of such mutated cells does occur (155). The net result of exposures that are sufficient to induce a chronic inflammatory response and enhanced cellular proliferation in the lung would be an increased likelihood of genetic mutation and, ultimately, tumor formation. Implicit in this concept is the existence of a threshold; toxicant exposures that do not elicit inflammatory and proliferative responses that are of sufficient magnitude to overwhelm the defense mechanisms intrinsic to the lung (DNA repair and antioxidant status) would not be expected to pose an increased risk of tumorigenesis (156).

Virtually every cell in the lung can secrete inflammatory mediators such as leukotrienes, ROS, proteolytic enzymes, cytokines, and growth factors (157). Indeed, a characteristic of activated inflammatory cells (macrophages and polymorphonuclear leukocytes) is the increased generation of potentially genotoxic ROS and reactive nitrogen species (RNS), including hydrogen peroxide, superoxide anion, nitric oxide, and hydroxyl radical (154). Many of these inflammatory mediators can be considered to be multifunctional, performing a number of roles within the lung environment. For example, cytokines have been shown to participate in the stimulation of cellular proliferation; expression of cell adhesion molecules and subsequent attraction, adhesion and activation of polymorphonuclear leukocytes; and induction of apoptosis and respiratory tissue remodeling (158). Such effects could contribute to the process of secondary genotoxicity via amplification of ROS secretion by polymorphonuclear leukocytes and the modulation of cellular proliferation and apoptosis.

In addition to the possible genotoxic effects exerted by inflammatory ROS/RNS, several studies have reported that oxidants produced under conditions of inflammation have the potential to bioactivate chemical carcinogens. *In vitro* experiments have demonstrated that polymorphonuclear leukocytes can bioactivate pharmaceuticals including acetaminophen, chlorpromazine, and *p*-aminobenzoic acid, in addition to the chemical carcinogens benzene and PAHs (159). Chemical intermediates resulting from inflammatory-mediated bioactivation could be directly genotoxic to proximal lung cells. Such a model of secondary genotoxicity has been applied to at least partially explain the biological activity of several inert dusts, which can be characterized by the absence of direct genotoxicity and poor solubility in the lung environment. These particles, which would include titanium dioxide, talc, and carbon black, have been established as rat lung tumorigens following chronic exposure to doses capable of inducing sustained inflammation and enhanced epithelial cell proliferation (160). The persistent inflammatory response to inert dust exposure has been shown to parallel *in vivo* lung epithelial cell mutations at the HPRT gene locus, implying that inflammation is at least partly responsible for the mutagenic effects of toxicants that are not direct genotoxicants (161). The magnitude of the inflammatory response and its duration appear to be key factors for elicitation of the secondary genotoxicity phenomenon (160).

The failure of tobacco smoke to induce significant numbers of lung tumors in purportedly susceptible rodent models is suggestive of an indirect or secondary mechanism of tumorigenicity. It would appear that tobacco-smoke-exposed animals do exhibit an increased hyperplastic and inflammatory response relative to air-exposed controls (66,138). A reasonable correlation exists for CSC-induced hyperplasia and inflammation in mouse skin when compared with eventual tumor development (92). With particular regard to tobacco smoke exposures, results from rodent inhalation studies suggest that the induced hyperplasia represents a transient response that may not continue with repeated inhalation exposure (138). Thus, it can be hypothesized that, in the absence of exposures sufficient to induce sustained inflammation and hyperplasia in the lung epithelia, secondary genotoxicity and, ultimately, lung tumorigenesis may not occur.

Although the dissimilar tumor responses exhibited by A/J and rasH2 Tg mice (65) have, for the present time, been attributed by our laboratory to be related to the sensitivity of the former to tobacco-smoke-induced apoptosis (60,66), it remains a curiosity that nose-only exposure of rasH2 failed to increase lung tumor formation in a manner consistent with whole-body dosing. In the absence of readily available dosimetry markers of actual smoke exposure of the target tissue, it remains a possibility that nose-only exposure constitutes a means of toxicant delivery that is considerably more effective in terms of constituent delivery and, hence, possesses a greater potential for inducing apoptotic signaling within the lung. Although it is possible that secondary exposure associated with whole-body dosing (through transdermal and gastrointestinal routes) contributed to the observed differences, our laboratory is not currently pursuing this possibility. The potential for secondary exposure through whole-body, as well as nose-only dosing was affirmed by Chen et al. (162) during studies employing radiolabeled cigarette smoke. Pelt contamination was found to be fourfold greater after whole-body dosing as compared to nose-only dosing, with radioactivity in the lungs of whole-body exposed rats approximately double that of nose-only exposed animals.

Another plausible explanation for the disparate results obtained with whole-body vs. nose-only MTS inhalation exposure of rasH2 Tg mice is the possibility that secondary (indirect) genotoxicity occurred to a greater extent as a consequence of the longer exposure periods (i.e., 6-h vs. 3-h dosing, respectively). Excessive and persistent formation of ROS from inflammatory cells (macrophages and polymorphonuclear leukocytes) during particle-induced lung inflammation is considered to be a hallmark of secondary genotoxicity (163). Similarly observed is the production of inflammatory cell-derived growth factors responsible for epithelial cell proliferating events that possess the potential to promote tumor development. Results from *in vitro* studies have implicated ROS as genotoxic agents based on their ability to induce the oxidation of DNA bases, DNA strand breaks, and/or lipid peroxidation-mediated DNA adducts (155,164–167). Important to the current discussion, lung inflammation is reported to occur and persist only at exposures sufficient for the induction of tumorigenesis (168–170), supporting the concept that these events exhibit a threshold effect. In contrast to direct-acting DNA-reactive carcinogens, an exposure and dose that results in a lung burden without adverse indirect effects would constitute a no-effect exposure level in terms of tumorigenic potential.

Current understanding regarding the toxicokinetics of inhaled, poorly soluble, low-toxicity particles and their corresponding adverse effects appears to implicate two threshold effects (171). A dosimetric threshold exists related to pulmonary clearance capacity; this threshold is defined by the particle deposition rate being greater or lower than the pulmonary clearance rate, with the former situation leading to lung overload and chronic inflammation. Separately, a mechanistic threshold exists that is defined by antioxidant defenses and DNA repair capacity. Accordingly, the presence of persistent inflammation and release of ROS would eventually overwhelm antioxidant defenses and lead to the development of lung lesions. Chronic inhalation exposures to toxicants that rely largely on secondary genotoxicity to elicit a carcinogenic response but, in this instance, fail to induce significant inflammation, are presumably less likely to exhibit a detectable lung tumor response.

Our laboratory has developed a working hypothesis that takes into account the contribution of dysregulated cellular proliferation and apoptosis as modulated by the sequential alteration of oncogenes and/or tumor suppressor genes, the impact of an adaptive response that ultimately leads to progressive cell damage in the absence of apoptosis, and the

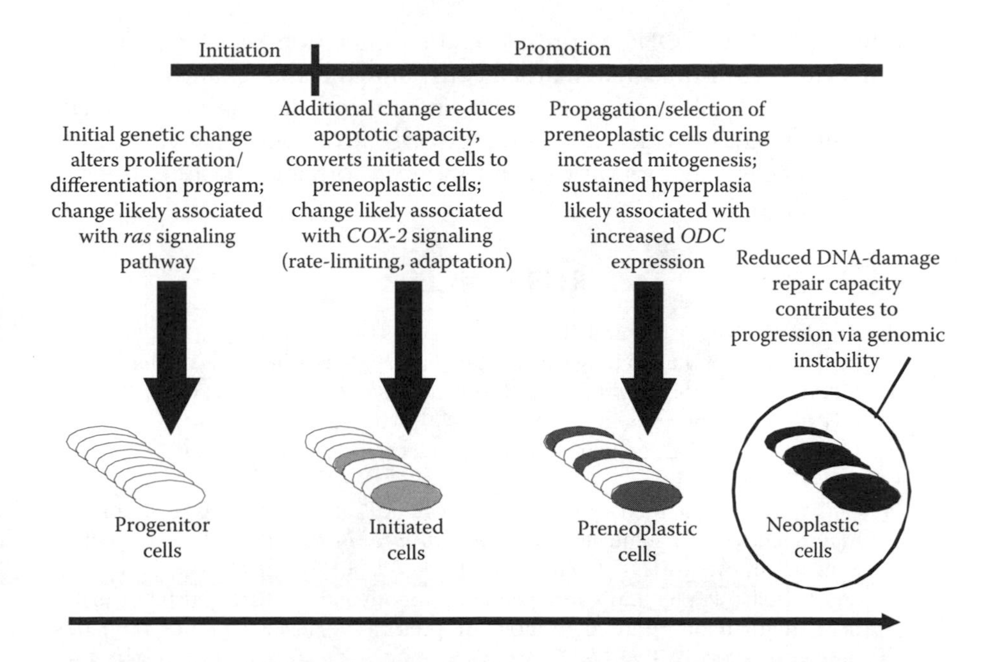

Figure 15.3 Hypothesis regarding the molecular pathology of lung tumor development, expressed in terms of the multistage model for tumorigenesis. Conceptually, initial genetic damage (presumably *Ras* mutation) alters proliferative signaling and leads to the generation of an initiated cell population. Excessive and persistent inflammation exceeds a threshold to induce secondary genotoxicity, ultimately leading to the modulation of COX-2 function (reduced apoptotic capacity associated with adaptation and the generation of preneoplastic cells) and increased cell proliferation (selection of preneoplastic cells to neoplasia).

consideration that secondary genotoxicity likely drives both altered cell signaling and adaptation (refer to Figure 15.3). Specifically, an initial genetic change would define the initiation step, resulting in the generation of a putative stem cell population that has been reprogrammed in terms of proliferation and/or differentiation potential. Altered *ras* signaling, a likely genetic change associated with initiation, could effectively increase expression of COX-2, which in turn would depress apoptotic signaling and increase cellular proliferation. A persistent and excessive inflammatory response would serve to elevate ROS production, ultimately resulting in additional genetic change (secondary genotoxicity) associated with conversion of the initiated cell population to preneoplastic or dormant tumor cells. This conversion step is proposed to be associated with enhanced COX-2 expression and a significant attenuation of apoptotic signaling in the presence of progressive cellular damage, thereby constituting adaptation. The propagation of preneoplastic cells would then be driven by elevated COX-2 and ODC expression, the generation of growth factors and/or cytokines from inflammatory cells, and the moderate tumor-promoting potential inherent to tobacco smoke constituents. Collectively, these events may provide a selective growth advantage relative to nontransformed cells and support toxicant-induced formation of lung tumors.

REFERENCES

1. Sekido, Y., Fong, K.M., and Minna, J.D. Progress in understanding the molecular pathogenesis of human lung cancer. *Biochem Biophys Acta* 1998; 1378: F21–F59.
2. Wittinghofer, A., Scheffzek, K., and Ahmadian, M.R. The interaction of Ras with GTPase-activating proteins. *FEBS Lett* 1997; 410: 63–67.
3. Grandori, C. and Eisenman, R.N. Myc target genes. *Trends Biochem Sci* 1997; 22: 177–181.
4. Yin, C., Knudson, C.M., Korsmeyer, S.J., and Van Dyke, T. Bax suppresses tumorigenesis and stimulates apoptosis *in vivo*. *Nature* 1997; 385: 637–640.
5. Sundaresan, V., Ganly, P., Hasleton, P., Rudd, R., Sinha, G., Bleehen, N.M., and Rabbitts, P. p53 and chromosome 3 abnormalities, characteristic of malignant lung tumours, are detectable in preinvasive lesions of the bronchus. *Oncogene* 1992; 7: 1989–1997.
6. Chung, G.T., Sundaresan, V., Hasleton, P., Rudd, R., Taylor, R., and Rabbitts, P.H. Sequential molecular genetic changes in lung cancer development. *Oncogene* 1995; 11: 2591–2598.
7. Hung, J., Kishimoto, Y., Sugio, K., Virmani, A., McIntire, D.D., Minna, J.D., and Gazdar, A.F. Allele-specific chromosome 3p deletions occur at an early stage in the pathogenesis of lung carcinoma. *JAMA* 1995; 273: 558–563.
8. Kishimoto, Y., Sugio, K., Hung, J.Y., Virmani, A.K., McIntire, D.D., Minna, J.D., and Gazdar, A.F. Allele-specific loss in chromosome 9p loci in preneoplastic lesions accompanying non-small-cell lung cancers. *J Natl Cancer Inst* 1995; 87: 1224–1229.
9. Sugio, K., Kishimoto, Y., Virmani, A.K., Hung, J.Y., and Gazdar, A.F. K-*ras* mutations are a relatively late event in the pathogenesis of lung carcinomas. *Cancer Res* 1994; 54: 5811–5815.
10. Sagawa, M., Saito, Y., Fujimura, S., and Linnoila, R.I. K-*ras* point mutation occurs in the early stage of carcinogenesis in lung cancer. *Br J Cancer* 1998; 77: 720–723.
11. Brambilla, E. and Brambilla, C. p53 and lung cancer. *Pathol Biol* 1997; 45: 852–863.
12. Keohavong, P., DeMichele, M.A., Melacrinos, A.C., Landreneau, R.J., Weyant, R.J., and Siegfried, J.M. Detection of K-*ras* mutations in lung carcinogenesis: relationship to prognosis. *Clin Cancer Res* 1996; 2: 411–418.
13. Li, Z.H., Zheng, J., Weiss, L.M., and Shibata, D. c-k-*ras* and p53 mutations occur very early in adenocarcinoma of the lung. *Am J Pathol* 1994; 144: 303–309.
14. Cooper, C.A., Carey, F.A., Bubb, V.J., Lamb, D., Kerr, K.M., and Wyllie, A.H. The pattern of K-*ras* mutation in pulmonary adenocarcinoma defines a new pathway of tumour development in the human lung. *J Pathol* 1997; 181: 401–404.

15. Westra, W.H., Baas, I.O., Hruban, R.H., Askin, F.B., Wilson, K., Offerhaus, G.J., and Slebos, R.J. K-*ras* oncogene activation in atypical alveolar hyperplasias of the human lung. *Cancer Res* 1996; 56: 2224–2228.
16. Slaughter, D.P., Southwick, H.W., and Smejkal, W. “Field cancerization” in oral stratified squamous epithelium—clinical implications of multicentric origin. *Cancer* 1953; 6: 963–968.
17. SENCAR. Mice: Proceedings of an EPA workshop on their use in toxicological testing. Published in, *Environ Health Perspect* 1986; 68: 3–151.
18. NTP Technical Report: Comparative Initiation/Promotion Skin Paint Studies of B6C3F$_1$ Mice, Swiss (CD-1) Mice, and SENCAR Mice. National Toxicology Program, U.S. Department of Health and Human Services; 1996.
19. Frame, S., Crombie, R., Liddell, J., Stuart, D., Linardopoulos, S., Nagase, H., Portella, G., Brown, K., Street, A., Akhurst, R., and Balmain, A. Epithelial carcinogenesis in the mouse: correlating the genetics and the biology. *Philos Trans R Soc Lond B Biol Sci* 1998; 353: 839–845.
20. Greenhalgh, D.A., Wang, X.-J., Donehower, L.A., and Roop, D.R. Paradoxical tumor inhibitory effect of p53 loss in transgenic mice expressing epidermal-targeted v-ras^{Ha}, v-fos, or human transforming growth factor α. *Cancer Res* 1996; 56: 4413–4423.
21. Yuspa, S.H., Dlugosz, A.A., Cheng, C.K., Denning, M.F., Tennenbaum, T., Glick, A.B., and Weinberg, W.C. Role of oncogenes and tumor suppressor genes in multistage carcinogenesis. *J Invest Dermatol* 1994; 103: 90S–95S.
22. Nelson, M.A., Futscher, B.W., Kinsella, T., Wymer, J., and Bowden, G.T. Detection of mutant Ha-ras genes in chemically initiated mouse skin epidermis before the development of benign tumors. *Proc Natl Acad Sci USA* 1992; 89: 6398–6402.
23 Strickland, J.E., Greenhalgh, D.A., Koceva-Chyla, A., Hennings, H., Restrepo, C., Balaschak, M., and Yuspa, S.H. Development of murine epidermal cell lines which contain an activated ras^{Ha} oncogene and form papillomas in skin grafts on athymic nude mouse hosts. *Cancer Res* 1988; 48: 165–169.
24. Bizub, D., Wood, A.W., and Skalka, A.M. Mutagenesis of the Ha-ras oncogene in mouse skin tumors induced by polycyclic aromatic hydrocarbons. *Proc Natl Acad Sci USA* 1986; 83: 6048–6052.
25. Brown, K., Quintanilla, M., Ramsden, M., Kerr, I.B., Young, S., and Balmain, A. v-ras genes from Harvey and BALB murine sarcoma viruses can act as initiators of two stage mouse skin carcinogenesis. *Cell* 1986; 46: 447–456.
26. Quintanilla, M., Brown, K., Ramsden, M., and Balmain, A. Carcinogen-specific mutation and amplification of Ha-ras during mouse skin carcinogenesis. *Nature* 1986; 322: 78–80.
27. Roop, D.R., Lowy, D.R., Tambourin, P.E., Strickland, J., Harper, J.R., Balaschak, M., Spangler, E.F., and Yuspa, S.H. An activated Harvey ras oncogene produces benign tumours on mouse epidermal tissue. *Nature* 1986; 323: 822–824.
28. Balmain, A. and Pragnell, I.B. Mouse skin carcinomas induced *in vivo* by chemical carcinogens have a transforming Harvey ras oncogene. *Nature* 1983; 303: 72–74.
29. Yuspa, S.H., Vass, W., and Scolnick, E. Altered growth and differentiation of cultured mouse epidermal cells infected with oncogenic retrovirus: contrasting effects of viruses and chemicals. *Cancer Res* 1983; 43: 6021–6030.
30. Goodman, J.I. The traditional toxicologic paradigm is correct: dose influences mechanism. *Environ Health Perspect* 1998; 106: 285–288.

31. Belinsky, S.A., Nikula, K.J., Palmisano, W.A., Michels, R., Saccomanno, G., Gabrielson, E., Baylin, S.B., and Herman, J.G. Aberrant methylation of p16(INK4a) is an early event in lung cancer and a potential biomarker for early diagnosis. *Proc Natl Acad Sci USA* 1998; 95: 11891–11896.
32. Fujishita, T., Mizushima, Y., Kashii, T., and Kobayashi, M. Coincident alterations of p16INK4A/CDKN2 and other genes in human lung cancer cell lines. *Anticancer Res* 1998; 18: 1537–1542.
33. Owens, D.M., Spalding, J.W., Tennant, R.W., and Smart, R.C. Genetic alterations cooperate with v-Ha-ras to accelerate multistage carcinogenesis in Tg. AC transgenic mouse skin. *Cancer Res* 1995; 55: 3171–3178.
34. Owens, D.M., Wei, S.-J.C., and Smart, R.C. A multihit, multistage model of chemical carcinogenesis. *Carcinogenesis* 1999; 20: 1837–1844.
35. Boutwell, R.K. Some biological aspects of skin carcinogenesis. *Prog Exp Tumor Res* 1964; 19: 207–250.
36. Hecker, E. Co-carcinogenic principles from the seed oil of Croton tiglium and from other Euphorbiaceae. *Cancer Res* 1968; 28: 2338–2349.
37. Furstenberger, G., Sorg, B., and Marks, F. Tumor promotion by phorbol esters in skin: evidence for a memory effect. *Science* 1983; 220: 89–91.
38. Furstenberger, G., Kinzel, V., Schwarz, M., and Marks, F. Partial inversion of multistage tumorigenesis in the skin of NMRI mice. *Science* 1985; 230: 76–78.
39. Lankas, G.R., Baxter, C.S., and Christian, R.T. Effect of tumor-promoting agents on mutation frequencies in cultured V79 Chinese hamster cells. *Mutat Res* 1977; 45: 153–156.
40. Birnboim, H.C. DNA strand breakage in human leukocytes exposed to a tumor promoter, phorbol myristate acetate. *Science* 1982; 215: 1247–1249.
41. Dzarlieva-Petrusevska, R.T. and Fusenig, N.E. Tumor promoter 12-*O*-tetradecanolyphorbol-13-acetate (TPA)-induced chromosome aberrations in mouse keratinocyte cell lines: a possible genetic mechanism of tumor promotion. *Carcinogenesis* 1985; 6: 1447–1456.
42. Petrusevska, R.T., Furstenberger, G., Marks, F., and Fusenig, N.E. Cytogenetic effects caused by phorbol ester tumor promotion in primary mouse keratinocyte cultures: correlation with the convertogenic activity of TPA in multistage skin carcinogenesis. *Carcinogenesis* 1988; 9: 1207–1215.
43. Furstenberger, G., Schurich, B., Kaina, B., Petrusevska, R.T., Fusenig, N.E., and Marks, F. Tumor induction in initiated mouse skin by phorbol esters and methyl methanesulfonate: correlation between chromosome damage and conversion (stage 1 of tumor promotion) *in vivo*. *Carcinogenesis* 1989; 10: 749–752.
44. Warren, B.S., Naylor, M.F., Winberg, L.D., Yoshimi, N., Volpe, J.P., Gimenez-Conti, I., and Slaga, T.J. Induction and inhibition of tumor progression. *Proc Soc Exp Biol Med* 1993; 202: 9–15.
45. Bianchi, A.B., Navone, N.M., Aldaz, C.M., and Conti, C.J. Overlapping loss of heterozygosity by mitotic recombination on mouse chromosome 7F1-ter in skin carcinogenesis. *Proc Natl Acad Sci USA* 1991; 88: 7590–7594.
46. Bianchi, A.B., Aldaz, C.M., and Conti, C.J. Non-random duplication of the chromosome bearing a mutated Ha-ras-1 allele in mouse skin tumors. *Proc Natl Acad Sci USA* 1990; 87: 6902–6906.
47. Aldaz, C.M., Trono, D., Larcher, F., Slaga, T.J., and Conti, C.J. Sequential trisomization of chromosomes 6 and 7 in mouse skin premalignant lesions. *Mol Carcinog* 1989; 2: 22–26.

48. Aldaz, C.M., Conti, C.J., Klein-Szanto, A.J.P., and Slaga, T.J. Progressive dysplasia and aneuploidy are hallmarks of mouse skin papillomas: relevance to malignancy. *Proc Natl Acad Sci USA* 1987; 84: 2029–2032.
49. Slaga, T.J., O'Connell, J., Rotstein, J., Patskan, G., Morris, R., Aldaz, C.M., and Conti, C.J. Critical genetic determinants and molecular events in multistage skin carcinogenesis. *Symp Fundam Cancer Res* 1987; 39: 31–44.
50. U.S. Surgeon General. The Health Consequences of Smoking: Cancer, 1982. DHHS/PHS 82-50179. U.S. Public Health Service, Office on Smoking and Health; Rockville, MD.
51. Coggins, C.R.E. A review of chronic inhalation studies with mainstream cigarette smoke in rats and mice. *Toxicol Pathol* 1998; 26: 307–314.
52. Witschi, H., Espiritu, I., Peake, J.L., Wu, K., Maronpot, R.R., and Pinkerton, K.E. The carcinogenicity of environmental tobacco smoke. *Carcinogenesis* 1997; 18: 575–586.
53. Witschi, H., Espiritu, I., Maronpot, R.R., Pinkerton, K.E., and Jones, A.D. The carcinogenic potential of the gas phase of environmental tobacco smoke. *Carcinogenesis* 1997; 18: 2035–2042.
54. Witschi, H. Tobacco smoke as a mouse lung carcinogen. *Exp Lung Res* 1998; 24: 385–394.
55. Witschi, H., Espiritu, I., Yu, M., and Willits, N.H. The effects of phenethyl isothiocyanate, N-acetylcysteine, and green tea on tobacco-smoke-induced lung tumors in strain A/J mice. *Carcinogenesis* 1998; 19: 1789–1794.
56. Witschi, H., Espiritu, I., and Uyeminami, D. Chemoprevention of tobacco-smoke-induced lung tumors in A/J strain mice with dietary myoinositol and dexamethasone. *Carcinogenesis* 1999; 20: 1375–1378.
57. Witschi, H., Uyeminami, D., Moran, D., and Espiritu, I. Chemoprevention of tobacco-smoke lung carcinogenesis in mice after cessation of smoke exposure. *Carcinogenesis* 2000; 21: 977–982.
58. Bogen, K.T. and Witschi, H. Lung tumors in A/J mice exposed to environmental tobacco smoke: estimated potency and implied human risk. *Carcinogenesis* 2002; 23: 511–519.
59. D'Agostini, F., Balansky, R.M., Bennicelli, C., Lubet, R.A., Kelloff, G.J., and De Flora, S. Pilot studies evaluating the lung tumor yield in cigarette-smoke-exposed mice. *Int J Oncol* 2001; 18: 607–615.
60. De Flora, S., Balansky, R.M., D'Agostini, F., Izzotti, A., Camoirano, A., Bennicelli, C., Zhang, Z., Wang, Y., Lubet, R.A., and You, M. Molecular alterations and lung tumors in p53 mutant mice exposed to cigarette smoke. *Cancer Res* 2003; 63: 793–800.
61. Finch, G.L., Nikula, J.K.J., Belinsky, S.A., Barr, E.B., Stoner, G.D., and Lechner, J.F. Failure of cigarette smoke to induce or promote lung cancer in the A/J Mouse. *Cancer Lett* 1996; 99: 161–167.
62. Jones-Bolin, S.E., Johansson, E., Palmisano, W.A., Anderson, M.W., Wiest, J.S., and Belinsky, S.A. Effect of promoter and intron 2 polymorphisms on murine lung K-*ras* gene expression. *Carcinogenesis* 1998; 19: 1503–1508.
63. Lin, L., Festing, M.F.W., Devereux, T.R., Crist, K.A., Christiansen, S.C., Wang, Y., Yang, A., Svenson, K., Paigen, B., Malkinson, A.M., and You, M. Additional evidence that the K-*ras* proto-oncogene is a candidate for the major mouse pulmonary adenoma susceptibility (*Pas*-1) gene. *Exp Lung Res* 1998; 24: 481–497.

64. Saitoh, A., Kimura, M., Takahashi, R., Yokoyama, M., Nomura, T., Izawa, M., Sekiya, T., Nishimura, S., and Katsuki, M. Most tumors in transgenic mice with human c-Ha-*ras* gene contained somatically activated transgenes. *Oncogene* 1990; 5: 1195–1200.
65. Curtin, G.M., Higuchi, M.A., Ayres, P.H., Swauger, J.E., and Mosberg, A.T. Lung tumorigenicity in A/J and rasH2 transgenic mice following mainstream tobacco smoke inhalation. *Toxicol Sci* 2004; 81: 26–34.
66. D'Agostini, F., Balansky, R.M., Izzotti, A., Lubet, R.A., Kelloff, G.J., and De Flora, S. Modulation of apoptosis by cigarette smoke and cancer preventive agents in the respiratory tract of rats. *Carcinogenesis* 2001; 22: 375–380.
67. Institute of Medicine. *Clearing the Smoke: Assessing the Science Base for Tobacco Harm Reduction.* National Academy Press, Washington D.C., 2001.
68. Enzmann, H., Iatropoulos, M., Brunnemann, K.D., Bomhard, E., Ahr, H.J., Schlueter, G., and Williams, G.M. Short- and intermediate-term carcinogenicity testing—a review. Part 2: Available experimental models. *Food Chem Toxicol* 1998; 36: 997–1013.
69. Wagner, B.M. Letter to the editor: response to Dr. Witschi's letter. *Inhal Toxicol* 2001; 13: 731–735.
70. Meckley, D.R., Hayes, J.R., Van Kampen, K.R., Mosberg, A.T., and Swauger, J.E. A responsive, sensitive, and reproducible dermal tumor promotion assay for the comparative evaluation of cigarette smoke condensates. *Regul Toxicol Pharmacol* 2004; 39: 135–149.
71. Meckley, D.R., Hayes, J.R., Van Kampen, K.R., Ayres, P.H., Mosberg, A.T., and Swauger, J.E. Comparative study of smoke condensates from 1R4F cigarettes that burn tobacco versus ECLIPSE cigarettes that primarily heat tobacco in the SENCAR mouse dermal tumor promotion assay. *Food Chem Toxicol* 2004; 42: 851–863.
72. Toward Less Hazardous Cigarettes; National Cancer Institute Smoking and Health Program; Gori, GB, Deputy Director of the Division of Cancer Cause and Prevention, National Cancer Institute; Bethesda, MD (1976–1980).
73. Wynder, E.L. and Hoffmann, D. A study of tobacco carcinogenesis. X. Tumor-promoting activity. *Cancer* 1969; 24: 289–301.
74. Hoffmann, D. and Wynder, E.L. A study of tobacco carcinogenesis. XI. Tumor initiators, tumor accelerators, and tumor-promoting activity of condensate fractions. *Cancer* 1971; 27: 848–864.
75. Wynder, E.L. and Wright, G. A study of tobacco carcinogenesis. I. The primary fractions. *Cancer* 1957; 10: 255–271.
76. Wynder, E.L., Kopf, P., and Ziegler, H. A study of tobacco carcinogenesis. II. Dose-response studies. *Cancer* 1957; 10: 1193–1200.
77. Wynder, E.L., Fritz, L., and Furth, N. Effect of concentrations of benzopyrene in skin carcinogenesis. *J Natl Cancer Inst* 1957; 19: 361–370.
78. Wynder, E.L. and Hoffmann, D. A study of tobacco carcinogenesis. VII. The study of the higher polycyclic hydrocarbons. *Cancer* 1959; 12: 1079–1086.
79. Wynder, E.L. and Hoffmann, D. A study of tobacco carcinogenesis. VIII. The role of the acidic fractions as promoters. *Cancer* 1961; 14: 1306–1315.
80. Roe, F.J.C., Salaman, M.H., and Cohen, J. Incomplete carcinogens in cigarette smoke condensate: tumour-promotion by a phenolic fraction. *Br J Cancer* 1959; 13: 623–633.
81. Gellhorn, A. The cocarcinogenic activity of cigarette tobacco tar. *Cancer Res* 1958; 18: 510–517.

82. Van Duuren, B.L., Sivak, A., Segal, A., Orris, L., and Langseth, L. The tumor-promoting agents of tobacco leaf and tobacco smoke condensate. *J Natl Cancer Inst* 1966; 37: 519–526.
83. Van Duuren, B.L., Sivak, A., Goldschmidt, B.M., Katz, C., and Melchionne, S. Initiating action of aromatic hydrocarbons in two-stage carcinogenesis. *J Natl Cancer Inst* 1970; 44: 1167–1173.
84. Van Duuren, B.L. and Goldschmidt, B.M. Cocarcinogenic and tumor promoting agents in tobacco carcinogenesis. *J Natl Cancer Inst* 1976; 56: 1237–1242.
85. Rubin, H. Synergistic mechanisms in carcinogenesis by polycyclic aromatic hydrocarbons and by tobacco smoke: a biohistorical perspective with updates. *Carcinogenesis* 2001; 22: 1903–1930.
86. Rubin, H. Selective clonal expansion and microenvironmental permissiveness in tobacco carcinogenesis. *Oncogene* 2002; 21: 7392–7411.
87. Ilgren, E.B., Ed. *Initiation and Promotion in Skin or Liver Neoplasia: A 65-year Annotated Bibliography of International Literature*. CRC Press, Boca Raton, FL, 1992.
88. Slaga, T., Triplett, L., and Nesnow, S. Mutagenic and carcinogenic potency of extracts of diesel and related environmental emissions: two-stage carcinogenesis in skin tumor sensitive mice (SENCAR). *Environ Int* 1981; 5: 417–423.
89. Nesnow, S., Triplett, L.L., and Slaga, T.J. Studies on the tumor initiating, tumor promoting, and tumor co-initiating properties of respiratory carcinogens. *Carcinog Compr Surv* 1985; 8: 257–277.
90. Nesnow, S. Mouse skin tumors as predictors of human lung cancer for complex emissions: an overview, skin carcinogenesis: Mechanisms of human relevance. *Prog Clin Biol Res* 1989; 298: 347–361.
91. Slaga, T.J., Budunova, I.V., Gimenez-Conti, I.B., and Aldaz, C.M. The mouse skin carcinogenesis model. *J Invest Dermatol Symp Proc* 1996; 1: 151–156.
92. Curtin, G.M., Hanausek, M., Walaszek, Z., Mosberg, A.T., and Slaga, T.J. Short-term *in vitro* and *in vivo* analyses for assessing the tumor-promoting potentials of cigarette smoke condensates. *Toxicol Sci* 2004; 81: 14–25.
93. Clifford, A., Morgan, D., Yuspa, S.H., Soler, A.P., and Gilmour, S. Role of ornithine decarboxylase in epidermal tumorigenesis. *Cancer Res* 1995; 55: 1680–1686.
94. Feith, D.J., Shantz, L.M., and Pegg, A.E. Targeted antizyme expression in the skin of transgenic mice reduces tumor promoter induction of ornithine decarboxylase and decreases sensitivity to chemical carcinogenesis. *Cancer Res* 2001; 61: 6073–6081.
95. Gilmour, S.K., Robertson, F.M., Megosh, L., O'Connell, S.M., Mitchell, J., and O'Brien, T.G. Induction of ornithine decarboxylase in specific subpopulations of murine epidermal cells following multiple exposures to 12-*O*-tetradecanoylphorbol-13-acetate, mezerein and ethyl phenylpropriolate. *Carcinogenesis* 1992; 13: 51–56.
96. O'Brien, T.G., Megosh, L.C., Gilliard, G., and Soler, A.P. Ornithine decarboxylase over-expression is a sufficient condition for tumor promotion in mouse skin. *Cancer Res* 1997; 57: 2630–2637.
97. Smith, M.K., Trempus, C.S., and Gilmour, S.K. Cooperation between follicular ornithine decarboxylase and v-Ha-ras induces spontaneous papillomas and malignant conversion in transgenic skin. *Carcinogenesis* 1998; 19: 1409–1415.

98. Soler, A.P., Gilliard, G., Megosh, L., George, K., and O'Brien, T.G. Polyamines regulate expression of the neoplastic phenotype in mouse skin. *Cancer Res* 1998; 58: 1654–1659.
99. Argyris, T.S. Tumor promotion by abrasion induced epidermal hyperplasia in the skin of mice. *J Invest Dermatol* 1980; 75: 360–362.
100. Argyris, T.S. and Slaga, T.J. Promotion of carcinomas by repeated abrasion in initiated skin of mice. *Cancer Res* 1981; 41: 5193–5195.
101. Morris, R.J., Coulter, K, Tryson, K., and Steinberg, S.R. Evidence that cutaneous carcinogen-initiated epithelial cells from mice are quiescent rather than actively cycling. *Cancer Res* 1997; 57: 3436–3443.
102. International Agency for Research on Cancer. Beryllium, cadmium, mercury and exposures in the glass manufacturing industry. *IARC Sci Pub* 1993; 58: 119–238.
103. Waalkes, M.P. Cadmium carcinogenesis in review. *J Inorg Biochem* 2000; 79: 241–244.
104. Takenaka, S., Oldiges, H., Konig, H., Hochrainer, D., and Oberdorster, G. Carcinogenicity of cadmium chloride aerosols in Wistar rats. *J Natl Cancer Inst* 1983; 70: 367–373.
105. Waalkes, M.P. Cadmium carcinogenesis. *Mutat Res* 2003; 533: 107–120.
106. Yang, C.-F., Shen, H.-M., Shen, Y., Zhuang, Z.-X., and Ong, C.-N. Cadmium-induced oxidative cellular damage in human fetal lung fibroblasts (MRC-5 cells). *Environ Health Perspect* 1997; 105: 712–716.
107. Hartmann, A. and Speit, G. Effect of arsenic and cadmium on the persistence of mutagen-induced DNA lesions in human cells. *Environ Mol Mutat* 1996; 27: 98–104.
108. Hartmann, M. and Hartwig, A. Disturbance of DNA damage recognition after UV-irradiation by nickel (II) and cadmium (II) in mammalian cells. *Carcinogenesis* 1998; 19: 617–621.
109. Dally, H. and Hartwig, A. Induction and repair inhibition of oxidative DNA damage by nickel (II) and cadmium (II) in mammalian cells. *Carcinogenesis* 1997; 18: 1021–1026.
110. Spruill, M.D., Song, B., Whong, W.-Z., and Ong, T. Proto-oncogene amplification and overexpression in cadmium-induced cell transformation. *J Toxicol Environ Health* 2002; 65: 2131–2144.
111. Joseph, P., Muchnok, T.K., Klishis, M.L., Roberts, J.R., Antonini, J.M., Whong, W.Z., and Ong, T. Cadmium-induced cell transformation and tumorigenesis are associated with transcriptional activation of *c-fos, c-jun,* and *c-myc* proto-oncogenes: role of cellular calcium and reactive oxygen species. *Toxicol Sci* 2001; 61: 295–303.
112. Achanzar, W.E., Achanzar, K.B., Lewis, J.G., Webber, M.M., and Waalkes, M.P. Cadmium induces *c-myc*, p53, and *c-jun* expression in normal human prostate epithelial cells as a prelude to apoptosis. *Toxicol Appl Pharmacol* 2000; 164: 291–300.
113. Abshire, M.K., Busard, G.S., Shiraishi, N., and Waalkes, M.P. Induction of c-*myc* and c-*jun* proto-oncogene expression in rat L6 myoblasts by cadmium is inhibited by zinc preinduction of the metallothionein gene. *J Toxicol Environ Health* 1996; 48: 359–377.
114. Hechtenberg, S., Schafer, T., Benters, J., and Beyersmann, D. Effects of cadmium on cellular proto-oncogene expression. *Ann Clin Lab Sci* 1996; 26: 512–521.

115. Zheng, H., Liu, J., Choo, K.H., Michalska, A.E., and Klaassen, C.D. Metallothionein-I and II knockout mice are sensitive to cadmium-induced liver mRNA expression of *c-jun* and p53. *Toxicol Appl Pharmacol* 1996; 136: 229–235.
116. Huang, C., Zhang, Q., Li, J., Shi, X., Castranova, V., Ju, G., Costa, M., and Dong, Z. Involvement of Erks activation in cadmium-induced AP-1 transactivation in vitro and in vivo. *Mol Cell Biochem* 2001; 222: 141–147.
117. Takiguchi, M., Achanzar, W.E., Qu, W., Li, G., and Waalkes, M.P. Effects of cadmium on DNA-(cytosine-5) methyltransferase activity and DNA methylation status during cadmium-induced cellular transformation. *Exp Cell Res* 2003; 286: 355–365.
118. Meplan, C., Mann, K., and Hainaut, P. Cadmium induces conformational modifications of wild-type p53 and suppresses p53 response to DNA damage in cultured cells. *J Biol Chem* 1999; 274: 31663–31670.
119. Yuan, C., Kadiiska, M.B., Achanzar, W.E., Mason, R., and Waalkes, M.P. Studies on the mechanisms of cadmium-induced blockage of apoptosis: possible role of caspase-3 inhibition. *Toxicol Appl Pharmacol* 2000; 164: 321–329.
120. Shimada, H., Shiao, Y.H., Shibata, M.A., and Waalkes, M.P. Cadmium suppresses apoptosis induced by chromium. *J Toxicol Environ Health* 1998; 54: 159–168.
121. Achanzar, W.E., Webber, M.M., and Waalkes, M.P. Altered apoptotic gene expression and acquired apoptotic resistance in cadmium-transformed human prostate epithelial cells. *Prostate* 2002; 52: 236–244.
122. Briehl, M.M. and Baker, A.F. Modulation of the antioxidant defense as a factor in apoptosis. *Cell Death Different* 1996; 3: 63–70.
123. Deng, D.X., Chakrabarti, S., Waalkes, M.P., and Cherian, M.G. Metallothionein and apoptosis in primary human hepatocellular carcinoma and metastatic adenocarcinoma. *Histopathology* 1998; 32: 340–347.
124. Manna, S.K., Zhang, H.J., Yan, T., Oberley, L.W., and Aggarwal, B.B. Overexpression of manganese superoxide dismutase suppresses tumor necrosis factor-induced apoptosis and activation of nuclear transcription factor-κ-B and activated protein-1. *J Biol Chem* 1998; 273: 13245–13254.
125. Kondo, Y., Rusnak, J.M., Hoyt, D.G., Settineri, C.E., Pitt, B.R., and Lazo, J.S. Enhanced apoptosis in metallothionein null cells. *Mol Pharmacol* 1997; 52: 195–201.
126. Kayanoki, Y., Fujii, J., Islam, K.N., Suzuki, K., Kawata, S., Matsuzawa, Y., and Taniguchi, N. The protective role of glutathione peroxidase in apoptosis induced by reactive oxygen species. *J Biochem* 1996; 19: 817–822.
127. Slater, A.F., Stefan, C., Nobel, I., van den Dobbelsteen, D.J., and Orrenius, S. Intracellular redox changes during apoptosis. *Cell Death Different* 1996; 3: 57–62.
128. Hart, B.A., Voss, G.W., Shatos, M., and Doherty, J. Cross-tolerance to hyperoxia following cadmium aerosol pretreatment. *Toxicol Appl Pharmacol* 1990; 103: 255–270.
129. Hart, B.A., Voss, G.W., and Willean, C.L. Pulmonary tolerance to cadmium following cadmium aerosol pretreatment. *Toxicol Appl Pharmacol* 1989; 101: 447–460.
130. Hart, B.A. Cellular and biochemical response of the rat lung to repeated inhalation of cadmium aerosols. *Toxicol Appl Pharmacol* 1986; 82: 281–291.

131. Hart, B.A., Eneman, J.D., Gong, Q., and Durieux-Lu, C.C. Increased oxidant resistance of alveolar epithelial type II cells isolated from rats following repeated exposure to cadmium aerosols. *Toxicol Lett* 1995; 81: 131–139.
132. Hart, B.A., Gong, Q., Eneman, J.D., and Durieux-Lu, C.C. In vivo expression of metallothionein in rat alveolar macrophages and type II epithelial cells following repeated cadmium aerosol exposures. *Toxicol Appl Pharmacol* 1995; 133: 82–90.
133. Cerutti, P., Ghosh, R., Oya, Y., and Amstad, P. The role of cellular antioxidant defense in oxidant carcinogenesis. *Environ Health Perspect* 1994; 102: 123–129.
134. Halliwell, B. Free radicals, antioxidants, and human disease: curiosity, cause, or consequence. *The Lancet* 1994; 344: 721–724.
135. Eneman, J.D., Potts, R.J., Osier, M., Shukla, G.S., Lee, C.H., Chiu, J.-F., and Hart, B.A. Suppressed oxidant-induced apoptosis in cadmium adapted alveolar epithelial cells and its potential involvement in cadmium carcinogenesis. *Toxicol* 2000; 147: 215–228.
136. Voehringer, D.W., McConkey, D.J., McDonnel, T.J., Brisbay, S., and Meyn, R.E. Bcl-2 expression causes redistribution of glutathione to the nucleus. *Proc Natl Acad Sci USA* 1998; 95: 2956–2960.
137. Hart, B.A., Potts, R.J., and Watkin, R.D. Cadmium adaptation in the lung—a double-edged sword? *Toxicology* 2001; 160: 65–70.
138. Witschi, H., Oreffo, V.I.C., and Pinkerton, K.E. Six-month exposure of strain A/J mice to cigarette sidestream smoke: cell kinetics and lung tumor data. *Fundam Appl Toxicol* 1995; 26: 32–40.
139. Witschi, H., Espiritu, I., Dance, S.T., and Miller, M.S. A mouse lung tumor model of tobacco smoke carcinogenesis. *Toxicol Sci* 2002; 68: 322–330.
140. De Flora, S., D'Agostini, F., Balansky, R., Camoirano, A., Bennicelli, C., Bagnasco, M., Cartiglia, C., Tampa, E., Longobardi, M.G., Lubet, R.A., and Izzotti, A. Modulation of cigarette-smoke-related end points in mutagenesis and carcinogenesis. *Mutat Res* 2003; 523–524: 237–252.
141. Vane, J.R., Bakhle, Y.S., and Botting, R.M. Cyclo-oxygenases 1 and 2. *Annu Rev Pharmacol Toxicol* 1998; 38: 97–120.
142. Dannenberg, A.J., Altorki, N.K., Boyle, J.O., Dang, C., Howe, L.R., Weksler, B.B., and Subbaramaiah, K. Cyclo-oxygenase 2: A pharmacological target for the prevention of cancer. *Lancet Oncol* 2001; 2: 544–551.
143. Wardlaw, S.A., March, T.H., and Belinsky, S.A. Cyclo-oxygenase-2 expression is abundant in alveolar type II cells in lung-cancer-sensitive mouse strains and in premalignant lesions. *Carcinogenesis* 2000; 21: 1371–1377.
144. Bauer, A.K., Dwyer-Nield, L.D., and Malkinson, A.M. High cyclo-oxygenase 1 (COX-1) and cyclo-oxygenase-2 (COX-2) contents in mouse lung tumors. *Carcinogenesis* 2000; 21: 543–550.
145. Muller-Decker, K., Neufang, G., Berger, I., Neumann, M., Marks, F., and Furstenberger, G. Transgenic cyclo-oxygenase-2 overexpression sensitizes mouse skin for carcinogenesis. *Proc Natl Acad Sci USA* 2002; 99: 12483–12488.
146. Umemura, T., Kodama, Y., Hioki, K., Nomura, T., Nishikawa, A., Hirose, M., and Kurokawa, Y. The mouse rasH2/BHT model as an in vivo rapid assay for lung carcinogens. *Jpn J Cancer Res* 2002; 93: 861–866.
147. Imaoka, M., Kashida, Y., Watanabe, T., Ueda, M., Onodera, H., Hirose, M., and Mitsumori, K. Tumor-promoting effect of phenolphthalein on development of lung tumors induced by N-ethyl-N-nitrosourea in transgenic mice carrying human prototype c-Ha-*ras* gene. *J Vet Med Sci* 2002; 64: 489–493.

148. Heasley, L.E., Thaler, S., Nicks, M., Price, B., Skorecki, K., and Nemenoff, R.A. Induction of cytosolic phospholipase A2 by oncogenic *ras* in human non-small-cell lung cancer. *J Biol Chem* 1997; 272: 14501–14504.
149. Sheng, H., Willians, C.S., Shao, J., Liang, P., DuBois, R.N., and Beauchamp, R.D. Induction of cyclooxygenase-2 by activated Ha-*ras* oncogene in Rat-1 fibroblasts and the role of mitogen-activated protein kinase pathway. *J Biol Chem* 1998; 273: 22120–22127.
150. Han, J.A., Kim, J.I., Ongusaha, P.P., Hwang, D.H., Ballou, L.R., Mahale, A., Aaronson, S.A., and Lee, S.W. P53-mediated induction of COX-2 counteracts p53- or genotoxic stress-induced apoptosis. *EMBO J* 2002; 21: 5635–5644.
151. Shin, V.Y., Liu, E.S., Ye, Y.N., Koo, M.W., Chu, K.M., and Cho, C.H. A mechanistic study of cigarette smoke and cyclooxygenase-2 on proliferation of gastric cancer cells. *Toxicol Appl Pharmacol* 2004; 195: 103–112.
152. Oberdorster, G. and Yu, C.P. The carcinogenic potential of inhaled diesel exhaust: a particle effect? *J Aerosol Sci* 1990; 21: S397–401.
153. Janssen, Y.M.W., Heintz, N.H., Marsh, J.P., Borm, P.J.A., and Mossman, B.T. Induction of *c-fos* and *c-jun* proto-oncogenes in target cells of the lung and pleura by carcinogenic fibers. *Am J Respir Cell Mol Biol* 1994; 11: 522–530.
154. Janssen, Y.M.W., Borm, P.J.A., Van Houten, B., and Mossman, B.T. Cell and tissue responses to oxidative damage. *Lab Invest* 1993; 69: 261–274.
155. Ames, B.N. and Gold, L.S. Endogenous mutagens and the causes of aging and cancer. *Mutat Res* 1991; 250: 3–16.
156. Butterworth, B.E., Conolly, R.B., and Morgan, K.T. A strategy for establishing mode of action of chemical carcinogens as a guide for approaches to risk assessments. *Cancer Lett* 1995; 93: 129–146.
157. Borm, P.J.A., and Driscoll, K.E. Particles, inflammation and respiratory tract carcinogenesis. *Toxicol Lett* 1996; 88: 109–113.
158. Gauldie, J., Jordana, M., and Cox, G. Cytokines and pulmonary fibrosis. *Thorax* 1993; 48: 931–935.
159. Kehrer, J.P., Mossman, B.T., Sevanian, A., Trush, M.A., and Smith, M.T. Free radical mechanisms in chemical pathogenesis. *Toxicol Appl Pharmacol* 1988; 95: 349–362.
160. Driscoll, K.E. Role of inflammation in the development of rat lung tumors in response to chronic particle exposure. *Inhal Toxicol* 1996; 8: 139–153.
161. Driscoll, K.E., Deyo, L.C., Howard, B.W., Poynter, J., and Carter, J.M. Characterizing mutagenesis in the hprt gene of rat alveolar epithelial cells. *Exp Lung Res* 1995; 21: 941–956.
162. Chen, B.T., Benz, J.V., Finch, G.L., Mauderly, J.L., Sabourin, P.J., Yeh, H.C., and Snipes, M.B. Effect of exposure mode on amounts of radiolabeled cigarette particles in lungs and gastrointestinal tracts of F344 rats. *Inhal Toxicol* 1995; 7: 1095–1108.
163. Schins, R.P.F. Mechanisms of genotoxicity of particles and fibers. *Inhal Toxicol* 2002; 14: 57–78.
164. Kasai, H., Crain, P.F., Kuchino, Y., Nishimura, S., Octsuyama, A., and Tanocka, H. Formation of 8-hydroxyguanine moiety in cellular DNA by agents producing oxygen radicals and evidence for its repair. *Carcinogenesis* 1986; 7: 1849–1851.
165. Schraufstatter, I., Hyslop, P.A., Jackson, J.H., and Cochrane, C.G. Oxidant-induced DNA damage of target cells. *J Clin Invest* 1988; 82: 1040–1050.

166. Aruoma, O.I., Halliwell, B., and Dizdaroglu, N. Iron ion-dependent modification of bases in DNA by the super-oxide radical-generating system hypoxanthine/xanthine oxidase. *J Biol Chem* 1989; 264: 13024–13028.
167. Trush, M.A. and Kensler, T.W. An overview of the relationship between oxidative stress and chemical carcinogenesis. *Free Radic Biol Med* 1991; 10: 201–209.
168. Donaldson, K., Bolton, R.E., Jones, A.D., Brown, G.M., Robertson, M.D., Slight, J., Cowie, H., and Davis, J.M.G. Kinetics of the bronchoalveolar leukocyte response in rats during exposure to equal airborne mass concentrations of quartz, chrysoltile asbestos or titanium dioxide. *Thorax* 1988; 43: 525–533.
169. Donaldson, K., Brown, G.M., Brown, D.M., Robertson, M.D., Slight, J., Cowie, H., Jones, A.D., Bolton, R.E., and Davis, J.M.G. Contrasting bronchoalveolar leukocyte responses in rats inhaling coal mine dust, quartz, or titanium dioxide: Effects of coal rank, airborne mass concentration, and cessation of exposure. *Environ Res* 1990; 52: 62–76.
170. Driscoll, K.E., Lindenschmidt, R.C., Maurer, J.K., Higgins, J.M., and Ridder, G. Pulmonary response to silica or titanium dioxide: inflammatory cells, alveolar macrophage-derived cytokines, and histopathology. *Am J Respir Cell Mol Biol* 1990; 2: 381–390.
171. Greim, H., Borm, P., Schins, R., Donaldson, K., Driscoll, K., Hartwig, A., Kuempel, E., Oberdorster, G., and Speit, G. Toxicity of fibers and particles—Report of the workshop held in Munich, Germany, October 26–27, 2000. *Inhal Toxicol* 2001; 13: 737–754.

16

TOBACCO SMOKING

Hanspeter Witschi

CONTENTS

16.1 INTRODUCTION

The smoking of tobacco products is one of the major public health problems of our time. Its most dire consequence is lung cancer. A rare disease a century ago, it is now responsible for most cancer deaths worldwide, some 900,000 new cases annually in men and 300,000 in women (Stewart and Kleihues, 2003). In the U.S., lung cancer accounts for more deaths annually than colon, breast, and prostate cancer combined (Szabo, 2001). Although during the last few years the incidence of lung cancer in the U.S. seems to have decreased or at least plateaued out to some extent, there remains, nevertheless, some concern that it might continue to increase in certain selected subpopulations, such as teenage girls (Wingo et al., 1999). Worldwide, the disease seems to be destined to increase further, particularly in Asia, and is likely to become a massive public health burden with considerable costs to society (Lam et al., 2001; Pisani et al., 1999). It is a sobering thought that we now know that the main causative agent of lung cancer, tobacco smoke, is the only known human carcinogen for which it would be possible to accomplish zero exposure, and yet exposure continues to grow in large sections of the population.

16.2 HISTORICAL BACKGROUND

Lung cancer is a rather interesting human disease. It was first discovered in the 18th century in England that occupational exposures to chemical mixtures, such as soot, could produce cancer at the site of its contact with skin (e.g., scrotal cancer in chimney sweeps). Later on, it was established that exposure to chemicals manufactured in industry or handled in certain occupations did cause cancer (bladder cancer in the production of aniline dyes and lung cancer in mining of radioactive ore). This made it possible to establish cause-and-effect relationships involving exposure to a few defined chemical agents and their handling in certain professions. It proved to be much more elusive to establish similar correlations between defined agents and the most frequently seen cancers in the general population, such as cancers of the colon, breast, and prostate, urogenital cancers, and lymphomas or leukemias. Today, lung cancer is the only widespread cancer in the general population for which we know with certainty the cause in the majority of cases: the smoking of tobacco products, mostly cigarettes.

Some rather unusual circumstances are responsible for this. In the 19th century, lung cancer was an extremely rare disease which, according to then available statistics, accounted for less than 0.5% of all cancers seen at autopsy in major clinical centers. During the following decades, this percentage rose gradually and by 1925 was about 14%—more than a 20-fold increase in just a few decades. At the time, it was not clear what the causes were—the 1918 influenza pandemic, increased motor traffic, and the use of asphalt for covering the roads were all suspected. It was around 1930 that, for the first time, suspicion was directed toward tobacco. Cigarette smoking had become popular worldwide shortly before and during World War I, not least because the invention of machines had made the large-scale manufacturing of cigarettes feasible and their price affordable. Epidemiologic studies performed in Germany, England, and the U.S. between 1930 and 1950 soon established a clear causal relationship between smoking and lung cancer (Doll, 1998). It still took a while before the truth was fully accepted because smokers, including many physicians, could not or would not believe that a socially perfectly acceptable habit was detrimental to their health. In 1964, the surgeon general's report drew public attention to the problem, although it was not until the end of the 20th century that it became generally accepted, including by the tobacco industry, that lung cancer was caused by smoking and that it was a preventable disease.

16.3 CHEMICALS IN TOBACCO SMOKE

The burning of tobacco produces a complex mixture of gases and fine particles. Tobacco smoke contains more than 4000 individual chemicals (Hoffmann and Wynder, 1999). Among them are many agents known to cause cancer in animals and in humans (Table 16.1). When drawn through an absolute filter, tobacco smoke can be roughly separated into a "tar" phase and a "gas" phase. The gas phase contains mainly nitrogen, carbon monoxide, and carbon dioxide, as well as some other volatile agents such as acetaldehyde and formaldehyde, 1,3-butadiene, acrolein, volatile nitrosamines, and cyanide. In the tar or particle phase, nicotine is the key component. It is a stimulant of the ganglia of the autonomous nervous system and neuromuscular junctions. Central nervous system stimulation provides a temporary "rush" or feeling of well-being. In the surgeon general's report of 1964, nicotine was called a "habit-forming" substance (U.S. DHEW, 1964); most would nowadays agree that it is a powerful addicting agent (U.S. Surgeon General, 1988). Nicotine is the reason why many smokers are unable or unwilling to quit. Other ingredients of the particle phase are polycyclic aromatic hydrocarbons such as benzo(a)pyrene, benzo(a) anthracene, aromatic amines, naphthalene, fluorene, and tobacco-smoke-specific nitrosamines. Several inorganic agents such as nickel, cadmium, arsenic, lead, and polonium-210 are also present.

The International Agency for Research on Cancer (IARC) routinely evaluates human and animal data on chemicals for carcinogenicity. According

Table 16.1 Classes of Potentially Carcinogenic Chemicals Found in Tobacco Smoke

Chemical Class	*Examples*
Aliphatic hydrocarbons	Butadiene, ethylene
Monocyclic aromatic hydrocarbons	Benzene, styrene
Di- and polycyclic aromatic hydrocarbons	Anthracene, benzo(a)pyrene, dibenz(a,h) anthracene, 5-methylchrysene
Phenols and phenol esters	Catechol, hydroquinone
Aldehydes	Acetaldehyde, acrolein, formaldehyde
Lactones	Coumarin
Nitrogen compounds	*N*–nitrosodimethylamine, 4-(methylnitrosamino) -1-(3-pyridyl)-1-butanone (NNK)
Agricultural chemicals and derivatives	Captan, DDT, malathion
Halogen compounds	Vinyl chloride
Inorganic elements	As, Cd, Cr, Ni, Pb, Se

to current information, nine individual tobacco smoke constituents, as well as the complete smoke itself, including environmental tobacco smoke (ETS), are known human carcinogens. Many other chemicals found in tobacco smoke are classified as probable or possible human carcinogens; examples are given in Table 16.2. Interestingly, the two most thoroughly studied carcinogens present in tobacco smoke, benzo(a)pyrene and 4-(methylnitrosamino)-1-(3-pyridyl)-1-butanone (NNK), have been classified only as probable or possible carcinogens (Table 16.2). In epidemiological investigations, mixtures containing benzo(a)pyrene, such as cigarette smoke, coke oven emissions, and fumes emanating from roofing tars, have been associated with development of lung cancer in man. Nitrosamines found in tobacco smoke are potent animal carcinogens that react with critical targets in many organs. Evidence for such interactions has also been found in human cells and tissues. Nevertheless, some doubts have been offered on occasion about whether the comparatively small amounts of either benzo(a)pyrene or NNK in cigarette smoke are sufficient to account exclusively for its carcinogenic activity (Wynder, 1961; Druckrey, 1961; Brown et al., 2003).

A number of possible and highly plausible mechanisms are thought to be relevant to the explanation of the carcinogenic action of tobacco smoke.

Table 16.2 Known, Probable, and Possible Human Carcinogens in Tobacco Smoke

IARC Group 1: Known Human Carcinogens	*IARC Group 2A: Probably Human Carcinogens*	*IARC Group 2B: Possibly Human Carcinogens*
2-Naphthylamine	1,3-butadiene	Acetaldehyde
4-Aminobiphenyl	Benzo(a)anthracene	Acrylonitrile
Arsenic	Benzo(a)pyrene	Orthoanisidine
Benzene	Dibenz(a,h)anthracene	Acetamide
Cadmium	Formaldehyde	Benzo(b)fluoranthene
Nickel	*N*-nitrosodiethylamine	Benzo(j)fluoranthene
N-nitrosodiethylamine	*N*-nitrosodimethylamine	Benzo(k)fluoranthene
Chromium VI		Dibenzo(a,e)pyrene
Vinyl chloride	Orthotoluidine	Dibenzo(a,h)pyrene
Full mixture of tobacco smoke		DDT
		Hydrazine
Environmental tobacco smoke		Lead
		Naphthalene
		NNK
		N-nitrosomorpholine
		2-nitropropane
		Styrene
		Urethane

Benzo(a)pyrene is metabolically converted to the reactive (–)-benzo(a) pyrene-7,8-diol and then reacts with guanine in DNA; formation of adducts following metabolic activation and subsequent induction of genetic changes, often in critical "hotspots," may be one such key event (Denissenko et al., 1996; Hecht, 1999). Alkyldiazonium ions, derived from nitrosamines, form O^6-alkyldeoxyguanosine. DNA adducts may then become repaired or, if miscoding persists, may be responsible for mutations in multiple genes, such as the tumor suppressor gene p53, RAS, myc, p16, RB, or FHIT. However, adduct formation does not necessarily imply initiation, but often remains only a sign of exposure. Possible mechanisms of carcinogenesis are not limited to the metabolic activation of specific carcinogens. Oxidant stress placed on lung tissue in smokers may yield superoxide anion, which can then be converted to hydrogen peroxide. In the presence of iron, a tobacco smoke constituent, the Haber–Weiss reaction (Kehrer, 2000) may then result in the formation of hydroxyl radicals that can cause DNA strand breaks. The release of hydrogen peroxide by activated macrophages and lipid peroxidation through reactive aldehydes may place additional oxidant stress on the lung and, thus, eventually result in malignant transformation of the epithelia lining the airways.

16.4 LUNG CANCER IN HUMANS

The smoking of tobacco accounts for approximately 90% of lung cancers found in humans, but it is not the only agent known to cause the disease. It has been known for centuries that, in certain mining operations in Europe, workers developed an unusually high incidence—sometimes up to 100%—of lung cancer, caused by the radioactive gas radon (*Bergkkrankheit* in the mines of Joachimsthal). The active mining of uranium on the Colorado plateau during World War II, required for the development of the atomic bomb, also led to an increased lung cancer incidence in the miners (Proctor, 1995). In ore-processing and smelting operations, inhalation of metal fumes and dusts such as cadmium, nickel, or chromium may cause cancer of the airways. Asbestos, besides causing pulmonary fibrosis and the rare tumor mesothelioma, can also cause lung cancer, and the risk of developing the disease is substantially increased in smokers. At one time it was thought that acute exposure to the chemical warfare agent mustard gas, as occurred in World War I on a large scale, would cause lung cancer, but a clear association could not be established. Eventually, mustard gas was found to be a human carcinogen during World War II in Japanese production workers. Another known human lung carcinogen is chloromethylether. Mixtures of volatile organic chemicals, containing such polycyclic aromatic hydrocarbons as benzo(a)pyrene, are thought to be responsible for the development of lung cancers in coke oven workers and roofers who inhale tar fumes. Some widely encountered agents such as welding fumes, silica dust, or formaldehyde (though not proved directly to be human carcinogens), as well as air pollution in general, may contribute to increased incidence of the disease. It has been observed for some time that in urban areas with heavy pollution more lung cancers can be found than in rural areas (Beeson et al., 1998). Emissions originating from coal-fired power plants or from diesel engines are thought to be major contributing factors in heavily polluted air.

One characteristic feature of lung cancer in humans is a very long latency period, generally assumed to be between 20 and 40 yr. This often makes it difficult to attribute lung cancer unequivocally to some exposure that might have occurred decades ago. Human lung cancers are of four major types: squamous-cell carcinoma, adenocarcinoma (these two tumor types each account for approximately 30% of all human lung cancers), small-cell lung cancer (20 to 30%), and large-cell carcinoma. Most of these tumors originate from the mucosa lining the conducting airways; hence, lung cancer in humans is often called *bronchogenic carcinoma.* A fifth type, bronchoalveolar carcinoma, is a tumor that originates in the most distal airways and the alveolar region. Besides aggressively invading adjacent tissues, lung cancers frequently metastasize to distant organs, such as the brain and liver. In general, the prognosis for clinically diagnosed lung cancer is poor.

About 50 yr ago, most lung cancers found in smokers were classified as squamous-cell carcinomas. During the last few decades, a shift has occurred, and these days adenocarcinomas are more and more frequently found (Thun et al., 1997). This has been attributed to the "changing cigarette" (Hoffmann et al., 1991). High-tar, nonfilter cigarettes rich in nicotine are increasingly being replaced by low-nicotine cigarettes with effective filters designed to remove much of the tar. The tar phase (the tobacco smoke particulate phase excluding nicotine and water) contains mostly polycyclic aromatic hydrocarbons, thought to be responsible for lung cancer initiation and promotion. Because these newer cigarettes also deliver less nicotine per puff, it is believed that smokers, in order to fulfill their craving for nicotine, drag more frequently on their cigarettes and also take the less irritating smoke deeper into their lungs. Thus, many carcinogens of the gas phase may find their way deeper into the lungs. Such a hypothesis would agree with animal data in which benzo(a)pyrene has been found to produce squamous-cell lung cancer, whereas the nitrosamine NNK produces adenocarcinomas.

In active smokers, the risk of developing lung cancer increases with the number of cigarettes smoked and the duration of smoking (Doll, 1998). In former smokers, regardless of the age at which they quit, the risk no longer increases, and the younger the age at quitting, the greater the benefit. It is less clear to what extent the introduction of filter cigarettes that deliver less tar and nicotine has helped in reducing the overall incidence of lung cancer. Epidemiologic studies had suggested only little benefit (Tang et al., 1995), and in 1996 the American Thoracic Society concluded that there has been little if any beneficial effect (American Thoracic Society, 1996). A review of the available data in 2001 led to the conclusion that changes in the design and composition of cigarettes had had no impact on public health and that lower-tar cigarettes failed to substantially modify lung cancer incidence (NCI, 2001). These conclusions have not gone unchallenged. It was argued, how in Great Britain, the smoking of low-tar cigarettes by younger individuals and the avoidance of plain, of hand-rolled, and of black-tobacco cigarettes had substantially reduced lung cancer risk (Lee, 2001). It was observed that over the past decades lung cancer rates had decreased in Britain, but increased in the U.S. (Lee and Forey, 2003). Because reduction in tar and differences in tobacco composition could not explain this divergent trend, it was hypothesized that a decrease in air pollution in Great Britain since

1956 had had beneficial effects. On the other hand, in the U.S., exposure to radon, asbestos, smoking of marijuana, and increased consumption of dietary fat might have added to lung cancer risk. Currently, it is hoped that new tobacco products, such as cigarettes that heat rather than burn tobacco (Wagner et al., 2000) or nicotine replacement products, will eventually help to reduce the lung cancer burden imposed by smoking. A review of the available evidence concluded that all tobacco products, including modified tobacco "potential reduced exposure products (PREP) and cigarette-like PREPs are toxic and poisonous," but further research was recommended (Stratton et al., 2001).

16.5 ANIMAL MODELS OF TOBACCO SMOKE CARCINOGENESIS

Animal studies on tobacco smoke carcinogenesis were initiated as soon as it became reasonably clear that the smoking of cigarettes was associated with development of lung cancer in humans. The results of most studies were disappointing. Although it was evident that cigarette smoke condensates readily produced tumors when painted on the skin of rabbits and mice, it proved next to impossible to produce lung tumors in experimental animals. When the animal studies were summarized in 2004, it was found that in only about half of the studies had cigarette smoke caused lung tumors in experimental animals (IARC, 2004). Results of some of these studies are listed in Table 16.3. The most convincing evidence that tobacco smoke could cause signs of malignant tissue changes was found in the larynx of hamsters. Although the incidence of lesions was, with the exception of one strain, generally below 10%, the data nevertheless identified tobacco smoke as an animal carcinogen. The cancerous lesions were limited to the larynx and included squamous-cell metaplasia and microinvasive carcinomas. No tumors were found in the lower respiratory tract of hamsters. Nevertheless, the Syrian golden hamster has been deemed to be a most appropriate model to study tobacco smoke carcinogenesis (Hecht, 1999).

IARC discusses five rat studies, out of which two showed positive results. In the Dalbey et al. (1980) experiment, 80 rats were exposed nose-only to cigarette smoke. Although the number of tumor-bearing animals was not statistically significantly higher than in controls, the total number of respiratory tract tumors (nose and lung) was. In the Mauderly et al. (2004) experiment, several hundred rats were exposed to comparatively high concentrations of tobacco mainstream smoke (100 and 250 mg/m^3 of total particulate matter). A significantly higher lung tumor incidence

Table 16.3 Lung Tumor Incidence in Laboratory Rodents Exposed to Tobacco Smoke

Species	*Strain*	*Sex*	*Tobacco-Smoke Exposed*	*Controls*	*Reference*
Mouse[a]	C57Bl	M	9/162 (5.6%)	3/160 (1.9%)	Harris et al., 1974
		F	7/164 (4.3%)	1/159 (0.6%)	
		M + F	16/362 (5%)[b]	4/319 (1%)	
	Snells	M	18/143	7/120 (5.8%)	
		F	(12.6%)	6/102 (5.9%)	
		M + F	12/11 (10.8%) 30/254 (11.2%)	13/(222) (5.9%)	
	SWR	M	6/31 (19.4%)	1/26 (3.8%)	Witschi et al., 2002; de Flora et al., 2003
		F	9/21 (42.9%)[b]	2/22 (9.0%)	
		M + F	15/52 (29%)[b]	3/45 (7%)	
	Balb/c	M	9/27 (33.3%)	6/30 (20.0%)	Witschi et al., 2002
	CByB6F1	M	6/30 (20.0%)	4/30 (13.3%)	Unpublished
Rat[c]	F 344	F	7/68 (10.3%)	2/78 (2.6%)	Dalbey et al., 1980
Rat[d]	F344	M	11/260 (4.2%)	4/118 (3.4%)	Mauderly et al., 2004
		F	21/256 (8.2%)[b]	0/119 (0%)	
		M + F	32/484 (7%)[b]	4/233 (2%)	
Syrian Golden Hamster[e]	Outbred	M	21/240 (8.8%)[b]	0/400 (0%)	Dontenwill et al., 1973
		F	8/240 (3.3%)[b]	0/400 (0%)	
		M + F	29/480 (6%)[b]	0/800 (0%)	
	BioRad 15.16	M	9/48 (18.8%)[b]	0/60 (0%)	Bernfeld et al., 1974
	BioRad 87.20	M	2/45 (4.4%)[b]	0/60 (0%)	Bernfeld et al., 1974

Note: Data are given as number of respiratory-tract-tumor-carrying animals over total number of animals at risk (percentages of lung tumor incidence in parenthesis).

[a] Significantly higher than controls ($p < .05$, Fisher's exact test).

[b] Lung tumors in strains other than A/J.

[c] Respiratory tract tumors (nose and pulmonary).

[d] Benign and malignant neoplasms combined.

[e] Laryngeal lesions only.

was observed only in the female rats (14% in the high-dose group and 6% in the low-dose group, vs. 0% in the controls). Approximately one third of the lung lesions were classified as bronchioloalveolar carcinomas, the remainder being adenomas. Both these studies clearly show that it is possible to induce lung tumors in rats with tobacco smoke, but that rather large numbers of animals are needed to detect this comparatively weak carcinogenic response. Furthermore, in none of these studies were lesions observed that resembled squamous-cell carcinoma or adenocarcinoma of bronchial origin. Dogs inhaling cigarette smoke through a tracheostomy also showed lesions resembling tumors in the bronchoalveolar region (Auerbach et al., 1970). However, it was not certain to what extent the observed lesions represented truly neoplastic changes.

There have also been several studies with mice. The largest one, involving close to 4000 mice, failed to provide evidence for increased tumor incidence, although it indicated that tobacco smoke could accelerate the time to tumor development (Henry and Kouri, 1986). Evidence for significantly increased lung tumor incidence was obtained when the data for male and female C57Bl, Snells, and SWR mice were combined. In the first two strains, incidence was again comparatively low (around 5% in C57Bl and 13% in Snell mice), and large numbers of animals had to be exposed to produce a statistically significant response. A better response with fewer animals was observed in SWR mice, an outbred strain prone to developing lung tumors upon challenge with a carcinogen.

The comparatively low incidence of tobacco-smoke-induced lung tumors and the large number of animals usually required to obtain a positive response make it difficult to pinpoint the relevant mechanisms of carcinogenesis or to evaluate the effects of interventions designed to reduce tobacco-smoke-induced lung tumorigenesis. More recently, several laboratories reported mouse experiments that look somewhat more promising (Witschi, Espiritu, Peake et al., 1997; D'Agostini et al., 2001; Stinn et al., 2002, Curtin et al., 2004). The most frequently used strain was the A mouse, although some positive responses were also obtained in SWR mice. Strain A mice spontaneously develop a high incidence of lung tumors. Exposure to carcinogens greatly increases the number of tumors found in the lung. It is now generally accepted that strain A mice readily develop tumors following exposure to polycyclic aromatic hydrocarbons, carbamates, hydrazines, or nitrosamines, whereas they respond less readily, if at all, to other carcinogens such as aromatic amines, aminofluorenes, aromatic halides, or metals (Maronpot et al., 1986).

Experiments designed to study the carcinogenicity of inhaled tobacco smoke in A/J mice often follow an exposure–recovery protocol. The mice are placed in a whole-body inhalation chamber into which a mixture of tobacco sidestream smoke and mainstream smoke is drawn. The animals are usually exposed for 6 h a day, 5 d a week. Smoke concentrations in the chamber are typically from 50 to 160 mg/m^3 of TSP (total suspended particulate matter), 10 to 20 mg/m^3 of nicotine, and 200 to 400 ppm of CO; this atmosphere is well tolerated by the animals. In most experiments, the mice are exposed for 5 months to the tobacco smoke. While in smoke, strain A mice (but not SWR mice) fail to gain weight at the same rate as do control animals, presumably due to the stress of being exposed to a vile atmosphere. When the animals are removed into air and given a recovery period of 4 months, the animals gain weight and tumors develop rapidly (Witschi, 1997; Stinn et al., 2004). Results are usually expressed as tumor multiplicity, i.e., the average number of tumors per lung, including the lungs of non-tumor-bearing animals. Data from various studies are summarized in Table 16.4. The data show a dose effect with higher multiplicities observed at the higher concentrations of tobacco smoke. However, a typical dose–effect curve is not readily apparent and, although the slope of the regression line is significantly different from zero, the curve appears to be rather flat (Figure 16.1). In general, the results from these studies seem to confirm what could be derived from the earlier experiments: inhaled tobacco smoke is a comparatively weak animal carcinogen. The average number of tumors in tobacco-smoke-exposed mice, even at the highest concentration, is usually below 2.5 tumors/lung. For comparative purposes, it must be remembered that it is easily possible, with the injection of potent carcinogens, such as polycyclic aromatic hydrocarbons or nitrosamines, to obtain tumor multiplicities of up to 50 or more. As far as agents other than tobacco smoke are concerned, only a few inhalation studies with A/J mice are available. For bis(chloromethyl)ether, 1,2-dibromoethane, urethane, vinyl chloride, ozone, and nitrogen dioxide the assays were positive, whereas inhalation of naphthalene or diesel exhaust was not followed by sign of increased tumorigenesis (Adkins et al., 1986; Leong et al., 1971; Pepelko and Peirano, 1983; Hassett et al., 1985).

The carcinogenicity of tobacco smoke condensate (or tar) was established by skin-painting studies in mice; inhalation studies were largely unsuccessful. A recent review discussed the possibility that the benzo(a)pyrene concentrations in cigarette smoke condensate are too low to act by themselves as initiators of skin tumors (Rubin, 2001). Rather, benzo(a)pyrene and other polycyclic aromatic hydrocarbons might act as promoters. In 1974, Leuchtenberger and Leuchtenberger had already shown that the gas phase of tobacco smoke, practically devoid of polycyclic aromatic hydrocarbons, was as active as full tobacco smoke in producing multiple lung tumors in Snell mice. This observation was

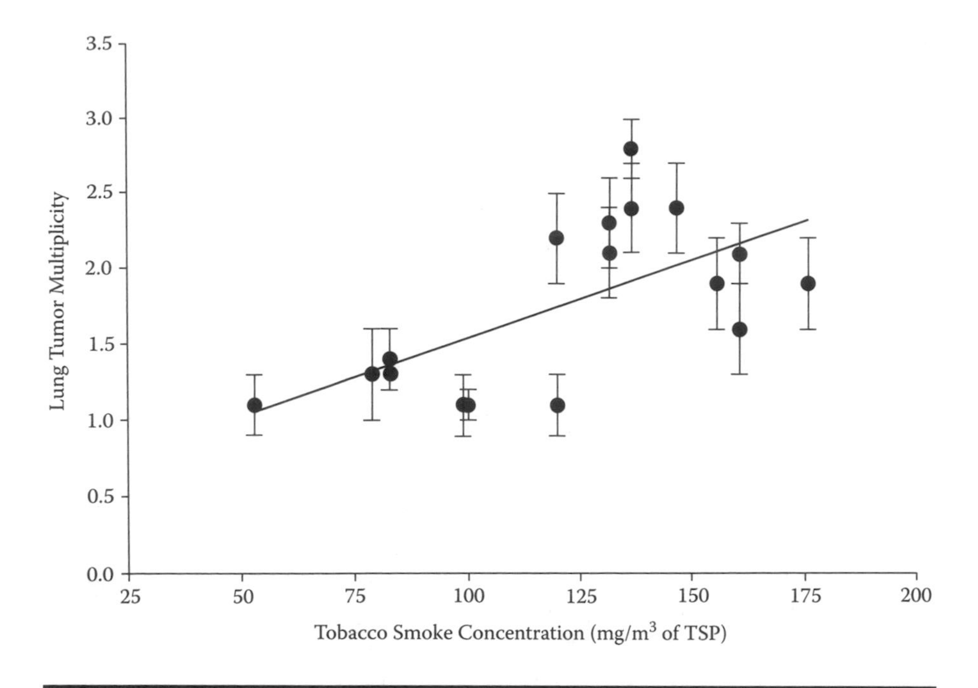

Figure 16.1 Lung tumor multiplicities in strain A mice exposed for 5 months to tobacco smoke, followed by 4 months in tobacco-smoke-free air, plotted in relation to tobacco smoke concentration (Table 16.4). Data are plotted as mean ± SE. The slope of the regression line is significantly different from zero ($p = .006$).

confirmed when strain A/J mice were exposed to unfiltered and filtered tobacco smoke (Witschi, et al., 1997; Witschi, 2005a). Analysis of some contents of the chamber atmospheres showed that several carcinogens were not present in high enough concentrations to account for tumor development. The notion that gas-phase constituents are at least as important as the agents found in the tar fraction might explain why the introduction of filter cigarettes had only a negligible effect on overall lung cancer incidence in man (American Thoracic Society 1996; Tang et al., 1995; NCI 2001). It should be mentioned, however, that in hamsters only exposure to full tobacco smoke, but not to the gas phase alone, has been found to produce carcinoma of the larynx (Dontenwill et al., 1973).

Several concerns have been raised with regard to using the induction of lung tumors in A/J mice by tobacco smoke as a model for human lung cancer. They are discussed in the following subsections.

16.5.1 Genetic Predisposition to Developing Lung Tumors

Strain A mice are genetically predisposed to developing lung tumors, making it often impossible to detect differences between treatment groups in tumor incidence, the generally accepted criteria for carcinogenicity. As a short-term bioassay widely used for hazard assessment, including measurement of the carcinogenic potential of cigarette smoke condensates (Wagner et al., 2000), skin-painting studies with SENCAR (sensitive to carcinogens) mice rely also on tumor multiplicity rather than tumor incidence as measurement of effect. As far as A/J mice are concerned, Table 16.4 shows that out of 12 experiments in which TSP concentrations were 100 mg/m^3 or higher, tumor multiplicities were significantly higher in 11 out of 12 experiments (92%). In the same studies referenced in Table 16.4, lung tumor incidences were significantly higher in only 8 out the 12 experiments (67%). The criterion for what constitutes a positive result in the strain A mouse lung tumor assay has always been tumor multiplicity (Shimkin and Stoner, 1975). Given a weak carcinogenic response, lung tumor incidence is usually a less reliable indicator than multiplicity, due to the high incidence of spontaneously occurring lung tumors in strain A mice. If animals are treated with a strong carcinogen, such as urethane, benzo(a)pyrene of NNK, the number of tumors per lung will be strictly dose dependent, but all animals will develop lung tumors. Under these experimental conditions, the efficacy of chemopreventive agents is invariably measured only in terms of reduction in tumor multiplicities.

16.5.2 Absence of Resemblance to Human Bronchogenic Carcinoma

In their discussion of the strain A/J model, IARC (2004) clearly states that both A/J and SWR mice are highly susceptible to the induction of peripheral lung tumors that originate primarily from type II alveolar cells. These cells are precursors for only about 5 to 10% of human adenocarcinomas (i.e., bronchioloalveolar carcinomas). Murine lung tumors thus resemble only a small fraction of human adenocarcinomas. On the other hand, similarities in molecular biology and signaling pathways suggest that they may be representative of human adenocarcinomas (Malkinson, 1998; Malkinson, 2001). It also must be noted that tobacco smoke has never been found to produce carcinomatous changes below the larynx in hamsters. In rats, all malignant and benign tumors observed following tobacco smoke inhalation are either in the nose or have been classified as bronchioloalveolar adenomas or carcinomas. In no animal model has cigarette smoke so far reproducibly induced the equivalent of human lung cancer.

Table 16.4 Table Lung Tumor Multiplicities in Strain A Mice Exposed to Tobacco Smoke

TSP[a]	*Smoke-Exposed*	*Controls*	*TS/Air*[b]	*Reference*
53	1.1 ± 0.2 (21)	0.7 ± 0.2 (23)	1.6	Witschi, et al., 1997a
79	1.3 ± 0.3 (26)[c]	0.5 ± 0.1 (24)	1.6	Witschi, Espiritu, et al., 1997b
83	1.3 ± 0.1 (33)[c]	0.9 ± 0.2 (29)	1.4	Witschi et al., 1998
83	1.1 ± 0.3 (9)[c]	0.22 ± 0.2 (9)	5.1	D'Agostini et al., 2001
87	1.4 ± 0.2 (24)[c]	0.5 ± 0.2 (24)	2.6	Witschi et al., 1997
99	1.1 ± 0.2 (25)	0.7 ± 0.2 (25)	1.6	Witschi, et al., 2004
100	1.1 ± 0.1[c]	0.5 ± 0.1[c]	2.1	Stinn et al., 2002
120	2.2 ± 0.3 (25)[c]	0.7 ± 0.2 (25)	3.1	Witschi, et al., 2004a
120	1.1 ± 0.2 (20)[c]	0.3 ± 0.1 (20)	3.7	D'Agostini et al., 2001
132	2.3 ± 0.3 (26)[c]	1.2 ± 0.2 (25)	1.9	Obermueller-Jevic et al., 2002
132	2.1 ± 0.3 (35)[c]	0.6 ± 0.1 (30)	3.5	Witschi et al., 1999
137	2.4 ± 0.3 (28)[c]	1.0 ± 0.1 (54)	2.4	Witschi et al., 2000
137	2.8 ± 0.2 (38)[c]	0.9 ± 0.2 (30)	3.1	Witschi et al., 2000
147	2.4 ± 0.3 (29)[c]	1.6 ± 0.2 (34)	1.5	Witschi and Espiritu, 2002
156	1.9 ± 0.3 (25)[c]	1.0 ± 0.2 (24)	1.9	Obermueller-Jevic et al., 2002
161	1.6 ± 0.3 (22)	1.1 ± 0.3 (15)	1.5	Witschi, et al., 2004b
161	2.1 ± 0.2 (52)[c]	0.9 ± 0.2 (47)	2.3	Witschi, Espiritu, Ly et al., 2004
176	1.9 ± 0.3 (22)[c]	0.7 ± 0.2 (25)	2.7	Witschi, et al., 2004a

Note: Data for lung tumor multiplicity are given as mean ± SE with number of animals in parenthesis.

[a] Average concentration of TSP (mg/m^3) in inhalation chambers.

[b] Ratio of lung tumor multiplicity in tobacco-smoke-exposed animals/controls.

[c] significantly different ($p < .05$) from controls.

16.5.3 Absence of Malignancy and Metastasis in Mouse Lung Tumors

Lung tumors in mice are mostly adenomas and thus lack two hallmarks of human lung cancer: malignancy and metastasis. In general, lung tumors in experimental animals (rodents or dogs) only rarely metastasize to distant organs. In mice, lung tumors evolve as small hyperplastic foci, which then grow into adenomas. When the tumors are between 5 and 10 months old, the great majority show an adenomatous pattern, with occasional ones showing foci of cells with malignant features. With time, the percentage of adenomas with carcinomatous foci increases, thus showing that the tumors can progress to malignancy (Foley et al., 1991). Metastatic spread of tumor cells into adjacent tissues or distant organs is only rarely found, although it does occur. As summarized by Shimkin and Stoner (1975), researchers found that only in 3.6% of the more than 5000 mice with lung tumors did distant metastases develop. In tobacco-smoke-exposed strain A mice, which are usually killed when they are 1 yr old, lesions in which tumor cells invade adjacent tissue or blood vessels are found only rarely. Lung tumors in mice thus represent an early stage of progression from hyperplasia to adenoma to adenocarcinoma. In humans, the diagnosis of lung cancer is usually only made in the terminal, most advanced stages of the disease process.

16.5.4 Dose Response

When the A/J lung tumor data from three different laboratories are plotted, a dose response becomes apparent (Figure 16.1). The curve is rather flat, although the slope is significantly different from zero. When three different doses were used in the same experiment, a dose response was apparent (Witschi, Espiritu, Uyeminami et al., 2004).

16.5.5 Reduced Weight Gain and Tumor Response

The reduced weight gain during tobacco smoke exposure suggest that the maximum tolerated dose has been exceeded and tumor response may have been compromised. If anything, reduced weight gain is known to decrease tumor formation, thus leading to underestimation of the response. In carcinogenicity assays, reduced weight gain is generally considered to be a sign of unwanted toxicity. In mice exposed to tobacco smoke, toxicity is fully reversible, because the animals' weights return to the control weight during the 4-month recovery period. It has now been shown that reduced weight gain has no effect on tumor development: animals were exposed to tobacco smoke and control animals were kept in tobacco-smoke-free air with restricted food so that their reduced weight gain approximated the one seen in the smoke-exposed animals. After the recovery period, only tobacco smoke was found to produce increased tumor multiplicity, but not the preceding food restriction. Initially reduced body weight gain is thus inconsequential for lung tumor development (Stinn et al., 2004). A plausible explanation offered for the initial inhibition of tumorigenesis during decreased weight gain is the increased blood levels of corticosterone, a stress response. Finally, it has been found that SWR mice exposed to tobacco smoke gain weight at the same rate as the controls and still develop significantly more lung tumors (Witschi et al., 2002).

16.5.6 Difficulty in Detection of Effects of Chemopreventive Agents

The comparatively low tumor multiplicities can make it difficult to detect small effects of chemopreventive agents. Given the usual tumor response of 1.8 to 2.8 (Table 16.4), power calculations show that to detect tumor-reducing effects of between 20 to 40% would require impracticably high numbers of animals. In a clinical trial, a reduction of lung cancers by 20% would be considered to be an encouraging finding. To document such an effect in the strain A mouse model of tobacco carcinogenesis is difficult, if not impossible. The average differences in lung tumor multiplicities in chemoprevention assays in which a 20% efficiency is observed lies usually between 0.2 and 0.5. Given the commonly seen standard deviations for multiplicity, it might under these circumstances easily take more than 150 animals per group to show a significant difference at $p < .05$ in the range of 0.7 to 0.9 of power. This number of animals would not only be impracticable, but also unacceptable from an animal welfare standpoint. On the other hand, in animals treated with NNK or other carcinogens, many more tumors are produced, and differences of 30% or more between controls and treated animals can usually be found with 20 to 25 animals per group.

16.6 CHEMOPREVENTION

Most lung cancers could be prevented by the simple expedient of quitting smoking. Unfortunately, there is a large segment of smokers who are either unable or unwilling to quit. The treatment of lung cancer still has a very poor success rate; less than 5% of patients treated with surgery, radiation, and chemotherapy survive for more than 5 yr. Chemoprevention of lung cancer seems to be an attractive strategy to decrease lung cancer rates (Goodmann, 2000). Chemoprevention is commonly understood to imply the intake of certain chemicals or, more often, natural products that will eventually interfere with the activation of potentially carcinogenic compounds to ultimate carcinogens; such agents are commonly called *blocking agents*. Chemopreventive agents also may delay, or stop altogether, the progression of precancerous lesions to the full disease; such *suppressing agents* might be particularly useful in helping to further reduce lung cancer risk in people who have successfully quit smoking.

About 20 yr ago, it was found that the antioxidant beta carotene would prevent cell layers from developing into hyperplastic and, eventually, dysplastic structures, events believed to be key in the development of squamous-cell cancer in the airways. Epidemiological data suggested that high intake of beta carotene seemed to protect to a certain degree against the development of cancer in humans (Peto et al., 1981). Taken together, the evidence was convincing enough to start three major clinical trials. Results were not as expected; in two of the three trials, patients treated with beta carotene and other supplements developed an increased lung cancer risk rather than a decrease; one of the trials had to be halted prematurely (Omenn, 1998). Another agent considered to be capable of interfering with the carcinogenic process was the antioxidant NAC (*N*-acetylcysteine). In a series of animal studies, it had been shown convincingly that NAC can prevent the development of many molecular changes that are thought to be important in lung cancer development (De Flora et al., 2003). However, clinical trials again failed to show an effect in humans (van Zandwjik et al., 2000). For the time being, chemoprevention of lung cancer in humans appears to be a much more difficult endeavor than chemoprevention of some other common human cancers, such as colon or breast cancer, in which impressive advances have already been made (Omenn, 2000).

Preclinical evaluation of potentially useful chemotherapeutic agents continues at a fast pace. Most often, these agents are evaluated in the strain A mouse lung tumor model. Strain A mice, when injected with a chemical carcinogen, readily develop multiple lung tumors. Counting their numbers in animals treated with a chemopreventive agent and comparison with the number of lung tumors in control animals then shows to what extent a given chemical may either prevent the development of tumors (when given before or during carcinogen administration) or interfere with the progression of hyperplastic or adenomatous lung lesions and dedifferentiation into carcinomatous lesions (Stoner et al., 1997). Against certain tobacco-smoke-specific carcinogens, particularly benzo(a)pyrene or NNK, agents of several chemical classes have been found to be highly effective. They include isothiocyanates, organoselenium compounds, glucocorticoids, naturally occurring agents such as tea polyphenols, d-limonene, and myoinositol, and nonsteroidal anti-inflammatory drugs or COX inhibitors (Hecht, 2002). However, it should also be pointed out that, on occasion, chemopreventive agents that work well against specific carcinogens are not effective when evaluated against the full complex mixture of tobacco smoke (Witschi, 2000). Treatment with mixtures of chemopreventive agents rather than with single chemicals appears to be a promising approach (Hecht et al., 2002; Mukhtar and Ahmad, 1999).

16.7 ENVIRONMENTAL TOBACCO SMOKE

After it was generally accepted that smoking of tobacco products poses a substantial risk of lung cancer, concerns arose about whether tobacco smoke also could harm "bystanders," i.e., people who were not actively smoking, but lived or worked in atmospheres that contained the so-called ETS. ETS is the mixture of smoke an active smoker exhales and sidestream smoke, which is greatly diluted by air. Exhaled mainstream smoke is partially scrubbed of many agents that cross the air–blood barrier such as nicotine, CO, CN, and others. Sidestream smoke is generated at lower burning temperatures than is mainstream smoke, and its chemical composition differs somewhat. It contains higher concentrations of certain carcinogenic agents, such as volatile nitrosamines, 2-naphthylamine, and 4-aminobiphenyl than does mainstream smoke, and it also has smaller particle size (Guerin et al., 1992).

After the surgeon general's report in 1986 had called attention to the potential dangers of ETS, the U.S. EPA released a first document on ETS in 1992; a more complete summary of all available evidence became available in 1999 (NCI, 1999). In its recent evaluation of ETS, the IARC (IARC, 2004) concluded that an excess risk of 20% for women and of 30% for men exists in nonsmoking spouses of smokers. For nonsmokers who work in places where they are exposed to ETS, the risk is statistically significant and of the order of 16 to 19%. The IARC concluded that there was sufficient evidence to consider ETS a human carcinogen. It is estimated that currently in the U.S. there are about 3000 lung cancer deaths annually that can be attributed to ETS exposure. The preponderance of evidence from human studies indicates that ETS is an important health hazard.

In experimental animals, inhalation of a mixture of cigarette sidestream (89%) and mainstream smoke (11%), a surrogate for ETS, has consistently been found to increase lung tumor multiplicities in mice (Witschi, 2005b). It must be pointed out, however, that the ETS concentrations required to obtain a tumorigenic response are orders of magnitude higher than what may be encountered in an environment in which humans are exposed to ETS. Sidestream smoke condensate produces more tumors when painted onto mouse skin than does mainstream condensate, providing some evidence that at lower burning temperatures more potent carcinogens are formed (Mohtashamipur et al., 1990). Implanted into the lungs of rats, it causes tumors. Because data were available for only one species, the mouse, the evidence for inhaled ETS being an animal carcinogen was deemed to be limited, but that for sidestream smoke condensate was deemed sufficient (IARC 2004).

There are health risks other than those for lung cancer from inhalation of ETS. It has been estimated that within the U.S. there are annually 35,000 to 62,000 deaths from ischemic heart disease, caused by exposure to ETS (NCI, 1999). There is a plausible biological mechanism: tobacco smoke, whether full or ETS, can cause increased adhesiveness ("stickiness") of blood platelets and lead to hemodynamic changes that may result in acute myocardial infarction. In an epidemiological study, it was found that in a small town in the U.S., where an ordinance had been introduced that prohibited smoking at work and in public places, hospital admissions for myocardial infarction decreased noticeably (Sargent et al., 2004). A prospective epidemiologic study came to the conclusion that measurements of urinary cotinine gave a better estimation of the risk of developing coronary heart disease than information on possible ETS exposure alone (Whincup et al., 2004).

A second concern relates to the health of children who live in houses with smokers (NCI, 1999). It has been known for a long time that active smoking by a pregnant women causes a substantial reduction in birth weight. It is plausible that inhalation of ETS might have a similar effect. Sudden infant death syndrome (SIDS) or "crib death" is associated with maternal smoking; odds ratios in several epidemiological studies range from 1.4 to 5. In infants exposed to ETS, there is an increased susceptibility to middle ear infections and bronchitis. A disquieting development is the increasing numbers of children with asthma. Inhalation of ETS may greatly enhance the risk of developing asthmatic attacks, and it is felt that this does occur in the U.S., where 400,000 to 1,000,000 children develop exacerbation of asthma because of ETS, with 8,000 to 26,000 new cases added annually. Thus, children seem to suffer the most from "involuntary smoking." Bans on smoking in public places or at workplaces have greatly helped in reducing health risk from ETS exposure. Such bans are not enforceable where, in all likelihood, exposure to ETS has the most deleterious effects, the home. Studies in animal models have consistently shown that exposure to a mixture of mainstream and sidestream smoke increases airway reactivity (Witschi, et al., 1997c). Some other evidence has shown that inhalation of ETS may render the lungs more susceptible to the effects of a subsequent inhalation of the common air pollutant ozone (Yu et al., 2002).

16.8 OTHER DISEASES CAUSED BY INHALATION OF TOBACCO SMOKE

From a global perspective, tobacco smoking threatens to become, in the next 20 yr or so, by far the largest cause of disease burden, particularly in underdeveloped countries where smoking prevalences are very high. Smoking may cause diseases of the respiratory tract as well as some diseases in other organs. The risk of lung cancer is not the only one to be increased by smoking. Associations have also been found to exist between smoking and cancer of the bladder; certain aromatic amines found in cigarettes smoke, such as 2-naphthylamine, are potent human bladder carcinogens. Other organs that may develop cancer following tobacco inhalation are the pancreas, colon, prostate, and cervix of the uterus. Epidemiological studies also have provided some supportive, or at least equivocal, evidence that tobacco smoke may be linked to cancer of the breast, brain, and stomach, leukemia, and lymphomas or non-Hodgkin's lymphoma. A recent epidemiological study led to the conclusion that for African-American men in the U.S., who have the highest cancer burden, smoking is the most important single risk factor (Leistikow, 2004). While smoking was on the rise between 1969 and 1990, not only lung cancer but cancers in all sites increased. Smoking declined after 1990 and so did all cancer-related death rates in this particular population.

Other noncancerous diseases that can be caused, or at least aggravated, by smoking are coronary heart disease and peripheral vascular disease (thrombangitis obliterans or Buerger's disease). In peripheral vascular disease, inflammation of the arteries, most often in the legs, and formation of intravascular blood clots (thrombi) lead to a progressive narrowing of the vascular lumen; cessation of smoking may halt the process. As far as coronary disease is concerned, smokers have twice the risk of developing myocardial infarction as nonsmokers (Doll, 1998).

Chronic obstructive pulmonary disease (COPD) is another condition associated with chronic morbidity and mortality. In 1990, COPD ranked 12th among disease burdens worldwide. It is projected that by the year 2020 it will become the third most common cause of death and the fifth most common cause of disability in the world (Lopez and Murray, 1998). It has been known for a long time that active smoking causes chronic bronchitis (*smoker's cough*). The pathology of this lesion is characterized by mucus hypersecretion, mucous gland hyperplasia, and an influx of inflammatory cells, such as neutrophils, macrophages, and lymphocytes, into the airways. It is possible that these changes will eventually favor the development of lung cancer at the chronically inflamed sites. Another consequence of chronic airway inflammation is the development of emphysema, an abnormal and permanent enlargement of the alveolar airspaces distal to the terminal bronchioles, with a reduction in the tissues forming and supporting the alveolar septa. It is thought that increased elastolytic activity is responsible for airspace destruction. Neutrophils may release elastases. Tobacco smoke is also thought to destroy the pulmonary α_1-antiproteinase activity, thus clearing the way for progressive tissue destruction. Emphysematous lesions have been produced in guinea pigs exposed to tobacco smoke (Wright and Churg, 2002). In mice that respond with decreased antioxidant defenses during inhalation of tobacco smoke, lung elastin decreases and airspaces develop emphysematous changes. In mice deficient in α_1-antiproteinase, cigarette smoke accelerates development of emphysema (Cavarra et al., 2001).

In conclusion, the available data seem to clearly indicate that smoking of tobacco products causes pulmonary diseases such as cancer and COPD, as well as pathologies in several extrapulmonary organ systems. It is encouraging to note that in many places active smoking is on the decrease and many tobacco-associated diseases will thus be prevented.

REFERENCES

Adkins, B.J., Van Stee, E.W., Simmons, J.E., and Eustis, S.L. (1986). Oncogenic response of strain A/J mice to inhaled chemicals. *J Toxicol Environ Health* 17, 311–322.

American Thoracic Society (1996). Cigarette smoking and health. *Am J Respir Crit Care Med* 153, 861–865.

Auerbach, O., Hammond, E.C., Kirman, D., and Garfinkel, L. (1970). Effects of cigarette smoking on dogs. II. Pulmonary neoplasms. *Arch Environ Health* 21, 754–768.

Beeson, W.L., Abbey, D.E., and Knutsen, S.F. (1998). Long-term concentrations of ambient air pollutants and incident lung cancer in California adults: results from the ASHMOG study. *Environ Health Perspect* 106, 813–823.

Bernfeld, P., Homburger, F., and Russfield, A.B. (1974). Strain differences in the response of inbred Syrian hamsters to cigarette smoke inhalation. *J Natl Cancer Inst* 53, 1141–1157.

Brown, B.G., Borschke A.J., and Doolittle, D.J. (2003). An analysis of the role of tobacco-specific nitrosamines in the carcinogenicity of tobacco smoke. *Nonlinearity Biol Chem Med* 2, 179–198.

Cavarra, E., Bartalesi, B., Lucattelli, M., Fineschi, S., Lunghi, B., Gambelli, F., Ortiz, L.A., Martorana, P.A., and Lungarella, G. (2001). Effects of cigarette smoke in mice with different levels of alpha(1)-proteinase inhibitor and sensitivity to oxidants. *Am J Respir Crit Care Med* 164, 886–890.

Curtin, G.M., Higuchi, M.A., Ayres, P.H., Swauger, J.E., and Mosberg, A.T. (2004). Lung tumorigenicity in A/J and rasH2 transgenic mice following mainstream tobacco smoke inhalation. *Toxicol Sci* 81, 26–34.

D'Agostini, F., Balansky, R.M., Bennicelli, C., Lubet, R.A., Kelloff, G.J., and De Flora, S. (2001). Pilot studies evaluating the lung tumor yield in cigarette-smoke-exposed mice. *Int J Oncol* 18, 607–615.

De Flora, S., D'Agostini, F., Balansky, R., Camoirano, A., Bennicelli, C., Bagnasco, M., Cartiglia, C., Tampa, E., Longobardi, M.G., Lubet, R.A., and Izzotti, A. (2003). Modulation of cigarette-smoke-related end points in mutagenesis and carcinogenesis. *Mutat Res* 523–524, 237–252.

Dalbey, W.E., Nettesheim, P., Griesemer, R., Caton, J.E., and Guerin, M.R. (1980). Chronic inhalation of cigarette smoke by F344 rats. *J Natl Cancer Inst* 64, 383–390.

Denissenko, M.F., Pao, A., Tang, M., and Pfeifer, G.P. (1996). Preferential formation of benzo[a]pyrene adducts at lung cancer mutational hotspots in P53. *Science* 274, 430–432.

Doll, R. (1998). Uncovering the effects of smoking: historical perspective. *Stat Methods Med Res* 7, 87–117.

Dontenwill, W., Chevalier, H.J., Harke, H.P., Lafrenz, U., Reckzeh, G., and Schneider, B. (1973). Investigations on the effects of chronic cigarette-smoke inhalation in Syrian golden hamsters. *J Natl Cancer Inst* 51, 1781–1832.

Druckrey, H. (1961). Experimental investigations on the possible carcinogenic effects of tobacco smoking. *Acta Med Scand* (Suppl. 369), 24–41.

Foley, J.F., Anderson, M.W., Stoner, G.D., Gaul, B.W., Hardisty, J.F., and Maronpot, R.R. (1991). Proliferative lesions of the mouse lung: progression studies in strain A mice. *Exp Lung Res* 17, 157–168.

Goodman, G.E. (2000). Prevention of lung cancer. *Crit Rev Oncol Hematol* 33, 187–197.

Guerin, M.R., Jenkins, R.A., and Tomkins, B.A. (1992). *The Chemistry of Environmental Tobacco Smoke: Composition and Measurement. Indoor Air Research Series.* Lewis, Boca Raton, FL.

Harris, R.J., Negroni, G., Ludgate, S., Pick, C.R., Chesterman, F.C., and Maidment, B.J. (1974). The incidence of lung tumours in c57bl mice exposed to cigarette smoke: air mixtures for prolonged periods. *Int J Cancer* 14, 130–136.

Hassett, C., Mustafa, M.G., Coulson, W.F., and Elashoff, R.M. (1985). Murine lung carcinogenesis following exposure to ambient ozone concentrations. *J Natl Cancer Inst* 75, 771–777. [published erratum appears in JNCI October 1986 77(4), 991].

Hecht, S.S. (1999). Tobacco smoke carcinogens and lung cancer. *J Natl Cancer Inst* 91, 1194–1210.

Hecht, S.S. (2002). Cigarette smoking and lung cancer: chemical mechanisms and approaches to prevention. *Lancet Oncol.* 3, 461–469.

Hecht, S.S., Upadhyaya, P., Wang, M., Bliss, R.L., McIntee, E.J., and Kenney, P.M. (2002). Inhibition of lung tumorigenesis in A/J mice by N-acetyl-S-(N-2-phenethylthiocarbamoyl)-L-cysteine and myo-inositol, individually and in combination. *Carcinogenesis* 23, 1455–1461.

Henry, C.J. and Kouri, R.E. (1986). Chronic inhalation studies in mice. II. Effects of long-term exposure to 2R1 cigarette smoke on (C57BL/Cum x C3H/AnfCum)F1 mice. *J Natl Cancer Inst* 77, 203–212.

Hoffmann, D., Hoffmann, I., and Wynder, E.L. (1991). Lung cancer and the changing cigarette. *IARC Sci Publ* 105, 449–459.

Hoffmann, D. and Wynder, E.L. (1999). Active and passive smoking. In *Toxicology* (Marquardt, H., Schaefer, S.G., McClellan, R., and Welsch, F., Eds.), Academic Press, New York, pp. 879–898.

IARC (2004). *Monographs on the Evaluation of the Carcinogenic Risk of Chemicals to Humans.* Vol. 83: Tobacco smoke and involuntary smoking. (WHO IARC, Ed.), IARC, Lyon.

Kehrer, J.P. (2000). The Haber-Weiss reaction and mechanisms of toxicity. *Toxicology* 149, 43–50.

Lam, T.H., Ho, S.Y., Hedley, A.J., Mak, K.H., and Peto, R. (2001). Mortality and smoking in Hong Kong: case-control study of all adult deaths in 1998. *Br Med J* 323, 361.

Lee, P.N. (2001). Lung cancer and type of cigarette smoked. *Inhal Toxicol* 13, 951–976.

Lee, P.N. and Forey, B.A. (2003). Why are lung cancer rate trends so different in the United States and United Kingdom? *Inhal Toxicol* 15, 909–949.

Leistikow, B. (2004). Lung cancer rates as an index of tobacco smoke exposures: validation against black male approximately nonlung cancer death rates, 1969–2000. *Prev Med* 38, 511–515.

Leong, B.K., MacFarland, H.N., and Reese, W.H., Jr. (1971). Induction of lung adenomas by chronic inhalation of bis (chloromethyl) ether. *Arch Environ Health* 22, 663–666.

Leuchtenberger, C. and Leuchtenberger, R. (1974). Differential response of Snell's and C57 black mice to chronic inhalation of cigarette smoke. Pulmonary carcinogenesis and vascular alterations in lung and heart. *Oncology* 29, 122–138.

Lopez, A.D. and Murray, C.C. (1998). The global burden of disease, 1990–2020. *Nat Med* 4, 1241–1243.

Malkinson, A.M. (1998). Molecular comparison of human and mouse pulmonary adenocarcinomas. *Exp Lung Res* 24, 541–555.

Malkinson, A.M. (2001). Primary lung tumors in mice as an aid for understanding, preventing, and treating human adenocarcinoma of the lung. *Lung Cancer* 32, 265–279.

Maronpot, R.R., Shimkin, M.B., Witschi, H.P., Smith, L.H., and Cline, J.M. (1986). Strain A mouse pulmonary tumor test results for chemicals previously tested in the National Cancer Institute carcinogenicity tests. *J Natl Cancer Inst* 76, 1101–1112.

Mauderly, J., Gigliotti, A.P., Barr, E.B., Bechtold, W.E., Belinsky, S.A, Hahn, F.F., Hobbs, C.A., March, T.H., Seilkop, S.K., and Fich, G.L. (2004). Chronic inhalation exposure to mainstream cigarette smoke increases lung and nasal tumor incidence in rats. *Toxicol Sci* 81, 280–292.

Mohtashamipur, E., Mohtashamipur, A., Germann, P.G., Ernst, H., Norpoth, K., and Mohr, U. (1990). Comparative carcinogenicity of cigarette mainstream and sidestream smoke condensates on the mouse skin. *J Cancer Res Clin Oncol* 116, 604–608.

Mukhtar, H. and Ahmad, N. (1999). Cancer chemoprevention: future holds in multiple agents. *Toxicol Appl Pharmacol* 158, 207–210.

NCI: National Cancer Institute (1999). Health Effects of Exposure to Environmental Tobacco Smoke: The Report of the California Environmental Protection Agency. NIH Pub. No. 99–4645, Bethesda MD.

NCI: National Cancer Institute (2001). Smoking and Tobacco Control Monographs. No. 13: Risks Associated with Smoking Cigarettes with Low Tar Machine Measured Yield of Tar and Nicotine. U.S. National Institutes of Health, Washington, D.C.

Obermueller-Jevic, U.C., Espiritu, I., Corbacho, A.M., Cross, C.E., and Witschi, H. (2002). Lung tumor development in mice exposed to tobacco smoke and fed beta-carotene diets. *Toxicol Sci* 69, 23–29.

Omenn, G.S. (1998). Chemoprevention of lung cancer: the rise and demise of beta-carotene. *Annu Rev Public Health* 19, 73–99.

Omenn, G.S. (2000). Chemoprevention of lung cancer is proving difficult and frustrating, requiring new approaches. *J Natl Cancer Inst* 92, 959–960.

Pepelko, W.E. and Peirano, W.B. (1983). Health effects of exposure to diesel exhaust emissions. *J Am Coll Toxicol* 2, 253–306.

Peto, R., Doll, R., Buckley, J.D., and Sporn, M.B. (1981). Can dietary beta-carotene materially reduce human cancer rates? *Nature* 290, 201–208.

Pisani, P., Parkin, D.M., Bray, F., and Ferlay, J. (1999). Estimates of the worldwide mortality from 25 cancers in 1990. *Int J Cancer* 83, 18–29.

Proctor, R.N. (1995). *Cancer Wars. How Politics Shapes What We Know About Cancer.* Basic Books, New York, 1995.

Rubin, H. (2001). Synergistic mechanisms in carcinogenesis by polycyclic aromatic hydrocarbons and by tobacco smoke: a biohistorical perspective with updates. *Carcinogenesis* 22, 1903–1930.

Sargent, R.P., Shepard, R.M., and Glantz, S.A. (2004). Reduced incidence of admissions for myocardial infarction associated with public smoking ban: before and after study. *Br Med J* 328, 977–980.

Shimkin, M.B. and Stoner, G.D. (1975). Lung tumors in mice: application to carcinogenesis bioassay. *Adv Cancer Res* 21, 1–58.

Stewart, B.W. and Kleihues, P. *World Cancer Report.* 2003. Lyon, IARC Press.

Stinn, W., Teredesai, A., Kuhl, P., Knörr-Wittmann, C., Kinot, R., Coggins, C.R.E., and Haussmann, H-J. (2004). Mechanisms involved in A/J mouse lung tumorigensis induced by inhalation of an environmental tobacco smoke surrogate. Inhal Tox. 17: 263–276.

Stoner, G.D., Morse, M.A., and Kelloff, G.J. (1997). Perspectives in cancer chemoprevention. *Environ Health Perspect* 105(Suppl. 4), 945–954.

Stratton, K., Shetty, P., Wallace, R., and Bondurant, S. (2001). *Clearing the Smoke. Assessing the Science Base for Tobacco Harm Reduction.* National Academy Press, Washington D.C.

Szabo, E. (2001). Lung epithelial proliferation: a biomarker for chemoprevention trials? *J Natl Cancer Inst* 93, 1042–1043.

Tang, J.L., Morris, J.K., Wald, N.J., Hole, D., Shipley, M., and Tunstall-Pedoe, H. (1995). Mortality in relation to tar yield of cigarettes: a prospective study of four cohorts. *Br Med J* 311, 1530–1533.

Thun, M.J., Lally, C.A., Flannery, J.T., Calle, E.E., Flanders, W.D., and Heath, C.W., Jr. (1997). Cigarette smoking and changes in the histopathology of lung cancer. *J Natl Cancer Inst* 89, 1580–1586.

U.S. DHEW (1964). *Smoking and Health. Report of the Advisory Committee to the Surgeon General of the Public Health Service.* U.S. Department of Health, Education and Welfare, Washington D.C., PHS Pub. 1103.

U.S. Surgeon General (1988). The Health Consequences of Smoking: Nicotine Addiction. Washington D.C., U.S. Government.

van Zandwijk, N., Dalesio, O., Pastorino, U., De Vries, N., and van Tinteren, H. (2000). Euroscan, a randomized trial of vitamin A and N-acetylcysteine in patients with head and neck cancer or lung cancer. *J Natl Cancer Inst* 92, 977–986.

Wagner, B.M., Cline, M.E., Dungworth, D.L., Fischer, T.H., Gardner, D.E., Pryor, W.A., Rennard, S.I., and Slaga, T.J. (2000). A safer cigarette? A comparative study. A consensus report. *Inhal Toxicol* 12(Suppl. 5), 1–48.

Whincup, P.H., Gilg, J.A., Emberson, J.R., Jarvis, M.J., Feyerabend, C., Bryant, A., Walker, M., and Cook, D.G. (2004). Passive smoking and risk of coronary heart disease and stroke: prospective study with cotinine measurement. *Br Med J* 329, 200–205.

Wingo, P.A., Ries, L.A., Giovino, G.A., Miller, D.S., Rosenberg, H.M., Shopland, D.R., Thun, M.J., and Edwards, B.K. (1999). Annual report to the nation on the status of cancer, with a special section on lung cancer and tobacco smoking. *J Natl Cancer Inst* 91, 1973–1996, 675–690.

Witschi, H. (2000). Successful and not so successful chemoprevention of tobacco-smoke-induced lung tumors. *Exp Lung Res* 26, 743–756.

Witschi, H. (2001). A short history of lung cancer. *Toxicol Sci* 64, 4–6.

Witschi, H.P. (2005a). *Toxicol Sci* 84: 81–87.

Witschi, H.P. (2005b). A/J Mouse as a model for lung tumorigenesis caused by tobacco smoke: strengths and weaknesses. *Exp Lung Res* 31: 3–18.

Witschi, H. and Espiritu, I. (2002). Development of tobacco-smoke-induced lung tumors in mice fed Bowman-Birk protease inhibitor concentrate (BBIC). *Cancer Lett* 183, 141–146.

Witschi, H.P., Espiritu, I., Peake, J.L., Wu, K., Maronpot, R.R., and Pinkerton, K.E. (1997a). The carcinogenicity of environmental tobacco smoke. *Carcinogenesis* 18, 575–586.

Witschi, H.P., Espiritu, I., Maronpot, R.R., Pinkerton, K.E., and Jones, A.D. (1997b). The carcinogenic potential of the gas phase of environmental tobacco smoke. *Carcinogenesis* 18, 2035–2042.

Witschi, H.P., Joad, J.P., and Pinkerton, K.E. (1997c). The toxicology of environmental tobacco smoke. *Annu Rev Pharmacol Toxicol* 37, 29–52.

Witschi, H., Espiritu, I., Yu, M., and Willits, N.H. (1998). The effects of phenethyl isothiocyanate, N-acetylcysteine and green tea on tobacco-smoke-induced lung tumors in strain A/J mice. *Carcinogenesis* 19, 1789–1794.

Witschi, H., Espiritu, I., and Uyeminami, D. (1999). Chemoprevention of tobacco-smoke-induced lung tumors in A/J strain mice with dietary myoinositol and dexamethasone. *Carcinogenesis* 20, 1375–1378.

Witschi, H., Uyeminami, D., Moran, D., and Espiritu, I. (2000). Chemoprevention of tobacco-smoke lung carcinogenesis in mice after cessation of smoke exposure. *Carcinogenesis* 21, 977–982.

Witschi, H., Espiritu, I., Dance, S.T., and Miller, M.S. (2002). A mouse lung tumor model of tobacco-smoke carcinogenesis. *Toxicol Sci* 68, 322–330.

Witschi, H., Espiritu, I., Uyeminami, D., Suffia, M., and Pinkerton, K. B. (2004a). Lung tumor response in strain a mice exposed to tobacco smoke: some dose-effect relationships. *Inhal Toxicol* 16, 27–32.

Witschi, H., Espiritu, I., Ly, M., and Uyeminami, D. (2004b). The effects of dietary myoinositol on lung tumor development in tobacco smoke-exposed mice. *Inhal Toxicol* 16, 195–201.

Wright, J.L. and Churg, A. (2002). Animal models of cigarette-smoke-induced COPD. *Chest* 122, 301S–306S.

Wynder, E.L. (1961). Laboratory contributions to the tobacco–cancer problem. *Acta Med Scand* (Suppl. 369), 63–95.

Yu, M., Pinkerton, K.E., and Witschi, H. (2002). Short-term exposure to aged and diluted sidestream cigarette smoke enhances ozone-induced lung injury in B6C3F1 mice. *Toxicol Sci* 65, 99–106.

INDEX

B

D

E

F

G

I

J

K

L

M

N

O

P

Q

R

S

T

U

V

W

X

Y

Z